6th International Symposium on Quantum Theory and Symmetries 2009

(QTS6)

Journal of Physics: Conference Series Volume 462

Lexington, Kentucky, USA
20 – 25 July 2009

ISBN: 978-1-62993-936-0
ISSN: 1742-6588

Printed from e-media with permission by:

Curran Associates, Inc.
57 Morehouse Lane
Red Hook, NY 12571

Some format issues inherent in the e-media version may also appear in this print version.

Printed by Curran Associates, Inc. (2014)

For permission requests, please contact the Institute of Physics
at the address below.

Institute of Physics
Dirac House, Temple Back
Bristol BS1 6BE UK

Phone: 44 1 17 929 7481
Fax: 44 1 17 920 0979

techtracking@iop.org

Additional copies of this publication are available from:

Curran Associates, Inc.
57 Morehouse Lane
Red Hook, NY 12571 USA
Phone: 845-758-0400
Fax: 845-758-2634
Email: curran@proceedings.com
Web: www.proceedings.com

TABLE OF CONTENTS

Journal of Physics: Conference Series **462** (2013) 011001 doi:10.1088/1742-6596/462/1/011001

Preface

The Symposia on Quantum Theory and Symmetries form a biennial international series devoted to applications of symmetry principles in theoretical and mathematical physics. The QTS symposia cut across a wide spectrum of subfields of physics and mathematics, emphasizing common themes and interdisciplinary approaches. In an era of increasing technical specialization, QTS is practically unique in its diversity, and provides valuable opportunities for researchers from different backgrounds to interact. Since its founding in 1999, QTS Symposia have been hosted in Goslar (Germany), Krakow (Poland), Cincinnati (USA), Varna (Bulgaria), and Valladolid (Spain).

The Sixth International Symposium on Quantum Theory and Symmetries (QTS6) was held from 20–25 July 2009 in Lexington, Kentucky, on the campus of the University of Kentucky. The symposium drew more than 200 researchers from 22 countries, and featured talks in atomic, nuclear, and condensed matter physics, cosmology and gravitation, quantum computing and information theory, string theory and other areas of mathematical physics. To provide coherence to an unusually diverse set of talks, the Organizing Committee structured the program around a few broad themes, such as holography and entanglement, which intersected with a large subset of the talks.

In keeping with the interdisciplinary nature of the Symposium, we have organized these Proceedings somewhat unconventionally. The contributions are listed alphabetically by presenting author, with flags indicating their relevance to some of the main themes and subject areas of the symposium. Supplementary resources are available online at http://www.pa.uky.edu/qts6, including the complete conference program and videos of the plenary talks.

QTS6 was jointly organized by the Department of Physics and Astronomy, the Department of Mathematics, and the Center for Computational Sciences of the University of Kentucky. We are grateful for the financial and practical support of each of these entities, as well as for generous support from the College of Arts and Sciences, the Office of the Vice President for Research of the University of Kentucky, and the Office of Science of the Department of Energy. The Singletary Center for the Arts provided conference facilities for our plenary sessions, and the Teaching and Academic Support Center provided technical support. We offer our deepest thanks to all the staff and students who assisted with the organization of the conference, especially Eva Ellis, Archisman Ghosh, Jae-Hyuk Oh, and Joshua Qualls.

Sumit R Das
Alfred D Shapere
Co-chairs, QTS6 Organizing Committee

Journal of Physics: Conference Series **462** (2013) 011001 doi:10.1088/1742-6596/462/1/011001

Journal of Physics: Conference Series **462** (2013) 011001 doi:10.1088/1742-6596/462/1/011001

Local Organizing Committee

Sumit R Das, co-chair
Alfred D Shapere, co-chair
Philip Argyres
John Connolly
Peter Hislop
Keh-Fei Liu
Ganpathy Murthy
Uwe Nagel
Rohana Wijewardhana

Advisory Committee

L Alvarez-Gaume (CERN)
M V Berry (Bristol)
S Brodsky (Stanford)
M Dijkgraaf (Amsterdam)
M Douglas (Rutgers)
T Eguchi (YITP)
M P A Fisher (Santa Barbara)
G Gibbons (Cambridge)
P Goddard (IAS)
M Henneaux (Brussels)
H Kawai (Kyoto)
K Lee (KIAS)
A Morozov (ITEP)
Y Nambu (Chicago)
J Preskill (Caltech)
E Rabinovici (Jerusalem)
S Randjbar-Daemi (ICTP)
S Sachdev (Harvard)
A Sen (Allahabad)
T Senthil (MIT)
R Shankar (Yale)
C Taubes (Harvard)
N Turok (Perimeter)
C Vafa (Harvard)
S Wadia (TIFR)
F Wilczek (MIT)

Conference Board

S T Ali (Montreal)
L L Boyle (Canterbury)
M A del Olmo (Valladolid)
V K Dobrev (Sofia)
H D Doebner (Clausthal), chair
E Kapuscik (Cracow)
V I Man'ko (Moscow)
G Marmo (Naples)
G S Pogosyan (Yerevan and Dubna)
T H Seligman (Cuernavaca)
A I Solomon (Paris)
P Suranyi (Cincinnati)
L C R Wijewardhana (Cincinnati)

Journal of Physics: Conference Series **462** (2013) 011002 doi:10.1088/1742-6596/462/1/011002

Peer review statement

All papers published in this volume of *Journal of Physics: Conference Series* have been peer reviewed through processes administered by the proceedings Editors. Reviews were conducted by expert referees to the professional and scientific standards expected of a proceedings journal published by IOP Publishing.

Journal of Physics: Conference Series **462** (2013) 012003 doi:10.1088/1742-6596/462/1/012003

Opening Remarks

Heinz-Dietrich Doebner
Technical University of Clausthal, Clausthal, Germany
E-mail: doebner@t-online.de

Abstract. This is the opening address given by the Chairman of the Conference Board of the symposium series, to the Sixth International Symposium on Quantum Theory and Symmetries.

Ladies and Gentlemen, dear Friends and Colleagues:

In the name of the Conference Board it is a pleasure and an honour to welcome you to the 6th International Symposium 'Quantum Theory and Symmetries' – QTS 6 – in Lexington, Kentucky.

The symposia of this series are biannual events. The first symposium was organized 1999 in Goslar, Germany, through the Arnold Sommerfeld Institute at the Technical University in Clausthal; the following editions were arranged in Cracow, Poland in 2001; in Cincinnati, US in 2003; in Varna, Bulgaria in 2005; in Valladolid, Spain in 2007; and in 2009 the symposium has come to Lexington. The series has been accepted by our community and is considered to be a very useful and active meeting place for scientists working in our field. The number of participants ranges between 150 and 220.

The Symposia are centered on selected subjects. They present discussions and results on active topics in our field including activities of the organizing institutions. Plenaries, parallel lectures, posters and proceedings articles reflect recent developments in physics, especially quantum physics. Included are a wide range of subjects, from axiomatic formulations to quantum information, that can be understood and analyzed through models which are based on the notion of symmetry. Aspects of chemistry, material science, mathematical biology, engineering and related sciences are also included in the range of subjects covered by this series.

Symmetry is understood as a collection of mathematical views and techniques, drawn from group theory, topology, algebra, differential geometry, noncommutative geometry, functional analysis, approximation methods, and numerical methods. If you ask me for a more detailed characterization, an indirect answer is: Collect the titles and the contributions of the proceedings of QTS 1 – 6 and get a characterization of symmetries in their widest sense, as related to quantum physics, general physics, and other sciences. I think that the experience of our community in modeling physical systems based on mathematical methods connected with symmetries can be applied also to the solution of problems that appear in other contexts, like economy and ecology.

The Conference Board is the organizing body of the QTS series. The board presently has 13 members working in 10 countries. The board encourages scientists and Institutions to present proposals for a QTS symposium, decides on the proposals for a given year and is prepared to assist the local organizers in matters of organization.

Journal of Physics: Conference Series **462** (2013) 012003 doi:10.1088/1742-6596/462/1/012003

The conference Board decided in the past year that QTS 7 in 2011 will be organized at the Czech Technical University in Prague, Czech Republic. The chair of the organizing Committee will be Cestmir Burdik. QTS 8 will be hosted by UNAM in Mexico City in August 2013.

In the last years the economic situation of scientific institutions has been difficult all over the world. In my country, Germany, there was and is a tendency to channel funds for development and research to activities with direct applications, preferably in cooperation with industrial projects, e.g., in energy research and the life sciences. Proposals of small research groups with a more fundamental character were and are not in the focus of interest and have had difficulties to get support. I understand and accept this. If you plot the time dependence of research activity in natural science and mathematics in Universities and smaller research institutions, excluding projects in 'Big Science', against different "research types" on a scale between industrial, applied, pure and fundamental problems, you find a wave-like curve; in some periods you observe a maximum around the region "fundamental problems", in other periods a maximum around "applications". However, with the 'economic crisis' it is more and more difficult to get support from our universities or from third parties for fundamental research, especially for workshops and conferences. I regret this, and think that the situation in other countries is similar.

With this background I acknowledge the successful activity of Al Shapere and Sumit Das. They managed in a difficult period very successfully to get support for an international QTS Symposium. This is remarkable. In the name of all of you I express my appreciation and my sincere thanks to the members of the Local Organizing Committee of QTS 6, especially to Al Shapere and Sumit Das.

The programme is excellent, it covers recent new developments of our field and interesting related topics. I noticed that the Symposium in Lexington has also attracted active young scientists representing possible future developments of our field. The organization was and is really efficient. The local organizers have provided excellent conditions for a very interesting, successful and smooth QTS Symposium.

Thank you again!

H. - D. Doebner

Journal of Physics: Conference Series **462** (2013) 012001 doi:10.1088/1742-6596/462/1/012001

Mass deformations of four-dimensional, rank 1, N=2 superconformal field theories

Philip C. Argyres

Physics Dept., U. Cinicinnati, Cincinnati OH 45221-0011, USA

E-mail: philip.argyres@gmail.com

Abstract. Turning on N=2 supersymmetry preserving relevant operators in a 4-dimensional N=2 superconformal field theory (SCFT) corresponds to a complex deformation compatible with the rigid special Kahler (RSK) geometry encoded in the low energy effective action. Field theoretic consistency arguments indicate that there should be many distinct such relevant deformations of each SCFT fixed point. Some new RSK-compatible complex deformations are constructed of isolated rank 1 SCFTs. We also make predictions for the dimensions of Higgs branches for some rank 1 SCFTs.

1. Introduction

This talk focuses on the construction of N=2 supersymmetric mass deformations of rank 1 SCFTs. Here by "rank" we mean the complex dimension of the Coulomb branch of the SCFT. It corresponds to the rank of the gauge group for Lagrangian theories, and generalizes that concept to the non-Lagrangian theories that will be our focus. After reviewing the classification of rank 1 N=2 SCFTs in section 2, we will see that even at rank 1, the classification and construction of all possible N=2 mass deformations is a difficult and open algebraic problem. In section 3 we will describe a direct approach to the problem and illustrate its difficulties, then turn in section 4 to a less systematic but more fruitful approach. We will be able to construct a new N=2 mass deformation of the E_6 conformal curve which has a G_2 global flavor symmetry. In sections 5 and 6 we will compare this result to predictions following from strong coupling dualities of Lagrangian N=2 SCFTs. In particular, in section 6, we will refine some earlier predictions by including information about the dimension of Higgs branches at N=2 fixed points.

2. Review of rank 1 N=2 SCFTs

Although a classification of rank 1 N=2 SCFTs is known [1, 2], a complete list of the possible mass deformations of those theories is not known. We start by reviewing the classification of the rank 1 N=2 SCFTs. This is done through the construction of their Seiberg-Witten curves [3]. These scale-invariant curves are the starting point for turning on mass deformations which break the scale invariance. A general picture that has emerged [4, 5] and is a useful orientation for discussing the curves of strongly-coupled (non-Lagrangian) N=2 supersymmetric fixed-point theories is that strongly coupled conformal fixed points have a similar structure as Lagrangian theories:

- the singularity of the Seiberg-Witten curve plays a role analogous to the gauge algebra ($\mathfrak{g}$) of the Lagrangian theory, and

- different complex structure deformations of the singular curve that preserve N=2 supersymmetry are analogous to adding matter hypermultiplets in different representations ($\mathbf{r}$) of $\mathfrak{g}$ that preserve conformal invariance (in the limit of zero masses).

2.1. Rank 1 Seiberg-Witten theory

The low-energy physics on the Coulomb branch can be encoded by a family of elliptic curves [3],

$$y^2 \;=\; x^3 \;+\; f(u, m_i)x \;+\; g(u, m_i) \tag{1}$$

depending on complex parameters $\{u, m_i\}$, and a meromorphic 1-form λ_{SW} with residues m_i and satisfying

$$\partial_u \lambda_{SW} \;=\; \frac{\mathrm{d}x}{y} \;+\; \partial_x(\star)\mathrm{d}x. \tag{2}$$

Here u is the a global complex coordinate on the Coulomb branch with scaling dimension $D(u)$, and m_i, $i = 1, \ldots, h$, are mass parameters, all of scaling dimension $D(m_i) = 1$.

When the $m_i = 0$, the theory is scale invariant. In particular, the scale-invariant vacuum corresponds to the origin, $u = 0$, of the Coulomb branch. In vacua with $u \neq 0$ the scale invariance is spontaneously broken. The scale-invariant theory has a global internal symmetry algebra $\mathfrak{u}(2)_R \oplus \mathfrak{h}$. Here $\mathfrak{u}(2)_R$ is the R-symmetry and $\mathfrak{h}$ is the global flavor symmetry. Turning on masses m_i explicitly breaks the flavor symmetry. The masses transform as weights in the adjoint representation of $\mathfrak{h}$, so turning on generic m_i breaks $\mathfrak{h} \to \oplus_{i=1}^{h} \mathfrak{u}(1)_i$, where $h = \mathrm{rank}(\mathfrak{h})$. Charges, n_i, of states under these $\mathfrak{u}(1)$'s are called quark numbers.

The complex structure of the torus described by the curve is the low energy $\mathfrak{u}(1)_{em}$ gauge coupling. The electric and magnetic $\mathfrak{u}(1)_{em}$ charges (n_e, n_m) and quark numbers n_i of a BPS state determine the homology class of a cycle, $\gamma(n_e, n_m) \;=\; n_e[\alpha] \;+\; n_m[\beta] + n_i[\delta_i]$, on the torus, which determines the central charge (and BPS mass) of these states by $Z \;=\; \oint_\gamma \lambda_{SW}$. Here α and β are a canonical basis of 1-cycles on the torus, and the δ_i are a basis of cycles around the poles of of λ_{SW}.

The elliptic curve is singular at values of u corresponding to the zeros of the discriminant

$$\Delta \;\equiv\; 4 \cdot f^3 \;+\; 27 \cdot g^2 \;=\; 0. \tag{3}$$

These singularities physically correspond to a breakdown of the low-energy description in terms of massless neutral modes on the Coulomb branch since charged states are becoming massless at these points in moduli space. For a curve representing a scale-invariant theory (with masses $m_i = 0$), then $\Delta \sim u^n$ for some n. When mass parameters are turned on they appear in the curve in the form of invariants of the Weyl group of the flavor symmetry,

$$\Delta = u^n + ... + P_{kD(u)}(\{m_i\})u^{n-k} + \cdots, \tag{4}$$

where P_n is a homogeneous polynomial of degree n. Since the order of Δ is independent of the flavor symmetry, this implies that different flavor symmetries correspond to different patterns of orders of zeros of Δ in u.

2.2. Kodaira classification and mass deformations

The possible scale-invariant singularities of rank-1 curves coincides with Kodaira's classification [6] of the degenerations of holomorphic families of elliptic curves over one variable [2]. The result is two infinite series and six "exceptional" curves.

$$\begin{array}{lllll} \text{name} & \text{curve} & \Delta & D(u) & \mathfrak{g} \\ \hline E_8 & y^2 = x^3 + 2u^5 & u^{10} & 6 & - \\ E_7 & y^2 = x^3 + u^3 x & u^9 & 4 & - \\ E_6 & y^2 = x^3 + u^4 & u^8 & 3 & - \\ D_4 & y^2 = x^3 + 3\tau u^2 x + 2u^3 & u^6 & 2 & \mathfrak{su}(2) \\ H_3 & y^2 = x^3 + u^2 & u^4 & 3/2 & - \\ H_2 & y^2 = x^3 + ux & u^3 & 4/3 & - \\ H_1 & y^2 = x^3 + u & u^2 & 6/5 & - \\ D_{n>4} & y^2 = x^3 + 3ux^2 + 4\Lambda^{-2(n-4)}u^{n-1} & u^{n+2} & 2 & \mathfrak{su}(2) \\ A_{n\geq 0} & y^2 = (x-1)(x^2 + \Lambda^{-(n+1)}u^{n+1}) & u^{n+1} & 1 & \mathfrak{u}(1) \end{array} \tag{5}$$

The "name" is a conventional name of each singular family of curves. Here $\mathfrak{g}$ is the gauge group for the curves in the list corresponding to Lagrangian theories, and Λ is the UV strong coupling scale of the IR-free CFTs and τ is the marginal gauge coupling. The other six curves correspond to strongly interacting fixed points. The $H_{1,2,3}$ series were found by tuning parameters in $\mathfrak{su}(2)$ and $\mathfrak{su}(3)$ Lagrangian field theories in [7, 1], while the $E_{6,7,8}$ curves were constructed in [2].

The maximal mass deformations of these curves correspond to the complex structure deformations

$$\begin{array}{lll} E_8 & y^2 = x^3 + x(M_2u^3 + M_8u^2 + M_{14}u + M_{20}) + (2u^5 + M_{12}u^3 + M_{18}u^2 + M_{24}u + M_{30}) & \mathfrak{e}_8 \\ E_7 & y^2 = x^3 + x(u^3 + M_8u + M_{12}) + (M_2u^4 + M_6u^3 + M_{10}u^2 + M_{14}u + M_{18}) & \mathfrak{e}_7 \\ E_6 & y^2 = x^3 + x(M_2u^2 + M_5u + M_8) + (u^4 + M_6u^2 + M_9u + M_{12}) & \mathfrak{e}_6 \\ D_4 & y^2 = x^3 + x(3\tau u^2 + M_2u + M_4) + (2u^3 + \widetilde{M}_4u + M_6) & \mathfrak{so}(8) \\ H_3 & y^2 = x^3 + x(M_{1/2}u + M_2) + (u^2 + M_3) & \mathfrak{u}(3) \\ H_2 & y^2 = x^3 + x(u) + (M_{2/3}u + M_2) & \mathfrak{u}(2) \\ H_1 & y^2 = x^3 + x(M_{4/5}) + (u) & \mathfrak{u}(1) \\ D_{n>4} & y^2 = x^3 + 3ux^2 + \Lambda^{-(n-4)}\widetilde{M}_n x + 4\Lambda^{-2(n-4)}(u^{n-1} + M_2u^{n-2} + \cdots + M_{2n-2}) & \mathfrak{so}(2n) \\ A_{n\geq 0} & y^2 = (x-1)(x^2 + \Lambda^{-(n+1)}[u^{n+1} + M_2u^{n-1} + M_3u^{n-2} + \cdots + M_{n+1}]) & \mathfrak{su}(n+1) \end{array} \tag{6}$$

Here the Lie algebra on the right denotes the flavor algebra, $\mathfrak{h}$. Note that for the A-D-E series of curves, the maximal flavor algebras are the corresponding Lie algebras. However, this is, in the following sense, a coincidence.

As an example, consider the maximal mass deformation of the A_n series corresponding to a $\mathfrak{u}(1)$ gauge theory with $n+1$ hypermultiplets of charge 1, contributing $b = n+1$ to the beta function. The total contribution to the beta function from the hypermultiplets determines the form of the singularity in an IR-free theory, but there are many inequivalent ways of contributing a given amount. For example, n_a hypermultiplets of charge $\pm r_a$ give $b = \sum_a n_a r_a^2$, and a $\oplus_a \mathfrak{u}(n_a)$ flavor symmetry, smaller than the $\mathfrak{su}(n+1)$ flavor symmetry of the maximal mass deformation. Thus there should be many inequivalent mass deformations of the A_n singularities preserving N=2 supersymmetry.

The same is true for the other Lagrangian curves, the D_n series which are IR-free $\mathfrak{su}(2)$ theories for $n > 4$, and the conformal $\mathfrak{su}(2)$ theory for $n = 4$. For the D_4 theory we can have $b = 0$ with 4 fundamental hypermultiplets or with 1 adjoint hypermultiplet. The first is the maximal mass deformation and has an $\mathfrak{so}(8)$ flavor symmetry, while the second has an $\mathfrak{sp}(1)$ flavor symmetry, with curve with a sub-maximal mass deformation [3]

$$y^2 = \prod_i (x - e_i u - e_i^2 M_2). \tag{7}$$

Journal of Physics: Conference Series **462** (2013) 012001 doi:10.1088/1742-6596/462/1/012001

The mass deformation of the $D_{n>4}$ curves in (6) corresponds to an $\mathfrak{su}(2)$ gauge theory with $2n$ half-hypermultiplets in the fundamental representation contributing $b = 2(n-4)$. (The total $b > 0$ so all these theories are IR free.) There are many other ways of adding matter hypermultiplets to contribute the same b. There are two classes of representations for $\mathfrak{su}(2)$, the real $\mathbf{2r+1}$ and the pseudoreal $\mathbf{2s}$ (denoting representations by their dimensions). To avoid anomalies [8] we must have $2n_r$ of each real representation and any number m_s of the pseudoreal such that their contribution to the appropriately normalized quadratic casimir, $\frac{1}{3}\Sigma_s m_s s(4s^2-1)$, is even. Then $b = \frac{4}{3}\Sigma_r n_r r(r+1)(2r+1) + \frac{1}{3}\Sigma_s m_s s(4s^2-1) - 8$ and the flavor symmetry that corresponds to this value of the beta function is $\oplus_r \mathfrak{sp}(n_r) \oplus_s \mathfrak{so}(m_s)$. Again, we see that there are many inequivalent sub-maximal mass deformations of the $D_{n>4}$ singularities that preserve N=2 supersymmetry.

3. Constructing new curves

So it is natural to try to construct other sub-maximal mass deformations of the non-Lagragian singularities. Note that evidence for the existence of new mass deformations of the $E_{6,7,8}$ singularities was found in [5] and will be reviewed in section 5. The maximal mass deformations of the $H_{1,2,3}$ singularities were found in [7, 1] by taking appropriate scaling limits of known Lagrangian curves, while the maximal mass deformations of the $E_{6,7,8}$ curves were worked out by Minahan and Nemeschansky in [2] just from guessing the flavor symmetry and imposing the conditions of N=2 supersymmetry. We will try to copy that procedure below.

To illustrate the problem, suppose we look for a sub-maximal mass deformation of the E_6 singularity with flavor symmetry $\mathfrak{su}(3)$. It then follows from arguments in [9] (extended in section 6, below) that the discriminant will have only 4 zeros for generic m_i instead of the maximal 8, which can only be achieved by one of the following possible inequivalent factorizations of the discriminant:

$$\begin{aligned}\Delta \sim\ & (u+...)^5(u^3+...) \quad \text{or} \quad (u+...)^4(u+...)^2(u^2+...) \quad \text{or} \quad (u+...)^3(u+...)(u^2+...)^2 \\ & \text{or} \quad (u^2+...)^3(u^2+...) \quad \text{or} \quad (u^4+...)^2. \end{aligned} \tag{8}$$

A laborious computer search of parameterizations of complex deformations of the E_6 singularity reveals 2 consistent factorization solutions. The first $\mathfrak{su}(3)$ factorization solution is actually a 1-parameter (ν) family:

$$\begin{aligned} y^2 &= x^3 + 3N_2x[u^2 + (1+\nu)N_2^3 + N_3^2] + [u^4 + u^2((1+2\nu)N_2^3 + 2N_3^2) \\ &\quad + \nu(1+\nu)N_2^6 + (1+2\nu)N_2^3N_3^2 + N_3^4], \\ \Delta &= -27[u^2 + (1+\nu)N_2^3 + N_3^2]^2[u^2 + (2+\nu)N_2^3 + N_3^2]^2 \ \sim\ (u^4+\ldots)^2, \end{aligned} \tag{9}$$

where $N_{2,3}$ are the $\mathfrak{su}(3)$ Weyl invariants. However, there does not exist a Seiberg-Witten 1-form for this curve for any ν, so these deformations are not compatible with $N=2$ supersymmetry. The second $\mathfrak{su}(3)$ factorization solution is

$$\begin{aligned} y^2 &= x^3 + u[3N_2x(u-4N_3) + u^3 - 12u^2N_3 - u(N_2^3 - 48N_3^2) - 64N_3^3] \\ \Delta &= -27u^2[u^3 - 12u^2N_3 + u(N_2^3 + 48N_3^2) - 64N_3^3]^2 \end{aligned}$$

We have not been able to construct or rule out a Seiberg-Witten 1-form for this curve.

As another example, we can search for a sub-maximal deformation of the E_7 singularity with flavor symmetry $\mathfrak{sp}(3) \oplus \mathfrak{su}(2)$ (as predicted to exist in [5]). This requires factoring a 9th-order polynomial into 6 generic zeros:

$$\Delta \sim (u+...)^4(u^5+...) \quad \text{or} \quad (u+...)^3(u+...)^2(u^4+...) \quad \text{or} \quad (u^3+...)^2(u^3+...). \tag{10}$$

A systematic search reduces the problem to solving on the order of 800 polynomial relationships among 160 unknowns. Though highly over-constrained, it is computationally very difficult to determine whether there are any solutions.

Similarly, a sub-maximal deformation of the E_8 singularity with flavor symmetry $\mathfrak{sp}(5)$ requires factoring a 10th-order polynomial into 7 generic zeros:

$$\Delta \sim (u+...)^4(u^6+...) \quad \text{or} \quad (u+...)^3(u+...)^2(u^5+...) \quad \text{or} \quad (u^3+...)^2(u^4+...). \tag{11}$$

A systematic search in this case is much too difficult. Clearly some other method is needed.

4. Isogenies: a non-systematic approach

An n-isogeny is an n-to-1 holomorphic map of a curve to itself that preserves the holomorphic 1-form. The existence of such a map implies that some of the zeros of the discriminant will have multiplicity n. Though there are, in principle, n-isogenies for any n, only certain 2- and 3-isogenies seem to have known closed-form expressions for general parameters. In particular, there are three traditional presentations of elliptic curves which are related by simple isogenies:

$$\begin{aligned} \text{Legendre:} \quad & y^2 = x^3 + fx + g, \\ \text{Jacobi:} \quad & \widetilde{y}^2 = \widetilde{x}^4 + \alpha\widetilde{x}^2 + \beta, \\ \text{Hessian:} \quad & \gamma = \widetilde{y}^3 + \delta\widetilde{x}\widetilde{y} + \widetilde{x}^3, \end{aligned} \tag{12}$$

where f, g, α, β, γ, and δ are all functions of u. We have been presenting our elliptic curves in the Legendre form so far. The map between the Jacobi and Legendre forms is a 2-isogeny, while that between the Hessian and Legendre forms is a 3-isogeny.

4.1. 2-isogenies

We can map the Jacobi form to the Legendre form by $\widetilde{x} = x^{\frac{1}{2}}$, $\widetilde{y} = yx^{-\frac{1}{2}}$. It then easily follows that the condition for a curve to have a Jacobi 2-isogenous form is that $D(\beta) = kD(u)$ where $k \in \mathbb{Z}^+$. From the scale dimensions of u and x that can be deduced from (5), it follows that the H_2, D_4, and E_7 curves have a 2-isogenous curve of the form

$$y^2 = x^3 + x\left(\beta - \frac{1}{3}\alpha^2\right) + \alpha\left(\frac{2}{27}\alpha^2 - \frac{1}{2}\beta\right). \tag{13}$$

- The H_2 isogenous curve is (13) with $\alpha = M_{2/3}$ and $\beta = u + M_{2/3}^2/3$. It can only have a $\mathfrak{u}(1)$ flavor symmetry. It would be interesting to identify this sub-maximal mass deformation of the H_2 theory by tuning parameters in an asymptotically free Lagrangian N=2 theory.
- The D_4 curve with a 2-isogeny has the form (13) with $\alpha = \tau u + M_2$, $\beta = u^2 + M_4$, and discriminant
$$\Delta = (u^2 + M_4)^2((\tau^2 - 4)u^2 + 2\tau M_2 u + (M_2^2 - 4M_4)). \tag{14}$$
If we take the special case $M_4 = (4-\tau^2)^{-1}M_2^2$ then we get the $\mathfrak{sp}(1)$ (adjoint hypermultiplet) sub-maximal deformation of the D_4 singularity found in [3]. Presumably demanding the existence of a Seiberg-Witten one-form enforces the $M_4 \propto M_2^2$ identification, though this has not been checked.
- The E_7 curve with a 2-isogeny is (13) with $\alpha = M_2 u + M_6$, $\beta = u^3 + M_8 u + M_{12}$. The mass parameter dimensions correspond to the dimensions of the Weyl invariants of the exceptional F_4 flavor symmetry. A systematic search for the SW 1-form for this curve is in progress [10].

4.2. 3-isogenies

A 3-isogenous map from the Hessian form to the Legendre form of the elliptic curve is given by $\tilde{x} = -xy^{-\frac{1}{3}}$, $\tilde{y} = y^{\frac{1}{3}}$. The resulting condition for a curve to have a Hessian 3-isogenous form is $D(\gamma) = kD(u)$ where $k \in \mathbb{Z}^+$. The scaling in (5) then shows that the H_3 and E_6 curves have a 3-isogenous deformation of the form

$$y^2 = x^3 - x\delta\left(\gamma + \frac{1}{12}\delta^3\right) + \left(-\frac{1}{108}\delta^6 - \frac{1}{6}\gamma\delta^3 + \gamma^2\right). \qquad (15)$$

- The H_3 isogenous curve has $\gamma = u + M_{1/2}^3/12$ and $\delta = M_{1/2}$, and can only have a $\mathfrak{u}(1)$ flavor symmetry. It would be interesting to identify this sub-maximal mass deformation of the H_3 theory by tuning parameters in an asymptotically free Lagrangian N=2 theory.
- The E_6 curve with a 3-isogeny is (15) with $\delta = M_2$, $\gamma = u^2 + M_6$ and discriminant

$$(u^2 + M_2^3 + M_6)^3(9u^2 + 5M_2^3 + 9M_6). \qquad (16)$$

We have explicitly constructed a SW 1-form for this curve [10], following the method of [2]. We find that the flavor symmetry of this curve is the G_2 exceptional algebra.

5. Predictions from S-duality

To summarize so far, we have succeeded in constructing an N=2 supersymmetric sub-maximal mass deformation of the E_6 singularity with flavor algebra $\mathfrak{h} = G_2$, have not been able to rule out the possibility of another such deformation with $\mathfrak{h} = \mathfrak{su}(3)$, and have slight evidence that a sub-maximal deformation of the E_7 singularity exists with $\mathfrak{h} = F_4$.

We would now like to compare these results with the predictions made in [5] for sub-maximal mass deformations of rank-1 SCFTs. These predictions were based on the fact that N=2 SCFTs can arise as decoupled factors of the strong-coupling limit of certain Lagrangian SCFTs. In particular, it was argued in [4] that the physics at infinite coupling of an N=2 Lagrangian SCFT with a gauge algebra $\mathfrak{g}$ of rank r is a weakly coupled scale-invariant gauge theory with a gauge algebra $\tilde{\mathfrak{g}}$ with smaller rank s which is coupled to an isolated rank-$(r-s)$ N=2 SCFT. The coupling between $\tilde{\mathfrak{g}}$ and the SCFT is the standard gauge coupling: $\tilde{\mathfrak{g}}$ gauges a subalgebra of the global symmetry algebra, $\mathfrak{h}$, of the SCFT.

If one knows the Seiberg-Witten curve and one-form of the $\mathfrak{g}$ Lagrangian SCFT, then one can derive the curve and one-form of the (mass-deformed) isolated SCFT simply by taking the infinite-coupling limit. However, this information (the low energy effective theory on the Coulomb branch) is not know for many theories.[1] In [5] we instead accumulated evidence for the existence of isolated SCFTs and their flavor algebras $\mathfrak{h}$ by assuming a strong-coupling limit of the form described above and testing it with some simple algebraic consistency checks. In particular, for each potential strong-coupling duality we demanded that on both sides of the duality the spectrum of dimensions of Coulomb branch vevs matched, the flavor symmetry algebras matched, and the number of marginal couplings matched. We then computed for the isolated SCFT from the presumed duality the flavor algebra central charge $k_{\mathfrak{h}}$, the $\mathfrak{u}(1)_R$ central charge k_R, the conformal anomaly a, and the existence of a global $\mathbb{Z}_2$-obstruction to gauging the flavor symmetry. Finally, we asked whether a given such isolated SCFT was predicted in more than one case.

After examining many examples with one or two marginal couplings, we predicted three new rank 1 sub-maximal mass deformations:

[1] In light of the recent solution by D. Gaiotto in [11], and its extension in many following papers, of a very large class of N=2 theories, it may be that sub-maximal mass deformations of some isolated SCFTs may now be able to be derived in this straightforward way.

- The E_8 singular curve should have a mass deformation with flavor algebra $\mathfrak{h} = \mathfrak{sp}(5)$, a Z_2 obstruction, and central charges $k_{\mathfrak{h}} = 7$, $(3/2)k_R = 98$, $48a = 164$.
- The E_7 singular curve should have a mass deformation with flavor algebra $\mathfrak{h} = \mathfrak{sp}(3) \oplus \mathfrak{su}(2)$, a Z_2 obstruction in the $\mathfrak{sp}(3)$ factor, and central charges $k_{\mathfrak{sp}(3)} = 5$, $k_{\mathfrak{su}(2)} = 8$, $(3/2)k_R = 58$, $48a = 100$.
- The E_6 singular curve should have a mass deformation with flavor algebra $\mathfrak{h} \neq E_6$ with $r = \text{rank}(\mathfrak{h})$ between $2 \leq r \leq 6$, and central charges $k_{\mathfrak{h}} = (8-n)/\text{I}$, $(3/2)k_R = 38 - 2n$, $48a = 68 - 2n$, where $n \in \{0, 1, 2\}$ and I is the (positive integer) Dynkin index of embedding of $\mathfrak{su}(2)$ in the maximal embedding $\mathfrak{h} \supset \mathfrak{su}(2) \oplus \mathfrak{u}(1)^p$ with $p \in \{1, 2\}$. If $p = 1$ then $n = 2$, and if $p = 2$ then $n = 0$ or 1. If $r = 2$ then $p = 1$, and if $r > 2$ then $p = 1$ or 2.

The predicted E_7 sub-maximal deformation does not match the possible $\mathfrak{h} = F_4$ deformation found as a 2-isogeny. If would be interesting to see if such a sub-maximal deformation could be located as a factor in the strong-coupling limit of a Lagrangian SCFT by the above methods.

The predicted E_6 sub-maximal deformation is loose enough to be compatible with the $\mathfrak{h} = G_2$ deformation constructed as a 3-isogeny.

6. Central charges and curves

We can sharpen the above predictions from N=2 S-dualities by a slight refinement [10] of the arguments of [9]. These arguments show how topologically twisted Seiberg-Witten theory computes the central charges ($k_{\mathfrak{h}}$, k_R, a) in terms of the low-energy data (d, h, Z) where

- $d = D(u)$ is the dimension of the Coulomb branch vev u,
- h is the number of $\mathfrak{u}(1)_{em}$-neutral hypermultiplets at a generic point on the Coulomb branch (*i.e.*, the quaternionic dimension of the Higgs branch above a generic point on the Coulomb branch), and
- Z is the number of singular points on the Coulomb branch of the curve at generic masses (*i.e.*, the number of zeros of the discriminant Δ counted *without* multiplicities).

In particular, for rank 1 SCFTs (*i.e.*, with one-dimensional Coulomb branches), the relations are $k_{\mathfrak{h}} = 2d - h$, $(3/2)k_R = 2Zd + 4 + 2h$, and $48a = 12d + 2Zd - 2 + 2h$. The difference between the second two relations gives a known identity, while the new relations can be inverted as

$$h = 2d - k_{\mathfrak{h}}, \qquad \text{and} \qquad Z = d^{-1}\left(k_{\mathfrak{h}} + \frac{3}{4}k_R - 2d - 2\right), \tag{17}$$

to determine the Higgs branch dimension, h, and the number of zeros of the discriminant, Z.

Since any massless hypermultiplets on the Lagrangian side of an N=2 S-duality necessarily transform under the global flavor symmetry $\mathfrak{f}$, this implies that the h hypermultiplets of the dual isolated SCFT factor must be charged appropriately under its flavor symmetry $\mathfrak{h}$ so that when the relevant $\widetilde{\mathfrak{g}}$ subalgebra of $\mathfrak{h} \supset \widetilde{\mathfrak{g}} \oplus \mathfrak{f}$ is gauged, the hypermultiplets split into the observed gauge-charged hypermultiplet plus gauge-singlet hypermultiplets which transform under $\mathfrak{f}$.

For example, the $\mathfrak{h} = \mathfrak{sp}(5)$ sub-maximal deformation of the E_8 singularity appears as a factor in the N=2 S-dual pair:

$$G_2 \text{ with } 8 \cdot \mathbf{7} \simeq \mathfrak{su}(2) \text{ with } (\mathbf{2} \oplus \text{SCFT}[6 : \mathfrak{sp}(5)]), \tag{18}$$

(where SCFT$[6 : \mathfrak{sp}(5)]$) denotes the $\mathfrak{sp}(5)$ sub-maximally deformed CFT). The values of the central charges imply by (17) that the number of half-hypermultiplets is $2h = 10$. This fills out the pseudoreal **10** irrep of $\mathfrak{h} = \mathfrak{sp}(5)$. Upon gauging $\widetilde{\mathfrak{g}} = \mathfrak{su}(2)$, $\mathfrak{h}$ is broken as $\mathfrak{sp}(5) \supset \mathfrak{su}(2) \oplus \mathfrak{sp}(4)$, under which the **10** decomposes as $\mathbf{10} = (\mathbf{2}, \mathbf{1}) \oplus (\mathbf{1}, \mathbf{8})$. This fits perfectly: the $(\mathbf{2}, \mathbf{1})$ is the doublet half-hypermultiplet on the right side of (18), while the

$(\mathbf{1}, \mathbf{8})$ are the 8 "quark" half-hypermultiplets transforming under the $\mathfrak{f} = \mathfrak{sp}(4)$ global symmetry apparent on the left side of (18). This also agrees with the predicted $\mathbb{Z}_2$-obstruction to gauging the $\mathfrak{h} = \mathfrak{sp}(5)$ flavor symmetry of the isolated SCFT: is has a single half-hypermultiplet in a pseudoreal representation.

Similar remarks apply to the $\mathfrak{h} = \mathfrak{sp}(3) \oplus \mathfrak{su}(2)$ sub-maximal deformation of the E_7 singularity, which has $2h = 6$ half-hypermultiplets transforming in the **6** of $\mathfrak{sp}(3)$, and none transforming under the $\mathfrak{su}(2)$.

Equation (17) in these two cases also implies that $Z = 7$ and 6, respectively. This was used in (11) and (10) above to constrain the possible pattern of factorizations of the discriminants of the sub-maximally deformed curves.

Now we use (17) to constrain the possible sub-maximal deformation of the E_6 curve mentioned in section 5. The relevant S-duality in this case is

$$\mathfrak{su}(3) \text{ with } \mathbf{3} \oplus \overline{\mathbf{3}} \oplus \mathbf{6} \oplus \overline{\mathbf{6}} \ \simeq\ \mathfrak{su}(2) \text{ with } (n \cdot \mathbf{2} \oplus \text{SCFT}[3 : \mathfrak{h}]). \tag{19}$$

Equation (17) together with the requirements that h and Z be non-negative integers, then gives only 4 solutions:

$$(n, p, I, Z, h) = (0, 2, 2, 5, 2) \text{ or } (0, 2, 8, 4, 5) \text{ or } (2, 1, 1, 5, 0) \text{ or } (2, 1, 2, 4, 3), \tag{20}$$

where p and I are parameters describing the embedding of $\mathfrak{su}(2) \subset \mathfrak{h}$, defined in section 5. Though this data is still not enough to determine $\mathfrak{h}$, it does put strong constraints on it. (The list of allowed $\mathfrak{h}$ is under investigation [10].) For example, it rules out the $\mathfrak{h} = G_2$ 3-isogeny deformation, since $\text{rank}(G_2) = 2$, so $p = 1$, and $Z = 5$ by (16), picking out the third of the possible solutions in (20). But there is no index $I = 1$ $\mathfrak{su}(2)$ subgroup of G_2 with commutant $\mathfrak{u}(1)$, so this possibility is ruled out. It would be interesting to see if the G_2 sub-maximally deformed SCFT appeared as a factor in other N=2 S-dualities.

Acknowledgments

This talk reports on work done in collaboration with John R. Wittig [10]. Also, it is a pleasure to thank D. Gaiotto, A. Shapere, and Y. Tachikawa for helpful discussions.

References

[1] Argyres P, Plesser R, Seiberg N and Witten E 1996 *Nucl. Phys.* B **461** 71
[2] Minahan J and Nemeschansky D 1996 *Nucl. Phys.* B **482** 142
[3] Seiberg N and Witten E 1994 *Nucl. Phys.* B **431** 484
[4] Argyres P and Seiberg N 2007 *JHEP* **0712** 088
[5] Argyres P and Wittig J 2008 *JHEP* **0801** 074
[6] Kodaira K 1963 *Ann. Math.* **77** 563
[7] Argyres P and Douglas M 1995 *Nucl. Phys.* B **448** 93
[8] Witten E 1982 *Phys. Lett.* B **117** 324
[9] Shapere A and Tachikawa Y 2008 *JHEP* **0809** 109
[10] Argyres P and Wittig J 2010 to appear
[11] Gaiotto D 2009 N=2 dualities *Preprint* arXiv:0904.2715

Journal of Physics: Conference Series **462** (2013) 012002 doi:10.1088/1742-6596/462/1/012002

Thermalization and Black Holes

Dongsu Bak

Physics Department, University of Seoul, Seoul 130-743 Korea

E-mail: dsbak@uos.ac.kr

Abstract. In the framework of AdS/CFT correspondence, we discuss how the thermalization in a Yang-Mills plasma is described by the exact solution of the time dependent black hole in the dual gravity side. We describe the prescription to match generic initial thermal states from the both sides. We also discuss implications of our solution on general non-equilibrium phenomena. The thermalization involves a loss of information leading to an increase of entropy, which is only possible in the thermodynamic limit of field theories. We argue that the large N limit plays a role of this thermodynamic limit in our AdS/CFT correspondence.

1. Introduction

In the advent of the AdS/CFT correspondence or more generally gauge/gravity duality, there has been great progress in our understanding of the nature of gravity and dual field theories in a strongly coupled regime[1]. The conformal gauge theories are well defined quantum mechanically and this can be used to define quantum gravity using the correspondence. One of the main trouble in this definition of quantum gravity is that one needs a detailed dictionary of correspondence beyond the current level of our understanding.

A well known example of such correspondence is the AdS_5/CFT_4 correspondence. The $\mathcal{N} = 4$ SU(N) super-Yang-Mills (SYM) theory is dual to the type IIB superstring theory on $AdS_5 \times S^5$ background[1]. The $\mathcal{N} = 4$ SYM theory is parametrized by the number of colour N and the gauge coupling g_{YM} and its large N planar limit is defined by sending $N \to \infty$ and $g_{YM} \to 0$ while holding the 't Hooft coupling $\lambda = g_{YM}^2 N$ fixed. In this limit, the string coupling $g_s = \frac{g_{YM}^2}{4\pi}$ responsible for the joining and splitting of strings becomes zero and thus the type IIB supergravity description becomes precise. If the AdS radius $L = \lambda^{1/4} l_s$ is large with l_s denoting the string length scale, the supergravity is weakly coupled whereas the corresponding SYM theory becomes strongly coupled. Below we shall work in this limited framework of the correspondence where the classical type IIB supergravity is dual to the planar $\mathcal{N} = 4$ SYM theory in the strongly coupled regime of large λ.

The main topic of this note is how the thermal nature of gravity arises in the context of the gauge/gravity correspondence. We shall first present a detailed example of the time dependent black hole solution[2, 3] describing thermalization of a finite-size perturbation above an equilibrium SYM state. In the gauge theory side, what is expected is quite clear. The perturbations decay away in time with characteristic time scales called relaxation time scales. Of course the local perturbation is also spatially decaying, which is due to the thermal screening effects. Among the spatial scales, the true mass gap and the Debye screening mass are two important characteristic scales of plasma at finite temperature[4, 5]. The relaxation in time eventually leads to a new equilibrium with an increased entropy. The total energy of the system is conserved in this process. In fact the relevant $\mathcal{N}=4$ SYM theory is defined over a compact space Σ which will be specified below. Thus the volume of the system is fixed in time. Then the first law of thermodynamics dictates that there is no flow of heat into the system. Except the strongly coupled aspect, the above is typical and generic in any thermal system in the thermodynamic limit.

The time dependent black hole is precisely describing the same physics of thermalization from the gravity side. At an initial time $t=0$, the perturbation is turned on deriving the time dependence of the horizon area caused by the (scalar) matter flux through the horizon. The entropy changes as the horizon area changes while the initial perturbation is relaxing away. Since there is no energy flux through the boundary, the energy and the volume of our system do not change through the processes which is simply following from the energy momentum conservation of gravity theories.

In fact one may be more specific about the correspondence. Namely using the generalized prescription of matching the initial states of the two sides based on the thermofield dynamics[6, 7, 8, 9], we shall show that the one-point function $\langle \mathcal{L}(t)\rangle$ of the field theory Lagrange density is decaying exponentially in time. The functional behaviours computed from both sides agree with each other precisely.

In this note we shall extend our discussion about detailed structure of the geometry. This will include the structure of the Euclidean geometry, from which one can generalize our prescription of matching initial states from both sides. We shall also discuss physical implications of the above correspondence. The first one is the thermalization time scale in the strong coupling limit. At high enough temperature, the thermalization happens very fast, which has a potential application to the understanding of the fast thermalization observed in the RHIC experiment. The second is how the information loss is occurring in the planar limit of the SYM theory via the change of the entropy. Our exact solution is particularly suited for the discussion of these aspects. We shall study the first law and its implication on what the gravity theory is describing. Our solution demonstrates clearly that the gravity includes dynamical effects that cannot be described by the change of entropy in time. This follows from a direct application of our first law of the thermodynamics. How to probe the time dependent entropy using the method of entanglement entropy will be discussed. We shall finish this note with a few final comments.

2. Time dependent black hole and correspondence

In this section we review the properties of time dependent black hole and the correspondence to the SYM system. We start with a simple scalar Einstein gravity system,

$$I=\frac{1}{16\pi G}\int d^d x\sqrt{g}\left(R-\frac{1}{2}g^{ab}\partial_a\phi\partial_b\phi+(d-1)(d-2)\right)\,, \tag{1}$$

where the spacetime dimension d is greater than or equals to three. Any solution of the above can be embedded into the type IIB supergravity for $d=5$ and $d=3$. The $d=5$ ($d=3$) solution describes the deformation of $AdS_5\times S^5$ ($AdS_3\times S^3$) geometry[10, 3]. Note that in five dimensions, ϕ is the IIB dilaton whereas $\phi/\sqrt{2}$ corresponds to the dilaton for $d=3$. The AdS radius denoted by L is set to be unity and we shall focus on the 5d case though all the results are extended to the three dimensions.

The metric for the time dependent black hole is given by

$$ds^2=f(\mu)(d\mu^2-d\tau^2+\cos^2\tau\, ds^2_\Sigma)\,, \tag{2}$$

where ds^2_Σ is describing the compact, smooth, finite volume Einstein space metric in $(d-2)$ dimensions satisfying $R^\Sigma_{kl}=-(d-3)g^\Sigma_{kl}$. The coordinate τ is ranged over $[-\pi/2,\pi/2]$. For $d=3$

case, the Σ space corresponds to a circle S^1 and, for higher dimensions, the space can be given by the quotient of the hyperbolic space H_{d-2} by a discrete subgroup of the hyperbolic symmetry group, $SO(1,d-2)$. The function $f(\mu)$ is given by the integral[10, 2]

$$\mu_0 \pm \mu = \int_f^\infty \frac{dx}{2\sqrt{x^3 - x^2 + \frac{\gamma^2}{(d-1)(d-2)}x^{4-d}}}, \tag{3}$$

where μ_0 is chosen such that $\mu = 0$ at the turning point. Then $\mu \in [-\mu_0, \mu_0]$ with $\mu_0 \geq \pi/2$.

The dilaton field is given implicitly by the integral

$$\phi(\mu) = \int_{-\mu_0}^{\mu} dx \frac{\sqrt{2}\gamma}{f^{\frac{d-2}{2}}(x)} + \phi_- . \tag{4}$$

and we introduce $\phi_\pm$ by $\phi_\pm = \phi(\pm\mu_0)$. This time dependent black hole solution is closely related to Janus solution dual to the interface conformal theory[10, 11]. If the deformation parameter γ vanishes, the solution corresponds to the BTZ black hole in 3d[12] and the topological black holes in higher dimensions[13, 14, 15, 16].

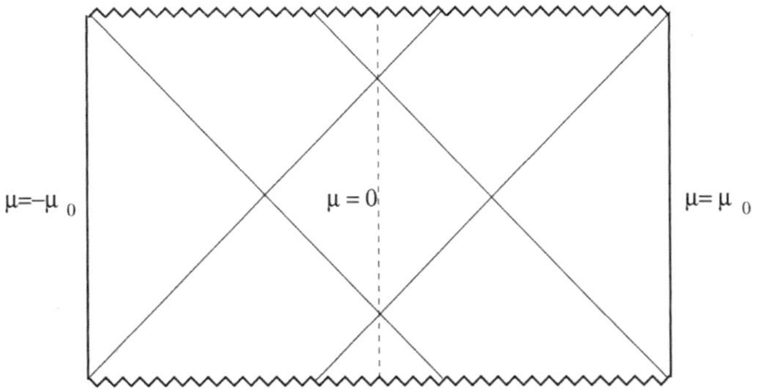

Figure 1. Penrose diagram for the time dependent black hole. The τ ($\in [-\pi/2, \pi/2]$) coordinate runs vertically upward and μ ($\in [-\mu_0, \mu_0]$) to the right horizontally.

The above form of the solution is particularly suited for drawing the Penrose diagram representing the global structure of the spacetime. As in Figure 1, the Penrose diagram is elongated horizontally. The vertical coordinate τ runs from $-\pi/2$ to $\pi/2$ and the horizontal coordinate μ from $-\mu_0$ to $+\mu_0$ where μ_0 is greater than or equal to $\pi/2$. In fact μ_0 increases monotonically as the integration constant γ is increasing. As γ approaching γ_c from below with

$$\gamma_c^2 = (d-2)\left(\frac{d-2}{d-1}\right)^{d-2}, \tag{5}$$

the μ_0 becomes infinitely large. Hence the horizontal size becomes larger and larger as γ is approaching γ_c from below. The black hole solution can have a physical meaning beyond the critical value of γ but the spacetime no longer involve two timelike boundaries. We shall restrict our discussion for $\gamma < \gamma_c$ below. The dilaton runs from ϕ_- to ϕ_+ from left boundary to the right boundary. Hence for $\gamma < \gamma_c$ there are two boundaries as the usual black hole in the AdS spacetime but the boundary coupling becomes different as a result of the deformation with the deformation parameter γ. There are two future and two past horizons as drawn in Figure 1 by 45 degree lines from the future and past infinities. For the future horizon, the horizon area starts from zero at $\tau = -\pi/2$ and monotonically increases reaching an equilibrium value

$$S_{final} = \frac{\mathcal{V}_\Sigma}{4G}. \tag{6}$$

By studying the near boundary behavior of the metric tensor, one can obtain the expectation value of the boundary energy momentum tensor:

$$\begin{aligned}\mathcal{E} &= \frac{1}{8\pi G}\left(\frac{d-2}{d-1}\right)\left(\frac{d-3}{d-1}\right)^{\frac{d-3}{2}} \\ p &= \frac{1}{8\pi G}\left(\frac{1}{d-1}\right)\left(\frac{d-3}{d-1}\right)^{\frac{d-3}{2}},\end{aligned} \tag{7}$$

where $\mathcal{E}$ is the energy density in the boundary field theory. The total energy of the system $E = \mathcal{E}\mathcal{V}_\Sigma$ agrees with the equilibrium value. The result is consistent with the tracelessness condition of the energy momentum tensor, i.e. $T^\mu{}_\mu = 0$, which is due to the conformal symmetry. We also note that the operator dual to the dilaton is the CFT Lagrange density $\mathcal{L}$. From the study of the boundary behaviours of the dilaton field, one finds

$$\langle \mathcal{L} \rangle = \frac{\gamma}{8\pi G}\frac{1}{\cosh^{d-1} t}, \tag{8}$$

where t is the boundary time related to the conformal time τ by $\tanh^{-1}(\sin\tau)$.

By choosing the conformal factor $h^2 = \cos^2\tau / f(\mu)$, the boundary metric for the CFT is given by

$$ds^2_{B\pm} = -dt^2 + ds^2_\Sigma\,. \tag{9}$$

There are two separated boundaries at $\mu = \pm\mu_0$ and the boundary time t $(\in (-\infty,\infty))$.

For $d = 5$ case, the $\mathcal{N} = 4$ SYM theory on the above boundary spacetime is the corresponding dual system. Since the values of the dilaton on the two boundaries are different from each other, the corresponding SYM theories of the two boundaries now become different as a result of the deformation. The number of colors N agrees with each other while the 't Hooft couplings $\lambda_\pm = 4\pi\, e^{\phi_\pm} N$ become different as a result of deformation.

To talk about the correspondence further, one has to specify the initial density matrix of the boundary theory. Here we encounter a new situation where there is no standard prescription to find the initial density matrix. Our proposal is based on a generalization of the thermofield dynamics which is a mathematical technique to describe the real time thermodynamics.

We note first that the time dependent black hole solution allows an analytic continuation, $\tau = -i\tau_E$, leading to the Euclidean geometry,

$$ds^2_E = f(\mu)(d\mu^2 + d\tau_E^2 + \cosh^2\tau_E ds^2_\Sigma) \tag{10}$$

with the scalar field $\phi(\mu)$ unchanged. The Euclidean coordinate τ_E is ranged over $(-\infty,\infty)$. The Euclidean geometry is smooth everywhere and has a boundary. The conformal shape of the (μ,τ_E) space is disk shaped. Below we shall provide the physical interpretation of the above Euclidean geometry in terms of thermofield dynamics[7, 8, 9].

We begin with the field theory instanton and its role as an initial state. For the $4d$ Poincarè-invariant theories, their instanton solution has $O(4)$ invariance and let us take $t_E = 0$ as the fixed point of the Z_2 symmetry $t_E \to -t_E$ where t_E is the Euclidean time. Thus, at $t_E = 0$, the time derivative of fields vanish due to the Z_2 symmetry. This $t_E = 0$ field configuration can then be interpreted as an initial configuration from which the Lorentzian dynamics follows. The subsequent Lorentzian dynamics for $t \geq 0$ can be obtained from the instanton solution by analytic continuation[17]. At $t_E = t = 0$, the Lorentzian and the Euclidean configurations agree with each other and the time derivatives (velocities) of both fields vanish, which helps them join smoothly. Thus at least semi-classically we conclude that the Euclidean solution provides an initial state for the Lorentzian time evolution.

In case of geometry, this corresponds to the so called Hartle-Hawking construction of the wave function of universe[18]. In our problem we shall basically follow the proposal of Ref. [6] to construct the thermofield initial state. We patch the half of Euclidean geometry sliced at $\tau_E = 0$ to the upper half of the Lorentzian solution from $\tau = 0$. Since the geometry involves two boundaries, the corresponding Hilbert space consists of $\mathcal{H} = \mathcal{H}_+ \times \mathcal{H}_-$. Unlike the conventional thermofield formalism, the two Hamiltonians H_+ and H_- differ from each other. Following the proposal in Ref. [6], the Euclidean geometry defines a boundary Hamiltonian and allows a Euclidean boundary time evolution, which will determine the initial thermofield state by

$$|\Psi\rangle = \frac{1}{\sqrt{Z}} \sum_{mn} \langle E_m^+ | U | E_n^- \rangle \; |E_m^+\rangle \otimes |E_n^-\rangle \tag{11}$$

where Z is the normalization factor. The Euclidean evolution operator U is given by

$$U = T \exp\Big[- \int_{t_-^E}^{t_+^E} dt^E \, H(t^E) \Big] \tag{12}$$

where t^E is the boundary Euclidean time with $t_\pm^E$ denoting the two boundary times at $\tau_E = 0$. Since the two boundary Hamiltonians differ from each other, $H(t^E)$ becomes time dependent. In this respect, the above is a slight generalization of the Maldacena's proposal but this naturally follows from the fact that our Euclidean boundary Hamiltonian is now time dependent. The conformal shape of (μ, τ_E) space is depicted in Figure 2.

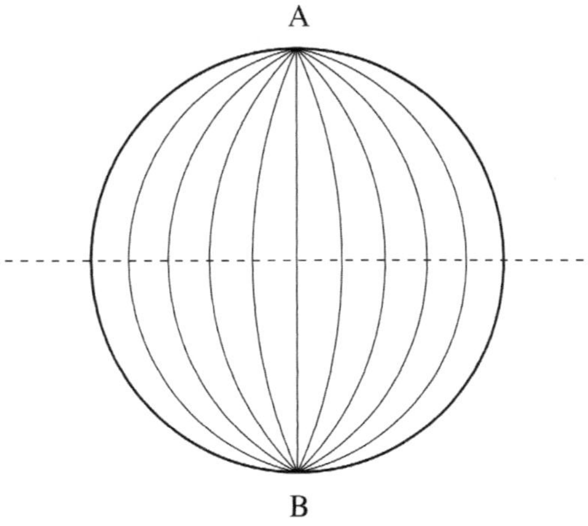

Figure 2. The conformal diagram of the Euclidean solution in $(\mu,\, \tau_E)$ space. The curves represent constant μ coordinate. On the curve, τ_E runs from $-\infty$ at the south pole to $+\infty$ at the north pole. The dotted line is the $\tau_E = 0$ line and the lower half is used to construct the thermofield initial state at $t = 0$.

This procedure can be extended to a more general case, once we restrict ourselves to a class of gravity solutions which possess a time reflection symmetry under $\tau \to -\tau$. This corresponds to the case where one has zero velocity at $\tau = 0$. By the same patching of the Euclidean solution, the thermofield state determined from the boundary evolution can serve as an initial thermofield state. Of course we assume that the resulting geometry remains asymptotically AdS at both boundaries.

The density matrix ρ_+ for the boundary system on the right hand side is given by

$$\rho_+(t) = \mathrm{tr}_- |\Psi(t,t')\rangle\langle\Psi(t,t')| \tag{13}$$

where $\text{tr}_{\pm}$ denotes the trace over the $\mathcal{H}_{\pm}$ Hilbert space. The time evolution for the thermofield state for the $+$ and $-$ boundary states are respectively by H_+ and H_- with time t and t'. If O_+ is any operator defined in the $\mathcal{H}_+$ space, the thermal expectation values are defined by

$$\langle O_+(t) \rangle = \text{tr}_+ \rho_+(t) O_+(t) \,. \tag{14}$$

This description of the system by the density matrix provides us with the single boundary view.

The density matrix ρ_+ is no longer commuting with H_+ and, hence, time dependent. Then the expectation value $\langle O_+(t) \rangle$ is in general time dependent, which is consistent with our result of the previous section for the operator $\mathcal{L}(t)$.

Using this identification of the density matrix, one can show for example the expectation values of energy E is time independent, which agrees with the gravity computation. Furthermore, using the so called conformal perturbation theory, one may show that the gravity evaluation of $\langle \mathcal{L}(t) \rangle$ can be recovered to the linear order in γ, which supports our identification of the initial density matrix[3].

3. Thermalization

The exponential decay of the one point function describes a thermalization occurring in our boundary SYM system. The initial difference of the kinetic minus potential energy densities at $t = 0$ is the finite perturbation above an equilibrium state. As the time goes by, the initial difference relaxes away exponentially reaching a new equilibrium state at $t = \infty$. Since there is no external perturbation, the total energy cannot be changing due to the energy conservation. This is a typical situation when one has a thermalization or a relaxation of finite-size perturbation. Namely, a system beginning with a non equilibrium state approaches a new equilibrium state as time goes by, whose time scale is called the relaxation time scale. From the general grounds, the relaxation time scale depends on operator causing the perturbation, the coupling and the final temperature T of the system. In conformal field theories, operators of definite scale dimension Δ are relevant from general grounds.

Recovering the temperature, one finds the thermalization time scale is given by

$$\tau_{relax} = 1/(2\pi T) \,. \tag{15}$$

This agrees with the value occurring in other studies[19, 20, 21] where it played the role of a limiting value.

The thermalization happens very fast at high enough temperatures. We are looking at the strongly coupled quark gluon plasma. So one can compare this time scale with those thermalization time scale of the plasma produced at the RHIC experiment. In a split second less than 0.6 fermi/c, the fireballs are thermalized and the hydrodynamic description works pretty well explaining the final experimental outcomes. Since the size of the fireball is about 5 fermi and the time scale of 0.6 fermi/c even does not allow any collisions in the average sense, such fast thermalization time scale has been quite puzzling. Note that the temperature keeps changing as the expansion goes on. We use the typical temperature T=300MeV to estimate the thermalization time scales, which is certainly an extreme temperature from any standard. This leads to the thermalization time scale 0.1 $fermi/c$ or 0.5 $fermi/c$ for the 99% thermalization, which explains the actual thermalization time scale pretty well.

4. Entropy and Information loss

In this section, we focus on the implication of the above result to the gravity theory and the strongly coupled SYM theory. We begin with the first law of the thermodynamics, which is given by

$$dQ = dE + p \, d\mathcal{V}_\Sigma \,, \tag{16}$$

for any thermodynamic processes. This is basically statement on the conservation of energy of system including the heat flow. For our case, we have shown $dE = 0$ and $d\mathcal{V}_\Sigma = 0$ since the volume of the compact space of the boundary is fixed in time. In addition, there is no inflow of heat as we have explained before, *i.e.* $dQ = 0$. On the other hand, we know that the entropy changes in time along the thermalization process. Therefore,

$$S_{final} > S_{initial} \,. \tag{17}$$

This implies that the relation $dQ = TdS$ does not hold along our time dependent processes. This simply implies that our system is not in any quasi-equilibrium states during the relaxation process. This is in contrast with the claim that the classical gravity may follow from the consideration of entropic force[22]. Our solution and the correspondence clearly shows that the classical gravity includes thermalization effects for which the concept of entropic force is not applicable.

Let us now turn to the discussion of the problem of entropy and the information loss. Our result on the exponential decay of the one-point function implies that the corresponding information of initial perturbation is permanently lost at the future infinity. This loss of information is of course not consistent with the unitarity in the time evolution of the system. Without any truncation, the full SYM theory defined on a compact space is of course unitary. In our case, the system involves a truncation of the planar limit where one ignores any subleading contributions in $1/N$: This is precisely what the classical supergravity is describing. Therefore we conclude that the source of the information loss is basically from the large N planar limit.

To talk about the information loss quantitatively, one needs to define entropy of the system. If it is not defined for all time, we need at least the definition of the initial entropy $S_{initial}$ and the final entropy S_{final}. This is not a simple matter. For instance one may try the following definition.

$$\mathcal{S}(t) = -\mathrm{tr}_+\, \rho_+(t) \ln \rho_+(t) \,. \tag{18}$$

But by a simple manipulation, one may show that $\mathcal{S}$ is time independent since the time evolution of the density matrix is simply given by a similarity transformation of the form $\rho_+(t) = e^{-iH_+ t} \rho_+(0)\, e^{iH_+ t}$ and the similarity transformation inside the trace has no effect. Hence above definition is not working. We do not know how to proceed for the thermal field theory, but it appears that one needs a some properly defined coarse graining to see the effect of changing entropy.

In the gravity side, one does not have any proper definition for the time-dependent entropy. The changing area along the future event horizon could be one candidate for the entropy. But the geometry has the Z_2 reflection symmetry along $\tau = t = 0$. Now if one is following the future horizon in the backward direction starting from the future infinity, the area keeps decreasing beyond the symmetric $t = 0$ point. This is not satisfactory. For this time dependent case, the apparent horizon and the event horizon do not agree with each other. Then which horizon is to be used to measure the area is not clear. There is no well defined definition of the time-dependent temperature. The usual method of the Hartle-Hawking prescription is not working for our case.

We wonder if the study of the entanglement entropy[23] for our background may help our understanding of the time dependence of the entropy. The generalization of our solutions to various cases will be quite interesting in understanding nature of thermalization in strongly coupled gauge theories. In particular, our ansatz preserves locally $SO(d-2,2)$ conformal symmetries. One obvious generalization will be the case of $SO(d-2) \times SO(2)$ symmetry where $SO(2)$ corresponds to the time translation. This leads to the non linearly coupled partial differential equations of two variables. The treatment this coupled system may be complicated due to its nonlinear nature. However, expanding in powers of the radial coordinate of S^{d-2}, the equation for the remaining variable becomes linear differential equation except the leading part whose solution is already known. Therefore a systematic study in this expansion seems feasible, which may lead to rich solutions for time dependent black holes including formation of black hole. Further study in this direction is required.

Acknowledgements

I am grateful to Sumit Das and Alfred Shapere for providing stimulating atmosphere and warm hospitality. This work is supported in part by 2009 UOS Academic Research Grant.

References

[1] J. M. Maldacena, Adv. Theor. Math. Phys. **2**, 231 (1998) [Int. J. Theor. Phys. **38**, 1113 (1999)] [arXiv:hep-th/9711200].
[2] D. Bak, M. Gutperle and S. Hirano, JHEP **0702**, 068 (2007) [arXiv:hep-th/0701108].
[3] D. Bak, M. Gutperle and A. Karch, JHEP **0712**, 034 (2007) [arXiv:0708.3691 [hep-th]].
[4] P. B. Arnold and L. G. Yaffe, Phys. Rev. D **52**, 7208 (1995) [arXiv:hep-ph/9508280].
[5] D. Bak, A. Karch and L. G. Yaffe, JHEP **0708**, 049 (2007) [arXiv:0705.0994 [hep-th]]; D. Bak, K. B. Fadafan and H. Min, arXiv:1003.5227 [hep-th].
[6] J. M. Maldacena, JHEP **0304**, 021 (2003) [arXiv:hep-th/0106112].
[7] J. S. Schwinger, J. Math. Phys. **2**, 407 (1961).
[8] L. V. Keldysh, Zh. Eksp. Teor. Fiz. **47** (1964) 1515 [Sov. Phys. JETP **20** (1965) 1018].
[9] Y. Takahasi and H. Umezawa, Collect. Phenom. **2**, 55 (1975).
[10] D. Bak, M. Gutperle and S. Hirano, JHEP **0305**, 072 (2003) [arXiv:hep-th/0304129].
[11] A. B. Clark, D. Z. Freedman, A. Karch and M. Schnabl, Phys. Rev. D **71**, 066003 (2005) [arXiv:hep-th/0407073].
[12] M. Banados, C. Teitelboim and J. Zanelli, Phys. Rev. Lett. **69**, 1849 (1992) [arXiv:hep-th/9204099].
[13] R. B. Mann, arXiv:gr-qc/9709039.
[14] J. D. E. Creighton and R. B. Mann, Phys. Rev. D **58**, 024013 (1998) [arXiv:gr-qc/9710042].
[15] D. Birmingham, Class. Quant. Grav. **16**, 1197 (1999) [arXiv:hep-th/9808032].
[16] R. Emparan, JHEP **9906**, 036 (1999) [arXiv:hep-th/9906040].
[17] C. G. . Callan and S. R. Coleman, Phys. Rev. D **16**, 1762 (1977).
[18] J. B. Hartle and S. W. Hawking, Phys. Rev. D **28**, 2960 (1983).
[19] P. K. Kovtun and A. O. Starinets, Phys. Rev. D **72**, 086009 (2005) [arXiv:hep-th/0506184].
[20] J. J. Friess, S. S. Gubser, G. Michalogiorgakis and S. S. Pufu, JHEP **0704**, 080 (2007) [arXiv:hep-th/0611005].
[21] C. P. Herzog, A. Karch, P. Kovtun, C. Kozcaz and L. G. Yaffe, JHEP **0607**, 013 (2006) [arXiv:hep-th/0605158].
[22] E. P. Verlinde, arXiv:1001.0785 [hep-th].
[23] S. Ryu and T. Takayanagi, Phys. Rev. Lett. **96**, 181602 (2006) [arXiv:hep-th/0603001]; S. Ryu and T. Takayanagi, JHEP **0608**, 045 (2006) [arXiv:hep-th/0605073].

Journal of Physics: Conference Series **462** (2013) 012003 doi:10.1088/1742-6596/462/1/012003

Some New Results in AdS_4/CFT_3 Duality[1]

Oren Bergman[a], **Shinji Hirano**[b], **Gilad Lifschytz**[c]

[a] Department of Physics, Technion, Israel Institute of Technology, Haifa 32000, Israel
[b] The Niels Bohr Institute, Blegdamsvej 17, DK-2100 Copenhagen, Denmark
[c] Department of Mathematics and Physics and CCMSC, University of Haifa at Oranim, Tivon 36006, Israel

E-mail: [a]`bergman@physics.technion.ac.il`, [b]`hirano@nbi.dk`, [c]`giladl@research.haifa.ac.il`

Abstract. In this talk I will present two new results related to the duality of AdS_4 backgrounds of M theory or Type IIA string theory and three-dimensional Chern-Simons-Matter theories. The first result is a $1/\lambda$ correction to the supergravity/CFT relation between R and λ in the original ABJM and ABJ models. This part is based on work with Shinji Hirano, that appeared in [1]. The second result concerns the generalization of the ABJM model to unequal CS levels, which has been conjectured to be dual to an AdS_4 solution of massive Type IIA supergravity. I will present a simple brane construction that supports this conjecture. This part is based on work with Gilad Lifschytz, that appeared in [2].

1. Introduction

The AdS/CFT correspondence relates quantum gravity in an Anti-de-Sitter (AdS) spacetime background to a conformal field theory in one lower dimension, which one can regard as living at the spatial boundary of the AdS spacetime. The examples that are known explicitly are derived from string theory using D-branes, which, in a specific large N and low energy limit, can simultaneously be described either in terms of the supergravity background they create, or in terms of the gauge field theory on their worldvolume. In its most symmetric form, this duality involves an AdS spacetime and a conformal field theory (CFT). The simplest example comes from D3-branes in Type IIB string theory, which lead to a duality between the maximally supersymmetric ($\mathcal{N} = 4$) Yang-Mills theory in four dimensions and the $AdS_5 \times S^5$ solution of Type IIB supergravity [3]. The parameters N and $\lambda = g_{YM}^2 N$ of the gauge theory translate, respectively, into the inverse coupling constant and the curvature radius of the gravity theory. This means that the low energy supergravity description is valid when the gauge theory is strongly coupled, and a perturbative gauge theory description is valid when the background is highly curved. This correspondence has been generalized and tested in many ways, and there is now an impressive collection of AdS_5/CFT_4 dual pairs derived from D3-branes in various non-trivial backgrounds.

The M2-brane and M5-brane of M theory also give rise to AdS backgrounds, however the corresponding CFTs cannot be derived directly, since the underlying degrees of freedom of M theory are not known. M2-branes in flat space, in particular, lead to an $AdS_4 \times S^7$ solution, and should therefore be described by a three-dimensional CFT with $\mathcal{N} = 8$ supersymmetry and

[1] Talk presented at QTS6 by Oren Bergman.

an $SO(8)$ global R symmetry. A formal definition of this CFT is provided by the reduction to Type IIA string theory; it can be identified as the IR fixed point of three-dimensional $\mathcal{N} = 8$ Super-Yang-Mills theory, corresponding to the worldvolume theory on a collection of D2-branes. However, the fixed point is strongly coupled, and this does not lead to a Lagrangian description. Faced with the failure to derive the required CFT, an obvious alternative is to guess its form. In [4], an attempt was made using an $SU(N)$ gauge field with a pure Chern-Simons (CS) action together with matter fields. The CS action is classically conformal invariant, and with a suitable matter content could be made into a superconformal theory. This did not quite work, since the maximal supersymmetry in this case is $\mathcal{N} = 3$, but it pointed in the right direction. The first breakthrough was made by Bagger and Lambert, who presented an explicit $\mathcal{N} = 8$ CFT which had the correct degrees of freedom to describe M2-branes [5]. Initially this model was formulated as a new kind of gauge theory based on a 3-algebra, but subsequently it was reformulated as an ordinary CS gauge theory with an $SU(2) \times SU(2)$ gauge group with opposite CS coefficients k and $-k$, and with matter in the bi-fundamental representation [6]. This theory was conjectured to describe two M2-branes in certain orbifold backgrounds of M theory [7]. However, it left a number of questions open, most importantly how to describe M2-branes in flat space, and how to generalize to N M2-branes, and thereby make contact with the supergravity dual.

These questions were answered by the ABJM model [8]. This model describes a collection of N coincident M2-branes in M theory at the singularity of an orbifold $\mathbb{C}^4/\mathbb{Z}_k$. The field theory in this case is an $\mathcal{N} = 6$ $U(N)_k \times U(N)_{-k}$ CS theory, with matter consisting of two hyper-multiplets in the bi-fundamental representation. At large N, the effective coupling constant is the 't Hooft coupling $\lambda = N/k$, so the field theory is weakly coupled if $k \gg N$. If $k \ll N$, there is a weakly curved dual supergravity description, given either by M theory on $AdS_4 \times S^7/\mathbb{Z}_k$ with N units of 4-form flux in AdS_4 if $k \ll N^{1/5}$, or by Type IIA string theory on $AdS_4 \times \mathbb{C}P^3$ with N units of RR 4-form flux in AdS_4 and k units of RR 2-form flux in $\mathbb{C}P^1 \subset \mathbb{C}P^3$ if $k \gg N^{1/5}$. For $k = 1$ this theory describes M2-branes in flat space, with the M theory dual $AdS_4 \times S^7$. In this case (and also for $k = 2$) the supersymmetry is enhanced to $\mathcal{N} = 8$. However, both the field theory and the Type IIA dual are strongly coupled in this case, and the enhancement of the supersymmetry is due to non-perturbative states [8, 9].

A simple extension of this model can be made by considering gauge groups with unequal ranks $U(N + l)_k \times U(N)_{-k}$, corresponding to adding l "fractional" M2-branes [10]. For $l \leq k$, the theory remains an $\mathcal{N} = 6$ CFT, with the same dual geometry, but with an additional flux turned on. In the M theory picture this is described by "discrete torsion", which is just a discrete holonomy of the C field on the torsion 3-cycle $S^3/\mathbb{Z}_k \subset S^7/\mathbb{Z}_k$, $\oint C_3/(2\pi) = l/k$. In the Type IIA description, this becomes a B field holonomy on $\mathbb{C}P^1 \subset \mathbb{C}P^3$, $b = \oint B_2/(2\pi)^2 = l/k$. This also implies an additional RR flux $\int_{\mathbb{C}P^2} F_4/(2\pi) = l$, corresponding to l D4-branes wrapped on the 2-cycle $\mathbb{C}P^1 \subset \mathbb{C}P^3$.

There has been a considerable amount of activity generalizing this model to theories with less supersymmetry, and there are now many explicit examples of AdS_4/CFT_3 dual pairs. In this talk I will describe two new results related to AdS_4/CFT_3 duality. The first is a correction to the duality relation due to higher curvature effects. The second result is related to the deformation of the model to unequal CS levels, which has been conjectured to be dual to a massive Type IIA supergravity AdS_4 background. I will describe a brane configuration that supports this conjecture, and show explicitly how it leads to different CS levels for the two gauge groups.

2. Anomalous corrections in the ABJ(M) model

In the $AdS_5 \times S^5$ case, the duality dictionary is exact to all orders in g_s and α' [11]. In particular, the relation between the parameters $R^4_{IIB}/(\alpha')^2 = 4\pi\lambda$ does not receive corrections beyond the low energy classical supergravity approximation. However, as we will now demonstrate, the analogous relation for $AdS_4 \times S^7/\mathbb{Z}_k$, $R^4_{IIA}/(\alpha')^2 = R^6_M/(kl^6_M) = 32\pi^2\lambda$, receives corrections due to higher curvature and flux couplings. In particular, these corrections become relevant at two loops in the string worldsheet sigma model, and are important for the strong coupling test of the all loop Bethe ansatz proposed in [12].

2.1. M theory description

In the M theory description, these corrections correspond to additional contributions to the M2-brane charge from the gravitational anomaly term and the Chern-Simons term in the low energy eleven dimensional supergravity theory,

$$S_{11} = \frac{1}{2\kappa_{11}^2}\left[\int d^{11}x\sqrt{-G}\left(R - \frac{1}{2}|G_4|^2\right) - \frac{1}{6}\int C_3 \wedge G_4 \wedge G_4 + (2\pi)^2\int C_3 \wedge I_8\right]. \quad (1)$$

The 8-form anomaly polynomial I_8 is related to the Euler class of the space transverse to the M2-brane, such that $\int I_8 = -\chi_{bulk}/24$, where χ_{bulk} is the contribution to the Euler characteristic from the bulk [13]. There are therefore three potential contributions to the M2-brane charge [14],

$$Q_{M2} = N - \frac{1}{2(2\pi)^2}\int_{\mathcal{M}_8} G_4 \wedge G_4 - \frac{\chi_{bulk}}{24}, \quad (2)$$

corresponding, respectively, to sources, flux, and the geometry of the transverse space.

Our transverse space is $\mathbb{C}^4/\mathbb{Z}_k$, which has a total Euler characteristic $\chi(\mathbb{C}^4/\mathbb{Z}_k) = k$ [15]. This has a bulk contribution from the singularity, and a contribution from the boundary $S^7/\mathbb{Z}_k$. The boundary contribution is easily computed by realizing that $\mathbb{Z}_k$ acts freely on S^7, which is the boundary of $\mathbb{C}^4$, and by the fact that $\chi_{bnd}(\mathbb{C}^4) = \chi(\mathbb{C}^4) = \chi(point) = 1$. The $\mathbb{Z}_k$ action gives $\chi_{bnd}(\mathbb{C}^4/\mathbb{Z}_k) = 1/k$ and therefore $\chi_{bulk}(\mathbb{C}^4/\mathbb{Z}_k) = k - 1/k$. The fixed point of the orbifold therefore carries an M2-brane charge given by

$$Q_{M2}(\mathbb{C}^4/\mathbb{Z}_k) = -\frac{\chi_{bulk}}{24} = -\frac{1}{24}\left(k - \frac{1}{k}\right). \quad (3)$$

This generalizes the known result for the so-called $OM2^-$ plane, $Q_{M2}(\mathbb{R}^8/\mathbb{Z}_2) = -1/16$ [16]. In this case there is an additional consistency check which comes from compactifying one of the coordinates of the $\mathbb{R}^8$, and reducing to Type IIA string theory. There are two $OM2^-$ planes in this case, that become a single orientifold plane $O2^-$ in Type IIA string theory. The D2-brane charge of the orientifold plane can be computed independently using string theory, and the result is $-1/8$, percisely twice the charge of the $OM2^-$ plane. There is no analogous simple Type IIA reduction for $k > 2$.

To compute the flux contribution in (2) we need to find a representative of the torsion class of G_4. We can do this by generalizing the construction of [16]. Start with the bundle $\mathcal{O}(-k)$ (the k-fold tensor product of the natural line bundle $\mathcal{O}(-1)$) over $\mathbb{C}P^3$. Taking a disk $|w| \leq 1$ in each fiber defines a smooth 8-dimensional manifold $\mathcal{M}$, whose boundary is $S^7/\mathbb{Z}_k$. The M2-brane charge can therefore be evaluated as

$$Q_{M2}(\text{flux}) = -\frac{1}{2}\int_{\mathcal{M}} \frac{G_4}{2\pi} \wedge \frac{G_4}{2\pi}. \quad (4)$$

The torsion class is represented by a 4-dimensional submanifold $\mathcal{W} \subset \mathcal{M}$, whose boundary is $S^3/\mathbb{Z}_k \subset S^7/\mathbb{Z}_k$:

$$\int_{\mathcal{W}} \frac{G_4}{2\pi} = \int_{S^3/\mathbb{Z}_k} \frac{C_3}{2\pi} = \frac{l}{k}. \quad (5)$$

Journal of Physics: Conference Series **462** (2013) 012003 doi:10.1088/1742-6596/462/1/012003

A representative for the class of G_4 may then be constructed using the Poincare dual of the base $\mathbb{C}P^3$, which is a 2-form X satisfying

$$\int_{\mathcal{W}} X \wedge X = -k \,. \tag{6}$$

We can therefore identify $G_4/(2\pi) = -(l/k^2) X \wedge X$, and the the M2-brane charge is given by

$$Q_{M2}(\text{flux}) = -\frac{l^2}{2k^4} \int_{\mathcal{M}} X \wedge X \wedge X \wedge X = \frac{l^2}{2k} \,. \tag{7}$$

This again generalizes the $OM2$ plane case. In that case there is only one choice of discrete torsion $l = 1$, corresponding to an $OM2^+$ plane which carries an M2-brane charge of $3/16$. Upon compactifying one of the $\mathbb{R}^8$ directions one then obtains one of the four variants of orientifold 2-planes in Type IIA string theory, $O2^-$, $O2^+$, $\widetilde{O2}^-$, or $\widetilde{O2}^+$, depending on which $OM2$-planes are placed at opposite points on the circle. Again, a similar consistency check by reduction to ten dimensions cannot be made for $k > 2$.

Adding the two corrections, the total M2-brane charge in the orbifold background is given by

$$Q_{M2} = N - \frac{1}{24}\left(k - \frac{1}{k}\right) + \frac{l^2}{2k} \,. \tag{8}$$

In the near-horizon M theory dual description $R_M^6/l_M^6 = 32\pi^2 Q_{M2}$, and therefore

$$\frac{R_{IIA}^4}{(\alpha')^2} = \frac{R_M^6}{k l_M^6} = 32\pi^2 \left(\lambda - \frac{1}{24}\left(1 - \frac{1}{k^2}\right) + \frac{l^2}{2k^2}\right) . \tag{9}$$

Note that the correction vanishes for $k = 1$, as it should. Note also that the correction is $\mathcal{O}(1/\lambda)$ relative to the classical low-energy supergravity result, which means it corresponds to a two-loop worldsheet sigma model effect.

2.2. Type IIA description and a puzzle

In the Type IIA description, the above correction corresponds to a shift in the D2-brane charge. We can compute this shift by considering the effects of other branes added to the near-horizon $AdS_4 \times \mathbb{C}P^3$ background. Specifically, we consider a D$(2n)$-brane which forms a domain wall in AdS_4 and wraps $\mathbb{C}P^{n-1} \subset \mathbb{C}P^3$, with $n = 1, 2$ and 3. On the one hand, each such brane produces a jump in one of the field theory parameters N, l or k, and on the other hand, each can carry charges of lower dimensional branes due to its worldvolume CS and curvature couplings. By comparing these two effects one can derive a map between the field theory parameters and the D-brane charges. More precisely, this relates the field theory parameters to the D-brane *Page* charges, which are defined by the modified RR field strengths $\hat{F} = \tilde{F} \wedge e^{-B_2}$, where $\tilde{F}$ are the gauge-invariant field strengths. Since the modified field strengths satisfy the ordinary Bianchi identity, $d\hat{F} = 0$, Page charge is conserved and quantized, although it is not gauge invariant. However, for the domain wall D-branes, the Page charge is well-defined up to large gauge transformations that shift the B-field holonomy b by an integer. The explicit map is [17][2]:

$$Q_{D2}^P = N + \frac{k}{12} \;,\quad Q_{D4}^P = l - \frac{k}{2} \;,\quad Q_{D6}^P = k \,. \tag{10}$$

[2] This was generalized in [2] to include D8-branes, and the associated parameter q corresponding to the difference between the two CS levels. This parameter will be discussed in the next section.

In particular, the shift in the D4-brane charge relative to l comes from the CS term on a D6-brane which wraps $\mathbb{C}P^2$. The quantization of the worldvolume magnetic flux is shifted by $1/2$ in this case due to an anomaly associated with non-spin manifolds [18]. Therefore each D6-brane domain wall, which shifts k by one, also changes the D4-brane charge by $1/2$. The shift in the D2-brane charge is a sum of the contribution from a CS term and a curvature term on the D6-brane domain wall.

While the Page charges are naturally related to the integer field theory parameters, the quantity that is naturally related to the curvature radius R_{IIA} is the Maxwell D2-brane charge, which is defined by the gauge invariant field strength $\tilde{F}_6$. This is related to the Page charges by

$$Q^M_{D2} = Q^P_{D2} + bQ^P_{D4} + \frac{1}{2}b^2 Q^P_{D6} \,. \tag{11}$$

The B-field holonomy b is given by

$$b = -\frac{l}{k} + \frac{1}{2} \,, \tag{12}$$

where the $1/2$ shift relative to the value claimed in [10] is again due to the anomaly of [18], this time applied to a D4-brane wrapped on $\mathbb{C}P^2$ [17]. The $1/2$ shift in the worldvolume flux must be supplemented by a $1/2$ shift in b to ensure an integer tadpole, which can be cancelled by including an integer number of strings. The sign convention for the B-field is also different here. Substituting (10) into (11) then gives

$$Q^M_{D2} = N + \frac{k}{12} - \frac{\left(l - \frac{k}{2}\right)^2}{2k} \,. \tag{13}$$

Note that this result is only valid at large k, where the weakly coupled Type IIA description holds. In particular, there could be an $\mathcal{O}(1/k)$ correction. Indeed, to get the known results for $k = 1$ and $k = 2$ requires adding an amount $1/(24k)$. We are led to conjecture that the total D2-brane Maxwell charge is

$$Q^M_{D2} = N - \frac{1}{24}\left(k - \frac{1}{k}\right) - \frac{l(l-k)}{2k} \,. \tag{14}$$

However, this raises a puzzle. Although both (8) and (14) reproduce the same known results for $k = 1$ and $k = 2$, they differ for $k > 2$. The resolution of this discrepency requires more work, but there is an indication that something is missing in the M theory calculation. Namely, the shifted quantization of Page charge for D4-branes wrapping $\mathbb{C}P^1$ requires a similar shift in the quantization of charge for M5-branes wrapping $S^3/\mathbb{Z}_k$. However, the mechanism for the shift must be different here, since the anomaly of [18] doesn't apply. We would like to argue that the M5-brane charge quantization is shifted due to an M2-brane parity anomaly, of the type described in [19]. In the usual argument for Dirac quantization we consider a Euclidean M2-brane wrapped on an S^3 in the S^4 surrounding an M5-brane. The M2-brane path integral acquires a phase $\exp(i\int_{S^3} C_3)$, which is single valued only if $\exp(i\int_{S^4} G_4) = 1$. The period of G_4 would therefore have to be an integer multiple of 2π. However, there is another phase factor coming from the path integral over the worldvolume fermions of the M2-brane. This is the parity anomaly. The condition of single-valuedness is shifted to

$$(-1)^{P_1(N(S^4))/2} e^{i\int_{S^4} G_4} = 1 \,, \tag{15}$$

where $N(S^4)$ is the normal bundle to S^4 in the total space. In our case the relevant part of the space is $S^7/\mathbb{Z}_k$, which can be described as an $SU(2)/\mathbb{Z}_k$ instanton bundle over S^4. Therefore $P_1(N(S^4)) = -2k$, and we obtain the required M5-brane charge shift by $-k/2$. This explains the shift from l to $l - k$ in comparing (8) with (14), but it does not completely resolve their discrepancy.

3. Branes and massive Type IIA duals

Massive Type IIA supergravity is a variant of Type IIA supergravity, in which the NSNS 2-form eats the RR 1-form and becomes massive [20]. This also gives rise to a cosmological constant given by the mass-squared. This theory was subsequently interpreted in terms of a background RR 0-form field strength F_0, which is associated with D8-branes in Type IIA string theory [21]. However it is not yet known whether D8-branes lift to M theory, and therefore whether massive Type IIA supergravity is part of M theory. On the other hand, a variety of supersymmetric $AdS_4 \times M_6$ solutions with $F_0 \neq 0$ are known [22, 23, 24], and this raises the question of what their dual 3d CFTs are.

In a number of simple cases, corresponding to deformations of the $\mathcal{N}=6$ solution with $F_0=0$, it was argued that the dual CFTs are deformations of the ABJM model to $U(N)_{k_1} \times U(N)_{k_2}$ Chern-Simons-Matter (CSM) theories with $k_1 + k_2 = F_0$, and additional superpotential terms for the matter fields [25]. The simplest example is a non-supersymmetric deformation of the field theory that preserves the $SO(6)$ global symmetry, in which only the sum of the CS levels is changed. This $\mathcal{N}=0$ CFT was conjectured to be dual to a non-supersymmetric $AdS_4 \times \mathbb{C}P^3$ solution of massive Type IIA supergravity with the $SO(6)$ invariant metric on $\mathbb{C}P^3$. The main evidence for this conjecture comes from considering the properties of D-branes in the $F_0 \neq 0$ background. For example, tadpole cancellation on the D0-brane requires F_0 strings to end on it, which agrees with the fact that the dual di-monopole operator of the field theory has extra gauge indices when $k_1 \neq -k_2$, which must be saturated by $|k_1 + k_2|$ semi-infinite Wilson lines.

In principle, the condition $k_1 + k_2 = F_0$ can be satisfied in many ways. However there should be a unique field theory dual to a given background. One of the questions we would like to address is how precisely the F_0 background affects each of the two CS levels. To answer this question, and also to gain more insight into the AdS_4/CFT_3 duality in the massive IIA case, we will construct a brane realization of these theories.

We start with Type IIA string theory in the $\mathcal{N}=6$ $AdS_4 \times \mathbb{C}P^3$ background. We then deform this background by nucleating a D8-brane that fills AdS_4, and wraps a 5-dimensional cycle in $\mathbb{C}P^3$. We can do this because the cycle is topologically trivial, and therefore the D8-brane carries no conserved charge. On the other hand it forms a domain wall in $\mathbb{C}P^3$, across which F_0 changes by one unit (Fig. 1a). Since this configuration breaks all the supersymmetry, the deformation is not flat, but we can still do it. Now imagine pulling the D8-brane all the way from one pole of the $\mathbb{C}P^3$, where it has a vanishing size, to the other pole, where it again has a vanishing size. This deforms the original $\mathcal{N}=6$ solution with $F_0=0$, to a new $\mathcal{N}=0$ solution with $F_0=1$. We can repeat the process and produce backgrounds with higher values of F_0. Furthermore, as long we keep to a small number of D8-branes relative to N and k, we can use the probe approximation, and the background metric remains the same. Namely, the $SO(6)$ symmetry is preserved, and the solution we obtain is the $\mathcal{N}=0$ $AdS_4 \times \mathbb{C}P^3$ solution.

Now let us turn to Type IIB string theory, and describe the T-dual brane configuration. The brane configuration used in the ABJM model consists of an NS5-brane, a $(1,k)$5-brane (a bound state of an NS5-brane and k D5-branes), and a number of D3-branes arranged as follows [26]:

	0	1	2	3	4	5	6	7	8	9
NS5	•	•	•	•	•	•				
$(1,k)5$	•	•	•	$\cos\theta$	$\cos\theta$	$\cos\theta$		$\sin\theta$	$\sin\theta$	$\sin\theta$
D3	•	•	•				•			

The angle θ is the relative orientation of the two 5-branes in the 3-7, 4-8 and 5-9 planes, and is related to k as $\tan\theta = k$ (for $g_s = 1$ and $C_0 = 0$). The coordinate x^6 is compact, and the D3-branes can either wind around it, or be suspended between the two 5-branes. This describes a three-dimensional $\mathcal{N}=3$ Yang-Mills-Chern-Simons theory with a gauge group $U(N+l)_k \times U(N)_{-k}$, and two bi-fundamental hyper-multiplets. The two ranks $N+l, N$ are given by the number of D3-branes on either side of the circle separated by the two 5-branes. For $l \leq k$ this theory flows in the IR to the $\mathcal{N} = 6$ superconformal CSM theory with the same gauge group.

The object dual to the D8-brane described above is a D7-brane, which is oriented as follows:

	0	1	2	3	4	5	6	7	8	9
D7	•	•	•	•	•			•	•	•

The new configuration breaks supersymmetry, and the D7-brane is repelled from the D3-branes in the x^5 direction. The position of the D7-brane in x^5 corresponds to the position of the D8-brane in the $\mathbb{C}P^3$, and the instability of the D3-D7 configuration corresponds to a "slipping" mode of the D8-brane. The D8-brane deformation described above corresponds to moving the D7-brane from $x^5 \to -\infty$ to $x^5 \to +\infty$ across the D3-branes (Fig. 1b). The key property of the D7-brane is that it sources a monodromy for the RR scalar potential in the 5-6 plane $C_0 \to C_0 + 2\pi$, which one can regard as occuring across a branch cut emanating from the D7-brane [27]. As the D7-brane is taken to $x^5 \to \infty$ we are left with a piecewise constant C_0 background that jumps by 2π across the cut. This leads to an additional CS term on the D3-brane that the cut intersects, that comes from the 4d RR coupling:

$$\int_{R^{1,2}} \int_{x^6} C_0 \mathrm{Tr}\,(F \wedge F) = 2\pi S_{CS}\,. \tag{16}$$

Depending on the position of the D7-brane on the circle, the theory will be either $U(N+l)_k \times U(N)_{-k+1}$ or $U(N+l)_{k+1} \times U(N)_{-k}$. More generally, for q D7-branes the resulting theory is $U(N+l)_{k_1} \times U(N)_{k_2}$, with $k_1 + k_2 = q$. There are $q+1$ distinct theories, corresponding to the different ways of distributing the D7-branes between the two halves of the circle. Alternatively, we can describe the distinct theories in terms of configurations with all the D7-branes on the same side, but with additional D5-branes attached. Start with the configuration describing $U(N+l)_k \times U(N)_{-k+1}$ (Fig. 1b), and move the D7-brane across the NS5-brane. This leads to the creation of a D5-brane [28], and therefore changes the part of the NS5-brane below the D7-brane to a $(1,1)$5-brane (Fig. 1c). The resulting theory is $U(N+l)_{(k-1)+1} \times U(N)_{-(k-1)}$, and is identical to the original theory. We can therefore put all the D7-branes on one side, and describe the different configurations by the number of D5-branes. In the Type IIA picture these different configurations correspond to D8-branes with additional D6-branes ending on them. We will fix our convention by identifying the D8-brane embedding without attached D6-branes with the D7-brane to the *left* of the NS5-brane without attached D5-branes. This then tells us that the background with $F_0 = q$ and $F_2 = k$ is dual to the unique CFT with $U(N+l)_k \times U(N)_{-k+q}$.

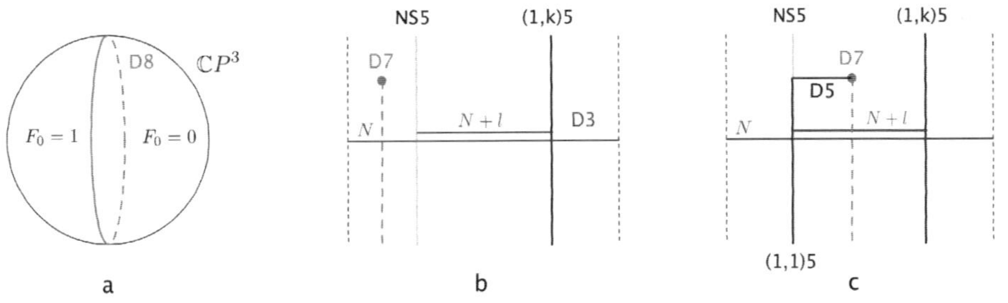

Figure 1. The D8-brane deformation and the dual Type IIB brane configuration.

Acknowledgments
OB thanks the organizers of the QTS6 conference for the opportunity to present this work. OB and GL also thank the Aspen Center for Physics, where some of this work was done. The work of OB and GL is supported in part by the Israel Science Foundation under grant no. 392/09. OB is also supported in part the US-Israel Binational Science Foundation under grant no. 2008-072.

References
[1] O. Bergman and S. Hirano, JHEP **0907**, 016 (2009) [arXiv:0902.1743 [hep-th]].
[2] O. Bergman and G. Lifschytz, arXiv:1001.0394 [hep-th].
[3] J. M. Maldacena, Adv. Theor. Math. Phys. **2**, 231 (1998) [Int. J. Theor. Phys. **38**, 1113 (1999)] [arXiv:hep-th/9711200].
[4] J. H. Schwarz, JHEP **0411**, 078 (2004) [arXiv:hep-th/0411077].
[5] J. Bagger and N. Lambert, Phys. Rev. D **77**, 065008 (2008) [arXiv:0711.0955 [hep-th]].
[6] M. Van Raamsdonk, JHEP **0805**, 105 (2008) [arXiv:0803.3803 [hep-th]].
[7] N. Lambert and D. Tong, Phys. Rev. Lett. **101**, 041602 (2008) [arXiv:0804.1114 [hep-th]]; J. Distler, S. Mukhi, C. Papageorgakis and M. Van Raamsdonk, JHEP **0805**, 038 (2008) [arXiv:0804.1256 [hep-th]].
[8] O. Aharony, O. Bergman, D. L. Jafferis and J. Maldacena, JHEP **0810**, 091 (2008) [arXiv:0806.1218 [hep-th]].
[9] A. Gustavsson and S. J. Rey, arXiv:0906.3568 [hep-th]; O. K. Kwon, P. Oh and J. Sohn, JHEP **0908**, 093 (2009) [arXiv:0906.4333 [hep-th]].
[10] O. Aharony, O. Bergman and D. L. Jafferis, JHEP **0811**, 043 (2008) [arXiv:0807.4924 [hep-th]].
[11] R. R. Metsaev and A. A. Tseytlin, "Type IIB superstring action in AdS(5) x S(5) background," Nucl. Phys. B **533**, 109 (1998) [arXiv:hep-th/9805028].
[12]
[12] N. Gromov and P. Vieira, JHEP **0901**, 016 (2009) [arXiv:0807.0777 [hep-th]]; C. Ahn and R. I. Nepomechie, "N=6 super Chern-Simons theory S-matrix and all-loop Bethe ansatz equations," JHEP **0809**, 010 (2008) [arXiv:0807.1924 [hep-th]];
[13] L. Alvarez-Gaume and E. Witten, "Gravitational Anomalies," Nucl. Phys. B **234**, 269 (1984).
[14]
[14] S. Sethi, C. Vafa and E. Witten, Nucl. Phys. B **480**, 213 (1996) [arXiv:hep-th/9606122].
[15] K. Mohri, "D-branes and quotient singularities of Calabi-Yau fourfolds," Nucl. Phys. B **521**, 161 (1998) [arXiv:hep-th/9707012].
[16] S. Sethi, "A relation between N = 8 gauge theories in three dimensions," JHEP **9811**, 003 (1998) [arXiv:hep-th/9809162].
[17] O. Aharony, A. Hashimoto, S. Hirano and P. Ouyang, arXiv:0906.2390 [hep-th].
[18] D. S. Freed and E. Witten, arXiv:hep-th/9907189.
[19] E. Witten, J. Geom. Phys. **22**, 1 (1997) [arXiv:hep-th/9609122].
[20] L. J. Romans, Phys. Lett. B **169**, 374 (1986).
[21] J. Polchinski, Phys. Rev. Lett. **75**, 4724 (1995) [arXiv:hep-th/9510017].
[22] K. Behrndt and M. Cvetic, Phys. Rev. Lett. **95**, 021601 (2005) [arXiv:hep-th/0403049]; K. Behrndt and M. Cvetic, Nucl. Phys. B **708**, 45 (2005) [arXiv:hep-th/0407263].
[23] A. Tomasiello, Phys. Rev. D **78**, 046007 (2008) [arXiv:0712.1396 [hep-th]].
[24] D. Lust and D. Tsimpis, JHEP **0502**, 027 (2005) [arXiv:hep-th/0412250]; J. Bovy, D. Lust and D. Tsimpis, JHEP **0508**, 056 (2005) [arXiv:hep-th/0506160]; C. Kounnas, D. Lust, P. M. Petropoulos and D. Tsimpis, JHEP **0709**, 051 (2007) [arXiv:0707.4270 [hep-th]]; P. Koerber, D. Lust and D. Tsimpis, JHEP **0807**, 017 (2008) [arXiv:0804.0614 [hep-th]]; C. Caviezel, P. Koerber, S. Kors, D. Lust, D. Tsimpis and M. Zagermann, Class. Quant. Grav. **26**, 025014 (2009) [arXiv:0806.3458 [hep-th]]; D. Lust and D. Tsimpis, JHEP **0904**, 111 (2009) [arXiv:0901.4474 [hep-th]].
[25] D. Gaiotto and A. Tomasiello, JHEP **1001**, 015 (2010) [arXiv:0901.0969 [hep-th]].
[26] T. Kitao, K. Ohta and N. Ohta, Nucl. Phys. B **539**, 79 (1999) [arXiv:hep-th/9808111]; O. Bergman, A. Hanany, A. Karch and B. Kol, JHEP **9910**, 036 (1999) [arXiv:hep-th/9908075].
[27] B. R. Greene, A. D. Shapere, C. Vafa and S. T. Yau, Nucl. Phys. B **337**, 1 (1990).
[28] A. Hanany and E. Witten, Nucl. Phys. B **492**, 152 (1997) [arXiv:hep-th/9611230].
[29] E. Witten, arXiv:hep-th/9903005.

Dimensional confinement and Yang-Baxter integrability in cold-atom systems via optical lattices

C. J. Bolech[1,3] and P. Kakashvili[2,3]

[1]Department of Physics, University of Cincinnati, Ohio 45221-0011, USA
[2]Nordic Institute for Theoretical Physics (NORDITA), 106 91 Stockholm, Sweden
[3]Physics & Astronomy Department, Rice University, Houston, Texas 77005, USA

Abstract. Using two-dimensional optical lattices, it has recently become possible to split up clouds of ultra cold atoms into collections of parallel one-dimensional tubes in which the dynamics is restricted to the direction along their axis. Since the Feshbach-resonance interactions among atoms are short ranged, they can be approximated as contact interactions. It is thus possible to realize optical-trap versions of some classic integrable systems first studied by Lieb, Gaudin, Yang and others. Such systems can display exotic physics like spin-charge separation and unusual paired states. We focus on the study of systems of fermions in one-dimension. At finite temperatures, regimes such as the spin-coherent and the spin-incoherent Luttinger liquids can be realized by tuning the inter-atomic interaction strength and trap parameters. We identify the noise correlation of density fluctuations as a robust observable to discriminate between these two regimes. We also address the effects of spin imbalance for the case of attractive interactions, and study the one-dimensional analog of spatially modulated (FFLO) superconducting states. Our focus is on how the temperature affects the density profiles that are being measured in the experiments. The theoretical study of temperature effects in these systems is of experimental relevance for the problem of thermometry.

1. Introduction

Recent developments in atomic and optical physics allow experimentalists to trap and cool down clouds of atomic gases beyond the limits of quantum degeneracy. This capability combined with the use of Feshbach resonances [1] and optical lattices [2], allows to engineer systems in which the interactions and the dimensionality are controllable. Thus, new opportunities open up to realize experimentally model systems with interesting quantum dynamics. One such class of systems of particular importance are integrable one-dimensional models. These models are special because an infinite number of conserved charges (or symmetries) allows to find closed non-perturbative solutions with non-trivial ground states and excitations.

The experiments that achieve the dimensional reduction to 1d, rely on multiple laser beams used to create a 2d optical lattice, say in the x-y plane, resulting in a configuration consisting of a square array of tubes as depicted in Fig. 1. Each tube constitutes an elongated quasi-1d trap if the tunneling between the tubes is negligible and the transverse trapping frequency is large compared to all other scales (including the atom-atom interactions). Further, if the transverse confinement is strong, then a single transverse band can be considered (*i.e.* no extra band quantum numbers are required). Such controlled 1d-trap arrays provided the first successful demonstrations of the 1d interacting boson gas or Tonks-Girardeau gas [3, 4] and played a key

Figure 1. Sketch of the 3d cigar-shaped cloud (light blue halo) and the array of 1d tubes formed by a 2d optical lattice potential.

role in the early theoretical proposals for the realization of the fermionic counterparts using cold-atom gases [5, 6].

In our work, we addressed in-depth the case of fermioninc 1d interacting gases. These are also integrable systems and were first discussed as such by Lieb, Gaudin, Yang and others [7, 8]. The main complication that arises in connecting the idealized theories with the realities of the cold-atom systems is due to the presence of the confining (or trapping) potential. The resulting inhomogeneous realizations of the models are no longer integrable and have to be treated in an approximate way that exploits the detailed knowledge available for the homogeneous cases.

The most studied example is the case of two-component fermions with repulsive interactions. At large densities and in the low-energy limit, the homogeneous system is usually referred to as *Luttinger liquid* and can be conveniently studied using bosonization. We developed a scheme to study the inhomogeneous system by combining a non-linear conformal mapping and finite-size bosonization [9, 10]. Our results stress the qualitative differences that arise when the fermion densities are not large and the system can be considered, instead, a realization of the so called *spin-incoherent Luttinger liquid* [11] (in our context, spin refers to the extra label used to distinguish between the two types of fermions – in practice, two different hyperfine states of the atoms). We showed that correlations of density fluctuations are able to qualitatively distinguish between different regimes and provide an ideal probe in the cold-atom context. While the off-diagonal correlators are positive for the (usual) spin-coherent Luttinger liquid, they are negative for its spin-incoherent counterpart. A result that does not depend on the details of the trap and should be easy to access experimentally.

One of the notable aspects of the experiments with cold atoms is that one can, unlike for other systems, turn the interactions to be attractive, rather than repulsive. For 1d systems, this is interesting since it allows to realize the one-dimensional analog of superfluid/superconducting phases. Combined with the additional experimental ability of controlling at will the population balance of the two fermionic species, one can realize exotic polarized superfluid phases. One such state is the Ferrell-Fulde-Larkin-Ovchinikov state (FFLO) in which fermions pair up with non-zero center-of-mass momentum [12, 13]. This modified superconducting states are not seen in usual materials, but are thought to be relevant in the context of heavy fermions [14] and organic conductors [15]. Not only that, they are also probably crucial to the understanding of the expected superfluid states of nuclear and quark matter [16].

In the 1d case, the scattering is factorizable in the Yang-Baxter sense and these models can be studied using Bethe Ansatz. A number of recent papers have looked at the homogeneous case [17], (there exists also some bosonization-based calculations [18]), but only numerical studies of the inhomogeneous case were available [19, 20]. In our presentation during the symposium, we reported about our recent contribution in which we extended the integrability-based calculations to the inhomogeneous case, and did so at finite temperatures using the so called *Thermodynamic Bethe Ansatz* (TBA) equations [21].

2. Model

We start from the Hamiltonian of a single 1d tube in a harmonic trap

$$H = -\frac{\hbar^2}{2m}\sum_i \frac{\partial^2}{\partial z_i^2} + g_{1\mathrm{d}}\sum_{i<j}\delta(z_i - z_j) + \frac{m\omega_z^2}{2}\sum_i z_i^2, \tag{1}$$

where m is the mass of the atoms, ω_z is the trap frequency and $g_{1\mathrm{d}} = \frac{2\hbar^2 a_{3\mathrm{d}}}{m a_\perp^2}\frac{1}{1 - C a_{3\mathrm{d}}/a_\perp}$ [with $C = |\zeta(1/2)|/\sqrt{2} \simeq 1.0326$] is the effective interaction strength between particles, which can be tuned using a Feshbach resonance. The parameter $a_{3\mathrm{d}}$ is the 3d scattering length and $a_\perp$ is the transverse oscillator length. If we measure lengths in units of the harmonic oscillator length $(\hbar/m\omega_z)^{1/2}$ and energies in units of $\hbar\omega_z/2$, the resulting dimensionless Hamiltonian is

$$H = -\sum \frac{\partial^2}{\partial z_i^2} - 4\Delta\sum \delta(z_i - z_j) + \sum z_i^2, \tag{2}$$

where $\Delta = -\frac{g_{1\mathrm{D}}}{2\hbar\omega_z}\sqrt{\frac{m\omega_z}{\hbar}} > 0$ (attractive interactions). We start from the solution of the uniform system and take the effects of the trap into account via a local density, Thomas-Fermi, approximation (widely used in studying trapped gases, though its violation has been observed in highly elongated 3d clouds [22] due to interface interaction effects [23, 24]). For the case of attractive fermions in 1d, one has smooth crossovers and comparisons with Quantum Monte Carlo calculations [19] indicate the Thomas-Fermi scheme gives good results for density profiles as soon as the densities are not too small.

3. TBA analysis and results

The corresponding TBA equations are given by:

$$\begin{aligned}
&f_u - \bar{f}_u = G * f_b - G * f_{s1},\\
&f_b - \bar{f}_b = 2[k^2 - \Delta^2 - \mu(z)]/T + K_1 * \bar{f}_u + K_2 * \bar{f}_b,\\
&f_{sn} - \bar{f}_{sn} = \delta_{n,1} G * \bar{f}_u + G * (f_{sn+1} + \hat{\delta}_{n,1} f_{sn-1}),\\
&\lim_{n\to\infty}(K_{n+1} * f_{sn} - K_n * f_{sn+1}) = -2h/T,
\end{aligned}$$

as reported elsewhere [21]. These are integral equations that can be solved numerically for different values of the temperature and interaction strength (cf. Ref. [25]). Density profiles obtained combining the solutions of the TBA and a finite-temperature Thomas-Fermi approximation are plotted in Fig. 2. The two plots illustrate two qualitatively different behaviors that are realized depending on the degree of global polarization, $P = (N_\uparrow - N_\downarrow)/(N_\uparrow + N_\downarrow)$. In one case, the edge of the cloud is a balanced paired superfluid while in the other is a fully polarized Fermi gas. In both cases, the inner core of the cloud is formed by a partially polarized 1d-superfluid with FFLO-like features. We have shown elsewhere that the different regions remain identifiable up to temperatures of about 10% of the Fermi temperature [21].

We stress that our results can also be used for thermometry. Indeed, the information from the particle-density trap profiles was used in the analysis of the experimental data obtained recently by the cold-atom experimental group at Rice University [26]. Those results show good agreement with the theory and the reported temperatures are about 10% of the Fermi temperature with some deviations towards larger values at the tails of the cloud (notice the experimental error bars also increase near the edges).

A more recent, and previously unreported, result is about the profiles for entropy density in the trap; these are shown in Fig. 3 for two different temperatures. One observes that the curves are peaked at the interface between the inner FFLO-like region and the outer cloud, and

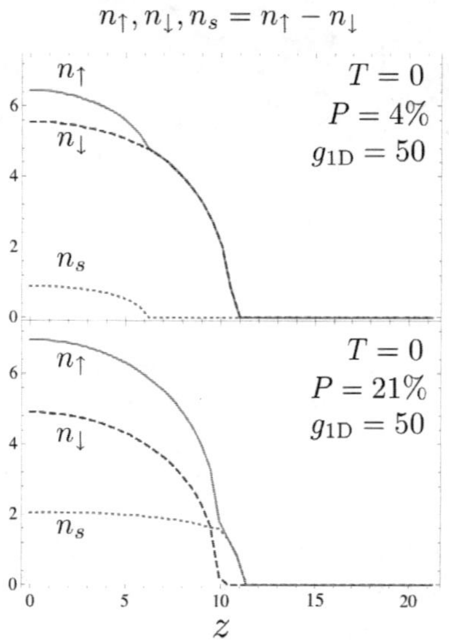

Figure 2. Trap density profiles for $N = 200$ and $T = 0$. Two different global polarizations, $P = 4\%$ and 21% are shown. The plots show spin-up ($n_\uparrow$, solid), spin-down ($n_\downarrow$, dashed) particle densities and their difference (n_s, dotted). The critical polarization at which all the densities go to zero together at the edge of the cloud is $P \approx 15\%$ (in that case the 1d-FFLO state spans the whole cloud). These behaviors were recently demonstrated experimentally [26]. (At the time of the QTS6 symposium, only preliminary experimental data was available.)

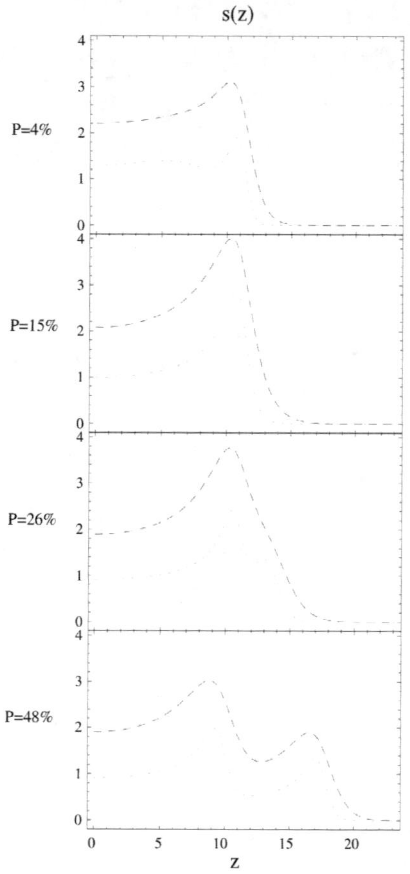

Figure 3. Trap density profiles of the entropy density for different total polarizations of the atomic cloud (indicated at the left of each plot). The parameters for the plots are: $\Delta = 25$, $T = 0.06T_F$ (green dotted line) and $0.12T_F$ (red dashed line).

then again near the edge of the outer cloud. One can understand the result in terms of atomic densities: when they become small, the phase space for arranging particles is larger (even if interactions are strong) and the entropy per particle is larger as a result. Similar observations were made before for lattice gases (bosonic and fermionic Hubbard models) and there have been suggestions of exploiting this effect for cooling schemes [27]. The idea is to somehow let the particles near the edge of the clouds fly out and take with them a comparatively large share of the entropy, thus lowering the total (and also average) entropy of the system. The process is different from the usual evaporative cooling technique in that the shape of the trap would be engineered so that there is no substantial reduction of the particle density at the trap center (this step would be experimentally challenging, but is supposed to produce big gains in terms of reduction of the system temperature as a fraction of the Fermi temperature).

4. Conclusions

To summarize, we have studied attractive spin-imbalanced ultracold atomic clouds of fermions in 1d. Combining the Bethe Ansatz method and the Thomas-Fermi approximation we have shown how to calculate trap profiles of different observables. We reported, in particular, on the particle density, magnetization density and entropy density profiles. Our finite-temperature calculations serve also to indicate how much one needs to cool in order to be able to observe exotic pairing states and shed light on the possibility of applying recently proposed cooling schemes to this 1d systems. During the symposium (and also in Ref. [21]) we discussed about ways of mapping out the 'phase diagram' of the trapped system by reading it out from the temperature profiles. Recent experiments did make use of those ideas but ultimately followed a different protocol for analyzing the data. The problem of dealing with a bundle of tubes, rather than a single one, introduces additional complications. The experiments used a method of reconstructing the total-number profiles per species across tubes by using an inverse Abel transform of their data after first projecting onto the plane perpendicular to the tube's axis in order to improve the signal to noise ratio (following our advise). Then our analysis of a single tube was repeated and combined for many tubes in order to produce predictions for the whole bundle – rather than trying to extract single-tube information from the experiments.

A similar TBA analysis can be also performed for repulsive interactions between particles, where the coexistence of different Luttinger-liquid regimes has been proposed as discussed in the introduction [9]. The approach is well suited to explore interesting crossover effects, which are challenging to capture with other methods. Such systems would also be within the reach of the present generation of experiments by working on the other side of the Feshbach resonance. Conversely, our inhomogeneous bosonization scheme could be applied to the attractive case in order to answer questions about dynamics that are currently still open. Efforts in this direction are underway.

It is extremely useful that the experiments are able to realized models that are close to integrable ones. This not only has intellectual appeal in terms of motivating further studies of these unique models, but also helps to develop the experimental tools required to probe non-trivial strongly correlated quantum systems. Such tools could later be applied to a wider class of systems and, along the way, perhaps improve our understanding of the origins of integrability.

Acknowledgments

We would like to thank the conference organizers for giving us the opportunity to present this work during the 6th edition of the international symposium on *Quantum Theory and Symmetries* (QTS6). This work was supported in part by DARPA-ARO (grant no. W911NF-07-1-0464) and by the W. M. Keck and Welch foundations (C-1681).

References

[1] Feshbach H 1958 *Ann. Phys.* **5** 357
[2] Bloch I 2005 *Nat. Phys.* **1** 23–30
[3] Paredes B *et al* 2004 *Nature* **429** 277–281
[4] Kinoshita T, Wenger T and Weiss D S 2004 *Science* **305** 1125
[5] Recati A *et al* 2003 *Phys. Rev. Lett.* **90** 020401 ; *ibid.* 2003 J. Opt. B **5** S55
[6] Kecke L, Grabert H and Hausler W 2005 *Phys. Rev. Lett.* **94** 176802
[7] Yang C N 1967 *Phys. Rev. Lett.* **19** 1312 ; Lieb E H and Wu F Y 1968 *Phys. Rev. Lett.* **20** 1445
[8] Gaudin M 1983 *La fonction d'onde de Bethe* (Paris: Masson)
[9] Kakashvili P, Bhongale S G, Pu H and Bolech C J 2008 *Phys. Rev. A* **78** 041602(R)
[10] Kakashvili P, Bhongale S, Pu H and Bolech C 2009 *Physica B: Condensed Matter* **404** 3320–3323
[11] Fiete G A 2007 *Rev. Mod. Phys.* **79** 801–20
[12] Fulde P and Ferrell R A 1964 *Phys. Rev.* **135** A550
[13] Larkin A I and Ovchinnikov Y N 1964 *Zh. Eksp. Teor. Fiz.* **47** 1136 ; Sov. Phys. JETP **20**, 762 (1965)
[14] Radovan H A *et al* 2003 *Nature* **425** 51 ; Miclea C F *et al* 2006 Phys. Rev. Lett. **96** 117001

[15] Singleton J *et al* 2000 *J. Phys.: Condens. Matter* **12** L641 ; Yonezawa S *et al* 2008 Phys. Rev. Lett. **100**, 117002. See also : Bolech C J and Giamarchi T 2004 Phys. Rev. Lett. **92** 127001 ; Bolech C J and Giamarchi T 2005 Phys. Rev. B **71** 024517
[16] Sedrakian A, Clark J W and Alford M G (eds) 2006 *Pairing in Fermionic Systems: Basic Concepts and Modern Applications* (Singapore: World Scientific)
[17] Orso G 2007 *Phys. Rev. Lett.* **98** 070402 ; Hu H, Liu X -J and Drummond P D 2007 *Phys. Rev. Lett.* **98** 070403 ; Guan X W *et al* 2007 *Phys. Rev. B* **76** 085120
[18] Yang K 2001 *Phys. Rev. B* **63** 140511 ; Zhao E and Liu W V 2008 *Phys. Rev. A* **78** 063605
[19] Casula M, Ceperley D M and Mueller E J 2008 *Phys. Rev. A* **78** 033607
[20] Feiguin A E and Heidrich-Meisner F 2007 *Phys. Rev. B* **76** 220508
[21] Kakashvili P and Bolech C J 2009 *Phys. Rev. A* **79** 041603(R)
[22] Partridge G B *et al* 2006 *Science* **311** 503 ; Partridge G B *et al.* 2006 *Phys. Rev. Lett.* **97**, 190407
[23] De Silva T N and Mueller E J 2006 *Phys. Rev. Lett.* **97** 070402
[24] Imambekov A, Bolech C J, Lukin M and Demler E 2006 *Phys. Rev. A* **74** 053626. See also : Baksmaty L O, Lu H, Bolech C J and Pu H 2010 *arXiv:1003.4488*
[25] Bolech C J and Andrei N 2005 *Phys. Rev. B* **71** 205104 ; *ibid.* 2002 *Phys. Rev. Lett.* **88** 237206
[26] Liao Y, Rittner A S C, Paprotta T, Li W, Partridge G B, Hulet R G, Baur S K and Mueller E J 2009 *arXiv:0912.0092*
[27] Ho T and Zhou Q 2009 *Proc. Nac. Acad. Sci.* **106** 6916–6920

Journal of Physics: Conference Series **462** (2013) 012005 doi:10.1088/1742-6596/462/1/012005

The Freudenthal triple classification of 3-qubit entanglement and STU black hole entropy

L Borsten

Blackett Laboratory, Imperial College, Prince Consort Road, London SW7 2BZ, UK

E-mail: `leron.borsten@imperial.ic.ac.uk`

Abstract. We show that the 3-qubit entanglement classification is naturally encoded by that elegant algebraic structure, well known from supergravity, the *Freudenthal triple system*. In particular, we show that the six entanglement classes: (1) Separable *A*-*B*-*C*, (2a) Biseparable *A*-*BC*, (2b) Biseparable *B*-*CA*, (2c) Biseparable *C*-*AB*, (3) W and (4) Greenberger-Horne-Zeilinger correspond respectively to the Freudenthal triple ranks 1, 2a, 2b, 2c, 3, 4. Based on work done in collaboration with D. Dahanayake, M. J. Duff, H. Ebrahim and W. Rubens.

1. Introduction

The interdisciplinary nature of Quantum Theory and Symmetries 6 made for a truly unique and rewarding experience - it was certainly a privilege to be counted amongst its participants. Moreover, as the week went by, it became apparent that not only is this union of disciplines exciting, it is unequivocally of our time. We are living at the dawn of the (quantum) information age whilst, hand-in-hand, pushing the boundaries of fundamental particle physics as we begin in earnest to probe new unchartered energy scales. In such a light it seems nothing but natural to bring together the frontiers of quantum information science and high energy physics.

While this is a tall order to live up to, the present work, at least in some limited sense, fits the bill. On the one hand, its primary focus centres on the classification of quantum entanglement and, in particular, the 3-qubit system. Entanglement lies at the heart of quantum information theory and its many diverse applications, such as quantum computation, cryptography and communication. On the other hand, the mathematical structures employed in this analysis are imported from the seemingly disparate field of M-theory and, in particular, the study of stringy black holes, which have played, and continue to play, an essential role in understanding the non-perturbative aspects of quantum gravity.

Indeed, while logically independent, this work has its provenance in the black-hole/qubit correspondence, reviewed in [1]. In 2006 [2] it was observed that the measure of 3-qubit entanglement, the so called *3-tangle* [3, 4], and the entropy of the STU black hole [5, 6] are both given by the same mathematical object, *Cayley's hyperdeterminant* [7]. Further work [8, 9, 10, 11, 12, 13, 14, 15, 16, 17, 18, 19, 1, 20, 21, 22] has led to a more complete dictionary translating a variety of phenomena in one language to those in the other. In fact, the black holes of the maximally supersymmetric $D = 4, \mathcal{N} = 8$ supergravity have been related to a very special tripartite entanglement of seven qubits [11, 12]. It is well known that $\mathcal{N} = 8$ supergravity is intimately related to the exceptional Jordan algebra of 3×3 split-octonionic Hermitian matrices and the *Freudenthal triple system* (FTS) [23]. Since the STU model may be consistently embedded in the $D = 4, \mathcal{N} = 8$ theory it is natural to study the application of the Jordan algebras and the FTS to both three qubits and STU black holes [8, 24, 11, 12, 13, 14, 17, 18, 1, 25].

Accordingly, in this contribution we briefly review the FTS classification of 3-qubit entanglement as presented in [26]. Unfortunately, the corresponding treatment of the STU black hole will only be touched upon due to limited space. We begin by recalling the conventional classification of 3-qubit entanglement as first obtained by Dür *et al.* using the paradigm of *stochastic local operations and classical communication* (SLOCC) [27]. We then set up the FTS framework before illustrating how it reproduces the correct entanglement classification. The entanglement orbits obtained in [26] via the FTS are also presented. We conclude by summarising the associated analysis of the STU black hole.

2. Conventional three-qubit entanglement classification

The proper qualitative and quantitative characterisation of multipartite entanglement remains one of the longest standing open problems in quantum information science. While the complete picture still eludes us, substantial progress has been made using the SLOCC framework, as first proposed in [28]. Two quantum states of a composite system are considered SLOCC-equivalent if and only if they may be probabilistically interrelated using purely local operations, i.e. operations restricted to act on constituent parts independently, supplemented by the classical communication of any relevant information between said constituents. The essential observation is that while such operations may be used to establish arbitrary classical correlations, they are manifestly unable to create genuine quantum correlations. Hence, two SLOCC-equivalent states possess the same entanglement. For more details see [29, 30] and the references therein.

In [27] it was argued that two states of an n-qubit system are SLOCC-equivalent if and only if they are related by $[\mathrm{SL}(2,\mathbb{C})]^{\otimes n}$, which we will refer to as the *SLOCC-equivalence group*. It may be usefully thought of as the "gauge" group of n-qubit entanglement. In particular, the 3-qubit (Alice, Bob and Charlie) state vector is given by,

$$|\psi\rangle = a_{ABC}|ABC\rangle, \qquad A, B, C = 0, 1, \tag{1}$$

where a_{ABC} transforms as a $(\mathbf{2}, \mathbf{2}, \mathbf{2})$ under the 3-qubit SLOCC-equivalence group $\mathrm{SL}_A(2,\mathbb{C}) \times \mathrm{SL}_B(2,\mathbb{C}) \times \mathrm{SL}_C(2,\mathbb{C})$.

Dür *et al.* [27] used simple arguments involving the ranks and ranges of the reduced density matrices to show that there are six SLOCC equivalence classes, representing six physically distinct forms of entanglement: (1) Separable A-B-C, (2a) Biseparable A-BC, (2b) Biseparable B-CA, (2c) Biseparable C-AB, (3) W and (4) Greenberger-Horne-Zeilinger (GHZ). These six classes may be distinguished [27] by the three local entropies,

$$S_A = \det \rho_A, \quad S_B = \det \rho_B, \quad S_C = \det \rho_C, \tag{2}$$

($\rho_A = \mathrm{tr}_{BC}|\psi\rangle\langle\psi|$ and similarly for ρ_B and ρ_C) and Cayley's hyperdeterminant [7, 4],

$$\mathrm{Det}\, a_{ABC} = -\tfrac{1}{2}\epsilon^{A_1A_2}\epsilon^{A_3A_4}\epsilon^{B_1B_2}\epsilon^{B_3B_4}\epsilon^{C_1C_3}\epsilon^{C_3C_4}a_{A_1B_1C_1}a_{A_2B_2C_2}a_{A_3B_3C_3}a_{A_4B_4C_4}, \tag{3}$$

which is the unique $[\mathrm{SL}(2,\mathbb{C})]^3$ quartic invariant also displaying a permutation symmetry under the interchange of Alice, Bob and Charlie. This classification of classes is summarised in Table 1. It is only the W and GHZ classes that possess genuine tripartite entanglement - "three qubits can be entangled in two inequivalent ways" [27]. The GHZ class is distinguish by non-vanishing hyperdeterminant and has the highest order of entanglement in the sense that it maximally violates the principle of local realism [31, 32]. The non-invertible SLOCC operations (lying outside the SLOCC-equivalence group) may be used to descend through these classes, creating the entanglement hierarchy as depicted on the right of Figure 1. Let us now turn our attention to FTS analysis of three qubits.

Table 1: The values of the local entropies S_A, S_B, and S_C and the hyperdeterminant $\operatorname{Det} a$ are used to partition three-qubit states into entanglement classes.

Class	Representative	Condition				
		ψ	S_A	S_B	S_C	$\operatorname{Det} a$
A-B-C	$\|000\rangle$	$\neq 0$	$= 0$	$= 0$	$= 0$	$= 0$
A-BC	$\|010\rangle + \|001\rangle$	$\neq 0$	$= 0$	$\neq 0$	$\neq 0$	$= 0$
B-CA	$\|100\rangle + \|001\rangle$	$\neq 0$	$\neq 0$	$= 0$	$\neq 0$	$= 0$
C-AB	$\|010\rangle + \|100\rangle$	$\neq 0$	$\neq 0$	$\neq 0$	$= 0$	$= 0$
W	$\|100\rangle + \|010\rangle + \|001\rangle$	$\neq 0$	$\neq 0$	$\neq 0$	$\neq 0$	$= 0$
GHZ	$\|000\rangle + \sqrt{\operatorname{Det} a}\|111\rangle$	$\neq 0$	$\neq 0$	$\neq 0$	$\neq 0$	$\neq 0$

Figure 1: Left: Classification of $\mathcal{N} = 8$ black holes according to susy. N denotes the number of intersecting D-branes in the microscopic picture. Right: The entanglement classification of 3-qubits. The arrows represent the removal of a D-brane or a non-invertible SLOCC operation.

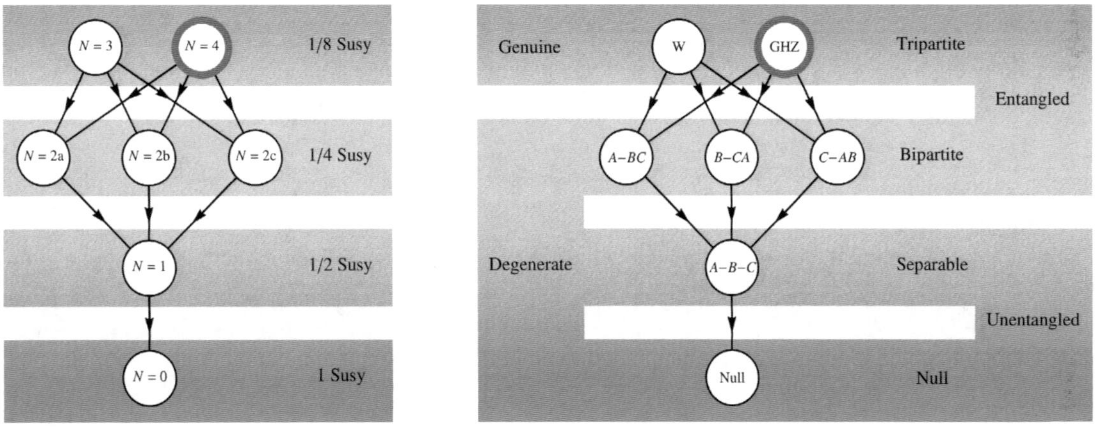

3. The FTS classification of qubit entanglement

3.1. Cubic Jordan algebras and the Freudenthal triple system

A Jordan algebra $\mathfrak{J}$ is vector space defined over a ground field $\mathbb{F}$ equipped with a bilinear product satisfying

$$X \circ Y = Y \circ X, \quad X^2 \circ (X \circ Y) = X \circ (X^2 \circ Y), \quad \forall\ X, Y \in \mathfrak{J}. \tag{4}$$

The class of *cubic* Jordan algebras are constructed as follows. Let V be a vector space equipped with a cubic norm, i.e. a homogeneous map of degree three, $N : V \to \mathbb{F}$ s.t. $N(\lambda X) = \lambda^3 N(X)$, $\forall \lambda \in \mathbb{F}, X \in V$, with trilinear linearisation

$$N(X, Y, Z) := \frac{1}{6}[N(X+Y+Z) - N(X+Y) - N(X+Z) - N(Y+Z) + N(X) + N(Y) + N(Z)]. \tag{5}$$

If V further contains a base point $N(c) = 1, c \in V$ one may define the following three maps,

$$\begin{aligned} \operatorname{Tr} : V \to \mathbb{F}; &\quad X \mapsto 3N(c, c, X), \\ S : V \times V \to \mathbb{F}; &\quad (X, Y) \mapsto 6N(X, Y, c), \\ \operatorname{Tr} : V \times V \to \mathbb{F}; &\quad (X, Y) \mapsto \operatorname{Tr}(X)\operatorname{Tr}(Y) - S(X, Y). \end{aligned} \tag{6}$$

A cubic Jordan algebra $\mathfrak{J}$, with multiplicative identity $\mathbb{1} = c$, may be derived from any such vector space if N is *Jordan cubic*. That is: (i) the trace bilinear form (6) is non-degenerate (ii) the quadratic adjoint map, $\sharp\colon \mathfrak{J} \to \mathfrak{J}$, uniquely defined by $\mathrm{Tr}(X^\sharp, Y) = 3N(X,X,Y)$, satisfies $(X^\sharp)^\sharp = N(X)X$, $\forall X \in \mathfrak{J}$. The Jordan product is then given by

$$X \circ Y = \tfrac{1}{2}\big(X \times Y + \mathrm{Tr}(X)Y + \mathrm{Tr}(Y)X - S(X,Y)\mathbb{1}\big), \tag{7}$$

where, $X \times Y$ is the linearisation of the quadratic adjoint, $X \times Y = (X+Y)^\sharp - X^\sharp - Y^\sharp$.

Given a cubic Jordan algebra $\mathfrak{J}$ defined over a field $\mathbb{F}$, one is able to construct an FTS by defining the vector space $\mathfrak{F}(\mathfrak{J})(=\mathfrak{F})$,

$$\mathfrak{F}(\mathfrak{J}) = \mathbb{F} \oplus \mathbb{F} \oplus \mathfrak{J} \oplus \mathfrak{J}. \tag{8}$$

An arbitrary element $x \in \mathfrak{F}(\mathfrak{J})$ may be written as a "2×2 matrix",

$$x = \begin{pmatrix} \alpha & X \\ Y & \beta \end{pmatrix} \quad \text{where } \alpha, \beta \in \mathbb{F} \quad \text{and} \quad X, Y \in \mathfrak{J}. \tag{9}$$

The FTS comes equipped with a non-degenerate bilinear antisymmetric quadratic form, a quartic form and a trilinear triple product [33, 34, 35, 36, 37]:

(i) Quadratic form $\{x, y\}\colon \mathfrak{F} \times \mathfrak{F} \to \mathbb{F}$

$$\{x, y\} = \alpha\delta - \beta\gamma + \mathrm{Tr}(X, W) - \mathrm{Tr}(Y, Z), \quad \text{where} \quad x = \begin{pmatrix} \alpha & X \\ Y & \beta \end{pmatrix}, \; y = \begin{pmatrix} \gamma & Z \\ W & \delta \end{pmatrix}. \tag{10a}$$

(ii) Quartic form $q : \mathfrak{F} \to \mathbb{F}$

$$q(x) = -2[\alpha\beta - \mathrm{Tr}(X, Y)]^2 - 8[\alpha N(X) + \beta N(Y) - \mathrm{Tr}(X^\sharp, Y^\sharp)]. \tag{10b}$$

(iii) Triple product $T : \mathfrak{F} \times \mathfrak{F} \times \mathfrak{F} \to \mathfrak{F}$ which is uniquely defined by

$$\{T(x, y, w), z\} = q(x, y, w, z) \tag{10c}$$

where $q(x, y, w, z)$ is the full linearisation of $q(x)$ such that $q(x, x, x, x) = q(x)$.

The *automorphism* group is given by the set of invertible $\mathbb{F}$-linear transformations preserving the quadratic and quartic forms [33, 34],

$$\mathrm{Aut}(\mathfrak{F}) = \{\sigma \in \mathrm{Iso}_{\mathbb{F}}(\mathfrak{F}) | q(\sigma x) = q(x), \{\sigma x, \sigma y\} = \{x, y\}, \ \forall x, y \in \mathfrak{F}\}. \tag{11}$$

The automorphism group corresponds to the U-duality group of a variety 4-dimensional supergravities (see for example [38, 39] and the references therein). The conventional concept of matrix rank may be generalised to Freudenthal triple systems in a natural and $\mathrm{Aut}(\mathfrak{F})$ invariant manner. The rank of an arbitrary element $x \in \mathfrak{F}$ is uniquely defined by [36, 37]:

$$\begin{aligned}
&\mathrm{Rank}x = 1 \Leftrightarrow 3T(x, x, y) + x\{x, y\}x = 0 \ \forall y;\\
&\mathrm{Rank}x = 2 \Leftrightarrow \exists y \text{ s.t. } 3T(x, x, y) + x\{x, y\}x \neq 0, \ T(x, x, x) = 0;\\
&\mathrm{Rank}x = 3 \Leftrightarrow T(x, x, x) \neq 0, \ q(x) = 0;\\
&\mathrm{Rank}x = 4 \Leftrightarrow q(x) \neq 0.
\end{aligned} \tag{12}$$

3.2. FTS representation of three-qubits
To define the 3-qubit FTS we use the particularly simple cubic Jordan algebra $\mathfrak{J}_{ABC} := \mathbb{C} \oplus \mathbb{C} \oplus \mathbb{C}$ with cubic form

$$N(X) = X_1 X_2 X_3, \quad \text{where} \quad X = (X_1, X_2, X_3) \in \mathfrak{J}_{ABC}. \tag{13}$$

It is not difficult to verify that this defines a *bona fide* cubic Jordan algebra [40, 26].

We are now in a position to define our 3-qubit FTS,

$$\mathfrak{F}_{ABC} := \mathfrak{F}(\mathfrak{J}_{ABC}) = \mathbb{C} \oplus \mathbb{C} \oplus \mathfrak{J}_{ABC} \oplus \mathfrak{J}_{ABC}, \tag{14}$$

as the representation space of three qubits. We identify three qubit wavefunctions $|\psi\rangle, |\phi\rangle \ldots$ with elements $\Psi, \Phi, \ldots$ of the 8-dimensional complex vector space $\mathfrak{F}_{ABC}$ via,

$$|\psi\rangle = a_{ABC}|ABC\rangle \quad \longleftrightarrow \quad \Psi = \begin{pmatrix} a_{111} & (a_{001}, a_{010}, a_{100}) \\ (a_{110}, a_{101}, a_{011}) & a_{000} \end{pmatrix}, \tag{15}$$

The powerful machinery of the FTS may now be applied to the 3-qubit system.

First, we note that the automorphism group is the SLOCC-equivalence group, $\text{Aut}(\mathfrak{F}_{ABC}) = \text{SL}_A(2, \mathbb{C}) \times \text{SL}_B(2, \mathbb{C}) \times \text{SL}_C(2, \mathbb{C})$. Using (10b) one finds that the quartic norm $q(\Psi)$ is related to Cayley's hyperdeterminant by

$$q(\Psi) = \{T(\Psi, \Psi, \Psi), \Psi\} = 2 \det \gamma^A = 2 \det \gamma^B = 2 \det \gamma^C = -2 \, \text{Det} \, a_{ABC}, \tag{16}$$

where, following [41], we have defined the matrix γ^A by

$$(\gamma^A)_{A_1 A_2} = \varepsilon^{B_1 B_2} \varepsilon^{C_1 C_2} a_{A_1 B_1 C_1} a_{A_2 B_2 C_2}, \tag{17}$$

and similar for $\gamma^{B,C}$, which transform respectively as $(\mathbf{3}, \mathbf{1}, \mathbf{1}), (\mathbf{1}, \mathbf{3}, \mathbf{1}), (\mathbf{1}, \mathbf{1}, \mathbf{3})$ under $[\text{SL}(2, \mathbb{C})]^3$. The γ's are related to the local entropies of section 2 by

$$S_A = 4\Big[\operatorname{tr} \gamma^{B\dagger} \gamma^B + \operatorname{tr} \gamma^{C\dagger} \gamma^C\Big], \tag{18}$$

and its cyclic permutations. The triple product maps a state Ψ, which transforms as a $(\mathbf{2}, \mathbf{2}, \mathbf{2})$ of $[\text{SL}(2, \mathbb{C})]^3$, to another state $T(\Psi, \Psi, \Psi)$, cubic in the state vector coefficients, also transforming as a $(\mathbf{2}, \mathbf{2}, \mathbf{2})$. Explicitly, $T(\Psi, \Psi, \Psi)$ may be written as

$$T(\Psi, \Psi, \Psi) = T_{ABC}|ABC\rangle, \quad \text{where} \quad T_{ABC} = \varepsilon^{A'A''} a_{A'BC} (\gamma^A)_{A''A}. \tag{19}$$

Note, T_{ABC} may also be written using $\gamma^{B,C}$ in the obvious manner. We may now assign an abstract FTS rank to an arbitrary 3-qubit state Ψ using (12).

3.3. The FTS rank entanglement classes
Rank 1 and the class of separable states: A non-zero state Ψ is rank 1 iff

$$\Upsilon := 3T(\Psi, \Psi, \Phi) + \{\Psi, \Phi\}\Psi = 0, \quad \forall \, \Phi, \tag{20}$$

In particular, (20) implies $T(\Psi, \Psi, \Psi) = 0$, from which we deduce that at most one γ is non-vanishing since

$$(\gamma^A)_{A_1 A_2} (\gamma^C)_{C_1 C_2} = \varepsilon^{B_2 B_1} a_{A_1 B_1 C_1} T_{A_2 B_2 C_2} + \varepsilon^{B_1 B_2} a_{A_2 B_2 C_1} T_{A_1 B_1 C_2} \tag{21}$$

Table 2: The entanglement classification of three qubits as according to the FTS rank system.

Class	Rank	FTS rank condition	
		vanishing	non-vanishing
Null	0	Ψ	–
A-B-C	1	$3T(\Psi,\Psi,\Phi)+\{\Psi,\Phi\}\Psi$	Ψ
A-BC	2a	$T(\Psi,\Psi,\Psi)$	γ^A
B-CA	2b	$T(\Psi,\Psi,\Psi)$	γ^B
C-AB	2c	$T(\Psi,\Psi,\Psi)$	γ^C
W	3	$q(\Psi)$	$T(\Psi,\Psi,\Psi)$
GHZ	4	–	$q(\Psi)$

and similarly for $(\gamma^B)_{B_1B_2}(\gamma^A)_{A_1A_2}$ and $(\gamma^C)_{C_1C_2}(\gamma^B)_{B_1B_2}$. Then, using

$$-\Upsilon_{A_1B_1C_1} = \varepsilon^{A_2A_3}b_{A_3B_1C_1}(\gamma^A)_{A_1A_2} + \varepsilon^{B_2B_3}b_{A_1B_3C_1}(\gamma^B)_{B_1B_2} + \varepsilon^{C_2C_3}b_{A_1B_1C_3}(\gamma^C)_{C_1C_2} \tag{22}$$

where

$$|\phi\rangle = b_{ABC}|ABC\rangle \quad \longleftrightarrow \quad \Phi = \begin{pmatrix} b_{111} & (b_{001}, b_{010}, b_{100}) \\ (b_{110}, b_{101}, b_{011}) & b_{000} \end{pmatrix}, \tag{23}$$

we observe that (20) implies all three γ's must vanish. Using (18) it is then clear that all three local entropies vanish.

Conversely, from (18), $S_A = S_B = S_C = 0$ implies that each of the three γ's vanish and the rank 1 condition is satisfied.

Rank 2 and the class of biseparable states: A non-zero state Ψ is rank 2 or less if and only if $T(\Psi,\Psi,\Psi)=0$. To not be rank 1 there must exist some Φ such that $3T(\Psi,\Psi,\Phi)+\{\Psi,\Phi\}\Psi \neq 0$. It was shown above that this is equivalent to one and only one non-vanishing γ matrix.

Using (18) it is clear that the choices $\gamma^A \neq 0$ or $\gamma^B \neq 0$ or $\gamma^C \neq 0$ give $S_A = 0, S_{B,C} \neq 0$ or $S_B = 0, S_{C,A} \neq 0$ or $S_C = 0, S_{A,B}, \neq 0$, respectively. These are precisely the conditions for the biseparable class *A-BC* or *B-CA* or *C-AB* presented in Table 1.

Conversely, using (18) and the fact that both the local entropies and $\mathrm{tr}(\gamma^\dagger\gamma)$ are positive semidefinite, we find that all states in the biseparable class are rank 2, the particular subdivision being given by the corresponding non-zero γ.

Rank 3 and the class of W-states: A non-zero state Ψ is rank 3 if $q(\Psi) = -2\,\mathrm{Det}\,a = 0$ but $T(\Psi,\Psi,\Psi) \neq 0$. From (19) all three γ's are then non-zero but from (16) all have vanishing determinant. In this case (18) implies that all three local entropies are non-zero but $\mathrm{Det}\,a = 0$. So all rank 3 Ψ belong to the W-class.

Conversely, from (18) it is clear that no two γ's may simultaneously vanish when all three S's are non-zero (> 0). $T(\Psi,\Psi,\Psi) = 0$ implies at least two of the γ's vanish. Consequently, for all W-states $T(\Psi,\Psi,\Psi) \neq 0$ and, therefore, all W-states are rank 3.

Rank 4 and the class of GHZ-states: The rank 4 condition is given by $q(\Psi) \neq 0$ and, since for the 3-qubit FTS $q(\Psi) = -2\,\mathrm{Det}\,a$, we immediately see that the set of rank 4 states is equivalent to the GHZ class of genuine tripartite entanglement as in Table 2. Note, the GHZ class is a one complex dimensional space of orbits parameterised the quartic norm q.

These results are summarised in Table 2. The stability groups H for each orbit coset $[\mathrm{SL}(2,\mathbb{C})]^3/H$, as obtained using the FTS framework in [26], are also presented in Table 3.

Journal of Physics: Conference Series **462** (2013) 012005 doi:10.1088/1742-6596/462/1/012005

Table 3: Stability groups and complex dimensions of the 3-qubit entanglement orbits

Class	FTS Rank	Stabiliser	dim	Projective stabiliser	dim
Separable	1	$[\mathrm{SO}(2,\mathbb{C})]^2 \ltimes \mathbb{C}^3$	4	$[\mathrm{SO}(2,\mathbb{C}) \ltimes \mathbb{C}]^3$	3
Biseparable	2	$\mathrm{SO}(3,\mathbb{C}) \times \mathbb{C}$	5	$\mathrm{SO}(3,\mathbb{C}) \times (\mathrm{SO}(2,\mathbb{C}) \ltimes \mathbb{C})$	4
W	3	$\mathbb{C}^2$	7	$\mathrm{SO}(2,\mathbb{C}) \ltimes \mathbb{C}^2$	6
GHZ	4	$[\mathrm{SO}(2,\mathbb{C})]^2$	7	$[\mathrm{SO}(2,\mathbb{C})]^2$	7

Table 4: Orbits stabilisers of STU black holes in the real case $\mathfrak{J}_{STU} = \mathbb{R} \oplus \mathbb{R} \oplus \mathbb{R}$.

Class	FTS Rank	$q(\Psi)$	Stabiliser	dim	Black hole	Susy
Separable	1	$= 0$	$[\mathrm{SO}(1,1)]^2 \ltimes \mathbb{R}^3$	4	Small	1/2
Biseparable	2	$= 0$	$\mathrm{SO}(2,1) \times \mathbb{R}$	5	Small	1/2
W	3	$= 0$	$\mathbb{R}^2$	7	Small	1/2
GHZ	4	> 0	$[\mathrm{U}(1)]^2$	7	Large	1/2
GHZ	4	> 0	$[\mathrm{U}(1)]^2$	7	Large	0
GHZ	4	< 0	$[\mathrm{SO}(1,1)]^2$	7	Large	0

Note, we consider the states as both points in the Hilbert space and rays in its projectivisation. The 3-qubit separable projective coset is just a direct product of three individual qubit cosets $\mathrm{SL}(2,\mathbb{C})/\,\mathrm{SO}(2,\mathbb{C})\ltimes\mathbb{C}$. The biseparable projective coset is just the direct product of an entangled 2-qubit coset $[\mathrm{SL}(2,\mathbb{C})]^2/\,\mathrm{SO}(3,\mathbb{C})$ with a 1-qubit coset as one would expect.

4. Black holes

As emphasized in section 1 there is a growing dictionary relating various features of qubit systems to aspects of black holes in supergravity. Using the dictionary relating the STU black hole to the 3-qubit system it is clear that this FTS analysis carries through. The results are summarised in Table 4. We see that the physically distinct forms of 3-qubit entanglement are directly related to physically distinct STU black hole solutions [8]. The entangled GHZ class corresponds to "large" black holes with non-vanishing classical horizon, while the remaining classes are related to "small" black holes with vanishing classical horizon. However, there are differences. The black hole charges are real valued (with Jordan algebra $\mathfrak{J}_{STU} = \mathbb{R} \oplus \mathbb{R} \oplus \mathbb{R}$) and really correspond to the case of real qubits or "rebits' which are, as noted in [42, 9], qualitatively different from the complex case. This feature manifests itself in the different real forms appearing in the GHZ class as in Table 4. For details see [23, 38, 43, 44, 1, 26]. A microscopic interpretation of this correspondence may be given by embedding the STU model in type II string theory compactified on a 6-torus [16]. The 3-qubit states are given by the possible wrapping configurations of intersecting D-branes around the three 2-tori of the compact space, explaining the origin of their binary nature. In this case the entanglement classes are related to the amount of supersymmetry preserved by the corresponding intersecting D-brane configuration as in Figure 1. For more details the interested reader is referred to [16, 1].

Acknowledgments

It is a pleasure to thank all the organisers for a fantastic conference. I would also like to extend my gratitude to Sergio Ferrara, Peter Levay and Alessio Marrani for interesting and useful discussions. Finally, a warm thank you to my collaborators, Duminda Dahanayake, Hajar Ebrahim, William Rubens and Michael J. Duff for his ongoing support and encouragement.

References

[1] Borsten L, Dahanayake D, Duff M J, Ebrahim H and Rubens W 2009 *Phys. Rep.* **471** 113–219 (*Preprint* 0809.4685)
[2] Duff M J 2007 *Phys. Rev.* **D76** 025017 (*Preprint* hep-th/0601134)
[3] Coffman V, Kundu J and Wootters W K 2000 *Phys. Rev.* **A61** 052306 (*Preprint* quant-ph/9907047)
[4] Miyake A and Wadati M 2002 *Quant. Info. Comp.* **2 (Special)** 540–555 (*Preprint* quant-ph/0212146)
[5] Duff M J, Liu J T and Rahmfeld J 1996 *Nucl. Phys.* **B459** 125–159 (*Preprint* hep-th/9508094)
[6] Behrndt K, Kallosh R, Rahmfeld J, Shmakova M and Wong W K 1996 *Phys. Rev.* **D54** 6293–6301 (*Preprint* hep-th/9608059)
[7] Cayley A On the theory of linear transformations *Camb. Math. J.* **4** (1845) 193–209
[8] Kallosh R and Linde A 2006 *Phys. Rev.* **D73** 104033 (*Preprint* hep-th/0602061)
[9] Lévay P 2006 *Phys. Rev.* **D74** 024030 (*Preprint* hep-th/0603136)
[10] Ferrara S and Kallosh R 2006 *Phys. Rev.* **D73** 125005 (*Preprint* hep-th/0603247)
[11] Duff M J and Ferrara S 2007 *Phys. Rev.* **D76** 025018 (*Preprint* quant-ph/0609227)
[12] Lévay P 2007 *Phys. Rev.* **D75** 024024 (*Preprint* hep-th/0610314)
[13] Duff M J and Ferrara S 2008 *Lect. Notes Phys.* **755** 93–114 (*Preprint* hep-th/0612036)
[14] Duff M J and Ferrara S 2007 *Phys. Rev.* **D76** 124023 (*Preprint* 0704.0507)
[15] Lévay P 2007 *Phys. Rev.* **D76** 106011 (*Preprint* 0708.2799)
[16] Borsten L, Dahanayake D, Duff M J, Rubens W and Ebrahim H 2008 *Phys. Rev. Lett.* **100** 251602 (*Preprint* 0802.0840)
[17] Borsten L 2008 *Fortschr. Phys.* **56** 842–848
[18] Lévay P and Vrana P 2008 *Phys. Rev.* **A78** 022329 (*Preprint* 0806.4076)
[19] Lévay P, Saniga M and Vrana P 2008 *Phys. Rev.* **D78** 124022 (*Preprint* 0808.3849)
[20] Levay P, Saniga M, Vrana P and Pracna P 2009 *Phys. Rev.* **D79** 084036 (*Preprint* 0903.0541)
[21] Borsten L, Dahanayake D, Duff M J and Rubens W 2009 (*Preprint* 0908.0706)
[22] Levay P and Szalay S 2010 (*Preprint* 1004.2346)
[23] Ferrara S and Günaydin M 1998 *Int. J. Mod. Phys.* **A13** 2075–2088 (*Preprint* hep-th/9708025)
[24] Pioline B 2006 *Class. Quant. Grav.* **23** S981 (*Preprint* hep-th/0607227)
[25] Borsten L, Dahanayake D, Duff M J and Rubens W 2009 *Phys. Rev.* **D80** 026003 (*Preprint* 0903.5517)
[26] Borsten L, Dahanayake D, Duff M J, Rubens W and Ebrahim H 2009 *Phys. Rev.* **A80** 032326 (*Preprint* 0812.3322)
[27] Dür W, Vidal G and Cirac J I 2000 *Phys. Rev.* **A62** 062314 (*Preprint* quant-ph/0005115)
[28] Bennett C H, Popescu S, Rohrlich D, Smolin J A and Thapliyal A V 2000 *Phys. Rev.* **A63** 012307 (*Preprint* quant-ph/9908073)
[29] Plenio M B and Virmani S 2007 *Quant. Inf. Comp.* **7** 1 (*Preprint* quant-ph/0504163)
[30] Horodecki R, Horodecki P, Horodecki M and Horodecki K 2009 *Rev. Mod. Phys.* **81** 865–942 (*Preprint* quant-ph/0702225)
[31] Greenberger D M, Horne M and Zeilinger A 1989 *Bell's Theorem, Quantum Theory and Conceptions of the Universe* (Dordrecht: Kluwer Academic) ISBN 0-7923-0496-9
[32] Greenberger D M, Horne M A, Shimony A and Zeilinger A 1990 *American Journal of Physics* **58** 1131–1143 URL http://link.aip.org/link/?AJP/58/1131/1
[33] Freudenthal H 1954 *Nederl. Akad. Wetensch. Proc. Ser.* **57** 218–230
[34] Brown R B 1969 *J. Reine Angew. Math.* **236** 79–102
[35] Faulkner J R A Construction of Lie Algebras from a Class of Ternary Algebras *Trans. Amer. Math. Soc.* **155** (1971) no. 2, 397–408
[36] Ferrar C J Strictly Regular Elements in Freudenthal Triple Systems *Trans. Amer. Math. Soc.* **174** (1972) 313–331
[37] Krutelevich S 2007 *J. Algebra* **314** 924–977 (*Preprint* math/0411104)
[38] Bellucci S, Ferrara S, Günaydin M and Marrani A 2006 *Int. J. Mod. Phys.* **A21** 5043–5098 (*Preprint* hep-th/0606209)
[39] Borsten L *et al.* 2010 (*Preprint* 1002.4223)
[40] McCrimmon K 2004 *A Taste of Jordan Algebras* (New York: Springer-Verlag New York Inc.) ISBN 0-387-95447-3
[41] Duff M J 2006 *Phys. Lett.* **B641** 335–337 (*Preprint* hep-th/0602160)
[42] Acín A, Andrianov A, Costa L, Jané E, Latorre J I and Tarrach R 2000 *Phys. Rev. Lett.* **85** 1560–1563 (*Preprint* quant-ph/0003050)
[43] Bellucci S, Ferrara S and Marrani A 2007 Contribution to the Proceedings of the XVII SIGRAV Conference, Turin, Italy, 4–7 Sep 2006 (*Preprint* hep-th/0702019)
[44] Bellucci S, Ferrara S, Marrani A and Yeranyan A 2008 *Entropy* **10** 507–555 (*Preprint* 0807.3503)

Supersymmetric lattices[1]

Simon Catterall

Department of Physics, Syracuse University, Syracuse, New York, 13244 USA

E-mail: `smc@physics.syr.edu`

Abstract. Discretization of supersymmetric theories is an old problem in lattice field theory. It has resisted solution until quite recently when new ideas drawn from orbifold constructions and topological field theory have been brought to bear on the question. The result has been the creation of a new class of lattice gauge theory in which the lattice action is invariant under one or more supersymmetries. The resultant theories are local and free of doublers and in the case of Yang-Mills theories also possess exact gauge invariance. In principle they form the basis for a truly non-perturbative definition of the continuum supersymmetric field theory. In this talk these ideas are reviewed with particular emphasis being placed on $\mathcal{N} = 4$ super Yang-Mills theory.

1. Introduction

The problem of formulating supersymmetric theories on lattices has a long history going back to the earliest days of lattice gauge theory. However, after initial efforts failed to produce useful supersymmetric lattice actions the topic languished for many years. Indeed a folklore developed that supersymmetry and the lattice were mutually incompatible. However, recently, the problem has been re-examined using new tools and ideas such as topological twisting, orbifold projection and deconstruction and a class of lattice models have been constructed which maintain one or more supersymmetries exactly at non-zero lattice spacing.

While in low dimensions there are many continuum supersymmetric theories that can be discretized this way, in four dimensions there appears to a unique solution to the constraints – $\mathcal{N} = 4$ super Yang-Mills. The availability of a supersymmetric lattice construction for this theory is clearly very exciting from the point of view of exploring the connection between gauge theories and string/gravitational theories. But even without this connection to string theory it is clearly of great importance to be able to give a non-perturbative formulation of a supersymmetric theory via a lattice path integral in the same way that one can formally define QCD as a limit of lattice QCD as the lattice spacing goes to zero and the box size to infinity. From a practical point of view one can also hope that some of the technology of lattice field theory such as strong coupling expansions and Monte Carlo simulation can be brought to bear on such supersymmetric theories.

In this talk I will outline some of the key ingredients that go into these constructions, the kinds of applications that have been considered so far and highlight the remaining difficulties.

First, let me explain why discretization of supersymmetric theories resisted solution for so long. The central problem is that naive discretizations of continuum supersymmetric

[1] Contributed to 6th International Symposium on Quantum Theory and Symmetries (QTS6), Lexington, Kentucky, USA, 20-25 July 2009

theories break supersymmetry completely and radiative effects lead to a profusion of relevant supersymmetry breaking counterterms in the renormalized lattice action. The coefficients to these counterterms must then be carefully fine tuned as the lattice spacing is sent to zero in order to arrive at a supersymmetric theory in the continuum limit. In most cases this is both unnatural and practically impossible – particularly if the theory contains scalar fields. Of course, one might have expected problems – the supersymmetry algebra is an extension of the Poincaré algebra which is explicitly broken on the lattice. Specifically, there are no infinitesimal translation generators on a discrete spacetime so that the algebra $\{Q,\overline{Q}\} = \gamma_\mu p_\mu$ is already broken at the classical level. Equivalently it is a straightforward exercise to show that a naive supersymmetry variation of a naively discretized supersymmetric theory fails to yield zero as a consequence of the failure of the Leibniz rule when applied to lattice difference operators[2]. In the last five years or so this problem has been revisited using new theoretical tools and ideas and a set of lattice models have been constructed which retain exactly some of the continuum supersymmetry at non-zero lattice spacing. The basic idea is to maintain a particular subalgebra of the full supersymmetry algebra in the lattice theory. The hope is that this exact symmetry will constrain the effective lattice action and protect the theory from dangerous susy violating counterterms.

Two approaches have been pursued to produce such supersymmetric actions; one based on ideas drawn from the field of topological field theory [7, 8, 9, 10] and another pioneered by David B. Kaplan and collaborators using ideas of orbifolding and deconstruction [4, 5, 6]. Remarkably these two seemingly independent approaches lead to the same lattice theories – see [11, 12, 13, 14] and the recent reviews [15, 16]. This convergence of two seemingly completely different approaches leads one to suspect that the final lattice theories may represent essentially unique solutions to the simultaneous requirements of locality, gauge invariance and at least one exact supersymmetry. We will only have time to discuss the approach via topological twisting in this talk.

2. Topological twisting

Perhaps the simplest way to understand how this subalgebra emerges is to reformulate the target theory in terms of "twisted fields". The basic idea of twisting goes back to Witten in his seminal paper on topological field theory [17] but actually had been anticipated in earlier work on staggered fermions [18]. In our context the idea is decompose the fields of the theory in terms of representations not of the original (Euclidean) rotational symmetry $SO_{\rm rot}(D)$ but a twisted rotational symmetry which is the diagonal subgroup of this symmetry and an $SO_{\rm R}(D)$ subgroup of the R-symmetry of the theory.

$$SO(D)' = {\rm diag}(SO_{\rm Lorentz}(D) \times SO_{\rm R}(D)) \quad (1)$$

To be explicit consider the case where the total number of supersymmetries is $Q = 2^D$. In this case I can treat the supercharges of the twisted theory as a $2^{D/2} \times 2^{D/2}$ matrix q. This matrix can be expanded on products of gamma matrices

$$q = \mathcal{Q}I + \mathcal{Q}_\mu\gamma_\mu + \mathcal{Q}_{\mu\nu}\gamma_\mu\gamma_n u + \ldots \quad (2)$$

The 2^D antisymmetric tensor components that arise in this basis are the twisted supercharges and satisfy a corresponding supersymmetry algebra following from the original algebra

$$\mathcal{Q}^2 = 0 \quad (3)$$

[2] Significant work has gone into generalizing the Leibniz rule to finite difference operators in the context of non-commutative models using the techniques of Hopf algebras see [1, 2, 3]. This approach will not be discussed in this talk

$$\{\mathcal{Q}, \mathcal{Q}_\mu\} = p_\mu \quad (4)$$
$$\cdots \quad (5)$$

The presence of the nilpotent scalar supercharge $\mathcal{Q}$ is most important; it is the algebra of this charge that we can hope to translate to the lattice. The second piece of the algebra expresses the fact that the momentum is the $\mathcal{Q}$-variation of something which makes plausible the statement that the energy-momentum tensor and hence the entire action can be written in $\mathcal{Q}$-exact form[3]. Notice that an action written in such a $\mathcal{Q}$-exact form is trivially invariant under the scalar supersymmetry provided the latter remains nilpotent under discretization.

The rewriting of the supercharges in terms of twisted variables can be repeated for the fermions of the theory and yields a set of antisymmetric tensors $(\eta, \psi_\mu, \chi_{\mu\nu}, \ldots)$ which for the case of $Q = 2^D$ matches the number of components of a real Kähler-Dirac field. This repackaging of the fermions of the theory into a Kähler-Dirac field is at the heart of how the discrete theory avoids fermion doubling as was shown by Becher, Joos and Rabin in the early days of lattice gauge theory [19, 20].

It is important to recognize that the transformation to twisted variables corresponds to a simple change of variables in flat space – one more suitable to discretization. A true topological field theory only results when the scalar charge is treated as a true BRST charge and attention is restricted to states annihilated by this charge. In the language of the supersymmetric parent theory such a restriction corresponds to a projection to the vacua of the theory. It is *not* employed in these lattice constructions.

3. An example: 2D super Yang-Mills

This theory satisfies our requirements for supersymmetric latticization; its R-symmetry possesses an $SO(2)$ subgroup corresponding to rotations of the its two degenerate Majorana fermions into each other. Explicitly the theory can be written in twisted form as

$$S = \frac{1}{g^2}\mathcal{Q}\int \mathrm{Tr}\left(\chi_{\mu\nu}\mathcal{F}_{\mu\nu} + \eta[\overline{\mathcal{D}}_\mu, \mathcal{D}_\mu] - \frac{1}{2}\eta d\right) \quad (6)$$

The degrees of freedom are just the twisted fermions $(\eta, \psi_\mu, \chi_{\mu\nu})$ previously described and a complex gauge field $\mathcal{A}_\mu$. The latter is built from the usual gauge field and the two scalars present in the untwisted theory $\mathcal{A}_\mu = A_\mu + iB_\mu$ with corresponding complexified field strength $\mathcal{F}_{\mu\nu}$.

Notice that the original scalar fields transform as vectors under the original R-symmetry and hence become vectors under the twisted rotation group while the gauge fields are singlets under the R-symmetry and so remain vectors under twisted rotations. This structure makes possible the appearance of a complex gauge field in the twisted theory. Notice though,that the theory is only invariant under the usual $U(N)$ gauge symmetry and not its complexified cousin.

The nilpotent transformations associated with $\mathcal{Q}$ are given explicitly by

$$\begin{aligned}
\mathcal{Q}\,\mathcal{A}_\mu &= \psi_\mu \\
\mathcal{Q}\,\psi_\mu &= 0 \\
\mathcal{Q}\,\overline{\mathcal{A}}_\mu &= 0 \\
\mathcal{Q}\,\chi_{\mu\nu} &= -\overline{\mathcal{F}}_{\mu\nu} \\
\mathcal{Q}\,\eta &= d \\
\mathcal{Q}\,d &= 0
\end{aligned}$$

[3] Actually in the case of $\mathcal{N} = 4$ there is an additional $\mathcal{Q}$-closed term needed

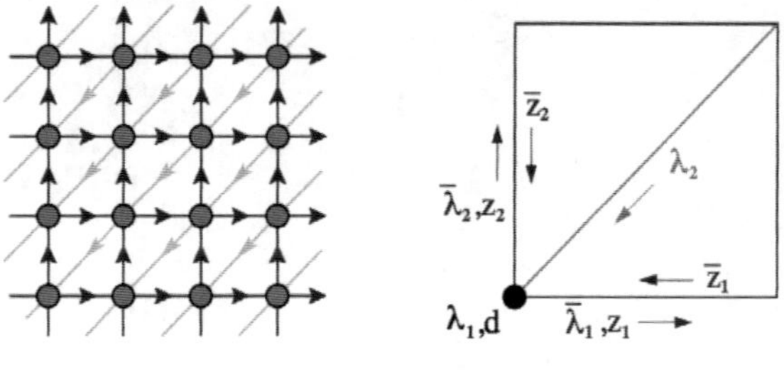

Figure 1. 2d four supercharge lattice

Performing the $\mathcal{Q}$-variation and integrating out the auxiliary field d yields

$$S = \frac{1}{g^2}\int \mathrm{Tr}\left(-\overline{\mathcal{F}}_{\mu\nu}\mathcal{F}_{\mu\nu} + \frac{1}{2}[\overline{\mathcal{D}}_\mu, \mathcal{D}_\mu]^2 - \chi_{\mu\nu}\mathcal{D}_{[\mu}\psi_{\nu]} - \eta\overline{\mathcal{D}}_\mu\psi_\mu\right) \tag{7}$$

To untwist the theory and verify that indeed in flat space it just corresponds to the usual theory one can do a further integration by parts to produce

$$S = \frac{1}{g^2}\int \mathrm{Tr}\left(-F_{\mu\nu}^2 + 2B_\mu D_\nu D_\nu B_\mu - [B_\mu, B_\nu]^2 + L_F\right) \tag{8}$$

where $F_{\mu\nu}$ is the usual Yang-Mills term. It is now clear that the imaginary parts of the gauge fields B_μ can now be given an interpretation as the scalar fields of the usual formulation. Similarly one can build spinors out of the twisted fermions and write the action in the manifestly Dirac form

$$L_F = \begin{pmatrix} \chi_{12} & \frac{\eta}{2} \end{pmatrix} \begin{pmatrix} -D_2 - iB_2 & D_1 + iB_1 \\ D_1 - iB_1 & D_2 - iB_2 \end{pmatrix} \begin{pmatrix} \psi_1 \\ \psi_2 \end{pmatrix} \tag{9}$$

4. Discretization

The prescription for discretization is somewhat natural. (Complex) gauge fields are represented as complexified Wilson gauge links $\mathcal{U}_\mu(x) = e^{\mathcal{A}}_\mu(x)$ living on links of a lattice which for the moment we can think of as hypercubic. These transform in the usual way under $U(N)$ lattice gauge transformations

$$\mathcal{U}_\mu(x) \rightarrow G(x)\mathcal{U}_\mu(x)G^\dagger(x) \tag{10}$$

Supersymmetric invariance then implies that $\psi_\mu(x)$ live on the same links and transform identically. The scalar fermion $\eta(x)$ is clearly most naturally associated with a site and transforms accordingly

$$\eta(x) \rightarrow G(x)\eta(x)G^\dagger(x) \tag{11}$$

The field $\chi_{\mu\nu}$ is slightly more difficult. Naturally as a 2-form it should be associated with a plaquette. In practice we introduce diagonal links running through the center of the plaquette and choose $\chi_{\mu\nu}$ to lie *with opposite orientation* along those diagonal links. This choice of orientation will be necessary to ensure gauge invariance. Figure 1. shows the resultant lattice theory (with $\psi_\mu \rightarrow \overline{\lambda}_\mu$ and $\eta \rightarrow \lambda_1$ and $\chi_{12} \rightarrow \lambda_2$)

To complete the discretization we need to describe how continuum derivatives are to be replaced by difference operators. A natural technology for accomplishing this in the case of adjoint fields was developed many years ago and yields expressions for the derivative operator applied to arbitrary lattice p-forms [21]. In the case discussed here we need just two derivatives given by the expressions

$$\mathcal{D}_\mu^{(+)} f_\nu \;=\; \mathcal{U}_\mu(x) f_\nu(x+\mu) - f_\nu(x)\mathcal{U}_\mu(x+\nu) \tag{12}$$

$$\overline{\mathcal{D}}_\mu^{(-)} f_\mu \;=\; f_\mu(x)\overline{\mathcal{U}}_\mu(x) - \overline{\mathcal{U}}_\mu(x-\mu) f_\mu(x-\mu) \tag{13}$$

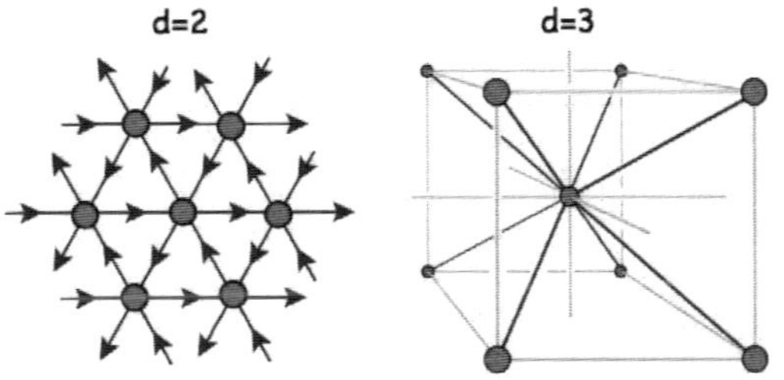

Figure 2. Other $Q = 8$ supersymmetric lattices

The lattice field strength is then given by the gauged forward difference $\mathcal{F}_{\mu\nu} = D^{(+)}_\mu \mathcal{U}_\nu$ and is automatically antisymmetric in its indices. Furthermore it transforms like a lattice 2-form and yields a gauge invariant loop on the lattice when contracted with $\chi_{\mu\nu}$. Similarly the covariant backward difference appearing in $\overline{\mathcal{D}}_\mu \mathcal{U}_\mu$ transforms as a 0-form or site field and hence can be contracted with the site field η to yield a gauge invariant expression.

This use of forward and backward difference operators guarantees that the solutions of the theory map one-to-one with the solutions of the continuum theory and hence fermion doubling problems are evaded [19]. Indeed, by introducing a lattice with half the lattice spacing one can map this Kähler-Dirac fermion action into the action for staggered fermions [22]. Notice that, unlike the case of QCD, there is no rooting problem in this supersymmetric construction since the additional fermion degeneracy is already required by the continuum theory.

Many other examples of supersymmetric lattices exist. Figure 2. shows two such lattices arising in the case of eight supercharges – a two dimensional triangular lattice and a generalized hypercubic lattice (including body and face links) in three dimensions. Notice that in all cases almost all fields live on links with the exception of a small number of fermion site fields – the number of those corrresponding to the number of exact supersymmetries preserved in the lattice theory. Furthermore, in all cases the number of fermions exactly fills out multiples of a basic Kähler-Dirac field in the corresponding number of dimensions.

5. Twisted $\mathcal{N} = 4$ super Yang-Mills

In four dimensions the constraint that the target theory possess 16 supercharges singles out a single theory for which this construction can be undertaken – $\mathcal{N} = 4$ SYM.

The continuum twist of $\mathcal{N} = 4$ that is the starting point of the twisted lattice construction was first written down by Marcus in 1995 [23] although it now plays a important role in the Geometric-Langlands program and is hence sometimes called the GL-twist [24]. This four dimensional twisted theory is most compactly expressed as the dimensional reduction of a five dimensional theory in which the ten (one gauge field and six scalars) bosonic fields are realized as the components of a complexified five dimensional gauge field while the 16 twisted fermions naturally span one of the two Kähler-Dirac fields needed in five dimensions. Remarkably, the action of this theory contains a $\mathcal{Q}$-exact piece of precisely the same form as the two dimensional theory given in eqn. 6 provided one extends the field labels to run now from one to five. In addition the Marcus twist requires a new $\mathcal{Q}$-closed term which was not possible in the two dimensional theory.

$$S_{\text{closed}} = -\frac{1}{8} \int \text{Tr}\, \epsilon_{mnpqr} \chi_{qr} \overline{\mathcal{D}}_p \chi_{mn} \tag{14}$$

The supersymmetric invariance of this term then relies on the Bianchi identity $\epsilon mnpqr \mathcal{D}_p \mathcal{F}_{qr} = 0$.

The four dimensional lattice that emerges from examining the moduli space of the resulting discrete theory is called A_4^* and is constructed from the set of five basis vectors v_a pointing out from the center of a four dimensional equilateral simplex out to its vertices together with their inverses $-v_a$. It is the four dimensional analog of the 2D triangular lattice. Complexified Wilson gauge link variables $\mathcal{U}_a$ are placed on these links together with their $\mathcal{Q}$-superpartners ψ_a. Another 10 fermions are associated with the diagonal links $v_a + v_b$ with $a > b$. Finally, the exact scalar supersymmetry implies the existence of a single fermion for every lattice site. The lattice action corresponds to a discretization of the Marcus twist on this A_4^* lattice and can be represented as a set of traced closed bosonic and fermionic loops. It is invariant under the exact $\mathcal{Q}$ scalar susy, lattice gauge transformations and a global permutation symmetry S^5 and can be proven free of fermion doubling problems as discussed before. The $\mathcal{Q}$-exact part of the lattice action is again given by eqn. 7 where the indices μ, ν now correspond to the indices labeling the five basis vectors of A_4^*.

While the supersymmetric invariance of this $\mathcal{Q}$-exact term is manifest in the lattice theory it is not clear how to discretize the continuum $\mathcal{Q}$-closed term. Remarkably, it is possible to discretize eqn. 14 in such a way that it is indeed exactly invariant under the twisted supersymmetry!

$$S_{\text{closed}} = -\frac{1}{8}\sum_{\mathbf{x}} \text{Tr}\, \epsilon_{mnpqr}\chi_{qr}(\mathbf{x}+\mu_m+\mu_n+\mu_p)\overline{\mathcal{D}}_p^{(-)}\chi_{mn}(\mathbf{x}+\mu_p) \tag{15}$$

and can be seen to be supersymmetric since the lattice field strength satisfies an exact Bianchi identity [21].

$$\epsilon_{mnpqr}\mathcal{D}_p^{(+)}\mathcal{F}_{qr} = 0 \tag{16}$$

6. Numerical results

The lattice theory may be simulated using standard algorithms and techniques drawn from lattice QCD. The first step is to integrate out the twisted fermions which results in a Pfaffian of the twisted fermion kinetic operator $\text{Pf}\left(M(\mathcal{U},\overline{\mathcal{U}})\right)$. Assuming that the phase of this operator can be neglected[4] this Pfaffian can be represented by the integral

$$\text{Pf}M(\mathcal{U},\overline{\mathcal{U}}) = \int \mathcal{D}F\mathcal{D}F^\dagger e^{-F^\dagger\left(M^\dagger M\right)^{\frac{1}{4}}F} \tag{17}$$

where the *bosonic* pseudofermion field F carries the same labels as its fermion counterpart. The fractional inverse power is then approximated to whatever accuracy is desired by a partial fraction expansion whose coefficients are determined using the remez algorithm to minimize the error over some interval in the spectrum. A standard RHMC algorithm can then be used to simulate the full partition function of the resultant theory.

6.1. *Exact supersymmetry*

One of the first things to check is the of course the exact supersymmetry. Table 1. compares the Monte Carlo measurements of the bosonic action against the exact value computed assuming the exact supersymmetry (the bosonic action can be derived using an exact $\mathcal{Q}$ Ward identity). Results are shown for $SU(2)$ and several lattice couplings κ for both the two dimensional model with 4 supercharges and $\mathcal{N}=4$ super Yang-Mills in four dimensions with 16 supercharges [25].

These results confirm rather convincingly that the lattice theory does indeed possess exact supersymmetry.

[4] Numerical and analytical arguments suggest that this is the case

κ	κS_B	exact
1.0	4.40(2)	4.5
10.0	4.47(2)	4.5
100.0	4.49(1)	4.5

$\mathcal{Q} = 4$

κ	κS_B	exact
1.0	13.67(4)	13.5
10.0	13.52(2)	13.5
100.0	13.48(2)	13.5

$\mathcal{Q} = 16$

Table 1. Tests of exact SUSY - bosonic action for $SU(2)$ theory at various lattice couplings

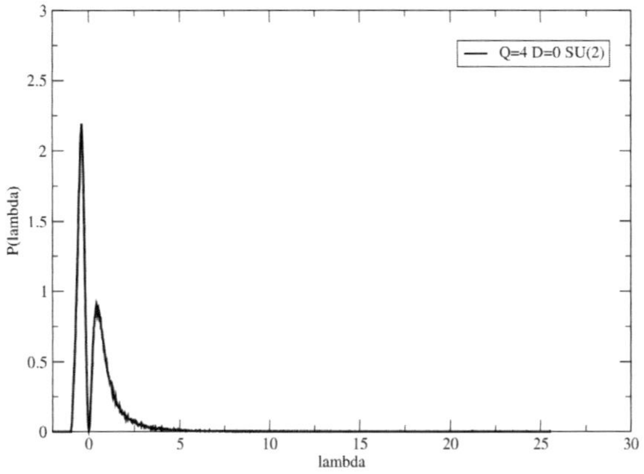

Figure 3. Distribution of eigenvalues of $\overline{\mathcal{U}}_\mu \mathcal{U}_\mu - I$ for $Q = 4$ SU(2) theory

6.2. Moduli space

Next we turn to the moduli space. In the continuum and on the lattice both the 4 and 16 supercharge theories possesses an infinite set of classical vacua corresponding to constant commuting $\mathcal{U}$ matrices. Furthermore, since the complex fields are in principle unbounded from above one might worry that the partition function would be unbounded. In the usual formulation the usual statement is that the scalar fields can run off to infinity along these flat directions. In practice we find that the partition function and low moments of the scalar field distribution are well defined in these theories. Figures 3. and 4. show the distribution of eigenvalues of the scalar fields (given by $\overline{\mathcal{U}}_\mu \mathcal{U}_\mu - 1$) for the 4 and 16 supercharge $SU(2)$ models. While one observes long power law tails particularly in the 4 supercharge case the fields remain localized around the origin in the moduli space, the partition function is finite and no instability is seen in the simulations.

6.3. Fermion eigenvalues

It is interesting to examine the eigenvalues of the fermion operator accumulated over the same set of Monte Carlo configurations used in the moduli space analysis. These are shown in figures 5. and 6. for 4 and 16 supercharges respectively.

Notice these picture show only the region close to the origin – the full spectrum extends to larger values and shows a clustering along the imaginary (vertical) axis corresponding to eigenvalues of the derivative operator appearing in the fermion kinetic terms. Nevertheless, these pictures reveal one strong difference between the 4 and 16 supercharge cases; the relative scarcity of eigenvalues close to the origin in the 16 supercharge case. The absence of a zero

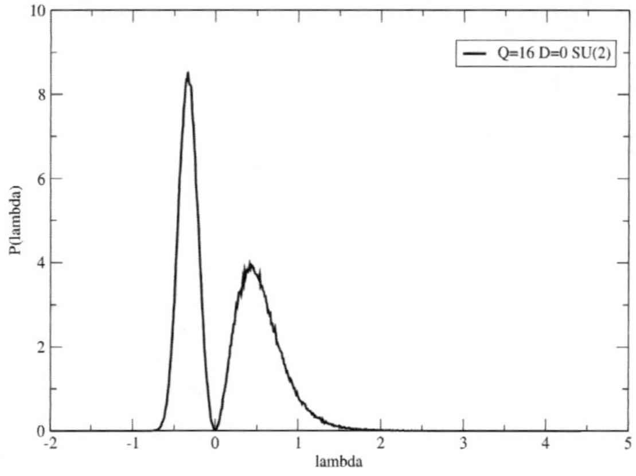

Figure 4. Distribution of eigenvalues of $\overline{\mathcal{U}}_\mu \mathcal{U}_\mu - I$ for $Q = 16$ SU(2) theory

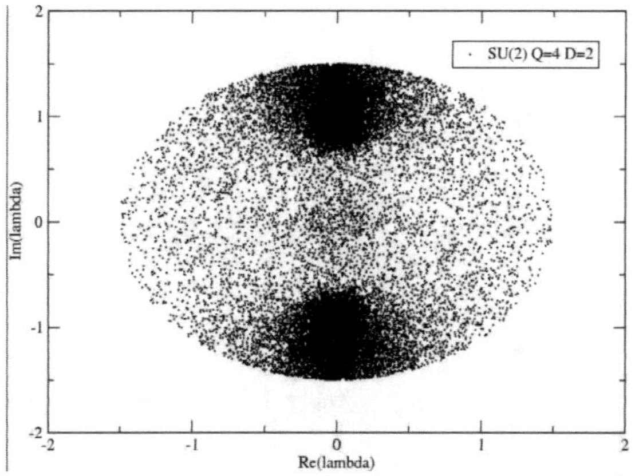

Figure 5. Scatter plot of real and imaginary parts of fermion eigenvalues for $Q = 4$ $SU(2)$ model on 2×2 lattice

eigenvalue in the fermion spectrum indicates by supersymmetry a corresponding absence of boson zero modes – those associated with the classical flat directions. In the context of the Monte Carlo this is consistent with the observation that the scalars do not penetrate far down the classical flat directions but rather correspond to a wavefunction which localizes close to the origin of moduli space. In contrast the scalars in the 4 supercharge theory can be found at relatively large distance from the origin (the tail of the eigenvalue distribution falls only as a single inverse power of the eigenvalue and thus the variance of the eigenvalues actually diverges logarithmically). At such large distances there is a near bosonic zero corresponding to translations along the flat directions and corresponding a near zero fermion eigenvalue as is seen in the spectrum. Indeed, the presence of a nearly massless fermion mode in the spectrum is a necessary condition for supersymmetry breaking where it would play the role of a Goldstino. An order parameter for supersymmetry breaking is the Witten index. The measured value of W is close to zero in this model (it can be measured via the phase of the Pfaffian in the

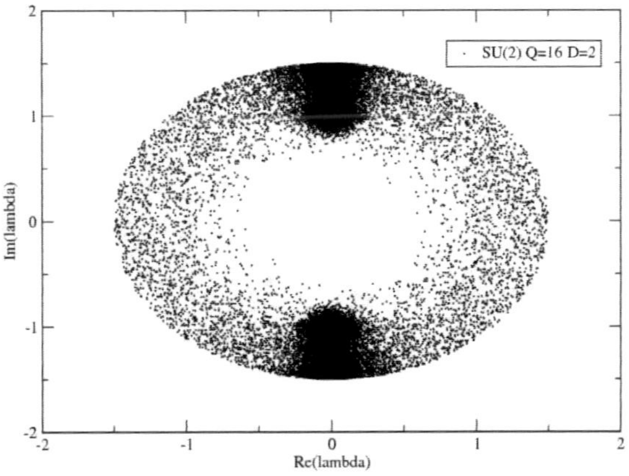

Figure 6. Scatter plot of real and imaginary parts of fermion eigenvalues for $Q = 16$ $SU(2)$ model on 2×2 lattice

phase quenched theory we simulate). Thus we conclude that our simulations yield tantalizing evidence for dynamical supersymmetry breaking in the two dimensional model. No such breaking apparently occurs for the four dimensional theory in line with expectations.

7. Prospects

One of the key issues that still remains to be explored is the question of how much residual fine tuning will be required to achieve a continuum limit in which full supersymmetry is restored. This is controlled by the flows in all relevant operators which could be induced in the effective action as a result of quantum corrections. We have used the exact lattice symmetries together with power counting to enumerate the possible set of such lattice operators.

Only four terms appear in this counter term analysis and three of these correspond to wavefunction renormalizations of kinetic terms already present in the bare lattice action. There is one additional term of the form

$$\mathcal{Q}(\eta\mathcal{U}_\mu\overline{\mathcal{U}}_\mu) \tag{18}$$

which leads to supersymmetric mass terms for the fermions and scalars. However, such a term would lead to a lifting of the classical moduli space which does not happen in the continuum theory. We have computed the effective action to one loop in the lattice theory and find that it vanishes as a consequence of exact $\mathcal{Q}$ supersymmetry just as in the continuum. Indeed, this result can be generalized to any finite order of perturbation theory using the $\mathcal{Q}$-exact property of the lattice action and indicates that such a dangerous radiative correction is *not* induced to all orders of perturbation theory [26]. This is a strong and useful result as it indicates at worst a logarithmic tuning of the other terms in the action will be needed.

Having eliminated all such mass terms the remaining question concerning the restoration of full supersymmetry then rests on whether the ratios of the coefficients to the renormalized kinetic operators flow away from their classical values as the lattice spacing is decreased. A one loop calculation is in progress which should shed light on this issue.

Beyond this issue it would be very interesting to use Monte Carlo simulation to test aspects of the AdS/CFT conjecture. Parallel code has now been developed to study $\mathcal{N} = 4$ super Yang-Mills and a program of numerical investigations of this theory is ongoing. At finite temperature this theory and its dimensional reductions and deformations should be dual to a variety of black hole solution in supergravity – see the recent work reported in [27, 28]. One would hope that the results of such simulations could be useful in the quest to understand how aspects of the quantum geometry can be understood in terms of the dual gauge theory.

8. Acknowledgments
S.C would like to thanks Joel Giedt, Anosh Joseph, Mithat Unsal and Toby Wiseman for useful discussions. Numerical results shown in this talk were obtained using USQCD resources at Fermilab. SMC is supported in part by DOE under grant number DE-FG02-85ER40237.

References
[1] A. D'Adda, I. Kanamori, N. Kawamoto, K. Nagata, Twisted superspace on a lattice, Nucl. Phys. B707 (2005) 100–144. http://arxiv.org/abs/hep-lat/0406029
[2] A. D'Adda, I. Kanamori, N. Kawamoto, K. Nagata, Exact extended supersymmetry on a lattice: Twisted n = 2 super yang-mills in two dimensions, Phys. Lett. B633 (2006) 645–652. http://arxiv.org/abs/hep-lat/0507029
[3] A. D'Adda, I. Kanamori, N. Kawamoto, K. Nagata, Exact Extended Supersymmetry on a Lattice: Twisted N=4 Super Yang-Mills in Three Dimensions, Nucl. Phys. B798 (2008) 168–183. http://arxiv.org/abs/0707.3533
[4] A. G. Cohen, D. B. Kaplan, E. Katz, M. Unsal, Supersymmetry on a euclidean spacetime lattice. i: A target theory with four supercharges, JHEP 08 (2003) 024. http://arxiv.org/abs/hep-lat/0302017
[5] A. G. Cohen, D. B. Kaplan, E. Katz, M. Unsal, Supersymmetry on a euclidean spacetime lattice. ii: Target theories with eight supercharges, JHEP 12 (2003) 031. http://arxiv.org/abs/hep-lat/0307012
[6] D. B. Kaplan, M. Unsal, A euclidean lattice construction of supersymmetric yang- mills theories with sixteen supercharges, JHEP 09 (2005) 042. http://arxiv.org/abs/hep-lat/0503039
[7] S. Catterall, Lattice supersymmetry and topological field theory, JHEP 05 (2003) 038. http://arxiv.org/abs/hep-lat/0301028
[8] F. Sugino, A lattice formulation of super yang-mills theories with exact supersymmetry, JHEP 01 (2004) 015. http://arxiv.org/abs/hep-lat/0311021
[9] S. Catterall, S. Ghadab, Lattice sigma models with exact supersymmetry, JHEP 05 (2004) 044. http://arxiv.org/abs/hep-lat/0311042
[10] S. Catterall, A geometrical approach to n = 2 super yang-mills theory on the two dimensional lattice, JHEP 11 (2004) 006. http://arxiv.org/abs/hep-lat/0410052
[11] P. H. Damgaard, S. Matsuura, Lattice supersymmetry: Equivalence between the link approach and orbifolding, JHEP 09 (2007) 097. http://arxiv.org/abs/0708.4129 [hep-lat]
[12] P. H. Damgaard, S. Matsuura, Classification of Supersymmetric Lattice Gauge Theories by Orbifolding http://arxiv.org/abs/0704.2696 [hep-lat]
[13] S. Catterall, From Twisted Supersymmetry to Orbifold Lattices, JHEP 01 (2008) 048. http://arxiv.org/abs/0712.2532
[14] M. Unsal, Twisted supersymmetric gauge theories and orbifold lattices, JHEP **0610**, 089 (2006) [arXiv:hep-th/0603046].
[15] J. Giedt, Progress in four-dimensional lattice supersymmetry http://arxiv.org/abs/0903.2443
[16] Simon Catterall, David B. Kaplan and Mithat Ünsal, Exact lattice supersymmetry Phys.Rept.484:71-130,2009. http://arxiv.org/abs/0903.4881
[17] E. Witten, Topological quantum field theory, Commun. Math. Phys. 117 (1988) 353.
[18] S. Elitzur, E. Rabinovici, A. Schwimmer, Supersymmetric models on the lattice, Phys. Lett. B119 (1982) 165.
[19] J. M. Rabin, Homology theory of lattice fermion doubling, Nucl. Phys. B201 (1982) 315.
[20] P. Becher, H. Joos, The dirac-kahler equation and fermions on the lattice, Zeit. Phys. C15 (1982) 343.
[21] H. Aratyn, M. Goto, A. H. Zimmerman, A lattice gauge theory for fields in the adjoint representation, Nuovo Cim. A84 (1984) 255.
[22] Tom Banks, Y. Dothan, D. Horn, Geometric Fermions. Phys.Lett.B117:413,1982.
[23] N. Marcus, The other topological twisting of n=4 yang-mills, Nucl. Phys. B452 (1995) 331–345. http://arxiv.org/abs/hep-th/9506002
[24] A. Kapustin, E. Witten, Electric-magnetic duality and the geometric langlands program http://arxiv.org/abs/hep-th/0604151

[25] S. Catterall, First results from simulations of supersymmetric lattices, JHEP 01 (2009) 040. http://arxiv.org/abs/0811.1203

[26] S. Catterall, E. Dzeimkowski, J. Giedt and A. Joseph, in preparation.

[27] Extracting black hole physics from the lattice. Simon Catterall, Toby Wiseman arXiv:0909.4947 [hep-th]
Black hole thermodynamics from simulations of lattice Yang-Mills theory. Simon Catterall, Toby Wiseman, Phys.Rev.D78:041502,2008.
Towards lattice simulation of the gauge theory duals to black holes and hot strings. Simon Catterall, Toby Wiseman, JHEP 0712:104,2007.

[28] First Results from Lattice Simulation of the PWMM. Simon Catterall, Greg van Anders, arXiv:1003.4952 [hep-th]

Journal of Physics: Conference Series **462** (2013) 012007 doi:10.1088/1742-6596/462/1/012007

Associativity, Jacobi, Bremner, and All That

Thomas Curtright

Abstract. I discuss various aspects of multi-linear algebras related to associativity.

Department of Physics, University of Miami, Coral Gables, FL 33124 USA

E-mail: curtright@miami.edu

1. Introduction

Nambu introduced a multilinear operator N-bracket in the context of a novel formulation of mechanics [15]:

$$[A_1 A_2 \cdots A_N] = \sum_{\sigma \in S_N} \operatorname{sgn}(\sigma)\ A_{\sigma_1} \cdots A_{\sigma_N}\ , \tag{1}$$

where the sum is over all $N!$ permutations of the operators. For example, the operator 3-bracket is

$$[ABC] = ABC - ACB + BCA - BAC + CAB - CBA\ . \tag{2}$$

The operator product here is assumed to be *associative*.

The same construction independently appeared in the mathematical literature more than 50 years ago [12, 13]. The theory of such multi-operator products, as well as their "classical limits" in terms of multivariable Jacobians, has been studied extensively [1, 2, 3, 5, 6, 7, 8, 9, 10, 11, 14, 16, 17, 18, 19, 20].

From an algebraic point of view, it is natural to seek *the analogue of the Jacobi identity* for operator N-brackets. For the case of even N-brackets, the obvious generalization where one N-bracket acts on another leads to a true identity. However, for odd N-brackets this usually does *not* work. For instance, it is almost always true that

$$[[ABC]\,DE] - [[ADE]\,BC] - [A\,[BDE]\,C] - [AB\,[CDE]] \neq 0\ . \tag{3}$$

That is to say, the so-called FI ("fundamental identity") *fails*. There is one especially notable exception: $su\,(2)$, as described by Nambu [15].

Fortunately, even brackets are not odd. They need only act twice to yield an identity. Namely [6, 11],

$$[\ B_1 \cdots B_{N-1}\ [B_N \cdots B_{2N-1}]\ ] = 0 \quad \text{for } N \text{ even.} \tag{4}$$

Here, total antisymmetrization of all the Bs is understood. When $N = 2$ this is the familiar Jacobi identity. The proof is by direct calculation and follows as a consequence of associativity.

Unfortunately, an odd N-bracket acting on just one other odd N-bracket does *not* vanish even when totally antisymmetrized over all entries, but rather produces a $(2N-1)$-bracket [6, 5]. Therefore the simplest identity obeyed by odd brackets of only one type, that does not introduce higher-order brackets, requires that they act at least *thrice*.

Journal of Physics: Conference Series **462** (2013) 012007 doi:10.1088/1742-6596/462/1/012007

Bremner Identity and GBIs

Bremner [1] proved an identity (henceforth the "BI") for associative operator 3-brackets acting thrice.

$$[\ [A\ [bcd]\ e]\ f\ g] = [\ [Abc]\ [def]\ g]\ , \tag{5}$$

where it is understood that all *lower* case entries are totally antisymmetrized by implicitly summing over all $6! = 720$ *signed* permutations of them.

The BI can be proven through a resolution of both LHS and RHS as a series of canonically ordered words. By direct calculation we find

$$\begin{aligned}[][A\,[bcd]\,e]\,fg] &= 24\ Abcdefg - 36\ bAcdefg + 36\ bcAdefg \\ &\quad - 24\ bcdAefg + 36\ bcdeAfg - 36\ bcdefAg + 24\ bcdefgA\ .\end{aligned} \tag{6}$$

The same expansion holds for $[[Abc]\,[def]\,g]$, again by direct calculation. That is to say, both $[[A\,[bcd]\,e]\,fg]$ and $[[Abc]\,[def]\,g]$ can be rendered as a 7-bracket plus another 3-bracket containing 3-brackets:

$$[[A\,[bcd]\,e]\,fg] = \frac{1}{20}\ [Abcdefg] - \frac{1}{6}\ [A\,[bcd]\,[efg]] = [[Abc]\,[def]\,g]\ . \tag{7}$$

Thus the BI amounts to the combinatorial statement that there are two distinct ways to write a 7-bracket in terms of nested 3-brackets.

Xiang Jin, Luca Mezincescu, and I proved that a similar identity holds for any odd-order bracket acting thrice [4]. For odd $N = 2L + 1$, this generalized BI ("GBI") is

$$\begin{aligned}&[\ [AB_1 \cdots B_{2L}]\ [B_{2L+1} \cdots B_{4L+1}]\ B_{4L+2} \cdots B_{6L}\ ] \\ &= [\ [A\ [B_1 \cdots B_{2L+1}]\ B_{2L+2} \cdots B_{4L}]\ B_{4L+1} \cdots B_{6L}\ ]\ .\end{aligned} \tag{8}$$

Again, this identity is a consequence of only associativity. Thus all odd brackets built from associative products of operators need only act thrice to yield an identity. Given an hypothesized closed algebra of odd N-brackets, the GBI provides the simplest test for consistency with an underlying associative product.

To prove the GBI, we again expanded in terms of canonically ordered words. By direct calculation,

$$[[A\,[B_1 \cdots B_{2L+1}]\ B_{2L+2} \cdots B_{4L}]\ B_{4L+1} \cdots B_{6L}\ ] = \sum_{n=0}^{6L} (-1)^n\, m_n\ B_1 \cdots B_n\ A\ B_{n+1} \cdots B_{6L}\ , \tag{9}$$

where it is implicit that one is to totally antisymmetrize over all the Bs. All the coefficients in the resolution are integers. Explicitly,

$$\begin{aligned} m_n &= (2L+1)!\,(2L)!\,(2L-1)! \times c_n\ , \\ c_n &= \begin{cases} (n+1)\,(4L-n)\,/2 & \text{for} \quad 0 \le n \le 2L \\ 10L^2 - 6Ln + L + n^2 & \text{for} \quad 2L+1 \le n \le 3L \\ c_{6L-n} & \text{for} \quad 3L+1 \le n \le 6L \end{cases} \ . \end{aligned} \tag{10}$$

The same expansion holds for $[[AB_1 \cdots B_{2L}]\ [B_{2L+1} \cdots B_{4L+1}]\ B_{4L+2} \cdots B_{6L}]$, again by direct calculation. Hence the GBI.

For example, $L = 1$ gives the previous coefficients (6), while $L = 2$ gives

$$\begin{pmatrix} m_0 \\ m_1 \\ m_2 \\ m_3 \\ m_4 \\ m_5 \\ m_6 \\ m_7 \\ m_8 \\ m_9 \\ m_{10} \\ m_{11} \\ m_{12} \end{pmatrix} = 5!4!3! \times \begin{pmatrix} c_0 \\ c_1 \\ c_2 \\ c_3 \\ c_4 \\ c_5 \\ c_6 \\ c_7 \\ c_8 \\ c_9 \\ c_{10} \\ c_{11} \\ c_{12} \end{pmatrix} = 5!4!3! \times \begin{pmatrix} 4 \\ 7 \\ 9 \\ 10 \\ 10 \\ 7 \\ 6 \\ 7 \\ 10 \\ 10 \\ 9 \\ 7 \\ 4 \end{pmatrix} = \begin{pmatrix} 69\,120 \\ 120\,960 \\ 155\,520 \\ 172\,800 \\ 172\,800 \\ 120\,960 \\ 103\,680 \\ 120\,960 \\ 172\,800 \\ 172\,800 \\ 155\,520 \\ 120\,960 \\ 69\,120 \end{pmatrix} . \tag{11}$$

For simplicity, I emphasize the 3-bracket case in the following.

3-Algebras
For a 3-algebra with linearly independent operators T_a that obey

$$[T_a T_b T_c] = i\ F_{abc}{}^{d}\ T_d\ , \tag{12}$$

the BI becomes, with implicit total antisymmetrization of the six b_j indices,

$$F_{ab_1b_2}{}^{x} F_{b_3b_4b_5}{}^{y} F_{xyb_6}{}^{z} = F_{b_1b_2b_3}{}^{x} F_{axb_4}{}^{y} F_{yb_5b_6}{}^{z}\ . \tag{13}$$

Alternatively, after renaming and cycling indices,

$$F_{b_1b_2b_3}{}^{x} \left(F_{axb_4}{}^{y} F_{yb_5b_6}{}^{z} - F_{ab_4b_5}{}^{y} F_{yxb_6}{}^{z} \right) = 0\ . \tag{14}$$

This *trilinear relation* is a condition on the structure constants required by an underlying associativity for any posited 3-algebra.

Exercise (25 points; show all details; due Friday)**:** Use (14) to prove a classification theorem for 3-algebras.

To be more specific, consider now any *closed bilinear algebra* where all commutators and anticommutators are also elements of the algebra, as given by

$$[T_a T_b] = i f_{ab}{}^{c}\ T_c\ , \quad \{T_a T_b\} = g_{ab}{}^{c}\ T_c\ . \tag{15}$$

For example, for $u\,(N)$ with the T_a given by $N \times N$ matrices, the second RHS involves the well-known $d_{ab}{}^{c}$ symbol, as well as Kronecker delta terms. Or, with a bit of freedom of interpretation, one may think of the operator product expansion of any CFT in this way.

For a bilinear algebra of this form, the corresponding 3-algebra is also completely determined, or "induced." This follows from

$$2 \times [ABC] = \{[AB]\,C\} + \{[BC]\,A\} + \{[CA]\,B\}\ . \tag{16}$$

The induced 3-algebra structure constants are given in terms of the f and g symbols by

$$2\ F_{abc}{}^{x} = f_{ab}{}^{u}\ g_{uc}{}^{x} + f_{bc}{}^{u}\ g_{ua}{}^{x} + f_{ca}{}^{u}\ g_{ub}{}^{x}\ . \tag{17}$$

Thus the BI conditions on the induced 3-algebra structure constants can be re-expressed in terms of f and g. Again with implicit antisymmetrizations, the BI conditions become

$$f_{b_1 b_2}{}^{u}\, g_{ub_3}{}^{x} \left(F_{axb_4}{}^{y} F_{yb_5b_6}{}^{z} - F_{ab_4b_5}{}^{y} F_{yxb_6}{}^{z} \right) = 0\ . \tag{18}$$

These conditions are indeed obeyed when the f and g symbols satisfy the conditions wrought by associativity.

The Jacobi identity (JI),

$$[A\,[BC]] + [B\,[CA]] + [C\,[AB]] = 0\ , \tag{19}$$

is a consequence of associativity but it is *not* equivalent to it, even when augmented with the super Jacobi identity (SJI),

$$[\{AB\}\,C] = \{A\,[BC]\} + \{B\,[AC]\}\ . \tag{20}$$

Here, we have used the usual (anti)commutator notation, sans commas, $\{AB\} = AB + BA$ and $[AB] = AB - BA$.

However, there is *another* trilinear identity which, when paired with the SJI, *is* equivalent to associativity. Namely,

$$[A\,[BC]] = \{\{AB\}\,C\} - \{\{AC\}\,B\}\ . \tag{21}$$

For want of a more compelling name, we will refer to this third relation as the "super-duper Jacobi identity" (SDJI)[1]. Note that the JI follows from the SDJI.

For a closed bilinear algebra the SJI and SDJI identities require the following conditions to be obeyed by the structure constants:

$$g_{ab}{}^{u}\, f_{uc}{}^{x} = g_{au}{}^{x}\, f_{bc}{}^{u} + g_{bu}{}^{x}\, f_{ac}{}^{u}\ , \qquad f_{bc}{}^{u}\, f_{au}{}^{x} = g_{ab}{}^{u}\, g_{uc}{}^{x} - g_{ac}{}^{u}\, g_{ub}{}^{x}\ . \tag{22}$$

The more familiar JI conditions on the bilinear algebra structure constants follow from the second of these.

As I mentioned already, the BI conditions for the induced 3-algebra structure constants, expressed in terms of f and g, follow from these two conditions. Moreover, the structure constants F for *any induced N-bracket algebra* can be expressed in terms of f and g for such closed bilinear algebras, and it can be shown that the conditions on F imposed by associativity are indeed satisfied as a consequence of these same two conditions on f and g.

As an *aside*, there are deformed versions of these identities involving the "quommutators"

$$[AB]_\lambda \equiv \lambda AB - \lambda^{-1} BA\ . \tag{23}$$

These naturally lead to 3-brackets. For example, the "Jaquobi identity":

$$[[AB]_\lambda\,C]_\mu + [[BC]_\lambda\,A]_\mu + [[CA]_\lambda\,B]_\mu = \tfrac{1}{2}\left(\lambda + \frac{1}{\lambda}\right)\left(\mu - \frac{1}{\mu}\right)[ABC] + \tfrac{1}{2}\left(\lambda - \frac{1}{\lambda}\right)\left(\mu - \frac{1}{\mu}\right)\{ABC\}\ . \tag{24}$$

Etc.[2] Here, we have also used the totally symmetrized 3-bracket: $\{ABC\} = ABC + ACB + BCA + BAC + CAB + CBA$.

[1] Apologies to Irving Berlin.

[2] The *ultimate Jaquobi identity* would involve six complex parameters, λ_j and μ_j, $j = 1, 2, 3$, as in: $\left[[AB]_{\lambda_1} C\right]_{\mu_1} + \left[[BC]_{\lambda_2} A\right]_{\mu_2} + \left[[CA]_{\lambda_3} B\right]_{\mu_3}$. Requiring that this vanish gives six equations for the parameters. Assuming $\mu_1 \neq 0$, $\lambda_3 \neq 0$, $\mu_3 \neq 0$, the *generic* solution is: $\lambda_1 = \mu_1 \lambda_3 \mu_3$, $\lambda_2 = \mu_1 \lambda_3$, and $\mu_2 = \frac{1}{\mu_1 \mu_3}$. So the solution manifold has complex dimension three, including an overall complex scale. For fixed scale, it is in fact a geometrically *ruled surface*, and it must contain the usual Jacobi, the super Jacobi, and the super-duper Jacobi identities. Indeed, it does. The symmetric group is a symmetry of the solution manifold, so other solutions are given by permutations of 1, 2, 3. There is also parity: Another solution is obtained from the generic one just by flipping the signs of all the parameters.

Classical Manifolds

There are also interesting questions for classical manifold theory that arise in this context. A *classical* 3-bracket is defined by

$$[A, B, C] = \omega^{abc}\ \partial_a A\ \partial_b B\ \partial_c C\ , \tag{25}$$

with antisymmetric but otherwise arbitrary 3-tensor ω^{abc}. The combination that constitutes the so-called FI is

$$\begin{aligned} &[E, F, [A, B, C]] - [[E, F, A], B, C] - [A, [E, F, B], C] - [A, B, [E, F, C]] \\ &= \left(\omega^{abc}\omega^{def} - \omega^{dbc}\omega^{aef} - \omega^{adc}\omega^{bef} - \omega^{abd}\omega^{cef}\right)\ \partial_d\left(\partial_a A\ \partial_b B\ \partial_c C\ \partial_e E\ \partial_f F\right) \\ &+ \left(\partial_a A\ \partial_b B\ \partial_c C\ \partial_e E\ \partial_f F\right)\left(\omega^{def}\partial_d\omega^{abc} - \omega^{dbc}\partial_d\omega^{aef} - \omega^{adc}\partial_d\omega^{bef} - \omega^{abd}\partial_d\omega^{cef}\right)\ . \end{aligned} \tag{26}$$

In the literature, when ω is such that this vanishes, this is called a *Nambu-Poisson manifold.* This gives two types of bilinear constraints on ω, obviously.

But, to conclude this talk, it seems more reasonable (to me at least) that one should impose, instead of the FI, a classical analogue of the BI. For a classical N-bracket involving $n \geq N$ (odd) independent variables, with antisymmetric but otherwise arbitrary N-tensor $\omega^{a_1 \cdots a_N}$,

$$[B_{i_1}, B_{i_2}, \cdots, B_{i_N}] = \omega^{a_1 \cdots a_N}\ \partial_{a_1} B_{i_1}\ \cdots\ \partial_{a_N} B_{i_N}\ , \tag{27}$$

we define a *Bremner-Poisson manifold* as one for which the BI holds. This leads to requirements on the ω tensor that differ from those imposed by the FI. So defined, Bremner-Poisson and Nambu-Poisson manifolds are different, in general. We will discuss this in more detail elsewhere.

Perhaps N-brackets and algebras have an important role to play in physics, as originally suggested by Nambu. Recently there has been considerable interest in N-brackets, especially 3-brackets, as expressed in the physics literature (see [2] and references therein). These ideas await further development.

Acknowledgments *I thank the organizers of this conference for the opportunity to talk about this work, and I thank my collaborators, David Fairlie, Xiang Jin, Luca Mezincescu, and Cosmas Zachos, for sharing their thoughts about Nambu brackets. This work was supported by NSF Awards 0555603 and 0855386.*

[1] M R Bremner, "Identities for the ternary commutator" J Algebra **206** (1998) 615–623; M R Bremner and L A Peresi, "Ternary analogues of Lie and Malcev algebras" Linear Algebra and its Applications **414** (2006) 1-18. (Similar results were independently found by J Nuyts, *unpublished*, 2008.)
[2] T Curtright, D Fairlie, X Jin, L Mezincescu, and C Zachos, "Classical and Quantal Ternary Algebras" Phys Lett **B675** (2009) 387-392 [arXiv:0903.4889 [hep-th]].
[3] T Curtright, D Fairlie, and C Zachos, "Ternary Virasoro-Witt Algebra" Phys Lett **B666** (2008) 386-390 [arXiv:0806.3515 [hep-th]].
[4] T Curtright,X Jin, and L Mezincescu, "Multi-operator brackets acting thrice" J Phys A **42** (2009) 462001, [arXiv:0905.2759 [math-ph]].
[5] T L Curtright and C K Zachos, "Classical and quantum Nambu mechanics" Phys Rev **D68** (2003) 085001 [arXiv:hep-th/0212267].
[6] J A de Azcárraga, and J C Pérez Bueno, "Higher-order simple Lie algebras" Commun Math Phys 184 (1997) 669-681 [arXiv:hep-th/9605213].
[7] C Devchand, D Fairlie, J Nuyts, and G Weingart, "Ternutator Identities" J Phys A **42** (2009) 475209, [arXiv:0908.1738 [hep-th]].
[8] G Dito and M Flato, "Generalized Abelian Deformations: Application to Nambu Mechanics" Lett Math Phys **39** (1997) 107-125 [arXiv:hep-th/9609114].
[9] V T Filippov, "n-Lie Algebras" Sib Math Journal **26** (1986) 879-891; "On n-Lie Algebra of Jacobians" Sib Math Journal **39** (1998) 573-581.
[10] P Gautheron, "Some Remarks Concerning Nambu Mechanics" Lett Math Phys **37** (1996) 103-116.
[11] P Hanlon and M Wachs, "On Lie k-Algebras" Adv Math **113** (1995) 206-236.

[12] P Higgins, "Groups with multiple operators" Proc London Math Soc **6** (1956) 366-416.
[13] A G Kurosh, "Multioperator rings and algebras" Russian Math Surveys **24** (1969) 1-13.
[14] T Lada and J Stasheff, "Introduction to SH Lie algebras for physicists" Int J Theor Phys **32** (1993) 1087-1103.
[15] Y Nambu, "Generalized Hamiltonian Dynamics" Phys Rev **D7** (1973) 2405-2412.
[16] A P Pojidaev, "Enveloping Algebras of Filippov Algebras" Comm Algebra **31** (2003) 883-900.
[17] M Schlesinger and J D Stasheff, "The Lie algebra structure of tangent cohomology and deformation theory" J Pure Appl Algebra **38** (1985) 313-322.
[18] L Takhtajan, "On foundation of the generalized Nambu mechanics" Comm Math Phys **160** (1994) 295-315 [arXiv:hep-th/9301111].
[19] L Vainerman and R Kerner, "On special classes of n-algebras" J Math Physics **37** (1996) 2553-2565.
[20] I Vaisman, "A survey on Nambu–Poisson brackets" Acta Math Univ Comenianae **LXVIII**, 2 (1999) 213-241 [arXiv:math/9901047].

Jordan duality and Freudenthal duality

Duminda Dahanayake

Theoretical Physics, Blackett Laboratory, Imperial College London, London SW7 2AZ, United Kingdom

E-mail: duminda.dahanayake@imperial.ac.uk

Abstract. We define a *Jordan duality* $A \to A^\star$ on the quantised charges A of five dimensional black holes and strings and an analogous *Freudenthal duality* $x \to \tilde{x}$ on the quantised charges x of four dimensional stringy black holes, both of which are distinct from U-duality but leave the lowest order entropy invariant. We discuss the action of these maps on discrete U-duality invariants. Based on joint work with L. Borsten, M. J. Duff, and W. Rubens.

1. Introduction

The quantised charges A of five dimensional black strings and quantised charges B of five dimensional black holes are assigned to elements of an *integral Jordan algebra* $\mathfrak{J}$ whose cubic norm N determines the lowest order entropy

$$S_{5(\text{black string})} = 2\pi\sqrt{N(A)}, \qquad S_{5(\text{black holes})} = 2\pi\sqrt{N(B)}, \tag{1}$$

and whose *reduced structure group* $\mathrm{Str}_0(\mathfrak{J})$ is the U-duality group. Integral cubic Jordan algebras are defined in section 2.1 and the reduced structure group is described in section 2.2. We proceed to define the Jordan duality operation in section 2.3.

Similarly the quantised charges x of black holes of the four dimensional supergravities arising from string and M-theory are assigned to elements of an *integral Freudenthal triple system* (FTS) $\mathfrak{M}(\mathfrak{J})$ (where $\mathfrak{J}$ is the integral cubic Jordan algebra underlying the corresponding 5D supergravity [1, 2, 3, 4, 5, 6, 7]) whose quartic form $\Delta(x)$ determines the lowest order entropy

$$S_4 = \pi\sqrt{|\Delta(x)|}, \tag{2}$$

and whose *automorphism group* $\mathrm{Aut}(\mathfrak{M}(\mathfrak{J}))$ is the U-duality group. FTSs are defined in section 3.1 and the automorphism group is described in section 3.2. We proceed to define the Freudenthal duality operation in section 3.3.

Examples of integral Jordan algebras and FTSs with the corresponding U-duality groups $\mathrm{Str}_0(\mathfrak{J})$ and $\mathrm{Aut}\,\mathfrak{M}(\mathfrak{J})$ are given in Table 1. In particular this includes the cases $\mathcal{N} = 2$ *STU*, $\mathcal{N} = 2$ coupled to n vector multiplets, magic $\mathcal{N} = 2$, and $\mathcal{N} = 8$ [8, 9, 1, 10, 11, 12, 13, 14, 15, 16]. The $\mathcal{N} = 4$ heterotic string with $\mathrm{SL}_2^{\mathbb{Z}} \times \mathrm{SO}_{6,22}^{\mathbb{Z}}$ U-duality may also be included using $\mathfrak{J} = \mathbb{Z} \oplus Q_{5,21}$ [15, 17]. The notation $J_3^{\mathbb{A}}$ denotes sets of 3×3 Hermitian matrices defined over the four division algebras $\mathbb{A} = \mathbb{R}, \mathbb{C}, \mathbb{H}$ or $\mathbb{O}$ (or their split signature cousins). The notation $\mathbb{Z} \oplus Q_n$ denotes the infinite sequence of *spin factors* $\mathbb{Z} \oplus Q_n$, where Q_n is an n-dimensional vector space over $\mathbb{Z}$ [18, 19, 7, 20, 21].

Table 1. The automorphism group $\mathrm{Aut}(\mathfrak{M}(\mathfrak{J}))$ and the dimension of its representation $\dim \mathfrak{M}(\mathfrak{J})$ given by the Freudenthal construction defined over the integral cubic Jordan algebra $\mathfrak{J}$ with dimension $\dim \mathfrak{J}$ and reduced structure group $\mathrm{Str}_0(\mathfrak{J})$. The quantised $\mathcal{N}=8$ theories in 5 and 4 dimensions have U-duality groups $E^{\mathbb{Z}}_{6(6)}$ and $E^{\mathbb{Z}}_{7(7)}$ respectively.

Jordan algebra $\mathfrak{J}$	$\mathrm{Str}_0(\mathfrak{J})$	$\dim \mathfrak{J}$	$\mathrm{Aut}(\mathfrak{M}(\mathfrak{J}))$	$\dim \mathfrak{M}(\mathfrak{J})$
$\mathbb{Z}$	–	1	$\mathrm{SL}_2^{\mathbb{Z}}$	4
$\mathbb{Z}\oplus\mathbb{Z}$	$\mathrm{SO}^{\mathbb{Z}}_{1,1}$	2	$\mathrm{SL}_2^{\mathbb{Z}}\times\mathrm{SL}_2^{\mathbb{Z}}$	6
$\mathbb{Z}\oplus\mathbb{Z}\oplus\mathbb{Z}$	$\mathrm{SO}^{\mathbb{Z}}_{1,1}\times\mathrm{SO}^{\mathbb{Z}}_{1,1}$	3	$\mathrm{SL}_2^{\mathbb{Z}}\times\mathrm{SL}_2^{\mathbb{Z}}\times\mathrm{SL}_2^{\mathbb{Z}}$	8
$\mathbb{Z}\oplus Q_n$	$\mathrm{SO}^{\mathbb{Z}}_{1,1}\times\mathrm{SO}^{\mathbb{Z}}_{n-1,1}$	$n+1$	$\mathrm{SL}_2^{\mathbb{Z}}\times\mathrm{SO}^{\mathbb{Z}}_{2,n}$	$2n+4$
$J_3^{\mathbb{Z}}$	$\mathrm{SL}_3^{\mathbb{Z}}$	6	$\mathrm{Sp}_6^{\mathbb{Z}}$	14
$J_3^{\mathbb{C}}$	$\mathrm{SL}_3^{\mathbb{Z}}$	9	$\mathrm{SU}^{\mathbb{Z}}_{3,3}$	20
$J_3^{\mathbb{H}}$	$\mathrm{SU}_6^{\star\mathbb{Z}}$	15	$\mathrm{SO}_{12}^{\star\ \mathbb{Z}}$	32
$J_3^{\mathbb{O}}$	$E^{\mathbb{Z}}_{6(-26)}$	27	$E^{\mathbb{Z}}_{7(-25)}$	56
$J_3^{\mathbb{O}^s}$	$E^{\mathbb{Z}}_{6(6)}$	27	$E^{\mathbb{Z}}_{7(7)}$	56

The Bekenstein-Hawking entropy and other macroscopic physical quantities are invariant under the continuous U-duality group of the underlying low energy supergravity action, but in the full quantum theory the symmetry is broken to a discrete subgroup due to the Dirac-Schwinger quantisation conditions. As a result, physical quantities in the quantised theory can depend on previously absent discrete invariants. Typically the discrete invariants are given by the gcds of particular dyon charge combinations [22, 23, 24, 25, 26]. These are described for 5D in section 2.4 and for 4D in section 3.4 along with their behaviour under J/F-duality.

2. 5D Black strings/holes and the Jordan dual

2.1. Cubic Jordan algebras

The bilinear product $\circ$, of a Jordan algebra $\mathfrak{J}$ [27, 28, 29, 18, 19], satisfies

$$X\circ Y = Y\circ X, \qquad X^2\circ(X\circ Y) = X\circ(X^2\circ Y), \qquad \forall\ X,Y\in\mathfrak{J}. \tag{3}$$

Of particular interest to us is the class of integral cubic Jordan algebras, which possess a cubic map $N:\mathfrak{J}\to\mathbb{Z}$ homogeneous of degree three

$$N(\lambda X) = \lambda^3 N(X), \quad \forall\ \lambda\in\mathbb{Z},\ X\in\mathfrak{J}. \tag{4}$$

With the aid of a base point $\mathbb{1}\in\mathfrak{J}$ satisfying $N(\mathbb{1})=1$ one may determine the properties of $\mathfrak{J}$ from the cubic form through a construction due to Freudenthal, Springer and Tits [30, 20, 31]. First the cubic form is polarised to a trilinear map

$$N(X,Y,Z) := \tfrac{1}{6}\big[N(X+Y+Z)-N(X+Y)-N(X+Z)-N(Y+Z)+N(X)+N(Y)+N(Z)\big]. \tag{5}$$

Second the trilinear map defines the following four maps

$$\begin{aligned}
&\text{trace} & \mathrm{Tr}(X) &:= 3N(X,\mathbb{1},\mathbb{1}), && (6a)\\
&\text{spur quadratic map} & S(X) &:= 3N(X,X,\mathbb{1}), && (6b)\\
&\text{spur bilinear map} & S(X,Y) &:= 6N(X,Y,\mathbb{1}), && (6c)\\
&\text{trace bilinear form} & \mathrm{Tr}(X,Y) &:= \mathrm{Tr}(X)\,\mathrm{Tr}(Y) - S(X,Y). && (6d)
\end{aligned}$$

Third the trace bilinear form uniquely defines the quadratic adjoint map $^{\#}: \mathfrak{J} \to \mathfrak{J}$ and its polarisation

$$\mathrm{Tr}(X^{\#}, Y) := 3N(X, X, Y), \tag{7a}$$
$$X\#Y := (X+Y)^{\#} - X^{\#} - Y^{\#}. \tag{7b}$$

Finally the Jordan product is defined as

$$X \circ Y := \tfrac{1}{2}\big(X\#Y + \mathrm{Tr}(X)Y + \mathrm{Tr}(Y)X - S(X,Y)\mathbb{1}\big). \tag{8}$$

The result is a cubic Jordan algebra provided the cubic form is *Jordan cubic*, to wit

(i) The trace bilinear form (6d) is non-degenerate.
(ii) The quadratic adjoint map (7a) satisfies

$$(X^{\#})^{\#} = N(X)X, \qquad \forall X \in \mathfrak{J}. \tag{9}$$

For example, in the $J_3^{\mathbb{A}}$ case the Jordan product is $X \circ Y = \frac{1}{2}(XY + YX)$, where XY is just the conventional matrix product. See [19] for a comprehensive account. In all cases, one defines the Jordan triple product as

$$\{X, Y, Z\} := (X \circ Y) \circ Z + X \circ (Y \circ Z) - (X \circ Z) \circ Y. \tag{10}$$

In general an integral Jordan algebra is not closed under the Jordan product, but the cubic norm and trace bilinear form are integer valued, which are the crucial properties for our purposes. Furthermore $\mathfrak{J}$ is closed under the (un)polarised quadratic adjoint map as required.

2.2. Jordan ranks

The structure group, $\mathrm{Str}(\mathfrak{J})$, is composed of all linear bijections on $\mathfrak{J}$ that leave the cubic norm N invariant up to a fixed scalar factor,

$$N(g(X)) = \lambda N(X), \quad \forall g \in \mathrm{Str}(\mathfrak{J}). \tag{11}$$

The reduced structure group $\mathrm{Str}_0(\mathfrak{J})$ leaves the cubic norm invariant and therefore consists of those elements in $\mathrm{Str}(\mathfrak{J})$ for which $\lambda = 1$ [32, 19, 9]. The usual concept of matrix rank may be generalised to cubic Jordan algebras and is invariant under both $\mathrm{Str}(\mathfrak{J})$ and $\mathrm{Str}_0(\mathfrak{J})$ [18, 7]. The ranks are specified by the vanishing or not of three rank polynomials linear, quadratic, and cubic in A (resp. B) as shown in Table 2. Large BPS black holes and strings correspond to rank 3 with $N(A), N(B) \neq 0$ and small BPS correspond to ranks 1 and 2 with $N(A), N(B) = 0$. In Table 2 we have listed the fraction of unbroken supersymmetry for the $\mathcal{N} = 8$ case.

Table 2. Partition of the space $\mathfrak{J}$ into four orbits of $\mathrm{Str}_0(\mathfrak{J})$ or ranks.

Rank	Condition			$\mathcal{N} = 8$ BPS
	A	$A^{\#}$	$N(A)$	
0	$= 0$	$= 0$	$= 0$	-
1	$\neq 0$	$= 0$	$= 0$	1/2
2	$\neq 0$	$\neq 0$	$= 0$	1/4
3	$\neq 0$	$\neq 0$	$\neq 0$	1/8

2.3. Jordan dual

Given a black string with charges A or black hole with charges B, we define its Jordan dual by

$$A^\star := A^\# N(A)^{-1/3}, \qquad B^\star := B^\# N(B)^{-1/3}. \tag{12}$$

J-duality is well defined for large rank 3 strings for which both $A^\#$ and $N(A)$ are nonzero and large rank 3 holes for which both $B^\#$ and $N(B)$ are nonzero. It can be shown [33] that the Jordan dual leaves the cubic form invariant $N(A) = N(A^\star)$ and satisfies $A^{\star\star} = A$. For a valid dual $A^\star$, we require that $N(A)$ is a perfect cube. Despite the non-polynomial nature of the transformation, the J-dual scales linearly in the sense

$$A^\star(nA) = nA^\star(A), \qquad B^\star(nB) = nB^\star(B), \qquad n \in \mathbb{Z}. \tag{13}$$

The U-duality integral invariants $\mathrm{Tr}(X, Y)$ and $N(X, Y, Z)$ are not generally invariant under Jordan duality while $\mathrm{Tr}(A^\star, A)$ and $N(A)$, and hence the lowest-order black hole entropy are. However, higher order corrections to the black hole entropy depend on some of the discrete U-duality invariants, to which we now turn.

2.4. Discrete U-invariants

J-duality commutes with U-duality in the sense that $A^\star$ transforms contragredient to A. This follows from the property that a linear transformation s belongs to the norm preserving group if and only if

$$s(A)\#s(B) = s'(A\#B) \tag{14}$$

where s' is given by

$$\mathrm{Tr}(s(A), s'(B)) = \mathrm{Tr}(A, B) \tag{15}$$

and always belongs to the norm preserving group if s itself does [34]. This implies

$$(s(A))^\star = s'(A^\star). \tag{16}$$

The gcd of a collection of not all zero integral Jordan algebra elements is defined to be the greatest integer that divides them. By definition gcd is positive. The gcd may be used to define the following set of discrete U-duality invariants [6]:

$$d_1(A) = \gcd(A), \qquad d_2(A) = \gcd(A^\#), \qquad d_3(A) = |N(A)|, \tag{17}$$

which are the gcds of the rank polynomials. Clearly $d_3(A)$ is conserved as expected, but this is not necessarily the case for $d_1(A)$ and $d_2(A)$. Nevertheless the product $d_1(A)d_2(A)$ is preserved.

While we require that $N(A)$ is a perfect cube for a valid J-dual this is not a sufficient condition because we further require that

$$d_3(A) = \left[\frac{d_2(A)}{d_1(A^\star)}\right]^3 = \left[\frac{d_2(A^\star)}{d_1(A)}\right]^3 = d_3(A^\star). \tag{18}$$

For $\mathfrak{J} = J_3^{\mathbb{A}}$ with $\mathbb{A} \in \{\mathbb{C}^s, \mathbb{H}^s, \mathbb{O}^s\}$ the orbit representatives of all black strings (holes) have been fully classified [6] and it can be shown that the three d_i uniquely determine the representative.

3. 4D black holes and the Freudenthal dual

3.1. Freudenthal triple systems

Given an integral cubic Jordan algebra $\mathfrak{J}$, one is able to construct an integral FTS by defining the vector space $\mathfrak{M}(\mathfrak{J})$,

$$\mathfrak{M}(\mathfrak{J}) = \mathbb{Z} \oplus \mathbb{Z} \oplus \mathfrak{J} \oplus \mathfrak{J}. \tag{19}$$

with elements $x \in \mathfrak{M}(\mathfrak{J})$ written

$$x = (\alpha, \beta, A, B), \quad \alpha, \beta \in \mathbb{Z}, \quad A, B \in \mathfrak{J}. \tag{20}$$

For convenience we identify the quadratic quantity

$$\kappa(x) := \tfrac{1}{2}(\alpha\beta - \mathrm{Tr}(A, B)). \tag{21}$$

The FTS comes equipped with a non-degenerate bilinear antisymmetric quadratic form, a quartic form and a trilinear triple product [8, 9, 35, 36, 7]:

(i) Quadratic form $\{x, y\}$: $\mathfrak{M}(\mathfrak{J}) \times \mathfrak{M}(\mathfrak{J}) \to \mathbb{Z}$

$$\begin{aligned} \{x, y\} &= \alpha\delta - \beta\gamma + \mathrm{Tr}(A, D) - \mathrm{Tr}(B, C), \\ \text{where} \quad x &= (\alpha, \beta, A, B), \qquad y = (\gamma, \delta, C, D). \end{aligned} \tag{22a}$$

(ii) Quartic form $\Delta : \mathfrak{M}(\mathfrak{J}) \to \mathbb{Z}$

$$\Delta(x) = -4[\kappa(x)^2 + (\alpha N(A) + \beta N(B) - \mathrm{Tr}(A^\#, B^\#))]. \tag{22b}$$

The quartic norm $\Delta(x)$ is either $4k$ or $4k + 1$ for some $k \in \mathbb{Z}$.

(iii) Triple product $T : \mathfrak{M}(\mathfrak{J}) \times \mathfrak{M}(\mathfrak{J}) \times \mathfrak{M}(\mathfrak{J}) \to \mathfrak{M}(\mathfrak{J})$ which is uniquely defined by

$$\{T(w, x, y), z\} = 2\Delta(w, x, y, z), \tag{22c}$$

where $\Delta(w, x, y, z)$ is the fully polarised quartic form such that $\Delta(x, x, x, x) = \Delta(x)$.

Note that all the necessary definitions, such as the cubic and trace bilinear forms, are inherited from the underlying Jordan algebra $\mathfrak{J}$.

3.2. FTS ranks

The automorphism group $\mathrm{Aut}(\mathfrak{M}(\mathfrak{J}))$ is composed of all invertible $\mathbb{Z}$-linear transformations that leave both the antisymmetric bilinear form $\{x, y\}$ and the quartic form $\Delta(x)$ invariant [9]. The conventional concept of matrix rank may be generalised to Freudenthal triple systems in a natural and $\mathrm{Aut}(\mathfrak{M}(\mathfrak{J}))$ invariant manner. The rank of an arbitrary element $x \in \mathfrak{M}(\mathfrak{J})$ is uniquely defined by the vanishing or not of four rank polynomials linear, quadratic (in essence), cubic, and quartic in x as shown in Table 3 [36, 7]. Large BPS and large non-BPS black holes correspond to rank 4 with $\Delta(x) > 0$ and $\Delta(x) < 0$, respectively. Small BPS black holes correspond to ranks 1, 2 and 3 with $\Delta(x) = 0$. In Table 3 we have listed the fraction of unbroken supersymmetry for the $\mathcal{N} = 8$ case.

3.3. Freudenthal dual

Given a black hole with charges x, we define its Freudenthal dual by

$$\tilde{x} := T(x)|\Delta(x)|^{-1/2}, \tag{23}$$

where $T(x) \equiv T(x, x, x) \in \mathfrak{M}(\mathfrak{J})$. F-duality is well defined for large rank 4 black holes for which both $T(x)$ and $\Delta(x)$ are nonzero. It can be shown that the Freudenthal dual leaves the quartic

Table 3. Partition of the space $\mathfrak{M}(\mathfrak{J})$ into five orbits of $\mathrm{Aut}(\mathfrak{M}(\mathfrak{J}))$ or ranks.

Rank	Condition				$\mathcal{N}=8$ BPS
	x	$3T(x,x,y)+\{x,y\}x$	$T(x,x,x)$	$\Delta(x)$	
0	$=0$	$=0\ \forall y$	$=0$	$=0$	-
1	$\neq 0$	$=0\ \forall y$	$=0$	$=0$	1/2
2	$\neq 0$	$\neq 0$	$=0$	$=0$	1/4
3	$\neq 0$	$\neq 0$	$\neq 0$	$=0$	1/8
4	$\neq 0$	$\neq 0$	$\neq 0$	>0	1/8
4	$\neq 0$	$\neq 0$	$\neq 0$	<0	0

form invariant $\Delta(x)=\Delta(\tilde{x})$ and satisfies $\tilde{\tilde{x}}=-x$. For a valid $\tilde{x}$ we require that $\Delta(x)$is a perfect square. Despite the non-polynomial nature of the transformation, the F-dual scales linearly in the sense

$$\tilde{x}(nx)=n\tilde{x}(x), \qquad n\in\mathbb{Z}. \tag{24}$$

The U-duality integral invariants $\{x,y\}$ and $\Delta(x,y,z,w)$ are not generally invariant under Freudenthal duality while $\{\tilde{x},x\}$, $\Delta(x)$, and hence the lowest-order black hole entropy, are invariant. However, higher order corrections to the black hole entropy depend on some of the discrete U-duality invariants, to which we now turn.

3.4. Discrete U-invariants

We make the important observation that since

$$T(\sigma(x),\sigma(y),\sigma(z))=\sigma(T(x,y,z)), \qquad \forall\sigma\in\mathrm{Aut}\,\mathfrak{M}(\mathfrak{J}), \tag{25}$$

F-duality commutes with U-duality

$$\widetilde{\sigma(x)}=\sigma(\tilde{x}). \tag{26}$$

The gcd of a collection of not all zero integral FTS elements is defined to be the greatest integer that divides them. By definition gcd is positive. The gcd may be used to define the following[1] set of discrete U-duality invariants [7, 26]:

$$\begin{aligned} &d_1(x)=\gcd(x), && d_3(x)=\gcd(T(x,x,x)),\\ &d_2(x)=\gcd(3T(x,x,y)+\{x,y\}x)\quad\forall y, && d_4(x)=|\Delta(x)|,\\ &d_2'(x)=\gcd(B^{\#}-\alpha A,\ A^{\#}-\beta B,\ 2\kappa(x)C+2\{A,B,C\})\quad\forall C, && d_4'(x)=\gcd(x\wedge T(x)), \end{aligned} \tag{27}$$

where $\wedge$ denotes the antisymmetric tensor product. As in the 5D case, the unprimed invariants are gcds of the rank polynomials. Clearly $d_4(x)$ is conserved as expected and it can be shown [33] that $d_4'(x)$, $d_2(x)$, and $d_2'(x)$ are also invariant, but this is not necessarily the case for $d_1(x)$ and $d_3(x)$. Nevertheless the product $d_1(x)d_3(x)$ is preserved.

Typically, the literature on exact 4D black hole degeneracies [37, 38, 39, 40, 41, 42, 43, 15, 22, 23, 44, 24, 45, 46, 25, 26] deals only with *primitive* black holes $d_1(x)=1$. We are not required to impose this condition and generically do not do so. More generally a quantity is termed primitive if it has unit gcd. A related simplifying concept is *projectivity*, wherein a charge vector for the cases $\mathfrak{J}=\mathbb{Z}\oplus\mathbb{Z}\oplus\mathbb{Z}, J_3^{\mathbb{C}^s}, J_3^{\mathbb{H}^s}, J_3^{\mathbb{O}^s}$ is projective if the components of it's quadratic rank tensor are primitive.

[1] In the $\mathcal{N}=8$ case Sen [26] denotes $d_2'(x)$ by ψ, and $d_4'(x)$ by χ.

While we require that $|\Delta(x)|$ is a perfect square for a valid F-dual this is not a sufficient condition because we further require that

$$d_4(x) = \left[\frac{d_3(x)}{d_1(\tilde{x})}\right]^2 = \left[\frac{d_3(\tilde{x})}{d_1(x)}\right]^2 = d_4(\tilde{x}). \tag{28}$$

Unlike in 5D the invariants (27) are insufficient to uniquely determine the orbit representatives for the $\mathfrak{J} = J_3^{\mathbb{A}}$ with $\mathbb{A} \in \{\mathbb{C}^s, \mathbb{H}^s, \mathbb{O}^s\}$ cases.

4. Conclusions

A subset of 5D black holes admit a Jordan dual $A^\star$ preserving some, but not all, discrete U-invariants. Similarly, a subset of 4D black holes admit a Freudenthal dual $\tilde{x}$ preserving some, but not all, discrete U-invariants. In both the 4D and 5D cases, if the discrete invariant d_1 is preserved by the J/F-duality map, then all the listed discrete invariants are preserved. When d_1 isn't conserved the J/F-dual is not U-related to the original charge vector. In the simpler 5D case the preservation of d_1 ensures that $A^\star$ is U-related to A, but in 4D the analogous conclusion only holds in the projective case. For non-projective 4D black holes the situation is complicated by the absence of a complete orbit classification and uncertainty regarding what invariants are relevant to the higher-order corrections. This situation is summarised in Table 4. It would be interesting if in the non-projective case there were configurations with the same precision entropy that are F-related but not U-related.

Table 4. Are F or J duals related by U-duality?

Duality	d_1 conserved ?		U-dual ?
F-dual	Yes	Projective	Yes
		Non-projective	?
F-dual	No		No
J-dual	Yes		Yes
J-dual	No		No

From a physical standpoint F-duality and J-duality are defined on charge vectors rather than component fields of the lowest order action so their microscopic stringy interpretation remains unclear, but we remark that two black holes related by F-duality in 4D are related by J-duality when lifted to 5D. The 4D/5D lift [47] relates the entropy of non-rotating 4D black holes to the entropy of rotating 5D black holes and it can be shown [33] that the lift of the F-dual is related to the J-dual of the lift thus:

$$\begin{array}{ccc} \text{4D black hole } x & \xrightarrow{\text{4D/5D lift}} & \text{5D black string } A \sim \tilde{B}^\star \\ {\scriptstyle \text{Freudenthal dual}}\big\downarrow & & \big\downarrow{\scriptstyle \text{Jordan dual}} \\ \text{dual 4D black hole } \tilde{x} & \xrightarrow[\text{4D/5D lift}]{} & \text{dual 5D black hole } \tilde{B} \sim A^\star \end{array} . \tag{29}$$

Acknowledgments

The author wishes to thank L. Borsten, M. J. Duff, and W. Rubens, with whom this work was done.

Journal of Physics: Conference Series **462** (2013) 012008 doi:10.1088/1742-6596/462/1/012008

References

[1] Günaydin M, Sierra G and Townsend P K 1983 *Phys. Lett.* **B133** 72
[2] Günaydin M, Sierra G and Townsend P K 1984 *Nucl. Phys.* **B242** 244
[3] Günaydin M, Sierra G and Townsend P K 1985 *Nucl. Phys.* **B253** 573
[4] Elkies N and Gross B H 1996 *Internat. Math. Res. Notices* **14** 665–698
[5] Gross B H 1996 *Invent. Math.* **124** 263–279
[6] Krutelevich S 2002 *J. Algebra* **253** 276–295
[7] Krutelevich S 2007 *J. Algebra* **314** 924–977 (*Preprint* math/0411104)
[8] Freudenthal H 1954 *Nederl. Akad. Wetensch. Proc. Ser.* **57** 218–230
[9] Brown R B 1969 *J. Reine Angew. Math.* **236** 79–102
[10] Ferrara S and Maldacena J M 1998 *Class. Quant. Grav.* **15** 749–758 (*Preprint* hep-th/9706097)
[11] Ferrara S and Günaydin M 1998 *Int. J. Mod. Phys.* **A13** 2075–2088 (*Preprint* hep-th/9708025)
[12] Günaydin M 2005 *AIP Conf. Proc.* **767** 268–287 (*Preprint* hep-th/0502235)
[13] Bellucci S, Ferrara S, Günaydin M and Marrani A 2006 *Int. J. Mod. Phys.* **A21** 5043–5098 (*Preprint* hep-th/0606209)
[14] Ferrara S, Gimon E G and Kallosh R 2006 *Phys. Rev.* **D74** 125018 (*Preprint* hep-th/0606211)
[15] Pioline B 2006 *Class. Quant. Grav.* **23** S981 (*Preprint* hep-th/0607227)
[16] Borsten L, Dahanayake D, Duff M J, Ebrahim H and Rubens W 2009 *Phys. Rep.* **471** 113–219 (*Preprint* 0809.4685)
[17] Gunaydin M and Pavlyk O 2009 (*Preprint* 0901.1646)
[18] Jacobson N 1961 *J. Reine Angew. Math.* **207** 61–85
[19] Jacobson N 1968 *Structure and Representations of Jordan Algebras* vol 39 (Providence, Rhode Island: American Mathematical Society Colloquium Publications)
[20] McCrimmon K The Freudenthal-Springer-Tits construction of exceptional Jordan algebras *Trans. Amer. Math. Soc.* **139** (1969) 495–510
[21] Baez J C 2001 *Bull. Amer. Math. Soc.* **39** 145–205 (*Preprint* math/0105155)
[22] Sen A 2008 *Gen. Rel. Grav.* **40** 2249–2431 (*Preprint* 0708.1270)
[23] Banerjee S and Sen A 2008 *JHEP* **03** 022 (*Preprint* 0712.0043)
[24] Banerjee S and Sen A 2008 *JHEP* **04** 012 (*Preprint* 0801.0149)
[25] Sen A 2008 *JHEP* **07** 118 (*Preprint* 0803.1014)
[26] Sen A 2008 *JHEP* **08** 037 (*Preprint* 0804.0651)
[27] Jordan P 1933 *Zschr. f. Phys.* **80** 285
[28] Jordan P 1933 *Nachr. Ges. Wiss. Göttingen II* **39** 209–214
[29] Jordan P, von Neumann J and Wigner E P On an algebraic generalization of the quantum mechanical formalism *Ann. Math.* **35** (1934) no. 1, 29–64
[30] Springer T A 1962 *Nederl. Akad. Wetensch. Proc. Ser. A* **24** 259–265
[31] McCrimmon K 2004 *A Taste of Jordan Algebras* (New York: Springer-Verlag New York Inc.) ISBN 0-387-95447-3
[32] Schafer R 1966 *Introduction to Nonassociative Algebras* (New York: Academic Press Inc.) ISBN 0-1262-2450-1
[33] Borsten L, Dahanayake D, Duff M J and Rubens W 2009 *Phys. Rev.* **D80** 026003 (*Preprint* 0903.5517)
[34] Springer T A and Veldkamp F D 2000 *Octonions, Jordan Algebras and Exceptional Groups* (Berlin, Heidelberg, New York: Springer-Verlag) ISBN 3-540-66337-1
[35] Faulkner J R A Construction of Lie Algebras from a Class of Ternary Algebras *Trans. Amer. Math. Soc.* **155** (1971) no. 2, 397–408
[36] Ferrar C J Strictly Regular Elements in Freudenthal Triple Systems *Trans. Amer. Math. Soc.* **174** (1972) 313–331
[37] Strominger A and Vafa C 1996 *Phys. Lett.* **B379** 99–104 (*Preprint* hep-th/9601029)
[38] Dijkgraaf R, Verlinde E P and Verlinde H L 1997 *Nucl. Phys.* **B484** 543–561 (*Preprint* hep-th/9607026)
[39] Maldacena J M, Moore G W and Strominger A 1999 (*Preprint* hep-th/9903163)
[40] Pioline B 2005 *JHEP* **08** 071 (*Preprint* hep-th/0506228)
[41] Dabholkar A, Denef F, Moore G W and Pioline B 2005 *JHEP* **10** 096 (*Preprint* hep-th/0507014)
[42] Shih D, Strominger A and Yin X 2006 *JHEP* **06** 037 (*Preprint* hep-th/0506151)
[43] Shih D, Strominger A and Yin X 2006 *JHEP* **10** 087 (*Preprint* hep-th/0505094)
[44] Banerjee S, Sen A and Srivastava Y K 2008 *JHEP* **05** 098 (*Preprint* 0802.1556)
[45] Banerjee S, Sen A and Srivastava Y K 2008 *JHEP* **05** 101 (*Preprint* 0802.0544)
[46] Sen A 2008 *JHEP* **07** 078 (*Preprint* 0803.3857)
[47] Gaiotto D, Strominger A and Yin X 2006 *JHEP* **02** 024 (*Preprint* hep-th/0503217)

Journal of Physics: Conference Series **462** (2013) 012009 doi:10.1088/1742-6596/462/1/012009

The Double Life of Thermal QCD

Mohammed Mia, Keshav Dasgupta, Charles Gale, Sangyong Jeon

Department of Physics, McGill University, 3600 rue University, Montréal, QC, Canada H3A 2T8

Abstract. We study the gravity dual of a thermal gauge theory whose behavior parallels that of thermal QCD in the far IR. The UV of our theory has infinite degrees of freedom. We holographically renormalize the supergravity action to compute the stress tensor of the dual gauge theory with fundamental flavors incorporating the logarithmic running of the gauge coupling. From the stress tensor we obtain the shear viscosity and the entropy of the medium at a temperature T, and investigate the violation of the bound for the viscosity to the entropy ratio. This is a shortened and simplified companion paper to hep-th/0902.1540, and is based on the talks given by M. Mia at the "Strong and Electroweak Matter 2008" workshop and the McGill workshop on "AdS/CFT, Condensed Matter and QCD" in the fall of 2008; and K. Dasgupta at the "Sixth International Symposium on Quantum Theory and Symmetries (QTS6)", July 2009.

1. Introduction

Strongly coupled Quark Gluon Plasma (QGP) poses theoretically challenging yet experimentally accessible questions. The formation of QGP at RHIC is an example where theoretical descriptions are completely lacking at low energies because our perturbative techniques fail at strong couplings. However probing the nonperturbative regime of large N gauge theory through a *gravity* dual has led to some interesting results for the physics of the quark gluon plasma. A popular approach so far has been the AdS/CFT correspondence [1], even though QCD is not a conformal field theory at UV. However, for certain gauge theories with running couplings, there exist gravity duals. At zero temperatures these gravity duals are studied with [2] and without [3, 4] fundamental flavors. On the other hand, at high temperatures, there are examples of gravity duals without fundamental flavors in the literature [7]. In this paper we give an example of a gravity dual for a specific thermal gauge theory with fundamental flavors and logarithmically running coupling constants (see also [8]). Our aim here is to study quantities such as shear viscosity, entropy and the viscosity to entropy ratio with our gravity dual.

2. The Duality

The gauge theory that we consider arises from stacking N D3 branes at the tip of a six dimensional conifold with base $T^{1,1} = S^2 \times S^3$, then placing M D5 branes in such a way that it wraps the two cycle S^2 on the base of the conifold alongwith N_f $D7$ branes embedded via so-called Ouyang embedding [2] to introduce fundamental matter [5]. Finally by euclideanising and periodically identifying the time coordinate we can introduce temperature in the gauge theory. However the high temperature, i.e temperature above the deconfinement temperature that we would be interested in, will be generated by inserting a black hole in the dual gravitational backgound [5]. The D3 brane world volume and the unwrapped directions of

D5 and D7 branes extends in the four Minkowski directions. Strings can end on any of these branes and the excitation of the strings are described by an $SU(N+M) \times SU(N)$ gauge group with N_f flavors. If g_1, g_2 denote the gauge couplings of $SU(N+M)$ and $SU(N)$, then they have nontrivial beta functions (at zero temperature) $\frac{\partial}{\partial \log \Lambda}\left[\frac{4\pi^2}{g_1^2}+\frac{4\pi^2}{g_2^2}\right] = -\frac{3N_f}{4}$ and $\frac{\partial}{\partial \log \Lambda}\left[\frac{4\pi^2}{g_1^2}-\frac{4\pi^2}{g_2^2}\right] = 3M\left(1+\frac{3g_sN_f}{2\pi}\log\Lambda\right)$. We see that the two gauge couplings run in opposite directions and in the IR $SU(N+M)$ flows to strong coupling. Performing a Seiberg duality transformation, we identify the strongly coupled $SU(N+M)$ with a weakly coupled $SU(N-(M-N_f))$ at the IR. We see that not only the number of colors are reduced, but the difference of the size of the gauge group now decreases from M in [3] to $M - N_f$. This difference will decrease by the increments of N_f until it is smaller than or equal to N_f. Then there are two possible end points: (a) if N is still greater than zero then we will have an approximately conformal theory, or (b) if N decreases to zero but with finite M left over then we will have a $SU(M)$ theory with N_f flavors that confines in the far IR (see [2] for more details). The latter theory, or more particularly the high temperature limit of the latter theory, is what we are interested in and henceforth we will only consider that[1].

At zero temperature the dual description (assuming of course that in this limit the gauge theory decouples from gravity) involves D7 branes, deformed conifold and fluxes [2]. Due to the underlying RG flows the supergravity description *does not* capture the choppy cascading nature of the gauge theories! Rather the supergravity description only captures the smooth RG flows in the theory. This would also mean that the actual *number* of colors in the theory is rather subtle to define. These details are explored in [4, 5].

Once we switch on a temperature in the gauge theory, the dual gravity description loses its simplicity. We can no longer claim that the fluxes, warp factor etc would remain unchanged. Even the internal manifold cannot remain a simple warped deformed conifold any more. All the internal spheres would get squashed, and at $r = 0$ there could be both resolution as well as deformation of the two and three cycles respectively. Due to this complicated nature of our background our entire analysis in [5] is based on the following limit:

$$\left(g_s,\ g_sN_f,\ g_s^2MN_f,\ \frac{g_sM^2}{N}\right) \rightarrow 0, \qquad (g_sN,\ g_sM) \rightarrow \infty \tag{1}$$

In [5] we presented our results to $\mathcal{O}(g_sN_f, g_sM^2/N)$, and discuss how to extend this to higher orders in g_sN_f and g_sM^2/N. In the limit where the deformation parameter is small, we show that to $\mathcal{O}(g_sN_f, g_sM^2/N)$ we can analytically derive the background taking a resolved conifold geometry. The resolution parameter a depends on $g_sN_f, g_sM^2/N$ as well as on the horizon radius r_h i.e $a = a_0 + \mathcal{O}(g_sN_f, g_sM^2/N, r_h)$ with a_0 constant. Our background then has the

[1] For a review of Seiberg dualities and cascading theories, see [4]. For a review on brane constructions for cascading theories see [6].

following form:

$$\begin{aligned}
ds^2 &= \frac{1}{\sqrt{h}}\Big[-g_1(r)dt^2+dx^2+dy^2+dz^2\Big]+\sqrt{h}\Big[g_2(r)^{-1}dr^2+d\mathcal{M}_5^2\Big]\\
\widetilde{F}_3 &= 2M\mathbf{A_1}\left(1+\frac{3g_sN_f}{2\pi}\log r\right)e_\psi\wedge\frac{1}{2}(\sin\theta_1\,d\theta_1\wedge d\phi_1-\mathbf{B_1}\sin\theta_2\,d\theta_2\wedge d\phi_2)\\
&-\frac{3g_sMN_f}{4\pi}\mathbf{A_2}\frac{dr}{r}\wedge e_\psi\wedge\left(\cot\frac{\theta_2}{2}\sin\theta_2\,d\phi_2-\mathbf{B_2}\cot\frac{\theta_1}{2}\sin\theta_1\,d\phi_1\right)\\
&-\frac{3g_sMN_f}{8\pi}\mathbf{A_3}\sin\theta_1\sin\theta_2\left(\cot\frac{\theta_2}{2}d\theta_1+\mathbf{B_3}\cot\frac{\theta_1}{2}d\theta_2\right)\wedge d\phi_1\wedge d\phi_2\\
H_3 &= 6g_s\mathbf{A_4}M\left(1+\frac{9g_sN_f}{4\pi}\log r+\frac{g_sN_f}{2\pi}\log\sin\frac{\theta_1}{2}\sin\frac{\theta_2}{2}\right)\frac{dr}{r}\wedge\frac{1}{2}\Big(\sin\theta_1\,d\theta_1\wedge d\phi_1\\
&-\mathbf{B_4}\sin\theta_2\,d\theta_2\wedge d\phi_2\Big)+\frac{3g_s^2MN_f}{8\pi}\mathbf{A_5}\mathcal{D}(r,\psi)\wedge\left(\cot\frac{\theta_2}{2}d\theta_2-\mathbf{B_5}\cot\frac{\theta_1}{2}d\theta_1\right)\\
C_0 &= \frac{N_f}{4\pi}(\psi-\phi_1-\phi_2),\qquad F_5=\frac{1}{g_s}\left[d^4x\wedge dh^{-1}+*(d^4x\wedge dh^{-1})\right]\\
e^{-\Phi} &= \frac{1}{2g_s}\left[\frac{1}{r^{\epsilon_a}}-\frac{3\epsilon_a a^2}{2r^2}-\frac{g_sN_f}{2\pi}\log\left(\sin\frac{\theta_1}{2}\sin\frac{\theta_2}{2}\right)+\text{constant}\right] \qquad (2)
\end{aligned}$$

where $\mathcal{D}(r,\psi)\equiv\frac{dr}{r}\wedge e_\psi-\frac{1}{2}de_\psi$; g_i are the black hole factors[2]; $\epsilon_a=\frac{3g_sN_f}{4\pi}$; $\mathbf{A}_i,\mathbf{B}_i$ are $\mathcal{O}(a^2,g_sN_f)$ corrections that take us away from the zero temperature Ouyang background [2], and are worked out in details in [5]; and $\mathcal{M}_5$ is a warped *resolved* conifold with the warp factor:

$$h=\frac{L^4}{r^{4-\epsilon_1}}+\frac{L^4}{r^{4-2\epsilon_2}}-\frac{2L^4}{r^{4-\epsilon_2}}+\frac{L^4}{r^{4-r^{\epsilon_2^2/2}}}\equiv\sum_{\alpha=1}^{4}\frac{L^4_{(\alpha)}}{r^4_{(\alpha)}} \qquad (3)$$

where $\epsilon_i, r_{(\alpha)}$ etc are defined as:

$$\begin{aligned}
\epsilon_1 &= \frac{3g_sM^2}{2\pi N}+\frac{g_s^2M^2N_f}{8\pi^2N}+\frac{3g_s^2M^2N_f}{8\pi N}\log\left(\sin\frac{\theta_1}{2}\sin\frac{\theta_2}{2}\right),\quad \epsilon_2=\frac{g_sM}{\pi}\sqrt{\frac{2N_f}{N}}\\
&r_{(\alpha)}=r^{1-\epsilon_{(\alpha)}},\quad \epsilon_{(1)}=\frac{\epsilon_1}{4},\quad \epsilon_{(2)}=\frac{\epsilon_2}{2},\quad \epsilon_{(3)}=\frac{\epsilon_2}{4},\quad \epsilon_{(4)}=\frac{\epsilon_2^2}{8}\\
&r_{(\pm\alpha)}=r^{1\mp\epsilon_{(\alpha)}},\quad L_{(1)}=L_{(2)}=L_{(4)}=L^4,\quad L_{(3)}=-2L^4 \qquad (4)
\end{aligned}$$

and L^4 is defined in the same way as in the AdS/CFT correspondence. The D7 branes are embedded via Ouyang embedding in this background. Note that our analysis to this order is analytic, but beyond this order we loose all analytic control and we can only find the background numerically (see also [7] for the background without flavor D7 branes and [9] for the background with flavor branes). Furthermore the log r dependences of all the fluxes, warp factor and dilaton is only for regions close to the D7 brane. To get a good behavior at far infinity we need to embed the whole system in F-theory [10]. That would mean inserting extra $24-N_f$ seven branes at infinity; and then all the fluxes would go as inverse powers of r at large r. In fact in [5] we made a concrete prediction for the large r warp factor: it is given by (3) except that the sum over α now ranges from $-\infty$ to $+\infty$. This way we can get rid of possible Landau poles (or naked singularities) in the background (see also [11]).

[2] Note that they are generically unequal.

Once we know the dual background we can say that the Hilbert space of the gauge theory can be obtained from the Hilbert space of the string theory on this geometry. In fact the dual supergravity background captures the strongly coupled gauge theory where we expect the RG flow to be smooth without any choppy Seiberg dualities, although we might be able to interpret the strongly coupled gauge theory also as some kind of approximate cascading theory.

As we now know very well, the UV completion of cascading type theories require *infinite* degrees of freedom. Once we have infinite degrees of freedom at the UV, we no longer expect a finite boundary action from supergravity analysis! What we need is to regularise and renormalise the supergravity boundary action so that finite correlation functions could be extracted. This would also mean that the usual Witten type proposal [12] for the AdS/CFT correspondence can be re-expressed in terms of the boundary variables to give us the complete picture. Therefore we can rewrite the ansatze proposed by Witten *et al* [12] for our background geometry to take the following Wilsonian form:

$$\begin{aligned} \mathcal{Z}_{\rm QCD}[\phi_0] &\equiv \langle \exp \int_{M^4} \phi_0 \mathcal{O} \rangle = \mathcal{Z}_{\rm total}[\phi_0] \\ &\equiv \exp(S_{\rm total}[\phi_0] + S_{\rm GH} + S_{\rm counterterm}) \equiv \exp S_{\rm ren}[\phi_0] \end{aligned} \tag{5}$$

where M^4 is Minkowski manifold, $S_{\rm total}$ is the low energy type IIB action defined in the string frame[3], ϕ_0 is the fluctuation over a given background, $S_{\rm GH}$ is the Gibbons-Hawking boundary term [14] and $S_{\rm counterterm}$ is the counter-term action added to renormalise the action.

The interesting part now is that we can define *two* classes of theories in this kind of backgrounds by introducing a cut-off at $r = r_c$:

• The first class is to analyse the theory right at the usual boundary where $r_c \to \infty$. This is the standard picture where there are infinite degrees of freedom at the boundary, and the theory has a smooth RG flow from UV to IR till it confines (at least from the weakly coupled gravity dual).

• The second class is to analyse theories by specifying the degrees of freedom at generic r_c and then defining the theories at the boundary. All these theories would meet the cascading theory at certain scales under RG flows. The gravity duals of these theories are the usual resolved-deformed conifold geometries cut-off at various r_c with appropriate UV caps added (or, alternatively, degrees of freedom specified). These UV caps are non-trivial geometries added to the resolved-deformed conifold geometry from $r = r_c$ to $r = \infty$[4]. In addition to that, both these classes of theories ought to have appropriate number of seven branes so that all the Landau poles could be removed.

We can also say that the boundary degrees of freedom are now given by $\mathcal{N}_{\rm uv}$ where $\mathcal{N}_{\rm uv} = \epsilon^{-n}, \epsilon \to 0$ and $n >> 1$ specifies the parent cascading theory. For all other theories that we define at the boundary $r = \infty$, by cutting off the geometry at $r = r_c$, will have $\mathcal{N}_{\rm uv}$ going as $\mathcal{N}_{\rm uv} = \epsilon^{-n}$ but now $n \geq 1$. Yet another interesting thing about all these theories is that *any* thermodynamical quantitities defined in these theories will be completely independent of

[3] There are subtleties involved in using $S^{\rm total}_{\rm string} \simeq S^{\rm total}_{\rm SUGRA}$. For detailed discussions, consult [5].

[4] For example we could add AdS geometries from $r = r_c$ to $r = \infty$. This is like adding $\mathcal{N} = 4$ or 2 degrees of freedom at the cut-off scale to UV complete the gauge theories. One issue here is the connection to the work of [13]. The question is can we write a boundary theory at $r = r_c$ itself instead of going to the actual $r = \infty$ boundary? As shown in [13] this is possible in AdS case because the theory on the surface of a ball in AdS space doesn't have to be a local quantum field theory. The non-local behavior in such a "boundary" theory is completely captured by the Wilsonian effective action at the so called boundary. As detailed in [5], this is tricky in the Klebanov-Strassler model precisely because there is no unique strongly coupled gauge theory here. Again, as in the RG flow pictures of [5], when the dual gravity theory is weakly coupled we have a smooth RG flow in the gauge theory side, but the theory at any given scale can be given by infinite number of representative gauge theories none of which completely capture the full dynamics. Thus it makes more sense to define the theories at $r = \infty$ boundary only and not at any generic $r = r_c$.

Journal of Physics: Conference Series **462** (2013) 012009 doi:10.1088/1742-6596/462/1/012009

the cut-off r_c and only depend on the temperature T (with an exponentially small dependences on the boundary degrees of freedom)! This will be briefly demonstrated in the next section (see [5] for the full exposition).

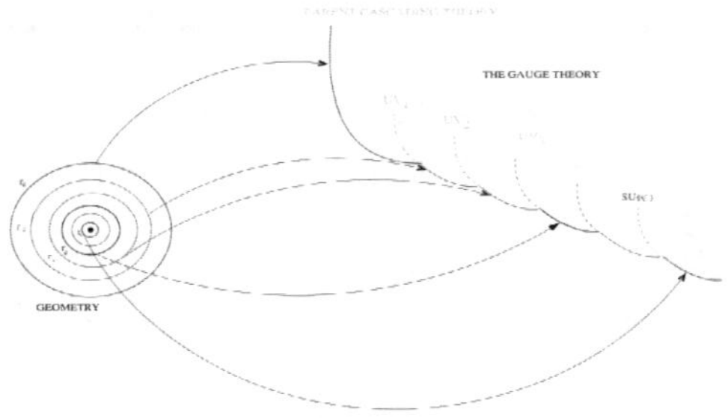

Figure 1. A very approximate way to denote all the UV completed theories that we could study in our framework by cutting off the geometry at various $r = r_c$. For a more correct illustration of the situation see [5]. The outermost blue curves on both sides denote the parent cascading theory.

3. The Stress Tensor

One of the important thermodynamical quantity to extract from our background would be the stress tensor. As the stress tensor couples to the metric, it's expectation value is obtained from the renormalised action $S_{\rm ren}$ via $\langle T^{ij}(\Lambda_c)\rangle = \frac{\delta S_{\rm ren}[l_{ij}]}{\delta l_{ij}}\Big|_{l_{ij}=0}$ where l_{ij} is the four dimensional effective metric induced by the ten dimensional metric of the form (2) by approximating the metric first by a five dimensional slice and then taking the required pull-back (see [5] for details on the procedure). At every fixed r the five dimensional metric induces a four dimensional metric and with appropriate rescaling of the time coordinate, we can view this effective four dimensional metric as a flat Minkowski metric (with possible perturbations). If O_{ij} denote the perturbations on the background by quark strings that stretch from the D7 branes to the horizon of the black hole, then writing the supergravity action up to quadratic order in O_{ij}; using integration by parts together with appropriate Gibbons-Hawking terms; and holographically renormalising the subsequent action, we obtain:

$$T^{mm}_{\rm medium+quark} = \int \frac{d^4q}{(2\pi)^4}\sum_{\alpha,\beta}\Bigg\{(H^{mn}_{|\alpha|} + H^{nm}_{|\alpha|})s^{(4)[\beta]}_{nn} - 4(K^{mn}_{|\alpha|} + K^{nm}_{|\alpha|})s^{(4)[\beta]}_{nn}$$
$$+ (K^{mn}_{|\alpha|} + K^{nm}_{|\alpha|})s^{(5)[\beta]}_{nn} + \sum_{j=0}^{\infty} \hat{b}^{(\alpha)}_{n(j)}\widetilde{J}^n\delta_{nm}e^{-j\mathcal{N}_{\rm uv}} + \mathcal{O}(\mathcal{T}e^{-\mathcal{N}_{uv}})\Bigg\} \tag{6}$$

where $(\hat{b}^{(\alpha)}_{n(j)}, \mathcal{N}_{\rm uv})$ together will specify the full boundary theory for a specific UV complete theory. The quantities $H^{mn}, \widetilde{J}^m, K^{mn}$ etc. with $m, n = 0, .., 3$ are completely independent of the radial coordinate r and depend on the 3+1 dimensional spacetime coordinates. Note also that the stress tensor depends only on the temperature T and is independent of any cut-offs. The dependence on the UV degrees of freedom is exponentially suppressed, and in the limit $\mathcal{N}_{\rm uv} = \epsilon^{-n}, n >> 1$ we reproduce the result for the parent cascading theory.

4. Results and Discussion

With the general formulation of stress tensor (6), which is similar to the AdS/CFT results [16] in the limiting case $M = N_f = 0$, we can compute the wake a moving quark creates in a plasma with O_{ij} being the metric perturbation due to string (see [17] for an equivalent AdS calculation). Furthermore we can compute the shear viscosity η from the Kubo formula with the propagator obtained from the dual partition function (5) and the entropy density s using Wald's formula. The result for the ratio η/s is given by:

$$\frac{\eta}{s} = \frac{1+\sum_{k=1}^{\infty}\alpha_k e^{-4k\mathcal{N}_{\rm uv}}}{4\pi + \frac{1}{\pi}\log^2\left(1-T^4e^{-4\mathcal{N}_{uv}}\right)} - \frac{c_3\kappa}{3L^2\left(1-T^4e^{-4\mathcal{N}_{uv}}\right)^{3/2}}\left[\frac{B_o(4\pi^2-\log^2 C_o)+4\pi A_o\log C_o}{\left(4\pi^2-\log^2 C_o\right)^2+16\pi^2\log^2 C_o}\right] \tag{7}$$

where $(A_o, B_o, C_o, \alpha_k)$ are constants that depend on the temperature T and $e^{-\mathcal{N}_{\rm uv}}$; and c_3 is the coefficient of the Riemann square term coming from the backreactions of the D7 branes in the background. These have been explicitly worked out in [5]. The c_3 dependence of the η/s ratio first appeared in [18].

With certain choices of parameters ($g_s = 0.01, N = 1000, M = 100, N_f = 3, c_3 = 0.001$) and ignoring $\mathcal{O}(g_sN_f, g_sM^2/N)$ corrections to the black hole parameters $g_i(r)$ the result for η/s is shown in figure 2. It is easy to see that we are violating the celebrated KSS bound [19][5]. The

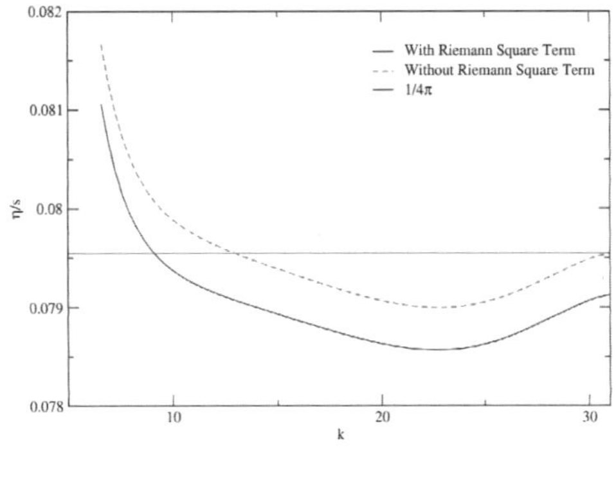

Figure 2. Plot of η/s versus the UV degrees of freedom.

plot in figure 2 is shown for η/s with and without Riemann Square term. The x-axis is defined as $k = \frac{e^{\mathcal{N}_{\rm uv}}}{T}$ where $\mathcal{N}_{\rm uv}$ is the UV degrees of freedom and T is the temperature of the cascading theories. For the parent cascading theory, and their corresponding family of theories, $k \to \infty$ and we see a violation of the bound (the solid blue line). As k decreases (assuming this is possible!) the red dashed line dips slightly below the $1/4\pi$ axis, but the solid blue line remains considerably below the $1/4\pi$ axis. For k sufficiently small the bound is not violated. However all the models that we studied here and in [5] can *only* realise the infinite k limit, so that the small k limit depicted in Fig. 2 above is just an extrapolation of (7).

[5] Recent papers dealing with the violation of the KSS bound are [18, 20].

5. Conclusion

We have analyzed the gravity dual of a thermal gauge theory with logarithmic running coupling which may resemble QCD in the far IR. It appears that the KSS bound for η/s may be violated in certain gauge theories if one carefully takes into account both the running of the couplings as well as the backreactions of the flavor branes in the dual gravity picture, for a range of parameter values (see details in [5]).

It is of course important to test and verify the robustness of this limit, and we view the current work as contributing to this effort. More work is needed in order to identify the size and extent of this violation, at the moment this violation is only parametric and our parameters need to have better defined physical origins. In the end, one may need to rely on the empirical identification of key quantities, like transport coefficients for example [21]. In this regard, the role of heavy ion experiments at RHIC and at the LHC can't be overestimated.

Note added: In our recent paper [22], we have given a concrete example of a UV complete theory that consistently proves all the statements that we made here and in [5]. Our new geometry has a UV cap given by an asymptotic AdS-Schwarzchild geometry. The IR dynamics of our theory is captured by the OKS-BH geometry. The deformation of the OKS-BH geometry at the boundary is indeed captured by an *interpolating* geometry that ties smoothly, in the presence of sources and fluxes, to the asymptotic AdS-Schwarzchild geometry. All the relevant details like changes in stress-tensor, entropy etc due to the UV cap can now be calculated. In addition to that we have been able to study zero temperature linear confinement as well as high temperature quarkonium suppression and melting from our UV complete theory. Our results are also consistent with many of the earlier predictions made in the literature using completely different techniques than ours.

6. Acknowledgments

We would like to thank Alex Buchel, Paul Chesler, Andrew Frey, Jaume Gomis, Evgeny Kats, Ingo Kirsch, Rob Myers, Peter Ouyang, Omid Saremi, Aninda Sinha, Diana Vaman, Larry Yaffe and especially Ofer Aharony and Matt Strassler for many helpful discussions and correspondence. This work is supported in part by the Natural Sciences and Engineering Research Council (NSERC) of Canada, and in part by McGill University.

References

[1] J. M. Maldacena, "The large N limit of superconformal field theories and supergravity," Adv. Theor. Math. Phys. **2**, 231 (1998) [Int. J. Theor. Phys. **38**, 1113 (1999)] [arXiv:hep-th/9711200].

[2] P. Ouyang, "Holomorphic D7-branes and flavored N = 1 gauge theories," Nucl. Phys. B **699**, 207 (2004) [arXiv:hep-th/0311084]; H. Y. Chen, P. Ouyang and G. Shiu, "On Supersymmetric D7-branes in the Warped Deformed Conifold," arXiv:0807.2428 [hep-th]; K. Dasgupta, P. Franche, A. Knauf and J. Sully, "D-terms on the resolved conifold," arXiv:0802.0202 [hep-th].

[3] I. R. Klebanov and M. J. Strassler, "Supergravity and a confining gauge theory: Duality cascades and χ_{SB}-resolution of naked singularities," JHEP **0008**, 052 (2000) [arXiv:hep-th/0007191].

[4] M. J. Strassler, "The duality cascade," arXiv:hep-th/0505153; "An unorthodox introduction to supersymmetric gauge theory," arXiv:hep-th/0309149.

[5] M. Mia, K. Dasgupta, C. Gale and S. Jeon, "Five easy pieces: The dynamics of quarks in strongly coupled plasmas," arXiv:0902.1540 [hep-th].

[6] K. Dasgupta and S. Mukhi, "Brane Constructions, Conifolds and M-Theory," Nucl. Phys. B **551**, 204 (1999) [arXiv:hep-th/9811139]; "Brane constructions, fractional branes and anti-de Sitter domain walls," JHEP **9907**, 008 (1999) [arXiv:hep-th/9904131]; K. Dasgupta, K. Oh and R. Tatar, "Geometric transition, large N dualities and MQCD dynamics," Nucl. Phys. B **610**, 331 (2001) [arXiv:hep-th/0105066]; "Open/closed string dualities and Seiberg duality from geometric transitions in M-theory," JHEP **0208**, 026 (2002) [arXiv:hep-th/0106040]; K. Dasgupta, K. h. Oh, J. Park and R. Tatar, "Geometric transition versus cascading solution," JHEP **0201**, 031 (2002) [arXiv:hep-th/0110050].

[7] O. Aharony, A. Buchel and A. Yarom, "Holographic renormalization of cascading gauge theories," Phys. Rev. D **72**, 066003 (2005) [arXiv:hep-th/0506002]; "Short distance properties of cascading gauge theories," JHEP **0611**, 069 (2006) [arXiv:hep-th/0608209]; O. Aharony, A. Buchel and P. Kerner, "The black hole in the throat - thermodynamics of strongly coupled cascading gauge theories," Phys. Rev. D **76**, 086005 (2007) [arXiv:0706.1768 [hep-th]].

[8] R. Casero, C. Nunez and A. Paredes, "Towards the string dual of N = 1 SQCD-like theories," Phys. Rev. D **73**, 086005 (2006) [arXiv:hep-th/0602027]; F. Bigazzi, A. L. Cotrone and A. Paredes, "Klebanov-Witten theory with massive dynamical flavors," JHEP **0809**, 048 (2008) [arXiv:0807.0298 [hep-th]]; F. Bigazzi, A. L. Cotrone, A. Paredes and A. Ramallo, "Non chiral dynamical flavors and screening on the conifold," arXiv:0810.5220 [hep-th]; "The Klebanov-Strassler model with massive dynamical flavors," arXiv:0812.3399 [hep-th]; G. Bertoldi, F. Bigazzi, A. L. Cotrone and J. D. Edelstein, "Holography and Unquenched Quark-Gluon Plasmas," Phys. Rev. D **76**, 065007 (2007) [arXiv:hep-th/0702225]; A. L. Cotrone, J. M. Pons and P. Talavera, "Notes on a SQCD-like plasma dual and holographic renormalization," JHEP **0711**, 034 (2007) [arXiv:0706.2766 [hep-th]].

[9] N. Borodatchenkova, M. Haack and W. Muck, "Towards Holographic Renormalization of Fake Supergravity," arXiv:0811.3191 [hep-th].

[10] C. Vafa, "Evidence for F-Theory," Nucl. Phys. B **469**, 403 (1996) [arXiv:hep-th/9602022]; A. Sen, "F-theory and Orientifolds," Nucl. Phys. B **475**, 562 (1996) [arXiv:hep-th/9605150]; K. Dasgupta and S. Mukhi, "F-theory at constant coupling," Phys. Lett. B **385**, 125 (1996) [arXiv:hep-th/9606044].

[11] M. Grana and J. Polchinski, "Gauge / gravity duals with holomorphic dilaton," Phys. Rev. D **65**, 126005 (2002) [arXiv:hep-th/0106014]; I. Kirsch and D. Vaman, "The D3/D7 background and flavor dependence of Regge trajectories," Phys. Rev. D **72**, 026007 (2005) [arXiv:hep-th/0505164]; J. Erdmenger, N. Evans, I. Kirsch and E. Threlfall, "Mesons in Gauge/Gravity Duals - A Review," Eur. Phys. J. A **35**, 81 (2008) [arXiv:0711.4467 [hep-th]].

[12] E. Witten, "Anti-de Sitter space and holography," Adv. Theor. Math. Phys. **2**, 253 (1998) [arXiv:hep-th/9802150]; S. S. Gubser, I. R. Klebanov and A. M. Polyakov, "Gauge theory correlators from non-critical string theory," Phys. Lett. B **428**, 105 (1998) [arXiv:hep-th/9802109].

[13] V. Balasubramanian and P. Kraus, "Spacetime and the holographic renormalization group," Phys. Rev. Lett. **83**, 3605 (1999) [arXiv:hep-th/9903190].

[14] G. W. Gibbons and S. W. Hawking, "Action Integrals And Partition Functions In Quantum Gravity," Phys. Rev. D **15**, 2752 (1977).

[15] A. Buchel, J. T. Liu and A. O. Starinets, "Coupling constant dependence of the shear viscosity in N=4 supersymmetric Yang-Mills theory," Nucl. Phys. B **707**, 56 (2005) [arXiv:hep-th/0406264]; R. C. Myers, M. F. Paulos and A. Sinha, "Quantum corrections to eta/s," Phys. Rev. D **79**, 041901 (2009) [arXiv:0806.2156 [hep-th]].

[16] M. Henningson and K. Skenderis, "The holographic Weyl anomaly," JHEP **9807**, 023 (1998) [arXiv:hep-th/9806087]; K. Skenderis, "Lecture notes on holographic renormalization," Class. Quant. Grav. **19**, 5849 (2002) [arXiv:hep-th/0209067]; S. de Haro, S. N. Solodukhin and K. Skenderis, "Holographic reconstruction

of spacetime and renormalization in the AdS/CFT correspondence," Commun. Math. Phys. **217**, 595 (2001) [arXiv:hep-th/0002230]; K. Skenderis, "Asymptotically anti-de Sitter spacetimes and their stress energy tensor," Int. J. Mod. Phys. A **16**, 740 (2001) [arXiv:hep-th/0010138]; A. Karch, A. O'Bannon and K. Skenderis, "Holographic renormalization of probe D-branes in AdS/CFT," JHEP **0604**, 015 (2006) [arXiv:hep-th/0512125].

[17] P. M. Chesler and L. G. Yaffe, "The wake of a quark moving through a strongly-coupled $\mathcal{N} = 4$ supersymmetric Yang-Mills plasma," Phys. Rev. Lett. **99**, 152001 (2007) [arXiv:0706.0368 [hep-th]]; "The stress-energy tensor of a quark moving through a strongly-coupled N=4 supersymmetric Yang-Mills plasma: comparing hydrodynamics and AdS/CFT," Phys. Rev. D **78**, 045013 (2008) [arXiv:0712.0050 [hep-th]]; P. M. Chesler, K. Jensen, A. Karch and L. G. Yaffe, "Light quark energy loss in strongly-coupled N = 4 supersymmetric Yang-Mills plasma," arXiv:0810.1985 [hep-th].

[18] Y. Kats and P. Petrov, "Effect of curvature squared corrections in AdS on the viscosity of the dual gauge theory," JHEP **0901**, 044 (2009), arXiv:0712.0743 [hep-th].

[19] P. Kovtun, D. T. Son and A. O. Starinets, "Viscosity in strongly interacting quantum field theories from black hole physics," Phys. Rev. Lett. **94**, 111601 (2005) [arXiv:hep-th/0405231].

[20] M. Brigante, H. Liu, R. C. Myers, S. Shenker and S. Yaida, "Viscosity Bound Violation in Higher Derivative Gravity," Phys. Rev. D **77**, 126006 (2008) [arXiv:0712.0805 [hep-th]]; M. Brigante, H. Liu, R. C. Myers, S. Shenker and S. Yaida, "The Viscosity Bound and Causality Violation," Phys. Rev. Lett. **100**, 191601 (2008) [arXiv:0802.3318 [hep-th]]; A. Dobado, F. J. Llanes-Estrada and J. M. T. Rincon, "The status of the KSS bound and its possible violations (How perfect can a fluid be?)," AIP Conf. Proc. **1031**, 221 (2008) [arXiv:0804.2601 [hep-ph]]; A. Buchel, R. C. Myers, M. F. Paulos and A. Sinha, "Universal holographic hydrodynamics at finite coupling," Phys. Lett. B **669**, 364 (2008) [arXiv:0808.1837 [hep-th]]; X. H. Ge, Y. Matsuo, F. W. Shu, S. J. Sin and T. Tsukioka, "Viscosity Bound, Causality Violation and Instability with Stringy Correction and Charge," JHEP **0810**, 009 (2008) [arXiv:0808.2354 [hep-th]]; A. Adams, A. Maloney, A. Sinha and S. E. Vazquez, "1/N Effects in Non-Relativistic Gauge-Gravity Duality," arXiv:0812.0166 [hep-th]; A. Buchel, R. C. Myers and A. Sinha, "Beyond $\frac{\eta}{s} = \frac{1}{4\pi}$," arXiv:0812.2521 [hep-th].

[21] M. Luzum and P. Romatschke, "Conformal Relativistic Viscous Hydrodynamics: Applications to RHIC results at $\sqrt{s_{NN}} = 200$ GeV," Phys. Rev. C **78**, 034915 (2008).

[22] M. Mia, K. Dasgupta, C. Gale and S. Jeon, "Toward Large N Thermal QCD from Dual Gravity: The Heavy Quarkonium Potential," arXiv:1004.0387 [hep-th].

Journal of Physics: Conference Series **462** (2013) 012010 doi:10.1088/1742-6596/462/1/012010

Testing Quantum Erasure and Reversible Quantum Measurement with Holladay's Simple Experiment

Michael Devereux

Los Alamos National Laboratory (Retired)

1373 B 40th St. Los Alamos, NM 87544

E-mail: dbar_x@cybermesa.com

Abstract. Schrödinger's continuous differential equation, which, as far as we know, describes the time development of all physical systems, from microscopic to cosmological, is time-reversal invariant. It is the process of measurement of a system that von Neumann found to be irreversible, and may so account for the distinction between past and future. Developed over the last thirty years, quantum eraser theory has claimed that some quantum measurements are reversible. Holladay proposed a very simple double-slit quantum eraser experiment, hitherto never performed, which he said would refute von Neumann's measurement description and support the quantum erasure thesis. Now, that experiment can actually be implemented with inexpensive, easily accessible equipment. Remarkably, results of the experiment refute quantum eraser theory and confirm von Neumann's measurement reduction process instead.

1. Introduction

Quantum erasure (QE) has been an influential thesis of the interference phenomena associated with quantum measurement for nearly thirty years. Its advocates say that interference may be eliminated by acquisition (or, in some cases, just availability) of the information indicating which of several alternative paths a *single* system's quantum wavefunction has followed. They assert that erasing that information will restore the interference. [1] Many research groups have now claimed to confirm the quantum eraser phenomenon observationally.

About ten years ago, in response to a criticism of quantum erasure by Mohrhoff [2] some of its proponents have re-construed QE as a more ordinary sorting of the ensemble of measured quantum systems into correlated sub-ensembles. [3] But the preeminent QE experimentalist for many years, Paul Kwiat, has confirmed, with a very recent article, [4] that information erasure resulting in restoration of interference from a simple two-slit screen would indeed be a demonstration of quantum erasure, were one able to insure that just one coherent laser photon were passing through the screen at any time. I'll show here that a very simple incoherent light source and a two-slit screen, much in the spirit of Thomas Young's original experiments, can be used to refute that quantum erasure thesis of interference restoration due to information erasure.

It's well known that the wavefunction evolution of all non-relativistic physical systems, of whatever size, can be described by Schrödinger's differential equation. Such wavefunction development is time-reversal invariant and incapable of the distinction between past and future that is indicated, for instance, by the monotonic increase in entropy (and decrease in information) mandated by Planck's formulation of the second law of thermodynamics. [5] Von Neumann, however, disputed Schrödinger's well-known, avid contention [6] that all of wavefunction development is continuous and reversible. Seconded by Bohr and Heisenburg, among many others, even today, he argued that there is a disparate, discontinuous event which transforms the wavefunction at measurement. [7] And he found this distinct development, named "process 1", to be irreversible.

In contrast, during the last thirty years, the seminal, theoretical development of quantum erasure has become the most persistent and influential defense of continuous Schrödinger evolution during quantum measurement. Marlan Scully and his collaborators claimed to show that the wavefunction of a spinning particle in a Stern-Gerlach magnet can evolve continuously, even as which path it takes through the magnet is measured. [8] In fact, employing continuous wavefunction evolution, they calculated a nearly diagonalized density matrix beyond the magnet for this historic prototype of quantum measurement.

But in spite of many experiments now claiming to confirm the quantum eraser hypothesis, that theory, and continuous wavefunction development during measurement, has not been established scientifically. It's obvious, in fact, from the myriad of contradictory theses, that even today, eighty years after its advent, there exists no consistent, comprehensive theory of measurement in quantum mechanics.Ballentime has said that "It is now widely understood that von Neumann's 'reduction' is not a real physical process (which would contradict the equations of motion for the combined system plus apparatus, and so make the theory hopelessly inconsistent)." [9] While Holladay, on the other hand, has labeled von Neumann's reduction process the "orthodox" view. [10] Indeed, I believe it is the lack of an understanding of the *process* of quantum measurement which is the ascendant source of the current confusion and controversy in quantum theory.

Below, in section 2, I show that Scully's theoretical analyses of Stern and Gerlach's experiment, upon which all of quantum erasure is based, is clearly not credible. Holladay suggested a very simple, optical, double-sit experiment which he said would confirm quantum erasure and refute von Neumann's quantum measurement reduction. [11] I devote section 3 to its description and implementation. I've explained how to use a small, incoherent (random phase) light to actually perform the experiment. Remarkably, that experiment refutes quantum erasure and supports von Neumann's explanation. Section 4 includes related comments and conclusions. A forthcoming article examines other, more sophisticated, optical experiments also purporting to confirm quantum erasure. Careful, critical analysis of those experiments shows them to also, simultaneously, confirm hypotheses incompatible with quantum erasure.

2. Theoretical Test

Quantum eraser theory began with the mathematical analysis of a modified Stern-Gerlach experiment developed by Marlan Scully, R. Shea, and J. D. McCullen (SSM) in 1978. [12] The authors installed a bi-level atom along one of the two possible paths of a heavy, spin one-half molecule traversing an inhomogeneous magnetic field. The molecule is said to invariably kick the atom from its ground to its excited state if it travels the route occupied by that atomic detector. This would, according to our usual Stern-Gerlach analyses, measure the spin direction of the molecule from its deflection by the magnetic field. Crucially, SSM supposed that a detector atom much lighter than the spinning molecule would transfer negligible momentum to the molecule at collision, and so result in no effective change of the molecule's quantum wavefunction. Employing this assumption, they derived a nearly diagonalized density matrix for the atom-molecule system after detection. Since SSM assumed that time-reversible, continuous Schrödinger evolution describes the entire passage of the molecule through the magnetic field, this would seem to satisfy Schrödinger's ardent conviction that there exist no quantum jumps, even at measurement. According to SSM, "This 'state reduction' (i.e. the loss of off-diagonality of the density matrix of the system) takes place even if the system is not 'messed up' or perturbed in the process of 'looking'." [13]

Note, however, that SSM ignored the quantized energy, always equal to the difference between the atom's ground and excited level, transferred from the molecule to the atom at collision. And, that every unique energy eigenvalue of the molecule is associated with a different, linearly-independent eigenfunction, which specifies a different measured molecular state. So, one is justified in denying the continuous development of the molecular wavefunction at detection, essential to the SSM analysis.

Nevertheless, Scully suggested that the SSM analysis might describe a real physical phenomenon called quantum erasure. He and his colleagues had written that if "we reverse time and pass our spins back through the apparatus, allowing them to again interact with the detector, the interaction could take an excited detector atom back to the ground state and return the total system to a pure case…." [14] This could imply that if the wavefunction of the measured object evolved continuously when information was provided to the detector (bi-level atom), then erasing that information should reverse the effects of measurement on the object. And, Scully and Walter reiterated this understanding some twenty years later "…if we put a *WelcherWeg* detector in place (so we lose interference even if we don't look at the detector) and then erase the which way information after the particles have passed through…such a 'quantum eraser' process (would) restore the interference fringes." [15] And, there appears to be no useful way to interpret this, the original, foundational description of quantum erasure, as an effective method for re-sorting an ensemble of incident particles.

Early on, Scully and his collaborators proposed several devices that might be used to confirm the reality of quantum erasure, but they initially proved impractical. [16] More recently, many groups claim to have confirmed quantum erasure experimentally. But, such confirmation may not be completely convincing. We shall see that hypotheses other than quantum erasure can often explain those results. Attempted refutation of a scientific hypothesis, rather than repeated confirmation, is the appropriate test.

3. The Holladay experiment

Holladay proposed a double-slit screen illuminated by a monochromatic light source (presumably, a "point" source,) as a quantum erasure devise, Figure 1. [17] He explained that a Young-type interference pattern would be produced on a far-field screen by each incoherent photon's wavefunction that passed through both slits. Then each slit is covered with a Polaroid filter, one horizontal, and the other vertical. Holladay wrote that the crossed Polaroids "label" each possible path which a photon may travel. And, of course, observation shows that those crossed Polaroids do eliminate the interference pattern.

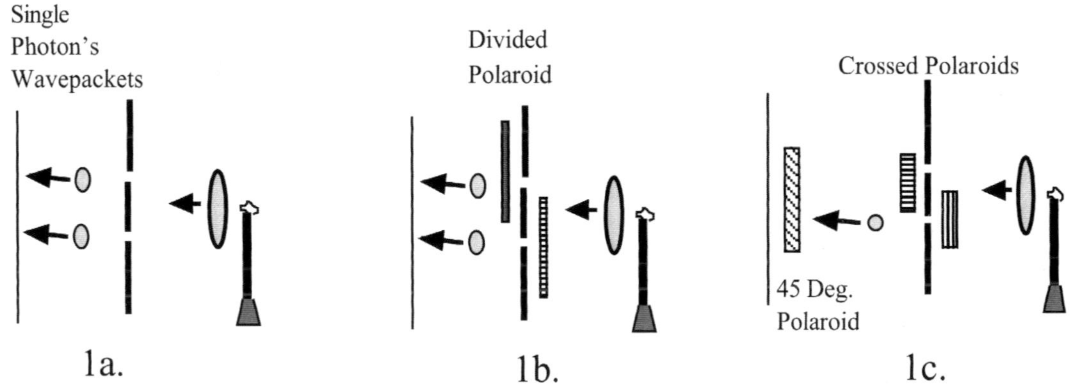

Figure 1. A minimaglight flashlight filament acts as a point source of incoherent, white light. When positioned about a meter from a two-slit screen it can produce a Young's interference pattern on a viewing screen. Polaroid filters project the wavefunction on to just one of many polarization eigenfunctions.

Journal of Physics: Conference Series **462** (2013) 012010 doi:10.1088/1742-6596/462/1/012010

To confirm quantum erasure, Holladay would cover both slits and their crossed polarizers with another Polaroid filter, this one at forty-five degrees from horizontal or vertical, Figure 1c. The Young's interference pattern, albeit at a reduced intensity of 25 percent, is supposed to reappear. According to the standard quantum erasure hypothesis, the final Polaroid is said to erase the which-way polarization information carried by a probability amplitude along each path. That last Polaroid is at forty-five degrees to each orthogonal polarizer, and the probability that a photon at the viewing screen traveled either path is one half. So, it's not possible to specify which way a photon went. Such an information erasure in the two amplitudes is believed by quantum eraser proponents to account for the putative, restored interference. Significantly, that interference revival was not tested by observation.

If we suppose, as Holladay did, that the final polarizer would restore interference, then "when the path of the photon is labeled by a definite state of polarization..., so that the photon with that particular polarization transits definitely one (or the other) of the paths, it does not make sense to say that the wave 'collapses' along that path and that the amplitude along the other path vanishes. If this phenomenon occurred, it is very difficult to see how the insertion of the quantum eraser Polaroid could restore the interference of amplitudes *from both paths.* Thus, these experiments call into serious question the orthodox position of von Neumann and his followers that posit a dual dynamics in quantum mechanics – the normal unitary change in the wave function given by the Schrödinger time dependent equation *and* the non-unitary change given by von Neumann's projection postulate....In these quantum eraser experiments apparently the photon is definitely traversing one path (or the other) but the wave, with one path 'empty', continues along both paths as manifested by the restoration of interference by the quantum eraser." [18] But, Holladay only presumed that interference would return. Until very recently, the actual observation had not been attempted.

Within the last few years, Paul Kwiat, the most prominent of the QE experimentalists, and his student, Rachel Hillmer, described a do-it-yourself quantum eraser demonstration. [19] It's almost identical to Holladay's proposal, so confirming the suitability of that simple experiment as a real test of the QE theory. They replace the double slits with a thin, straight wire to generate an interference pattern, and employ a small laser, like a laser pointer, rather than an incoherent light source. Then, Polaroid filters in the usual orientations are arranged as a demonstration of the quantum eraser phenomenon. They show how two orthogonal polarizers, one on either side of the wire, will provide the information to distinguish which way a photon traveled, destroying interference. Then, an additional forty-five degree Polaroid, covering the crossed polarizers, acts as an "eraser..., something that can erase the information indicating which path each particle has followed, thereby restoring the indistinguishability of the alternatives and restoring interference." [20] This last Polaroid does revive interference.

But, the authors recognize that only if one photon at a time passes the thin wire and polarizers can one claim genuine observational confirmation of quantum erasure. Otherwise, because the laser creates innumerable, coherent photons, the interference pattern, originally produced and then revived, could result from two particle-like photons passing simultaneously to either side of the wire. (Dirac's admonition that a photon only interferes with itself [21] was incorrect. Unfortunately, in spite of Glauber's efforts, [22] many physicists remain in the dark about this.) . Quantum erasure is supposed to revive interference due to a *single* system's wavefunction that still travels both ways, even after the which-way path has been detected. Kwiat emphasizes this point: "...you could only truly prove that...individual photons that make up the light wave are indeed doing the full quantum dance...by sending the photons through the apparatus and detecting them one at a time." [23] Or, by using incoherent photons so that only those individual photon wavefunctions that transit both slits contribute to interference.

4. Performing the experiment

Wonnell demonstrated the use of a tiny, Minimaglight, incandescent, flashlight filament, 1.3mm long and 0.25mm wide, to produce a two-slit, Young's interference pattern. [24] Such an incandescent light source is very unusual today when inexpensive, versatile lasers are conveniently available. Moreover, the chaotic source generates white-light photons that are bunched together [25] and have an extremely short coherence length. Nevertheless, unlike a laser, it does possess the essential property that interference can be constructed from only those photons' wavefunctions that pass through both of the two slits. What it lacks in modern sophistication it supplies in efficient functionality, and can actually be used to test quantum erasure. With the filament exposed by removing the lens cover and set up about one meter from an appropriate two-slit film, like a Cornell Slitfilm Demonstrator, that filament acts as a point source of incoherent, white light. With a two-slit film of about 0.1 mm width, and about 0.3 mm separation I implemented Holladay's experiment.

I recorded the photograph of Figure 2a with one slit covered by opaque material. A small (4 megapixel) digital camera, focused on infinity, was secured a few centimeters from the slit film, and automatically adjusted the timed exposure to a few seconds. The resulting Fraunhofer diffraction pattern takes the form, $\frac{I(\theta)}{I_0} = \frac{\sin^2\beta}{\beta^2}$, with $\beta = \frac{\pi a \sin\theta}{\lambda}$, where I is the light intensity, I_0 is intensity at the center of the pattern (q = 0), l is the wavelength, and $a \approx 0.1$ mm, in this instance, is the slit width. The same arrangement, now with both slits open, was used to take the photograph of Figure 2b. Young's interference, characterized by equally spaced maxima, is plainly seen superimposed on the Fraunhofer image. The pattern is a product of Fraunhofer diffraction and Young's interference, $\frac{I(\theta)}{I_0} = \frac{\sin^2\beta}{\beta^2}\cos^2\gamma$, with $\gamma = \frac{\pi d \sin\theta}{\lambda}$, where $d \approx 0.3$ mm was the slit separation. There is, of course, some color aberration, due to the white light, but not so much that the patterns are obscured.

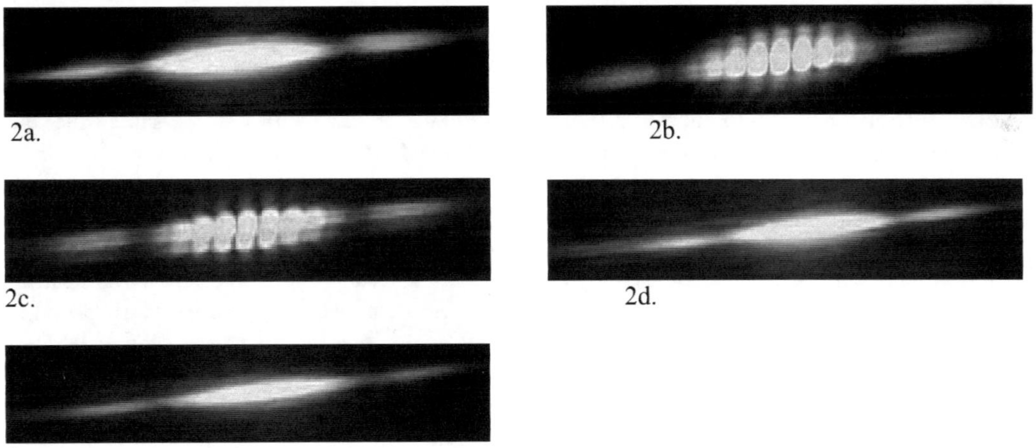

Figure 2. The first photograph, 2a, shows the Fraunhofer pattern produced by a single slit. Figure 2b is the Young's interference pattern resulting from two slits, each identical to the single slit of the first photograph. The next photograph, 2c, shows that two separate polarizing filters, one over each slit, if aligned, still result in interference at the viewing screen. Figure 2d is the photograph when the slits are covered by crossed Polaroids. Figure 2e is the resulting photograph, indicating no restored interference, if a forty-five degree Polaroid covers the crossed polarizers. Color, and much improved resolution, may be available on-line.

Journal of Physics: Conference Series **462** (2013) 012010 doi:10.1088/1742-6596/462/1/012010

Wonnell has measured the very short coherence length of photons from the Minimaglight filament at about ten microns. [26] Following Holladay's prescription for a quantum eraser experiment, it's necessary to cover each slit with a piece of polarizing film. Unless those two, crossed Polaroids have nearly identical film thickness, tiny differences in optical path length, rather than orthogonal polarization through each slit, could prevent photon overlap, eliminating interference. So, I placed one piece of polarizing film over each slit, with each polarization axis aligned horizontally. The photograph, Fig 2c shows that Young's interference persists. And, the two pieces of Polaroid (cut from the same sheet of film) must be of very nearly identical thickness, since wavetrains from the flashlight continue to overlap at the viewing screen.

Next, I oriented these two polarizing films ninety degrees from each other. The photograph, Figure 2d, shows that Young's interference is eliminated by the orthogonal polarizers. But not eliminated by a difference in optical path length through the two films, since they have essentially uniform thickness. Finally, as prescribed by quantum eraser theory, I placed a Polaroid film with axis forty-five degrees from vertical over the two slits and crossed Polaroids, Figure 1c. The photograph, Figure 2e, shows clearly that the Young's interference pattern is not restored as quantum erasure requires. Moreover, the use of a collimated sodium arc lamp, for instance, with the same arrangement, would, no doubt, reinforce this conclusion with nearly monochromatic photons of substantially longer coherence length.

Though it's often forgotten or ignored, even some prominent supporters of the QE theory recognize that a Polaroid filter acts as a von Neumann-type projection operator in the mathematical formalism of quantum mechanics. [27] The wavevector of an incident photon approaching the double-slit film with crossed polarizers is in a superposition, $|\Psi(\mathbf{r},t)\rangle = \frac{1}{\sqrt{2}}(|\psi_1(\mathbf{r}_1,t)\rangle|\chi_{\phi+90}\rangle + |\psi_2(\mathbf{r}_2,t)\rangle|\chi_\phi\rangle)$, where either of two possible polarization directions, ϕ or $\phi + 90$ degrees, are possible when measured. Schrödinger time development of the wavefunction is unitary and continuous and it cannot evolve a superposition of polarizations into just one value. But, as von Neumann explained, the Polaroids *project* just one of these two eigenfunctions, and they erase phase information in the wavefunction. That is a *measurement* of each photon's polarization, so the photon wavefunction can survive on only one path from the two slits. Therefore there can be no possibility of resurrecting interference after the polarization measurement, as predicted by quantum erasure.

Incandescent photons are more likely to be bunched together than those from a laser, so a minimaglight filament might generate more coincident, overlapping photons at the two-slit screen than a laser. If two such incoherent photons reach the crossed Polaroids and slit film at the same time, the probability is one-half that each will traverse a different slit, even after projection to just one path by a polarization measurement. But, crucially, the forty-five degree polarizer cannot restore the interference, it cannot reverse measurement, because those two photons have random phases. Two coincident, coherent laser photons (as in the Kwiat-Hilmer demonstration) only seem to reverse the polarization measurement because they are in phase, and so they can restore interference even after transiting different slits. It is essential that a real QE experiment demonstrate measurement reversal (as in interference restoration) in single quantum systems.

An experiment that claims to have done this was recently reported by Schneider and La Puma. [28] Their experiment is very similar to Holiday's proposal. Instead of a double-slit screen, they used a Mach-Zehnder interferometer, which also can divide a single photon's wavefunction along two indistinguishable paths. A well-defined interference pattern of many light and dark fringes is recorded in a simple digital camera if a small laser provides photons to their interferometer. They place crossed Polaroid filters in each arm of the interferometer and this is supposed to encode the photon's path in its polarization. As they show, it does wipe out the interference pattern. Then, they place a third Polaroid, oriented at forty-five degrees, over the output of the interferometer. This is said to erase the which-way encoding. It is clearly shown to revive the photographic image of many light and dark fringes in the interference pattern.

Schneider and LaPuma are aware, of course, that even a small laser (they don't tell us its power or wavelength) produces a prodigious number of photons per second. And, an hypothesis of two simultaneous coherent photons traveling separate interferometer paths, with overlying wavetrains at the camera, would also account for their observed result, instead of confirming the quantum erasure of a single system. So, they attenuate their laser by a factor of about 10^9, which is supposed to eliminate the possibility of overlapping photons. This produces a resurrected, wavering, and very tenuous, single dark fringe over a light background in the observed pattern. [29]

Without specifying the coherence length of photons from their laser, the authors say that the distance between photons, on average, is about ten times their coherence length. This means that there is about a ten percent chance that one photon (emitted at random) in the attenuated beam will overlie at least half the wavetrain of another. This may be sufficient to account for the single, revived dark fringe, when each of those two photons passes simultaneously along a different arm of the interferometer. Even in an undergraduate laboratory, one could easily check the source of that single dark fringe by comparing attenuation of the laser beam with a corresponding lengthening of exposure time at the camera. Attenuation would reduce overlap, and beam intensity too, of course, but lengthened exposure time would only increase the resolution at the interference pattern.

5. Conclusion

Erwin Schrödinger was adamant that all time evolution of the quantum wavefunction is continuous, even during measurement. For over thirty years the most influential support for Schrödinger's view has been supplied by quantum eraser theory. That theory is founded on an analysis of Stern and Gerlach's experiment by Marlan Scully and his collaborators from 1978. But a critical evaluation of the Scully, Stern-Gerlach analysis shows that it is not credible. Though many experiments now claim to confirm quantum erasure, it's easy to refute that hypothesis with an incoherent light source in Holladay's experimental arrangement. Even a very simple flashlight filament, double-slit film, and several Polaroids are adequate to test the QE proposition. Unlike many other QE experiments, which claim to confirm that theory, the incoherent light source, rather than a laser, can be essential for actually testing quantum erasure.

We also need to recognize that observations which corroborate predictions of one theory may simultaneously support other, contradictory hypotheses. And that the more reliable test of a proposed scientific theory, like quantum erasure, is attempted refutation, not confirmation. Still, I've briefly addressed some other QE experiments in a forthcoming article. [30] Careful analyses show that those experiments, too, also support hypotheses incompatible with quantum erasure.

References

[1] Herzog, T, Kwiat P, Weinfurter H and Zeilinger A 1995 *Phys. Rev. Lett.* **75**, 3034-37 and Kim Y-h, Yu R, Kulik S, Shih Y and Scully M 2000 *Phys. Rev. Lett.* **84**, 1-5
[2] Mohrhoff U 1996 *Am. J. Phys.* **64**, 1468-75
[3] Englert B-G, Scully M and Walter H 1999 *Am. J. Phys.* **67**, 325-29
[4] Kwiat P and Hillmer R, May 2007 *Sci. Am.* 90-95
[5] Planck M 1926 *Treatise on Thermodynamics* (New York; Longman) p 103
[6] Schrödinger E 1952 *The British Journal for the Philosophy of Science 3,* 109-23
[7] Von Neumann J 1955 *Mathematical Foundations of Quantum Mechanics* (Princeton; NJ: Princeton U. Press)
[8] Scully M, Shea R and McCullen J D 1978 *Phys. Rep.* **43,** 485-98
[9] Ballentine L 2003 *Am. J. Phys.* **71**, 639-40, p. 639.
[10] Holladay W 1993 Physics Letters A **183**, 280-82 p. 281

[11] Ibid.
[12] Ref. [8]
[13] Ref. [8], p 488
[14] Ref. [8], p 495
[15] Scully M and Walter H 1998 *Found. Phys.* **28**, 399-413
[16] Scully M and Druhl K 1982 *Phys. Rev.* A **25**, 2208-13
[17] Ref. [10]
[18] Ref. [10], pp 281-82
[19] Ref. [4]
[20] Ref. [4], p 94
[21] Dirac P A M 1930 *The Principles of Quantum Mechanics* (Oxford; Clarendon), p 15
[22] Glauber R 1995 *Am. J. Phys.* **6**, p 12
[23] Ref. [4], p 91
[24] Wonnell has shown how to use the minimaglight as incoherent light source at his web page: http://www.personal.psu.edu/faculty/r/e/ref//apparatus/2003competition/wonnel-DA-LC.htm
[25] Loudon R 1983 *The Quantum Theory of Light* (Oxford; Clarendon)
[26] Ref. [24]
[27] Kwiat P, Steinberg A and Chiao R 1992 *Phys. Rev.* A **45**, 7729-39 p 7737
[28] Schneider M and LaPuma I 2002 *Am. J. Phys.* **70**, 266-71
[29] Ref. [28] p 268
[30] To be published. Inquire to the author.

Spaces of Quantum Field Theories

M R Douglas
Simons Center for Geometry and Physics
Stony Brook University, Stony Brook NY 11794
IHES, Le Bois-Marie, Bures-sur-Yvette France 91440

E-mail: douglas@max2.physics.sunysb.edu

Abstract. The concept of a "space of quantum field theories" or "theory space" was set out in the 1970's in work of Wilson, Friedan and others. This structure should play an important role in organizing and classifying QFTs, and in the study of the string landscape, allowing us to say when two theories are connected by finite variations of the couplings or by RG flows, when a sequence of QFTs converges to another QFT, and bounding the amount of information needed to uniquely specify a QFT, enabling us to estimate their number. As yet we do not have any definition of theory space which can be used to make such arguments.

In this talk, we will describe various concepts and tools which should be developed for this purpose, inspired by the analogous mathematical problem of studying the space of Riemannian manifolds. We state two general conjectures about the space of two-dimensional conformal field theories, and we define a distance function on this space, which gives a distance between any pair of theories, whether or not they are connected by varying moduli.

1. Introduction

Quantum field theory (QFT) is a remarkably successful physical framework, describing particle physics, critical phenomena, and certain many-body systems. Superstring theory is also closely based on QFT. A central problem of theoretical physics is to classify the QFTs and understand the relations between them.

After more than fifty years of effort, we have no complete classification, nor is such a thing even on the horizon, except for a few special subclasses of theories. In the broadest terms, we can classify QFT's by the dimension of space-time, and the presence of continuous symmetry and/or supersymmetry. Within such a class, to get a well posed problem, we should place some upper bound on the total number of degrees of freedom, under some precise definition.

The most basic question of classification would then be: is there any sense in which the total set of QFTs in this class is finite, or countable, or compact, or something we can in any way put definite limits on? To put this a different way, can we make a list, say of Lagrangians for definiteness, in which every QFT is guaranteed a place on the list? In itself this need not be the final answer – an entry on the list might need to satisfy further consistency conditions to be a QFT, or several entries might describe the same QFT. Still, given such a starting point, we could continue by addressing these difficulties to systematically map out the space of QFTs. But, except for a few special classes, we have no such list, and no answer to this question.

Now, given sufficiently restrictive assumptions, it can happen that some finite subset of choices of field content and coupling constants fixes the entire theory, making the classification problem tractable. For example, if we are interested in four-dimensional theories with 16 supercharges,[1] there are good arguments that the only examples are the supersymmetric Yang-Mills theories. These are determined by a choice of gauge group G (a compact Lie group) and a complexified gauge coupling for each group factor, with certain known equivalences between theories (S-dualities).

Another class of QFTs which has been classified is the "minimal CFTs," the two-dimensional conformal field theories with central charge $c < 1$. Their classification is a bit more complicated, with an "algebraic" part (the representation theory of the Virasoro algebra) determining the spectrum of primary fields, and a "global" part which determines the list of closed operator algebras satisfying modular invariance (see [16, 7] for an overview of this classification).

While there are a few more results of this type, the general classification problem, even for cases of central interest such as $d = 4$, $N = 1$ supersymmetric theories, seems so distant that it is not much discussed. This is because we have no answer to the question we posed above. More specifically, because we do not sufficiently well understand the relation between the description of a QFT (such as a bare Lagrangian) and the actual observables, it is not known how to reduce the problem to finite terms.

To illustrate this point, let us consider what might appear to be a simple analog to the minimal CFTs, namely the 2d CFT's with $c < 2$. As is well known, the representation theory of the Virasoro algebra does not get us very far here; the number of primary fields grows very quickly with dimension (superpolynomially) and we do not have any way to reduce this large amount of data to a manageable subset. While one can consider larger algebras and rational CFTs, this amounts to looking under a rather well inspected lamppost; we know that there are nonrational CFTs, and that crucial aspects of the problem (such as the fact that CFTs can have moduli) are not respected by this simplification.

There is no reason why we must take an algebraic approach, and one might try to describe the space of $c < 2$ CFTs in other ways. For example, one could start by considering the Lagrangians with two scalar fields and a nontrivial potential. By the c-theorem, we know all such theories will flow to CFTs (often trivial) with $c < 2$. While these could be listed, there is no reason to think this includes all the $c < 2$ theories. An example of a $c < 2$ theory which probably cannot be obtained this way is the direct sum of an Ising model ($c = 1/2$) with two tricritical Ising models ($c = 7/10$).

This particular theory can be obtained using three scalar fields, and it is a reasonable hypothesis that any $c < 2$ theory could be obtained in a similar way, as a flow from a starting point with more scalar fields. If this were the case, we could imagine solving the classification problem by simply listing Lagrangians, but we would need one more ingredient for this list to have an end: an "inverse c-theorem" that tells us that all theories with $c < 2$ can be obtained as RG flows from a set of theories with at most N scalar fields. At present we do not know whether this is so, nor is there any conjecture for the N required for such a claim to hold.

This is an example of a fairly concrete approach to classification, which might or might not work. But one of the points I want to make in this lecture is that we (physicists) have too simple an idea of what a classification should be. I believe we need to broaden this idea to make progress, in ways analogous to those which were developed to address similar problems in mathematics.

Intuitively, one classifies a set of mathematical objects by listing them, and giving an algorithm which, given an object X, determines which one it is. The compact Lie groups are a paradigmatic example in which the objects are classified by discrete structures. As with the minimal CFTs, this classification has an algebraic part, the classification of Lie algebras (say in terms of Dynkin diagrams), and a global part, the classification of discrete subgroups which

[1] In this talk, we only discuss unitary field theories without gravity.

can appear as the fundamental group. In other cases, one has families of objects depending on parameters. One then needs to understand the moduli spaces in which these parameters take values; familiar examples are the moduli spaces of Riemann surfaces.

These are cases in which the classification program was successful. On the other hand, many other superficially analogous mathematical problems have not been solved in such concrete terms, and we will argue that this includes problems quite analogous to classification of QFT. This did not stop the mathematicians, but rather inspired the development of other ways to think about classification problems, more abstract than listing objects or defining moduli spaces. Some are familiar in physics, such as the idea of a topological invariant. Others are not, including most of the ideas of metric geometry. [18]

Before proceeding, let us recall the best existing "classification" of QFTs, which is based on perturbation theory. A large set of QFTs (perhaps all of them) can be defined by quantizing a classical action, say by using a regularized functional integral, computing in perturbation theory, and performing renormalization. For these theories, the classification problem has three parts. First, we need to enumerate actions and the couplings they depend on. We also need to identify equivalences between theories and redundant couplings. We then need to study the renormalization group and decide which operators are relevant and marginal, so that we can take the continuum limit. In perturbation theory, these steps are by now fairly well understood.

Finally, we need to understand the relation between the perturbation theory and the actual observables of the theory. In some cases, this is fairly clear – quantities computed in perturbation theory are asymptotic to the true correlation functions or S-matrix elements as the couplings go to zero. If the physics at finite coupling is qualitatively similar to the limiting free theory, one can essentially regard the coefficients of the perturbative expansion as physical observables. If these agree between two theories, they are the same; if they are "close" in some sense the theories are close, and so on.

Although this map from the action to the observables is somewhat complicated, as physicists we learn how to work with it, and this is our working definition of a "space of QFTs." Thus the stable particles correspond to fields, certain leading coefficients in scattering amplitudes or correlation functions correspond to couplings, and so on. The resulting moduli space of QFTs is a space with one component for each choice of field content, and with each component parameterized by a finite set of couplings in the Lagrangian.

One problem with this working definition is that it is based on perturbation theory, and there are many QFTs for which this is not a good description. Formally, we might distrust perturbation theory because the bare coupling is large, or because a coupling is relevant and becomes large at low energy. Now in itself this does not prove that the physics is qualitatively different from the picture given by perturbation theory; an example where it is not is the 2d Landau-Ginzburg models (*i.e.*, scalar fields with a potential with isolated critical points).

But in many other cases we know that it is. The perturbative excitations might be confined, so that only composite objects are observable. Even when the perturbative excitations are not confined, there might be solitons which are lighter, and in a physical sense are more fundamental. These thoughts lead to the concept of duality equivalences, and the idea that a large part of theory space might be described by patching together perturbative descriptions which are valid in different regimes.[2]

Of course, there might be other regions where no perturbative description is valid. Indeed, we do not know whether we yet know about all the types of QFT, or whether we have all of the concepts we will need to describe them. One test is that a local description of theory space must be complete, in the sense that any relevant or marginal perturbation keeps us within the space. One must also ask whether there are disconnected components of theory space, not obtained

[2] Perhaps the first significant use of this idea was in [22]. It was a common thread throughout the mid-90's, as illustrated by [23, 9].

by perturbing the theories we know about. To address this question, one's basic definitions probably must place the QFTs within some larger set with a simpler description.

These are attractive questions and ideas, which are part of the general study of the space of QFTs. This concept seems to have come out of statistical field theory; it was discussed in Wilson and Kogut's famous review [24] and many works since. At least as an intuition, it provides a framework for a good deal of work on QFT. Nevertheless, we do not know very much about it, as illustrated by the question of our introduction, or more precise questions such as these:

- Is theory space connected (by paths)? Given two QFTs X and Y, are they in the same component? If so, is the path between them "finite length" or "infinite length" ?
- How "big" is theory space? Can we show that the number of QFTs satisfying some conditions is finite, and estimate the number?
- How many parameters (or measurements) do we need to characterize a QFT? At weak coupling, we can answer this, but more generally?

In trying to make these ideas and questions precise, one runs up against unsolved problems. Certain necessary concepts and tools have not yet been developed, and another goal of the talk is to explain what these are.

In previous talks on this subject [12] I expressed this point with the phrase that "to study the space of Xs" (here meaning QFT), one must have a definition of an X. While I would still maintain this, having the definition is not enough; there are other equally important aspects of the problem. Let me list a few, which I will explain below:

- Topology and metric on theory space
- Embedding theorems
- Weak QFT

In explaining these ideas, we will follow the basic analogy between the classification of QFT and the classification of Riemannian manifolds, *i.e.* manifolds with metrics, which first emerged in Friedan's work on renormalization of 2d sigma models [14].

2. Topology and metric on theory space

The next point I want to make in this talk is that it is useful to make precise definitions of when two QFTs are "close." To give the basic idea, suppose we regard a QFT as defined by a list of all of the observables – S-matrix elements, or operator dimensions and o.p.e. coefficients. We then say that two theories are at a distance ϵ in "theory space" if the largest difference between these coefficients is ϵ. Of course, we could use this to define a distance in terms of the other definitions of QFT.

Now, one of the basic problems with regarding a QFT as defined by the list of all of its observables, is that since there are an infinite number of operators with an infinite number of correlation functions, there are an infinite set of observables, and this picture of theory space may seem rather unconstrained and useless. Of course, a first step in the discussion is to reduce this to simpler terms, and for 2d CFT it is known that we can derive all the other correlation functions from the three-point functions of primary fields. However, this is still an infinite amount of data.

This should not stop us. On reflection, one already has a milder version of the same sort of problem when thinking about spaces of ordinary functions, even for a real analytic function of a single real variable. Such a function is characterized by its Taylor expansion around a given point, an infinite set of coefficients. And, indeed, there is no *a priori* upper bound on the number of coefficients one needs to describe such a function. Nevertheless this point does not stop us; we have a good intuition for how to work with functions, and in "normal physics problems" we can

Journal of Physics: Conference Series **462** (2013) 012011 doi:10.1088/1742-6596/462/1/012011

find finite dimensional spaces of functions which are good enough approximations to describe the functions of interest.

Our goal is to develop the analogous intuition here, which tells us what finite dimensional subset of this data is needed to actually determine a particular QFT. Again, there is a simple conceptual answer to this question – if we are perturbing around a given theory, we need to know the couplings for the relevant and marginal operators, a finite subset. The problem is to say this in some way which, unlike the standard discussions, does not depend so drastically on which theory we are perturbing around. To do this, we need to formulate a few basic concepts, such as what it means to "perturb around a given theory," in a more general way.

Let us be a bit more concrete. Suppose we have a set of QFTs $T(\alpha)$, depending on various discrete or continuous parameters α. Each QFT has a space of local operators $\phi_i(x)$, and we grant that we know the n-point correlation functions for all the theories. The problem is to define a distance between pairs of theories, denoted

$$d(T_1, T_2), \tag{1}$$

which satisfies the axioms of a distance (or metric space[3]):

$$d(T_1, T_2) \geq 0 \qquad \text{(with equality only if} T_1 \cong T_2). \tag{2}$$

$$d(T_1, T_2) = d(T_2, T_1), \tag{3}$$

$$d(T_1, T_2) + d(T_2, T_3) \geq d(T_1, T_3) \qquad \text{(the triangle inequality)}. \tag{4}$$

One can keep some concrete sets of QFTs in mind which to test a definition. The first is the two-dimensional sigma models with target space a torus of dimension d, a flat metric g_{ij}, and constant B field. These QFTs are CFTs and their moduli space is well known to be

$$SO(d, d; \mathbb{Z})\backslash SO(d, d; \mathbb{R})/SO(d; \mathbb{R}) \times SO(d; \mathbb{R}) \tag{5}$$

The second is the $c = 1$ CFTs, which were classified in [17]. The moduli space of such theories consists of several disconnected components. One is made up of two half-lines parameterizing the radius of the target spaces S^1 and $S^1/\mathbb{Z}_2$, with the endpoint of the second half-line identified with a particular point on the first half-line. There are also three isolated "exceptional" theories which can be defined as quotients of the $SU(2)_1$ WZW model by the E-type discrete subgroups of $SU(2)$.

2.1. Zamolodchikov distance

There is a standard Riemannian metric on any moduli space on CFTs, the Zamolodchikov metric [26, 19]. It is defined as follows: a tangent vector t to the space of theories at a theory T corresponds to a marginal local operator O_t in the theory; the length of a tangent vector t is then

$$||t||^2_{Zam} \equiv \langle O_t(0) O_t(1) \rangle_T. \tag{6}$$

This corresponds to the metric on field space in a space-time effective action for string theory compactified on T.

A Riemannian metric can be used to define a standard distance function – the distance between T_1 and T_2 is the length of the shortest path connecting them. In the special case of the torus target spaces, the Zamolodchikov metric is homogeneous, so any pair of theories T_1 and T_2 is connected by a unique shortest geodesic which can easily be found explicitly. We then take

[3] In parts of the mathematics literature, such a distance function is referred to as a metric; the metric tensor familiar from general relativity is referred to as a Riemannian metric.

$d(T_1, T_2)$ to be the length of this geodesic; familiar geometric arguments show that this satisfies the stated axioms.

So far all this will probably be familiar, and the reader may be expecting us to propose the straightforward generalization of this. Any moduli space of CFTs will have a Zamolodchikov metric, and we can take the distance $d(T_1, T_2)$ to be the length of the shortest path between them. Again, this will satisfy the axioms.

While this metric is certainly very important in CFT and string theory, if we consider it as a candidate distance to use in defining theory space, it is not really what we want. The main problem with it is simply that it does not define a distance between every pair of CFTs (much less QFTs). For example, the $c < 1$ minimal models have no moduli and are thus isolated points in theory space. Consider the case $c = 4/5$; there are two unitary partition functions, one diagonal and the other that of the three-state Potts model. These are not connected within the space of CFTs, so they have no distance in this sense. But we do not want to say they are at infinite distance, as they can both be obtained as limits of spin systems related by simply varying bare couplings. Rather, a definition of the distance between them must be based on other concepts.

Disconnected components of the space of CFTs are common, such as the exceptional $c = 1$ theories. It is reasonable to expect many more examples, and our definition of theory space must be able to handle them.

2.2. Distance between CFTs

Let us propose a different definition of distance which can be applied to any pair of two-dimensional CFTs with the same central charge c. As explained in the founding paper [5], using the state-operator correspondence and the operator product expansion, all n-point functions can be expressed in terms of the spectrum of operator dimensions $(h_i, \bar{h}_i)$, and the 3-point functions

$$C_{ijk} = \langle \phi_i(z_1)\phi_j(z_2)\phi_k(z_3)\rangle \tag{7}$$

for some fixed z_1, z_2, z_3, or equivalently the o.p.e. coefficients.[4] Thus, the data we need to describe a CFT is this (admittedly infinite) list of numbers $(h_i, \bar{h}_i, C_{ijk})$. If we can define a distance purely in terms of this data, it will apply to any pair of CFTs.

Let us regard this data for a theory T as defining a vector $V(T)$ in the space of infinite sequences of real numbers; then we could define the distance between a pair of theories T_1 and T_2 as the distance between the corresponding vectors. Furthermore, since the space of sequences of real numbers is a linear space, we can define this distance in terms of a norm, so

$$d(T_1, T_2) = ||V(T_1) - V(T_2)||. \tag{8}$$

A standard class of such norms are the ℓ_p norms,

$$||V||_p = \left(\sum_i |v_i|^p\right)^{1/p}, \tag{9}$$

where $p \geq 1$ is a real number. One can also take the limit $p \to \infty$ to get

$$||V||_\infty = \max_i |v_i|. \tag{10}$$

[4] To simplify the discussion here, we normalize all the operators to have unit two-point function. This is actually not the best choice when considering a general family of theories, as it can lead to spurious divergences in the o.p.e. coefficients, but we will discuss this point elsewhere.

Thus, by choosing some p, we get a distance between T_1 and T_2.

We will also need to choose a normalization convention for the vectors V. While at first it might seem natural to take $||V|| = 1$, with this choice the maximum distance between any pair of points will be 2, which is not what we want. Later we will fix this choice by other considerations.

This style of definition has many disadvantages compared to the Zamolodchikov metric. It is less natural – we have to choose p, and we are combining the various CFT data in a somewhat arbitrary way. At first sight, it does not correspond to any observable in string theory. On the other hand, it has the great advantage that it defines a distance between **every** pair of CFTs, since every CFT corresponds to a vector $V(T)$.

2.3. Symmetries and dualities

Before discussing the appropriate choice of p, we need to deal with some problems with the definition as stated. One evident problem is that the vector $V(T)$ is not uniquely defined by the CFT, since it depends on a choice of basis. We might try to state a prescription for ordering the basis, such as taking the lowest dimension operators first. However, if there are operator degeneracies, and especially if the CFT has symmetries, such prescriptions will not lead to a unique choice of V.

A better way to deal with this problem is as follows. We consider the Hilbert spaces for the two theories $\mathcal{H}_1$ and $\mathcal{H}_2$, and the set of unitary transformations $U : \mathcal{H}_1 \rightarrow \mathcal{H}_2$. Any such unitary transformation determines a corresponding linear transformation on the Hamiltonian $L_0 + \bar{L}_0$, the o.p.e. coefficients C_{ijk}, and on their matrix elements in an explicit basis. Thus, it defines an action on the space of linear sequences $V(T)$. Using this, we take the previous definition and minimize over all possible choices of U,[5]

$$d(T_1, T_2) = \min_U ||U(V(T_1)) - V(T_2)||. \tag{11}$$

While this obviously satisfies the other axioms, we need to check the triangle inequality. This can be seen by writing Eq. (4) as

$$||U_{12}(V(T_1)) - V(T_2)|| + ||U_{32}(V(T_3)) - V(T_2)|| \geq \min_{U'} ||U' \cdot U_{12}(V(T_1)) - U_{32}(V(T_3))||. \tag{12}$$

where U_{12} and U_{32} are the choices minimizing Eq. (11). Since the U's all come from unitaries acting on $\mathcal{H}$, minimizing over U' in this expression is the same as minimizing over U in the expression Eq. (11) for $d(T_1, T_3)$. Now, already for $U' = 1$, this inequality is true, since it is just the ordinary triangle inequality. But the additional step of minimizing over U' can only decrease the right hand side, respecting the inequality.

In practice, if T_1 and T_2 are not very close, it may be difficult to find the minimizing U, and in this sense the definition is nonconstructive. Nevertheless it is a standard mathematical definition (used to define the Gromov-Hausdorff distance between manifolds) and it has a number of advantages. Most importantly, it takes account of symmetry and duality relations between theories. Suppose the theories T_1 and T_2 are T-dual; then there will exist a unitary transformation U such that the distance $d(T_1, T_2) = 0$, as it should be.

2.4. Problems with high dimension operators

While this definition is better than the first one, it still does not work. Consider the example of a free boson with radius R; one set of operators in the two theories are the winding operators

$$V_n =: \exp in(\phi_L - \phi_R) : , \qquad n \in \mathbb{Z} \tag{13}$$

[5] Note that U is not a general unitary matrix acting on the space of vectors V (in which case the minimum would be determined by the lengths $||V(T_1)||$ and $||V(T_2)||$), since it is constructed as a tensor product of unitaries acting on $\mathcal{H}$.

Journal of Physics: Conference Series **462** (2013) 012011 doi:10.1088/1742-6596/462/1/012011

with dimension $h = n^2R^2$. Suppose we consider a pair of theories related by a small variation of R. Clearly the operator V_n in the first theory will correspond to V_n in the second, so we will have

$$h^{(R_1)} - h^{(R_2)} = n^2 \left(R_1^2 - R_2^2\right). \tag{14}$$

Thus, for any nonzero $R_1 - R_2$, as n becomes large this difference will become arbitrarily large, and the distance defined this way (for any p) will be infinite.

Looking at other examples, this problem is very general. For any definition of this type to make sense, the effect of varying high dimension operators and their o.p.e.'s must be suppressed.

Conversely, if there were to exist pairs of CFTs which agreed in all the o.p.e.'s of low dimension operators, and only differed for operators with large h, we would not be able to define a space of CFTs. Intuitively, we do not expect this, because the structure of the theory is determined by the relevant and marginal operators, but we do not know of any precise claims to this effect.

One step towards making this precise would be to demonstrate the following

Conjecture 1 *There exists an H, growing linearly with c, such that if two CFT's agree in their spectrum and o.p.e. coefficients for all operators with $h \le H$ and $\bar{h} \le H$, they must be the same CFT.*

Although the intuition based on relevant and marginal operators suggests $H = d$, this need not be the case. A related question which comes up in gauge-gravity duality [25], is to find theories with a large gap Δ_1 in the low lying spectrum of primary fields. Clearly, if there exist two such theories, H must be at or below the gap. Simple arguments show that, for a chiral CFT, the gap cannot be larger than $\Delta_1 = c/24 + 1$, and this bound is saturated for $c = 24$. Although it is not known whether such "extremal CFTs" exist for $c > 24$, from works such as [15] it seems plausible that Δ_1 grows linearly with c. This suggests the growth of H proposed in the conjecture.

While some property like this seems necessary for a theory space as we are envisioning it to exist at all, in itself it would not answer our previous questions. For example, it would not directly imply any bound on the number of such operators in a given CFT. Nevertheless we believe this would be an important claim about CFT.

2.5. Convergence and relation to comparing partition functions

To fix the problem we discussed in the previous subsection, we can put in a convergence factor which suppresses the contributions of high dimension operators. Thus, instead of Eq. (8), we use

$$\begin{aligned} d(T_1, T_2) &= \Bigg(\sum_i |e^{-t2\Delta_i^{(1)}} \Delta_i^{(1)} - e^{-t2\Delta_i^{(2)}} \Delta_i^{(2)}|^p \\ &+ \sum_{i,j,k} |e^{-t(\Delta_i^{(1)}+\Delta_j^{(1)}+\Delta_k^{(1)})} C_{ijk}^{(1)} - e^{-t(\Delta_i^{(2)}+\Delta_j^{(2)}+\Delta_k^{(2)})} C_{ijk}^{(2)}|^p \Bigg)^{1/p} \end{aligned} \tag{15}$$

where $\Delta_i = h_i + \bar{h}_i$, and with some fixed $t > 0$. This particular choice of convergence factor is made to realize the axioms Eq. (3) and Eq. (4), and can be motivated in terms of the world-sheet origin of the data h_i and C_{ijk}. For example, C_{ijk} corresponds to an amplitude on a sphere with three punctures, and the regulated version is obtained by replacing each puncture with a boundary of length $2\pi t$ (obtaining the "pair of pants" diagram). Defining the operator

$$H \equiv \sum_a L_0^{(a)} + \bar{L}_0^{(a)} \tag{16}$$

where the sum over a represents a sum over the boundaries, we can write this as

$$d(T_1, T_2) = \min_U ||e^{-tH_1} U V(T_1) - e^{-tH_2} V(T_2)||, \tag{17}$$

where we reintroduced the unitary transformation of Eq. (11).

This type of definition could also be expressed in terms of differences between the correlation functions of the two CFTs. By expanding in a sum over intermediate states, we can write a four-point function as

$$\langle \phi_1(x_1)\phi_2(x_2)\phi_3(x_3)\phi_4(x_4)\rangle = f(x) \sum_i C^i_{12} C_{34i} e^{-th_i - \bar{t}\bar{h}_i}, \tag{18}$$

where $(t, \bar{t})$ are simple functions of the x_a (essentially the logarithms of cross-ratios), as is $f(x)$. At fixed t one gets a regulated sum of o.p.e. coefficients of the form we want, and clearly one could use the t dependence to extract the coefficients and construct an expression like Eq. (15).

This still leaves the need to minimize over unitaries U. One could reduce these issues by, instead of comparing individual correlation functions and then adding the results, taking a physical observable which already combines all the o.p.e. coefficients. Consider the genus two partition function; by thinking of this as obtained by attaching two "pairs of pants" one sees that it has an expansion of the form

$$Z_{g=2}(T;\ t_1, t_2, t_3) = \sum_{ijk} |C_{ijk}|^2 e^{-t_1\Delta_i - t_2\Delta_j - t_3\Delta_k} \tag{19}$$

and could also be used for this purpose. We will return to this idea shortly.

2.6. More physical definitions

Some of these mathematical prescriptions can be translated into more physical terms. A prototype for this is to consider boundary conditions in a single CFT. Given a pair of boundary conditions B_1 and B_2, a natural way to define the distance between them is in terms of the lowest dimension open string operator between them [10, 11, 20]. Given the partition function Z_{12} for such open strings, we define

$$d^2(B_1, B_2) = \frac{c}{24} - \lim_{t\to\infty} \log Z_{12}(t) \tag{20}$$

in conventions in which an operator of dimension h contributes $\exp(c/24-h)t$ to Z. The relation $d^2 = h$ is motivated by the standard formula for the dimension of a winding state in the large volume limit.

This will be zero if $B_1 \cong B_2$, since the identity operator will contribute. Naively it can also be zero if $B_1 \neq B_2$ but the branes intersect, but in a unitary CFT the only operator of dimension zero is the identity operator, which cannot exist in this Hilbert space. In fact there will be additional "twist" contributions to the ground state energy which raise the dimension. While this gives us two of the axioms, we do not know a proof of the triangle inequality for this definition (away from the large volume limit).

Let us now consider a distance between a pair of $d = 2$ CFTs. The direct analog of the preceding discussion would involve a hypothetical $d = 3$ theory which admitted $d = 2$ CFTs as boundary conditions. However another option, which could be applied to any pair of CFTs T_1 and T_2, would be to make some physical construction in the tensor product theory $T_1 \otimes T_2$.

We then need a physical analog of the unitary operator U which relates the two Hilbert spaces. The simplest thing to consider would be the boundary states in the product theory.

Equivalently, we can consider a defect line $L(U)$ which separates a region of the world-sheet carrying theory T_1 from one carrying theory T_2 [4]. Given any world-sheet definition in which the two theories are related by perturbations and flows, a precise definition of defect line can be made, as discussed in many works such as [6].

If $T_1 \cong T_2$, we can find a "topological boundary state" which can be freely moved without changing the energy; in this sense it has zero tension. This suggests that we define $d(T_1, T_2)$ in terms of the tension of the defect, minimized over all such defects.

Let us now return to Eq. (17). For the special case $p = 2$, we can write the norm $||V||$ as the square root of an inner product (V, V). Now, making the world-sheet identifications of the previous subsection, the inner product on the CFT Hilbert space $\mathcal{H}$, gives us an inner product on the space of coefficients V (for example, C_{ijk} lives in the dual of $\mathcal{H} \otimes \mathcal{H} \otimes \mathcal{H}$, etc.). Thus, we can rewrite the regulated difference between the o.p.e. coefficients appearing in Eq. (17) as

$$\begin{aligned} d_C(T_1, T_2)^2 &= \min_U (e^{-tH_1} UV(T_1) - e^{-tH_2} V(T_2), e^{-tH_1} UV(T_1) - e^{-tH_2} V(T_2)) \qquad (21) \\ &= ||e^{-tH_1} V(T_1)||^2 + ||e^{-tH_2} V(T_2)||^2 - 2 \max_U (e^{-tH_1} UV(T_1), e^{-tH_2} V(T_2)) \\ &= Z_{g=2}(T_1\ 2t, 2t, 2t) + Z_{g=2}(T_2\ 2t, 2t, 2t) - 2 \max_U Z_{g=2}(T_1|L(U)|T_2;\ 2t, 2t, 2t) \end{aligned}$$

where the last term

$$Z_{g=2}(T_1|L(U)|T_2\ 2t, 2t, 2t) = (e^{-tH_1} UV(T_1), e^{-tH_2} V(T_2)) \qquad (22)$$

is defined by these equations. Now, the point is that this term also has a world-sheet interpretation. Since the unitary U maps $\mathcal{H}_1$ to $\mathcal{H}_2$, we realize it as a defect line $L(U)$, separating two regions of the world-sheet carrying the two theories. Thus it is the partition function on a genus 2 Riemann surface, built by sewing together two pairs of pants, but with a the common defect line $L(U)$ along all three components of the junction.

A similar discussion shows that the regulated difference between the operator dimensions in the two theories can be written as the d/dt derivative (or equivalently an insertion of $L_0 + \bar{L}_0$) of a difference of genus one amplitudes,

$$d_h(T_1, T_2)^2 = -\frac{d}{dt} \left(Z_{g=1}(T_1\ 2t) + Z_{g=1}(T_1\ 2t) - 2 \max_U Z_{g=1}(T_1|L(U)|T_2;\ 2t) \right). \qquad (23)$$

Finally, the squared distance $d^2(T_1, T_2)$ is defined as the sum of Eq. (21) and Eq. (23),

$$d(T_1, T_2)^2 = \min_U c_h(t) d_h(T_1, T_2)^2 + c_C(t) d_C(T_1, T_2)^2 \qquad (24)$$

with the choice of defect line $L(U)$ which minimizes the total distance (so the argument around Eq. (12) will work) and where $c_h(t)$ and $c_C(t)$ are normalization factors chosen to cancel the leading divergences of the partition functions (for example, $c_h(t) \sim \exp ct/6$). By the previous formal arguments, this will satisfy the triangle inequality as well as the other axioms. Thus we have a physical interpretation of the distance Eq. (17) for the particular case $p = 2$.

There are some evident variations on this one could consider. For example, one could add higher genus terms, or even integrate over moduli space (in the critical c or with some suitable weight) to make this look more like a string theory observable. There are also a number of questions which we hope to clarify in subsequent work. One is whether to require that the defect lines be local on the world-sheet. If so, then for Eq. (17) and Eq. (24) to be the same, we would need to show that the minimizing U is local.

With some refinements, such as specifying the specific metric on the surfaces, the definition Eq. (24) might make sense for general 2d QFTs as well. This point is also under study.

2.7. *Topology and boundedness*

One role for a definition of distance between CFTs or QFTs is as a technical or mathematical tool, used to establish properties like the convergence of an approximate definition. We will discuss this idea in §4. But there are also interesting physical questions about it.

One property which is reasonable to ask, and which we would conjecture holds for the definition Eq. (17) or its physical counterpart Eq. (24), is that it is uniform with respect to the Zamolodchikov distance. In other words,

- A sequence of CFTs which converges in the Zamolodchikov metric, also converges in the metric Eq. (24). Similarly, if a sequence converges in Eq. (24), and all pairs of CFTs in the sequence have well-defined Zamodchikov distances, it will converge there as well.
- One can also talk about an unbounded sequence, meaning a sequence T_i for which the distances $d(T, T_n)$ to some reference theory T go to infinity for large n. This notion should also be the same.

Because the definition of Eq. (24) is essentially linear, questions of convergence and especially of boundedness are likely to be much simpler to study.

A more specific relation to the Zamolodchikov metric can be found by computing derivatives of Eq. (24) with respect to the moduli. It seems plausible that, with a suitable choice of the coefficients $c(t)$ in Eq. (24), this distance will be equal to or less than the Zamolodchikov distance. These points are being checked and explicit examples worked out [13].

The boundedness property allows us to restate a conjecture of Kontsevich and Soibelman (as cited in [1]). Loosely stated, it is that the only infinite distance limits in the space of CFTs are the large volume limits. More precisely, it is that a family of CFTs with fixed c and a lower bound on the nonzero operator dimensions h_i, is totally bounded (or precompact). Besides determining the infinite distance limits, this implies that the space can be completed, meaning that the limit of any convergent sequence of CFTs can be regarded as defining a point in a larger completion of the space. There are many examples of such limit points obtained by shrinking volumes of cycles to zero, whose interpretation in string theory involves nonperturbative effects such as new massless states, instanton corrections, or both. Such limit points need not satisfy all of the axioms of CFT, and it would be interesting to have a general theory of them.

As a final comment, although we arrived at a nice definition with a concrete world-sheet interpretation, in mathematics one often uses more than one definition of distance and/or norm, even in the same problem, as each can have its own advantages with respect to the other structures in the problem.

3. Embedding theorems

Any approach based on observables such as the above faces the problem that the consistency conditions for observables to come from a QFT are very complicated. On the other hand, the consistency conditions on a Lagrangian which can be used to define a QFT by a functional integral are fairly simple. This suggests that we base the classification on the Lagrangian approach, however we do not know whether all QFTs can be obtained this way.

A very direct way to address this question would be to give a prescription which, given some subset of the observables of a QFT, produces a Lagrangian which is guaranteed to reproduce the original QFT. While at first this may sound implausible, what I have in mind is a prescription which is highly redundant, and perhaps only useful to make this conceptual point. For example, we might imagine a prescription in which every operator up to some very high dimension is represented by its own field in the Lagrangian. Let us consider the Landau-Ginzburg theories with a single real field ϕ and the Lagrangian

$$\mathcal{L} = (\partial\phi)^2 - \sum_{i=1}^{n} t_i \phi^i, \tag{25}$$

then we would introduce operators $O_k =: \phi^k$ up to $k = n - 2$, and possibly further operators involving derivatives of ϕ. We would then introduce couplings which enforce the relations between these fields, such as Lagrange multipliers

$$\lambda(O_2 - C_{112}O_1^2) \tag{26}$$

where C_{112} is an o.p.e. coefficient, and so on. The idea would be to have so many couplings that we could tune as many dimensions and o.p.e. coefficients as we needed to reproduce the desired QFT.

Is this a reasonable goal? Even if it is possible, it might turn out to have the problem we mentioned in the introduction, that even after putting some upper bound on the number of degrees of freedom of our QFT, we might need an arbitrarily large number of fields in order to reproduce it. If so, this approach would probably not be worth the trouble to develop.

Restricting attention to 2d CFT, we are thus led to conjecture the "inverse c-theorem" of the introduction,

Conjecture 2 *All 2d CFTs of a given central charge c can be obtained as flows from the theory of N free bosons, for some definite N growing as the square of the central charge c.*

What reason is there to believe this, and what should we expect N to be? There are many directly analogous questions in mathematics which can guide us.

Suppose we want to classify the manifolds of a given dimension D. Might it be that all such manifolds can be obtained as submanifolds of $\mathbb{R}^N$, the Euclidean N-dimensional space? More explicitly, we are asking for an embedding, a function $X : M \hookrightarrow \mathbb{R}^N$ such that $X(p) = X(q)$ implies $p = q$, and the Jacobian $\partial_i X^I$ has maximal rank D.

This question was asked and answered early on; for smooth manifolds this is the Whitney Embedding Theorem, and one needs $N = 2D$. Thus, if all CFTs could be obtained as sigma models with target spaces of dimension equal to the central charge, and we did not need to tune the metric or couplings, the conjecture would hold with $N = 2c$.

This is of course a drastic oversimplification of the real situation. A closer analog might be to ask that we can reproduce the relevant and marginal couplings of the sigma model, meaning the potential, metric and B-field via the embedding. Denoting these as (V, g_{ij}, B_{ij}), the mathematical question is to find a function f on some $\mathbb{R}^N$ such that $f = 0$ on an embedding $X : M \hookrightarrow \mathbb{R}^N$ and is positive elsewhere, so that

$$g_{ij} = \partial_i X^I \partial_j X^J \delta_{IJ}; \qquad B_{ij} = \partial_i X^I \partial_j X^J \omega_{IJ}$$

on M. The total potential would then be $Cf + \tilde{V}$, where $\tilde{V}$ is a function on $\mathbb{R}^N$ which restricts to V on M, and C is some large constant.

Looking into the mathematics, finding f and $\tilde{V}$ is not hard.[6] To realize the metric as the restriction of the Euclidean metric on $\mathbb{R}^N$, and the B-field as the pullback of a constant two-form, since together these have D^2 coefficients, one might think that it would be possible given D^2 adjustable functions, so that $N = D^2$.

Showing that this can be done globally (*i.e.* over all of M) is much more difficult. Nevertheless, the Nash embedding theorem states that this is possible for a smooth metric; one needs $N = (D + 2)(D + 3)/2$ (see [2] for a recent review). While I don't know of a similar theorem for the combined g and B problem, there is no obstruction to realizing B as a pullback, so it seems plausible that this can also be done with N of order D^2.

This at least suggests that the conjecture could hold with N of order the square of the central charge c. Of course, one needs to discuss renormalization and many other issues to follow this

[6] One can take for f the sums of squares of functions which define M locally, and combine these with V using a partition of unity on $\mathbb{R}^N$.

geometric route. Clearly the problem will be far more constrained given supersymmetry. For the case of $(2,2)$ SCFT, the relevant analogy is to the Kodaira embedding theorem, which suggests a general prescription for taking data from the topological open sector of an SCFT, and constructing a gauged linear sigma model which flows back to the same SCFT [3].

4. Weak QFTs

Our final approach is to try to sit the problem of constructing and classifying QFTs, in some larger problem obtained by relaxing some of the many axioms and consistency conditions. The hope is of course that this problem will be easier to solve, and that we can then go on to enforce the additional conditions required to get actual or "honest" QFTs. In other words, we start by classifying "weak QFTs" which obey some but not all of the axioms of QFT.

The term[7] is by analogy to the idea in mathematics of a "weak solution" of a partial differential equation. This is a solution which need not be a function but might be a distribution or something else. Even if the goal is to find a solution which is a function, often it can be easier to first show that such a weak solution exists, and then show that it is continuous, smooth or satisfies other desired properties. The basic example is the solution of linear PDE's by Fourier transform, which does not naturally produce a function but rather an "L^2 function," which is only defined up to a set of measure zero.

When stated in such generality, of course this approach has been used extensively, one might even say universally, in the problem of defining QFTs. For example, a lattice theory or a regulated functional integral might be regarded as defining a "weak" form of a Euclidean or Lorentz invariant local QFT. Of course, in this case one has good reasons to expect the desired symmetry and locality properties to be recovered in the long distance limit.

The class of weak theories we have in mind here would be one in which we can define the weak QFT in terms of a finite amount of data, and then express the additional axioms needed for it to be an honest QFT as some finite set of equations which the data must satisfy.

A prototype for this is the classification of the minimal $c<1$ CFTs. Such a CFT is defined by giving a finite list of primary fields and the finite set of their o.p.e. coefficients. This data then must satisfy the BPZ (associativity) and Cardy conditions to define an honest CFT.

It would be interesting to be able to generalize this to $c>1$. The main problem is of course that we cannot work with the actual list of primary fields. We need to know something like Conjecture 1 to cut this down to size.

Suppose we knew the data of Conjecture 1; *e.g.* the complete list of primary fields and o.p.e. coefficients for all operators with dimension $h<H$ and $\bar{h}<H$. Could we use this to define a "weak QFT" ?

An interesting related idea appears in the work [21] of Rattazzi *et al.* They show that a CFT with a scalar operator ϕ of dimension h_ϕ, must contain an operator $:\phi^2:$ in the o.p.e.

$$\phi\,\phi \rightarrow 1 + :\phi^2: + \ldots \tag{27}$$

with an upper bound on its dimension $h_{\phi^2} \leq f(h_\phi) < 2h_\phi$. Interestingly, the argument does not use Virasoro representation theory, only the finite-dimensional conformal group $SO(d,2)$, and thus could work in any space-time dimension. One needs explicit expressions for the conformal blocks, which were found for $d=2$ and $d=4$ in [8].

The argument relies on associativity of the 4-point function, schematically

$$\sum_i \langle \phi(z_1)\phi(z_2)|i\rangle \, \langle i|\phi(z_3)\phi(z_4)\rangle = \sum_j \langle \phi(z_2)\phi(z_3)|j\rangle \, \langle j|\phi(z_4)\phi(z_1)\rangle. \tag{28}$$

[7] suggested to me by Edward Witten

It is important that every term on both sides is positive, by unitarity (reflection positivity). Using conformal symmetry, one can write this (in any d) as a relation between functions of one complex variable z, a generalization of the cross ratio in $d = 2$ which maps the coincident limits $z_1 \to z_2, z_3, z_4$ to $z \to 0, \infty, 1$ respectively. Then, while this is a functional relation, one focuses on the behavior at the symmetric point $z = 1/2$. Although it is not at all obvious, from numerical study it turns out that the difference of the conformal blocks entering the two sides of this equation, behaves differently at $z = 1/2$ for intermediate operators O_i with $h_i < f(h_\phi)$ and those with $h_i > f(h_\phi)$. This can be expressed as a sign of a second derivative which must vanish for Eq. (28) to hold, and thus both types of operators must be present in the sum.

In the terms we are using here, one can think of this as a "weak CFT" argument. One is considering a class of theories which has most of the structure of CFT, such as symmetry and positivity, but need not satisfy associativity. Within this class, one can write down four-point functions which match an assumed part of the spectrum and o.p.e. Eq. (27). One then imposes associativity and gets a constraint.

Let us return to our earlier question: how can we use the data of Conjecture 1 to define a "weak QFT" ? What we clearly can do, is construct approximate four-point functions by summing over intermediate operators with $h, \bar{h} \leq H$, such as

$$\sum_{m \text{ such that } h \leq H} \langle \phi_i(z_1)\phi_j(z_2)|m\rangle \langle m|\phi_k(z_3)\phi_l(z_4)\rangle \tag{29}$$

While these will reproduce the data, they will not be associative.

Then, we define a function U which expresses how far these approximate four-point functions are from associativity, schematically

$$U = \sum ||F - F'|| \tag{30}$$

where $F - F'$ is the difference between the two sides of Eq. (28), and the sum is over all correlation functions of operators with $h, \bar{h} \leq H$.

Of course, associativity of the four-point functions is not the only consistency condition on a CFT, there is also modular invariance at genus one, and perhaps others (since we did not impose associativity of all four-point functions). But we could go on and add other terms to measure the failure of these properties to hold.

We can then define a flow which decreases U. Given a choice of parameters to vary, say the dimensions and o.p.e. coefficients, and given a metric on this space, we have the gradient flow

$$\dot{V}^i = -G^{ij}\frac{\partial U}{\partial V^j}. \tag{31}$$

Thus, starting with a weak QFT, we flow to an honest QFT.

This scheme is admittedly somewhat complicated. Worse, it has two major problems:

- If we do not put in all the consistency conditions, we may end up at $U = 0$ with a set of correlation functions which do not correspond to any QFT.
- The function U might have local minima with $U > 0$. Indeed, as we put in more and more consistency conditions, the function U becomes complicated, and is likely to have more and more local minima.

So, we should probably not rush to implement this scheme numerically just yet. But the general idea might be a good one, analogous to constructing CFTs via RG flow but making different aspects of the problem manifest. Given enough analytic understanding of the space of CFTs one is trying to describe, one might be able to make good choices for U and the metric which avoid the problems we just raised.

References

[1] B. S. Acharya and M. R. Douglas, "A finite landscape?," arXiv:hep-th/0606212.
[2] B. Andrews, "Notes on the isometric embedding problem and the Nash-Moser implicit function theorem," 2002.
[3] Paul. S. Aspinwall and M. R. Douglas, work in progress.
[4] C. Bachas, J. de Boer, R. Dijkgraaf and H. Ooguri, "Permeable conformal walls and holography," JHEP **0206**, 027 (2002) [arXiv:hep-th/0111210].
[5] A. A. Belavin, A. M. Polyakov and A. B. Zamolodchikov, "Infinite conformal symmetry in two-dimensional quantum field theory," Nucl. Phys. B **241**, 333 (1984).
[6] I. Brunner and D. Roggenkamp, "Defects and Bulk Perturbations of Boundary Landau-Ginzburg Orbifolds," JHEP **0804**, 001 (2008) [arXiv:0712.0188 [hep-th]].
[7] A. Cappelli and J. B. Zuber, "A-D-E Classification of Conformal Field Theories," arXiv:0911.3242 [hep-th].
[8] F. A. Dolan and H. Osborn, "Conformal partial waves and the operator product expansion," Nucl. Phys. B **678**, 491 (2004) [arXiv:hep-th/0309180].
[9] M. R. Douglas, "Superstring dualities, Dirichlet branes and the small scale structure of space," arXiv:hep-th/9610041.
[10] M. R. Douglas, "D-branes and matrix theory in curved space," Nucl. Phys. Proc. Suppl. **68**, 381 (1998) [arXiv:hep-th/9707228].
[11] M. R. Douglas, "Two lectures on D-geometry and noncommutative geometry," arXiv:hep-th/9901146.
[12] M. R. Douglas, talk given at the Stony Brook Symposium in Honor of Chen Ning Yang and James H. Simons, March 27–28, 2008.
[13] M. R. Douglas, work in progress.
[14] D. H. Friedan, "Nonlinear Models In Two + Epsilon Dimensions," Annals Phys. **163**, 318 (1985).
[15] M. R. Gaberdiel, S. Gukov, C. A. Keller, G. W. Moore and H. Ooguri, "Extremal N=(2,2) 2D Conformal Field Theories and Constraints of Modularity," arXiv:0805.4216 [hep-th].
[16] T. Gannon, "Monstrous moonshine and the classification of CFT," arXiv:math/9906167.
[17] P. H. Ginsparg, "Curiosities at c = 1," Nucl. Phys. B **295**, 153 (1988).
[18] *Metric structures for Riemannian and non-Riemannian spaces*, M. Gromov, M. Katz, P. Pansu, and S. Semmes, 2001, Birkhäuser ISBN: 978-0-8176-4582-3.
[19] D. Kutasov, "Geometry on the space of conformal field theories and contact terms," Phys. Lett. B **220**, 153 (1989).
[20] G. W. Moore, "K-theory from a physical perspective," arXiv:hep-th/0304018.
[21] R. Rattazzi, V. S. Rychkov, E. Tonni and A. Vichi, "Bounding scalar operator dimensions in 4D CFT," JHEP **0812**, 031 (2008) [arXiv:0807.0004 [hep-th]].
[22] N. Seiberg and E. Witten, "Monopole Condensation, And Confinement In N=2 Supersymmetric Yang-Mills Theory," Nucl. Phys. B **426**, 19 (1994) [Erratum-ibid. B **430**, 485 (1994)] [arXiv:hep-th/9407087].
[23] C. Vafa, "Lectures on strings and dualities," arXiv:hep-th/9702201.
[24] K. G. Wilson and J. B. Kogut, "The Renormalization group and the epsilon expansion," Phys. Rept. **12**, 75 (1974).
[25] E. Witten, "Three-Dimensional Gravity Revisited," arXiv:0706.3359 [hep-th].
[26] A. B. Zamolodchikov, "Irreversibility of the Flux of the Renormalization Group in a 2D Field Theory," JETP Lett. **43**, 730 (1986)

Journal of Physics: Conference Series **462** (2013) 012012 doi:10.1088/1742-6596/462/1/012012

The black hole/qubit correspondence

M. J. Duff

Theoretical Physics, Blackett Laboratory, Imperial College London, London SW7 2AZ, United Kingdom

E-mail: m.duff@imperial.ac.uk

Abstract. We review recent progress in the black hole/qubit correspondence.

1. Introduction

The work connects two previously disparate areas of theoretical physics:

(i) Hawking black hole entropy.

(ii) Qubit entanglement in quantum information theory .

In the 1970s Hawking showed that quantum black holes can radiate energy and hence carry entropy. Although Hawking gave a macroscopic explanation, a microscopic understanding was provided only in 1996 by superstring theory. In 2006 it was noticed that the entropy formula for a particular 8-charge black hole appearing in string theory is given by the 'hyperdeterminant', a quantity introduced by the mathematician Cayley in 1845. Remarkably, this same quantity also appears in the completely different field of quantum information theory (QIT). A qubit is a two-state (up/down) quantum system; so three qubits (Alice, Bob and Charlie) can have 8 different states. These states can get entangled in a way that Einstein described as "Spooky" (Alice, Bob and Charlie could live in different galaxies) and Cayley's hyperdeterminant is the '3-tangle' which measures the degree of entanglement.

This turned out to be the tip of an iceberg and since then a two-way dictionary between black holes and qubits has been established and knowledge of string theory has been used to discover new things about QIT and vice-versa. These include (a) a microscopic interpretation of these 8 states in terms of 8 ways of wrapping D3-branes around the extra dimensions of string theory; (b) the three-way entanglement of seven qubits, suggesting that predictions based on the mathematics of octonions may be testable in the laboratory; (c) the discovery of "Freudenthal duality", a new symmetry of black holes; (d) a new way of classifying multi-qubit entanglement based on Jordan algebras.

Current goals include:

(i) To pursue the notion of "superqubits" which introduces the concept of bose-fermi symmetry into QIT and finds applications in condensed-matter physics.

(ii) To use intersecting D-branes to solve long-outstanding problems in the currently topical field of multi-qubit entanglement, thus providing falsifiable predictions of string theory.

(iii) To seek a physical origin of these black hole/qubit coincidences, which would have far-reaching implications for a unified theory of fundamental forces.

2. Black holes, string theory and M-theory

If current ideas are correct, a unified theory of all physical phenomena will require some radical ingredients:

2.1. Supersymmetry

For each known *boson* (integer spin 0, 1, 2 and so on), there is a *fermion* (half-integer spin 1/2, 3 /2, 5/2 and so on), and vice versa. The number of these supersymmetries is denoted $\mathcal{N}$ and ranges from 1 to 8 in four spacetime dimensions. The Large Hadron Collider in Geneva will be looking for these superparticles. Moreover, supersymmetry implies gravity: if Einstein had not already discovered general relativity, supersymmetry would force us to invent it. Many theorists favour supersymmetry because it provides a framework within which the weak, electromagnetic and strong forces may be united with gravity.

2.2. Extra dimensions

Curiously, supergravity places an upper limit of eleven on the dimension of spacetime. The familiar universe, of course, has three space dimensions and one time dimension. But in the 1920s Kaluza and Klein suggested that spacetime may have a hidden fifth dimension. This extra dimension would not be infinite, like the others; instead it would close in on itself, forming a circle. The kind of real, four-dimensional world supergravity ultimately predicts depends on how the extra seven dimensions are rolled up, a la Kaluza and Klein.

2.3. Strings

In 1984, however, 11-dimensional supergravity was knocked off its pedestal by superstring theory in 10 dimensions. There were five competing theories: the $E_8 \times E_8$ heterotic, the $SO(32)$ heterotic, the $SO(32)$ Type I, and the Type IIA and Type IIB strings. The $E_8 \times E_8$ seemed, at least in principle, capable of explaining the elementary particles and forces, including their handedness. And strings seemed to provide a theory of gravity consistent with quantum effects.

2.4. Branes and M-theory

Yet the spacetime of eleven dimensions allows for a membrane, which may take the form of a bubble or a two-dimensional sheet. In 1987 it was shown [1] that if one of the 11 dimensions is a circle, we can wrap the sheet around it once, pasting the edges together to form a tube. If the radius becomes sufficiently small, the rolled-up membrane ends up looking like a string in 10 dimensions; it yields precisely the Type IIA superstring. In a landmark talk at the University of Southern California in 1995, Witten [2] drew together all this work on strings, branes and 11 dimensions under the umbrella of M-theory in 11 dimensions.

2.5. Black holes

Such breakthroughs have led to a new interpretation of black holes as intersecting black-branes wrapped around seven curled dimensions. As a result, there are strong hints that M-theory may even clear up the paradoxes of black holes raised by Hawking. In 1974 Hawking showed that black holes are not entirely black but may radiate energy [3]. In that case, they must possess entropy, which measures the disorder by accounting for the number of quantum states available. Yet the microscopic origin of these states stayed a mystery. Using Polchinski's *D-branes*, Strominger and Vafa [4] were able to count the number of quantum states in black-branes. They found an entropy that agrees with Hawking's prediction, placing another feather in the cap of M-theory. Branes now occupy centre stage as the microscopic constituents of M-theory, as the higher-dimensional progenitors of black holes and as entire universes in their own right. Despite all these successes, physicists are glimpsing only small corners of M-theory; the big picture is still lacking. Over the next few years we hope to discover what M-theory really is. Understanding black holes will be an essential pre-requisite.

3. Entanglement and quantum information theory

In completely separate developments, exciting things were happening in the world of QIT. Quantum entanglement is a quantum mechanical phenomenon in which the states of two or more objects have to be described with reference to each other, even though the individual objects may be spatially separated. In their seminal 1935 paper, Einstein, Podolsky, and Rosen (EPR) demonstrated that quantum mechanics cannot provide a complete description of "local realism" for two spatially separated but quantum mechanically correlated particles. It was not until 1964, however, that CERN theorist John Bell [5] proposed a falsifiable experiment to test EPR and it was not until 1983 that Alain Aspect actually performed it. Quantum mechanics won and local realism lost. As quantum information science developed, the impact of entanglement went far beyond the testing of the conceptual foundations of quantum mechanics. Entanglement is now of central importance in QIT as a physical resource essential to numerous quantum information tasks including: quantum cryptography, superdense coding, and quantum computation.

Utilizing the paradigm of *stochastic local operations and classical communication* (SLOCC) it was established in [6] that three qubits (Alice, Bob and Charlie: $|\Psi\rangle = a_{ABC}|ABC\rangle$ with $A, B, C = 0, 1$) can be entangled in several physically distinct ways: tripartite GHZ (Greenberger-Horne-Zeilinger), tripartite W, biseparable A-BC, separable A-B-C and null:

$$\begin{array}{ccccccccc} & & & & \text{GHZ} & & & & \\ & & \text{W} & & & & \text{W} & & \\ A\text{-}BC & & & & A\text{-}BC & & & & A\text{-}BC \\ & & A\text{-}B\text{-}C & & & & A\text{-}B\text{-}C & & \\ & & & & \text{Null} & & & & \end{array} \tag{1}$$

The GHZ state is distinguished by a non-vanishing invariant known as the 3-tangle τ_{ABC}, given by Cayley's hyperdeterminant of a_{ABC}.

However, the crucial problem of classifying arbitrary n-qubit entanglement remains unsolved.

4. Black holes and qubits

Studying the entanglement/entropy correspondence, we have begun to see interesting and non-trivial links between the two fields of string theory and QIT. Further work may lead to the realisation that entanglement and entropy are dual descriptions of the same phenomenon. Successes so far include:

4.1. Black holes in 4D and three qubits

An explicit correspondence between the above 3-tangle and the entropy S of the 8-charge STU black hole of supergravity [8]:

$$S = \tfrac{\pi}{2}\sqrt{\tau_{ABC}}.$$

Moreover, the physically distinct forms of 3-qubit entanglement correspond directly to the physically distinct black hole solutions. Similar results relate $5D$ black holes to qutrits (3-state systems) [9].

4.2. Wrapped branes as qubits

We have demonstrated that the eight states of the 3-qubit system correspond to eight ways of wrapping four D3-branes around three 2-tori of Type II string theory compactified to four dimensions [10]. Similarly, the nine states of the 2-qutrit system correspond to nine ways of wrapping two M2-branes around two 3-tori of M-theory compactified to 5 dimensions.

Journal of Physics: Conference Series **462** (2013) 012012 doi:10.1088/1742-6596/462/1/012012

Table 1. A 4-charge black hole in $4D$ spacetime 0,1,2,3 from four D3-branes wrapping the six extra dimensions 4,5,6,7,8,9 yields a GHZ state. Each row represents a D3 brane and the first row represents a brane wrapping dimensions 4,6 and 8.

0	1	2	3	4	5	6	7	8	9	brane	$\lvert ABC\rangle$
x	o	o	o	x	o	x	o	x	o	D3	$\lvert 000\rangle$
x	o	o	o	x	o	o	x	o	x	D3	$\lvert 011\rangle$
x	o	o	o	o	x	x	o	o	x	D3	$\lvert 101\rangle$
x	o	o	o	o	x	o	x	x	o	D3	$\lvert 110\rangle$

4.3. Black holes, qubits and octonions

We have extended the $\mathcal{N} = 2$ STU model example to the most general case of black holes in $\mathcal{N} = 8$ supergravity where the global symmetry, referred to as U-duality, is $E_{7(7)}$, one of the exceptional Lie groups appearing in Cartan's classification. Noting that $E_{7(7)} \supset [SL(2)]^7$,we have shown that the corresponding system in quantum information theory is that of seven qubits (Alice, Bob, Charlie, Daisy, Emma, Fred and George), $[SL(2)]^7$ being the corresponding SLOCC equivalence group. However, the larger symmetry requires that they undergo at most a tripartite entanglement of a very specific kind. The entanglement measure will be given by the quartic Cartan $E_7(\mathbb{C})$ invariant. The entanglement may be represented by the Fano plane where the vertices A, B, C, D, E, F, G represent the seven qubits and the seven lines $ABD, BCE, CDF, DEG, EFA, FGB$ and GAC represent the tripartite entanglement. See Figure 1. The Fano plane also corresponds to the multiplication table of the imaginary octonions.

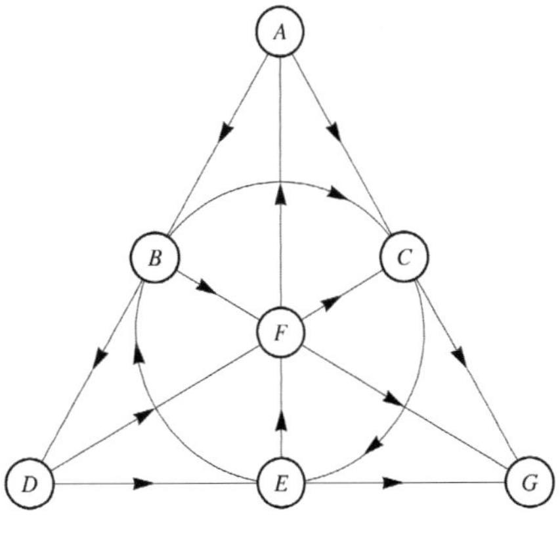

Figure 1. The Fano plane. The vertices A, B, C, D, E, F, G represent the seven qubits and the seven lines $ABD, BCE, CDF, DEG, EFA, FGB, GAC$ represent the tripartite entanglement.

From the point of view of M-theory, this suggests a whole new way of studying its symmetries based on the 7 imaginary octonions (completely different from the Jordan algebra approach that uses all 8 *split* octonions.) From the QIT point of view, seven qubits is just at the limit of what can be reached experimentally, so predictions based on the mathematics of octonions, described by the Fano Plane, may now be testable in the laboratory. Octonions have fascinated mathematicians and physicists for decades but have yet to find any physical application. In recent books both Roger Penrose and Ray Streater have characterised octonions as one of the

Journal of Physics: Conference Series **462** (2013) 012012 doi:10.1088/1742-6596/462/1/012012

great 'lost causes' in physics. Yet we hope that the tripartite entanglement of seven qubits will prove them wrong and provide away of seeing the effects of octonions in the laboratory [11, 12].

4.4. Freudenthal classification of three qubits
New methods to classify three qubits exploiting that elegant branch of mathematics involving Jordan algebras and the related Freudenthal triple system (FTS), familiar from black holes in string theory [13]. The FTS naturally reproduces the conventional 3-qubit entanglement classification (1). Moreover, using the FTS analysis we determined the mathematical *coset* characterisation of both the entanglement and black hole classes, hitherto unknown results.

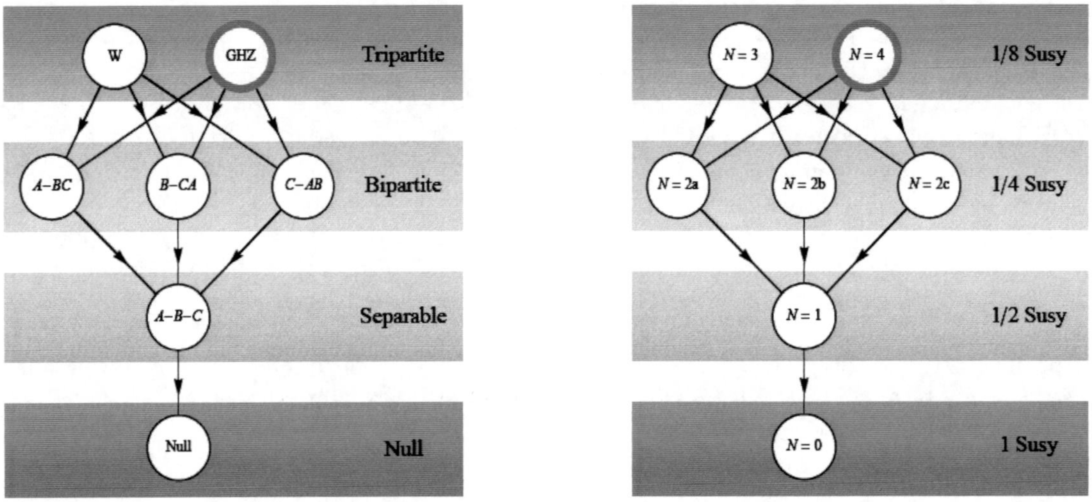

Figure 2: The classification of three qubits (left) exactly matches the classification of black holes from N wrapped branes (right). Only the GHZ state has a nonzero 3-tangle and only the $\mathcal{N} = 4$ black hole has nonzero entropy.

Figure 2.

4.5. Freudenthal duality
When the black hole charges are quantized the U-duality symmetry is broken to an infinite discrete subgroup and the square of the entropy becomes integer valued. Invoking recent important developments in number theory, we have introduced a new symmetry in the case when this integer is a perfect square. This "Freudenthal duality" is best understood in terms of the *integral* FTS and may relate quantized black holes that are not necessarily U-duality equivalent while preserving their lowest order entropy [14]. It would be interesting to understand their role in the physics of stringy black holes. For example, this symmetry rules out small black holes whose masslessness would spoil the conjectured finiteness of $\mathcal{N} = 8$ supergravity.

5. Current research
5.1. Superqubits
Although this topic grew out of our black hole/qubit correspondence, it is logically independent of it. In QIT n-qubit states lie in the fundamental representation of the SLOCC equivalence group $[SL(2, \mathbb{C})]^n$. We propose a supersymmetric generalisation of the qubit, the *superqubit*,

by extending this to the supergroup $[OSp(2|1)]^n$. For $n = 1$ a single superqubit forms a 3-dimensional representation of $OSp(2|1)$ consisting of two commuting "bosonic" components a_A and one anti-commutating "fermionic" component $a_\bullet$.

$$|\Psi\rangle = a_0|0\rangle + a_1|1\rangle + a_\bullet|\bullet\rangle$$

This mathematical construction seems a very natural one. For $n = 2$ and $n = 3$ we introduce

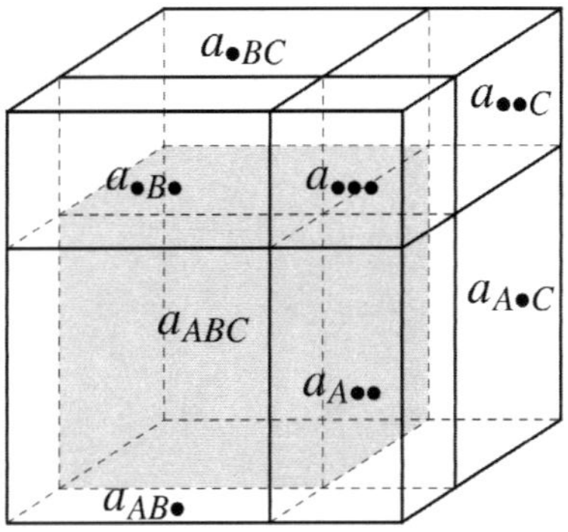

Figure 3. The $3 \times 3 \times 3$ cubic superhypermatrix

the appropriate supersymmetric generalisations of the conventional entanglement measures. In particular, super-Bell and super-GHZ states are characterised respectively by non-vanishing superdeterminant (distinct from the Berezinian) and superhyperdeterminant. We have already uncovered some unexpected consequences. For example, the two superqubit state

$$|\Psi\rangle = i|\bullet\bullet\rangle$$

looks naively separable but is entangled because the product of two Grassmann numbers $a_\bullet b_\bullet$ cannot yield a complex number. In fact it is *maximally* entangled.

From a physical point of view, the supergroup $OSp(2|1)$ is also used in some models of strongly correlated electrons in condensed matter physics. The t-J model is a specialisation of the Hubbard model where the effective field theory describes a system in which the charge and spin degrees of freedom separate and are free to move across the lattice. In certain regions of the parameter space the t-J model becomes supersymmetric and the chargeon and the spinon transform in the fundamental representation of $OSp(2|1)$ [15]. This means that they form a valid realisation of a superqubit (cf. how the two polarisations of a photon can be a realisation of a qubit). Long chains of these superqubits are described in the lattice and we can use our techniques to quantify any super-entanglement that may be present and, in particular, discover if the supersymmetry is able to control the decoherence times of these systems. The influence that this may have on quantum error correction and quantum computing should not be ignored. Already the role of superqubits in super quantum computing is being explored by Castellani et al [16].

Another example where our superqubits can be used is the *supersymmetric quantum Hall effect* [17], observed in two-dimensional electron systems subjected to low temperatures and strong magnetic fields, whereby conductivity is quantised.

5.2. D-branes and qubits: falsifiable predictions of string theory

The partial nature of our understanding of exactly how the extra dimensions of string/M-theory are curled up has so far prevented any kind of *smoking gun* experimental test. This has led some critics of string theory to suggest that it is not true science. This is easily refuted[1] by studying the history of scientific discovery; the thirty year time lag between EPR's germ of an idea and Bell's falsifiable prediction provides a nice example. Nevertheless it cannot be denied that such a prediction in string theory would be very welcome. One of our aims is to do just that, namely to use D-brane intersection rules to predict new results in multi-qubit entanglement.

Macroscopically, the entropy is just one quarter the area of the event horizon of the black hole. To give a microscopic derivation [4] we need to invoke ten-dimensional string theory whose associated Dp-branes wrapping around the six compact dimensions provide the string-theoretic interpretation of the black holes. A Dp-brane wrapped around a p-dimensional cycle of the compact directions $(x^4, x^5, x^6, x^7, x^8, x^9)$ looks like a D0-brane from the four-dimensional (x^0, x^1, x^2, x^3) perspective. In the string literature one may find *D-brane intersection rules* which tell us how N branes can intersect over one another and the fraction ν of supersymmetry that they preserve. Up to $N = 4$ the results are given by [18]

$$\begin{array}{ccccccc} & & & N=4,\nu=1/8 & & & \\ & N=3,\nu=1/8 & & & & N=3,\nu=1/8 & \\ N=2,\nu=1/4 & & & N=2,\nu=1/4 & & & N=2,\nu=1/4 \\ & N=1,\nu=1/2 & & & & N=1,\nu=1/2 & \\ & & & N=0,\nu=1 & & & \end{array} \qquad (2)$$

In our black hole/qubit correspondence, the microscopic description of a GHZ state is that of the $N = 4, \nu = 1/8$ case of D3-branes of Type IIB string theory wrapping the (469), (479), (569), (578) cycles of a six-torus and intersecting over a string. The wrapped circles are denoted by crosses and the unwrapped circles by noughts. $|0\rangle$ corresponds to xo and $|1\rangle$ to ox, as in Table 1. The number of qubits here is three because the number of extra dimensions is six. This also explains where the two-valuedness enters on black hole side. To wrap or not to wrap; that is the qubit.

Repeating the exercise for the $N < 4$ cases and using our dictionary, we see that string theory predicts the three qubit entanglement classification (1), in complete agreement with the standard results of QIT. Allowing for different p-branes wrapping tori of different dimensions, we can also describe qutrits and more generally qudits. Furthermore, for the well-documented cases of 2×2, 2×3, 3×3, $2\times 2\times 3$ and $2\times 2\times 4$, our D-brane intersection rules are also in complete agreement. However, for higher entanglements, such as $2 \times 2 \times 2 \times 2$, then the QIT results are partial or not known or else contradictory. This is currently a very active area of research in QIT because the experimentalists can now control entanglement with greater numbers of qubits. Our goal is to use the allowed wrapping configurations and D-brane intersection rules to predict new qubit entanglement classifications.

5.3. Physical underpinnings

Despite these mathematical coincidences, however, the underlying physical reason why black hole and qubits should be related is still a mystery. There are intriguing hints that the answer may lie in the above wrapping of D-branes around the extra dimensions of string theory. If so, this qubit interpretation will radically change the way we look at string and M-theory.

Quantum entanglement lies at the heart of quantum information theory, with applications to quantum computing, teleportation, cryptography and communication. So whether or not there is a physical underpinning, the esoteric techniques of string theory might yet have practical applications.

[1] See or listen to my 2007 debate with Lee Smolin at the Royal Society for the Arts `http://www.thersa.org/events/speakers-archive/d/michael-duff`

Acknowledgments
I would like to thank my collaborators L. Borsten, D. Dahanayake, H. Ebrahim, S. Ferrara and W. Rubens

References

[1] Duff M J, Howe P S, Inami T and Stelle K S 1987 *Phys. Lett.* **B191** 70
[2] Witten E 1995 *Nucl. Phys.* **B443** 85–126 (*Preprint* hep-th/9503124)
[3] Hawking S W 1975 *Commun. Math. Phys.* **43** 199–220
[4] Strominger A and Vafa C 1996 *Phys. Lett.* **B379** 99–104 (*Preprint* hep-th/9601029)
[5] Bell J S On the Einstein-Podolsky-Rosen paradox *Physics* **1** (1964) no. 3, 195
[6] Dür W, Vidal G and Cirac J I 2000 *Phys. Rev.* **A62** 062314 (*Preprint* quant-ph/0005115)
[7] Maldacena J M 1998 *Adv. Theor. Math. Phys.* **2** 231–252 (*Preprint* hep-th/9711200)
[8] Duff M J 2007 *Phys. Rev.* **D76** 025017 (*Preprint* hep-th/0601134)
[9] Duff M J and Ferrara S 2007 *Phys. Rev.* **D76** 124023 (*Preprint* 0704.0507)
[10] Borsten L, Dahanayake D, Duff M J, Rubens W and Ebrahim H 2008 *Phys. Rev. Lett.* **100** 251602 (*Preprint* 0802.0840)
[11] Duff M J and Ferrara S 2007 *Phys. Rev.* **D76** 025018 (*Preprint* quant-ph/0609227)
[12] Borsten L, Dahanayake D, Duff M J, Ebrahim H and Rubens W 2009 *Phys. Rep.* **471** 113–219 (*Preprint* 0809.4685)
[13] Borsten L, Dahanayake D, Duff M J, Rubens W and Ebrahim H 2009 *Phys. Rev.* **A80** 032326 (*Preprint* 0812.3322)
[14] Borsten L, Dahanayake D, Duff M J and Rubens W 2009 *Phys. Rev.* **D80** 026003 (*Preprint* 0903.5517)
[15] Wiegmann P B 1988 *Phys. Rev. Lett.* **60** 2445
[16] Castellani L, Grassi P A and Sommovigo L 2010 (*Preprint* 1001.3753)
[17] Hasebe K 2005 *Phys. Rev. Lett.* **94** 206802 (*Preprint* hep-th/0411137)
[18] Bergshoeff E, de Roo M, Eyras E, Janssen B and van der Schaar J P 1997 *Nucl. Phys.* **B494** 119–143 (*Preprint* hep-th/9612095)

6th International Symposium on Quantum Theory and Symmetries (QTS6) IOP Publishing
Journal of Physics: Conference Series **462** (2013) 012013 doi:10.1088/1742-6596/462/1/012013

Universal description of viscoelasticity with foliation preserving diffeomorphisms

Tatsuo Azeyanagi[1], **Masafumi Fukuma**[1], **Hikaru Kawai**[1,2] **and Kentaroh Yoshida**[1]

[1] Department of Physics, Kyoto University, Kyoto 606-8502, Japan
[2] Theoretical Physics Laboratory, Nishina Center, RIKEN, Saitama 351-0198, Japan

E-mail: aze@gauge.scphys.kyoto-u.ac.jp, fukuma@gauge.scphys.kyoto-u.ac.jp, hkawai@gauge.scphys.kyoto-u.ac.jp, kyoshida@gauge.scphys.kyoto-u.ac.jp

Abstract. We review our recent proposal for a universal description of generic single-component viscoelastic systems with a single relaxation time. Foliation preserving diffeomorphisms are introduced as an underlying symmetry which naturally interpolates between the two extreme characters of elasticity and fluidity. The symmetry is found to be powerful enough to determine the dynamics in the first order of strains.

1. Introduction

Viscoelasticity (see, e.g., [1, 2]) is a unifying notion which contains solid and fluid as limiting cases. It was defined by James Clerk Maxwell such as to describe the materials that behave as elastic solids at short times scales and as viscous fluids at long time scales [3, 4].

In order to grasp the idea of this definition, we consider a material (like a chewing gum) consisting of many particles bonding each other. They stay at their equilibrium positions in the absence of stresses (as in the leftmost of Fig. 1), and we now suppose that an external force is applied to deform the material. An internal stress is then produced in the body, and according to the definition, this internal stress can be treated as an elastic force at least during short intervals of time. However, if we keep the deformation much longer than some period of time (characteristic to each material and to be called the relaxation time), then the bonding structure changes to reduce the free energy, and the internal stress vanishes eventually. The point here is that the two configurations (the center and the rightmost of Fig. 1) exhibit the same shape (i.e. same configurations for the positions of particles) but have different bonding structures.

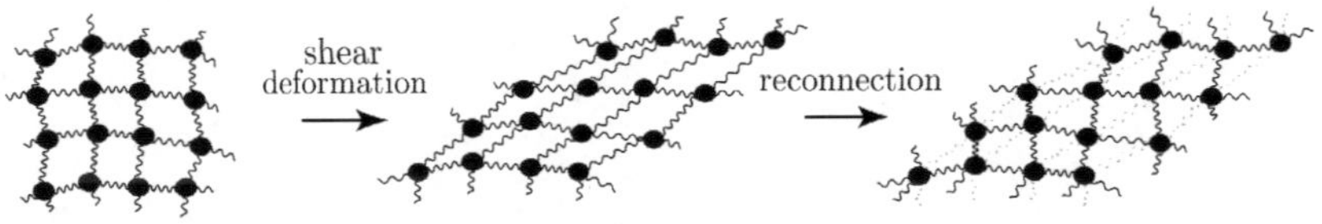

Figure 1. Shear deformation of viscoelastic material

We thus see that the bonding structure itself is dynamical for viscoelastic materials, and is lead to introduce a new dynamical variables (to be called *intrinsic metric* [5]) to specify the bonding structure. It was claimed in [5] that viscoelasticity may have a universal description if we define the intrinsic metric properly and assume the invariance of the system under *foliation preserving diffeomorphisms* (FPDs). In fact, it was shown there that FPDs naturally interpolate between the two extreme limits of elasticity and fluidity, and are powerful enough to determine the dynamics in the first order of strains.

The main purpose of the present article is to give a brief review of [5] in a concise way, and also to present some additional stuffs which were not mentioned in [5].

Our discussions will be constrained to single-component materials. Note that, even for this case, there can be two relaxation times, one being for shear deformations and another for bulk compressions (or expansions). In this article, we set the relaxation time for bulk compressions to be infinite, so that the bulk stress does not undergo relaxations (see Fig. 2). This ensures that the system exhibits fluidity at long intervals of time.

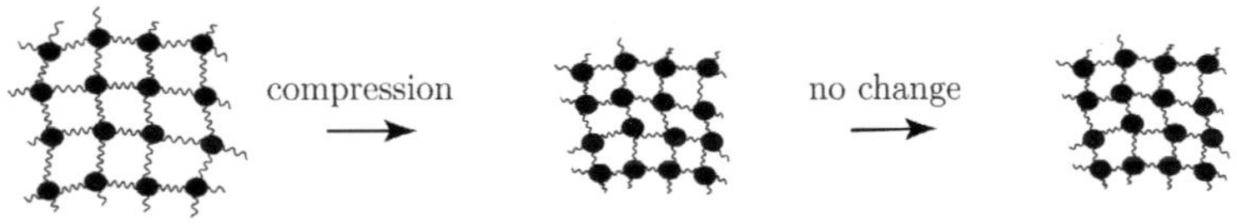

Figure 2. Compression of viscoelastic material

2. Geometrical setup

We describe a viscoelastic material within the framework of world-volume theory with the world-volume coordinates $\xi = (\xi^a)$ $(a = 1, 2, 3)$ (labeling the material particles of a body) and with the target space coordinates $X = (X^i)$ $(i = 1, 2, 3)$. The shape of the material at time t is then specified by giving the functions $X^i(\xi, t)$. In order to further describe the bonding structure, we introduce an intrinsic metric $\bar{h}_{ab}(\xi, t)$ as follows.

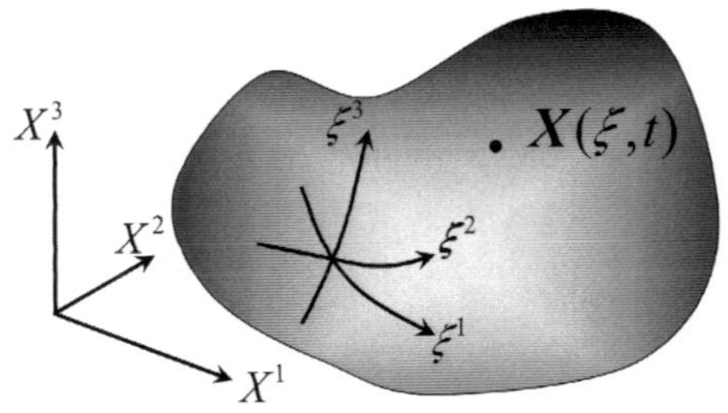

Figure 3. World-volume of a viscoelastic material

We first recall that our viscoelastic materials are defined to be elastic at least for a very short time. This means that for any (sufficiently small) portion in a material, its *natural shape* can be defined at each moment as the shape that would be taken when all stresses were virtually removed (see Fig. 4).

This observation enables us to define two kinds of metrics on the world-volume at each time t (see Fig. 5). The first one is the *induced metric* $h_{ab}(\xi, t)$ which measures the length in the real shape:

$$ds^2 = h_{ab}(\xi, t)\, d\xi^a d\xi^b\,, \tag{1}$$

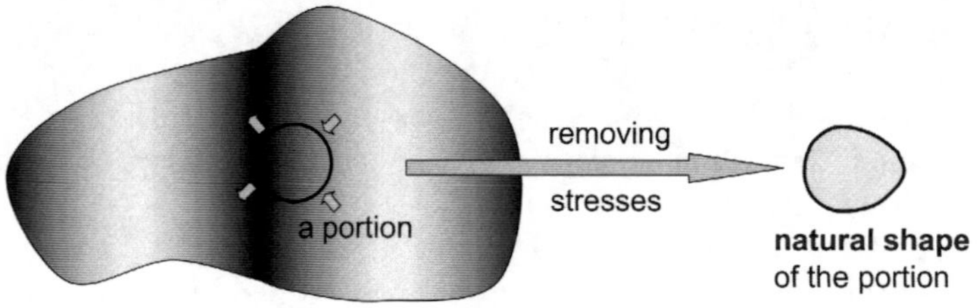

Figure 4. Definition of the natural shape for a small portion

where

$$h_{ab}(\xi,t) \equiv e^i_a(\xi,t)e^i_b(\xi,t) \qquad \big(e^i_a(\xi,t) \equiv \partial_a X^i(\xi,t)\ :\ \text{dreibein}\big)\,. \tag{2}$$

Another is the *intrinsic metric* $\bar{h}_{ab}(\xi,t)$ which measures the length in the natural shape:

$$d\bar{s}^2 = \bar{h}_{ab}(\xi,t)\,d\xi^a d\xi^b\,. \tag{3}$$

The difference between them defines the *strain tensor*,

$$\varepsilon_{ab}(\xi,t) \equiv \frac{1}{2}\big(h_{ab}(\xi,t) - \bar{h}_{ab}(\xi,t)\big). \tag{4}$$

In fact, for an elastic material at rest without stress, we can take the coordinates $\xi = (\xi^a)$ such that $\bar{h}_{ab} = \delta_{ab}$. Then, when the material is slightly deformed under stresses, $X^a(\xi,t) = \xi^a + u^a(\xi,t)$ (u^a: displacements), we have $h_{ab} = \partial_a X^i \partial_b X^i = \delta_{ab} + \partial_a u_b + \partial_b u_a + O(u^2)$, so that $\varepsilon_{ab} = (1/2)\big(\partial_a u_b + \partial_b u_a\big) + O(u^2)$.

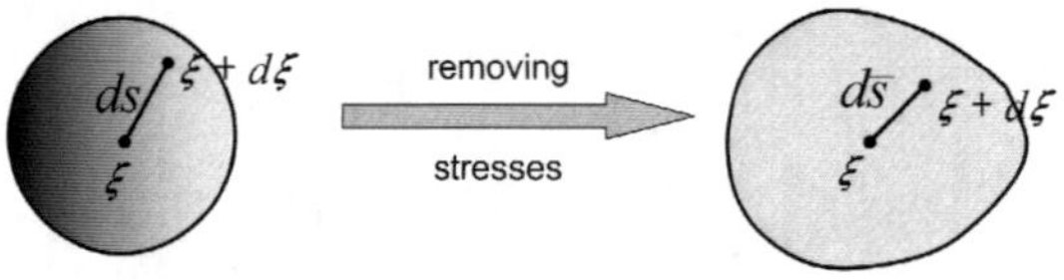

Figure 5. The induced and intrinsic metrics

We denote by ρ_0 the mass density in the absence of strains. In this article, we assume that ρ_0 is a constant independent of ξ and t. Then the mass contained in a volume element $d^3\xi = d\xi^1 d\xi^2 d\xi^3$ is given by $\rho_0\sqrt{\bar{h}}\,d^3\xi$. The mass density ρ in the real three-dimensional space is given by the mass conservation $\rho\sqrt{h}d^3\xi = \rho_0\sqrt{\bar{h}}d^3\xi$ as $\rho = \big(\sqrt{\bar{h}}/\sqrt{h}\big)\,\rho_0$.

Let us consider various deformations of a 2D material. Here (ξ^1,ξ^2) are comoving coordinates attached to material particles.

(a) shear deformations: Suppose that a material in equilibrium is deformed with an external force as in Fig. 6. The two metrics h_{ab} and $\bar{h}_{ab}$ agree before the deformation. When the material is deformed, the induced metric h_{ab} varies, expressing the change of the shape. In contrast, the intrinsic metric $\bar{h}_{ab}$ retains its original form just after the deformation. This is because the material should exhibit elasticity for a short period of time, and the natural shape at that moment is its original form in equilibrium. At sufficiently later times, $\bar{h}_{ab}$ gradually changes its values and comes to equal the induced metric.

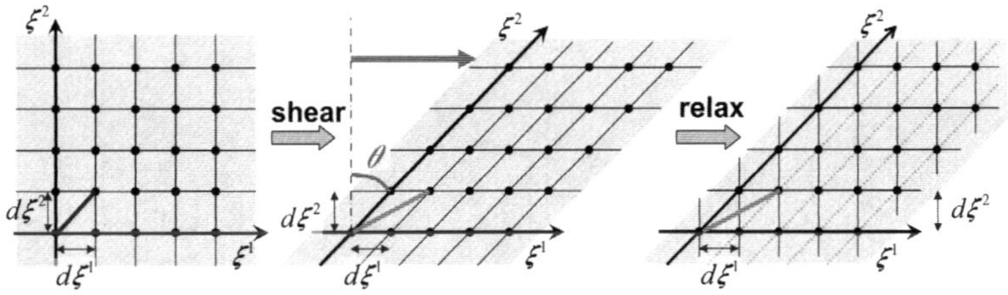

$$h_{ab} = \begin{pmatrix} 1 & 0 \\ 0 & 1 \end{pmatrix} \longrightarrow \begin{pmatrix} 1 & \tan\theta \\ \tan\theta & \sec^2\theta \end{pmatrix} \longrightarrow \begin{pmatrix} 1 & \tan\theta \\ \tan\theta & \sec^2\theta \end{pmatrix}$$

$$\bar{h}_{ab} = \begin{pmatrix} 1 & 0 \\ 0 & 1 \end{pmatrix} \longrightarrow \begin{pmatrix} 1 & 0 \\ 0 & 1 \end{pmatrix} \longrightarrow \begin{pmatrix} 1 & \tan\theta \\ \tan\theta & \sec^2\theta \end{pmatrix}$$

$$\rho = \rho_0 \longrightarrow \rho_0 \longrightarrow \rho_0$$

Figure 6. Shear deformation with the angle θ

(b) bulk compressions: Suppose that a material in equilibrium is compressed with a factor $\lambda\,(<1)$ as in Fig. 7. The two metrics h_{ab} and $\bar{h}_{ab}$ again agree before the deformation, and the induced metric h_{ab} changes after the deformation. However, according to our assumption that the bulk compression (or expansion) does not undergo relaxation, the intrinsic metric $\bar{h}_{ab}$ retains its original form.

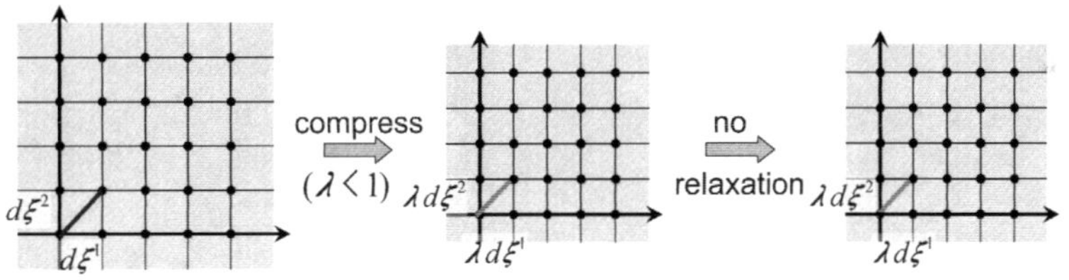

$$h_{ab} = \begin{pmatrix} 1 & 0 \\ 0 & 1 \end{pmatrix} \longrightarrow \begin{pmatrix} \lambda^2 & 0 \\ 0 & \lambda^2 \end{pmatrix} \longrightarrow \begin{pmatrix} \lambda^2 & 0 \\ 0 & \lambda^2 \end{pmatrix}$$

$$g_{ab} = \begin{pmatrix} 1 & 0 \\ 0 & 1 \end{pmatrix} \longrightarrow \begin{pmatrix} 1 & 0 \\ 0 & 1 \end{pmatrix} \longrightarrow \begin{pmatrix} 1 & 0 \\ 0 & 1 \end{pmatrix}$$

$$\rho = \rho_0 \longrightarrow \frac{\rho_0}{\lambda^2} \longrightarrow \frac{\rho_0}{\lambda^2}$$

Figure 7. Compression by the factor λ ($\lambda < 1$)

3. Foliation preserving diffeomorphisms

So far the world-volume coordinates ξ have been attached to a material, labeling the material particles. However, we can introduce arbitrary coordinates ξ at each time t, independently of the labeling of material particles. In this case we need to specify the relative motion of the material to the coordinates, but this can be carried out by introducing a vector field $\mathcal{N} = N^a(\xi,t)\partial_a$ (see Fig. 8) [5]. More precisely, the vector field is defined such that the material particle located at ξ at time t is supposed to move to the position $\xi^a + N^a(\xi,t)dt$ after the time interval dt.

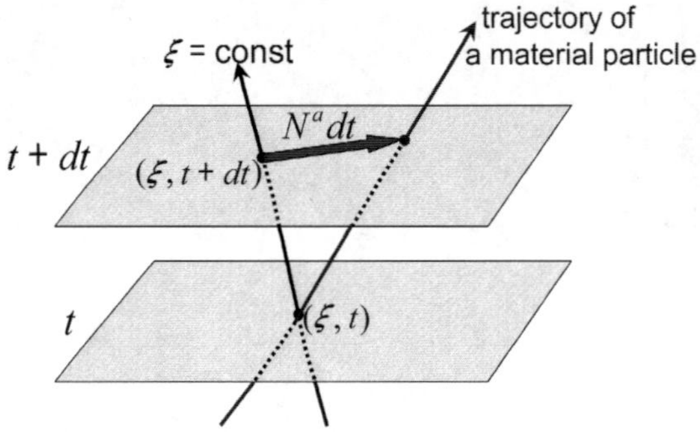

Figure 8. Definition of N^a

With $N^a(\xi,t)$, the system comes to have the gauge symmetry of *foliation preserving diffeomorphisms* (FPDs), reflecting the redundancy in the description. The set of FPDs consist of the $(3+1)$-dimensional reparametrizations that do not move the time slices, and is generated by the infinitesimal transformations of the form (see [5] for references of mathematical details):

$$\delta\xi^a(\xi,t)=\epsilon^a(\xi,t)\,,\quad \delta t(\xi,t)=0\,. \tag{5}$$

The fields X^i, h_{ab} and $\bar{h}_{ab}$ are intrinsically three-dimensional and thus transform covariantly as

$$\delta X^i=\epsilon^a\,\partial_a X^i\,, \tag{6}$$

$$\delta h_{ab}=\partial_a\epsilon^c\,h_{cb}+\partial_b\epsilon^c\,h_{ac}+\epsilon^c\,\partial_c h_{ab}\,, \tag{7}$$

$$\delta \bar{h}_{ab}=\partial_a\epsilon^c\,\bar{h}_{cb}+\partial_b\epsilon^c\,\bar{h}_{ac}+\epsilon^c\,\partial_c \bar{h}_{ab}\,, \tag{8}$$

while the transformation of the shift vector has an additional inhomogeneous term as

$$\delta N^a=-\partial_c\epsilon^a\,N^c+\epsilon^c\,\partial_c N^a-\dot{\epsilon}^a\,. \tag{9}$$

The time derivatives of a FPD-covariant tensor $T^{b_1b_2\cdots}_{a_1a_2\cdots}$ are no longer covariant, but can be made so by introducing the covariant derivative [5]:

$$\frac{D}{Dt}T^{b_1b_2\cdots}_{a_1a_2\cdots}\equiv\frac{\partial}{\partial t}T^{b_1b_2\cdots}_{a_1a_2\cdots}+\mathcal{L}_{\mathcal{N}}T^{b_1b_2\cdots}_{a_1a_2\cdots}\,, \tag{10}$$

where $\mathcal{L}_{\mathcal{N}}$ is the Lie derivative with respect to $\mathcal{N}=N^a\partial_a$.

For example, the velocity $v^i(\xi,t)$ and the acceleration $a^i(\xi,t)$ of the material particle located at (ξ,t) are given, respectively, by

$$v^i=\frac{DX^i}{Dt}=\dot{X}^i+N^a\partial_a X^i\,,\qquad a^i=\frac{Dv^i}{Dt}=\frac{D^2X^i}{Dt^2}=\dot{v}^i+N^a\partial_a v^i\,. \tag{11}$$

This can be easily confirmed by noting that the material particle at (ξ,t) moves after the time interval dt to the point $(\xi+N dt,\,t+dt)$:

$$X^i(\xi+N dt,\,t+dt)=X^i(\xi,t)+\big[\dot{X}^i(\xi,t)+N^a(\xi,t)\,\partial_a X^i(\xi,t)\big]dt\equiv X^i(\xi,t)+v^i(\xi,t)\,dt\,. \tag{12}$$

As for the metrics h_{ab} and $\bar{h}_{ab}$, their covariant derivatives are given, respectively, by

$$K_{ab}\equiv\frac{1}{2}\frac{Dh_{ab}}{Dt}=\frac{1}{2}\left(\dot{h}_{ab}+\partial_a N^c\,h_{cb}+\partial_b N^c\,h_{ac}+N^c\,\partial_c h_{ab}\right), \tag{13}$$

$$\bar{K}_{ab}\equiv\frac{1}{2}\frac{D\bar{h}_{ab}}{Dt}=\frac{1}{2}\left(\dot{\bar{h}}_{ab}+\partial_a N^c\,\bar{h}_{cb}+\partial_b N^c\,\bar{h}_{ac}+N^c\,\partial_c \bar{h}_{ab}\right), \tag{14}$$

and are called the extrinsic curvatures. Note that

$$K_{ab} - \bar{K}_{ab} = \frac{D\varepsilon_{ab}}{Dt}, \tag{15}$$

and

$$K_{ab} = \frac{1}{2}\big(\nabla_a v_b + \nabla_b v_a\big), \tag{16}$$

where $v_a \equiv e^i_a v^i$ and ∇_a is the covariant derivative with respect to h_{ab} [5].

As was stated above, the set of FPDs forms a gauge symmetry group of the system, and expresses a redundancy in the description of the system. This can be gauge-fixed arbitrarily according to convenience in describing the dynamics of a given system. Two of the useful gauge fixings are the followings:

[**Comoving frame**] This is useful in describing elastic materials and is defined by the condition $N^a(\xi,t) = 0$, which leads to $v^i(\xi,t) = \dot{X}^i(\xi,t)$.

[**Laboratory frame**] This is useful in describing fluids and is defined by the condition $X^a(\xi,t) \equiv \xi^a$, from which follow the equations $v^a = N^a$, $h_{ab} = \delta_{ab}$, $\rho = \rho_0\sqrt{\bar{h}}$, and $a^a = \dot{v}^a + v^b\,\partial_b v^a$. The last equation shows that the covariant derivative in this frame becomes the material (Lagrangian) derivative.

4. Fundamental equations

We now write down a set of equations which determine the time evolution of twelve variables X^i (three), $\bar{h}_{ab}$ (six) and N^a (three) up to FPDs (three). Our strategy is to first write down equations which hold in both of the elastic and fluid limits (this can be done in specific frames), and then to make the equations covariant under FPDs.

4.1. Equation of motion for $\bar{h}_{ab}(\xi,t)$

We first consider the limiting case of elasticity in the comoving frame ($N^a = 0$). We then have $\dot{\bar{h}}_{ab} = 0$ because the natural shape does not evolve in time for elastic materials. Its FPD-covariant expression is $\bar{K}_{ab} = 0$, and thus we expect that nonvanishing $\bar{K}_{ab}$ represents genuinely plastic (i.e. nonelastic) deformations. However, one can show that its trace part $\bar{K} \equiv \bar{h}^{ab}\bar{K}_{ab}$ always vanishes due to the mass conservation [5]:

$$\bar{K} = 0. \tag{17}$$

In fact, in the laboratory frame, $\bar{K}$ is expressed as $(1/\rho)\big(\dot{\rho} + \partial_a(\rho v^a)\big)$ [5] and thus must vanish.

The above consideration shows that only the traceless (shear) component of the extrinsic curvature $\bar{K}_{ab}$ expresses the changing rate of the bonding structure, which should be proportional to shear strains in the following form [5]:

$$\bar{K}_{ab} - \frac{1}{3}\bar{K}\,\bar{h}_{ab} = \frac{1}{\tau}\Big(\varepsilon_{ab} - \frac{1}{3}\big(\bar{h}^{cd}\varepsilon_{cd}\big)\bar{h}_{ab}\Big). \tag{18}$$

Note that the second term on the left-hand side can actually be omitted because $\bar{K} = 0$. The inverse of the proportional coefficient, τ, has the dimension of time, and should correspond to the relaxation time.

In order to check the consistency of this equation (named the *rheology equation* in [5]), we take two extreme limits. In the elastic limit ($\tau \to \infty$), we obtain $\bar{K}_{ab} = 0$, which again means that there occurs no time evolution of intrinsic metric. In contrast, in the fluid limit ($\tau \to 0$), we have $\varepsilon_{ab} \propto \bar{h}_{ab}$, or $h_{ab} \propto \bar{h}_{ab}$. This means that the material acquires only homogeneous compression around each point, which is actually the characteristic property defining fluid.

4.2. Equation of motion for $X^i(\xi,t)$

The dynamics of $X^i(\xi,t)$ should be expressed as Euler's equation

$$\rho\, a_a = -\nabla^b T_{ba}\,. \tag{19}$$

The leading form of the stress tensor T_{ab} in the derivative expansion can be determined by the following requirements:

- T_{ab} be symmetric and covariant under FPDs.
- T_{ab} be linear in the strain ε_{ab} and the spatial derivative of the velocity, $\nabla_a v_b$.

These requirements imply that T_{ab} is a linear combination of the irreducible components of ε_{ab} and $K_{ab} = (1/2)(\nabla_a v_b + \nabla_b v_a)$ (see (16)):

$$T_{ab} = -2\mu\,\varepsilon'_{ab} - \frac{1}{\kappa}\,\varepsilon\, h_{ab} - 2\gamma\, K'_{ab} - \zeta\, K h_{ab}\,, \tag{20}$$

where

$$\varepsilon \equiv h^{ab}\varepsilon_{ab}\,, \quad \varepsilon'_{ab} \equiv \varepsilon_{ab} - \frac{1}{3}\,\varepsilon\, h_{ab}\,, \quad K \equiv h^{ab}K_{ab}\,, \quad K'_{ab} \equiv K_{ab} - \frac{1}{3}\,K h_{ab}\,. \tag{21}$$

The set of equations (17), (18), (19) and (20) give nine independent equations, and thus can describe the time evolution of twelve dynamical variables up to FPDs. In order to see the meaning of the coefficients in (20), we consider two extreme limits of elasticity and fluidity. In the elastic limit, we have $\bar{K}_{ab} = 0$ and thus $K'_{ab} = \big(D\varepsilon_{ab}/Dt\big)\big(1 + O(\varepsilon)\big)$, $K = \big(D\varepsilon/Dt\big)\big(1 + O(\varepsilon)\big)$. Then the stress tensor becomes

$$T_{ab} = -2\mu\,\varepsilon'_{ab} - \frac{1}{\kappa}\,\varepsilon\, h_{ab} - 2\gamma\,\frac{D\varepsilon'_{ab}}{Dt} - \zeta\,\frac{D\varepsilon}{Dt} h_{ab} + O(\varepsilon^2)\,, \tag{22}$$

which shows that the four terms represent, respectively, shear and bulk moduli, and shear and bulk frictions. On the other hand, the fluid limit is realized by considering at time scales much longer than the relaxation time τ. By using the rheology equation in the form $\varepsilon'_{ab} = \tau\, K'_{ab}\big(1 + O(\varepsilon)\big)$, Eq. (20) becomes

$$T_{ab} = -2\eta\, K'_{ab} - \zeta\, K h_{ab} - \frac{1}{\kappa}\,\varepsilon\, h_{ab} + O(\varepsilon^2)\,, \tag{23}$$

where $\eta = \gamma + \mu\tau$ and ζ are shear and bulk viscosities. The last term expresses the pressure with κ being the bulk compressibility.

5. Dispersion relations

In order to get dispersion relations for the set of equations (17)–(20), we expand solutions around the hydrostatic solution in the laboratory frame ($X^a(\xi,t) = \xi^a$, $h_{ab} = \delta_{ab}$) as

$$\bar{h}_{ab} = \delta_{ab} + \Delta\bar{h}_{ab}\,, \quad N^a = 0 + \Delta N^a\,. \tag{24}$$

We consider plane waves propagating in the $X^3(=\xi^3)$-direction:

$$\Delta\bar{h}_{ab}(\xi,t) = \Delta\bar{h}_{ab}(k,\omega)e^{ik\xi^3 - i\omega t}\,, \quad \Delta N^a(\xi,t) = \Delta N^a(k,\omega)e^{ik\xi^3 - i\omega t}\,. \tag{25}$$

The equations can be separated according to the helicity, and we have the following equations:

- helicity 0 (shear modes) $\{\Delta N^1 \pm i\Delta N^2,\ \Delta\bar{h}_{31} \mp i\Delta\bar{h}_{32}\}$:

$$\omega^3 + i\Big[\frac{1}{\tau} + \frac{1}{\rho_0}\Big(\zeta + \frac{4}{3}\gamma\Big)k^2\Big]\omega^2 - \frac{1}{\rho_0}\Big[\frac{1}{\tau}\Big(\zeta + \frac{4}{3}\gamma\Big) + \Big(\frac{1}{\kappa} + \frac{4}{3}\mu\Big)\Big]k^2\omega - \frac{i}{\rho_0\kappa\tau}k^2 = 0 \tag{26}$$

$$\text{fluid limit } (\tau \to 0):\ \omega^2 + \frac{i}{\rho_0}\Big(\zeta + \frac{4}{3}\gamma\Big)k^2\omega - \frac{1}{\rho_0\kappa}k^2 = 0 \tag{27}$$

$$\text{elastic limit } (\tau \to \infty):\ \omega^2 + \frac{i}{\rho_0}\Big(\zeta + \frac{4}{3}\gamma\Big)k^2\omega - \frac{1}{\rho_0}\Big(\frac{1}{\kappa} + \frac{4}{3}\mu\Big)k^2 = 0 \tag{28}$$

- helicity 1 (sound modes) $\{\Delta N^3,\ \Delta\bar{h}_{33},\ \Delta\bar{h}_{11} + \Delta\bar{h}_{22}\}$:

$$\omega^2 + i\Big(\frac{1}{\tau} + \frac{\gamma}{\rho_0}k^2\Big)\omega - \frac{1}{\rho_0}\frac{\gamma + \mu\tau}{\tau}k^2 = 0 \tag{29}$$

$$\text{fluid limit } (\tau \to 0):\ i\omega - \frac{\eta}{\rho_0}k^2 = 0 \tag{30}$$

$$\text{elastic limit } (\tau \to \infty):\ \omega^2 + i\frac{\gamma}{\rho_0}k^2\omega - \frac{\mu}{\rho_0}k^2 = 0 \tag{31}$$

- helicity 2 (scalar modes) $\{\Delta\bar{h}_{11} - \Delta\bar{h}_{22} \mp 2i\,\Delta\bar{h}_{12}\}$:

$$\omega = -\frac{i}{\tau} \tag{32}$$

6. Conclusion and discussion

In this work we have shown that viscoelasticity can have a universal description by introducing an intrinsic metric to express bonding structures and by adopting FPDs as an underlying gauge symmetry. We have seen that FPDs naturally interpolate between elasticity and fluidity, and that covariance under FPDs uniquely determines the dynamics in the first order of strains.

The next step should be to extend our framework to the systems with more than one relaxation time and with multiple components, in order to describe more realistic materials.

We conclude with a comment that in the sound modes the diffusion equation becomes a telegraph-like equation when the relaxation time is taken finite (see Eq. (29)). This means that information propagates only inside the "light cone", and thus the causality problem (inherent in diffusion equations) disappears when the effect of elasticity is taken into account. This property should be taken over also in the "relativistic extension of viscoelasticity" [6], and would serve as a causal completion of relativistic fluid mechanics.

Acknowledgments

MF thanks the organizers of the QTS6 conference, especially Sumit Das and Al Shapere, for inviting him to give a talk. Elucidating discussions with Michael Berry and Yuho Sakatani are also appreciated. This work was supported by the Grant-in-Aid for the Global COE program "The Next Generation of Physics, Spun from Universality and Emergence" from the Ministry of Education, Culture, Sports, Science and Technology (MEXT) of Japan. This work was also supported by the Japan Society for the Promotion of Science (JSPS) (No. 20·892) and by MEXT (No. 19540288 and No. 18540264).

References

[1] Christensen R M 1971 *Theory of Viscoelasticity* (Academic Press)
[2] Bingham E C 2007 *Fluidity and Plasticity* (Kessinger Publishing)
[3] Landau L D and Lifshitz E M 1987 *Fluid Mechanics* (Butterworth-Heinemann)
[4] Landau L D and Lifshitz E M 1986 *Theory of Elasticity* (Butterworth-Heinemann)
[5] Azeyanagi T, Fukuma M, Kawai H and Yoshida K 2009 *Phys. Lett.* B **681** 290 (*Preprint* arXiv:0907.0656)
[6] Fukuma M and Sakatani Y (in progress)

Journal of Physics: Conference Series **462** (2013) 012014 doi:10.1088/1742-6596/462/1/012014

Asymptotically free $\mathcal{N}=2$ theories and irregular conformal blocks

Davide Gaiotto

School of Natural Sciences, Institute for Advanced Study, Princeton, NJ 08540, USA

E-mail: dgaiotto@ias.edu

Abstract. A surprising connection between $N=2$ gauge theory instanton partition functions and conformal blocks has been recently proposed. We illustrate through simple examples the generalization to asymptotically free $N=2$ gauge theories.

1. Introduction

In [1] evidence was presented for a direct correspondence between the instanton partition functions [2, 3] of $\mathcal{N}=2$ superconformal quivers of $SU(2)$ gauge groups ("A_1 superconformal quivers" in short) and Virasoro conformal blocks. At the root of the correspondence lies a specific realization [4, 5, 6] of the A_1 superconformal quivers as the twisted compactification of a six dimensional theory: the A_1 $(2,0)$ six dimensional SCFT. Asymptotically free quivers of $SU(2)$ gauge groups (A_1 quivers) admit a very similar realization [4, 5]. The only new ingredient is a larger class of codimension two defects in the six dimensional theory.

All theories in the A_1 class are associated to a meromorphic quadratic differential ϕ_2 on a Riemann surface, which is allowed to have poles at the location of the defects. The poles are always of degree 2 or smaller for A_1 superconformal quivers. (The terminology "regular", or "tame" could be used for these singularities). General theories in the A_1 class are associated to a quadratic differential with poles of higher order.(The terminology "irregular", or "wild" could be used for these singularities). This has a simple physical motivation. These asymptotically free theories are obtained from superconformal theories by tuning some mass parameter to be very large, while adjusting the marginal gauge coupling in the UV so that the running coupling in the IR remains finite. From a six-dimensional perspective, the limiting procedure brings two or more standard punctures together to produce a single puncture with a larger degree of divergence for ϕ_2.

It would be reasonable to carry over the same limiting procedure to the instanton partition function, and then the corresponding conformal block, in order to extend the results of [1] to asymptoticlly free theories. Instead, we will follow here an instructive shortcut. We will define directly some exotic states in the CFT which correspond to the irregular singularities of $\phi_2(z)$. "Irregular" conformal blocks with insertions of such states at punctures will reproduce the instanton partition functions of the asymptotically free theories. We refer to [7], [8], [9] for previous attempts to define such type of states in the $c=1$ theory.

We will use the standard dictionary of [1]: set a scale by $\epsilon_1\epsilon_2=1$ and identify $\epsilon_1=b$, $Q=b+\frac{1}{b}$, $c=1+6Q^2$. The expressions for the quadratic differential are taken from [5], possibly with trivial redefinitions/rescalings of the parameters. The form of the Seiberg-Witten curves $x^2 = \phi_2(z)$ may be unfamiliar to the reader, as it differs in form from the original expression [10, 11]. This form of the curve follows naturally from the brane construction in [4], and the relation to the spectral curve of a Hitchin system. The conventions about mass parameters in the instanton partition function are such that $m \to -m$ exchanges fundamental and antifundamental matter contributions. Masses are shifted by $Q/2$ with respect to the convention for which $m \to Q - m$ exchanges fundamental and antifundamental matter contributions

2. $SU(2)$ $N_f = 0$

This theory is associated to a sphere with two punctures, such that the quadratic differential ϕ_2 has a pole of degree 3 at each puncture. If the punctures are set at $z = 0, \infty$ we can take

$$\phi_2 = \frac{\Lambda^2}{z^3} + \frac{2u}{z^2} + \frac{\Lambda^2}{z}. \tag{1}$$

Here Λ is fixed, and coincides with the scale of the $SU(2)$ theory, while u parameterizes the Coulomb branch.

ϕ_2 has been identified in [1] with the semiclassical limit of the energy momentum tensor $T(z)$. We would like to consider a two dimensional "conformal block" with two special punctures on the sphere, i.e. the inner product

$$\langle \Delta, \Lambda^2 | \Delta, \Lambda^2 \rangle. \tag{2}$$

The state $|\Delta, \Lambda^2\rangle$ should live in the Verma module of a highest weight state of conformal dimension $\Delta = \frac{Q^2}{4} - a^2$. We hope to identify $\pm a$ with the eigenvalues of the vector multiplet scalar in the instanton partition function. To reproduce the singularity of $\phi_2(z)$ at $z = 0$ we are led to the requirements

$$L_1|\Delta, \Lambda^2\rangle = \Lambda^2|\Delta, \Lambda^2\rangle \qquad L_2|\Delta, \Lambda^2\rangle = 0. \tag{3}$$

Notice that the Virasoro commutation relations are sufficient to imply then $L_n|\Delta, \Lambda^2\rangle = 0$ for all $n > 2$.

We aim to define $|\Delta, \Lambda^2\rangle$ as a (possibly formal) power series in Λ^2, i.e,

$$|\Delta, \Lambda^2\rangle = v_0 + \Lambda^2 v_1 + \Lambda^4 v_2 + \cdots \tag{4}$$

Here v_0 is the highest weight vector $|\Delta\rangle$ and v_n is a level n descendant such that $L_1 v_n = \Lambda^2 v_{n-1}$ and $L_2 v_n = 0$. It is not fully clear to us why such vectors should exist. Surprisingly, we found that these equations can be recursively, and uniquely solved to as high a level n as we cared to check. (level 8)

The inner product

$$\langle \Delta, \Lambda^2 | \Delta, \Lambda^2 \rangle = \sum \Lambda^{4n} |v_n|^2 \tag{5}$$

coincides order by order (again, we only checked up to level 8) with the instanton partition function for $SU(2)$ $N_f = 0$, with the simple identification of instanton factor as $q = \Lambda^4$

For reference, we report here the first few v_n

$$v_0 = |\Delta\rangle \tag{6}$$

$$v_1 = \frac{1}{2\Delta} L_{-1}|\Delta\rangle \tag{7}$$

$$v_2 = \frac{1}{4\Delta(2c\Delta + c + 16\Delta^2 - 10\Delta)}((c + 8\Delta)L_{-1}^2 - 12\Delta L_{-2})|\Delta\rangle \tag{8}$$

$$v_3 = \frac{1}{24\Delta(2 + c - 7\Delta + c\Delta + 3\Delta^2)(c + 2c\Delta + 2\Delta(-5 + 8\Delta))} \tag{9}$$

$$(12\Delta(-3 - c + 7\Delta)L_{-3} - 12(c + 3c\Delta + \Delta(-7 + 9\Delta))L_{-2}L_{-1} + \tag{10}$$

$$+(c^2 + c(8 + 11\Delta) + 2\Delta(-13 + 12\Delta))L_{-1}^3)|\Delta\rangle. \tag{11}$$

3. $SU(2)$ $N_f = 1$
This theory is associated to a sphere with two punctures, such that the quadratic differential ϕ_2 has a pole of degree 3 at a puncture (say at $z = 0$) and a pole of degree 4 at the other puncture (say at $z = \infty$). We can make some convenient choices

$$\phi_2 = \frac{\Lambda^2}{2z^3} + \frac{2u}{z^2} - \frac{2\Lambda m}{z} - \Lambda^2. \tag{12}$$

Here Λ is fixed, and coincides with the scale of the $SU(2)$ theory. m is also fixed, and coincides with the mass of the single flavor hypermultiplet. u parameterizes the Coulomb branch. The minus signs and the $\frac{1}{2}$ factor are introduced to simplify later expressions.

We will consider the following inner product:

$$\langle \Delta, \Lambda, m | \Delta, \Lambda^2/2 \rangle. \tag{13}$$

Both states should live in the Verma module of the highest weight state of conformal dimension $\Delta = \frac{Q^2}{4} - a^2$. $|\Delta, \Lambda^2/2\rangle$ is the same state as in the previous section, with a trivial redefinition of Λ^2 . We require $|\Delta, \Lambda, m\rangle$ to satisfy

$$L_2|\Delta, \Lambda, m\rangle = -\Lambda^2|\Delta, \Lambda, m\rangle \qquad L_1|\Delta, \Lambda, m\rangle = -2m\Lambda|\Delta, \Lambda, m\rangle. \tag{14}$$

Notice that the Virasoro commutation relations are again sufficient to imply $L_n|\Delta, \Lambda, m\rangle = 0$ for all $n > 2$.

We aim to define $|\Delta, \Lambda, m\rangle$ as a (possibly formal) power series in Λ, i.e,

$$|\Delta, \Lambda, m\rangle = w_0 + \Lambda w_1 + \Lambda^2 w_2 + \cdots \tag{15}$$

Here w_0 is the highest weight vector $|\Delta\rangle$ and w_n is a level n descendant such that $L_1 w_n = -2m w_{n-1}$ and $L_2 w_n = -w_{n-2}$. Again, it is not fully clear to us why such vectors should exist, but we checked their existence and unicity for the first few n.

The inner product

$$\langle \Delta, \Lambda, m | \Delta, \Lambda^2/2 \rangle = \sum \Lambda^{3n} 2^{-n} \langle w_n | v_n \rangle \tag{16}$$

coincides order by order (again, we only checked the first few levels) with the instanton partition function for $SU(2)$ $N_f = 1$, with $q = \Lambda^3$ and with mass parameter m.

As a reference, we report here the first few w_n

$$w_0 = |\Delta\rangle \tag{17}$$
$$w_1 = -\frac{m}{\Delta} L_{-1}|\Delta\rangle \tag{18}$$
$$w_2 = \frac{1}{\Delta(c+2c\Delta+2\Delta(-5+8\Delta))}((cm^2 + \Delta(3+8m^2))L_{-1}^2 - 2\Delta(1+2\Delta+6m^2)L_{-2})|\Delta\rangle. \tag{19}$$

4. $SU(2)$ $N_f = 2$**, first realization**
This theory has two distinct realizations. The first realization is associated to a sphere with two punctures, such that the quadratic differential ϕ_2 has a pole of degree 4 at both punctures (say at $z = 0, \infty$). We can take some convenient choices

$$\phi_2 = -\frac{\Lambda^2}{z^4} - \frac{2\Lambda m_1}{z^3} + \frac{2u}{z^2} - \frac{2\Lambda m_2}{z} - \Lambda^2. \tag{20}$$

Here Λ is fixed, and coincides with the scale of the $SU(2)$ theory. $m_{1,2}$ are also fixed, and coincide with the masses of the two flavor hypermultiplets. u parameterizes the Coulomb branch.

We will consider the following inner product:

$$\langle \Delta, \Lambda, m_2 | \Delta, \Lambda, m_1 \rangle. \tag{21}$$

Here we meet a spurious "$U(1)$ factor". The inner product coincides order by order (as usual, we only checked the first few levels) with the instanton partition function for $SU(2)$ $N_f = 2$, with $q = 4\Lambda^2$, multiplied by an overall factor which depends on Λ only. The overall factor has a power expansion which we conjecture to coincide with $\exp 2\Lambda^2$. The appearance of this spurious factor is hardly a surprise. The spurious factor for $N_f = 4$ was some $(1-q)^{2\tilde{m}_0(Q-\tilde{m}_1)}$, where $\tilde{m}_{0,1}$ are appropriate combinations of the mass parameters. In the asymptotically free limit, $q \to 0$ as $(\Lambda/M)^{4-N_f}$, where M is the mass scale sent to infinity. The spurious factor goes to 1 if $N_f < 2$, but can have a finite limit otherwise. In this realization of the $N_f = 2$ theory, both $m_{0,1}$ scale as M, leading to a finite limit.

5. $SU(2)$ $N_f = 2$, second realization

There is a second way to realize the same four dimensional theory from a distinct six dimensional setup. The setup contains two regular punctures (say at $z = 0, 1$) and a puncture with a pole of rank 3 (say at $z = \infty$).

$$\phi_2 = -\frac{m_+^2}{z^2} - \frac{m_-^2}{(z-1)^2} + \frac{2\tilde{u}}{z(z-1)} + \frac{\Lambda^2}{z}. \tag{22}$$

We will consider a three point function, involving the highest weights of dimensions $\Delta_\pm = \frac{Q^2}{4} - m_\pm^2$ and $|\Delta, \Lambda^2\rangle$. The masses $m_\pm$ are such that the fundamental masses are $m_1 = m_+ + m_-$ and $m_2 = m_+ - m_-$The calculation is straightforward, and reproduces the instanton partition function right away, with no overall spurious factors and $q = \Lambda^2$

6. $SU(2)$ $N_f = 3$

Finally, to get a theory with three flavors we have to consider a setup which contains two regular punctures (say at $z = 0, 1$) and a puncture with a pole of rank 4 (say at $z = \infty$).

$$\phi_2 = -\frac{m_+^2}{z^2} - \frac{m_-^2}{(z-1)^2} + \frac{2\tilde{u}}{z(z-1)} - \frac{2m_3\Lambda}{z} - \Lambda^2. \tag{23}$$

We will consider a three point function, involving the highest weights of dimensions $\Delta_\pm = \frac{Q^2}{4} - m_\pm^2$ and $|\Delta, \Lambda, m_3\rangle$. The calculation is straightforward, and reproduces the instanton partition function, with $q = -2\Lambda$, multiplied by what appears to be $\exp(1/b + b - 2m_1)$. Again, $m_1 = m_+ + m_-$ and $m_2 = m_+ - m_-$. The spurious factor has a simple interpretation as the appropriate asymptotically free limit of the $N_f = 4$ spurious factor.

7. Concluding remarks

It appears that the special punctures required to realize the instanton partition function of asymptotically free theories as 2d CFT "conformal blocks" can be defined in a straightforward way, requiring them to be eigenvectors of a set of Virasoro generators which is in one-to-one correspondence with the coefficients of ϕ_2 which are fixed in gauge theory. Although the examples we provide only cover singularities up to z^{-4}, higher degrees of singularity should work in a similar fashion. Notice that the proper formulation for the boundary conditions of ϕ_2 at an irregular singularity involves fixing the singular part of $\sqrt{\phi_2}$. If the degree of singularity of ϕ_2 is $2N$ (or $2N-1$), the coefficients in the Laurent polynomial for ϕ_2 from z^{-2N} to z^{-N-1} are fixed. The corresponding Virasoro generators L_n with n from $2N-2$ to $N-1$ can act as a multiple of the identity when acting on a state killed by $L_{n'}$ with $n' > 2N-2$, without violating the Virasoro algebra. We conjecture that such a state always exist (at least as a formal, level-by-level sum) and is unique.

Acknowledgments
The author has benefited from discussions with Y.Tachikawa and L.F.Alday. We also want to thank Y.Tachikawa for sharing with us some powerful Mathematica notebooks devoted to instanton computations. D.G. is supported in part by the DOE grant DE-FG02-90ER40542. D.G. is supported in part by the Roger Dashen membership in the Institute for Advanced Study.

References
[1] L. F. Alday, D. Gaiotto, and Y. Tachikawa, "Liouville Correlation Functions from Four-Dimensional Gauge Theories," `arXiv:0906.3219 [hep-th]`.
[2] N. A. Nekrasov, "Seiberg-Witten Prepotential from Instanton Counting," *Adv. Theor. Math. Phys.* **7** (2004) 831–864, `arXiv:hep-th/0206161`.
[3] N. Nekrasov and A. Okounkov, "Seiberg-Witten theory and random partitions," `arXiv:hep-th/0306238`.
[4] E. Witten, "Solutions of four-dimensional field theories via M- theory," *Nucl. Phys.* **B500** (1997) 3–42, `arXiv:hep-th/9703166`.
[5] D. Gaiotto, G. W. Moore, and A. Neitzke, "Wall-crossing, Hitchin Systems, and the WKB Approximation," `arXiv:0907.3987 [hep-th]`.
[6] D. Gaiotto, "N=2 dualities," `arXiv:0904.2715 [hep-th]`.
[7] G. W. Moore, "Geometry of the string equations," *Commun. Math. Phys.* **133** (1990) 261–304.
[8] G. W. Moore, "Matrix models of 2-D gravity and isomonodromic deformation," *Prog. Theor. Phys. Suppl.* **102** (1990) 255–286.
[9] T. Miwa, "CLIFFORD OPERATORS AND RIEMANN'S MONODROMY PROBLEM," *Publ. Res. Inst. Math. Sci. Kyoto* **17** (1981) 665.
[10] N. Seiberg and E. Witten, "Monopole Condensation, And Confinement In N=2 Supersymmetric Yang-Mills Theory," *Nucl. Phys.* **B426** (1994) 19–52, `arXiv:hep-th/9407087`.
[11] N. Seiberg and E. Witten, "Monopoles, duality and chiral symmetry breaking in N=2 supersymmetric QCD," *Nucl. Phys.* **B431** (1994) 484–550, `arXiv:hep-th/9408099`.

Noncommutative teleparallel gravity from a consistent deformation of gauge theory

Ignacio Cortese and J Antonio García
Departamento de Física de Altas Energías,
Instituto de Ciencias Nucleares
Universidad Nacional Autónoma de México,
Apartado Postal 70-543, México D.F. 04510

E-mail: nachoc@nucleares.unam.mx, garcia@nucleares.unam.mx

Abstract.
Starting from a standard noncommutative gauge theory and using the Seiberg-Witten map we propose a new version of a noncommutative gravity. We use consistent deformation theory starting from a free gauge action and gauging a killing symmetry of the background metric to construct a deformation of the gauge theory that we can relate with gravity. The result of this consistent deformation of the gauge theory is nonpolynomial in A_μ. From here we can construct a version of noncommutative gravity that is simpler than previous attempts. Our proposal is consistent and is not plagued with the problems of other approaches like twist symmetries or gauging other groups.

1. Deformed gauge theory and Gravity

Consider the usual Yang-Mills action given by

$$S_{YM} = \int d^4x F^A_{\mu\nu} F^{\mu\nu}_A, \tag{1}$$

where the non abelian field is A^A_μ and its associated $F^A_{\mu\nu}$ is given by

$$F^A_{\mu\nu} = \partial_\mu A^A_\nu - \partial_\nu A^A_\mu + f_{BC}{}^A A^B_\mu A^C_\nu, \tag{2}$$

and the gauge symmetry

$$\delta A^A_\mu = D_\mu \omega^A, \tag{3}$$

where

$$D_\mu \omega^A = \partial_\mu \omega^A + A^B_\mu f_{BC}{}^A \omega^C. \tag{4}$$

Now suppose that we want to deform this action (using the consistent deformation approach given in [1, 2]) and construct the action that is invariant under the deformed transformations

$$\delta A^A_\mu = D_\mu \omega^A + \varepsilon^\nu \partial_\nu A^A_\mu + A^A_\nu \partial_\mu \varepsilon^\nu = D_\mu \omega^A + \mathcal{L}_\varepsilon A^A_\mu, \tag{5}$$

where

$$\varepsilon^\mu \equiv \omega^A \xi^\mu_A, \tag{6}$$

and ξ^μ_A a constant matrix. This deformation can be done and the result is (for details see [2, 3])

$$L = -\frac{1}{4}(1 + \xi^\rho_A A^A_\rho)\hat{F}^A_{\mu\nu}\hat{F}^{\mu\nu}_A, \tag{7}$$

where

$$\hat{F}^A_{\mu\nu} = E^\rho_\mu E^\sigma_\nu F^A_{\rho\sigma}, \tag{8}$$

and

$$E_\mu{}^\nu = \delta_\mu{}^\nu - \xi^\rho_A E^A_B A^B_\mu, \tag{9}$$

Here E^A_B are defined by

$$E_B{}^C(\delta^A_C + \xi^\mu_C A^A_\mu) = \delta^A_B, \tag{10}$$

that are nonpolynomial in the gauge fields A^A_μ. At first sight this theory looks very complicated because among the nonlinearities of the YM theory we have infinite coupling terms encoded in the non polynomial (in the gauge fields) matrices $E_\mu{}^\nu$. Nevertheless this action have very nice properties. First of all, it can be written in a "geometric form"

$$L = -\frac{1}{4}\sqrt{-g}g^{\mu\rho}g^{\nu\sigma}F_{\mu\nu}{}^A F_{\rho\sigma A}, \tag{11}$$

where

$$g_{\mu\nu} = \eta_{\alpha\beta}e^\alpha_\mu e^\beta_\nu = 1 + \mathcal{A}^s + \mathcal{A}\mathcal{A}^T = \eta_{\mu\nu} + \mathcal{A}_{\mu\nu} + \mathcal{A}_{\nu\mu} + \mathcal{A}_{\mu\rho}\mathcal{A}_\nu{}^\rho, \quad e_\mu{}^\nu = \delta_\mu{}^\nu + \mathcal{A}_\mu{}^\nu, \tag{12}$$

and

$$\mathcal{A}_\mu{}^\nu = \xi^\nu_A A^A_\mu. \tag{13}$$

Surprisingly enough, the transformation properties of this "metric" are as a usual space-time metric, *i.e.*, transform under the gauge transformations (5) as a symmetric two-tensor but now under the diffeomorphism generated by $\varepsilon^\mu = \omega^A \xi^\mu_A$

$$\delta g_{\mu\nu} = \mathcal{L}_\varepsilon g_{\mu\nu}. \tag{14}$$

The "vielbein" transforms as a space-time vector for the index μ

$$\delta e_\mu{}^\nu = \varepsilon^\rho\partial_\rho e_\mu{}^\nu + e_\rho{}^\nu\partial_\mu\varepsilon^\rho. \tag{15}$$

Notice that the metric that rise an lower spece-time indices is the flat Minkowski background $\eta_{\mu\nu}$ and

$$E_\mu{}^\nu e_\nu{}^\rho = \delta_\mu{}^\rho. \tag{16}$$

So our question is: Can we relate this theory with Einstein Gravity? The answer is yes and we will show how we can do that.

Using the field redefinition that interchanges the role of the gauge field A^A_μ with the vielbein (12) we can define a new "field strength" as

$$\mathcal{F}_{\mu\nu}{}^\rho = \xi^\rho_A F_{\mu\nu}{}^A = \partial_\mu\mathcal{A}^\rho_\nu - \partial_\nu\mathcal{A}^\rho_\mu = \partial_\mu e^\rho_\nu - \partial_\nu e^\rho_\mu, \tag{17}$$

where $F^A_{\mu\nu}$ is the usual Yang-Mills field strength given by (2). We are taking as our internal gauge group an abelian group and $\xi^A_\mu = \delta^A_\mu$ for simplicity. This imply that we are adding four copies of deformed Maxwell actions defined as the same deformed YM action with structure constants of the gauge group equal to zero.

By making the crucial observation that this redefined field strength (17) can be related to the Ricci rotation coefficients $\Omega_{\mu\nu}{}^{\rho}$ of the standard construction of Weitzenböck gravity [4, 5], we can now build from the deformation of the abelian Yang-Mills theory the corresponding action of General Relativity. Up to an irrelevant numerical factor the torsion tensor is $T_{\mu\nu}{}^{\rho} \sim \Omega_{\mu\nu}{}^{\rho}$. Now the explicit relation between the Ricci rotation coefficients and the field strength of the deformed gauge theory is

$$\Omega_{\sigma\kappa}{}^{\rho} = \hat{\mathcal{F}}_{\sigma\kappa}{}^{\rho} = E^{\mu}_{\sigma} E^{\nu}_{\kappa} \mathcal{F}_{\mu\nu}{}^{\rho}. \tag{18}$$

In terms of $\Omega_{\mu\nu}{}^{\rho}$ the deformed Yang-Mills Lagrangian (7) is

$$L = -\frac{1}{4} e\ \Omega_{\mu\nu}{}^{\rho} \Omega^{\mu\nu}{}_{\rho}, \tag{19}$$

where we have identified the factor $(1 + \mathrm{Tr}\,\mathcal{A})$ as the determinant of the vielbein[1]. Since there is no curvature tensor available in this construction the most general Lagrangian for this theory of gravity can be written as a combination of the so called Weitzenböck invariants

$$I_1 = \Omega_{\mu\nu\rho}\Omega^{\mu\nu\rho}, \quad I_2 = \Omega_{\mu\nu\rho}\Omega^{\rho\mu\nu}, \quad I_3 = \Omega_{\mu\rho}{}^{\rho}\Omega^{\mu}{}_{\sigma}{}^{\sigma}. \tag{20}$$

The Pellegrini-Plebańsky Lagrangian [6] is

$$L = ec^i I_i. \tag{21}$$

To fix the coefficients c^i in order to have Einstein gravity it is instructive to write the linearized action. Denoting the symmetric part of $\mathcal{A}_{\mu\nu}$ by $\mathcal{A}^S{}_{\mu\nu}$ and the antisymmetric part as $\mathcal{A}^A{}_{\mu\nu}$ and retaining terms up to second order in $\mathcal{A}$, using the field redefinition $e \to 1 + \mathcal{A}$ in the Pellegrini-Plebański Lagrangian (21), the result is [4], $c_1 = 1, c_2 = 2, c_3 = -4$. Of course we can construct the full nonlinear action by the use of our field redefinition to get

$$L = e(\Omega_{\mu\nu\rho}\Omega^{\mu\nu\rho} + 2\Omega_{\mu\nu\rho}\Omega^{\rho\mu\nu} - 4\Omega_{\mu\rho}{}^{\rho}\Omega^{\mu}{}_{\sigma}{}^{\sigma}). \tag{22}$$

This is the desired Hilbert-Einstein action. It is interesting to observe that all the invariance under diffeomorphisms is recovered only with this particular choice for the coefficients c^i. It is also worth noticing that this action is not only invariant under the local transformations (15) but also is actually invariant under the full local Poincaré transformations just because the Einstein-Hilbert action is invariant under this symmetry. With this form of the Einstein-Hilbert action in terms of the vielbein e^{ν}_{μ} we can check that the combination of terms given by (22) is such that the action can be written in terms of the fields $g_{\mu\nu}$ and its derivatives as the standard action of General Relativity. In the following section we will construct the noncommutative counterpart of this action.

2. Noncommutative Gravity

As we have four copies of the deformed Maxwell action we can start this section by deforming in the noncommutative sense (using the Seiberg-Witten map) the action of our previous section for the case of an abelian internal group.

This new deformation is a noncommutative deformation with parameter ϑ. The noncommutative deformation can be applied to the original gauge theory or to the nonpolynomial deformed one. We will try the noncommutative deformation of the original gauge theory because

[1] It is worth noticing that the Weiberg-Witten theorem [7] is not violated in our approach because we are not constructing the graviton directly from the field theory. To make contact with gravity we need to consider also other invariants (see below). We can mention also that the background of our noncommutative theory is not the Minkowski metric alone but we have also the background tensor $\theta^{\mu\nu}$.

this deformation is now well understood and under control [8, 9]. We will develop the deformation of the gauge theory using the Seiberg-Witten map. As is well known this map to first order in ϑ is

$$F^1_{\mu\nu} = -\frac{1}{2}\theta^{\sigma\rho}(\{A_\sigma, \partial_\rho F_{\mu\nu}\} - \{F_{\mu\sigma}, F_{\nu\rho}\}), \qquad F_{\mu\nu} \to F_{\mu\nu} + F^1_{\mu\nu}, \tag{23}$$

where $\{\cdot,\cdot\}$ is the anticommutator and $\theta^{\sigma\rho}$ is the noncommutative tensor defined by[2]

$$[x^\mu, x^\nu]_\star = i\vartheta\theta^{\mu\nu}. \tag{24}$$

The effect in the noncommutative Maxwell theory

$$L = -\frac{1}{4}\eta^{\mu\rho}\eta^{\nu\sigma}F_{\mu\nu} \star F_{\rho\sigma}, \tag{25}$$

of this field redefinition is to add a new effective vertex to the free Maxwell action. The result of this after the noncommutative deformation is (to first order in $\theta^{\mu\nu}$)

$$L = -\frac{1}{4}(1 + \xi^\rho A_\rho)\left(\hat{F}_{\mu\nu}\hat{F}^{\mu\nu} + 2\theta^{\mu\sigma}\hat{F}_{\mu\nu}\hat{T}^\nu_\sigma\right), \tag{26}$$

where $\hat{F}_{\mu\nu} = E^\rho_\mu E^\sigma_\nu F_{\rho\sigma}$ and $\hat{T}^\nu_\sigma = -\frac{1}{4}\delta^\nu_\sigma \hat{F}_{\alpha\beta}\hat{F}^{\alpha\beta} + \hat{F}_{\sigma\beta}\hat{F}^{\nu\beta}$. The deformed gauge symmetry is

$$\delta_\lambda A_\mu = \partial_\mu\lambda + \lambda\xi^\nu \hat{F}_{\nu\mu}. \tag{27}$$

To show that the NC deformation is consistent we can calculate the gauge variation of the Lagrangian (26) under the deformed transformation (27). Under the basic assumption that the tensor θ is invariant under the background metric killing vector that we want to gauge

$$\mathcal{L}_\xi \theta^{\mu\nu} = 0, \tag{28}$$

the result is

$$\delta L = -\frac{1}{4}\partial_\rho\left(\varepsilon^\rho(1 + \xi\cdot A)\left(\hat{F}_{\mu\nu}\hat{F}^{\mu\nu} + 2\theta^{\mu\sigma}\hat{F}_{\mu\lambda}\hat{T}^\lambda_\sigma\right)\right) = \partial_\rho(\varepsilon^\rho L) \tag{29}$$

So for example in the case of a translation $\xi^\mu = a^\mu$ with a^μ a constant, the condition (28) is automatically valid and the complete deformation of the action (26) and its gauge symmetry (27) is not obstructed.

2.1. NC Gravity from deformed NCYM Theory

The next step is to try the same trick but with the Deformed Yang-Mills theory. The generalization of our idea, presented in the previous section, to the case of the YM theory is straightforward. We start from the NCYM action in flat space-time

$$L = -\frac{1}{4}\eta^{\mu\rho}\eta^{\nu\sigma}F_{\mu\nu}{}^A \star F_{\rho\sigma\,A} \tag{30}$$

where

$$F^A_{\mu\nu} = \partial_\mu A_\nu - \partial_\nu A_\mu - [A_\mu, A_\nu]^A_\star \tag{31}$$

[2] We will take the not very extended interpretation that θ is a Lorentz tensor and that the $\star$ product is covariant under global Poincaré transformations. For details see [10]. We will absorb the deformation parameter ϑ in the definition of $\theta^{\mu\nu}$.

and the usual $\star$ gauge symmetry. Applying the Seiberg-Witten map (up to first order in the deformation parameter)

$$F^C_{\mu\nu} \to F^C_{\mu\nu} + \frac{1}{2}\theta^{\alpha\beta} d^{ABC}\left(F^A_{\mu\alpha}F^B_{\nu\beta} - A^A_\alpha \partial_\beta F^B_{\mu\nu} + \frac{1}{2} f^{BDE} A^A_\alpha A^E_\beta F^D_{\mu\nu}\right), \tag{32}$$

we obtain

$$L = -\frac{1}{4}\left(Tr(F_{\mu\nu}F^{\mu\nu}) + \theta^{\alpha\beta} d_{ABC} F^{A\mu\nu}(\frac{1}{4}F^B_{\beta\alpha}F^C_{\mu\nu} + F^B_{\mu\alpha}F^C_{\nu\beta})\right), \tag{33}$$

where we are taking the trace over the group indices as usual and the coefficients d_{ABC} are defined by

$$\{T_A, T_B\} = d_{ABC} T_C. \tag{34}$$

Notice that this theory is defined on the global gauge algebra of the Poincaré group that comes from the algebra of the killing vectors of the background metric. This theory is invariant under the usual nonabelian gauge symmetry

$$\delta A^A_\mu = D_\mu \omega^A, \tag{35}$$

and the global symmetry defined by

$$\delta_B A^A_\mu = \xi^\nu_B F^A_{\mu\nu}, \tag{36}$$

for each B. Now we perform the same steps to deform this action (to first order is ϑ) to all orders in κ. The new action is

$$L = -\frac{1}{4}(1 + \xi^\rho_A A^A_\rho)\left(Tr(\hat{F}_{\mu\nu}\hat{F}^{\mu\nu}) + \theta^{\alpha\beta} d_{ABC} \hat{F}^{A\mu\nu}(\frac{1}{4}\hat{F}^B_{\beta\alpha}\hat{F}^C_{\mu\nu} + \hat{F}^B_{\mu\alpha}\hat{F}^C_{\nu\beta})\right), \tag{37}$$

where $\hat{F}_{\mu\nu} = E^\rho_\mu E^\sigma_\nu F_{\rho\sigma}$. The deformed gauge symmetry is

$$\delta_w A^A_\mu = D_\mu \omega^A + \omega^B \xi^\nu_B \hat{F}^A_{\nu\mu}. \tag{38}$$

or

$$\delta A^A_\mu = D_\mu \omega^A + \varepsilon^\nu \partial_\nu A^A_\mu + A^A_\nu \partial_\mu \varepsilon^\nu = D_\mu \omega^A + \mathcal{L}_\varepsilon A^A_\mu, \tag{39}$$

where

$$\varepsilon^\mu = \omega^A \xi^\mu_A. \tag{40}$$

The action is invariant under the deformed symmetry (38) provided

$$\omega^A \mathcal{L}_{\xi_A} \theta^{\mu\nu} = 0. \tag{41}$$

The condition (41) do not restrict in any sense the definition of noncommutativity as stated in (24) because this noncommutation of the coordinates is invariant under global translations.

The content of the condition (41) is to gauge less symmetry than the deformation (27) is capable to gauge. This is a central point in the construction of our model for noncommutative gravity. The base gauge symmetry upon we are constructing gravity from the deformed NCYM action are only local translations. This versatility of the gauging procedure adopted here allow us to construct a model that in principle is free from the usual obstructions present in many models of noncommutative gravity[3].

[3] In the case of the gauge theory the generators of gauge symmetries are constructed from the global symmetries ξ^A with local parameters ω^A. For space-time symmetries the generators are local and given by ε^μ. To relate them we need to a map from each generator ξ^μ_A to ε^μ given by (40).

As a consequence of the condition (41) the transformation rule for the vielbein (15) becomes

$$\delta e_\mu{}^\nu = \varepsilon^\rho \partial_\rho e_\mu{}^\nu + e_\rho{}^\nu \partial_\mu \varepsilon^\rho. \tag{42}$$

In that case the transformation rule associated with the vielbein comes only trough its lower space-time index. In this sense we are turning off the local Lorentz rotations associated with the frames defined by the vielbeins. The frames are "rigid" and the associated theory of gravity that emerges from here is the teleparallel gravity [11]. This framework will be used here to build a noncommutative gravity theory. The reason why we need to impose the condition (41) is to preserve the gauge invariance of our noncommutative model.

As we will choose the translations as the only symmetries that will be gauged then we can use the same rules to construct noncommutative gravity as the ones that we have used to construct standard gravity from the commutative deformed action (7). Namely to identify the Ricci rotation coefficients Ω with $\hat{F}$

$$\Omega_{\sigma\kappa}{}^\rho = \hat{\mathcal{F}}_{\sigma\kappa}{}^\rho = E^\mu_\sigma E^\nu_\kappa \mathcal{F}_{\mu\nu}{}^\rho, \tag{43}$$

and

$$\mathcal{F}_{\mu\nu}{}^\rho = \xi^\rho_A F_{\mu\nu}{}^A = \partial_\mu \mathcal{A}^\rho_\nu - \partial_\nu \mathcal{A}^\rho_\mu = \partial_\mu e^\rho_\nu - \partial_\nu e^\rho_\mu. \tag{44}$$

The next move is to substitute this expressions in the noncommutative action (37) and build the analogs of the invariants I_i for the noncommutative case. In this way we will end with a version of a noncommutative Pellegrini-Plebanski Lagrangian that is the central result of our note

$$L_{NCG} = L_{PP} + L_{NC}, \tag{45}$$

where L_{PP} is the usual Pellegrini-Plebanski Lagrangian (with the coefficients adjusted in such a way to recover Einstein Gravity) and L_{NC} is the non commutative correction to gravity.

As we have essentially four copies of the Maxwell action the non commutative deformed YM action is (after the identification of the Ricci coefficients Ω with $\hat{F}$),

$$L_1 = e\left(\Omega_{\mu\nu}{}^\rho \Omega^{\mu\nu}{}_\rho + \theta^{\alpha\beta} d_{\rho\sigma\delta} \Omega^{\rho\mu\nu} (\frac{1}{4}\Omega^\sigma_{\beta\alpha}\Omega^\delta_{\mu\nu} + \Omega^\sigma_{\mu\alpha}\Omega^\delta_{\nu\beta})\right), \tag{46}$$

for some suitable coefficients $d_{\rho\sigma\delta} = 1$ ($d_{\rho\sigma\delta} = 1$ when the indices are equal and zero otherwise, just the sum of four Maxwell terms as stated above) and the identification of $\det e^\nu_\mu$ with the factor in front of the noncommutative Lagrangian (37).

This is a central result of this article. To identify the others invariants that correspond to the noncommutative version of the invariants I_i we read from (46) the replacement in terms of Ω in the form

$$\Omega_{\mu\nu}{}^\rho \rightarrow \Omega_{\mu\nu}{}^\rho + (\Omega_{NC})_{\mu\nu}{}^\rho, \tag{47}$$

and we found

$$(\Omega_{NC})_{\mu\nu}{}^\rho = \theta^{\alpha\beta} d_{\sigma\delta}{}^\rho (\frac{1}{4}\Omega^\sigma_{\beta\alpha}\Omega^\delta_{\mu\nu} + \Omega^\sigma_{\mu\alpha}\Omega^\delta_{\nu\beta}). \tag{48}$$

Using this dictionary we can write the corresponding noncommutative quadratic invariants

$$L_2 = e\left(\Omega_{\mu\nu\rho}\Omega^{\rho\mu\nu} + \Omega^{\rho\mu\nu}(\Omega_{NC})_{\mu\nu\rho} + \Omega^{\mu\nu\rho}(\Omega_{NC})_{\rho\mu\nu}\right). \tag{49}$$

Taking into account that the trace of Ω_{NC} is

$$(\Omega_{NC})_\mu = (\Omega_{NC})_{\mu\nu}{}^\nu = \theta^{\alpha\beta} d_{\sigma\delta}{}^\nu (\frac{1}{4}\Omega^\sigma_{\beta\alpha}\Omega^\delta_{\mu\nu} + \Omega^\sigma_{\mu\alpha}\Omega^\delta_{\nu\beta}), \tag{50}$$

the last noncommutative invariant is

$$L_3 = e\left(\Omega_\mu \Omega^\mu + 2\Omega^\mu (\Omega_{NC})_\mu\right). \tag{51}$$

The searched noncommutative gravity Lagrangian is then

$$L_{NCG} = L + 2e\left(\Omega^{\rho\mu\nu}(\Omega_{NC})_{\rho\mu\nu} + \Omega^{\rho\mu\nu}(\Omega_{NC})_{\mu\nu\rho} + \Omega^{\mu\nu\rho}(\Omega_{NC})_{\rho\mu\nu} - 4\Omega^\mu(\Omega_{NC})_\mu\right). \tag{52}$$

As a final comment let us consider the field content and symmetries of this NC gravity Lagrangian. The first piece in (52) is the standard Einstein Hilbert Lagrangian and can be written in terms of the curvature associated to the Levi-Civita connection or if we want using the spin connection as usual. The second piece is much more tricky. The field content of this NC correction terms is just the vielbein e_μ^ν and its derivatives. Our present understanding of these terms does not allow us to write them as $f(R)$ corrections to Einstein Gravity as we may expect if the theory were invariant under the full diffeomorphism symmetry. Here R is the curvature associated with the Levi-Civita connection . All that we can do is to write this piece in terms of the Torsion of the Weitzenböck connection. A possible implication of this fact is that this piece of the Lagrangian (52) breaks the complete diffeomorphism symmetry already present in the first piece (Einstein-Hilbert Lagrangian) to the local translation symmetry just as in teleparallel gravity. In turn this imply that the symmetry of the Lagrangian (52) is just the symmetry of the teleparallel gravity. Perhaps the reason behind this breakdown of the symmetry is the presence of the noncommutative parameter and its transformation properties under local Lorentz transformations[4]. This issue and the geometrical and phenomenological implications of this noncommutative correction to the Einstein Gravity need further study that we leave for a future work.

Acknowledgements
This work was partially supported by Mexico's National Council of Science and Technology (CONACyT) grant 50-155I, as well as by DGAPA-UNAM grant IN116408.

References

[1] G. Barnich and M. Henneaux, "Consistent couplings between fields with a gauge freedom and deformations of the master equation," Phys. Lett. B **311**, 123 (1993) [arXiv:hep-th/9304057].
[2] F. Brandt, "Gauge theories of spacetime symmetries," Phys. Rev. D **64**, 065025 (2001) [arXiv:hep-th/0105010].
[3] I. Cortese and J. A. Garcia, "Emergent Noncommutative gravity from a consistent deformation of gauge theory," arXiv:1001.4180 [hep-th].
[4] T. Ortin, "Gravity And Strings," *Cambridge University, Cambridge University Press, 2004*
[5] M. Blagojevic, "Gravitation and gauge symmetries," *Bristol, UK: IOP (2002) 522 p*
[6] C. Pellegrini and J. Plebanski, *Mat. Fys. Skr. Dan. Vid. Selsk.* **2** nr. 4 (1963).
[7] S. Weinberg and E. Witten, "Limits On Massless Particles," Phys. Lett. B **96**, 59 (1980). F. Loebbert, "The Weinberg-Witten theorem on massless particles: An Essay," Annalen Phys. **17**, 803 (2008).
[8] N. Seiberg and E. Witten, "String theory and noncommutative geometry," JHEP **9909**, 032 (1999) [arXiv:hep-th/9908142].
[9] G. Barnich, M. A. Grigoriev and M. Henneaux, "Seiberg-Witten maps from the point of view of consistent deformations of gauge theories," JHEP **0110**, 004 (2001) [arXiv:hep-th/0106188]. G. Barnich, F. Brandt and M. Grigoriev, "Seiberg-Witten maps in the context of the antifield formalism," Fortsch. Phys. **50**, 825 (2002) [arXiv:hep-th/0201139].

[4] Another fact that need a further investigation comes from string theory. The noncommutative leading order correction of Einstein Gravity that comes from string theory start at second order terms in the noncommutative parameter $\theta^{\mu\nu}$. We need to develop this second order correction terms in our formalism to see if we can write a Lagrangian with a complete diffeomorphism invariance.

[10] J. Antonio García and I. Cortese, "Space-time symmetries in Noncommutative field theories: Covariance vs. Invariance", *to appear*

[11] F. W. Hehl, J. D. McCrea, E. W. Mielke and Y. Ne'eman, "Metric-affine gauge theory of gravity: field equations, Noether identities, world spinors, and breaking of dilation invariance", Phys. Rep. **258** (1995) 1 [arXiv:gr-qc/9402012].

Journal of Physics: Conference Series **462** (2013) 012016 doi:10.1088/1742-6596/462/1/012016

String Theory as a Theory of Species

Cesar Gomez

Instituto de Física Teórica UAM-CSIC, C-XVI Universidad Autónoma de Madrid, Cantoblanco, 28049 Madrid, Spain

Abstract. In this talk I summarize my recent work with Gia Dvali on the dynamics of theories containing gravity and weakly coupled particle species.

1. Gravity and Species

What makes gravity special and different from any other form of interaction in Nature is that the scale setting the intensity of the interaction, namely the Planck scale, is also fixing the bound on information storage [1] [2]. This double role of the Planck scale is at the core of the holographic meaning of gravity. The existence of N weakly coupled particle species, where by that we will mean particles with decay width much less than their mass , entails a minimum amount of information of N bits. The species scale L_N is simply defined as the minimum size where this information can be stored setting a length scale below which physical resolution of species becomes gravitationally impossible [3]. In asymptotically flat space-time of dimension $D = 4 + d$ this scale is given by:

$$L_N = N^{\frac{1}{D-2}} L_D \tag{1}$$

with L_D the Planck length in D dimensions. In this sense a theory of N weakly coupled species is characterized by two length scales L_D and L_N related by (1). The aim of this talk is to show that a theory of species fits into a weakly coupled string theory with the species scale playing the role of the string scale [4].

2. Proving the Species Scale: Black Holes

The simplest way to prove the species scale is to consider the effect of species on black hole physics [5]. In particular on the evaporation of neutral black holes with negative specific heat. One immediate consequence of the existence of particle species is to increase the rate of evaporation of semiclassical black holes reducing the evaporation time $\tau = \frac{M_{bh}^3}{M_{pl}^4}$ by a factor of N. This means that the black hole stops to behave semi-classically at

$$\tau = N.R(M_{bh}) \tag{2}$$

with $R(M_{bh})$ the gravitational radius. The previous semi-classicality bound sets - as the minimal possible size for Einstenian black holes - the species scale L_N. Thus we can already conclude that quantum mechanically a theory of gravity with species possesses a natural UV cutoff L_N larger than the Planck scale beyond which Einstenian gravity should be modified.

Journal of Physics: Conference Series **462** (2013) 012016 doi:10.1088/1742-6596/462/1/012016

A hint on how to proceed in order to UV complete the theory between the Planck scale and the species scale comes from the expression of the mass and entropy of the black hole when it reaches the species scale, namely:

$$M_{bh} = NL_N^{-1} \tag{3}$$

and

$$S = N \tag{4}$$

Expression (3) is formally identical to the mass formula for a string state with number of oscillators $N_{osc} = N^2$ and string length $l_s = L_N$. Moreover , at least asymptotically , the entropy of such string state goes like $\sqrt{N_{osc}}$ in agreement with (4). However, in order to interpret this string entropy as a Bekenstein-Hawking entropy we should require

$$S = \sqrt{N_{osc}} = (\frac{l_s}{L_{Pl}})^8 \tag{5}$$

corresponding to a black hole of size equal to the string length. This leads to the well known black hole string correspondence [6] relation $\sqrt{N_{osc}} = \frac{1}{g^2}$ for g the string coupling.

Hence we can formally map the theory of gravity and N species into a string theory by the following rule of correspondences:

$$L_N \to l_s \tag{6}$$

and

$$N \to \frac{1}{g^2} \tag{7}$$

Under this form of correspondence the standard relation

$$L_{Pl} = g^{\frac{1}{4}} l_s \tag{8}$$

becomes the definition (1) of the species scale in ten dimensions. The correspondence (6) and (7) naturally maps some features of black hole physics in the presence of weakly coupled species into the dynamics of black holes once we complete gravity in the UV with string theory. In particular in the presence of N species the black hole entropy is bounded by the number of species:

$$S > N \tag{9}$$

corresponding to the bound on the black hole entropy in string theory

$$S > \frac{1}{g^2} \tag{10}$$

Similarly the maximum Hagedorn temperature $T = \frac{1}{l_s}$ corresponds to the Hawking temperature of the smallest possible black hole in the presence of species, namely $\frac{1}{L_N}$. In summary when in a theory of species a black hole reaches the species scale it becomes a string mode with number of oscillators given by $\sqrt{N_{osc}} = N$. It is important to stress that the existence of a length scale - different to the Planck scale - where gravity should be UV completed is a necessary requirement of any consistent theory of species coupled to gravity.

3. Chan Paton "species"

Probably the most natural frame for defining a theory with species and gravity is open string theory with Chan Paton factors attached at the end points of the string. For n different Chan

Paton factors the number of species is $N = n^2$. In this particular case the resolution scale for these species is just the string length. This leads to the bound

$$L_N > l_s \tag{11}$$

which obviously implies a bound on the number of D-9 branes we can pile up one on the top of each other. Since D-branes gravitate with a tension $T_9 = \frac{M_s^10}{g}$ the collective gravitational source created by a stuck of n of them is

$$R^{-2} = nT_9G_{10} = ngM_s^2 \tag{12}$$

The bound (11) appears as a consequence of requiring that R should not exceeds the string length. In fact for $R = l_s$ we get

$$N = n^2 = \frac{1}{g^2} \tag{13}$$

that leads to $L_N = l_s$.

The species scale plays a natural role in the definition of gravitational holographic duals of theories without gravity. In a theory without gravity and N species, the number of degrees of freedom for the regularized theory defined in a finite box of volume one, depends on the UV cutoff Λ - we have decided to choose - as $\frac{N}{\Lambda^3}$. This number must be equal to the number of holographic degrees of freedom of the five dimensional gravity dual theory i.e $\frac{A(\Lambda)}{G_5}$ for $A(\Lambda)$ the corresponding regularized boundary area [7]. However the number of species is given by the standard relation (1) so we get

$$A(\Lambda) = \frac{L_N^8.G_5}{G_{10}\Lambda^3} \tag{14}$$

leading to the well known AdS/CFT [8] result $A(\Lambda) = \frac{R^3}{\Lambda^3}$ for $R = L_N$ the species scale.

4. Strings at Strong Coupling

In the previous sections we have mapped theory of species into weakly coupled string theory. A natural question is of course what happens when we move into strongly coupled string theory. Consistency with the picture we have presented until now would require the existence of a weakly coupled dual description with g_d the dual string coupling with

$$N_d = \frac{1}{g_d^2} \tag{15}$$

the effective number of species of the dual description and with

$$L_N^d = N_d^{\frac{1}{8}} L_{10} \tag{16}$$

the dual species scale. As an example let us consider the case of M-theory as the strong coupling description of type IIA string theory. In this case the weakly coupled species in the dual description are the D-0 branes [9]. The number of those species is $n = \frac{R}{L_{11}}$ for $R = l_s.g$. Now it is easy to check that

$$L_{11} = n^{\frac{1}{8}} L_{10} \tag{17}$$

i.e the eleven dimensional Planck length becomes the species scale of the dual theory. Moreover we get

$$n = g^{\frac{2}{3}} \tag{18}$$

leading to define the dual coupling as

$$g_d = g^{-\frac{1}{3}} \tag{19}$$

which is the one we get when we move from the string frame into he M-theory frame: $X \to g^{\frac{1}{3}} X$, for X the matrix coordinates of D-0 branes. In summary the meaning of L_{11} is to set the species scale for the weakly coupled dual D-0 species.

5. Concluding Remarks

The conjecture we are putting forward in this talk is that a natural answer to the question: What is string theory? could be, the UV completion of a theory of species and gravity with the species scale as the string scale. It is important to stress that the need for an UV completion above the Planck scale is a direct and quite universal consequence of the holographic interplay between black hole dynamics and the underlying information encoded in the existence of particle species.

Acknowledgments

I would like to thank Sumit Das and Al Shapere for the excellent workshop and for giving me the opportunity to present this stuff in such a wonderful scientific setting. This work has been partially supported by the Spanish DGI contract FPA2003-02877 and the CM grant HEPHACOS S2009ESP-1473.

References

[1] G. 't Hooft, "Dimensional reduction in quantum gravity", gr-qc/9310026; L. Susskind, "The World As A Hologram", J. Math. Phys. **36**, 6377 (1995), hep-th/9409089.
[2] J.D. Bekenstein, "Do we understand black hole entropy?", arXiv:gr-qc/9409015.
[3] G. Dvali and C Gomez, "Quantum Information and Gravity Cutoff in Theories with Species", Phys. Lett. **B674** (2009) 303, arXiv:0812.1940 [hep-th] .
[4] G. Dvali and C Gomez, "Species and Strings", arXiv:1004.3744 [hep-th]
[5] G. Dvali, "Black Holes and Large N Species Solution to the Hierarchy Problem," arXiv:0706.2050 [hep-th]; G. Dvali, "Nature of Microscopic Black Holes and Gravity in Theories with Particle Species", arXiv:0806.3801 [hep-th];'
[6] G. T. Horowitz and J. Polchinski, A correspondence principle for black holes and strings, Phys. Rev. D 55, 6189 (1997), hep-th/9612146 G.T. Horowitz and J. Polchinski, Phys. Rev. D 57, 2557 (1998) [hep-th/9707170]; T. Damour and G. Veneziano, Nucl. Phys. B 568, 93 (2000) [hep-th/9907030]. L. Susskind, hep-th/9309145, E. Halyo, A. Ra jaraman, and L. Susskind, hep-th/9605112, E. Halyo, B. Kol, A. Ra jaraman, and L. Susskind, hep-th/9609075.
[7] L. Susskind and E. Witten, "The Holographic bound in anti-de Sitter space", arXiv:9805114 [hep-th]
[8] J. M. Maldacena, " The Large N limit of superconformal field theories and supergravity", Adv.Theor.Math.Phys. **2** (1998) 231, Int.J.Theor.Phys. **38** (1999) 1113, hep-th/9711200; S.S. Gubser, I.R. Klebanov, A. M. Polyakov, "Gauge theory correlators from noncritical string theory", Phys. Lett. **B428** (1998) 105, hep-th/9802109; E. Witten, "Anti-de Sitter space and holography", Adv.Theor.Math.Phys. **2** (1998) 253, hep-th/9802150
[9] T. Banks , W. Fischler , S.H. Shenker, L. Susskind, " M theory as a matrix model: A Conjecture", Phys.Rev.D55:5112-5128,1997. hep-th/9610043

6th International Symposium on Quantum Theory and Symmetries (QTS6) IOP Publishing
Journal of Physics: Conference Series **462** (2013) 012017 doi:10.1088/1742-6596/462/1/012017

Non-relativistic AdS/CFT and the GCA

Arjun Bagchi, Rajesh Gopakumar

Harish-Chandra Research Institute, Chhatnag Road, Jhusi. India 211019.

E-mail: arjun, gopakumr@hri.res.in

Abstract. We construct a non-relativistic limit of the AdS/CFT conjecture by taking, on the boundary side, a parametric group contraction of the relativistic conformal group. This leads to an algebra with the same number of generators called the Galilean Conformal Algebra (GCA). The GCA is to be contrasted with the more widely studied Schrodinger algebra which has fewer generators. The GCA, interestingly, can be given an infinite dimensional lift for any dimension of spacetime and this infinite algebra contains a Virasoro Kac-Moody sub-algebra. We comment briefly on potential realizations of this algebra in real-life systems. We also propose a somewhat unusual geometric structure for the bulk gravity dual to the GCA. This involves taking a Newton-Cartan like limit of Einstein's equations in anti de Sitter space which singles out an AdS_2 comprising of the time and radial direction. The infinite dimensional GCA arises out of the contraction of the bulk Killing vectors and is identified with the (asymptotic) isometries of this Newton-Cartan structure.

1. Introduction

Even after more than a decade, the AdS-CFT conjecture [1] continues to throw up rich, new avenues of investigation. One such recent direction has been to consider extensions of the conjecture from its original relativistic setting to a non-relativistic context. This opens the door to potential applications of the spirit of gauge-gravity duality to a variety of real-life strongly interacting systems. It was pointed out in [2] that the Schrodinger symmetry group [3, 4, 5], a non-relativistic version of conformal symmetry, is relevant to the study of cold atoms. A gravity dual possessing these symmetries was then proposed in [6, 7].

Instead of the Schrodinger group here we consider an *alternative* non-relativistic realization of conformal symmetry in the context of the AdS/CFT conjecture [8]. This symmetry will be obtained by considering the nonrelativistic group contraction of the relativistic conformal group $SO(d+1,2)$ in $d+1$ space-time dimensions. This Galilean conformal group is to be contrasted with the more studied Schrodinger group from which it differs in some crucial respects as we shall go on to outline.

However, the most interesting feature of the GCA seems to be its natural extension to an infinite dimensional symmetry algebra. This is somewhat analogous to the way in which the finite conformal algebra of $SL(2,C)$ in two dimensions extends to two copies of the Virasoro algebra. Our algebra contains one copy of a Virasoro together with an $SO(d)$ current algebra (on adding the appropriate central extension). We comment on the potential realization of this extended symmetry.

In addition to possible applications to non-relativistic systems, one of the motivations for studying the contracted $SO(d+1,2)$ conformal algebra is to examine the possibility of a new tractable limit of the parent AdS/CFT conjecture. In fact, the BMN limit [9] of the AdS/CFT conjecture is an instance where, as result of taking a particular scaling limit, one obtains a contraction of the original $SO(4,2) \times SO(6)$ (bosonic) global symmetry. In our case, the non-relativistic contraction is obtained by taking a similar scaling limit on the parent theory. Like in the BMN case, taking this limit would isolate a closed subsector of the full theory. There are, however, some important differences here from a BMN type limit. Normally the BMN type scaling leads to a Penrose limit of the geometry in the vicinity of some null geodesic. These are typically pp-wave like geometries whose isometry is the same as that of the contracted symmetry group on the boundary. The non-relativistic scaling limit that leads to the GCA on the boundary is at first sight more puzzling to implement in the bulk. This is because, under the corresponding scaling, the bulk metric degenerates in the spatial directions x_i. We propose a novel non-metric Newton-Cartan like formulation of the bulk theory, which has a geometry with an AdS_2 base and the spatial R^d fibred over it. As a check of this proposal, we will see that the infinite dimensional GCA symmetries are realized in this bulk geometry as asymptotic isometries. These generators will also be seen to reduce to the generators of the GCA on the boundary.

2. Non-Relativistic Conformal Symmetries

2.1. Schrodinger Symmetry

The Schrodinger symmetry group in $(d+1)$ dimensional spacetime, $Sch(d,1)$ has been studied as a non-relativistic analogue of conformal symmetry. Its name arises from being the group of symmetries of the free Schrodinger wave operator in $(d+1)$ dimensions. In other words, it is generated by those transformations that commute with the operator $S = i\partial_t + \frac{1}{2m}\partial_i^2$. However, this symmetry is also believed to be realized in interacting systems, most recently in cold atoms at criticality.

The symmetry group contains the usual Galilean group (denoted as $G(d,1)$) with its central extension.

$$\begin{aligned}
[J_{ij}, J_{rs}] &= so(d), \quad [J_{ij}, B_r] = -(B_i\delta_{jr} - B_j\delta_{ir}) \\
[J_{ij}, P_r] &= -(P_i\delta_{jr} - P_j\delta_{ir}), \quad [J_{ij}, H] = 0, \quad [P_i, P_j] = 0 \\
[B_i, B_j] &= 0, \quad [B_i, P_j] = m\delta_{ij} \quad [H, P_i] = 0, \quad [H, B_i] = -P_i.
\end{aligned} \tag{2.1}$$

Here J_{ij} $(i,j = 1 \dots d)$ are the usual $SO(d)$ generators of spatial rotations. P_r are the d generators of spatial translations and B_j those of boosts in these directions. Finally H is the generator of time translations. The parameter m is the central extension and has the interpretation as the non-relativistic mass (which also appears in the Schrodinger operator S).

As vector fields on the Galilean spacetime $R^{d,1}$, they have the realization (in the absence of the central term)

$$J_{ij} = -(x_i\partial_j - x_j\partial_i), \quad H = -\partial_t, \quad P_i = \partial_i, \quad B_i = t\partial_i \tag{2.2}$$

In addition to these Galilean generators there are *two* more generators which we will denote by $\tilde{K}, \tilde{D}$. $\tilde{D} = -(2t\partial_t + x_i\partial_i)$ is a dilatation operator, which unlike the relativistic case, scales time and space differently. ($x_i \to \lambda x_i, \quad t \to \lambda^2 t$.) $\tilde{K}$ acts something like the time component of special conformal transformations. It has the form $\tilde{K} = -(tx_i\partial_i + t^2\partial_t)$ and generates the finite transformations $x_i \to \frac{x_i}{(1+\mu t)}, \quad t \to \frac{t}{(1+\mu t)}$.

These two additional generators have non-zero commutators

$$\begin{aligned}
[\tilde{K}, P_i] &= B_i, \quad [\tilde{K}, B_i] = 0, \quad [\tilde{D}, B_i] = -B_i \\
[\tilde{D}, \tilde{K}] &= -2\tilde{K}, \quad [\tilde{K}, H] = -\tilde{D}, \quad [\tilde{D}, H] = 2H.
\end{aligned} \tag{2.3}$$

Note that there is no analogue in the Schrodinger algebra of the spatial components K_i of special conformal transformations. Thus we have a smaller group compared to the relativistic conformal group (12 generators +1 central extension in $(3+1)$ dimensions as opposed to 15.).

2.2. Contraction of the Relativistic Conformal Group

We know that the Galilean algebra $G(d,1)$ arises as a contraction of the Poincare algebra $ISO(d,1)$. Physically this comes from taking the non-relativistic scaling

$$t \to \epsilon^r t \qquad x_i \to \epsilon^{r+1} x_i \tag{2.4}$$

with $\epsilon \to 0$. This is equivalent to taking the velocities $v_i \sim \epsilon$ to zero (in units where $c = 1$). For the process of group contraction the parameter r will play no role apart from modifying an over all factor which is unimportant. Hence we will mostly take $r = 0$.

Starting with the expressions for the Poincare generators ($\mu, \nu = 0, 1 \ldots d$)

$$J_{\mu\nu} = -(x_\mu \partial_\nu - x_\nu \partial_\mu) \qquad P_\mu = \partial_\mu, \tag{2.5}$$

the above scaling gives us the Galilean vector field generators of (2.2)

$$\begin{aligned} J_{ij} &= -(x_i\partial_j - x_j\partial_i) \qquad P_0 = H = -\partial_t \\ P_i &= \partial_i \qquad J_{0i} = B_i = t\partial_i. \end{aligned} \tag{2.6}$$

They obey the commutation relations (without central extension) of (2.1). To obtain the Galilean Conformal Algebra, we simply extend the scaling (2.4) to the rest of the generators of the conformal group $SO(d+1,2)$. Namely to

$$D = -(x \cdot \partial) \qquad K_\mu = -(2x_\mu(x\cdot\partial) - (x\cdot x)\partial_\mu) \tag{2.7}$$

where D is the relativistic dilatation generator and K_μ are those of special conformal transformations. The non-relativistic scaling in (2.4) now gives (see also [10])

$$D = -(x_i\partial_i + t\partial_t), \;\; K = K_0 = -(2tx_i\partial_i + t^2\partial_t), \;\; K_i = t^2\partial_i. \tag{2.8}$$

Note that the dilatation generator D is the *same* as in the relativistic theory and scales space and time in the same way. Therefore it is different from the dilatation generator $\tilde{D}$ of the Schrodinger group. Similarly, the temporal special conformal generator K in (2.8) is different from $\tilde{K}$. Finally, we now have spatial special conformal transformations K_i which were not present in the Schrodinger algebra.

Since the usual Galilean algebra $G(d,1)$ for the generators (J_{ij}, P_i, H, B_i) is a subalgebra of the GCA, we will not write down their commutators. The other non-trivial commutators of the GCA are [10]

$$\begin{aligned} [K, K_i] &= 0, \quad [K, B_i] = K_i, \quad [K, P_i] = 2B_i \\ [J_{ij}, K_r] &= -(K_i\delta_{jr} - K_j\delta_{ir}), \quad [J_{ij}, K] = 0, \quad [J_{ij}, D] = 0 \\ [K_i, K_j] &= 0, \quad [K_i, B_j] = 0, \quad [K_i, P_j] = 0, \quad [H, K_i] = -2B_i, \\ [D, K_i] &= -K_i, \quad [D, B_i] = 0, \quad [D, P_i] = P_i, \\ [D, H] &= H, \quad [H, K] = -2D, \quad [D, K] = -K. \end{aligned} \tag{2.9}$$

We can compare the relevant commutators in (2.9) with those of (2.3) and we notice that they are different. Thus the Schrodinger algebra and the GCA only share a common Galilean subgroup and are otherwise different. In fact, one can verify using the Jacobi identities for (D, B_i, P_j) that the Galilean central extension in $[B_i, P_j]$ is *not* admissible in the GCA. This is another difference from the Schrodinger algebra, which as mentioned above, does allow for the central extension. Thus in some sense, the GCA is the symmetry of a "massless" (or gapless) nonrelativistic system.

3. The Infinite Dimensional Extended GCA

The most interesting feature of the GCA is that it admits a very natural extension to an infinite dimensional algebra of the Virasoro-Kac-Moody type. To see this we denote

$$L^{(-1,0,1)} = H, D, K, \quad M_i^{(-1,0,1)} = P_i, B_i, K_i. \tag{3.10}$$

The finite dimensional GCA which we had in the previous section can now be recast as

$$\begin{aligned}
[L^{(m)}, L^{(n)}] &= (m-n)L^{(m+n)}, \quad [L^{(m)}, M_i^{(n)}] = (m-n)M_i^{(m+n)} \\
[J_{ij}, M_k^{(m)}] &= -(M_i^{(m)}\delta_{jk} - M_j^{(m)}\delta_{ik}), \quad [M_i^{(m)}, M_j^{(n)}] = 0, \quad [J_{ij}, L^{(n)}] = 0.
\end{aligned} \tag{3.11}$$

The indices $m, n = 0, \pm 1$. We have made manifest the $SL(2, R)$ subalgebra with the generators $L^{(0)}, L^{(\pm 1)}$. In fact, we can define the vector fields

$$L^{(n)} = -(n+1)t^n x_i \partial_i - t^{n+1}\partial_t, \quad M_i^{(n)} = t^{n+1}\partial_i \tag{3.12}$$

with $n = 0, \pm 1$. These (together with J_{ij}) are then exactly the vector fields in (2.2) and (2.8) which generate the GCA (without central extension).

If we now consider the vector fields of (3.12) for *arbitrary* integer n, and also define

$$J_a^{(n)} \equiv J_{ij}^{(n)} = -t^n(x_i\partial_j - x_j\partial_i) \tag{3.13}$$

then we find that this collection obeys the current algebra

$$\begin{aligned}
[L^{(m)}, L^{(n)}] &= (m-n)L^{(m+n)} \qquad [L^{(m)}, J_a^{(n)}] = -nJ_a^{(m+n)} \\
[J_a^{(n)}, J_b^{(m)}] &= f_{abc}J_c^{(n+m)} \qquad [L^{(m)}, M_i^{(n)}] = (m-n)M_i^{(m+n)}.
\end{aligned} \tag{3.14}$$

The index a labels the generators of the spatial rotation group $SO(d)$ and f_{abc} are the corresponding structure constants. We see that the vector fields generate a $SO(d)$ Kac-Moody algebra without any central terms. In addition to the Virasoro and current generators we also have the commuting generators $M_i^{(n)}$ which function like generators of a global symmetry. We can, for instance, consistently set these generators to zero. The presence of these generators therefore do not spoil the ability of the Virasoro-Kac-Moody generators to admit the usual central terms in their commutators.

We wish to understand what these symmetries mean. There is a simple interpretation for the generators $M_i^{(n)}, L^{(n)}, J_a^{(n)}$. We know that $M_i^{(-1,0,1)}$ generate uniform spatial translations, velocity boosts and accelerations respectively. From (3.12) one can see that $M_i^{(n)}$ generate arbitrary time dependent (but spatially independent) accelerations. $x_i \to x_i + b_i(t)$. Similarly $J_{ij}^{(n)}$ generate arbitrary time dependent, space-independent rotations: $x_i \to R_{ij}(t)x_j$. These two set of generators together generate the Coriolis group: the biggest group of "isometries" of "flat" Galilean spacetime [15].

The action of generators $L^{(n)}$ can read this off from (3.12): $t \to f(t)$, $x_i \to \frac{df}{dt}x_i$. It amounts to a reparametrisation of the absolute time t. Under this reparametrisation the spatial coordinates x_i act as vectors (on the worldline t). It seems as if this is some kind of "conformal isometry" of the Galilean spacetime, rescaling coordinates by the arbitrary time dependent factor $\frac{df}{dt}$.

Given that the Galilean limit can be obtained by taking a definite scaling limit within a relativistic theory, we expect to see the GCA (and perhaps its extension) as a symmetry of some subsector within every relativistic conformal field theory. For instance, in the best studied case of $\mathcal{N} = 4$ Yang-Mills theory, we ought to be able to isolate a sector with this symmetry.

One clue is the presence of the $SL(2,R)$ symmetry together with the preservation of spatial rotational invariance. One might naively think this should be via some kind of conformal quantum mechanics obtained by considering only the spatially independent modes of the field theory. But this is probably not totally correct for the indirect reasons explained in the next paragraph.

Recently, the nonrelativistic limit of the relativistic conformal hydrodynamics, which describes the small fluctuations from thermal equilibrium, have been studied [11, 12, 13]. One recovers the non-relativistic incompressible Navier-Stokes equation in this limit. The symmetries of this equation were then studied by [12] (see also [13]). One finds that all the generators of the finite GCA are indeed symmetries except for the dilatation operator D .The generator K acts trivially. In particular it has the K_i as symmetries. It is not surprising that the choice of a temperature should break the scaling symmetry of D. The interesting point is that the arbitrary accelerations $M_i^{(n)}$ are also actually a symmetry [14] (generating the so-called the Milne group [15]). Thus we have a part of the extended GCA as a symmetry of the non-relativistic Navier-stokes equation which should presumably describe the hydrodynamics in every nonrelativistic field theory.

Coming back to the Navier-Stokes equation, if the viscosity is set to zero, one gets the incompressible Euler equations In this case one has the entire finite dimensional GCA being a symmetry since D is now also a symmetry. It is the viscous term which breaks the symmetry under equal scaling of space and time. This shows that one can readily realize "gapless" non-relativistic systems in which space and time scale in the same way!

4. The Bulk Dual

Now we wish to turn our attention to the string theory side. Here it should be possible to take a similar scaling limit along the lines of the non-relativistic limit studied in [16, 17, 18]. Below we will only consider features of this scaling limit when the parent bulk theory is well described by gravity. This will already involve some novel features. This has to with the fact that the usual pseudo-Riemannian metric degenerates when one takes a non-relativistic limit. Nevertheless, there is a well defined, albeit somewhat unfamiliar, geometric description of gravity in such a limit [19]. In the (asymptotically) flat space case this is known as the Newton-Cartan theory of gravity which captures Newtonian gravity in a geometric setting. This is a *non-metric* gravitational theory. One can generalise this to the case of a negative cosmological constant as well. A variant of this is what we propose below as the right framework for the gravity dual of systems with the GCA. In the next subsection we will briefly review features of the Newton-Cartan theory and then go onto describe the case with a negative cosmological constant.

4.1. Newton-Cartan Theory of Gravity and its modification in AdS

In the Newton-Cartan description of gravity, the $(d+1)$ dimensional spacetime M has a time function t on it which foliates the spacetime into d dimensional spatial slices. Stated more precisely (see for example [20]): one defines a contravariant tensor $\gamma = \gamma^{\mu\nu}\partial_\mu \otimes \partial_\nu$ $(\mu,\nu = 0\ldots d)$ such there is a time 1-form $\tau = \tau_\mu dx^\mu$ which is orthogonal to γ in the sense that $\gamma^{\mu\nu}\tau_\mu = 0$. The metric γ, which has three positive eigenvalues and one zero eigenvalue, will be the non-dynamical spatial metric on slices orthogonal to the worldlines defined by τ. There is no metric on the spacetime as a whole. In fact, its geometric structure is that of a fibre bundle with a one dimensional base (time) and the d dimensional spatial slices as fibres.

The dynamics is encoded in a torsion free affine connection $\Gamma^\mu_{\nu\lambda}$ on M. We will demand that this connection is compatible with both γ and τ i.e.

$$\nabla_\rho \gamma^{\mu\nu} = 0 \qquad \nabla_\rho \tau_\nu = 0. \tag{4.15}$$

This enables one to define a time function t ("absolute time") since we have $\tau_\mu = \nabla_\mu t$. Unlike the Christoffel connections which are determined by the spacetime metric in Einstein's theory, this Newton-Cartan connection is not fixed by just these conditions. One has to impose some additional relations. Defining $R^{\mu\nu}_{\lambda\sigma} = \gamma^{\nu\alpha} R^{\mu}_{\lambda\alpha\sigma}$, one can define a Newtonian connection as one which obeys the additional condition $R^{\mu\nu}_{\lambda\sigma} = R^{\nu\mu}_{\sigma\lambda}$.

Now, we would like to parametrically carry out the non-relativistic scaling on the bulk AdS_{d+2} which would capture the physics of the nonrelativistic limit in the $(d+1)$ dimensional boundary theory. In the next section we will describe the bulk scaling in more detail. Here we will simply motivate its qualitative features and give the resulting Newton-Cartan like description of the bulk geometry.

We know that the boundary metric degenerates in the nonrelativistic limit with the d spatial directions scaling as $x_i \propto \epsilon$ while $t \propto \epsilon^0$. We expect this feature to be shared by the bulk metric. One expects that the geometry on constant radial sections to have such a scaling. Since the radial direction of the AdS_{d+2} is an additional dimension, we have to fix its scaling. The radial direction is a measure of the energy scales in the boundary theory via the holographic correspondence. We therefore expect it to also scale like time i.e. as ϵ^0. This means that in the bulk the time and radial directions of the metric *both* survive when performing the scaling. Together these constitute an AdS_2 sitting inside the original AdS_{d+2}.

What this implies for the dynamics is that we should have a Newton-Cartan like description but with the special role of time being replaced by an AdS_2. The geometric structure, in analogy with that of the previous section, is that of a fibre bundle with AdS_2 base and the d dimensional spatial slices as fibres.

Accordingly, we will consider a ("spatial") metric $\gamma = \gamma^{\mu\nu}\partial_\mu \otimes \partial_\nu$ $(\mu, \nu = 0 \ldots d+1)$ which now has *two* zero eigenvalues corresponding to the time and radial directions. (In a canonical choice of coordinates these directions will correspond to $\mu = 0, d+1$). Mathematically the two null eigenvectors will be taken to span the space of left invariant 1-forms of AdS_2. These will also define the AdS_2 metric $g_{\alpha\beta}$ in the usual way (This is the analogue of the time metric defined in the previous subsection).

We will once again have dynamical, torsion free affine connections $\Gamma^{\mu}_{\nu\lambda}$ which are compatible with both the spatial and AdS_2 metrics

$$\nabla_\rho \gamma^{\mu\nu} = 0 \qquad \nabla_\rho g_{\alpha\beta} = 0. \tag{4.16}$$

There will also be Christoffel connections from the AdS_2 and spatial metrics which will not be dynamical if we do not allow these metrics, specifically $g_{\mu\nu}$, to fluctuate.

4.2. GCA in the Bulk

In this section we will carry out the non-relativistic scaling limit on the AdS_5 piece of the bulk. We will also do this for the $SO(4,2)$ isometries of AdS_5 and obtain the same contracted algebra as before. Consider the metric of AdS_5 in Poincare coordinates

$$ds^2 = \frac{1}{z'^2}(\eta_{\mu\nu}dx^\mu dx^\nu - dz'^2) = \frac{1}{z'^2}(dt'^2 - dz'^2 - dx_i^2) \tag{4.17}$$

The nonrelativistic scaling limit that we will be considering is, as motivated in the previous section $t', z' \to \epsilon^0 t', \epsilon^0 z'$ $x_i \to \epsilon^1 x_i$.

In this limit we see that only the components of the metric in the (t', z') directions survive to give the metric on an AdS_2. The d dimensional spatial slices parametrised by the x_i are fibred over this AdS_2. The Poincare patch has a horizon at $z' = \infty$ and to extend the coordinates beyond this we will choose to follow an infalling null geodesic, in an analogue of the Eddington-Finkelstein coordinates. Therefore define $z = z'$ and $t = t' + z'$. In these coordinates

$$ds^2 = \frac{1}{z^2}(-2dtdz + dt^2) = \frac{dt}{z^2}(dt - 2dz). \tag{4.18}$$

In the infalling Eddington-Finkelstein coordinates, the Killing vectors of AdS_5 read as

$$\begin{aligned} P_i &= \partial_i, \ B_i = (t-z)\partial_i - x_i\partial_t \\ K_i &= (t^2 - 2tz - x_j^2)\partial_i + 2tx_i\partial_t + 2zx_i\partial_z + 2x_ix_j\partial_j \\ J_{ij} &= -(x_i\partial_j - x_j\partial_i), \ D = -t\partial_t - z\partial_z - x_i\partial_i \\ H &= -\partial_t, \ K = -(t^2 + x_i^2)\partial_t - 2z(t-z)\partial_z - 2(t-z)x_i\partial_i \end{aligned} \tag{4.19}$$

Carrying out the scaling mentioned before we obtain the contracted Killing vectors

$$\begin{aligned} P_i &= \partial_i, \ B_i = (t-z)\partial_i, \quad K_i = (t^2 - 2tz)\partial_i, \quad J_{ij} = -(x_i\partial_j - x_j\partial_i) \\ H &= -\partial_t, \quad D = -t\partial_t - z\partial_z - x_i\partial_i, \quad K = -t^2\partial_t - 2(t-z)(z\partial_z + x_i\partial_i) \end{aligned} \tag{4.20}$$

We see that at the boundary $z = 0$ these reduce to the contracted Killing vectors of the relativistic conformal algebra. It can also be checked that these obey the same algebra as (2.1) and (2.9) or equivalently (3.11) after the relabeling of (3.10).

The interpretation of most of the generators is straightforward. We note that the H, K, D are scalars under the spatial $SO(d-1)$ and generate, as before, an $SL(2,R)$. We identify this as the isometry group of the AdS_2 base of our Newton-Cartan theory. We can again define an infinite family of vector fields in the bulk

$$\begin{aligned} M_i^{(m)} &= (t^{m+1} - (m+1)zt^m)\partial_i, \quad J_{ij}^{(n)} = -t^n(x_i\partial_j - x_j\partial_i) \\ L^{(n)} &= -t^{n+1}\partial_t - (n+1)(t^n - nzt^{n-1})(x_i\partial_i + z\partial_z) \end{aligned} \tag{4.21}$$

These vector fields reduce on the boundary to (3.12) and (3.13).

It is rather remarkable that these vector fields also obey the commutation relations of the Virasoro-Kac-Moody algebra, the same as in the boundary theory

$$\begin{aligned} [L^{(m)}, L^{(n)}] &= (m-n)L^{(m+n)} \qquad [L^{(m)}, J_a^{(n)}] = -nJ_a^{(m+n)} \\ [J_a^{(n)}, J_b^{(m)}] &= f_{abc}J_c^{(n+m)} \qquad [L^{(m)}, M_i^{(n)}] = (m-n)M_i^{(m+n)}. \end{aligned} \tag{4.22}$$

How do we interpret all these additional vector fields from the point of view of the bulk? Firstly, notice that the vector fields $M_i^{(n)}$ and $J_a^{(n)}$ only act on x_i. From the viewpoint of the fibre bundle structure, these are simply rotations and translations on the spatial slices dependent on time as well as z. These are the isometries of the spatial metric. They are also trivially isometries of the AdS_2 metric since they do not act on those coordinates. Now we come to the action of the Virasoro generators, $L^{(n)}$. These turn out to be the asymptotic isometries of the AdS_2 in the sense of [21] and ordinary symmetries of the flat fibres. Thus the $L^{(n)}, J_a^{(n)}, M_i^{(N)}$ together generate (asymptotic) isometries of the spatial and AdS_2 metrics γ^{ij} and g_{ab} [8]. Therefore it seems natural to consider the action of these generators on the Newton-Cartan like geometry.

5. Concluding Remarks

We have seen that the nonrelativistic conformal symmetry obtained as a scaling limit of the relativistic conformal symmetry has several novel features which make it a potentially interesting case for further study. The GCA, we have argued, is different from the Schrodinger group which has been studied recently. It also has the advantage of being embedded within the relativistic theory. Hence we ought to have realizations of the GCA in every interacting relativistic conformal field theory. The obvious question is to understand this sector in a particular case such as $\mathcal{N} = 4$ Super Yang-Mills theory and to see whether the infinite dimensional extension can be dynamically realized. We have provided indications why this might be the case generically.

Using this algebra on the boundary, we can construct representations and these turn out to be labelled by the boosts and the dilations [22]. One can, like in the case of the relativistic conformal theory, use the finite algebra to determine the two and three point correlation functions upto numerical factors. There has also been recent progress in understanding the quantum nature of GCA in two dimensions [23], where the infinite algebra arises out of contraction of linear combinations of the two copies of the Virasoro algebra. This indicates that the GCA might be thought of as a high spin sector of the relativistic theory. There is also a recent realization of the quantum two-dimensional GCA in a cosmological topologically massive three-dimensional gravity theory [24].

A straightforward generalization of the results in the above discussion would be to a supersymmetric extension of the Kac-Moody algebra. These have been investigated in [25, 26].

The bulk description in terms of a Newton-Cartan like geometry is somewhat unfamiliar and it would be good to understand it better. In particular, one needs a precise bulk-boundary dictionary to characterize the duality. At least implicitly this is determined by taking the parametric limit of the relativistic duality.

In the case of the Schrodinger symmetry the dual gravity theory is proposed to live in *two* higher dimensions than the field theory. This also provided the route for embedding the dual geometry in string theory. It is interesting to ask if there is something analogous in our case, whereby the GCA is realized as a standard isometry of a higher dimensional geometry (e.g. $(d+3)$ dimensional for a $(d+1)$ dimensional field theory).

Coming back to the boundary theory, it is interesting to ask whether there are intrinsically non-relativistic realizations of the GCA, perhaps in a real life system. It is encouraging in this context that the incompressible Euler equations concretely realize the GCA, providing an example of a gapless non-relativistic system.

References

[1] J. M. Maldacena, *Adv. Theor. Math. Phys.* **2**, 231 (1998)
[2] Y. Nishida and D. T. Son, *Phys. Rev. D* **76**, 086004 (2007)
[3] C. R. Hagen, *Phys. Rev. D* **5**, 377 (1972).
[4] U. Niederer, *Helv. Phys. Acta* **45**, 802 (1972).
[5] M. Henkel, *J. Statist. Phys.* **75**, 1023 (1994)
[6] D. T. Son, *Phys. Rev. D* **78**, 046003 (2008)
[7] K. Balasubramanian and J. McGreevy, *Phys. Rev. Lett.* **101**, 061601 (2008)
[8] A. Bagchi and R. Gopakumar, *J. High Energy Phys.* JHEP**0907**, 037 (2009)
[9] D. E. Berenstein, J. M. Maldacena and H. S. Nastase, *J. High Energy Phys.* JHEP**0204**, 013 (2002)
[10] J. Lukierski, P. C. Stichel and W. J. Zakrzewski, *Phys. Lett. A* **357**, 1 (2006)
[11] I. Fouxon and Y. Oz, *Phys. Rev. Lett.* **101**, 261602 (2008)
[12] S. Bhattacharyya, S. Minwalla and S. R. Wadia, *Preprint* arXiv:0810.1545 [hep-th].
[13] I. Fouxon and Y. Oz, *Preprint* arXiv:0812.1266 [hep-th].
[14] V. N. Gusyatnikova and V. A. Yumaguzhin, *Acta Applicandae Mathematicae* **15** (January, 1989) 65–81.
[15] C. Duval, *Class. Quant. Grav.* **10**, 2217 (1993).
[16] J. Gomis and H. Ooguri, *J. Math. Phys.* **42**, 3127 (2001)
[17] U. H. Danielsson, A. Guijosa and M. Kruczenski, *J. High Energy Phys.* JHEP**0103**, 041 (2001)
[18] J. Gomis, J. Gomis and K. Kamimura, *J. High Energy Phys.* JHEP**0512**, 024 (2005)
[19] C. W. Misner, K. S. Thorne and J. A. Wheeler, *Gravitation*, (San Francisco 1973)
[20] C. Ruede and N. Straumann, *Helv. Phys. Acta* **70**, 318 (1997)
[21] J. D. Brown and M. Henneaux, *Commun. Math. Phys.* **104**, 207 (1986).
[22] A. Bagchi and I. Mandal, *Phys. Lett. B* **675**, 393 (2009)
[23] A. Bagchi, R. Gopakumar, I. Mandal and A. Miwa, *Preprint* arXiv:0912.1090 [hep-th].
[24] K. Hotta, T. Kubota and T. Nishinaka, *Preprint* arXiv:1003.1203 [hep-th].
[25] A. Bagchi and I. Mandal, *Phys. Rev. D* **80**, 086011 (2009)
[26] M. Sakaguchi, *Preprint* arXiv:0905.0188 [hep-th]

Radiation Damping in a Non-Abelian Strongly-Coupled Gauge Theory

Mariano Chernicoff†, **J. Antonio García**⋆ **and Alberto Güijosa**⋆

† Departament de Física Fonamental, Universitat de Barcelona, Marti i Franquès 1, E-08028 Barcelona, Spain, Email: mchernicoff@ub.edu

⋆ Departamento de Física de Altas Energías, Instituto de Ciencias Nucleares Universidad Nacional Autónoma de México, Apdo. Postal 70-543, México D.F. 04510, México Email: garcia, alberto@nucleares.unam.mx

Abstract. We study a 'composite' or 'dressed' quark in strongly-coupled $\mathcal{N} = 4$ super-Yang-Mills (SYM), making use of the AdS/CFT correspondence. We show that the standard string dynamics nicely captures the physics of the quark and its surrounding non-Abelian field configuration, making it possible to derive a relativistic equation of motion that incorporates the effects of radiation damping. From this equation one can deduce a non-standard dispersion relation for the composite quark, as well as a Lorentz covariant formula for its rate of radiation.

1. Motivation

A considerable effort has been invested in the study of strongly-coupled thermal non-Abelian plasmas by means of the AdS/CFT correspondence [1, 2]. The jet quenching observed at RHIC indicates that partons crossing the quark-gluon plasma experience significant energy loss. This question has been studied in a strongly-coupled $\mathcal{N} = 4$ SYM plasma for a variety of probes, including quarks [3], mesons [4], baryons [5, 6], gluons [5, 7] and k-quarks [5]. Here we will analyze quark energy loss *in vacuum*, which for a strongly-coupled non-Abelian gauge theory is interesting already in itself, and is in addition helpful for developing intuition that might extend to the finite temperature case [8, 9].

We expect an accelerating quark to radiate, and to experience a damping force due to the emitted radiation. In the context of classical electrodynamics, and for a non-relativistic electron modeled as a vanishingly small charge distribution, this effect is incorporated in the classic Abraham-Lorentz equation [10, 11]

$$m\left(\frac{d^2\vec{x}}{dt^2} - t_e\frac{d^3\vec{x}}{dt^3}\right) = \vec{F}\ . \tag{1}$$

The second term in the left-hand side is the damping term, which is seen to be associated with a characteristic timescale $t_e \equiv 2e^2/3mc^3$, set by the classical electron radius. The search for a Lorentz-covariant version of (1) led to the (Abraham-)Lorentz-Dirac equation [12],

$$m\left(\frac{d^2x^\mu}{d\tau^2} - t_e\left[\frac{d^3x^\mu}{d\tau^3} - \frac{1}{c^2}\frac{d^2x_\nu}{d\tau^2}\frac{d^2x^\nu}{d\tau^2}\frac{dx^\mu}{d\tau}\right]\right) = \mathcal{F}^\mu\ , \tag{2}$$

with τ the proper time and $\mathcal{F}^{\mu} \equiv \gamma(\vec{F}\cdot\vec{v}/c, \vec{F})$ the 4-force. The second term within the square brackets is the negative of the rate at which 4-momentum is carried away from the charge by radiation (as given by the covariant Lienard formula), so it is only this term that can properly be called radiation reaction. The first term within the square brackets, usually called the Schott term, and whose spatial part yields the damping force of (1) in the non-relativistic limit, is known to arise from the effect of the charge's 'near' (as opposed to radiation) field [13, 14].

The appearance of a third-order term in (1) and (2) leads to unphysical behavior, including pre-accelerating and self-accelerating (or 'runaway') solutions. These deficiencies are known to originate from the assumption that the charge is pointlike. For a non-relativistic charge distribution of small but finite size ℓ, (1) is corrected by an infinite number of higher-derivative terms,

$$m\left(\frac{d^2\vec{x}}{dt^2} - t_e\left[\frac{d^3\vec{x}}{dt^3} - \sum_{n=1}^{\infty} b_n \frac{\ell^n}{c^n}\frac{d^{3+n}\vec{x}}{dt^{3+n}}\right]\right) = \vec{F} \tag{3}$$

(with b_n some numerical coefficients), and is physically sound as long as $\ell > ct_e$ [14]. Upon shifting attention to the quantum case, one intuitively expects the pointlike 'bare' electron to acquire an effective size of order the Compton wavelength $\lambda_C \equiv \hbar/mc$, due to its surrounding cloud of virtual particles. Given that $\lambda_C \gg ct_e$, there should then be no room for unphysical behavior. Indeed, in [15] it was shown that nonrelativistic QED leads to

$$m\left(\frac{d^2\vec{x}}{dt^2} - t_e\left[\frac{d^3\vec{x}}{dt^3} - \sum_{n=1}^{\infty} d_n \frac{\lambda_C^n}{c^n}\frac{d^{3+n}\vec{x}}{dt^{3+n}}\right]\right) = \vec{F}~, \tag{4}$$

which has no runaway solutions, and shows that the charge develops a characteristic size $\ell = \lambda_C$.

Going further to the quantum non-Abelian case is a serious challenge. Nevertheless, we will show here that the AdS/CFT correspondence [1] allows us to address this question rather easily in certain strongly-coupled non-Abelian gauge theories [16, 17]. We expect the basic story we will uncover to apply generally to all instances of the gauge/string duality (including cases with finite temperature or chemical potentials), but for simplicity we will concentrate on the specific example of $\mathcal{N}=4$ SYM. Besides the gauge field, this maximally supersymmetric and conformally invariant theory (CFT) contains 6 real scalar fields and 4 Weyl fermions, all in the *adjoint* representation of the gauge group. We will be able to derive a Lorentz covariant equation summarizing the dynamics of a quark in this strongly-coupled theory, which will turn out to be a nonlinear generalization of the Lorentz-Dirac equation (2), with a higher-derivative structure somewhat similar to (4).

2. Basic Setup

From this point on we work in natural units, $c = \hbar = 1$. It is by now well-known that $\mathcal{N}=4$ $SU(N_c)$ SYM with coupling g_{YM} is, despite appearances, completely equivalent [1] to Type IIB string theory on a background that asymptotically approaches the five-dimensional anti-de Sitter (AdS) geometry

$$ds^2 = G_{MN}dx^M dx^N = \frac{R^2}{z^2}\left(-dt^2 + d\vec{x}^2 + dz^2\right)~, \tag{5}$$

where $R^4/l_s^4 = g_{YM}^2 N_c \equiv \lambda$ denotes the 't Hooft coupling, and l_s is the string length. The radial direction z is mapped holographically into a variable length scale in the gauge theory, in such a way that $z \to 0$ and $z \to \infty$ are respectively the ultraviolet and infrared limits. The directions $x^{\mu} \equiv (t, \vec{x})$ are parallel to the AdS boundary $z = 0$ and are directly identified with the gauge theory directions. The state of IIB string theory described by the unperturbed metric (5) corresponds to the vacuum of the $\mathcal{N}=4$ SYM theory, and the closed string sector describing (small or large) fluctuations on top of it fully captures the gluonic (+ adjoint scalar and fermionic) physics. The string theory description is under calculational control only for small string coupling and low curvatures, which translates into $N_c \gg 1$, $\lambda \gg 1$.

It is also known that one can add to SYM N_f flavors of matter in the *fundamental* representation of the $SU(N_c)$ gauge group by introducing in the string theory setup an open string sector associated with a stack of N_f D7-branes [18]. We will refer to these degrees of freedom as 'quarks,' even though, being $\mathcal{N}=2$ supersymmetric, they include both spin 1/2 and spin 0 fields. The D7-brane embedding is chosen to be translationally invariant along the gauge theory directions x^μ, and extend in the radial direction from the boundary of AdS at $z=0$ up to a location

$$z_m = \frac{\sqrt{\lambda}}{2\pi m} \tag{6}$$

determined by the mass m of the quarks. For $N_f \ll N_c$, the backreaction of the D7-branes on the geometry can be neglected; in the field theory this corresponds to a 'quenched' approximation.

An isolated quark of mass m is dual to an open string that has one endpoint on the D7-branes at $z=z_m$, and extends radially all the way to the AdS horizon at $z\to\infty$. The string dynamics is governed by the Nambu-Goto action

$$S_{NG} = -\frac{1}{2\pi l_s^2}\int d^2\sigma\,\sqrt{-\det g_{ab}} \equiv \int d^2\sigma\,\mathcal{L}_{NG}\;, \tag{7}$$

where $g_{ab} \equiv \partial_a X^M \partial_b X^N G_{MN}(X)$ $(a,b=0,1)$ denotes the induced metric on the worldsheet. We can exert an external force $\vec{F}$ on the string endpoint by turning on an electric field $F_{0i}=F_i$ on the D7-branes. This amounts to adding to (7) the usual minimal coupling, which in terms of the endpoint/quark worldline $x^\mu(\tau)\equiv X^\mu(\tau,z_m)$ reads

$$S_F = \int d\tau\, A_\mu(x(\tau))\,\frac{dx^\mu(\tau)}{d\tau}\;. \tag{8}$$

Variation of $S_{NG}+S_F$ implies the standard Nambu-Goto equation of motion for all interior points of the string, plus the boundary condition

$$\Pi^z_\mu(\tau)|_{z=z_m} = \mathcal{F}_\mu(\tau) \quad \forall\,\tau\;, \tag{9}$$

where $\Pi^z_\mu \equiv \partial\mathcal{L}_{NG}/\partial(\partial_z X^\mu)$ is the worldsheet (Noether) current associated with spacetime momentum, and $\mathcal{F}_\mu = -F_{\nu\mu}\partial_\tau x^\nu = (-\gamma\vec{F}\cdot\vec{v},\gamma\vec{F})$ the Lorentz four-force.

Notice that the string is being described (as is customary) in first-quantized language, and, as long as it is sufficiently heavy, we are allowed to treat it semiclassically. In gauge theory language, then, we are coupling a first-quantized quark to the gluonic (+ other SYM) field(s), and then carrying out the full path integral over the strongly-coupled field(s) (the result of which is codified by the AdS spacetime), but treating the path integral over the quark trajectory $x^\mu(\tau)$ in a saddle-point approximation.[1]

In more detail, it is really the endpoint of the string that corresponds to the quark, while the body of the string codifies the profile of the (near and radiation) gluonic (+ other SYM) field(s) set up by the quark.[2] The latter can be mapped out explicitly by computing one-point functions of local operators ($\langle \mathrm{tr}\,F^2\rangle, \langle T_{\mu\nu}\rangle, \ldots$) in the presence of the quark, which, via the standard GKPW recipe [2], requires a determination of the near-boundary profile of the closed string fields (ϕ, $h_{\mu\nu}$, ...) generated by the macroscopic string.

[1] For a study of quantum fluctuations about this classical string configuration, see the very recent work [19].

[2] In other words, the string here is the $\mathcal{N}=4$ SYM analog of the 'QCD string', with the surprising twist that it lives not in 4 but in 5 (+5) dimensions.

For the interpretation of our results, it will be crucial to keep in mind that the quark described by this string is not 'bare' but 'composite' or 'dressed'. This can be seen most clearly by working out the expectation value of the gluonic field surrounding a static quark located at the origin [20],

$$\frac{1}{4g_{YM}^2}\langle \mathrm{tr}\, F^2(x)\rangle = \frac{\sqrt{\lambda}}{16\pi^2|\vec{x}|^4}\left[1-\frac{1+\frac{5}{2}\left(\frac{2\pi m|\vec{x}|}{\sqrt{\lambda}}\right)^2}{\left(1+\left(\frac{2\pi m|\vec{x}|}{\sqrt{\lambda}}\right)^2\right)^{5/2}}\right]. \qquad (10)$$

For $m \to \infty$ ($z_m \to 0$), this is just the Coulombic field expected (by conformal invariance) for a pointlike charge. For finite m, the profile is still Coulombic far away from the origin, but in fact becomes non-singular at the location of the quark,

$$\frac{1}{4g_{YM}^2}\langle \mathrm{tr}\, F^2(x)\rangle = \frac{\sqrt{\lambda}}{128\pi^2}\left[15\left(\frac{2\pi m}{\sqrt{\lambda}}\right)^4 - \frac{35}{|\vec{x}|^4}\left(\frac{2\pi m|\vec{x}|}{\sqrt{\lambda}}\right)^6 + \ldots\right] \quad \text{for } |\vec{x}| < \frac{\sqrt{\lambda}}{2\pi m}. \qquad (11)$$

As seen in these equations, the characteristic thickness of this non-Abelian charge distribution is precisely the length scale z_m defined in (6). This is then the size of the gluonic cloud that surrounds the quark, or in other words, the analog of the Compton wavelength for our non-Abelian source.

3. Generalized Lorentz-Dirac Equation for the Quark

By Lorentz invariance, given the description of the static quark, we know that a quark moving at constant velocity corresponds to a purely radial string, moving as a rigid vertical rod. When the quark accelerates, the body of the string will trail behind the endpoint, and will therefore exert a force on the latter. Remembering that the body of the string codifies the SYM fields sourced by the quark, we know that in the gauge theory this force is interpreted as the backreaction of the gluonic field on the quark. In other words, in the AdS/CFT context the quark has a 'tail', and it is this tail that is responsible for the damping effect we are after. This mechanism had been previously established in the computations of the drag force exerted on the quark by a thermal plasma, which is described in dual language in terms of a string living on a black hole geometry [21, 22]. Our analysis here will make it clear that the damping effect is equally present in the gauge theory vacuum [16, 17].

To flesh out this story, we need to determine the string profile corresponding to a given accelerated quark/endpoint trajectory, by solving the nonlinear equation of motion following from the Nambu-Goto action (7). Fortunately, for this task we can make use of the results obtained in a remarkable paper by Mikhailov [23], which we now briefly review (a more detailed explanation can be found in [8]). This author considered an infinitely massive quark ($z_m = 0$), and was able to solve the equation of motion for the dual string on AdS, for an *arbitrary* timelike trajectory of the string endpoint. In terms of the coordinates used in (5), his solution is

$$X^\mu(\tau, z) = z\frac{dx^\mu(\tau)}{d\tau} + x^\mu(\tau)\ , \qquad (12)$$

with $x^\mu(\tau)$ the worldline of the string endpoint at the AdS boundary— or, equivalently, the worldline of the dual, infinitely massive, quark— parametrized by its proper time τ. It is easy to check that the lines at constant τ are null with respect to the induced worldsheet metric, a fact that plays an important role in Mikhailov's construction.

The solution (12) is retarded. To see this, note that, parametrizing the quark worldline by $x^0(\tau)$ instead of τ, and using $d\tau = \sqrt{1-\vec{v}^2}dx^0$, where $\vec{v} \equiv d\vec{x}/dx^0$, the $\mu = 0$ component of (12) takes the form

$$t = z\frac{1}{\sqrt{1-\vec{v}^2}} + t_{\mathrm{ret}}\ , \qquad (13)$$

where $t_{\rm ret}$ denotes the value of the quark/endpoint coordinate time $x^0(\tau)$ at the proper time τ relevant to the (t,z) segment of the string under consideration, and the endpoint velocity $\vec{v}$ is meant to be evaluated at $t_{\rm ret}$. In these same terms, the spatial components of (12) can be formulated as

$$\vec{X}(t,z) = z\frac{\vec{v}}{\sqrt{1-\vec{v}^2}} + \vec{x}(t_{\rm ret}) = (t-t_{\rm ret})\vec{v} + \vec{x}(t_{\rm ret})\ . \tag{14}$$

We see here that the behavior at time $t = X^0(\tau,z)$ of the string segment located at radial position z (which, roughly speaking, codifies the gluonic field profile at length scale z and position $\vec{x} = \vec{X}(t,z)$ in the gauge theory) is completely determined by the behavior of the string endpoint at the *earlier* time $t_{\rm ret}(t,z)$ determined from (13), i.e., by projecting back toward the boundary along the null line at fixed τ. (Note the analogy with the construction of the Lienard-Wiechert fields in classical electromagnetism.)

An analogous advanced solution built upon the same endpoint/quark trajectory can be obtained by reversing the sign of the first term in the right-hand side of (12). In gauge theory language, this choice of sign corresponds to the choice between a purely outgoing or purely ingoing boundary condition for the waves in the gluonic field at spatial infinity. Both on the string and the gauge theory sides, more general configurations should of course exist, but obtaining them explicitly is difficult due to the highly nonlinear character of the system. Henceforth we will focus solely on the retarded solutions, which are the ones that capture the physics of present interest, with influences propagating outward from the quark to infinity.

Using (13) and (14), Mikhailov was able to rewrite the total string energy in the form

$$E(t) = \frac{\sqrt{\lambda}}{2\pi}\int_{-\infty}^{t} dt_{\rm ret}\frac{\vec{a}^2 - [\vec{v}\times\vec{a}]^2}{(1-\vec{v}^2)^3} + E_q(\vec{v}(t))\ , \tag{15}$$

where of course $\vec{a} \equiv d\vec{v}/dx^0$. The second term in the above equation arises from a total derivative on the string worldsheet, and gives the expected Lorentz-covariant expression for the energy intrinsic to the quark [8],

$$E_q(\vec{v}) = \frac{\sqrt{\lambda}}{2\pi}\left(\frac{1}{\sqrt{1-\vec{v}^2}}\frac{1}{z}\right)\Bigg|_{\infty}^{z_m=0} = \gamma m\ . \tag{16}$$

The first term in (15) must then represent the accumulated energy *lost* by the quark over all times prior to t. Surprisingly, the rate of energy loss is seen to have precisely the same form as the standard Lienard formula from classical electrodynamics. We therefore learn that in this non-Abelian, strongly-coupled theory, the energy loss of an infinitely massive (pointlike) quark depends *locally* on the quark worldline. For the spatial momentum, [23, 8] similarly find

$$\vec{P}(t) = \frac{\sqrt{\lambda}}{2\pi}\int_{-\infty}^{t} dt_{\rm ret}\frac{\vec{a}^2 - [\vec{v}\times\vec{a}]^2}{(1-\vec{v}^2)^3}\vec{v} + \vec{p}_q(\vec{v}(t))\ , \tag{17}$$

with

$$\vec{p}_q = \frac{\sqrt{\lambda}}{2\pi}\left(\frac{\vec{v}}{\sqrt{1-\vec{v}^2}}\frac{1}{z}\right)\Bigg|_{\infty}^{z_m=0} = \gamma m\vec{v}\ . \tag{18}$$

We see then that, in spite of the non-linear nature of the system, Mikhailov's procedure leads to a clean separation between the tip and the tail of the string, i.e., between the quark (including its near field) and its gluonic radiation field.

In fact, using (12), the standard string dynamics reduces to standard particle dynamics at the level of the action: plugging (12) back into (7)+(8), we can explicitly carry out the integral over z to obtain [17]

$$\begin{aligned} S_{\rm NG}+S_{\rm F} &= -\frac{R^2}{2\pi l_s^2}\int d\tau \int_{z_m\to 0}^{\infty}\frac{dz}{z^2}+\int d\tau\, A_\mu(x(\tau))\frac{dx^\mu}{d\tau}(\tau) \\ &= -m\int d\tau\ +\int d\tau\, A_\mu(x(\tau))\frac{dx^\mu}{d\tau}(\tau)\ , \end{aligned} \tag{19}$$

which is evidently the standard action for a pointlike externally forced relativistic particle (with mass $m\to\infty$). Notice that the associated equation of motion does *not* include a damping force, which is just as one would expect for an infinitely massive charge, because the coefficient $t_e \propto 1/m$ of the damping terms in (1) and (2) approaches zero as $m\to\infty$.

Let us now consider the more interesting case of a quark with finite mass, $z_m>0$, where there should be a noticeable damping effect. As we emphasized at the end of the previous section, in this case our non-Abelian source is no longer pointlike but has size z_m. On the string theory side, the string endpoint is now at $z=z_m$, and we must again require it to follow the given quark trajectory, $x^\mu(\tau)$. As before, this condition by itself does not pick out a unique string embedding. Just like we discussed for the infinitely massive case, we additionally require the solution to be retarded, in order to focus on the gluonic field causally set up by the quark. As in [8], we can inherit this structure by truncating a suitably selected retarded Mikhailov solution. The embeddings of interest to us can thus be regarded as the $z\geq z_m$ portions of the solutions (12). These are parametrized by data at the AdS boundary $z=0$, which are now merely auxiliary and will henceforth be denoted with tildes, to distinguish them from the actual physical data associated with the endpoint/quark at $z=z_m$. In this notation, the string embedding reads

$$X^\mu(\tilde\tau,z)=z\frac{d\tilde x^\mu(\tilde\tau)}{d\tilde\tau}+\tilde x^\mu(\tilde\tau)\ . \tag{20}$$

Differentiation with respect to $\tilde\tau$ and evaluation at $z=z_m$ (where we can read off the quark trajectory $x^\mu(\tilde\tau)\equiv X^\mu(\tilde\tau,z_m)$) leads to

$$\frac{dx^\mu}{d\tilde\tau}=z_m\frac{d^2\tilde x^\mu}{d\tilde\tau^2}+\frac{d\tilde x^\mu}{d\tilde\tau}\ , \tag{21}$$

which in turn implies

$$d\tau^2\equiv -dx^\mu dx_\mu = d\tilde\tau^2\left[1-z_m^2\left(\frac{d^2\tilde x}{d\tilde\tau^2}\right)^2\right] \tag{22}$$

and

$$\frac{d^2x^\mu}{d\tilde\tau^2}=z_m\frac{d^3\tilde x^\mu}{d\tilde\tau^3}+\frac{d^2\tilde x^\mu}{d\tilde\tau^2}\ . \tag{23}$$

We are now finally ready to derive the desired equation of motion for the quark. This must simply be dual to the equation of motion satisfied by the string endpoint, which we know to be given by the standard boundary condition (9). For the embeddings (20), this condition reads

$$\Pi^z_\mu(\tau)=\frac{\sqrt{\lambda}}{2\pi}\frac{d\tilde\tau}{d\tau}\left[\frac{1}{z_m}\frac{d^2\tilde x^\mu}{d\tilde\tau^2}+\left(\frac{d^2\tilde x^\mu}{d\tilde\tau^2}\right)^2\frac{d\tilde x^\mu}{d\tilde\tau}\right]=\mathcal{F}_\mu\ . \tag{24}$$

Using (22), (23) and carrying out some additional algebra (see [17] for details), this can be rewritten in the form

$$\frac{d}{d\tau}\left(\frac{m\frac{dx^\mu}{d\tau}-\frac{\sqrt{\lambda}}{2\pi m}\mathcal{F}^\mu}{\sqrt{1-\frac{\lambda}{4\pi^2m^4}\mathcal{F}^2}}\right)=\frac{\mathcal{F}^\mu-\frac{\sqrt{\lambda}}{2\pi m^2}\mathcal{F}^2\frac{dx^\mu}{d\tau}}{1-\frac{\lambda}{4\pi^2m^4}\mathcal{F}^2}\ , \tag{25}$$

which is our main result [16, 17].

Notice that the characteristic length scale appearing in (25) is precisely $z_m = \sqrt{\lambda}/2\pi m$, which as discussed below (11), is the quark Compton wavelength. Let us now examine the behavior of a quark that is sufficiently heavy, or is forced sufficiently softly, that the condition $\sqrt{\lambda}|\mathcal{F}^2|/2\pi m^2 \ll 1$ holds. It is then natural to expand the equation of motion in a power series in this small parameter. To zeroth order in this expansion, we correctly reproduce the pointlike result $m\partial_\tau^2 x^\mu = \mathcal{F}^\mu$. If we instead keep terms up to first order, we find

$$m\frac{d}{d\tau}\left(\frac{dx^\mu}{d\tau} - \frac{\sqrt{\lambda}}{2\pi m^2}\mathcal{F}^\mu\right) = \mathcal{F}^\mu - \frac{\sqrt{\lambda}}{2\pi m^2}\mathcal{F}^2\frac{dx^\mu}{d\tau}~.$$

In the $\mathcal{O}(\sqrt{\lambda})$ terms it is consistent, to this order, to replace $\mathcal{F}^\mu$ with its zeroth order value, thereby obtaining

$$m\left(\frac{d^2x^\mu}{d\tau^2} - \frac{\sqrt{\lambda}}{2\pi m}\frac{d^3x^\mu}{d\tau^3}\right) = \mathcal{F}^\mu - \frac{\sqrt{\lambda}}{2\pi}\frac{d^2x^\nu}{d\tau^2}\frac{d^2x_\nu}{d\tau^2}\frac{dx^\mu}{d\tau}~. \qquad (26)$$

Interestingly, this coincides *exactly* with the Lorentz-Dirac equation (2), with the Compton wavelength (6) playing the role of characteristic size t_e for the composite quark. This is indeed the natural quantum scale of the problem.

We can continue this expansion procedure to arbitrarily high order in $\sqrt{\lambda}|\mathcal{F}^2|/2\pi m^2$. Our full equation (25) is thus recognized as an extension of the Lorentz-Dirac equation that automatically incorporates the size z_m of our non-pointlike non-Abelian source. It is curious to note that (25), which incorporates the effect of radiation damping on the quark, has been obtained from (12), which does *not* include such damping for the string itself (the latter is $1/N_c^2$ suppressed).

The full physical content of (25) can be made transparent by rewriting it in the form

$$\frac{dP^\mu}{d\tau} \equiv \frac{dp_q^\mu}{d\tau} + \frac{dP^\mu_{rad}}{d\tau} = \mathcal{F}^\mu, \qquad (27)$$

recognizing P^μ as the total string (= quark + radiation) four-momentum,

$$p_q^\mu = \frac{m\frac{dx^\mu}{d\tau} - \frac{\sqrt{\lambda}}{2\pi m}\mathcal{F}^\mu}{\sqrt{1 - \frac{\lambda}{4\pi^2 m^4}\mathcal{F}^2}} \qquad (28)$$

as the intrisic momentum of the quark including the contribution of the near-field sourced by it (or, in quantum mechanical language, of the gluonic cloud surrounding the quark), and

$$\frac{dP^\mu_{\text{rad}}}{d\tau} = \frac{\sqrt{\lambda}\mathcal{F}^2}{2\pi m^2}\left(\frac{\frac{dx^\mu}{d\tau} - \frac{\sqrt{\lambda}}{2\pi m^2}\mathcal{F}^\mu}{1 - \frac{\lambda}{4\pi^2 m^4}\mathcal{F}^2}\right) \qquad (29)$$

as the rate at which momentum is carried away from the quark by chromo-electromagnetic radiation. Both (28) and (29) diverge when $\mathcal{F}^2 \to \mathcal{F}^2_{\text{crit}}$, where $\mathcal{F}^2_{\text{crit}} = 4\pi^2 m^4/\lambda$ is the critical value at which the force becomes strong enough to nucleate quark-antiquark pairs (or, in dual language, to create open strings) [24].

Unlike its classical electrodynamic counterpart (2), our dressed quark equation of motion has no self-accelerating (runaway) solutions: in the (continuous) absence of an external force, (25) uniquely predicts that the 4-acceleration of the quark must vanish. Interestingly, the converse to this last statement is not true: constant 4-velocity does not uniquely imply a vanishing force. E.g., for motion purely along one dimension, (25) reads

$$ma = \frac{F(1-v^2)^{3/2}}{\sqrt{1-\frac{\lambda}{4\pi^2 m^4}F^2}} + \frac{\frac{\sqrt{\lambda}}{2\pi m}\frac{dF}{dt}(1-v^2)}{1-\frac{\lambda}{4\pi^2 m^4}F^2}\,. \qquad (30)$$

For $a = 0$, we obtain a differential equation with solution $F(t) = \pm(2\pi m^2/\sqrt{\lambda})\,\mathrm{sech}[2\pi m(t - t_0)/\gamma\sqrt{\lambda}]$, with t_0 an integration constant [17]. The energy provided to the system by this particular $F(t)$ does not translate into an increase of the quark/endpoint velocity, but into a continuous modification of the string tail, or, in gauge theory language, a change of the gluonic field profile. This is again a consequence of the extended, and hence deformable, nature of the quark.

All in all, then, we have in (25) a physically sensible and interesting description of the dynamics of a composite quark in SYM. It is truly remarkable that the AdS/CFT correspondence grants us such direct access to this piece of strongly-coupled non-Abelian physics.

References

[1] J. M. Maldacena, Adv. Theor. Math. Phys. **2**, 231 (1998) [Int. J. Theor. Phys. **38**, 1113 (1999)] [arXiv:hep-th/9711200]; O. Aharony, S. S. Gubser, J. M. Maldacena, H. Ooguri and Y. Oz, Phys. Rept. **323**, 183 (2000) [arXiv:hep-th/9905111].
[2] S. S. Gubser, I. R. Klebanov and A. M. Polyakov, Phys. Lett. B **428**, 105 (1998) [arXiv:hep-th/9802109]; E. Witten, Adv. Theor. Math. Phys. **2**, 253 (1998) [arXiv:hep-th/9802150].
[3] C. P. Herzog, A. Karch, P. Kovtun, C. Kozcaz and L. G. Yaffe, JHEP **0607** (2006) 013 [arXiv:hep-th/0605158]; S. S. Gubser, Phys. Rev. D **74** (2006) 126005 [arXiv:hep-th/0605182]; J. Casalderrey-Solana and D. Teaney, Phys. Rev. D **74** (2006) 085012 [arXiv:hep-ph/0605199]; H. Liu, K. Rajagopal and U. A. Wiedemann, Phys. Rev. Lett. **97** (2006) 182301 [arXiv:hep-ph/0605178].
[4] K. Peeters, J. Sonnenschein and M. Zamaklar, Phys. Rev. D **74**, 106008 (2006) [arXiv:hep-th/0606195]; H. Liu, K. Rajagopal and U. A. Wiedemann, Phys. Rev. Lett. **98**, 182301 (2007) [arXiv:hep-ph/0607062]; M. Chernicoff, J. A. García and A. Güijosa, JHEP **0609**, 068 (2006) [arXiv:hep-th/0607089]; K. Dusling, J. Erdmenger, M. Kaminski, F. Rust, D. Teaney and C. Young, JHEP **0810**, 098 (2008) [arXiv:0808.0957 [hep-th]].
[5] M. Chernicoff and A. Güijosa, JHEP **0702** (2007) 084 [arXiv:hep-th/0611155].
[6] C. Athanasiou, H. Liu and K. Rajagopal, JHEP **0805**, 083 (2008) [arXiv:0801.1117 [hep-th]].
[7] S. S. Gubser, D. R. Gulotta, S. S. Pufu and F. D. Rocha, "Gluon energy loss in the gauge-string duality," JHEP **0810**, 052 (2008) [arXiv:0803.1470 [hep-th]].
[8] M. Chernicoff and A. Güijosa, JHEP **0806**, 005 (2008) [arXiv:0803.3070 [hep-th]].
[9] Y. Hatta, E. Iancu and A. H. Mueller, arXiv:0803.2481 [hep-th]; K. B. Fadafan, H. Liu, K. Rajagopal and U. A. Wiedemann, Eur. Phys. J. C **61**, 553 (2009) [arXiv:0809.2869 [hep-ph]].
[10] M. Abraham, Ann. Physik **10** (1903) 105.
[11] H. A. Lorentz, *The Theory of Electrons and Its Applications to the Phenomena of Light and Radiant Heat*, 2nd ed., (Dover, 1952).
[12] P. A. M. Dirac, Proc. Roy. Soc. Lond. A **167** (1938) 148.
[13] C. Teitelboim, Phys. Rev. D **1** (1970) 1572 [Erratum-ibid. D **2** (1970) 1763].
[14] F. Rohrlich, *Classical Charged Particles*, 2nd. ed. (Addison Wesley, Redwood City, California, 1990); Am. J. Phys. **65** (1997) 1051.
[15] E. J. Moniz and D. H. Sharp, Phys. Rev. D **15** (1977) 2850.
[16] M. Chernicoff, J. A. García and A. Güijosa, Phys. Rev. Lett. **102** (2009) 241601 [arXiv:0903.2047 [hep-th]].
[17] M. Chernicoff, J. A. García and A. Güijosa, JHEP **0909** (2009) 080 [arXiv:0906.1592 [hep-th]].
[18] A. Karch and E. Katz, JHEP **0206** (2002) 043 [arXiv:hep-th/0205236].
[19] E. Cáceres, M. Chernicoff, A. Güijosa and J. F .Pedraza, arXiv:1003.5332 [hep-th].
[20] J. L. Hovdebo, M. Kruczenski, D. Mateos, R. C. Myers and D. J. Winters, Int. J. Mod. Phys. A **20** (2005) 3428.
[21] C. P. Herzog, A. Karch, P. Kovtun, C. Kozcaz and L. G. Yaffe, JHEP **0607** (2006) 013 [arXiv:hep-th/0605158].
[22] S. S. Gubser, Phys. Rev. D **74** (2006) 126005 [arXiv:hep-th/0605182].
[23] A. Mikhailov, arXiv:hep-th/0305196.
[24] J. Casalderrey-Solana and D. Teaney, JHEP **0704** (2007) 039 [arXiv:hep-th/0701123].

Journal of Physics: Conference Series **462** (2013) 012019 doi:10.1088/1742-6596/462/1/012019

Energy functionals for Calabi-Yau metrics

M Headrick, A Nassar
Martin Fisher School of Physics, Brandeis University, Waltham MA 02454, USA
mph@brandeis.edu, anassar@brandeis.edu

Abstract. We identify a set of "energy" functionals on the space of metrics in a given Kähler class on a Calabi-Yau manifold, which are bounded below and minimized uniquely on the Ricci-flat metric in that class. Using these functionals, we recast the problem of numerically solving the Einstein equation as an optimization problem. We apply this strategy, using the "algebraic" metrics (metrics for which the Kähler potential is given in terms of a polynomial in the projective coordinates), to the Fermat quartic and to a one-parameter family of quintics that includes the Fermat and conifold quintics. We show that this method yields approximations to the Ricci-flat metric that are exponentially accurate in the degree of the polynomial (except at the conifold point, where the convergence is polynomial), and therefore orders of magnitude more accurate than the balanced metrics, previously studied as approximations to the Ricci-flat metric. The method is relatively fast and easy to implement. On the theoretical side, we also show that the functionals can be used to give a heuristic proof of Yau's theorem.

1. Introduction

Calabi-Yau (CY) manifolds play a pivotal role in complex differential geometry, algebraic geometry, and string theory. Part of the reason for their importance, for both mathematics and physics, lies in the Ricci-flat metrics they admit. The existence of these metrics is guaranteed by Yau's theorem; this theorem, however, is not constructive, and, outside of flat metrics and orbifolds thereof, no exact solutions to the Einstein equation are known on any CY manifold. This state of affairs naturally leads to the question of whether it is feasible to compute useful numerical approximations to the Ricci-flat metrics, a question that has given rise to a small literature in recent years [1, 2, 3, 4, 5, 6, 7]. A few different methods have been put forward, all of which make crucial use of the advantages offered by Kähler geometry compared to real geometry.[1] The purpose of this paper is to introduce a new method for this problem, that scores well on all three of the major considerations that make such a method useful: ease of implementation, scale of computational resources required, and accuracy of the approximations obtained. The method is based on a mathematical observation that may be of some interest in its own right.

The mathematical and physical motivations for solving the Einstein equation on CYs are discussed at length in the papers cited above, and we will not repeat them here. However, it is important to note that the scope of the potential applications depends strongly on the performance of the methods available. For example, for some problems it is sufficient to compute

[1] Related methods for solving the Einstein equation on toric manifolds were discussed in [8, 9, 10].

the metric at a single point in the CY moduli space, while for others it is necessary to scan over a part of the moduli space, requiring much faster methods.

Any numerical method for solving a PDE must address two issues. The first is how to represent, or store, the function being solved for—in this case, the Kähler potential—for example using a real-space lattice or a spectral representation. The second is, given that representation, how to solve the equation, or, more precisely, how to find the best approximation within that representation to the exact solution. For the problem at hand, the principal challenges in terms of the representation are the high dimensionality of the manifold (usually four or six real dimensions, so that even for moderate resolutions storage quickly becomes a limiting factor) and its typically very complicated topology. The principal challenge in terms of solving the Einstein equation is its nonlinearity.

The first methods developed for solving the Einstein equation on CYs, by Headrick and Wiseman [1], used a real-space lattice representation of the Kähler potential, and solved the equation by a Gauss-Seidel relaxation method. The method was straightforward, if not particularly elegant. One downside was the messiness of dealing with coordinate patches. The main factor limiting the accuracy of the metrics obtained was storage, due to the large number of lattice points required for high resolution in high dimensions (the method was tested on K3; storage would have severely limited the possible resolutions for threefolds).

Subsequently, Donaldson proposed a radically different set of methods based on a spectral (momentum-space) representation of the metric [2]. Specifically, he advocated the use of so-called algebraic metrics, which had been studied from a theoretical viewpoint by Tian and others. The salient properties for our present purposes are (1) they are based on polynomials, and therefore very easy to work with both theoretically and computationally; (2) they eliminate the messiness of coordinate patches; and (3) within a certain Kähler class they are capable of approximating any given smooth metric—including the Ricci-flat one—exponentially well in k, the degree of the polynomials (k is analogous to the highest mode number of a finite set of Fourier modes used to approximate a function, for example, on S^1). The latter property essentially promises a form of data compression, and, when applied to the Ricci-flat metric, has potential to solve the storage problem mentioned above.

As with any spectral representation, the challenge then comes in solving the PDE. Linear PDEs (especially homogeneous ones) become simpler when written in momentum space, as the different modes decouple from each other. For nonlinear ones such as the Einstein equation, on the other hand, every mode is coupled to every other mode, leading to a complicated system of nonlinear equations. Donaldson proposed two kinds of approximate solutions to the Einstein equation within the framework of the algebraic metrics. The first was the so-called balanced metrics. For each k the balanced metric is the unique solution to a certain integral equation, which can be found by iterating a certain integral map.[2] Donaldson showed that they approach the Ricci-flat metric as $k \to \infty$, but only as a power of k. Thus they do not achieve the exponential accuracy promised by the algebraic metrics. The second proposal, called "refined" metrics, were based on a Galerkin method: the error in solving the Einstein equation (more precisely, the equivalent Monge-Ampère equation) was required to be orthogonal to the full set of basis functions (in this case, the degree-k polynomials). The refined metrics are expected to convergence exponentially to the Ricci-flat metric. However, efficiently computing the refined metric proved difficult.

Now, it is well known in numerical analysis that it is far easier to minimize a well-behaved function than to solve a non-linear set of equations.[3] Here, by "well-behaved", we mean smooth, bounded below, and having no critical points other than a single global minimum.This is true

[2] The balanced metrics are of significant mathematical and possibly physical interest quite apart from the problem of solving the Einstein equation [2, 11].

[3] For a standard discussion of this point, see e.g. section 9.6.1 of [12].

Journal of Physics: Conference Series **462** (2013) 012019 doi:10.1088/1742-6596/462/1/012019

both from a theoretical viewpoint and in terms of actual available algorithms; while there exist several efficient and robust numerical optimization algorithms, for solving non-linear equations there is essentially only the Newton-Raphson method, which has poor global convergence properties, and a few variants on it. So it is nearly always advantageous, when possible, to convert the latter type of problem into the former. In the case of an elliptic PDE, a well-behaved functional whose variational derivative yields the PDE is usually called an energy functional. Note that most elliptic PDEs do not admit such a variational formulation. (The utility of such a formulation is also of course clear from the mathematical point of view, for example for proving the existence of solutions.) Given an energy functional and a representation scheme (real-space, spectral, or otherwise), one minimizes the energy functional within the corresponding finite-dimensional space of functions.[4] The prototype for this method is the Rayleigh-Ritz variational method for finding the smallest eigenvalue of a linear operator. Especially for spectral representations, this strategy can vastly simplify the problem of solving the PDE.

In the paper [13], we show how to follow this strategy for case of the Einstein equation on a CY manifold.[5] We define a family of energy functionals on the space of Kähler metrics on a given CY manifold, that are bounded below and minimized within each Kähler class precisely on the Ricci-flat metric.[6] The existence of these functionals depends crucially on the magic of Kähler geometry and should by no means be taken for granted—comparable functionals do not, as far as we know, exist in the Riemannian setting (for example, the Einstein-Hilbert action is unbounded below, while the integrated square of the Ricci tensor has many spurious local minima). Note that the functionals we define are not tied to the algebraic metrics, and could be used to solve the Einstein equation given any spectral or other representation scheme. We show, incidentally, that these functionals can also be used to give a heuristic proof of Yau's theorem. The argument has loopholes, as we discuss, but may nonetheless be useful as a quick way to understand why one should expect Yau's theorem to hold.[7] The simplest functional takes the following form:

$$E[J] = \int_X \Omega \wedge \bar{\Omega}(\eta - 1)^2 , \tag{1}$$

where Ω is the holomorphic $(n, 0)$ form on the CY X, and

$$\eta = \frac{J^n}{\Omega \wedge \bar{\Omega}} . \tag{2}$$

We then combine the algebraic metrics with one of our energy functionals to define "optimal" metrics. The figures show the results of applying our method to the Fermat quartic and to the following one-real-parameter family of quintics,

$$\sum_{a=1}^{5} (z^a)^5 - 5\psi \prod_{a=1}^{5} z^a = 0 \tag{3}$$

which includes the Fermat and conifold quintics (at $\psi = 0, 1$ respectively). The integer parameter k is the degree of the polynomial in the projective coordinates used to approximate the Kähler potential; the larger k is, the better the approximation. The method is relatively simple to

[4] Note that the error in solving the equation will necessarily be orthogonal, in the full function space, to the subspace in which one is working. Hence any method of this type automatically solves some Galerkin-type condition.

[5] The possibility of following a functional-minimization strategy was briefly mentioned in [4].

[6] These functionals share some of the properties of the Calabi energy [14], the Mabuchi K-energy [15], and its generalizations [9, 16], although they are somewhat simpler.

[7] The recent paper [16] uses a variational approach based on the Mabuchi K-energy [15] to prove a weak version of Yau's theorem.

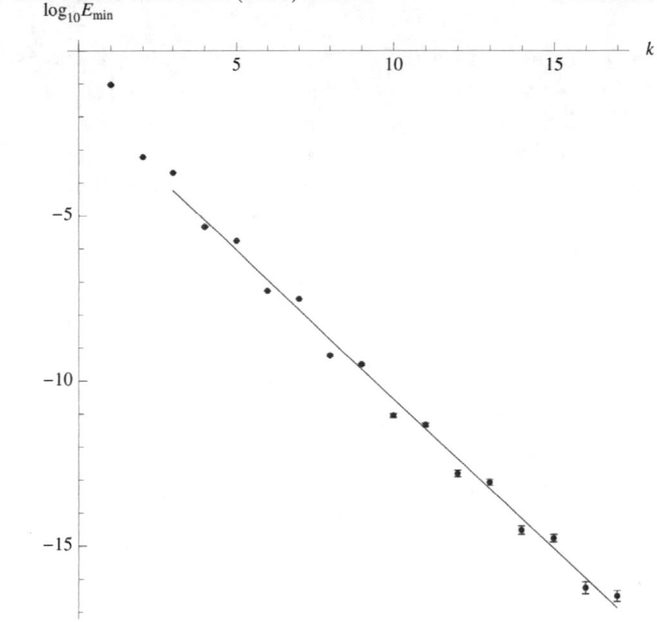

Figure 1. Value of E for the optimal algebraic metric on the Fermat quartic, versus k. The exponential decrease is clear, accompanied by a small even-odd modulation. The line is 0.03×8^{-k}, obtained by fitting to the points with $k = 3$ through 17.

implement (requiring just a few pages of *Mathematica* code), and very efficient (running for a few minutes on a laptop computer, it yields metrics that are likely accurate enough for most applications). In particular, the method achieves the exponential accuracy promised by the algebraic metrics. As a result, the metrics found are far more accurate, for a given value of k, than the balanced metrics. (The high-accuracy metrics we obtain on the compact conifold and deformed conifold, in particular, are perhaps relevant for certain models of string phenomenology; see [17] and references therein).

[1] Headrick M and Wiseman T 2005 *Class. Quant. Grav.* **22** 4931–4960 (*Preprint* `hep-th/0506129`)
[2] Donaldson S K 2009 *Pure Appl. Math. Q.* **5** 571–618 ISSN 1558-8599
[3] Douglas M R, Karp R L, Lukic S and Reinbacher R 2007 *JHEP* **12** 083 (*Preprint* `hep-th/0606261`)
[4] Douglas M R, Karp R L, Lukic S and Reinbacher R 2008 *J. Math. Phys.* **49** 032302 (*Preprint* `hep-th/0612075`)
[5] Braun V, Brelidze T, Douglas M R and Ovrut B A 2008 *JHEP* **05** 080 (*Preprint* `0712.3563`)
[6] Braun V, Brelidze T, Douglas M R and Ovrut B A 2008 *JHEP* **07** 120 (*Preprint* `0805.3689`)
[7] Keller J and Lukic S (*Preprint* `0907.1387 [math.DG]`)
[8] Doran C, Headrick M, Herzog C P, Kantor J and Wiseman T 2008 *Commun. Math. Phys.* **282** 357–393 (*Preprint* `hep-th/0703057`)
[9] Keller J 2009 *J. Inst. Math. Jussieu* **8** 743–768 ISSN 1474-7480 (*Preprint* `0709.1490`) URL `http://dx.doi.org/10.1017/S1474748009000103`
[10] Bunch R S and Donaldson S K 2008 *Handbook of geometric analysis. No. 1* (*Adv. Lect. Math. (ALM)* vol 7) (Int. Press, Somerville, MA) pp 1–28
[11] Douglas M R and Klevtsov S 2008 (*Preprint* `0811.0367`)
[12] Press W H, Teukolsky S A, Vetterling W T and Flannery B P 2007 *Numerical Recipes: The Art of Scientific Computing* 3rd ed (New York, NY, USA: Cambridge University Press) ISBN 0521880688, 9780521880688
[13] Headrick M and Nassar A 2009 (*Preprint* `0908.2635`)
[14] Calabi E 1982 *Seminar on Differential Geometry* (*Ann. of Math. Stud.* vol 102) (Princeton, N.J.: Princeton Univ. Press) pp 259–290
[15] Mabuchi T 1986 *Tohoku Math. J. (2)* **38** 575–593 ISSN 0040-8735

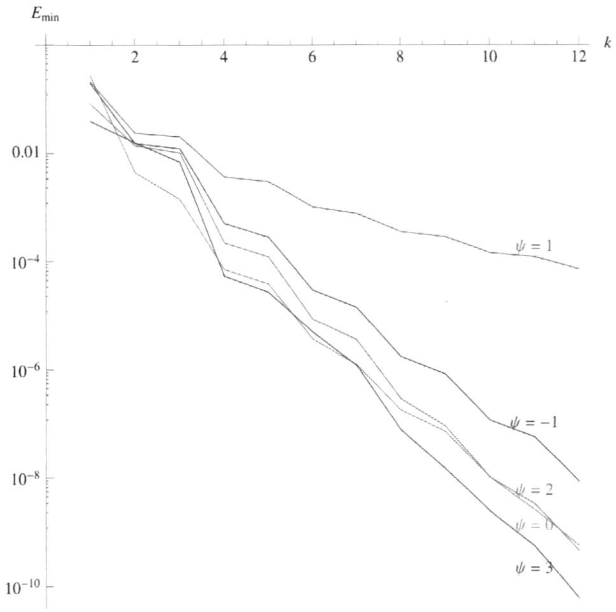

Figure 2. E_{min} versus k for algebraic metrics on the quintic (3) for the values of ψ indicated on the plot. The exponential decrease is clear for all values of ψ except $\psi = 1$, which is the conifold point in the quintic moduli space. A linear fit of $\log E_{\text{min}}$ gives $E_{\text{min}} \approx 0.07 \times 5^{-k}$ (not plotted) for $\psi = 0$ (the Fermat quintic).

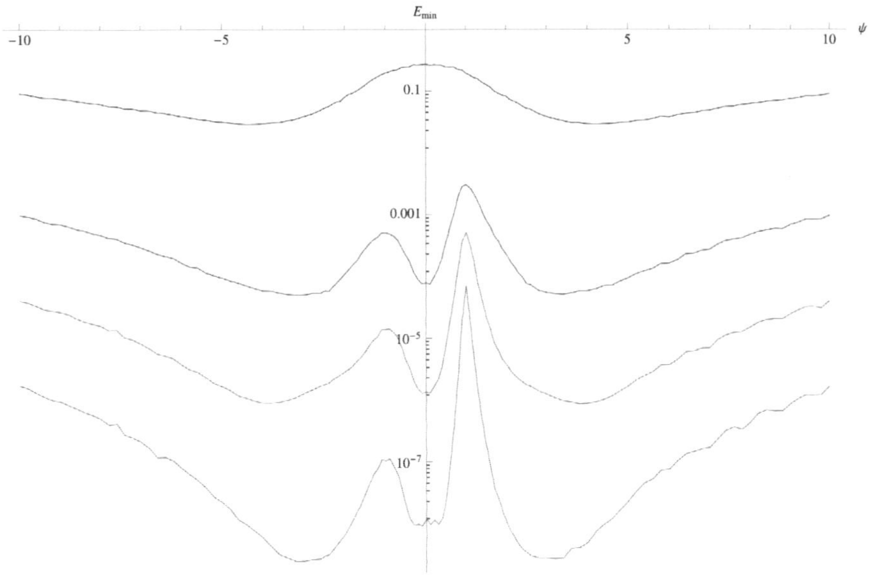

Figure 3. E_{min} versus ψ for (top to bottom) $k = 1, 4, 7, 10$ (the top curve represents the FS metric). (Error bars are not indicated to avoid cluttering the plot; however, the roughness of the curves, which is due to numerical errors, gives a good indication of their size.)

[16] Berman R J, Boucksom S, Guedj V and Zeriahi A 2000 (*Preprint* 0907.4490)
[17] Baumann D, Dymarsky A, Kachru S, Klebanov I R and McAllister L 2010 (*Preprint* 1001.5028)

Journal of Physics: Conference Series **462** (2013) 012020 doi:10.1088/1742-6596/462/1/012020

Six-dimensional $(2, 0)$ theory on tori

Måns Henningson

Department of Fundamental Physics, Chalmers University of Technology, S-412 96 Göteborg, Sweden

E-mail: mans@chalmers.se

Abstract. The six-dimensional $(2, 0)$ theories are a comparatively new and rather abstract type of quantum theory with important relations both to supersymmeric Yang-Mills theory in lower dimensions and to string- and M-theory in higher dimensions. After a short introduction to these theories, we focus on the case when they are considered on flat tori [1][2]. In particular, we give an example of how their ground state degeneracies can be computed, and also briefly discuss the spectrum of BPS-states. Finally, we comment on the automorphic transformation properties of the partition function of such a theory under the mapping class group of a six-torus.

The maximal dimension of a space(-time) which admits superconformal symmetry is $d = 1+5$. The symmetry algebra is then

$$osp(2, 6|2n) = so(2, 6) \oplus sp(2n) \oplus \text{ odd generators}$$

for some $n = 1, 2, \ldots$, where the first two terms are the conformal algebra in six-dimensions and the R-symmetry algebra [3].

Indications for the existence of a quantum theory with such symmetry (for $n = 2$) follows by considering type IIB string theory on a $(1 + 9)$-dimensional space-time with a codimension four singularity of some ADE-type. A self-consistent six-dimensional theory without dynamical gravity on the locus of the singularity then decouples from the bulk theory [4]. These so called $(2, 0)$ theories are highly unique: Apart from their ADE-type, they have no other discrete or continuous parameters.

Some reasons to study the $(2, 0)$ theories:

- They are quite different from other theories we know of, and still rather mysterious. Understanding them is likely to lead to much new mathematics and physics.
- They give a good opportunity to learn about important aspects of string theory without having to deal with quantum gravity.
- They are related to Yang-Mills theory in lower dimensions. In particular, they give a geometric understanding of S-duality of $N = 4$ super Yang-Mills theory in four dimensions.
- The study of $(2, 0)$ theory might be our best way towards a rigorous definition of quantum theory with infinitely many degrees of freedom.

A way to introduce a parameter in a $(2, 0)$ theory is to consider it on a space-time of the form $M^{1,5} = M^{1,4} \times S^1$ with a compact direction of radius R. At longer distances, the effective

Journal of Physics: Conference Series **462** (2013) 012020 doi:10.1088/1742-6596/462/1/012020

theory is then given by maximally supersymmetric Yang-Mills theory on $M^{1,4}$ with coupling constant $g = R^{1/2}$ and action

$$S = \frac{1}{R}\int_{M^{1,4}} \mathrm{Tr}\,(F \wedge *F + \ldots)\,.$$

The gauge group is of the form G/C, where G is simply connected with center subgroup C:

type	G	C
A_{n-1}	$SU(n)$	$\mathbb{Z}_n$
D_{2k}	$Spin(4k)$	$\mathbb{Z}_2 \times \mathbb{Z}_2$
D_{2k+1}	$Spin(4k+2)$	$\mathbb{Z}_4$
E_6	E_6	$\mathbb{Z}_3$
E_7	E_7	$\mathbb{Z}_2$
E_8	E_8	1.

In this way, $(2,0)$ theory can be seen as providing an ultra-violet completion of the Yang-Mills theory. The negative power of R indicates that $(2,0)$ theory has no Lagrangian description [5].

Let $M^{1,4} = \mathbb{R} \times M^4$, where the first factor denotes time. Quantum states of the Yang-Mills theory on this space are characterized by their magnetic 't Hooft flux

$$m \in H^2(M^4, C)$$

which determines the topological class of the gauge bundle over M^4, and their electric 't Hooft flux

$$e \in \mathrm{Hom}(H^1(M^4, C), U(1)) \simeq H^3(M^4, C)$$

which determines the transformation properties under 'large' gauge transformations [6]. From the perspective of $(2,0)$ theory on $M^{1,5} = M^{1,4} \times S^1$, we instead have a self-dual 't Hooft flux

$$f = e + m \in H^3(M^4 \times S^1, C).$$

A particularly interesting case is to consider $M^{1,5} = \mathbb{R} \times M^4 \times S^1$ with $M^4 = T^4$ a flat four-torus, since this preserves 16 supersymmetries. We can think of this as Yang-Mills theory on $\mathbb{R} \times T^4$ or as $(2,0)$ theory on $\mathbb{R} \times T^5$ with

$$T^5 = T^4 \times S^1.$$

The quantum states are characterized by their

- self-dual 't Hooft flux $f \in H^3(T^5, C)$
- energy $E \in \mathbb{R}^+$
- spatial momentum $p \in H^1(T^5, \mathbb{R})$
  (with the fifth component given by the Yang-Mills instanton number over T^4)
- $sp(4)$ R-symmetry representation R.

Supersymmetry implies the energy bound

$$E \geq |p|.$$

There are three classes of states:

- Vacuum states have $E = p = 0$ and are annihilated by all supercharges.
- BPS states have $E = |p| > 0$ and are annihilated by half of the supercharges.

- non-BPS states have $E > |p|$.

The spectra of vacua and BPS states are invariant under smooth deformations of the geometry of T^5, and can thus be followed from weak to strong coupling in the Yang-Mills perspective. We will discuss the computation of the spectrum of vacua from the Yang-Mills perspective. (Similar reasoning may be applied to the BPS-states, although we do not yet have any non-trivial checks on the results.)

At weak coupling, vacuum states are localized at orbifold singularities of the moduli space of flat connections over T^4. The low energy theory is given by maximally supersymmetric matrix quantum mechanics based on the subgroup S of the gauge group G/C left unbroken by the configuration at the singularity. This quantum mechanical model has a number n_S, depending on S, of normalizable ground states. Summing over the orbifold singularities gives the complete spectrum of vacua, which may be decomposed according to their electric and magnetic 't Hooft fluxes e and m.

Covariance under the $SL_4(\mathbb{Z})$ mapping class group of T^4 is manifest in the Yang-Mills formulation, but $(2,0)$ theory indicates covariance under the $SL_5(\mathbb{Z})$ mapping class group of $T^5 = T^4 \times S^1$. This leads to predictions that appear quite non-trivial from the Yang-Mills point of view.

As an example, we consider the D_{2k+1} $(2,0)$ theories. There are 6 $SL_5(\mathbb{Z})$ orbits of self-dual 't Hooft flux $f \in H^3(T^5, \mathbb{Z}_4)$. But a single orbit may be realized in different ways in the corresponding $Spin(4k+2)/\mathbb{Z}_4$ Yang-Mills theory. In this way we get alternative expressions for the generating functions

$$\mathcal{N}_f(q) = \sum_{k=0}^{\infty} N_f(D_{2k+1}) q^{4k+2}$$

of the number $N_f(D_{2k+1})$ of vacua with 't Hooft flux f. (Here q is a formal parameter.) E.g. for a certain $SL_5(\mathbb{Z})$ orbit of f, we have three alternative expressions (modulo q^{4k}-terms) for $\mathcal{N}_f(q)$:

$$\begin{aligned}
\mathcal{N}_f(q) &= \frac{1}{8}\left(P^8_{\text{even}}(q) + P^8_{\text{odd}}(q)\right) \\
&= \frac{1}{4} P^4_{\text{even}}(q) P^4_{\text{odd}}(q) \\
&= Q^4(q)\left(P^3_{\text{odd}}(q^2) P^9_{\text{even}}(q^2) + 3P^5_{\text{odd}}(q^2) P^7_{\text{even}}(q^2)\right. \\
&\quad \left. + 3P^7_{\text{odd}}(q^2) p^5_{\text{even}}(q^2) + P^9_{\text{odd}}(q^2) P^3_{\text{odd}}(q^2)\right) \\
&= q^6 + 10q^{10} + 67q^{14} + 350q^{18} + \ldots .
\end{aligned}$$

Here

$$\begin{aligned}
P_{\text{even}}(q) &= \frac{1}{2}\prod_{k=1}^{\infty}(1+q^{2k-1}) + \frac{1}{2}\prod_{k=1}^{\infty}(1-q^{2k-1}) \\
P_{\text{odd}}(q) &= \frac{1}{2}\prod_{k=1}^{\infty}(1+q^{2k-1}) - \frac{1}{2}\prod_{k=1}^{\infty}(1-q^{2k-1}) \\
Q(q) &= \prod_{k=1}^{\infty}(1+q^{2k}).
\end{aligned}$$

A promising approach to understand the complete spectrum of states is to consider the partition functions

$$Z_f = \mathrm{Tr}_{\mathcal{H}_f} \exp(-tE + ix \cdot P + iAR),$$

where $\mathcal{H}_f$ is the Hilbert space of states with self-dual 't Hooft flux $f \in H^3(T^5, C)$, and t, x, and A are some formal parameters. After continuation to Euclidean time, these partition functions can be seen as pertaining to a particular decomposition of a flat six-torus $T^6 = S^1 \times T^5$ defined by the T^5 geometry together with t and x, where the first factor denotes the 'time' direction. Twisting by R-symmetry in the spatial directions determines, together with the parameters A, a flat $sp(4)$ connection over this T^6.

The set of partition functions Z_f for $f \in H^3(T^5, C)$ should have automorphic properties under the $SL_6(\mathbb{Z})$ mapping class group of T^6. Indeed, these partition function can be regarded as components of an element Z of a certain vector space V. The space V furnishes an irreducible representation of a discrete Heisenberg algebra generated by elements Φ_v for $v \in H^3(T^6, C)$ subject to the relations

$$\Phi_v \Phi_w = \exp\left(2\pi i \int_{T^6} v \wedge w\right) \Phi_w \Phi_v.$$

To construct a basis of V, we break the covariance and decompose

$$H^3(T^6, C) \ni v = f + g \in H^3(T^5, C) \oplus H^2(T^5, C).$$

We then have a basis E_f, $f \in H^3(T^5, C)$ of V. This is such that

$$\begin{aligned} E_f &= \Phi_f E_0 \\ \Phi_g E_0 &= E_0. \end{aligned}$$

The Z_f are the components of Z relative to this basis.

E.g. under a continuous shift of the 'time' cycle of T^6 by an integer linear combination β of the spatial cycles, Z_f is multiplied by an f-dependent phase factor:

$$Z_f \mapsto Z_v \exp\left(\pi i \int_{T^5} f \wedge f[\beta]\right).$$

The Hamiltonian interpretation is that the spatial momentum $p \in H^1(T^5, \mathbb{R})$ obeys the shifted quantization law

$$p - f \cdot f \in H^1(T^5, \mathbb{Z}),$$

where $f \cdot f \in H^1(T^5, \mathbb{R}/\mathbb{Z})$. The best known example of this phenomenon is the possible non-integrality of the fifth component of p (i.e. the instanton number over T^4) for a non-trivial G/C bundle (i.e. of non-trivial magnetic 't Hooft flux m). Another example is the possible non-integrality of the four spatial components of p in situations where both the electric and magnetic 't Hooft fluxes e and m are non-trivial.

Acknowledgments

This research is supported by grants from the Swedish Research Council and the Göran Gustafsson foundation.

References

[1] M. Henningson, "The low-energy spectrum of (2,0) theory on $T^5 \times \mathbb{R}$," JHEP **0811**, 028 (2008) [arXiv:0809.4156 [hep-th]].
[2] M. Henningson, "BPS states in (2,0) theory on $\mathbb{R} \times T5$," JHEP **0903**, 021 (2009) [arXiv:0901.0785 [hep-th]].
[3] W. Nahm, "Supersymmetries and their representations," Nucl. Phys. B **135**, 149 (1978).
[4] E. Witten, "Some comments on string dynamics," arXiv:hep-th/9507121.
[5] E. Witten, "Conformal Field Theory In Four And Six Dimensions," arXiv:0712.0157 [math.RT].
[6] G. 't Hooft, "A Property Of Electric And Magnetic Flux In Nonabelian Gauge Theories," Nucl. Phys. B **153**, 141 (1979).

Journal of Physics: Conference Series **462** (2013) 012021 doi:10.1088/1742-6596/462/1/012021

Unsterile-Active Neutrino Mixing

Jimmy A. Hutasoit[1]

Department of Physics, Carnegie Mellon University, Pittsburgh, PA 15213, USA.

E-mail: `jhutasoi@andrew.cmu.edu`

Abstract. We consider a sterile neutrino to be an unparticle, namely an "unsterile" neutrino, and study its mixing with a canonical active neutrino via a see-saw mass matrix. There is no unitary transformation that diagonalizes the mixed propagator and a field redefinition is required. The unsterile-like propagating mode features a resonance for anomalous dimension between 0 and 1/3, but the complex pole disappears when the anomalous dimension is larger than 1/3. The active-like propagating mode is described by a stable pole, but inherits a non-vanishing spectral density above the unparticle threshold. We also find that the radiative decay width of the unsterile neutrino into the active neutrino (and a photon) via charged current loops is suppressed, and this suppression weakens the bound from the X-ray or soft gamma-ray background when one considers the unsterile neutrino to be a dark matter candidate.

1. Introduction

With the advancement of string theory, in particular with the advent of the much celebrated gauge/gravity duality, conformal invariance, along with supersymmetry, enjoys a lot of attention from the theoretical physics community. However, unlike supersymmetry, the pursuit of the phenomenological implications of having a conformal invariant sector in our universe had been lacking. This is not surprising, considering that in our everyday infrared life, size does matter and particles do have definite masses.

Recently, Georgi suggested an extension of the standard model in which particles couple to a *hidden* conformal sector [2]. At low energy, there emerges an effective interpolating field that features an anomalous scaling dimension. Georgi called this field the *unparticle* field as it has a phase space that looks like those of particles, but with a non-integer number of particles.

In this talk, we will consider the possibility of a fermionic unparticle that is not charged under any gauge groups of the standard model, but that is mixed with the standard model neutrinos, or the active neutrinos, via a see-saw type mass matrix. Henceforth, we will call this fermionic unparticle the *unsterile* neutrino.

2. The Set-up

Let us consider a simple set-up that consists of an unsterile and an active Dirac neutrino, mixed via a see-saw mass matrix. In momentum space, the Lagrangian density is given by

$$\mathcal{L} = \overline{\psi}_U \; (\not{p} - M) \; F(p) \; \psi_U + \overline{\nu}_a \not{p} \; \nu_a - m \; \left(\overline{\psi}_U \, \nu_a + \overline{\nu}_a \, \psi_U\right) , \qquad (1)$$

[1] This talk is based on a work with D. Boyanovsky and R. Holman [1].

where

$$F(p) = \left[\frac{-p^2 + M^2 - i\epsilon}{\Lambda^2}\right]^{-\eta}, \tag{2}$$

with $0 < \eta < 1$. Here, ψ_U and ν_a are the unsterile and the active neutrinos, respectively.

The kinetic term for the unsterile can be understood in many ways, but let us explain it in the language of RG resummation [3, 4, 5]. Let us consider the unsterile to be interacting with a field $\mathcal{A}$ of the conformal hidden sector by an interaction of the form

$$\mathcal{L}_{\text{int}} = g\,\overline{\psi}_U \mathcal{A} \psi_U, \tag{3}$$

where g is a small dimensionless coupling. To lowest order in g, the self-energy of ψ_U is given by

$$\Sigma(p) = -\eta\,(\not{p} - M)\,\ln\left[\frac{-p^2 + M^2 - i\epsilon}{\Lambda^2}\right], \tag{4}$$

where $\eta = cg^2$, with c is a constant whose value depends on the nature of the conformal field $\mathcal{A}$, and Λ is a renormalization scale. We have also subtracted the self-energy such that it vanishes at $\not{p} = M$. The Feynman diagram for the self-energy is depicted in Fig. 1.

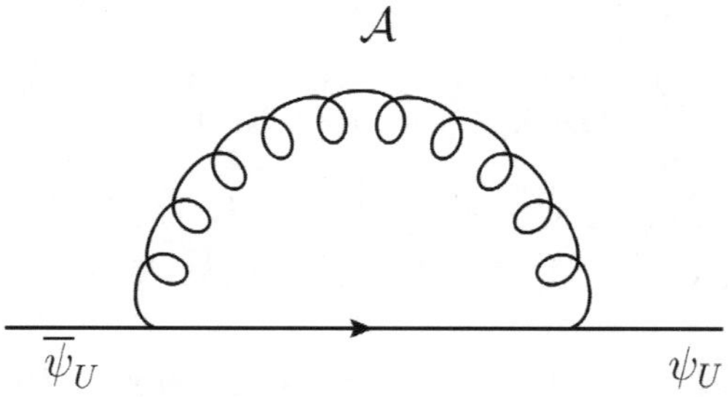

Figure 1. The leading order contribution to the self-energy.

Integrating out the conformal field $\mathcal{A}$ leads to the following effective action for ψ_U

$$\begin{aligned} \mathcal{L}_{\text{eff}} &= \overline{\psi}_U\,(\not{p} - M)\left[1 - \eta\ln\left(\frac{-p^2 + M^2 - i\epsilon}{\Lambda^2}\right)\right]\psi_U \\ &\approx \overline{\psi}_U\,(\not{p} - M)\left[\frac{-p^2 + M^2 - i\epsilon}{\Lambda^2}\right]^{-\eta}\psi_U, \end{aligned} \tag{5}$$

where in going to the last line we have invoked a renormalization group resummation of the infrared threshold divergences.

3. Diagonalization and the Results

By introducing the "flavor" doublet

$$\Psi = \begin{pmatrix} \nu_a \\ \psi_U \end{pmatrix}, \tag{6}$$

we can rewrite the Lagrangian density in a more compact form

$$\mathcal{L} = \overline{\Psi}(-p)\left[\not{p}\,\mathbb{F} - \mathbb{M}\right]\Psi(p), \tag{7}$$

where

$$\mathbb{F} = \begin{pmatrix} 1 & 0 \\ 0 & F(p) \end{pmatrix} \quad \text{and} \quad \mathbb{M} = \begin{pmatrix} 0 & m \\ m & MF(p) \end{pmatrix}. \tag{8}$$

To gain insight into how we can diagonalize the Lagrangian density, it is instructive to rewrite it in the helicity basis

$$\mathcal{L} = \sum_{h=\pm 1} \begin{pmatrix} \Psi_R^{h\,\dagger} & \Psi_L^{h\,\dagger} \end{pmatrix} \begin{pmatrix} (p^0 - h|\vec{p}|)\mathbb{F} & \mathbb{M} \\ \mathbb{M} & (p^0 + h|\vec{p}|)\mathbb{F} \end{pmatrix} \begin{pmatrix} \Psi_R^h \\ \Psi_L^h \end{pmatrix}. \tag{9}$$

As the action is manifestly Lorentz invariant, the transformation we need to perform in order to diagonalize the action must be such that the final result is still Lorentz invariant. In particular, Lorentz invariance does not allow us to mix the $(2,1)$-representation of the Lorentz group with the $(1,2)$-representation. Therefore, to diagonalize the action, we are going to perform the transformation $\Psi^h_{R,L} \to \mathcal{U}_{R,L}\,\Psi^h_{R,L}$, such that all the following matrices:

$$\mathcal{U}_R\left[(p^0 - h|\vec{p}|)\mathbb{F}\right]\mathcal{U}_R^{-1}, \quad \mathcal{U}_L\left[(p^0 + h|\vec{p}|)\mathbb{F}\right]\mathcal{U}_L^{-1}, \quad \mathcal{U}_R\,\mathbb{M}\,\mathcal{U}_L^{-1} \quad \text{and} \quad \mathcal{U}_L\,\mathbb{M}\,\mathcal{U}_R^{-1} \tag{10}$$

are all diagonal. Since $\mathbb{F}$ is diagonal, and yet not proportional to the identity, the only possible unitary transformations that diagonalize the first two are

$$\mathcal{U}_{R,L} = \begin{pmatrix} 1 & 0 \\ 0 & \pm 1 \end{pmatrix} \quad \text{or} \quad \begin{pmatrix} \pm 1 & 0 \\ 0 & 1 \end{pmatrix}. \tag{11}$$

However, none of the combinations of these possibilities diagonalize the last two matrices in 10. Therefore, there is no unitary transformation that diagonalizes the full propagator.

Instead, we are going to diagonalize the action by first rescaling the unsterile field, such that the flavor doublet is now given by

$$\nu = \begin{pmatrix} \nu_a \\ \nu_U \end{pmatrix}, \tag{12}$$

with $\nu_U = \sqrt{F(p)}\,\psi_U$, and the Lagrangian density becomes

$$\mathcal{L} = \bar{\nu}\left[\not{p}\,\mathbb{I} - \widetilde{\mathbb{M}}\right]\nu, \tag{13}$$

where

$$\widetilde{\mathbb{M}} = \frac{1}{\sqrt{\mathbb{F}}}\,\mathbb{M}\,\frac{1}{\sqrt{\mathbb{F}}} = \begin{pmatrix} 0 & \frac{m}{\sqrt{F(p)}} \\ \frac{m}{\sqrt{F(p)}} & M \end{pmatrix}, \tag{14}$$

and $\mathbb{I}$ is the identity in flavor space.

It is now pretty straightforward to diagonalize the action. The propagating or "mass" eigenstates are given by

$$\begin{pmatrix} \nu_a \\ \nu_U \end{pmatrix} = U(p)\begin{pmatrix} \nu_1 \\ \nu_2 \end{pmatrix}, \tag{15}$$

with

$$U(p) = \frac{1}{\sqrt{2}} \begin{pmatrix} \left[1+\widetilde{\mathcal{C}}(p)\right]^{\frac{1}{2}} & \left[1-\widetilde{\mathcal{C}}(p)\right]^{\frac{1}{2}} \\ -\left[1-\widetilde{\mathcal{C}}(p)\right]^{\frac{1}{2}} & \left[1+\widetilde{\mathcal{C}}(p)\right]^{\frac{1}{2}} \end{pmatrix} \quad \text{and} \quad \widetilde{\mathcal{C}}(p) = \left[1 + \frac{4m^2}{M^2 F(p)}\right]^{-\frac{1}{2}}. \tag{16}$$

The diagonalized "mass" matrix is then given by

$$\widetilde{\mathbb{M}}_d = U^{-1}(p)\, \widetilde{\mathbb{M}}\, U(p) = \begin{pmatrix} M_1(p) & 0 \\ 0 & M_2(p) \end{pmatrix}, \tag{17}$$

with

$$M_{1,2}(p) = \frac{M}{2}\left(1 - \sqrt{1 + \frac{4m^2}{M^2 F(p)}}\right). \tag{18}$$

The active-like "mass" eigenstate has an isolated pole below the unparticle threshold $p^2 = M^2$. This pole lies on the real axis and describes a massive and stable propagating mode. Near this pole, the propagator behaves like

$$\frac{1}{p^2 - M_1^2(p)} \approx \frac{Z_1}{p^2 - \mathcal{M}_1^2}, \quad \text{with} \quad \mathcal{M}_1^2 = \frac{m^4}{M^2}\left[\frac{M^2}{\Lambda^2}\right]^{2\eta} \quad \text{and} \quad Z_1^{-1} \approx 1 + 2\eta\, \frac{\mathcal{M}_1^2}{M^2}. \tag{19}$$

The active-like propagator also features an inherited spectral density

$$\rho_1(x) = \frac{\Theta(x)}{\pi}\, \frac{\frac{\Delta^2}{4}\, x^{2\eta} \sin(2\pi\eta)}{\left[x + 1 - \frac{\Delta^2}{4}\, x^{2\eta}\cos(2\pi\eta)\right]^2 + \left[\frac{\Delta^2}{4}\, x^{2\eta}\sin(2\pi\eta)\right]^2}, \tag{20}$$

where

$$x = \frac{p^2 - M^2}{M^2} \quad \text{and} \quad \Delta = 2\, \frac{m^2}{M^2}\left[\frac{M^2}{\Lambda^2}\right]^{\eta}. \tag{21}$$

This spectral density is depicted in Fig. 2. It vanishes at threshold $p^2 = M^2$, increases rapidly reaching a broad maximum and diminishes for increasing x.

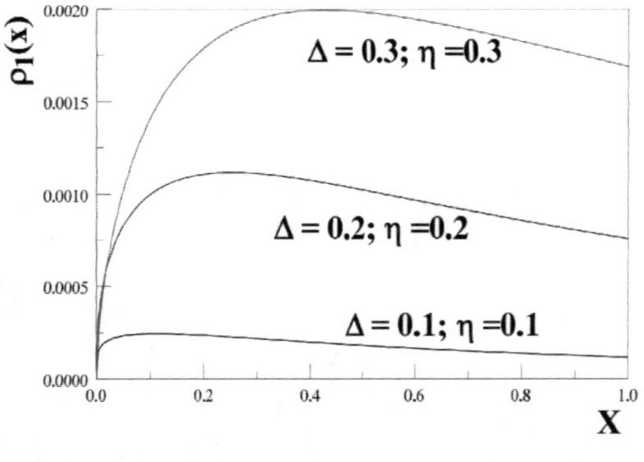

Figure 2. Spectral density for the active-like mode.

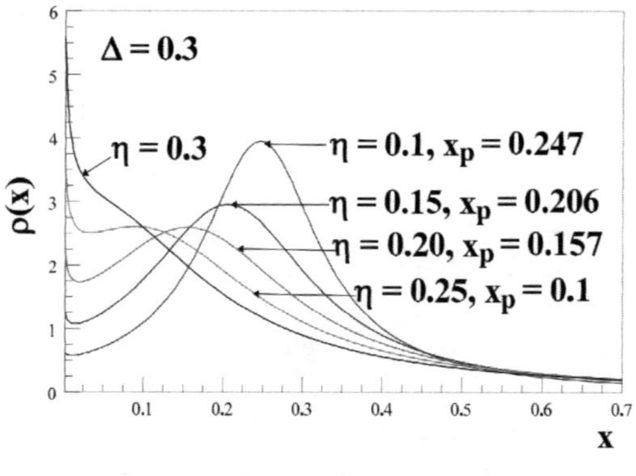

Figure 3. Spectral density for the unsterile-like mode.

The dispersion relation for the unsterile-like "mass" eigenstate has a solution only when $\mathrm{Re}(x) > 0$ and $0 \leq \eta < 1/3$. It describes a pole in the complex plane. Near this pole, the propagator behaves like

$$\frac{1}{p^2 - M_2^2(p)} \approx \frac{Z_2}{p^2 - \mathcal{M}_2^2 + i\mathcal{M}_2\Gamma}, \tag{22}$$

where

$$\mathcal{M}_2^2 = M^2\left[1 + \Delta^{\frac{1}{1-\eta}}\cos\left(\frac{\pi\eta}{1-\eta}\right)\right] \quad \text{and} \quad \Gamma = \frac{M^2}{\mathcal{M}_2}\Delta^{\frac{1}{1-\eta}}\sin\left(\frac{\pi\eta}{1-\eta}\right). \tag{23}$$

The spectral density for the unsterile-like mode is given by

$$\rho_2(x) = \frac{\Theta(x)}{\pi}\frac{\Delta\, x^\eta \sin(\pi\eta)}{\left[x - \Delta\, x^\eta\cos(\pi\eta)\right]^2 + \left[\Delta\, x^\eta \sin(\pi\eta)\right]^2}, \tag{24}$$

which is displayed in Fig. 3. For $0 \leq \eta < 1/3$, there is a resonance with a maximum confirmed to be given by

$$x_p = \Delta^{\frac{1}{1-\eta}}\cos\left(\frac{\pi\eta}{1-\eta}\right), \tag{25}$$

as obtained in Eq. 23.

In order to understand the significance of the imaginary part of the unsterile-like pole, let us revisit the RG resummation argument we used to derive the kinetic term of the unsterile field. To do that let us add

$$\mathcal{L}_\mathrm{m} = -\overline{\psi}_U\, m\, \nu_a + \mathrm{h.c.} \tag{26}$$

to Lagrangian density 3. It is straightforward to diagonalize this action. The mass eigenstates are given by

$$\begin{aligned} \psi_1 &= \nu_a \cos\theta_0 + \psi_U \sin\theta_0, && \text{with} && M_1 \approx m^2/M, \\ \psi_2 &= -\nu_a \sin\theta_0 + \psi_U\cos\theta_0, && \text{with} && M_2 \approx M + m^2/M, \end{aligned} \tag{27}$$

where

$$\cos\theta_0 \approx 1 \quad \text{and} \quad \sin\theta_0 \approx m/M. \tag{28}$$

In terms of these mass eigenstates, the interaction between the unsterile and the "hidden" conformal sector can be rewritten as

$$\mathcal{L}_\mathrm{int} = g\left(\overline{\psi}_2\cos\theta_0 - \overline{\psi}_1\sin\theta_0\right)\mathcal{A}\left(\psi_2\cos\theta_0 - \psi_1\sin\theta_0\right). \tag{29}$$

We can then obtain the decay rate for the process $\psi_2 \to \mathcal{A}\,\psi_1$ by applying Cutkosky cut to the self-energy diagram for the mass eigenstate ψ_2 as depicted in Fig. 4. It is given by

$$\Gamma_{\psi_2\to\mathcal{A}\psi_1} = 2\pi\eta\, M_2\sin^2\theta_0\cos^2\theta_0 \approx \pi\eta\frac{2m^2}{M}. \tag{30}$$

To lowest order in η and m/M, this result coincides with the non-perturbative imaginary part of the unsterile-like pole. Therefore, we see that this imaginary part Γ describes the *decay* of the unsterile-like mode into the active-like mode and fields in the "hidden" conformal sector.

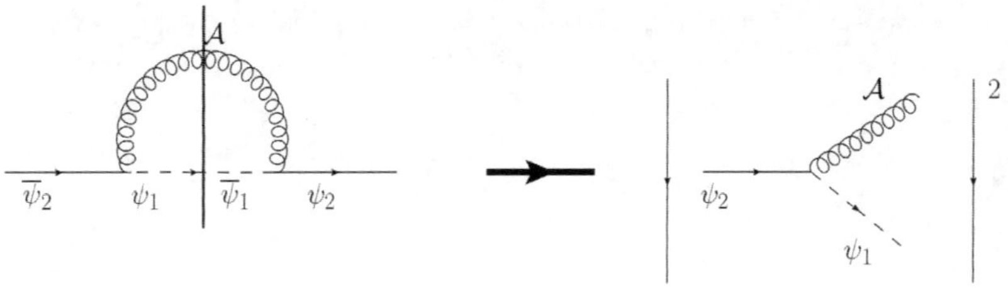

Figure 4. Cutkosky cut for the self-energy of mass eigenstate ψ_2.

4. A Warm Dark Matter Candidate

A motivation to consider sterile neutrino is that it is a potential warm dark matter candidate and could provide possible solutions to a host of astrophysical problems [6]. However, the radiative decay of a sterile-like neutrino mass eigenstate into an active-like mass eigenstate and a photon leads to a decay line that could be observable in the X-ray or soft gamma ray background. Thus, the non-observation of this line provides a constraint on the mass and mixing angle of sterile-like neutrinos.

Let us first consider the (canonical) sterile-active neutrino mixing

$$\begin{aligned} \nu_a &= \nu_1 \cos\theta_0 + \nu_2 \sin\theta_0, \\ \nu_s &= -\nu_1 \sin\theta_0 + \nu_2 \cos\theta_0, \end{aligned} \tag{31}$$

where ν_s is the canonical sterile neutrino. The charged current interaction then yields an interaction between the sterile-like neutrino and the charged lepton as follows

$$\mathcal{L}_{CC} = g\,\overline{\nu}_{aL}\,\not{W}\,l_L = g\left(\overline{\nu}_{1L}\,\cos\theta_0 + \overline{\nu}_{2L}\,\sin\theta_0\right)\,\not{W}\,l_L. \tag{32}$$

This interaction vertex leads to the radiative decay of the sterile-like neutrino $\nu_2 \to \nu_1\,\gamma$ and the diagrams that describe this process in unitary gauge are shown in Fig. 5. The radiative decay width is given by

$$\Gamma_{\nu_2\to\nu_1\gamma} \approx \frac{\alpha_{em}}{2}\left[\frac{3G_F}{32\pi^2}\right]^2 M_2^5\left[\frac{m_l}{M_W}\right]^4 \sin^2\theta_0\,\cos^2\theta_0. \tag{33}$$

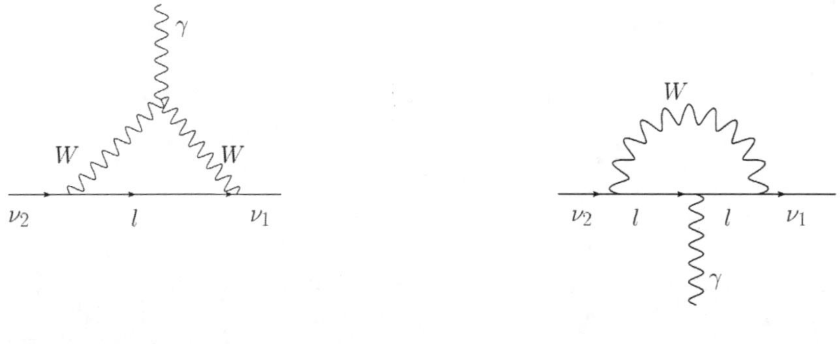

Figure 5. Contributions to radiative decay of a sterile-like neutrino in unitary gauge.

What about the radiative decay of the unsterile-like neutrino? In order to obtain the decay width for the unsterile case, we have to include the wave function renormalization $Z_{1,2}$. Furthermore, the mixing angles are accounted for by the following replacements

$$\begin{aligned} \cos^2(\theta_0) &\rightarrow \frac{1}{2}\left[1+\widetilde{\mathcal{C}}(p)\right]_{p^2=M_1^2}, \\ \sin^2(\theta_0) &\rightarrow \frac{1}{2}\left|1-\widetilde{\mathcal{C}}(p)\right|_{p^2=M_2^2}. \end{aligned} \tag{34}$$

The result is

$$\frac{\Gamma^U_{\nu_2\rightarrow\nu_1\gamma}}{\Gamma_{\nu_2\rightarrow\nu_1\gamma}} = \frac{\Delta^{\frac{\eta}{1-\eta}}\left[\frac{M^2}{\Lambda^2}\right]^\eta}{(1-\eta)\left(1+\eta\frac{\Delta^2}{2}\right)} \approx \frac{\left[2\,\frac{m^2}{\Lambda^2}\right]^{\frac{\eta}{1-\eta}}}{(1-\eta)\left(1+\eta\frac{\Delta^2}{2}\right)}. \tag{35}$$

We see that the unparticle nature of the sterile neutrino can lead to a *substantial suppression* of the radiative decay rate. As an example, taking $m/M \sim 10^{-5}$, $M \sim$ keV and $\Lambda \sim$ TeV, which are within the range of expectation for physics beyond the standard model, and taking $\eta \sim 0.1$, we find that the ratio 35 $\lesssim \mathcal{O}(10^{-3})$.

5. Summary and Outlook

In this talk, we considered the possibility that the $SU(2)$ singlet sterile neutrino might be an unparticle, an interpolating field whose correlation function feature an anomalous scaling dimension η as a consequence of coupling to a "hidden" conformal sector. We studied the consequences of its mixing with an active neutrino via a see-saw mass matrix by focusing on the simplest setting of one unsterile and one active Dirac neutrino. We found that there is no unitary transformation that diagonalizes the full propagator due to the non-canonical nature of the unsterile neutrino. This forces us to make a field redefinition for its complete diagonalization.

The active-like propagating mode corresponds to a stable particle, but inherits a non-trivial spectral density even in the absence of standard model interactions. The unsterile-like propagating mode is described by a complex pole above the unparticle threshold for $0 \leq \eta < 1/3$, featuring an "invisible width" that is a result from the decay of the unsterile-like mode into an active-like mode and fields in the conformal sector.

We found that there is a substantial suppression of the radiative decay line width, resulting in a weakening of the bounds from the X-ray and soft gamma ray backgrounds. This renders the unsterile neutrino as a good potential warm dark matter candidate. To further explore the possibility of an unsterile neutrino as a dark matter candidate, understanding its production process is necessary. Since an unsterile neutrino only interacts directly with the active one, the most effective dark matter production mechanism in this scenario is via unsterile-active neutrino oscillations. It would be interesting to study the implications of our results in the dark matter production mechanism along the line of Ref. [7].

We can also add another active neutrino into the mix, which results in the so-called $2+\tilde{1}$ model [8], and study the dynamics of the active-active neutrino oscillations. The complications that come from the momentum-dependent mixing angles and non-trivial spectral densities render a full quantum field theoretical treatment of this model necessary. A work studying the active-active neutrino oscillations dynamics in the $2+\tilde{1}$ scheme using the real time quantum field theoretical formalism as introduced in [9], in particular Section IV, is in progress. Perhaps the results will lead to the reconciliation of the LSND and MiniBooNE results.

References

[1] D. Boyanovsky, R. Holman and J. A. Hutasoit, "Unsterile-Active Neutrino Mixing: Consequences on Radiative Decay and Bounds from the X-ray Background," Phys. Rev. D **80**, 025012 (2009) [arXiv:0905.4729 [hep-ph]].

[2] H. Georgi, "Unparticle Physics," Phys. Rev. Lett. **98**, 221601 (2007) [arXiv:hep-ph/0703260].
[3] N.N.Bogoliubov, D.V.Shirkov, *Introduction to the theory of quantized fields*, (Interscience Publishers, N.Y. 1959).
[4] D. Boyanovsky and H. J. de Vega, "Dynamical renormalization group approach to relaxation in quantum field theory," Annals Phys. **307**, 335 (2003) [arXiv:hep-ph/0302055].
[5] M. Neubert, "Unparticle Physics with Jets," Phys. Lett. B **660**, 592 (2008) [arXiv:0708.0036 [hep-ph]].
[6] A. Kusenko, "Sterile neutrinos," AIP Conf. Proc. **917**, 58 (2007) [arXiv:hep-ph/0703116].
[7] S. Dodelson and L. M. Widrow, "Sterile Neutrinos as Dark Matter," Phys. Rev. Lett. **72**, 17 (1994) [arXiv:hep-ph/9303287].
[8] D. Boyanovsky, R. Holman and J. A. Hutasoit, "Oscillation dynamics of active-unsterile neutrino mixing in a $2+\tilde{1}$ mixing scheme," Phys. Rev. D **81**, 033009 (2010) [arXiv:0912.2093 [hep-ph]].
[9] J. Wu, J. A. Hutasoit, D. Boyanovsky and R. Holman, "Neutrino oscillations, entanglement and coherence: a quantum field theory study in real time," arXiv:1002.2649 [hep-ph].

Gauge theory duals of black hole - black string transitions of gravitational theories on a circle

Simon Catterall[1], **Anosh Joseph**[12] **and Toby Wiseman**[3]

[1] Department of Physics, Syracuse University, Syracuse, New York, 13244 USA
[3] Theoretical Physics Group, Blackett Laboratory, Imperial College, London SW7 2AZ, UK

E-mail: smc@physics.syr.edu, ajoseph@physics.syr.edu, t.wiseman@imperial.ac.uk

Abstract. We study the black hole - black string phase transitions of gravitational theories compactified on a circle using the holographic duality conjecture. The gauge theory duals of these theories are maximally supersymmetric and strongly coupled $1+1$ dimensional $SU(N)$ Yang-Mills theories compactified on a circle, in the large N limit. We perform the strongly coupled finite temperature gauge theory calculations on a lattice, using the recently developed exact lattice supersymmetry methods based on topological twisting and orbifolding. The spatial Polyakov line serves as relevant order parameter of the confinement - deconfinement phase transitions in the gauge theory duals.

1. Introduction

The rise of string theory as a candidate for a unified quantum theory including gravity has generated a resurgence of interest in black holes. In an appropriate limit, string theory reduces to supergravity. There are different types of black hole solution possible in supergravity and more interestingly, there are phase transitions possible between these black solutions. Examples of such transitions include the Hawking - Page transition [1] in AdS spaces and the Gregory - Laflamme transition [2] in $D1$-brane systems compactified on a circle [3]. These phase transitions are very difficult to tackle analytically in the gravitational theories. However, they have dual descriptions as thermal phase transitions in strongly coupled gauge theories. It is possible to analyze these phase transitions by studying the phase structure of their strongly coupled dual gauge theories. The dynamics of strongly coupled dual gauge theories can be studied on a lattice, using the recently developed exact lattice supersymmetry methods based on topological twisting and orbifolding [4].

The holographic duality conjecture in string theory, in an appropriate limit, relates the finite temperature low energy supergravity limit of string theory to strongly coupled finite temperature supersymmetric Yang-Mills theory. For a large number N of coincident Dp-branes in the "decoupling" limit, the dual gauge theory is $(1+p)$-dimensional, strongly coupled, maximally supersymmetric $SU(N)$ Yang-Mills theory taken at large N.

The decoupling limit is described as follows. String theory contains D-branes, which are solitonic objects. When we consider the full theory in the presence of these solitons we have modes that propagate in the bulk and modes that propagate on the solitons. The modes on the

[2] Speaker.

soliton interact with each other and with the bulk modes. It is possible, however, to define a limit of the full theory in which the bulk modes decouple from the modes living on the D-brane. This is typically a low energy limit, in which we tune the coupling constant so as to keep only the interactions among the modes living on the D-brane. In this limit the D-brane theory becomes super Yang-Mills theory (for the case $p \leq 3$).

The simplest case corresponds to $D0$-branes, where the dual theory is a finite temperature 16 supercharge Yang-Mills quantum mechanics with gauge group $SU(N)$ [5]. (See [6] for recent lattice studies of super Yang Mills quantum mechanics.) When $p = 3$, we have the case of $D3$-branes and the original AdS/CFT correspondence [7].

In this talk, we focus on the gauge theory duals of black hole - black string phase transitions due to Gregory - Laflamme instabilities developed in a system of N black $D1$-branes compactified on a circle. In the gauge theory dual interpretation, these transitions correspond to thermal phase transitions in $1+1$ dimensional maximally supersymmetric $SU(N)$ Yang-Mills theory on a circle at strong coupling [8].

2. Two dimensional super Yang-Mills and $D1$-branes

2.1. Maximally supersymmetric Yang-Mills on T^2

We consider the $1+1$ dimensional 16 supercharge strongly coupled SYM theory with gauge group $SU(N)$. It can be obtained from the dimensional reduction of $\mathcal{N} = 4$, $d = 4$ SYM theory into two dimensions. The gauge theory is at strong coupling $\lambda = g_{YM}^2 N \gg 1$, where λ is the 't Hooft coupling, g_{YM} is the Yang-Mills coupling and N is the number of colors.

It is natural to consider the case where the spatial dimension is compactified on a circle with circumference L. Since we are interested in studying the thermodynamics of this gauge theory on a lattice, we consider the Euclidean version of the theory. To study the field theory at a finite temperature we compactify the temporal direction on a circle with period β. That is $t \sim t + \beta$, where $\beta = 1/T$ is the inverse temperature. Thus the two dimensional Euclidean $SU(N)$ Yang-Mills theory is compactified on a torus T^2.

Euclidean $1+1$ dimensional $SU(N)$ SYM on a torus has two non-contractible Wilson loops (Polyakov lines), namely the Wilson loops running around the temporal (thermal) circle (P_t) and the spatial circle (P_x). The fermions are anti-periodic on the thermal circle and we impose periodic boundary conditions on the spatial circle. In this way, we can distinguish between spatial and temporal directions of the compactified theory.

The compactified theory on T^2 exhibits phase transitions as we change the temperature. The expectation value of the spatial Polyakov loop

$$P_x = \left\langle \mathrm{Tr}(P \exp(i \oint A_x)) \right\rangle , \tag{1}$$

where A_x is the gauge field along x-direction, in the Euclidean theory serves as order parameter for this phase transition.

The expectation value of P_x changes from zero to non-zero values as we increase the temperature. The expectation value of P_t is always non-zero at all temperatures. Thus upon heating, the SYM theory undergoes a transition from a "confined phase" ($P_x = 0$) to a "deconfined phase" ($P_x \neq 0$).

2.2. The gravitational theory on a circle

The dual gravitational theory corresponding to the $(1+1)$ dimensional maximally supersymmetric $SU(N)$ Yang-Mills theory describes the near-horizon geometry of the $D1$-brane metric of type IIB string theory [5].

The string theory contains a collection of N $D1$-branes. The decoupling limit considers keeping finite energy excitations of the $D1$-branes while taking the limit, $g_{YM}^2 = \frac{1}{2\pi}\frac{g_s}{\alpha'} =$ fixed, $\alpha' \to 0$, where g_s is the string coupling and α' is the slope parameter.

Since the dual gauge theory is at finite temperature T, we are interested in near extremal black $D1$-brane configurations of the gravitational theory in the above limit. We have compactified the spatial direction of the gauge theory on a circle of circumference L. This implies that we should impose a periodic identification of the spatial coordinate in the dual gravitational theory. Thus we have a space that is asymptotically $R^{8,1} \times S^1$.

We may write a IIB supergravity approximation for the string frame metric and dilaton (for a charged black string winding around S^1) as

$$ds^2 = \alpha'\Big\{\frac{u^3}{\sqrt{d_1\lambda'}}\Big[-\Big(1-\frac{u_0^6}{u^6}\Big)\frac{d\tau^2}{L^2}+\frac{d\theta^2}{(2\pi)^2}\Big]+\frac{\sqrt{d_1\lambda'}}{u^3\Big(1-\frac{u_0^6}{u^6}\Big)}du^2+u^{-1}\sqrt{d_1\lambda'}d\Omega_7^2\Big\}\,, \quad (2)$$

$$e^{\phi} = 2\pi\frac{\lambda'}{N}\sqrt{\frac{d_1\lambda'}{u^6}}\,, \quad (3)$$

where $\lambda' = \lambda L^2$, $d_1 = 2^6\pi^3$, and θ is the angular periodic coordinate describing the spatial S^1, so that $\theta \sim \theta + 2\pi$. From this supergravity solution we may derive the thermodynamics of the black hole from the region of the solution near the horizon using standard methods. One can compute that the constant of integration u_0 is related to the temperature of the solution as $u_0^2 = \frac{16\pi^{5/2}}{3}t\sqrt{\lambda'}$, where we have defined the natural dimensionless temperature $t = TL$. If one continues time to obtain a Euclidean solution, then there is a conformal boundary which is a torus T^2 whose cycles derive from the spatial circle and the Euclidean time circle [8].

Analysis of α' corrections implies that this supergravity solution near its horizon is a good approximation to the actual IIB string background provided that $t << \sqrt{\lambda'}$. The compact spatial circle means one must also ensure that winding string modes are not important corrections to the supergravity near the horizon, and this implies $1/\sqrt{\lambda'} << t$. Thus for large $\lambda' >> 1$ this supergravity solution correctly approximates the horizon region of the true string background for a range of temperatures. This solution is believed to be perturbatively stable within the supergravity approximation.

Below dimensionless temperatures $t \sim 1/\sqrt{\lambda'}$, the above background is strongly effected by string winding modes near the horizon which wrap over the spatial circle. We then cannot analyse the IIB string theory behaviour simply using IIB supergravity. However, we may employ the trick of T-duality, acting on the spatial circle, to map the IIB solution to one in IIA theory which in fact can be described in IIA supergravity. One finds the following IIA solution after T-duality,

$$ds^2 = \alpha'\Big\{-\frac{u^3}{\sqrt{d_1\lambda'}}\Big(1-\frac{u_0^6}{u^6}\Big)\frac{d\tau^2}{L^2}+\frac{\sqrt{d_1\lambda'}}{u^3}\left[\frac{du^2}{\Big(1-\frac{u_0^6}{u^6}\Big)}+(2\pi)^2d\tilde{\theta}^2\right]+u^{-1}\sqrt{d_1\lambda'}d\Omega_7^2\Big\}\,, \quad (4)$$

$$e^{\phi} = (2\pi)^2\frac{\lambda'}{N}\left(\frac{d_1\lambda'}{u^6}\right)^{3/4}\,, \quad (5)$$

now with $\tilde{\theta}$ as the angular coordinate on the (T-dual) spatial S^1, again with $\tilde{\theta} \sim \tilde{\theta} + 2\pi$. The requirement that the supergravity solution and its perturbations are good, gives the condition $\lambda' >> 1$ and $t << 1$ from requiring α' and winding corrections to be small near the horizon.

A gravity analysis now shows that in the T-dual IIA supergravity picture the solution above is perturbatively unstable to a gravity mode that breaks the translational invariance along the $\tilde{\theta}$

circle direction, and is localized near the horizon. This is an instability of the Gregory-Laflamme type. The analysis shows that it occurs below a temperature,

$$t_{GL} = \frac{3}{4\sqrt{\pi}} \frac{(2\pi a)^2}{\sqrt{\lambda'}} \tag{6}$$

where a is a constant determined numerically to be $a \approx 0.37$. We note that this is in the regime of validity of the IIA supergravity solution provided $\lambda' >> 1$.

In the T-dual IIA picture which allows us to describe the $D1$-branes using supergravity we see the instability as a perturbative graviton mode that occurs for $\lambda' >> 1$ at low temperatures t. Under T-duality the $D1$-branes wrapped on the S^1 become a density of $D0$-branes 'smeared' over the S^1. Physically this configuration becomes unstable to clumping of the $D0$ branes at low temperature. However, the interpretation in terms of the original $D1$-branes with $\lambda' >> 1$ is as a stringy winding mode instability about the spatial S^1, which occurs for $t \sim 1/\sqrt{\lambda'}$ (and hence can't be described using the IIB gravity solution) and is associated with symmetry breaking on the circle.

From a thermal Euclidean perspective the AdS/CFT correspondence therefore maps (the near-horizon limit of) near extremal solutions of the full type IIB string theory compactified on a circle to thermal phases of maximally supersymmetric Yang-Mills theory on $S^1 \times S^1$. In the gauge theory the natural interpretation of the above stringy instability is a transition from a "confining phase" ($P_x = 0$) to a "deconfining phase" ($P_x \neq 0$) as the dimensionless temperature t varies. The expectation value of P_t is always non-vanishing at all temperatures indicating the presence of event horizon in the dual gravitational theory, or from a Euclidean geometric perspective, the contractibility of the Euclidean time circle.

3. Strongly coupled $2D$ super Yang-Mills on a lattice

In recent years a new approach on how to put supersymmetric field theories on a lattice has been developed using ideas drawn from topological twisting [9] and orbifolding [10]. This approach indicates that a class of supersymmetric theories may be discretized by preserving just a fraction of the original supersymmetry. (See [4] for a review.)

The approach mentioned above, exposes a nilpotent supersymmetry. This supersymmetry say, $\mathcal{Q}$ is a Lorentz scalar and it does not produce translations on the lattice. The nilpotent property allows us to rewrite the two-dimensional super Yang-Mills action as a $\mathcal{Q}$-exact term. This action can then easily be discretized on a lattice.

3.1. Topological twisting

In this talk we discuss the results derived from the lattice simulations of topologically twisted two-dimensional (Euclidean) super Yang-Mills theory with gauge group $SU(N)$. The fields are all in the adjoint representation with an anti-hermitian basis for the generators. The two-dimensional 16 supercharge theory can be obtained from dimensional reduction of the four-dimensional $\mathcal{N} = 4$ SYM down to two dimensions.

We rewrite the original action of the four-dimensional theory in terms of "twisted fields." The basic idea of twisting is not new, it goes back to Witten in his seminal paper on topological field theory [11]. In the case of four dimensional $\mathcal{N} = 4$ SYM, the process of twisting is a decomposition of the fields in terms of representations of a twisted rotation group, which is a diagonal subgroup of the original (Euclidean) rotational symmetry $SO_{\rm Lorentz}(4)$ and an $SO_{\rm R}(4)$ subgroup of the R-symmetry of the theory. (The four-dimensional theory has $SO_{\rm R}(6)$ R-symmetry.) The twisted rotation group is:

$$SO(4)' = {\rm diag}\Big(SO_{\rm Lorentz}(4) \times SO_{\rm R}(4)\Big) . \tag{7}$$

We can treat the supercharges of the twisted four-dimensional $\mathcal{N}=4$ theory as a 4×4 matrix q. This matrix can be expanded on products of gamma matrices

$$q = \mathcal{Q}I + \mathcal{Q}_\mu\gamma_\mu + \mathcal{Q}_{\mu\nu}\gamma_\mu\gamma_\nu + \ldots \tag{8}$$

After the twist the original supersymmetry algebra becomes a twisted algebra

$$\mathcal{Q}^2 = 0 \tag{9}$$
$$\{\mathcal{Q}, \mathcal{Q}_\mu\} = p_\mu \tag{10}$$
$$\cdots \tag{11}$$

The first piece of the algebra shows that the nilpotent scalar supercharge $\mathcal{Q}$ does not produce translations on the lattice. That is, we can easily write down the twisted theory on a lattice. The second piece of the algebra expresses the fact that the momentum is the $\mathcal{Q}$-variation of something, which makes plausible the statement that the energy-momentum tensor and hence the entire action can be written in $\mathcal{Q}$-exact form. An action written in such a $\mathcal{Q}$-exact form is trivially invariant under the scalar supersymmetry provided the latter remains nilpotent under discretization.

The rewriting of the supercharges in terms of twisted variables can be repeated for the fermions of the theory and yields a set of p-forms $(\eta, \psi_\mu, \chi_{\mu\nu}, \ldots)$ which for the case of $Q=16$ matches the number of components of a real Kähler-Dirac field. This repackaging of the fermions of the theory into a Kähler-Dirac field is at the heart of how the discrete theory avoids fermion doubling as was shown by Becher, Joos and Rabin in the early days of lattice gauge theory [12, 13].

The twisted four-dimensional $\mathcal{N}=4$ theory can be written in a compact form involving five-dimensional fields [4]

$$S = \frac{1}{g^2}\int \mathrm{Tr}\,\mathcal{Q}\left(\chi_{ab}\mathcal{F}_{ab} + \eta[\overline{\mathcal{D}}_a, \mathcal{D}_a] - \frac{1}{2}\eta d\right) - \frac{1}{8g^2}\int \mathrm{Tr}\,\epsilon_{abcde}\chi_{de}\overline{\mathcal{D}}_c\chi_{ab}\,. \tag{12}$$

The fields are now in $SU(5)$ representations with the Roman indices a, b running from $1, \cdots, 5$. The sixteen fermionic degrees of freedom are packed in the $SU(5)$ decomposition $(1\oplus 5\oplus \overline{10}) \to (\eta, \psi_a, \chi_{ab})$. The ten bosonic degrees of freedom are encoded in a complex gauge fields $\mathcal{A}_a$ and $\overline{\mathcal{A}}_a$, decomposing as $5\oplus\overline{5}$ under $SU(5)$. We have $\mathcal{A}_a = A_a + iB_a$ with corresponding complexified field strength $\mathcal{F}_{ab}$. The field d is an auxiliary field introduced for the off-shell completion of the twisted algebra.

The nilpotent transformations associated with $\mathcal{Q}$ are given explicitly by

$$\begin{aligned} &\mathcal{Q}\,\mathcal{A}_a = \psi_a, && \mathcal{Q}\,\chi_{ab} = -\overline{\mathcal{F}}_{ab}, \\ &\mathcal{Q}\,\psi_a = 0, && \mathcal{Q}\,\eta = d, \\ &\mathcal{Q}\,\overline{\mathcal{A}}_a = 0, && \mathcal{Q}\,d = 0. \end{aligned}$$

The action (12), after $\mathcal{Q}$-variation, integrating out the auxiliary field d and dimensionally reducing along the 5th direction gives the four-dimensional action of the twisted $\mathcal{N}=4$ theory

$$\begin{aligned} S = {}& \frac{1}{g^2}\int \mathrm{Tr}\left(\overline{\mathcal{F}}_{\mu\nu}\mathcal{F}_{\mu\nu} + \frac{1}{2}[\overline{\mathcal{D}}_\mu, \mathcal{D}_\mu]^2 + \frac{1}{2}[\overline{\phi}, \phi]^2 + (\mathcal{D}_\mu\phi)^\dagger(\mathcal{D}_\mu\phi) - \chi_{\mu\nu}\mathcal{D}_{[\mu}\psi_{\nu]}\right. \\ & \left. -\overline{\psi}_\mu\mathcal{D}_\mu\overline{\eta} - \overline{\psi}[\phi, \psi_\mu] - \eta\overline{\mathcal{D}}_\mu\psi_\mu - \eta[\overline{\phi}, \overline{\eta}] - \chi^*_{\mu\nu}\overline{\mathcal{D}}_\mu\overline{\psi}_\nu - \chi^*_{\mu\nu}[\overline{\phi}, \chi_{\mu\nu}]\right), \end{aligned} \tag{13}$$

where the Greek indices run from $1, \cdots, 4$; the last two terms arise from the dimensional reduction of the $\mathcal{Q}$-closed term; and χ^* is the Hodge dual of χ: $\chi_{\mu\nu} = \frac{1}{2}\epsilon_{\mu\nu\rho\lambda}\chi_{\rho\lambda}$. (See [4] for more details.)

This theory can now be further dimensionally reduced to two dimensions.

3.2. Discretization of the twisted theory

We follow the geometric discretization scheme to write down the lattice version of the theory. The complex gauge fields are represented as complexified Wilson gauge links $\mathcal{U}_\mu(x) = e^{\mathcal{A}_\mu}(x)$ living on links of a lattice which for the moment we can think of a square lattice. These link fields transform in the usual way under $U(N)$ lattice gauge transformations

$$\mathcal{U}_\mu(x) \to G(x)\mathcal{U}_\mu(x)G^\dagger(x) \ . \tag{14}$$

Supersymmetric invariance then implies that $\psi_\mu(x)$ live on the same links and transform identically. The scalar fermion $\eta(x)$ transforms as a site field: $\eta(x) \to G(x)\eta(x)G^\dagger(x)$. The field $\chi_{\mu\nu}$ is a 2-form and we choose it to lie along a diagonal link.

The continuum derivatives are replaced by difference operators on the lattice. We use $\mathcal{D}_\mu^{(+)}$ for curl-like operations and $\mathcal{D}_\mu^{(-)}$ for divergence-like operations. (For more details on the lattice implementation we refer the reader to [4].)

4. Simulation results

We have simulated the two-dimensional thermal theory in the range of inverse temperature $0.02 < \beta < 1.0$. We have focused on two observables, the spatial and temporal Polyakov lines. For a lattice of linear size L these are given respectively by

$$P_x = \frac{1}{N}\left\langle \left| \operatorname{Tr} \Pi_{a_x=0}^{L-1} U_{a_x} \right| \right\rangle, \quad P_t = \frac{1}{N}\left\langle \left| \operatorname{Tr} \Pi_{a_t=0}^{L-1} U_{a_t} \right| \right\rangle \ . \tag{15}$$

We have performed the simulations on a 2×8 lattice for the values of color $N = 2, 3$ and 4, with values of infrared regulator $\mathrm{m} = 0.05$, 0.10 and 0.20. (To regulate the divergence with Monte Carlo simulations we have introduced a scalar mass term m.) The spatial Polyakov line undergoes a sudden transition from non-zero to zero values around $\beta \approx 0.2$ indicating that there is a phase transition in the gravitational theory. The temporal Polyakov loop does not undergo a transition indicating the presence of event horizons in the black solutions of the gravitational theory. We summarize the results in figures 1 and 2. More detailed results are to appear in [14].

4.1. Acknowledgments

SC and AJ are supported in part by the US Department of Energy under grant DE-FG02-85ER40237. TW is supported by a STFC advanced fellowship and a Halliday award. Simulations were performed using USQCD resources at Fermilab.

References

[1] S. W. Hawking and D. N. Page, "Thermodynamics Of Black Holes In Anti-De Sitter Space," Commun. Math. Phys. **87**, 577 (1983); E. Witten, "Anti-de Sitter space, thermal phase transition, and confinement in gauge theories," Adv. Theor. Math. Phys. **2**, 505 (1998) [arXiv:hep-th/9803131].

[2] R. Gregory and R. Laflamme, "Black strings and p-branes are unstable," Phys. Rev. Lett. **70**, 2837 (1993) [arXiv:hep-th/9301052]; R. Gregory and R. Laflamme, "The Instability of charged black strings and p-branes," Nucl. Phys. B **428**, 399 (1994) [arXiv:hep-th/9404071]; R. Gregory and R. Laflamme, "Evidence For Stability Of Extremal Black P-Branes," Phys. Rev. D **51**, 305 (1995) [arXiv:hep-th/9410050].

[3] M. Li, E. J. Martinec and V. Sahakian, "Black holes and the SYM phase diagram," Phys. Rev. D **59**, 044035 (1999) [arXiv:hep-th/9809061]; E. J. Martinec and V. Sahakian, "Black holes and the SYM phase diagram. II," Phys. Rev. D **59**, 124005 (1999) [arXiv:hep-th/9810224].

[4] S. Catterall, D. B. Kaplan and M. Unsal, "Exact lattice supersymmetry," Phys. Rept. **484**, 71 (2009) [arXiv:0903.4881 [hep-lat]]..

[5] N. Itzhaki, J. M. Maldacena, J. Sonnenschein and S. Yankielowicz, "Supergravity and the large N limit of theories with sixteen supercharges," Phys. Rev. D **58**, 046004 (1998) [arXiv:hep-th/9802042].

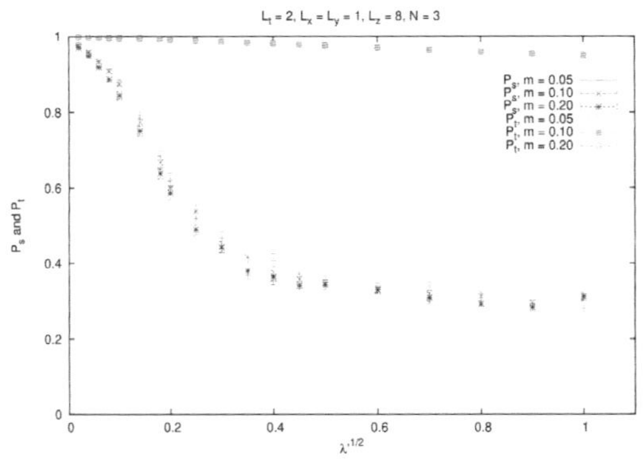

Figure 1. Spatial and temporal Polyakov lines (P_s and P_t) vs square root of the coupling ($\sqrt{\lambda'}$) for maximally supersymmetric $SU(3)$ Yang-Mills on a 2×8 lattice using different values of the infrared regulator m.

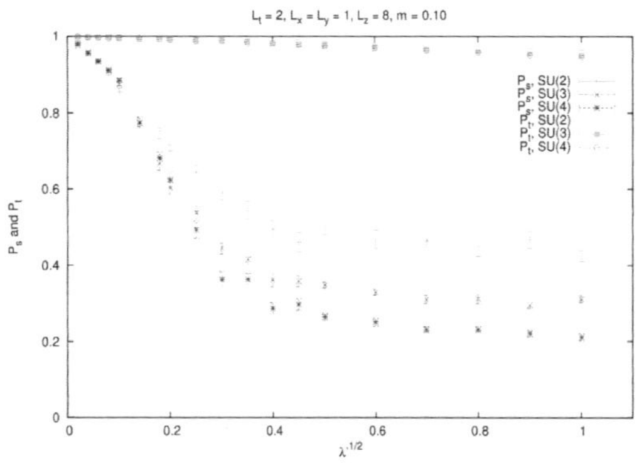

Figure 2. Spatial and temporal Polyakov lines (P_s and P_t) vs square root of the coupling ($\sqrt{\lambda'}$) for maximally supersymmetric $SU(N)$ Yang-Mills with $N = 2, 3, 4$ on a 2×8 lattice using the value $\mathrm{m} = 0.10$ for the infrared regulator.

[6] S. Catterall and T. Wiseman, "Towards lattice simulation of the gauge theory duals to black holes and hot strings," JHEP **0712**, 104 (2007) [arXiv:0706.3518 [hep-lat]]; S. Catterall and T. Wiseman, "Black hole thermodynamics from simulations of lattice Yang-Mills theory," Phys. Rev. D **78**, 041502 (2008) [arXiv:0803.4273 [hep-th]]; S. Catterall and A. Joseph, "Lattice Actions for Yang-Mills Quantum Mechanics with Exact Supersymmetry," Phys. Rev. D **77**, 094504 (2008) [arXiv:0712.3074 [hep-lat]].

[7] J. M. Maldacena, "The large N limit of superconformal field theories and supergravity," Adv. Theor. Math. Phys. **2**, 231 (1998) [Int. J. Theor. Phys. **38**, 1113 (1999)] [arXiv:hep-th/9711200].

[8] O. Aharony, J. Marsano, S. Minwalla and T. Wiseman, "Black hole-black string phase transitions in thermal 1+1-dimensional supersymmetric Yang-Mills theory on a circle," Class. Quant. Grav. **21**, 5169 (2004) [arXiv:hep-th/0406210]; O. Aharony, J. Marsano, S. Minwalla, K. Papadodimas, M. Van Raamsdonk and T. Wiseman, "The phase structure of low dimensional large N gauge theories on tori," JHEP **0601**, 140 (2006) [arXiv:hep-th/0508077].

[9] S. Catterall, "Lattice Supersymmetry and Topological Field Theory," JHEP **0305**, 038 (2003) [arXiv:hep-lat/0301028]; F. Sugino, "A lattice formulation of super Yang-Mills theories with exact supersymmetry," JHEP **0401**, 015 (2004) [arXiv:hep-lat/0311021]; S. Catterall, "A geometrical approach to N = 2 super Yang-Mills theory on the two dimensional lattice," JHEP **0411**, 006 (2004) [arXiv:hep-lat/0410052].

[10] A. G. Cohen, D. B. Kaplan, E. Katz and M. Unsal, "Supersymmetry on a Euclidean spacetime lattice. I: A target theory with four supercharges," JHEP **0308**, 024 (2003) [arXiv:hep-lat/0302017]; "Supersymmetry on a Euclidean spacetime lattice. II: Target theories with eight supercharges," JHEP **0312**, 031 (2003) [arXiv:hep-lat/0307012]; D. B. Kaplan and M. Unsal, "A Euclidean lattice construction of supersymmetric Yang-Mills theories with sixteen supercharges," JHEP **0509**, 042 (2005) [arXiv:hep-lat/0503039].

[11] E. Witten, "Topological quantum field theory," Commun. Math. Phys. **117** (1988) 353.

[12] J. M. Rabin, "Homology theory of lattice fermion doubling," Nucl. Phys. **B201** (1982) 315.

[13] P. Becher, H. Joos, "The dirac-kahler equation and fermions on the lattice," Zeit. Phys. **C15** (1982) 343.

[14] S. Catterall, A. Joseph and T. Wiseman, in preparation.

Journal of Physics: Conference Series **462** (2013) 012023 doi:10.1088/1742-6596/462/1/012023

Quantum Hall Effect, Bosonization and Chiral Actions in Higher Dimensions

Dimitra Karabali

Department of Physics and Astronomy, Lehman College, CUNY, Bronx, NY 10468, USA

E-mail: dimitra.karabali@lehman.cuny.edu

Abstract. We give a brief review of the Quantum Hall effect in higher dimensions and its relation to fuzzy spaces. For a quantum Hall system, the lowest Landau level dynamics is given by a one-dimensional matrix action. This can be used to write down a bosonized noncommutative field theory describing the interactions of higher dimensional nonrelativistic fermions with abelian or nonabelian gauge fields in the lowest Landau level. This general approach is applied explicitly to the case of QHE on $\mathbf{CP}^k$. It is shown that in the semiclassical limit the effective action contains a bulk Chern-Simons type term whose anomaly is exactly canceled by a boundary term given in terms of a chiral, gauged Wess-Zumino-Witten action suitably generalized to higher dimensions.

1. Introduction

Quantum Hall effect in two dimensions is a very important physical phenomenon [1]. In addition to a variety of fascinating experimental results it has provided a clear theoretical framework for exploring a number of field/string theory ideas such as bosonization, conformal invariance, topological field theories, noncommutative geometry, D-brane physics, etc. The study of QHE in more general contexts, such as higher dimensions and different geometries, has recently attracted a lot of attention [2]-[10], following the original work of Zhang and Hu who studied the QHE on S^4 with an $SU(2)$ background magnetic field [2].

In this brief review I will outline our work on the formulation of the QHE on arbitrary even dimensions with particular focus on the edge and bulk dynamics of the corresponding higher dimensional quantum Hall droplets in the presence of external gauge fields and the connection to noncommutative field theories [3]-[9]. Applications of these results beyond the QHE framework are mentioned at the end.

2. Matrix formulation of quantum Hall dynamics

The basic phenomenon of QHE refers to the dynamics of charged fermions on a manifold in the presence of a uniform magnetic field. At the single particle level, the energy eigenstates are grouped into the Landau levels. For high values of the magnetic field at low temperatures, the separation of levels is high compared to the available thermal excitation energy and the dynamics is confined to the lowest Landau level. Although our detailed formulation of higher dimensional QHE is based on the Landau problem on $\mathbf{CP}^k$, we start with a general matrix formulation of the dynamics of noninteracting fermions in the lowest Landau level, which eventually leads to a bosonization approach in terms of a noncommutative field theory.

Journal of Physics: Conference Series **462** (2013) 012023 doi:10.1088/1742-6596/462/1/012023

Let N denote the dimension of the one-particle Hilbert space corresponding to the states of the lowest Landau level, K of which are occupied by fermions. The spin degree of freedom is neglected, so each state can be occupied by a single fermion. In a physical sample, there is also a potential $\hat{V}$ which confines the fermions to within the sample. The fermions are localized around the minimum of the potential but, because of the exclusion principle, they spread over an area forming an incompressible droplet. The droplet is mathematically characterized by a diagonal density matrix $\hat{\rho}_0$ which is equal to 1 for occupied states and zero for unoccupied states. The most general fluctuations which preserve the LLL condition and the number of occupied states are unitary transformations of $\hat{\rho}_0$, namely $\hat{\rho}_0 \to \hat{\rho} = \hat{U}\hat{\rho}_0\hat{U}^\dagger$, where $\hat{U}$ is an $(N \times N)$ unitary matrix. The action which determines $\hat{U}$ is given by

$$S_0 = \int dt \ \mathrm{Tr} \left[i\hat{\rho}_0 \hat{U}^\dagger \partial_t \hat{U} \ - \ \hat{\rho}_0 \hat{U}^\dagger \hat{V} \hat{U} \right] \tag{1}$$

(The Hamiltonian is $\hat{V}$ up to an additive constant.) $\hat{U}$ can be thought of as a collective variable describing all the possible excitations within the LLL. The equation of motion resulting from (1) is the expected evolution equation for the density matrix $\hat{\rho}$, namely

$$i\frac{\partial \hat{\rho}}{\partial t} = [\hat{V}, \hat{\rho}] \tag{2}$$

The action S_0 can also be written as

$$S_0 = \frac{N}{N'} \int d\mu dt \ \mathrm{tr} \ \left[i(\rho_0 * U^\dagger * \partial_t U) \ - \ (\rho_0 * U^\dagger * V * U) \right] \tag{3}$$

where $d\mu$ is the volume measure of the space where QHE has been defined and ρ_0, U, V are the symbols of the corresponding matrices on this space. (The hatted expressions correspond to matrices and unhatted ones to the corresponding symbols, which are fields on the space where QHE is defined.) Equation (3) is written for the case of nonabelian fermions coupled to a background gauge field in some representation J' of dimension N'; the corresponding symbols are $(N' \times N')$ matrix valued functions. Our notation is such that "Tr" indicates trace over the N-dimensional LLL Hilbert space while "tr" indicates trace over the N'-dimensional representation J'. In the case of abelian fermions, $N' = 1$ and tr is trivial.

The general definitions of the symbol and star-product are as follows. If $\Psi_m(\vec{x})$, $m = 1, \cdots, N$, represent the correctly normalized LLL single-particle wavefunctions, then the definition of the symbol corresponding to a $(N \times N)$ matrix $\hat{O}$, with matrix elements O_{ml} is

$$O(\vec{x}, t) = \frac{1}{N} \sum_{m,l} \Psi_m(\vec{x}) O_{ml}(t) \Psi_l^*(\vec{x}) \tag{4}$$

The star-product is defined as

$$(\hat{O}_1 \hat{O}_2)_{symbol} = O_1(\vec{x}, t) * O_2(\vec{x}, t) \tag{5}$$

The action S_0 in (1), or equivalently (3), provides an exact bosonization for the original noninteracting fermion problem. It does not explicitly depend on the particular space and its dimensionality or the abelian or nonabelian nature of the underlying fermionic system. This information is encoded in equation (3) in the definition of the symbol, the star product and the measure.

This matrix formulation can be extended to include external, fluctuating gauge fields [7, 8] (in addition to the uniform background magnetic field which defines the Landau problem). Gauge interactions are described by a matrix action S which is the gauged version of S_0,

$$S = \int dt \ \mathrm{Tr} \left[i\hat{\rho}_0 \hat{U}^\dagger (\partial_t + i\hat{\mathcal{A}}) \hat{U} \ - \ \hat{\rho}_0 \hat{U}^\dagger \hat{V} \hat{U} \right] \tag{6}$$

Journal of Physics: Conference Series **462** (2013) 012023 doi:10.1088/1742-6596/462/1/012023

where $\hat{\mathcal{A}}$ is a matrix gauge potential. It can also be written in terms of the corresponding symbols as

$$S = \frac{N}{N'} \int dt\ d\mu\ \mathrm{tr}\ \left[i\rho_0 * U^\dagger * \partial_t U\ -\ \rho_0 * U^\dagger * V * U - \rho_0 * U^\dagger * \mathcal{A} * U \right] \tag{7}$$

The action (7) is now invariant under the infinitesimal transformations

$$\begin{aligned} \delta U &= -i\lambda * U \\ \delta \mathcal{A}(\vec{x}, t) &= \partial_t \lambda(\vec{x}, t) - i\left(\lambda * (V + \mathcal{A}) - (V + \mathcal{A}) * \lambda\right) \end{aligned} \tag{8}$$

The key physical question is how the field $\mathcal{A}$ is related to the gauge fields A_μ to which the original fermions couple. Once this is known, the action (7) can be expressed in terms of the usual gauge fields. Since S describes gauge interactions it has to be invariant under the usual gauge transformation

$$\begin{aligned} \delta A_\mu &= \partial_\mu \Lambda + i[\bar{A}_\mu + A_\mu,\ \Lambda] \\ \delta \bar{A}_\mu &= 0 \end{aligned} \tag{9}$$

Here Λ is the infinitesimal gauge parameter and $\bar{A}_\mu$ is a possible nonabelian background field. What we need is an expression for $\mathcal{A}$ in terms of A_μ such that when the gauge fields are transformed as in (9), the field $\mathcal{A}$ undergoes the transformation (8). In other words, the transformation (9) is induced by the transformation (8). Since $\mathcal{A}$ is the time-component of a noncommutative gauge field, the relation between $\mathcal{A}$ and the commutative gauge fields A_μ is essentially a Seiberg-Witten transformation [11, 12]. The bosonized action of the LLL fermionic system in the presence of gauge interactions then follows in a straightforward way [7, 8].

This approach, which is based on a matrix formulation, provides a very general way to construct the bosonic action describing the dynamics of the underlying LLL fermionic system in any space that admits a consistent formulation of QHE. This is a generalization of a method used by Sakita [13] to derive the electromagnetic interactions of LLL spinless electrons in the two-dimensional plane. The action S_0 in (1) was also used in the context of one-dimensional free fermions and their relation to $c = 1$ string theory [14]. Below we shall show how these general ideas get implemented in the case of $\nu = 1$ QHE on $\mathbf{CP}^k$, where ν is the filling fraction. We find, in the limit where $N \to \infty$ and the number of fermions is large, that the action S separates into a boundary term describing the coupling of the quantum Hall droplet to the external gauge field A_μ, and a purely A_μ-dependent bulk term, which is a Chern-Simons like term. An anomaly cancellation between these two terms renders the action gauge invariant, as expected.

3. Quantum Hall effect on $\mathbf{CP}^k$

Our analysis of the Landau problem on $\mathbf{CP}^k$ is based on the fact that this is a coset manifold, $\mathbf{CP}^k = SU(k+1)/U(k)$ [3]-[5]. Here we shall briefly describe the group theoretical method we have used to obtain the LLL wavefunctions entering the definition of the symbol and the star-product in (4, 5), necessary ingredients in deriving the action (7). Explicit details on the full Landau spectrum on $\mathbf{CP}^k$ are given in [5].

$\mathbf{CP}^k$ can be parametrized in terms of a $(k+1) \times (k+1)$ matrix g in the fundamental representation of $SU(k+1)$, by making the identification $g \sim gh$, where $h \in U(k)$. For $\mathbf{CP}^k$ one can have both abelian and nonabelian uniform magnetic fields. (The corresponding field strengths are proportional to the Riemannian curvature which is constant, proportional to the $U(k)$ structure constants, in the basis of the frame fields.) The $U(1)$ and $SU(k)$ background gauge fields are given by

$$\begin{aligned} a &= in\sqrt{\frac{2k}{k+1}}\mathrm{tr}(t_{k^2+2k}g^{-1}dg) \ , \qquad\qquad da = n\Omega \\ \bar{A}^a &= 2i\mathrm{tr}(t^a g^{-1}dg) \end{aligned} \tag{10}$$

The $U(1)$ field strength is proportional to the Kähler two-form Ω. The constant n can be expressed in terms of the radius R of $\mathbf{CP}^k$ and the $U(1)$ magnetic field B as, $n = 2BR^2$. Notice that $\bar{A}$ does not depend on n. t_A are the generators of $SU(k+1)$ as matrices in the fundamental representation, normalized so that $\mathrm{tr}(t_A t_B) = \frac{1}{2}\delta_{AB}$. They are classified into three groups: t_a, $a = 1,\ 2, \cdots,\ k^2-1$, corresponding to the $SU(k)$ part of $U(k) \subset SU(k+1)$; t_{k^2+2k}, the $U(1)$ direction of the subgroup $U(k)$ and the remaining coset generators t_α, $\alpha = 1, \cdots, 2k$. The latter ones can be further separated into the raising and lowering type $t_{\pm I} = t_{2I-1} \pm i t_{2I}$, $I = 1, \cdots, k$.

The Landau problem on $\mathbf{CP}^k$ is defined by the choice of "constant" magnetic fields (10). The corresponding wavefunctions form an $SU(k+1)$ representation and can be uniformly expressed in terms of the Wigner $\mathcal{D}$-functions which are the matrices corresponding to the group elements g in a particular representation J. Taking into account the proper normalization we have,

$$\Psi = \sqrt{N}\mathcal{D}^{(J)}_{L,R}(g) = \sqrt{N}\ \langle J, l_A |\ \hat{g}\ | J, r_A \rangle \tag{11}$$

where l_A, r_A label the states within the representation J.

We now define the right and left translation operators on g as

$$R_A\ g = g\ T_A \qquad\qquad L_A\ g = T_A\ g \tag{12}$$

where T_A are the $SU(k+1)$ generators in the representation to which g belongs. The $U(1)$ gauge field in (10) changes by a gauge transformation under a right $U(1)$ rotation of the form $g \to gh$, while it remains invariant under an $SU(k)$ right rotation. This implies that in the case where the fermions couple only to the abelian gauge field a, the corresponding single particle wavefunctions have a fixed $U(1)_R$ charge and are singlets under $SU(k)$ right rotations [4]. In particular the wavefunctions obey the condition

$$\begin{aligned} R_a\ \Psi_m &= 0, \qquad\qquad a = 1, \cdots, k^2-1 \\ R_{k^2+2k}\ \Psi_m &= -\frac{nk}{\sqrt{2k(k+1)}}\ \Psi_m \end{aligned} \tag{13}$$

On the other hand, the nonabelian gauge field $\bar{A}^a$ in (10) is invariant under right $U(1)$ rotations but noninvariant under right $SU(k)$ rotations. So in the case where the fermions have nonabelian degrees of freedom and couple to the full $U(k)$ background gauge field, the wavefunctions have the same fixed $U(1)_R$ charge as in (13) but under right rotations transform as a particular $SU(k)$ representation J' of dimension $N' = \mathrm{dim}J'$ [5]. In this case,

$$\begin{aligned} R_a\ \Psi_{m;a'} &= \Psi_{m;b'}\ (T_a)_{b'a'} \\ R_{k^2+2k}\ \Psi_{m;a'} &= -\frac{nk}{\sqrt{2k(k+1)}}\ \Psi_{m;a'} \end{aligned} \tag{14}$$

The indices $a', b' = 1, \cdots, N'$ label the states within the $SU(k)$ representation J' and can be thought of as the internal degrees of freedom of the nonabelian fermions coupled to the $U(k)$ background field. The matrices T_a are the $SU(k)$ generators in the representation J'.

The coset operators R_α correspond to covariant derivatives while the $SU(k+1)$ operators L_A correspond to magnetic translations. The Laplacian for the space is given by $-\nabla^2 =$

$R_{+I}R_{-I} + R_{-I}R_{+I} = 2R_{+I}R_{-I} +$ constant. The minimum of the Hamiltonian, and hence the lowest Landau level, is given by wavefunctions obeying

$$R_{-I}\ \Psi_{m;a'} = 0 \tag{15}$$

So, for the lowest Landau level in addition to the conditions (14), $|J, r\rangle$ must be a lowest weight state with $T_{-I}|J, r\rangle = 0$; we will denote these states as $|a', -n\rangle$. Once the representation J' is specified, one can identify representations J of $SU(k+1)$ which contain such a state. For example, if there is no $SU(k)$ field, the symmetric rank n representation of $SU(k+1)$ will contain the lowest weight state $|-n\rangle$, which is an $SU(k)$ singlet. The dimension of the J representation which defines the dimensionality of the LLL Hilbert space is then given by $N = (n+k)!/n!k!$. In the nonabelian case where the fermions couple to the $U(k)$ background field LLL wavefunctions form an irreducible $SU(k+1)$ representation J whose lowest weight state is an $SU(k)$ representation J'. Since the $U(1)$ charge is fixed in terms of n, there are constraints on the type of allowed J' $SU(k)$ representations [5]. The dimension N of the $SU(k+1)$ representation J depends now on the particular J' representation chosen, but for large n

$$N = \mathrm{dim}J \to \mathrm{dim}J'\ \frac{n^k}{k!} = N'\ \frac{n^k}{k!} \tag{16}$$

4. Edge and bulk effective action

We are now going to take the semi-clasical limit $N \gg K \gg 1$ (large-n limit) to extract the effective action S_0 describing the dynamics of the nonabelian LLL quantum Hall droplet. For this we need to derive expressions for the symbols of various operators and the star product. Using the definitions (4, 5) and the LLL wavefunctions obtained earlier we find

$$\begin{aligned} X * Y = XY &+ \frac{1}{n}P^{ij}D_iXD_jY - \frac{i}{n^2}P^{il}P^{kj}D_iX\bar{F}_{lk}D_jY \\ &+ \frac{1}{2n^2}P^{ik}P^{jl}\mathcal{D}_iD_jX\mathcal{D}_kD_lY + \mathrm{O}(1/n^3) \end{aligned} \tag{17}$$

where $\bar{F}_{lk} = \bar{F}^a_{lk}(T_a)^T$, $P^{ij} = g^{ij} + \frac{i}{2}(\Omega^{-1})^{ij}$ and $D_iX = \partial_iX + i[\bar{A}_i,\ X]$. $\mathcal{D}_i$ is the Levi-Civita covariant derivative for a curved space such as $\mathbf{CP}^k$, namely, $\mathcal{D}_iD_jX \equiv D_iD_jX - \Gamma^l_{ij}D_lX$, where Γ^l_{ij} is the Christoffel symbol for $\mathbf{CP}^k$. g_{ij} is the metric on $\mathbf{CP}^k$.

Equation (17) is valid for both abelian and nonabelian cases. In the abelian case, X, Y are commuting functions and $\bar{F}_{lk} \to 0$, so that $D_iX \to \partial_iX$ and $\mathcal{D}_iD_jX \to \partial_i\partial_jX - \Gamma^l_{ij}\partial_lX$.

The derivation for the action is simplified if we use an $SU(k)$ invariant confining potential V, so that the ground state has spherical symmetry [5]. In this case we find that the symbol corresponding to the density matrix is

$$(\rho_0)_{a'b'} = \rho_0(r^2)\delta_{a'b'}, \quad \rho_0(r^2) = \Theta\Big(1 - \frac{R^2r^2}{R_D^2}\Big) \tag{18}$$

where Θ is the step-function and R_D is the radius of the droplet. In the semiclassical limit the density matrix is constant over the phase volume occupied by the droplet and its derivative is essentially a delta-function with support at the droplet boundary. As a result the effective action S_0 in the absence of fluctuating gauge fields is a purely boundary action. It can be written in terms of a matrix valued field $G = e^{i\Phi} \in U(N')$, where Φ is the symbol corresponding to $\hat{\Phi}$ in $\hat{U} = e^{i\hat{\Phi}}$, as

$$S_0 \ = \ \frac{1}{4\pi}\mathrm{tr}\Big[\int_{\partial D} \ \Big[\Big(G^\dagger\dot{G} + \omega\ G^\dagger\mathcal{L}G\Big)\,G^\dagger\mathcal{L}G\Big]$$

$$+ \quad \frac{1}{4\pi}\int_D \mathrm{tr}\left[-d\left(i\bar{A}dGG^\dagger + i\bar{A}G^\dagger dG\right) + \frac{1}{3}\left(G^\dagger dG\right)^3\right] \wedge \left(\frac{n\Omega}{2\pi}\right)^{k-1} \frac{1}{(k-1)!} \quad (19)$$
$$\equiv \quad S_{\mathrm{WZW}}(A^L = A^R = \bar{A})$$

where, $\mathcal{L} = \frac{1}{n}(\Omega^{-1})^{ij}\hat{r}_j D_i$ is the component of the covariant derivative D perpendicular to the radial direction, along a special tangential direction on the droplet boundary; $\hat{r}_i$ is the radial unit vector normal to the boundary of the droplet and $\omega = (\frac{1}{n}\frac{\partial V}{\partial r^2})_{\mathrm{boundary}}$.

The action S_0 in (19) is a higher dimensional generalization of a chiral, Wess-Zumino-Witten action, vectorially gauged with respect to the time-independent background gauge field $\bar{A}$. The third term is a WZW-term written as an integral over a $(2k+1)$-dimensional region, corresponding to the droplet and time. The usual 3-form in the integrand of the WZW-term, $(G^\dagger dG)^3$, has now been augmented to the appropriate $(2k+1)$-form $(G^\dagger dG)^3 \wedge (\Omega)^{k-1}$. Since the WZW-term is the integral of a locally exact form, the whole action S_0 is a boundary action.

In the presence of gauge interactions the action describing the LLL dynamics is given by (7). A straightforward but rather tedious calculation [7, 8], following the idea outlined after eq.(8) determines $\mathcal{A}$ in terms of the gauge fields A_μ (up to additional gauge invariant terms),

$$\begin{aligned}
\mathcal{A} = A_0 \quad &- \frac{i}{2n}g^{ij}\left[A_i,\ 2D_iA_0 - \partial_0 A_i + i[A_i,\ A_0]\right] + \frac{1}{4n}(\Omega^{-1})^{ij}\{A_i, 2D_jA_0 - \partial_0 A_j + i[A_j,\ A_0]\} \\
&+ u^iA_i - \frac{i}{2n}g^{ij}\left[A_i,\ A_k\right]\partial_j u^k + \frac{1}{4n}(\omega_K^{-1})^{ij}\{A_i,\ A_k\}\partial_j u^k \\
&- \frac{i}{2n}g^{ij}\left[A_i,\ 2D_jA_k - D_kA_j + i[A_j,\ A_k]\ + 2\bar{F}_{jk}\ \right]u^k \\
&+ \frac{1}{4n}(\Omega^{-1})^{ij}\{A_i,\ 2D_jA_k - D_kA_j + i[A_j,\ A_k] + 2\bar{F}_{jk}\ \}u^k \\
&+ \frac{1}{2n^2}g^{ik}(\Omega^{-1})^{jl}\left(\mathcal{D}_iA_j + \mathcal{D}_jA_i\right)\nabla_k\partial_l V
\end{aligned} \quad (20)$$

where [,] and { , } indicate commutators and anticommutators respectively and $u^i = \frac{1}{n}(\Omega^{-1})^{ij}\partial_j V$. u^i is essentially the phase space velocity, if we think of the LLL as the phase space of a lower dimensional system, with symplectic structure $n\Omega$ and Hamiltonian V.

Substituting (20) into (7) we find that the effective action S splits into two pieces, a boundary action S_{edge} and a bulk one S_{bulk}. The boundary action is essentially a gauged version of the higher-dimensional WZW action encountered in (19). The gauging however appears in a left-right asymmetric way, indicating that the edge action by itself is not gauge-invariant. In particular,

$$S_{\mathrm{edge}} = S_{\mathrm{WZW}}(A^L, A^R) \quad (21)$$

where $A^L = A + \bar{A}$, $A^R = \bar{A}$.

The bulk contribution to the action is given by

$$\begin{aligned}
S_{\mathrm{bulk}} \quad &= \quad -\frac{N}{N'}\int dt d\mu\ \rho_0\ \mathrm{tr}\left(A_0 + u^kA_k\right) \\
&+ \quad \frac{kN}{4\pi nN'}\int dt\rho_0\left[\mathrm{tr}\left((A+\bar{A})d(A+\bar{A}) + \frac{2i}{3}(A+\bar{A})^3\right)\wedge\left(\frac{\Omega}{2\pi}\right)^{k-1}\right. \\
&- \quad \left.\frac{(k-1)}{2\pi}\ \mathrm{tr}\left[\left((A+\bar{A})d(A+\bar{A}) + \frac{2i}{3}(A+\bar{A})^3\right)dV\right]\wedge\left(\frac{\Omega}{2\pi}\right)^{k-2}\right] \\
&+ \quad \frac{N}{2nN'}\int dt d\mu\ \rho_0\ \mathrm{tr}\left[\nabla^i F_{ik} + (k+1)A_k\right]u^k
\end{aligned} \quad (22)$$

If we consider the approximation where R becomes large and the gradients of the external field are small compared to B, the metric-dependent terms in the last line of (22) can be neglected compared to the rest of the terms. In fact, with a little bit of algebra and using that $da = n\Omega$ and $N/N' = n^k/k!$ for large n, the "topological" part of (22) can be written as a single $(2k+1)$-dimensional Chern-Simons term

$$S_A^{\text{bulk}} = S_{CS}(\tilde{A}), \qquad \tilde{A} = \Big(A_0 + V, \; - a_i + \bar{A}_i + A_i\Big) \tag{23}$$

in agreement with [9].

The bulk action is not gauge-invariant. This has to do with the fact that a Chern-Simons action defined on a space with boundary is not gauge-invariant, the non-invariance given by a surface term. It is straightforward to check that the edge WZW action in (21) exactly cancels the gauge-anomaly of the Chern-Simons term rendering the total effective action gauge-invariant as expected. This is a higher dimensional and nonabelian analog of the anomaly cancellation between the edge and bulk actions, well known in the case of two-dimensional planar QHE [15].

5. Summary

An interesting observation which emerges from our analysis is that there is a universal one-dimensional matrix action describing the LLL dynamics, independent of the dimensionality and the abelian or nonabelian nature of the underlying fermionic system. The Hilbert space corresponding to the LLL states for the QHE on a space $\mathcal{M}$ defines a fuzzy version of $\mathcal{M}$. (In particular, the LLL of the example we have analyzed gives a definition of fuzzy $\mathbf{CP}^k$.) Using the star-product of this fuzzy $\mathcal{M}$ one derives the action of a noncommutative bosonic field theory, leading to an exact bosonization method at the level of the action. The semiclassical limit of this describes the dynamics of the corresponding $\nu = 1$ quantum Hall droplet and it naturally separates into a boundary contribution and a bulk contribution which are interesting higher dimensional generalizations of the Wess-Zumino-Witten and Chern-Simons action respectively. These ideas can be easily extended to describe the dynamics of the $\nu = n$ quantum Hall droplet, where n Landau levels are filled.

In two dimensions the corresponding WZW and Chern-Simons actions define conformal field theories [15]. The common origin of their higher dimensional counterparts, via the matrix formulation, suggests the possibility that these theories may share similar features, such as conformal symmetry and chiral algebra.

Although this presentation is in the context of QHE our analysis is relevant in the bosonization of a noninteracting fermionic system in higher dimensions by viewing it in phase space as a Landau problem with the symplectic structure being the magnetic field. Related work on phase space Hall droplets has been done in [16]. Other approaches towards bosonization in higher dimensions have appeared in [17]-[22].

A connection of our work to the Bergman kernel which can be used to approximate metrics on arbitrary Kähler manifolds is recently outlined in [23]. The Bergman kernel is essentially the density matrix ρ projected onto the LLL of the QHE defined on a manifold $\mathcal{M}$. The Bergman metric is defined by $g_B = \frac{1}{n}\partial\bar{\partial}\log\rho$ and provides, for large n, an approximation to Einstein metrics for Kähler manifolds embedded in $\mathbf{CP}^N$. In fact, an asymptotic expansion of ρ in $1/n$ is in terms of curvatures of these spaces [24]. In our case such an expansion can be obtained as the variation of the effective action S with respect to A_0. Our formulation provides a physical context to obtain this in the case of both abelian and nonabelian external gauge fields and compare with existing results in mathematical literature [24].

6. References

[1] For a recent review on QHE, see Girvin S 2000 in Comtet A *et al* (eds.) *Topological Aspects of Low Dimensional Systems* (Springer-Verlag, Berlin), and references therein (*Preprint* cond-mat/9906454)
[2] S.C. Zhang, J.P. Hu, *Science* **294** (2001) 823; J.P. Hu, S.C. Zhang, cond-mat/0112432.

[3] D. Karabali, V.P. Nair, *Nucl. Phys.* **B641** (2002) 533.

[4] D. Karabali and V.P. Nair, *Nucl. Phys.* **B679** (2004) 427.

[5] D. Karabali and V.P. Nair, *Nucl. Phys.* **B697** (2004) 513.

[6] D. Karabali, V.P. Nair and S. Randjbar-Daemi, hep-th/0407007, *Fuzzy spaces, the M(atrix) model and the quantum Hall effect*, in Ian Kogan memorial volume, "From Fields to Strings: Circumnavigating Theoretical Physics", ed. M. Shifman, A. Vainshtein and J. Wheater.

[7] D. Karabali, *Nucl. Phys.* **B726** (2005) 407; *Nucl. Phys.* **B750** (2006).

[8] D. Karabali and V.P. Nair, **J. Phys. A39** (2006) 12735.

[9] V.P. Nair, *Nucl. Phys.* **B750** (2006) 289.

[10] M. Fabinger, *JHEP* **0205** (2002) 037; Y.X. Chen, B.Y. Hou, B.Y. Hou, *Nucl. Phys.* **B638** (2002) 220; Y. Kimura, *Nucl. Phys.* **B637** (2002) 177; H. Elvang, J. Polchinski, hep-th/0209104; B.A. Bernevig, C.H. Chern, J.P. Hu, N. Toumbas, S.C. Zhang, *Ann. Phys.* **300** (2002) 185; B. A. Bernevig, J.P. Hu, N. Toumbas, S.C. Zhang, *Phys. Rev. Lett.* **91** (2003) 236803; S.C. Zhang, *Phys. Rev. Lett.* **90** (2003) 196801; B. Dolan, *JHEP* **0305** (2003) 18; G. Meng, *J. Phys.* **A36** (2003) 9415; S. Bellucci, P.Y. Casteill and A. Nersessian, *Phys. Lett.* **B574** (2003) 121; V.P. Nair and S. Randjbar-Daemi, *Nucl. Phys.* **B679** (2004) 447; A. Jellal, *Nucl. Phys.* **B725** (2005) 554; M Daoud and A. Jellal, *Nucl. Phys.* **B764** (2007) 109; K. Hasebe, *Phys. Rev.* **D78** (2008) 125024; *Phys. Rev.* **D81** (2010) 041702.

[11] N. Seiberg and E. Witten, *JHEP* **9909** (1999) 032.

[12] J. Madore, S. Schraml, P. Schupp and J. Wess *Eur. Phys. J.*.**C16** (2000) 161; B. Jurco, L. Moller, S. Schraml, P. Schupp and J. Wess *Eur. Phys. J.* **C21** (2001) 383; W. Behr and A. Sykora, *Nucl.Phys.* **B698** (2004) 473.

[13] B. Sakita, *Phys. Lett.* **B315** (1993) 124.

[14] S.R. Das, A. Dhar, G. Mandal and S.R. Wadia, *Int. J. Mod. Phys.* **A7** (1992) 5165; *Mod. Phys. Lett.* **A7** (1992) 71; A. Dhar, G. Mandal and S.R. Wadia, *Int. J. Mod. Phys.* **A8** (1993) 325; *Mod. Phys. Lett.* **A7** (1992) 3129; *Mod. Phys. Lett.* **A8** (1993) 3557; A. Dhar, *JHEP* **0507** (2005) 064.

[15] X.G. Wen, *Phys. Rev.* **B41** (1990) 12838; D.H. Lee and X.G. Wen, *Phys. Rev. Lett.* **66** (1991) 1765; M. Stone, *Phys. Rev.* **B42** (1990) 8399; *Ann. Phys.* (NY) **207** (1991) 38; J. Frohlich and T. Kepler, *Nucl. Phys.* **B354** (1991) 369.

[16] A.P. Polychronakos, *Nucl. Phys.* **B705** (2005) 457; *Nucl. Phys.* **B711** (2005) 505; *Phys. Rev. Lett.* **96** (2006) 186401.

[17] A. Luther, *Phys. Rev.* **B19** (1979) 320.

[18] F.D. Haldane, *Helv. Phys. Acta* **65** (1992) 152; cond-mat/0505529.

[19] A.H. Castro Neto and E. Fradkin, *Phys. Rev.* **B49** (1994) 10877.

[20] A. Houghton and B. Marston, *Phys. Rev.* **B48** (1993) 7790; H.J. Kwon, A. Houghton and B. Marston, *Phys. Rev.* **B52** (1995) 8002.

[21] P.W. Anderson and D. Khveshchenko, *Phys. Rev.*.**B52** (1995) 16415.

[22] A. Dhar, G. Mandal and N.V. Suryanarayana, *JHEP* **0601** (2006) 118; A. Dhar and G. Mandal, hep-th/0603154.

[23] M. Douglas, *Commun. Math. Phys.* **293** (2010) 205.

[24] G. Tian, *J. Differential Geom.* **32** (1990) 99; S. Zelditch, *Internat. Math. Res. Notices* **6** 317331 (1998); D. Catlin, in *Analysis and geometry in several complex variables* (Katata, 1997), Trends Math. p. 123, Birkhauser Boston, Boston, (1999); Z. Lu, *Amer. J. Math. 122* **2** (2000); X. Dai, K. Liu and X. Ma, *C. R. Acad. Sci. Paris*, Ser. I (2004) 339; X. Ma and G. Marinescu, *Adv. in Math.* **27** I. 4 (2008) 1756; X. Ma and G. Marinescu, *Progress in Mathematics, Birkhauser* **254** (2006).

Astrophysical Tests of Extra Dimensional Models

Michael Kavic
Department of Physics
The College of New Jersey
2000 Pennington Rd.
Ewing, NJ 08628
E-mail: kavicm@tcnj.edu

John H. Simonetti
Department of Physics, Virginia Tech
Blacksburg, VA 24061, U.S.A.
E-mail: jhs@vt.edu

Abstract. We investigate the possibility that astrophysical observations can be used to test models which predict the existence of extra spatial dimensions. The particular systems of an exploding primordial black hole in the presence of an extra dimension and a binary black hole/neutron star system in the Randall-Sundrum 2 model are considered.

1. INTRODUCTION

Early advances in understanding fundamental natural laws are inextricably linked to astrophysical observations. It is by no means a coincidence that three of the seminal figures in the early history of physics, Galileo Galilei, Johannes Kepler, and Sir Isaac Newton were all astrophysicists. It would seem that as physics matured as a science and terrestrial based experimentation became more prevalent astronomical observation was relegated to one voice in a chorus of sources for empirical data. However, upon more careful examination astronomical observations have continued to played a unique role in the investigation fundamental physics.

We need only consider the discovery of Hubble's law, or the detection of the cosmic microwave background to see the impact astrophysical observation had on physics in the twentieth century. Even more recently the finding of a non-zero neutrino mass from solar, cosmic ray, and supernova observations, and the discovery of dark matter and dark energy have again reminded us of the central role of astrophysics in fundamental physical inquiry.

Perhaps then the relationship between astronomy and fundamental physics extends to the realm of quantum gravity. Certainly black holes and the initial cosmological singularity (the big bang) serve as the two primary physical phenomena that require a theory of quantum gravity to fully describe. So we can ask whether there are extreme astrophysical phenomena which provide evidence of quantum gravity. We argue that such events are possible and we address two possible examples.

2. ASTROPHYSICAL SEARCHES FOR QUANTUM GRAVITY

While considering the best method to conduct an astrophysical search for quantum gravity we must bear in mind that a radically different approach may be required. We begin by noting that astronomical observations are often directed at a single target, and for as long as possible to obtain high precision measurements. However, high energy events may occur in seemingly random parts of the sky, over a short time scale, and could be missed if traditional astronomical methods are employed. Such transients are just the type of phenomena that could be related to quantum gravitational effects, and searches for these could provide a new arena in which to probe this elusive area of inquiry.

Observations of transient phenomena have already played a role in astrophysical exploration. The discovery of pulsars and gamma-ray bursts (GRBs) are prime examples. Moreover, a recently observed radio pulse of extragalactic origin has been found [1]. It has further been argued that this pulse could have been produced by a superconducting cosmic string, another phenomena which can provide evidence of quantum gravity [2].

We should also look beyond electromagnetic signals. Many explosive events that can produce short time scale electromagnetic pulses also produce a gravitational wave signature (i.e. supernova, cosmic strings, compact object mergers, etc.). Thus searches for coincident gravitational waves and electromagnetic pulses could be very profitable.

3. EXPLODING PRIMORIDAL BLACK HOLES & EXTRA DIMENSIONS

Now we consider an example of a transient event associated with the explosion of primordial black holes (PBHs), which depends upon two distinct quantum gravitational phenomena: Hawking radiation and the existence of an extra spatial dimension [3–5].

Rees noted that exploding primordial black holes could provide an observable coherent radio pulse that would be easier to detect than gamma-ray emissions [6].

Rees and Blandford [6; 7] describe the production of a coherent electromagnetic pulse by an explosive event in which the entire mass of the black hole is emitted. If significant numbers of electron-positron pairs are produced in the event, the relativistically expanding shell of these particles (a "fireball" of Lorentz factor γ_f) acts as a perfect conductor, reflecting and boosting the virtual photons of the interstellar magnetic field. An electromagnetic pulse results only for $\gamma_f \sim 10^5$ to 10^7, for typical interstellar magnetic flux densities and free electron densities. The energy of the electron-positron pairs is

$$kT \approx \frac{\gamma_f}{10^5} 0.1 \text{ TeV}. \tag{1}$$

Thus the energy associated with $\gamma_f \sim 10^5$ corresponds roughly to the electroweak scale.

There exists a remarkable relationship between the range of pulse-producing Lorentz factors for the emitted particles, and the TeV scale [3]. Since $\gamma_f \propto T$ at the time of the explosive burst, equation (1) yields

$$\frac{\gamma_f}{10^5} \approx \frac{10^{-19} \text{ m}}{R_s}, \tag{2}$$

where R_s is the Schwarzschild radius. Thus, the allowed range of Lorentz factors implies length scales $R_s \sim 10^{-19} - 10^{-21}$ m. Taking these as Compton wavelengths we find the associated energy scales to be

$$(R_s/\hbar c)^{-1} \sim 1 - 100 \text{ TeV}. \tag{3}$$

This relationship suggests that the production of an electromagnetic pulse by PBHs might be used to probe TeV-scale physics.

To make use of these interesting generic observations, a specific phenomenologically relevant explosive process is required. One such process, discussed by Kol [8], which connects quantum gravitational phenomena and the TeV scale, makes use of the possible existence of an extra dimension and relies on the physics of the black string/black hole phase transition.

A topological phase transition from the black string to the black hole occurs when an instability, the Gregory-LaFlamme point, is reached [9]. This transition is of first order, and results in a significant release of energy equivalent to a substantial increase in the luminosity of Hawking radiation [8].

The analysis of Rees and Blandford [6; 7] can be adapted to the topological phase transition scenario. Observing frequencies between ~ 1 GHz and 10^{15} Hz ($\gamma_f \sim 10^5$ to 10^7) samples the range of extra dimension size, $L \sim 10^{-18} - 10^{-20}$ m.

4. TRANSIENT PULSE SEARCHES

Searches for transient radio pulses from PBH explosions, cf. [10; 11], can probe for the existence of PBHs well below the limits established by observations of the diffuse γ-ray background [12; 13]. To date, these radio searches have utilized data collected for other purposes, or for limited times, all with negative results. A new generation of instruments, designed to operate at low radio frequencies, may be able to conduct extended searches for radio transients over wide fields of view (~ 1 steradian): the Long Wavelength Array (LWA) [14], Murchison Widefield Array (MWA) [15], and the Low Frequency Array (LOFAR) [16].

A continuous wide-field low-frequency radio transient search already underway uses the Eight-meter-wavelength Transient Array (ETA) [17–19] which operates at 38 MHz using 10 dual-polarization dipole antennas. ETA observations are most sensitive to $\gamma_f \approx 10^4$ to 10^5 ($L \approx 10^{-17}$ m to 10^{-18} m). A second array (ETA2) is under construction at a different site. Comparing the signals received at both sites will help mitigate radio interference — a technique that distinguishes all searches with distributed antenna arrays from single-antenna searches. This procedure enables the theoretical sensitivity to be attained. The sensitivity of a radio telescope to a pulse-producing source is dependent on the temporal broadening of an observed pulse due to interstellar scattering and due to dispersion across the finite-width frequency channels utilized in the observations. Taking account of these effects, the ETA is sensitive to transient pulses produced by black-string/black-hole phase transitions out to distances of about 300 pc.

5. EVOLUTION OF A BLACK-HOLE NEUTRON-STAR BINARY SYSTEM

Much work has been done to understand the nature of black holes in the braneworld scenario [20]. In particular, application of AdS/CFT indicates that the full classical 5D solution is equivalent to a 4D quantum corrected black hole. Moreover, this analysis has yielded surprising results which include a dramatically increased evaporation rate for large black holes due to the existence of conformal degrees of freedom [21]. Some observational implications of these results were explored in [22].

The evaporation rate of a black hole in the AdS/CFT braneworld scenario is given by

$$\dot{m}_{BH} = -2.8 \times 10^{-7} \left(\frac{m_{BH}}{M_\odot}\right)^{-2} \left(\frac{L}{10\ \mu\text{m}}\right)^{2} M_\odot\ \text{y}^{-1}, \tag{4}$$

where m_{BH} is the mass of the black hole and L is the AdS radius [21; 22]. The current bound on the AdS radius is $\sim 10\mu$m [23] which we take as a nominal value in this discussion.

To explore the observational implications of the above scenario, it is interesting to consider a black-hole-neutron-star (BH-NS) binary system. If the NS is a pulsar, it is possible to make precision observations that could reveal the effects of the AdS/CFT braneworld scenario. Precision measurements of the changing orbital period of this system would be key, as was the case in the demonstration of gravitational radiation effects in the binary pulsar system PSR 1913+16 (a NS-NS binary, where one of components is an observed pulsar).

Hadjidemetriou [24] discussed the detailed dynamical behavior of a binary system with isotropic mass loss from one or both components. In this situation the binary pair becomes less tightly bound while conserving angular momentum. Therefore the components must separate over time and the orbital period will increase. These results would apply directly to a BH-NS binary in the AdS/CFT braneworld scenario, if the mass-loss effect dominates the evolution of the orbital period in comparison with the effect of gravitational radiation.

Assuming mass loss dominates over energy loss by gravitational radiation, then to lowest order, the rates of change of the semi-major axis a and eccentricity e of the binary, averaged over an orbital period, are

$$\dot{a} = -\,a\frac{\dot{m}}{m}, \quad (5)$$

$$\dot{e} = 0. \quad (6)$$

where $\dot{m} = d/dt(m_{BH} + m_{NS}) = \dot{m}_{BH}$. Since the system obeys Kepler's third law at any moment, the corresponding rate of change of the orbital period is

$$\dot{P} = -2P\frac{\dot{m}}{m}. \quad (7)$$

The effect of mass loss would be to increase the semi-major axis and the orbital period. This is the opposite of the result produced by gravitational radiation. The observation of an increasing orbital period for a BH-NS binary system would yield dramatic evidence for an increased black-hole evaporation rate in an AdS/CFT braneworld scenario.

For a specific case with $P = 7.75$ hours and $e = 0.6$, motivated by PSR 1913+16, and $L = 10\mu$m the results due to mass-loss alone would be $\dot{a} = 16$ m y^{-1}, $\dot{P} = 0.40$ ms y^{-1}. In comparison, energy loss due to gravitational radiation, considered alone, would result in $\dot{a}_{GR} = -3.9$ m y^{-1}, and $\dot{P} = -0.12$ ms y^{-1}. To give an idea of how measurable the mass loss effect would be for $L = 10\mu$m, the observed rate of change of the orbital period for the binary pulsar PSR 1913+16 of $\dot{P} = (-2.4184 \pm 0.0009) \times 10^{-12}$ s/s [25] is equivalent to $\dot{P} = -0.076$ ms y^{-1} with an uncertainty about 2700 times smaller than the magnitude.

References

[1] Lorimer D R, Bailes M, McLaughlin M A, Narkevic D J and Crawford F 2007 *Science* **318** 777–780

[2] Vachaspati T 2008 (*Preprint* `arXiv:0802.0711 [astro-ph]`)

[3] Kavic M, Simonetti J H, Cutchin S E, Ellingson S W and Patterson C D 2008 *JCAP* **0811** 017 (*Preprint* `0801.4023`)

[4] Kavic M 2009 *Nucl. Phys. Proc. Suppl.* **192-193** 150–151

[5] Kavic M, Minic D and Simonetti J 2009 *Int. J. Mod. Phys.* **D17** 2495–2500 (*Preprint* `0805.2941`)

[6] Rees M J 1977 *Nature* **266** 333–334

[7] Blandford R D 1977 *Mon. Not. R. Astron. Soc.* **181** 489–498

[8] Kol B 2002 (*Preprint* `hep-ph/0207037`)

[9] Gregory R and Laflamme R 1993 *Phys. Rev. Lett.* **70** 2837–2840 (*Preprint* `hep-th/9301052`)

[10] Phinney S and Taylor J H 1979 *Nature* **277** 117–118

[11] Benz A O and Paesold G 1998 *Astron. Astrophys.* **329** 61–67

[12] Page D N and Hawking S W 1976 *Astrophys. J.* **206** 1–7

[13] Halzen F, Zas E, MacGibbon J H and Weekes T C 1991 *Nature* **353** 807–815

[14] Taylor G B 2006 *Long Wavelength Astrophysics, 26th meeting of the IAU, Joint Discussion 12, 21 August 2006, Prague, Czech Republic, JD12, #17* **12**

[15] Bowman J D *et al.* 2007 *Astron. J.* **133** 1505–1518 (*Preprint* astro-ph/0611751)

[16] Butcher H R 2004 *Proc. SPIE* **5489** 537–544

[17] Ellingson S W, Simonetti J H and Patterson C D 2007 *IEEE Trans. Antenn. Propag.* **55** 826–831

[18] Simonetti J H, Ellingson S W, Patterson C D, Taylor W, Venugopal V, Cutchin S and Boor Z 2006 *Bull. Amer. Astron. Soc.* **37** 1438–1438

[19] Patterson C D *et al.* 2009 *ACM Trans. Reconf. Tech. Syst.* **1** 1–19 (*Preprint* 0812.1255)

[20] Gregory R 2009 *Lect. Notes Phys.* **769** 259–298 (*Preprint* 0804.2595)

[21] Emparan R, Garcia-Bellido J and Kaloper N 2003 *JHEP* **01** 079 (*Preprint* hep-th/0212132)

[22] Emparan R, Fabbri A and Kaloper N 2002 *JHEP* **08** 043 (*Preprint* hep-th/0206155)

[23] Kapner D J *et al.* 2007 *Phys. Rev. Lett.* **98** 021101 (*Preprint* hep-ph/0611184)

[24] Hadjidemetriou J D 1963 *Icarus* **2** 440–451

[25] Weisberg J M and Taylor J H 2005 *Binary Radio Pulsars* (*Astronomical Society of the Pacific Conference Series* vol 328) ed F A Rasio & I H Stairs pp 25–+

Journal of Physics: Conference Series **462** (2013) 012025 doi:10.1088/1742-6596/462/1/012025

Knotting and higher order linking in physical systems

Roman V. Buniy[1], **Martha J. Holmes**[2] **and Thomas W. Kephart**[2,3]

[1] School of Earth and Space Exploration, and Physics Department, Arizona State University, Tempe, AZ 85287, USA

[2] Department of Physics and Astronomy, Vanderbilt University, Nashville, TN 37235, USA

E-mail: roman.buniy@gmail.com

E-mail: martha.j.holmes@vanderbilt.edu

E-mail: tom.kephart@gmail.com

Abstract. We discuss physical systems with topologies more complicated than simple gaussian linking. Our examples of these higher topologies are in non-relativistic quantum mechanics and in QCD.

1. Introduction

Knots occur in nature. Our focus will be on tight knots and links, and our main examples will come from quantum mechanics and quantum field theory, but we will begin with a mention of knots in plasma physics and biology, since these two areas have lead the way in applications of knot theory to physical systems [1].

2. Examples from classical physics

In plasma physics, magnetic flux tubes are under tension and tend to contract. One can observe plasma phenomena directly via images from the SOHO satellite where Sun spots and solar flares are clearly visible. The flares reach millions of miles above the surface of the sun. Pieces break off and are carried by the solar wind into the wider solar system. These free pieces of plasma have embedded magnetic fields, and some have non-vanishing magnetic helicity. As they interact with planet magnetospheres, they can exchange energy and magnetic helicity with them. The remainder is carried off to interstellar space. At a larger scale, our Milky Way and other galaxies have galactic winds due to all the stars they contain. Some of this plasma rushes off into intergalactic space. Hence it is thought that our Universe is full of plasma with trapped magnetic field, which in general has non-vanishing magnetic helicity.

3. Knots in biosystems

A plasma tends to expand when released into the vacuum, but if there are magnetic fields they can resist this expansion to some degree. In cases where there is magnetic helicity the expansion can be halted when an quasi-equilibrium is reached. In an ideal plasma magnetic helicity is a conserved quantity. There are minimal energy configurations determined by topology after

[3] speaker at Quantum Theory and Symmetries 6, University of Kentucky, July 2009

minimization of the energy holding magnetic helicity fixed. These configurations are known as Taylor states [2, 3, 4], and undisturbed plasma regions would tend to these states.

Tight knots were first discussed and some of their lengths estimated in [1]. However, most biosystems that have knotting do not have tight knots. DNA knots are a case in point. Many knot types have been discovered in DNA, but none are tight due to the persistence length of the DNA strand. Proteins can also be knotted, e.g., a computer model of the knotted protein in methanobacterium thermoautotrophicum has been studied by a group at Argonne National Laboratory. Even though proteins are compact objects, as knots they do not qualify as tight either.

4. Examples from quantum physics

Another example of a physical systems with linking and knotting is the Aharonov-Bohm effect [5] and its generalizations. The magnetic Aharonov-Bohm effect, see Fig. 1, results when a charged particle travels around a closed path in a region of vanishing magnetic field but non-vanishing vector potential. The wave function of the particle is affected by the vector potential and an interference pattern proportional to magnetic flux occurs at a detection screen. The conclusion one draws is that the vector potential is more fundamental than the magnetic field. Definitive experiments were done in the mid 1980s [6].

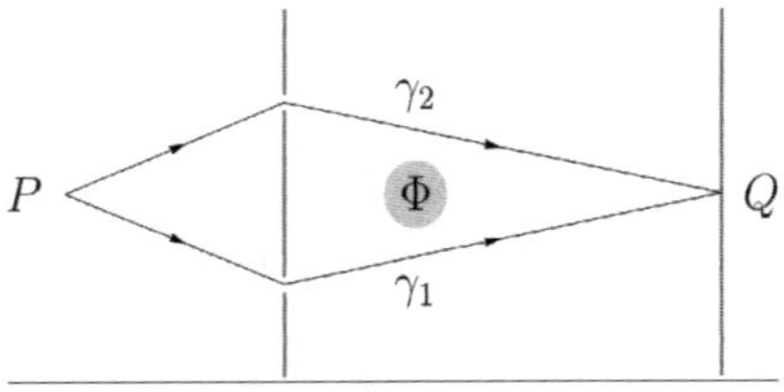

Figure 1. A plane projection of the standard magnetic Aharonov-Bohm effect apparatus.

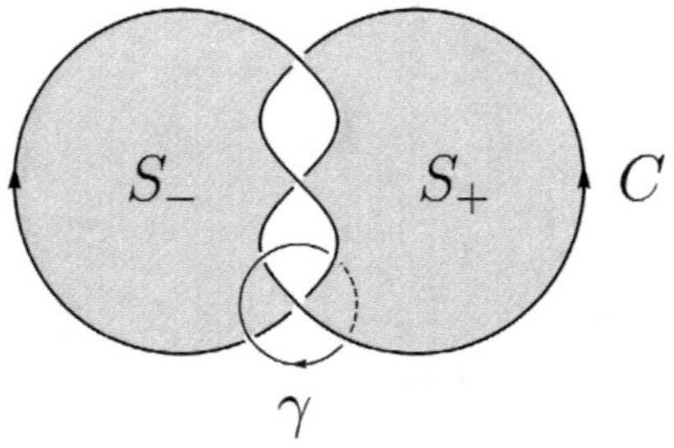

Figure 2. The trefoil knot C bounds an oriented surface $S = S_- + S_+$ called the Seifert surface. The curve γ passes through a hole in the Seifert surface.

In the standard Aharonov-Bohm effect, the two interfering paths of the electron and the solenoid are both copies of S^1 combined to form the Hopf link, topology of which is described by the fundamental group $\pi_1(S^3 - S^1) = \mathbb{Z}$. The resulting Gaussian linking is distinguished by a single winding number. Generalization of the Aharonov-Bohm effect to a trefoil knotted solenoids [7], where the fundamental group $\pi_1(S^3 - K) = \mathbb{Z} * \mathbb{Z}$ has a relation $\alpha^3 = \beta^2$ between its generators, leads to the following results.

Consider a solenoid along path C which bounds the surface $S = S_- + S_+$ called the Seifert surface as shown in Fig. 2, and a wave function along path C', not shown in Fig. 2. If the wave function along C' is a closed unknot that bounds the surface D' (also not shown) linked with a knotted solenoid C (i.e., if C' punches through S), then

$$\phi = \oint_{C'=\partial D'} A \cdot dx = \int_{D'} B \cdot dS = \frac{e\Phi}{\hbar c}.$$

On the other hand if the wave function is knotted and the solenoid is an unknot we find (interchanging paths C and C')

$$\phi = \oint_{C=\partial S} A \cdot dx = \int_S B \cdot dS = \frac{e\Phi}{\hbar c}.$$

In both cases the Gaussian linking is determined by the number of times wave function path passes through oriented surface that bounds the solenoid. Note that if the wave function is linked through one of the holes in the Siefert surface, see path γ in Fig. 2, then there is no Gaussian linking, and no standard Aharonov-Bohm effect, since there is no Gaussian linking. But the curves are nonetheless linked via higher order linking. It is rather difficult to explore higher order linking involving knots, so we will investigate a more tractable case. The simplest example of higher order linking is given by the Borromean rings. A generalization of the Aharonov-Bohm experiment of this type is shown in Fig. 3. After a careful choice of an appropriate gauge, one finds that the interference in the wave function along C_3 is proportional to the product of the fluxes in the two solenoids along C_1 and C_2,

$$\phi = \frac{e^2 \Phi_1 \Phi_2}{\hbar^2 c^2}.$$

The analysis of The Josephson effect [9] has a related generalization [10], shown in Fig. 4, which results in the current maximum

$$I = I_0 |\cos{(2\pi^2 \Phi_1 \Phi_2 / \Phi_0^2)}|$$

between points P and Q in the figure that is again proportional to the product of the fluxes in the two solenoids, where Φ_0 is the fluxoid.

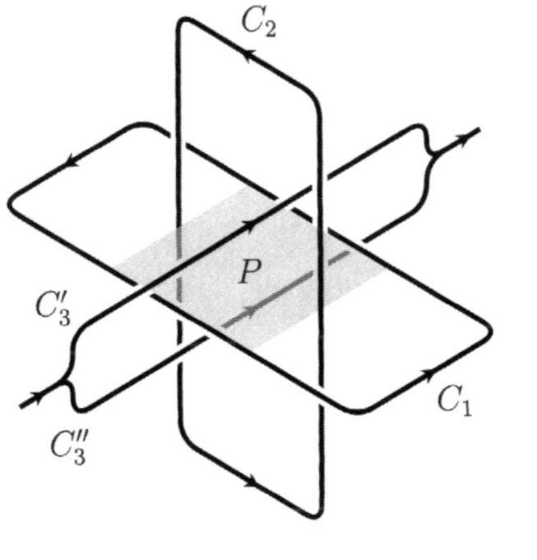

Figure 3. Schematic of a Borromean ring arrangement to detect the second order phase.

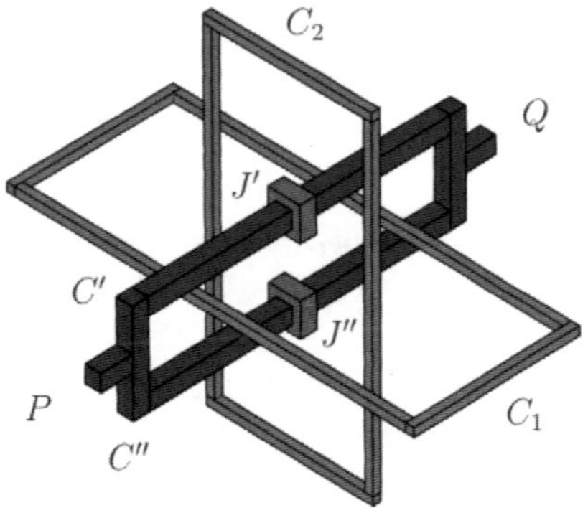

Figure 4. A schematic of a generalized magnetic Josephson effect apparatus. J' and J'' are Josephson junctions. C_1 and C_2 are solenoids.

5. Knots and links in quantum systems

There as many examples of quantum systems that support flux tubes, vortices, or other tube or line like structures. For instance, such systems include non-abelian gauge theories including

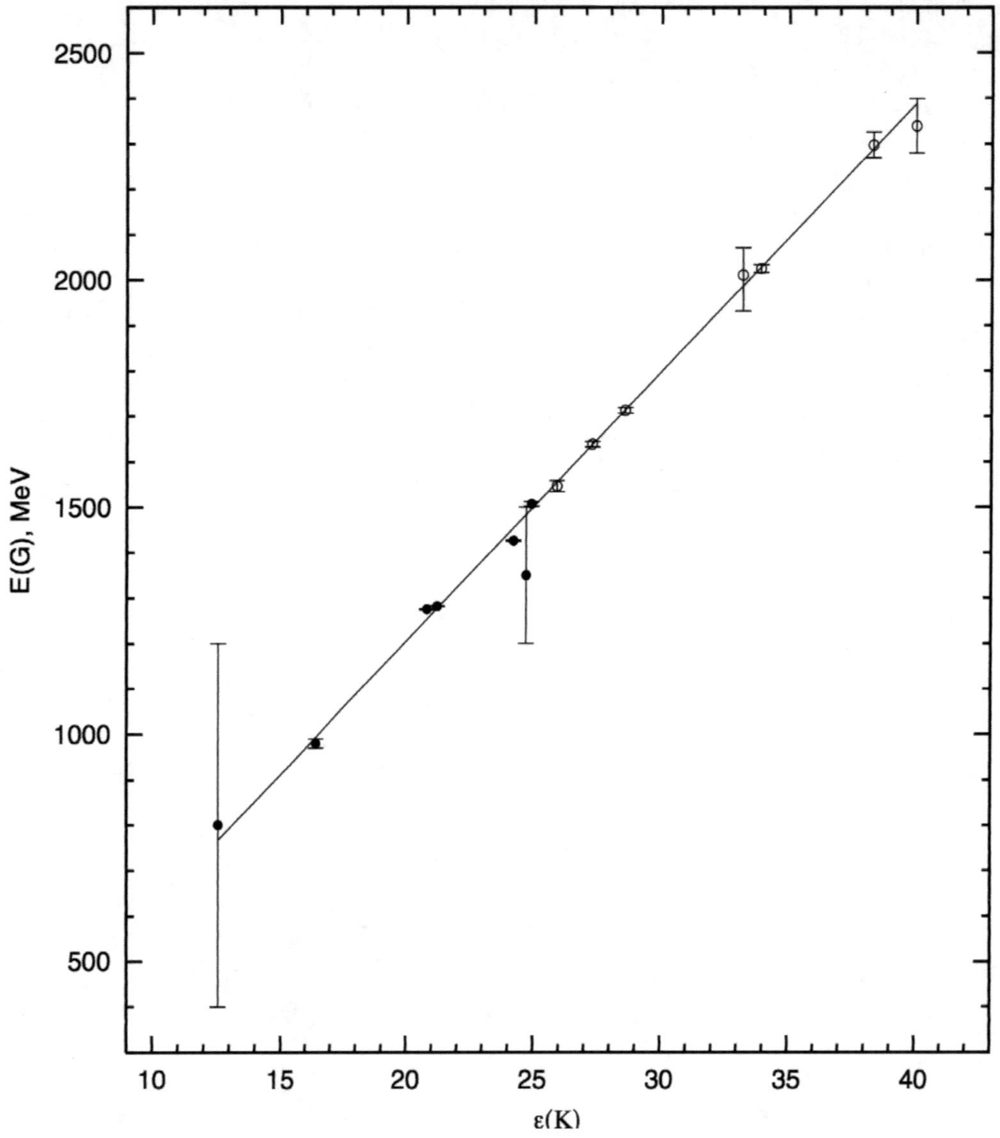

Figure 5. Relationship between the glueball spectrum E(G) and ratio of knot length to flux tube radius. Each point in this figure represents a glueball identified with a knot or link. The straight line is the model fit of f_J states to the tight knot/link spectrum from [12], which contains more discussion.

quantum chromodynamics, superconductors, superfluids, liquid crystals, and atomic condensates to name a few.

Knots were first discussed in the context of particle physics by Lord Kelvin who modeled elementrary atoms as knotted fluid vortices in the aether. While this model did not succeed due to insufficient scientific knowledge at the time, it did show that Kelvin was ahead of his time.

6. Tightly knotted QCD flux tubes as glueballs

The quark model is sufficient to describe most of the spectrum of hadronic bound states, but after filling the multiplets, a number of states remain and it has been suggested that at least one and perhaps several of these states are glueballs—states with no valance quarks. Two of us have suggested that the glueball spectrum of QCD is a result of tight knots and links of quantized chromo-electric flux [12]. This provides an infinite spectrum, up to stability, of new hadronic states, and predicts their energies. Identifying knot lengths with particle energies means the glueball spectrum is the same as the tight knot spectrum up to an overall scaling parameter. A fit matching knot/link lengths [13] with the presumed f_J glueball states [14] states, is shown in Fig. 5, see Ref. [12], where the lightest f_J state is identified with the shortest knot or link, the next lightest state with the next shortest knot or link, etc. E.g., we identify the $f_0(600)$ with the Hopf link, the $f_0(980)$ with the trefoil knot, etc. The fact that this is a one parameter fit means that any system of tightly knotted tubes will show the same universal behavior, with their energy/mass spectra corresponding to the length of knots and links up to one overall scaling parameter per system [12].

In the present model, glueball candidates of non-zero angular momentum, called f_J states, correspond to spinning knots and links. (These solitonic objects could also have intrinsic angular momentum, but that will not be discussed here.) To calculate rotational energies we need the moment of inertia tensor for each tight knot and link. We can get exact results for links with planar components. E.g., in its center of mass frame, the moment of inertia tensor for a Hopf link of uniform density is

$$I_{\text{Hopf}} = \begin{pmatrix} 21 & 0 & 0 \\ 0 & \frac{75}{2} & 0 \\ 0 & 0 & \frac{75}{2} \end{pmatrix} \pi^2 \rho\, a^5,$$

where one torus is in the xz-plane and the other is in the xz-plane. Note this inertia tensor corresponds to a prolate spheroid as do other straight chains of links with an even number of elements. Odd straight chains do not have so much symmetry. Moment of inertia tensors of other knots and links can be calculated by Monte Carlo methods [13]. For further discussion and a detailed analysis of both the zero and non-zero angular momentum f_J states including calculations of moment of inertia tensor see Ref. [15].

7. Conclusions

There are many classical systems that can be knotted or linked. Some of these knots can be tight, e.g., Taylor states in plasmas, but since the tubes can carry an arbitrary amount of flux, their radii are not fixed. In contrast, the flux in quantum systems is quantized leading to fixed radius tubes and hence quantized lengths for tight knots an links. This in turn leads to a quantized mass/energy spectrum. More generally, any quantum systems that supports quantized flux will have this same universal spectrum, i.e., one parameter per system—the slope.

Acknowledgments

RVB acknowledges support from a DoE grant at ASU and from the Arizona State Foundation. The work of MJH and TWK was supported by DoE grant number DE-FG05-85ER40226.

References

[1] V. Katritch, et al., Nature 384 (1996) 142; V. Katritch, et al., Nature 388 (1997) 148.
[2] L. Woltjer, PNAS, 44, 489 (1958).
[3] H. K. Moffatt, J. Fluid Mech. 35, 117 (1969).
[4] J. B. Taylor, Phys. Rev. Lett, 33, 1139 (1974).

[5] Y. Aharonov and D. Bohm, Phys. Rev. **115**, 485 (1959).
[6] A. Tonomura, N. Osakabe, T. Matsuda, T. Kawasaki, J. Endo, S. Yano, H. Yamada, Phys. Rev. Lett. 56, 792, (1986).
[7] R. V. Buniy and T. W. Kephart, Phys. Lett. A **373**, 919 (2009) [arXiv:0808.1891 [hep-th]].
[8] The relevant topology can be found in D. Rolfsen, *Knots and Links,* Publish or Perish, Berkeley, California (1976).
[9] B. D. Josephson, Physics Letters, 1, 251 (1962).
[10] R. V. Buniy and T. W. Kephart, J. Phys. A (2010), to appear; arXiv:0808.1892 [hep-th].
[11] R. V. Buniy and T. W. Kephart, Phys. Lett. A **372**, 2583 (2008) [arXiv:hep-th/0611334] and R. V. Buniy and T. W. Kephart, Phys. Lett. A **372**, 4775 (2008) [arXiv:hep-th/0611335].
[12] R. V. Buniy and T. W. Kephart, Phys. Lett. B **576**, 127 (2003) [arXiv:hep-ph/0209339]; R. V. Buniy and T. W. Kephart, Int. J. Mod. Phys. A **20**, 1252 (2005) [arXiv:hep-ph/0408027]; R. V. Buniy and T. W. Kephart, Published in Physical and Numerical Models in Knot Theory Including Applications to the Life Sciences. Edited by J. Calvo, K.C. Millett, E.J. Rawdon and A. Stasiak. World Scientific, 2005. p. 45. arXiv:hep-ph/0408025.
[13] T. Ashton, J. Cantarella, M. Piatek and E. Rawdon, arXiv:0508248 [math.DG].
[14] C. Amsler *et al.* [Particle Data Group], Phys. Lett. B **667**, 1 (2008).
[15] Martha J. Holmes, Ph.D. dissertation, Vanderbilt University (2009).

Journal of Physics: Conference Series **462** (2013) 012026 doi:10.1088/1742-6596/462/1/012026

Rigid Calabi-Yau threefolds, Picard Eisenstein series and instantons

L Bao[1], A Kleinschmidt[2], B E W Nilsson[3], D Persson[4] and B Pioline[5]

[1] Institut des Hautes Etudes Scientifiques, Le Bois-Marie 35, 91440 Bures-sur-Yvette, France
[2] Physique Théorique et Mathématique, Université Libre de Bruxelles and International Solvay Institutes, Campus Plaine CP 231, Boulevard du Triomphe, 1050 Brussels, Belgium
[3] Fundamental Physics, Chalmers University of Technology, SE-412 96, Göteborg, Sweden
[4] Institut für Theoretische Physik, Eidgenössische Technische Hochschule, CH-8093 Zürich, Switzerland
[5] Laboratoire de Physique Théorique et Hautes Energies, CNRS UMR 7589 and Université Pierre et Marie Curie - Paris 6, 4 place Jussieu, 75252 Paris cedex 05, France

E-mail: lingbao@ihes.fr, axel.kleinschmidt@ulb.ac.be, tfebn@chalmers.se, daniel.persson@itp.phys.ethz.ch, pioline@lpthe.jussieu.fr

Abstract. Type IIA string theory compactified on a rigid Calabi-Yau threefold gives rise to a classical moduli space that carries an isometric action of $U(2,1)$. Various quantum corrections break this continuous isometry to a discrete subgroup. Focussing on the case where the intermediate Jacobian of the Calabi-Yau admits complex multiplication by the ring of quadratic imaginary integers $\mathcal{O}_d$, we argue that the remaining quantum duality group is an arithmetic Picard modular group $PU(2,1;\mathcal{O}_d)$. Based on this proposal we construct an Eisenstein series invariant under this duality group and study its non-Abelian Fourier expansion. This allows the prediction of non-perturbative effects, notably the contribution of D2- and NS5-brane instantons. The present work extends our previous analysis in 0909.4299 which was restricted to the special case of the Gaussian integers $\mathcal{O}_1 = \mathbb{Z}[i]$.

1. Introduction

In this contribution we are interested in constructing the exact hypermultiplet moduli space metric of type IIA string theory compactified on a rigid Calabi-Yau threefold $\mathcal{X}$, admitting complex multiplication by the ring of quadratic imaginary integers $\mathcal{O}_d$. The rigidity condition means that there are no complex structure deformations ($h_{2,1}(\mathcal{X}) = 0$) and therefore the hypermultiplet sector of the resulting $\mathcal{N} = 2$ theory in $D = 3 + 1$ dimensions consists only of the so-called universal hypermultiplet, parameterized by four scalar fields. The complex multiplication property implies that the intermediate Jacobian of $\mathcal{X}$ takes the form

$$\mathcal{J}_d = H^3(\mathcal{X},\mathbb{C})/H^3(\mathcal{X},\mathbb{Z}) \simeq \mathbb{C}/\mathcal{O}_d\,, \tag{1}$$

where $\mathcal{O}_d$ is the set of algebraic integers in the quadratic number field $\mathbb{Q}(\sqrt{-d})$ where d is a square-free positive integer. For simplicity we restrict to the case where $\mathcal{O}_d$ is a principal ideal domain, hence admitting unique prime factorization. Remarkably, according to the Stark-Heegner theorem, there are only a finite number of values of d which satisfy this criterion:

$$d \in \{1, 2, 3, 7, 11, 19, 43, 67, 163\}\,. \tag{2}$$

The case $d = 1$ corresponds to the Gaussian integers $\mathcal{O}_1 = \mathbb{Z} + i\mathbb{Z}$ and was treated in detail in [1], where more details on the physical motivations behind the set-up studied in this contribution can be found. Here, we restrict exclusively to the case $d \equiv 3 \mod 4$ – which covers the entire Stark-Heegner sequence except $d = 1, 2$ – in which case

$$\mathcal{O}_d \equiv \mathbb{Z}[\omega_d] = \mathbb{Z} + \omega_d \mathbb{Z}, \qquad \text{where} \qquad \omega_d = \frac{-1 + i\sqrt{d}}{2}. \tag{3}$$

We parameterize the hypermultiplet moduli space $\mathcal{M}_{\text{UH}}$ by four scalars, namely the 4D dilaton ϕ and the axions $\zeta, \tilde{\zeta}, \sigma$. The scalars ζ and $\tilde{\zeta}$ are the periods of the Ramond-Ramond 3-form along a symplectic basis of $H_3(\mathcal{X}, \mathbb{Z})$, and the complex variable $\zeta + \omega_d \tilde{\zeta}$ thus parameterizes the intermediate Jacobian $\mathcal{J}_d$. The coordinate σ, on the other hand, corresponds to the $D = 4$ dual of the Neveu-Schwarz two-form, and parameterizes the fiber of a circle bundle over $\mathcal{J}_d$. The classical metric on $\mathcal{M}_{\text{UH}}$ can be written as

$$ds^2_{\mathcal{M}_{\text{UH}}} = d\phi^2 + \frac{1}{2} e^{2\phi} \frac{|d\zeta + \omega_d\, d\tilde{\zeta}|^2}{\Im(\omega_d)} + \frac{1}{4} e^{4\phi} \Big(d\sigma - \tilde{\zeta} d\zeta + \zeta d\tilde{\zeta} \Big)^2 , \tag{4}$$

which can be derived via the c-map from the quadratic prepotential $F(X) = \omega_d X^2/2$ on the type IIB side, with ω_d playing the role of the "period matrix". The metric (4) is locally isometric to the quaternion-Kähler symmetric space $(U(2) \times U(1)) \backslash U(2,1)$. Quantum corrections break the continuous isometry group $U(2,1)$ to a discrete subgroup Γ, while preserving the quaternion-Kähler property. Γ should contain at least the Heisenberg group

$$\zeta \longrightarrow \zeta + a\,, \quad \tilde{\zeta} \longrightarrow \tilde{\zeta} + b\,, \quad \sigma \longrightarrow \sigma + \big(2c + a + b + ab\big) - \big(a\tilde{\zeta} - b\zeta\big)\,, \tag{5}$$

for $a, b, c \in \mathbb{Z}$. These discrete periodicities are remnants of continuous (Peccei-Quinn) symmetries of the metric (4), as a result of D2 and NS5-brane instanton effects [5, 6]. Examining the list of generators of the Picard modular groups [2, 3, 4, 1], it is natural to conjecture that Γ is the *Picard modular group* $PU(2,1;\mathcal{O}_d) \equiv U(2,1) \cap PGL(3,\mathcal{O}_d)$, in the standard embedding. It contains (5) as a unipotent subgroup $N(\mathcal{O}_d) = N \cap \Gamma$, where N is associated to the Iwasawa decomposition $U(2,1) = KAN$. The additional generators in Γ can be identified physically as electric-magnetic duality and S-duality. Whereas the discrete symmetries (5) are well-established, the presence of electric-magnetic duality and S-duality at the quantum level is a strong assumption which merits further justification.[1] In [1] we have given some arguments in favour of this assertion and refer the reader to the discussion there.

Our aim in the remainder will be to construct a Γ-invariant automorphic form on the coset space $\mathcal{M}_{\text{UH}}$, namely the Eisenstein series $\mathcal{E}_s$ attached to a degenerate principal continuous representation of $U(2,1)$. The ultimate goal is to extract the *exact* metric on $\mathcal{M}_{\text{UH}}$ by using twistor techniques for quaternion-Kähler manifolds. The basic idea is that deformations of any quaternion-Kähler manifold $\mathcal{M}$ can be encoded in deformations of the complex contact structure of the associated twistor space $\mathcal{Z}_{\mathcal{M}}$, which is a $\mathbb{C}P^1$-bundle over $\mathcal{M}$ carrying a Kähler-Einstein metric [7, 8]. The Kähler potential $K_{\mathcal{Z}_{\mathcal{M}}}$ for the Kähler metric on $\mathcal{Z}_{\mathcal{M}}$ follows from the so called *contact potential* $\Phi(x, z)$, which depends on the collective coordinates x on the base $\mathcal{M}_{\text{UH}}$ and (holomorphically) on the fiber-coordinate $z \in \mathbb{C}P^1$, through the formula $K_{\mathcal{Z}} = \log\big[(1 + |z|^2)/|z|\big] + \Re(\Phi)$. The knowledge of the contact potential together with the so-called twistor lines – relating local Darboux coordinates on $\mathcal{Z}_{\mathcal{M}}$ to the pair $(x, z) \in \mathcal{M} \times \mathbb{C}P^1$

[1] It is debatable whether the correct quantum duality group ought to be $PU(2,1;\mathcal{O}_d)$ or $PSU(2,1;\mathcal{O}_d)$. Except for the $d = 1$ case discussed in [1], the latter group, unlike the former, does not include electric-magnetic duality. The simplest Eisenstein series constructed below is in fact invariant under the full $PU(2,1;\mathcal{O}_d)$.

Journal of Physics: Conference Series **462** (2013) 012026 doi:10.1088/1742-6596/462/1/012026

– then completely determines the metric on $\mathcal{M}_{\rm UH}$ [9, 10]. Our tentative proposal is that the exact contact potential at the 'north pole' $z = 0$ of the twistor space is given by

$$e^{\Phi} = e^{\phi}\, \mathcal{E}_{s=3/2}\,, \tag{6}$$

where ϕ on the r.h.s. is the dilaton and the order $s = 3/2$ of the Eisenstein series is fixed by studying the known perturbative terms. To completely specify the metric on $\mathcal{M}_{\rm UH}$, one should also determine the quantum corrections to the twistor lines, which were derived at the classical level in [11]. To motivate our proposal further we will now construct the Eisenstein series $\mathcal{E}_s$ and analyze its Fourier expansion with respect to the Heisenberg group (5).

2. Construction of Eisenstein series for $PU(2,1;\mathcal{O}_d)$

It will be useful to exploit the isomorphism between the coset space $(U(2)\times U(1))\backslash U(2,1)$ and complex hyperbolic space $\mathcal{U}$, where the latter is defined in terms of a pair of complex coordinates $\mathcal{Z} = (z_1, z_2)$ as follows

$$\mathcal{U} = \left\{ \mathcal{Z} = (z_1, z_2) \in \mathbb{C}^2 \,:\, \mathcal{F}(\mathcal{Z}) = -\Re(z_1) - \frac{1}{2}|z_2|^2 > 0 \right\}. \tag{7}$$

This space is preserved by the standard fractional linear action of the continuous $U(2,1)$, defined as the set of matrices leaving invariant the indefinite metric

$$\eta = \begin{pmatrix} 0 & 0 & 1 \\ 0 & 1 & 0 \\ 1 & 0 & 0 \end{pmatrix}, \qquad U(2,1) = \left\{ g \in GL(3,\mathbb{C}) \,:\, g^{\dagger}\eta g = \eta \right\}. \tag{8}$$

The Kähler metric on $\mathcal{U}$ can be derived from the Kähler potential $K = -\log(\mathcal{F})$, leading to

$$ds^2 = \frac{1}{4}\mathcal{F}^{-2}\big(dz_1 d\bar{z}_1 + z_2 dz_1 d\bar{z}_2 + \bar{z}_2 dz_2 d\bar{z}_1 - 2\Re(z_1) dz_2 d\bar{z}_2\big). \tag{9}$$

This becomes identical to the metric (4) upon identifying

$$z_1 \equiv i\Im(\omega_d)\sigma - \frac{1}{2}|\zeta + \omega_d\tilde{\zeta}|^2 - \Im(\omega_d)e^{-2\phi}, \quad z_2 \equiv \zeta + \omega_d\tilde{\zeta}, \tag{10}$$

so that $\mathcal{F} = \Im(\omega_d)\, e^{-2\phi}$. In the following it will prove convenient to work directly with the complex coordinates (z_1, z_2) rather than the real coordinates $(\phi, \zeta, \tilde{\zeta}, \sigma)$.

The metric (9) can alternatively be written as $ds^2 = -\frac{1}{8}\mathrm{Tr}(d\mathcal{K} d\mathcal{K}^{-1})$ in terms of the $U(2)\times U(1)$-invariant Hermitean coset representative

$$\mathcal{K} = \mathcal{F}^{-1}\begin{pmatrix} 1 & z_2 & z_1 \\ \bar{z}_2 & |z_2|^2 & \bar{z}_2 z_1 \\ \bar{z}_1 & \bar{z}_1 z_2 & |z_1|^2 \end{pmatrix} + \begin{pmatrix} 0 & 0 & 1 \\ 0 & 1 & 0 \\ 1 & 0 & 0 \end{pmatrix} \in SU(2,1)\,. \tag{11}$$

Defining also the singular matrix $\tilde{\mathcal{K}} = \mathcal{K} - \eta$, one sees that $\tilde{\mathcal{K}}$ can be written as

$$\tilde{\mathcal{K}} = \tilde{\mathcal{V}}^{\dagger}\tilde{\mathcal{V}} \qquad \text{with} \quad \tilde{\mathcal{V}} = \mathcal{F}^{-1/2}\begin{pmatrix} 1 & z_2 & z_1 \\ 0 & 0 & 0 \\ 0 & 0 & 0 \end{pmatrix}. \tag{12}$$

A Γ-invariant Eisenstein series $\mathcal{E}_s(\mathcal{Z})$ on $\mathcal{U}$ can then be defined as a (constrained) sum over non-zero lattice vectors $\vec{\Omega} = (\Omega_1, \Omega_2, \Omega_3)^T \in \mathcal{O}_d^3$ according to

$$\mathcal{E}_s(\mathcal{Z}) = \sum_{\substack{0\neq\vec{\Omega}\in\mathcal{O}_d^3 \\ \vec{\Omega}^\dagger\eta\vec{\Omega}=0}} (\vec{\Omega}^\dagger\mathcal{K}\vec{\Omega})^{-s} = \mathcal{F}^s \sum_{\substack{0\neq\vec{\Omega}\in\mathcal{O}_d^3 \\ \vec{\Omega}^\dagger\eta\vec{\Omega}=0}} |\Omega_3 z_1 + \Omega_2 z_2 + \Omega_1|^{-2s}\,. \tag{13}$$

The constraint

$$\vec{\Omega}^\dagger\,\eta\,\vec{\Omega} = |\Omega_2|^2 + 2\Re(\Omega_1\bar{\Omega}_3) = 0 \tag{14}$$

on the lattice vectors is necessary to ensure that $\mathcal{E}_s$ is an eigenfunction of the $PU(2,1)$ invariant Laplacian on $\mathcal{U}$. For vectors $\vec{\Omega}$ satisfying the constraint (14) one can replace $\mathcal{K}$ by $\tilde{\mathcal{K}}$ and then use the factorization (12) to rewrite the sum as shown in the last equality of (13).

When performing the Fourier expansion of $\mathcal{E}_s$ it is expedient to solve the constraint on $\vec{\Omega}$ explicitly. To this end we note that for fixed $\Omega_2, \Omega_3 \in \mathcal{O}_d$ with $\Omega_3 \neq 0$, solutions to (14) exist if and only if $|\Omega_2|^2/r \in \mathbb{Z}$, where r is defined as

$$r = \gcd\left(2p_1 - p_2, \frac{d+1}{2}p_2 - p_1\right)\,, \qquad \Omega_3 \equiv p_1 + \omega_d p_2\,, \qquad p_1, p_2 \in \mathbb{Z}\,. \tag{15}$$

In this case, the solutions of (14) can be parameterized as

$$\Omega_1 = -\frac{|\Omega_2|^2}{r}\Omega_1^{(0)} - i\sqrt{d}\frac{m}{r}\Omega_3 \qquad m \in \mathbb{Z}\,, \tag{16}$$

where $\Omega_1^{(0)} \in \mathcal{O}_d$ is a particular solution of $r = 2\Re(\Omega_1^{(0)}\bar{\Omega}_3)$, obtained via the Euclidean algorithm.

3. Abelian and non-Abelian Fourier expansion

Since the unipotent subgroup $N(\mathcal{O}_d) \subset \Gamma$ is a three-dimensional Heisenberg group, the general Fourier expansion of a Γ-invariant Eisenstein series consists of three distinct pieces: the so-called constant, Abelian and non-Abelian terms [12, 13]. The constant terms correspond to the zeroth Fourier coefficients, and hence do not depend on the axions; these terms can therefore be identified with perturbative contributions in the string coupling $g_s = e^\phi$. Due to the fact that the (restricted) Weyl group of $U(2,1)$ is of order 2, one expects two constant terms [14]. The axion-dependent terms (generic Fourier coefficients) further separate into Abelian and non-Abelian parts, respectively corresponding to the expansion along N/Z and Z, where Z is the center of N. We will now extract these pieces in turn by performing appropriate Poisson resummations.

To begin with, the Eisenstein series (13) can be separated into the terms where $\Omega_3 = 0$ (implying $\Omega_2 = 0$ by (14)) and those where $\Omega_3 \neq 0$ as

$$\mathcal{E}_s(\mathcal{Z}) = \mathcal{E}_s^{(0)}(\mathcal{F}) + \mathcal{A}_s(\mathcal{Z})\,. \tag{17}$$

The term $\mathcal{E}_s^{(0)}$ is the leading order contribution in the weak coupling limit $e^\phi \to 0$ and evaluates straightforwardly as

$$\mathcal{E}_s^{(0)} = \mathcal{F}^s \sum_{\Omega_1\neq 0} |\Omega_1|^{-2s} = \mathcal{F}^s \zeta_{\mathbb{Q}(i\sqrt{d})}(s)\,, \tag{18}$$

where we have introduced the Dedekind zeta function $\zeta_{\mathbb{Q}(i\sqrt{d})}(s)$. Using the structure of primes in the ring $\mathcal{O}_d$, it can be shown that

$$\zeta_{\mathbb{Q}(i\sqrt{d})}(s) = e_d\,\beta_d(s)\,\zeta(s) \tag{19}$$

Journal of Physics: Conference Series **462** (2013) 012026 doi:10.1088/1742-6596/462/1/012026

where e_d is the number of units in $\mathcal{O}_d$ (i.e. $e_3 = 6$ and $e_d = 2$ otherwise), $\beta_d(s) \equiv L\left(\left(\frac{\cdot}{-d}\right), s\right)$ is the Dirichlet L-function associated to the Legendre symbol and $\zeta(s)$ is the ordinary Riemann zeta function. $\mathcal{E}_s^{(0)}$ represents the first of the two constant terms.

Turning to the remainder term $\mathcal{A}_s$, we can now use the explicit solution of the constraint given in (16). Furthermore, we employ some standard tricks for the rewriting of powers as

$$M^{-s} = \frac{\pi^s}{\Gamma(s)} \int_0^\infty \frac{dt}{t^{s+1}} e^{-\frac{\pi}{t}M} \qquad \text{for } M > 0 \tag{20}$$

and for Poisson resummation (for real $x > 0$ and complex $a = a_1 + ia_2$)

$$\sum_{m\in\mathbb{Z}} e^{-\pi x|m+a|^2} = \frac{1}{\sqrt{x}} \sum_{\tilde{m}\in\mathbb{Z}} e^{-\frac{\pi}{x}\tilde{m}^2 - 2\pi i\tilde{m}a_1 - \pi x a_2^2}\,. \tag{21}$$

Using this leads to

$$\begin{aligned}
\mathcal{A}_s &= \mathcal{F}^s \sum_{\Omega_3\neq 0} \sum_{\substack{\Omega_2\in\mathcal{O}_d \\ |\Omega_2|^2\in r\mathbb{Z}}} \sum_{m\in\mathbb{Z}} \left| \Omega_3 z_1 + \Omega_2 z_2 - \frac{|\Omega_2|^2}{r}\Omega_1^{(0)} - i\sqrt{d}\frac{m}{r}\Omega_3 \right|^{-2s} \\
&= \mathcal{F}^s \frac{\pi^s}{\Gamma(s)} \sum_{\Omega_3\neq 0} \sum_{\substack{\Omega_2\in\mathcal{O}_d \\ |\Omega_2|^2\in r\mathbb{Z}}} \frac{r}{|\Omega_3|\sqrt{d}} \sum_{\tilde{m}\in\mathbb{Z}} \int_0^\infty \frac{dt}{t^{s+1/2}} e^{-\pi t \frac{r^2}{d|\Omega_3|^2}\tilde{m}^2 - 2\pi i\tilde{m}a_1 - \pi\frac{|\Omega_3|^2 d}{r^2 t}a_2^2}\,,
\end{aligned} \tag{22}$$

where

$$\begin{aligned}
a_1 &= -\frac{r}{\sqrt{d}|\Omega_3|^2}\left(|\Omega_3|^2\Im(z_1) + \Im(\Omega_2\bar{\Omega}_3 z_2) - \frac{|\Omega_2|^2}{r}\Im(\bar{\Omega}_3\Omega_1^{(0)}) \right), \\
a_2 &= -\frac{r}{\sqrt{d}|\Omega_3|^2}\left(|\Omega_3|^2\mathcal{F} + \frac{1}{2}|\Omega_2 - \Omega_3\bar{z}_2|^2 \right).
\end{aligned} \tag{23}$$

The second constant term and the Abelian Fourier coefficients can be extracted from the $\tilde{m} = 0$ terms, leading to

$$\begin{aligned}
\mathcal{A}_s^{(\tilde{m}=0)} &= \mathcal{F}^s \frac{\pi^s}{\Gamma(s)} \sum_{\Omega_3\neq 0} \sum_{\substack{\Omega_2\in\mathcal{O}_d \\ |\Omega_2|^2\in r\mathbb{Z}}} \frac{r}{|\Omega_3|\sqrt{d}} \int_0^\infty \frac{dt}{t^{s+1/2}} e^{-\pi\frac{|\Omega_3|^2 d}{r^2 t}a_2^2} \\
&= \mathcal{F}^s \frac{\pi^s\Gamma(s-1/2)\pi^{2s-1}}{\Gamma(s)\pi^{s-1/2}\Gamma(2s-1)} \sum_{\Omega_3\neq 0} \sum_{\substack{\Omega_2\in\mathcal{O}_d \\ |\Omega_2|^2\in r\mathbb{Z}}} \frac{r}{\sqrt{d}|\Omega_3|^{2-2s}} \int_0^\infty \frac{dt}{t^{2s}} e^{-\frac{\pi}{t}\left(|\Omega_3|^2\mathcal{F} + \frac{1}{2}|\Omega_2 - \Omega_3\bar{z}_2|^2\right)}\,,
\end{aligned} \tag{24}$$

where, in getting from the first to the second line, we have performed the integral over t and then used (20) to re-exponentiate part of the summand.

In the next step we have to treat the divisibility constraint $|\Omega_2|^2 \in r\mathbb{Z}$. This can be resolved by writing the solution as

$$\Omega_2 = \Omega_2^{(0)} + rv\,, \tag{25}$$

where $v \in \mathcal{O}_d$, and $\Omega_2^{(0)}$ runs over the set F_r of solutions to the equation $|\Omega_2^{(0)}|^2 = 0 \bmod r$ in a fundamental domain $\{n_1 + n_2\omega_d \,:\, 0 \leq n_1, n_2 < r\}$, with cardinality $N_d(r) \equiv \#F_r$. Now one can Poisson resum over $v \in \mathcal{O}_d$ using

$$\sum_{v\in\mathcal{O}_d} e^{-\pi x|v+a|^2} = \frac{2}{x\sqrt{d}} \sum_{u\in\mathcal{O}_d^*} e^{-\frac{\pi}{x}|u|^2 - 2\pi i\Re(u\bar{a})}\,, \tag{26}$$

where u runs over the dual lattice $\mathcal{O}_d^*$, spanned by $u_1 \equiv \frac{2}{d}\left(\frac{d+1}{2}+\omega_d\right) = 1+\frac{i}{\sqrt{d}}$ and $u_2 \equiv \frac{2}{d}(1+2\omega_d) = \frac{2i}{\sqrt{d}}$. The resulting Abelian terms are

$$\begin{aligned}\mathcal{A}_s^{(\tilde{m}=0)} &= \mathcal{F}^s \frac{4}{d}\frac{\pi^s\Gamma(s-1/2)\pi^{2s-1}}{\Gamma(s)\pi^{s-1/2}\Gamma(2s-1)} \sum_{\Omega_3\neq 0}\sum_{\Omega_2^{(0)}\in F_r} \frac{1}{r|\Omega_3|^{2-2s}} \\ &\times \sum_{u\in\mathcal{O}_d^*}\int_0^\infty \frac{dt}{t^{2s-1}} e^{-\frac{\pi}{t}|\Omega_3|^2\mathcal{F}-\frac{2\pi t}{r^2}|u|^2-2\pi i\frac{1}{r}\Re(u(\bar{\Omega}_2^{(0)}-\bar{\Omega}_3 z_2))}. \end{aligned} \tag{27}$$

The second constant term arises from the $u=0$ part of the sum:

$$\mathcal{A}_s^{(\tilde{m}=u=0)} = \mathcal{F}^{2-s}\frac{4}{d}\frac{\pi^s\Gamma(s-1/2)\pi^{2s-1}\Gamma(2s-2)}{\Gamma(s)\pi^{s-1/2}\Gamma(2s-1)\pi^{2s-2}} \sum_{\Omega_3\neq 0}\sum_{\Omega_2^{(0)}\in F_r}\frac{1}{r|\Omega_3|^{2s-2}}. \tag{28}$$

Noting that r divides $|\Omega_3|^2$, and using similar techniques as in [1] to sum up the Dirichlet series $\sum_{r>0} N_d(r) r^s$, one can rewrite the sums over Ω_3 and $\Omega_2^{(0)}$ as follows:

$$\sum_{\Omega_3\neq 0}\sum_{\Omega_2^{(0)}\in F_r}\frac{1}{r|\Omega_3|^{2s-2}} = \left(\sum_{r>0} N_d(r) r^{1-2s}\right)\frac{\zeta_{\mathbb{Q}(i\sqrt{d})}(s-1)}{\zeta(2s-2)} = \frac{\beta_d(2s-2)}{\beta_d(2s-1)}\zeta_{\mathbb{Q}(i\sqrt{d})}(s-1), \tag{29}$$

Orchestrating, the second constant term (28) becomes

$$\mathcal{A}_s^{(\tilde{m}=u=0)} = \mathcal{F}^{2-s}\zeta_{\mathbb{Q}(i\sqrt{d})}(s)\frac{\mathfrak{Z}_d(2-s)}{\mathfrak{Z}_d(s)}, \tag{30}$$

where we have defined the 'completed Picard function'

$$\mathfrak{Z}_d(s) = \zeta^*_{\mathbb{Q}(i\sqrt{d})}(s)\beta_d^*(2s-1) \tag{31}$$

in terms of the completed Dedekind zeta and Dirichlet L-functions,

$$\begin{aligned}\zeta^*_{\mathbb{Q}(i\sqrt{d})}(s) &= \left(\frac{d}{4}\right)^{s/2}\pi^{-s}\,\Gamma(s)\,\zeta_{\mathbb{Q}(i\sqrt{d})}(s), \\ \beta_d^*(s) &= \left(\frac{d}{4}\right)^{(s+1)/2}\left(\frac{\pi}{4}\right)^{-(s+1)/2}\Gamma\left(\frac{s+1}{2}\right)\beta_d(s), \end{aligned} \tag{32}$$

both being invariant under the interchange $s \leftrightarrow 1-s$. The two constant terms (18) and (30) can be now summarized as

$$\mathcal{E}_s^{(\mathrm{const})}(\mathcal{Z}) = \zeta_{\mathbb{Q}(i\sqrt{d})}(s)\left[\mathcal{F}^s + \frac{\mathfrak{Z}_d(2-s)}{\mathfrak{Z}_d(s)}\mathcal{F}^{2-s}\right]. \tag{33}$$

The Abelian terms correspond to the terms in (27) with $u\neq 0$. The integral over t produces a sum of modified Bessel functions,

$$\begin{aligned}\mathcal{A}_s^{(\tilde{m}=0,\,u\neq 0)} &= \mathcal{F}\frac{2^{s+2}}{d}\frac{\pi^s\Gamma(s-1/2)\pi^{2s-1}}{\Gamma(s)\pi^{s-1/2}\Gamma(2s-1)}\sum_{\Omega_3\neq 0}\sum_{\Omega_2^{(0)}\in F_r} r^{1-2s}\sum_{u\in\mathcal{O}_d^*}|u|^{2s-2}e^{-2\pi i\frac{1}{r}\Re(u\bar{\Omega}_2^{(0)})} \\ &\times e^{2\pi i\frac{1}{r}\Re(u\bar{\Omega}_3 z_2)}K_{2s-2}\left(2\pi\mathcal{F}^{1/2}\left|\frac{\sqrt{2}\bar{u}\Omega_3}{r}\right|\right). \end{aligned} \tag{34}$$

Holding $\Lambda \equiv \bar{u}\Omega_3/r$ fixed and carrying out the sum over $u \in \mathcal{O}_d^*$, $\Omega_3 \in \mathcal{O}_d$ and $\Omega_2^{(0)} \in F_r$, this may be rewritten using the variable change (10) as

$$\mathcal{A}_s^{(\tilde{m}=0,\, u\neq 0)} = e^{-2\phi}\frac{\zeta_{\mathbb{Q}(i\sqrt{d})}(s)}{\mathfrak{Z}_d(s)}\sum_{\Lambda\neq 0}\mu_s(\Lambda)\, e^{2\pi i(p\zeta - q\tilde{\zeta})}K_{2s-2}\left(2\pi\sqrt{2}\, e^{-\phi}\frac{|q+\omega_d\, p|}{\sqrt{\Im(\omega_d)}}\right), \tag{35}$$

for a certain summation measure $\mu_s(\Lambda) \equiv \mu_s(p,q)$, where p, q are the components of Λ on the basis (u_1, u_2) of $\mathcal{O}_d^*$, namely

$$\Lambda = \frac{\bar{u}\Omega_3}{r} \equiv p u_1 - q u_2 = -i\frac{(q+\omega_d\, p)}{\Im(\omega_d)}\ . \tag{36}$$

The fact that $\mu_s(p,q)$ has support on integer charges $p, q \in \mathbb{Z}$ is not obvious from their definition, but is required by the periodicity conditions (5), and should result from cancellations in the sum over $\Omega_2^{(0)}$. Using similar methods that were employed in [1] for the Gaussian integers $\mathcal{O}_1 = \mathbb{Z}[i]$, it is presumably possible to express $\mu_s(\Lambda)$ as a divisor sum, but we have not attempted to do so. At any rate, we stress that this method is only applicable when $\mathcal{O}_d$ admits unique prime factorization, i.e. when d belongs to the Stark-Heegner sequence.

Finally, the non-Abelian terms correspond to the contributions in (22) with $\tilde{m} \neq 0$. These can also be simplified along the lines of [1] but we refrain from displaying the end result here. In view of the constant terms (33), it is natural to conjecture that the Poincaré series $\mathcal{P}_s(\mathcal{Z}) \equiv \mathcal{E}_s(\mathcal{Z})/\zeta_{\mathbb{Q}(i\sqrt{d})}(s)$ for any d in the Stark-Heegner sequence (2) satisfies the functional relation

$$\mathfrak{Z}_d(s)\,\mathcal{P}_s(\mathcal{Z}) = \mathfrak{Z}_d(2-s)\,\mathcal{P}_{2-s}(\mathcal{Z}), \tag{37}$$

generalizing the conjecture put forward in [1]. This would require that $\mu_s(\Lambda) = \mu_{2-s}(\Lambda)$ for the instanton measure in (35).

4. Discussion of the results

We now turn to the discussion of the various terms. The constant terms, summarized in (33), are interpreted as corrections to the classical hypermultiplet metric (4) that arise from string perturbation theory. Two different powers of the string coupling $g_s = e^{\phi} = \mathcal{F}^{-1/2}\sqrt{\Im(\omega_d)}$ occur in this expansion. Referring back to our proposal for the contact potential (6) we see that for the value $s = 3/2$ of the order of the Eisenstein series $\mathcal{E}_s$, these two terms can be matched with string tree level $g_s^{-2} = e^{-2\phi}$ and one loop g_s^0 contributions. The relative coefficient between these two terms is known from explicit string theory computations (see for example [15]) and unfortunately it does not match the numerical value contained in the ratio of completed zeta functions in (33). (In particular, this ratio is negative for all d in the Stark-Heegner sequence.) This indicates that our proposal does not work as such. Despite this, we shall nevertheless see that the remaining terms in the Fourier expansion match remarkably well with the expected form of instanton contributions to the hypermultiplet metric.

Having fixed the value of $s = 3/2$ from the perturbative terms we can extract the weak-coupling ($e^{\phi} \to 0$ or $\mathcal{F} \to \infty$) limit from the Abelian terms. Examining (34) one deduces that these terms are weighted by $e^{-S_{\rm D2}}$, where ($u \neq 0$)

$$S_{\rm D2} = 2\pi\sqrt{2}\, e^{-\phi}\frac{|q+\omega_d p|}{\sqrt{\Im(\omega_d)}} - 2\pi i(p\zeta - q\tilde{\zeta})\,. \tag{38}$$

This is recognized as the instanton action for Euclidean D2-branes wrapped on an arbitrary 3-cycle $\Lambda \in H_3(\mathcal{X}, \mathbb{Z})$. The real part contains the expected $1/g_s$-suppression and the imaginary part exhibits the correct axionic "theta angles". The instanton measure $\mu_{3/2}(\Lambda)$ in (35) should count special Lagrangian 3-cycles in the rigid Calabi-Yau threefold $\mathcal{X}$.

A similar analysis can be performed for the non-Abelian terms. This we do for simplicity for vanishing $\Omega_2 = 0$ which corresponds to pure NS5-brane instantons. One may then infer from (22) that the leading contributions are weighted by $e^{-S_{\text{NS5}}}$, with (for $\tilde{m} \neq 0$)

$$S_{\text{NS5}} = \pi \, |r\tilde{m}| \left(e^{-2\phi} + \frac{1}{2} \frac{|\zeta + \omega_d \tilde{\zeta}|^2}{\Im(\omega_d)} \right) - \pi i r \tilde{m} \sigma \,, \tag{39}$$

which may be identified with the action of a charge $r\tilde{m}$ Euclidean NS5-brane [5]. We note in particular that the real part is suppressed by $1/g_s^2$ which is a characteristic feature of NS5-brane instantons. See [1] for a more detailed discussion when $d = 1$.

Although our qualitative results are on the right track, as mentioned above the numerical predictions are incorrect. This indicates that our assumptions may be reasonable but that the construction requires some modification. Drawing inspiration from related work on implementing $SL(2,\mathbb{Z})$-invariance on the type IIB hypermultplet moduli space [16], it is natural to propose that the correct $PU(2,1;\mathcal{O}_d)$-invariant automorphic form should be constructed directly in terms of the complex Darboux coordinates on the twistor space $\mathcal{Z}_{\mathcal{M}}$. Mathematically, this implies that the relevant automorphic form would be attached to the quaternionic discrete series of $U(2,1)$ (see [17, 11]), rather than the principal continuous series considered here. Steps towards identifying the NS5-brane instantons in twistor space were taken recently in [18]. If this endeavour is successful it would be the first example where the exact metric on the hypermultiplet moduli space of a type II string compactification is known, including all D- and NS5-instanton corrections. Irrespective of their correct physical interpretation, we have defined and analyzed Eisenstein series for various Picard modular groups and conjectured their functional relations.

Acknowledgments

AK would like to thank the organizers of QTS6 for organizing such a broad conference and the invitation to present these results. The authors are grateful to T. Finis, J. Rogawski and J. Stienstra for correspondence and discussions. AK is a Research Associate of the Fonds National de la Recherche Scientifique-FNRS, Belgium.

References

[1] Bao L, Kleinschmidt A, Nilsson B E W, Persson D and Pioline B 2010 *Commun. Num. Theor. Phys.* **4** (1)
[2] Falbel E and Parker J R 2006 *Duke Math. J.* **131** 249
[3] Falbel E, Francsics G, Lax P D and Parker J R 2009 *Preprint* arXiv:0911.1104
[4] Jiang Y, Wang J and Xie B 2010 *Preprint* arXiv:1003.2705
[5] Becker K, Becker M and Strominger A 1995 *Nucl. Phys.* B **456** 130
[6] Becker K and Becker M 1999 *Nucl. Phys.* B **551** 102
[7] Salamon S M 1982 *Invent. Math.* **67** 1
[8] LeBrun C 1995 *Internat. J. Math.* **6** 419
[9] Alexandrov S, Pioline B, Saueressig F and Vandoren S 2010 *Commun. Math. Phys.* **296** 353
[10] Alexandrov S, Pioline B, Saueressig F and Vandoren S 2009 *JHEP* **0903** 044
[11] Gunaydin M, Neitzke A, Pavlyk O and Pioline B 2008 *Commun. Math. Phys.* **283** 169
[12] Ishikawa Y 1999 *J. Math. Sci. Univ. Tokyo* **6** 477
[13] Pioline B and Persson D 2009 *Commun. Num. Theor. Phys.* **3** (4) 697
[14] Langlands R P 1976 *Springer Lecture Notes in Mathematics* **544** 1
[15] Antoniadis I, Minasian R, Theisen S and Vanhove P 2003 *Class. Quant. Grav.* **20** 5079
[16] Alexandrov S and Saueressig F 2009 *JHEP* **0909** 108
[17] Gross B H and Wallach N R 1996 *J. Reine. Angew. Math.* **481** 73
[18] Alexandrov S, Pioline B and Vandoren S 2009, to appear in *J. Math. Phys.*, *Preprint* arXiv:0912.3406 [hep-th]

Journal of Physics: Conference Series **462** (2013) 012027 doi:10.1088/1742-6596/462/1/012027

Spiky strings in the SL(2) Bethe Ansatz

L. Freyhult[1], **M. Kruczenski**[2] **and A. Tirziu**[2]

[1] Department of Physics and Astronomy, Uppsala University,
P.O. Box 803, S-75108, Uppsala, Sweden
and [2]Department of Physics, Purdue University,
W. Lafayette, IN 47907-2036, USA.

E-mail: `lisa.freyhult@physics.uu.se`, `markru@purdue.edu`, `atirziu@purdue.edu`

Abstract. We consider spiky strings in the context of the SL(2) Bethe ansatz equations. We obtain an asymmetric distribution of Bethe roots along one cut which determines the all loop anomalous dimension at leading and subleading orders in a large S expansion. At leading order in strong coupling (large λ) the energy of such states is given, in terms of the spin S and the number of spikes n by $E-S=\frac{n\sqrt{\lambda}}{2\pi}\left(\ln\frac{4\pi S}{\sqrt{\lambda}}+\ln\frac{4}{n}\sin\frac{\pi}{n}-1\right)+\mathcal{O}\left(\frac{\ln S}{S}\right)$. This result agrees perfectly with the same expansion obtained from the known spiky string classical solution. We also discuss a two cut spiky string Bethe root distribution at one-loop in the SL(2) Bethe ansatz.

1. Introduction

The study of the $SL(2)$ sector of $\mathcal{N}=4$ SYM gauge theory has lead to a remarkable progress in the understanding of the AdS/CFT correspondence. Much of this progress relates to twist two gauge theory operators of the type $\mathrm{tr}(\Phi D_+^S\Phi)$. On the string side they are described by the folded string solution [1] as can also be verified by an alternative computation in terms of the cusp anomaly [2]. In the field theory an all loop Bethe ansatz was proposed in [3] which can be used to compute their anomalous dimension at all loops in the planar approximation. Computations of quantum corrections to the folded string solution [4] at leading order in large S as well as its generalization to $AdS_3\times S^1$ [5, 6, 7] were recently used to successfully check such all loop asymptotic Bethe ansatz [8, 9, 10, 11]. The all loop asymptotic Bethe ansatz computation was further extended to the subleading orders in the large S expansion [11, 12, 13] and exact matching with the corresponding string theory computations [14] was found.

Beyond twist two, the $SL(2)$ sector contains operators of higher twist. For example operators of the type

$$\mathcal{O}=\mathrm{tr}\left(D_+^{\frac{S}{n}}\Phi\, D_+^{\frac{S}{n}}\Phi\dots D_+^{\frac{S}{n}}\Phi\right) \tag{1.1}$$

are described, on the string side, by the spiky string solutions [15]. By analogy with the twist two operators, the full understanding of these operators from the field theory point of view requires the construction of the Bethe ansatz solution that describes them. In this paper we provide a proposal to describe the spiky string solution in the all loop Bethe ansatz.

In [16, 17] and later in [18] such operators were described by a spin chain with a number of sites n being the same as the number of spikes. At leading order in large S the spiky string solution touches the boundary of AdS_3. It was shown [19] that at this order the spiky string solution can be mapped to the folded string solution by a conformal transformation, and thus the all loop energy at leading order in large S is given by

$$E - S = \frac{n}{2} f(\lambda) \ln S + \mathcal{O}(S^0, n) \tag{1.2}$$

where $f(\lambda)$ is a universal scaling function known as the cusp anomaly. Here we obtain this result from the field theory side by an all loop Bethe ansatz computation.

An important ingredient in the construction of the Bethe ansatz solution is a set of integers n_k which may be interpreted as the bosonic mode numbers of the waves propagating in the spin chain. The folded string solution in AdS_3 reduces [20] to the folded string solution in flat space in the limit $\frac{S}{\sqrt{\lambda}} \ll 1$. In flat space we can quantize the string exactly in terms of right and left moving waves propagating on the string. The folded string has the same number of right and left moving excitations, i.e. $n_L = n_R$, with the same wave numbers $k_L = k_R = 1$. In AdS_3, for all values of spin S and coupling λ, it was proposed that in a Bethe ansatz computation the folded string solution is described by two sets of modes, one with $n_k = -1$ and the other with $n_k = 1$. These numbers were used then as the input in the one loop SL(2) Bethe ansatz equations for two cuts (corresponding to $n_k = \pm 1$), and a symmetric Bethe root distribution was found to describe the folded string solution [21]. Similar bosonic quantum numbers were used later [22, 3] to find the corresponding solution to the all loop Bethe ansatz.

In our case, we look at the spiky string solution in flat space, find the left and right wave numbers and use them as an input to find the corresponding Bethe root equation that we then proceed to solve. It turns out that the bosonic quantum numbers that should be used in the Bethe ansatz equations are $n_L = -1$ and $n_R = n - 1$. Using this input we find an asymmetric distribution of Bethe roots spread over one cut along the real axis at one loop in weak coupling and leading order in the large S expansion. We then extend the computation to all loops and further obtain the all loop result also for the subleading order in large S. Remarkably, the result expanded at strong coupling precisely matches the result known from expanding the classical string solution [14], that is, at order $\ln S$ and also S^0.

We then proceed to consider a two cut solution to the one-loop Bethe ansatz equations. Again we use the "spiky string" bosonic quantum numbers for the integers n_k, i.e. one cut has $n_k = -1$ and the other $n_k = n - 1$ where n is the number of spikes. The solution has four real parameters $d < c < a < b$ which represent the position of the two cuts given by the segments $[d, c]$ and $[b, a]$. The solution obtained allows us to compute the energy E, spin S, R-charge J and number of spikes n in terms of these parameters, allowing to express the energy as a function of the other physical quantities $E = E(S, J, n)$.

2. Spiky strings in flat space

In flat space $ds^2 = -dt^2 + dx^2 + dy^2$ we can quantize the theory exactly, and it turns out that the spiky solutions are just a superpositions of a left and a right moving waves:

$$x = A \cos\left((n-1)\,\sigma_+\right) + A\,(n-1)\cos\left(\sigma_-\right) \tag{2.1}$$

$$y = A \sin\left((n-1)\,\sigma_+\right) + A\,(n-1)\sin\left(\sigma_-\right) \tag{2.2}$$

$$t = 2\,A\,(n-1)\,\tau = A\,(n-1)\,(\sigma_+ + \sigma_-) \tag{2.3}$$

where $\sigma_+ = \tau + \sigma$, $\sigma_- = \tau - \sigma$, n is the number of spikes and A is constant that determines the size of the string. Here, (τ, σ) parameterize the world-sheet of a string. The solutions are periodic in σ with period 2π and satisfy the equations of motion in conformal gauge, $(\partial_\tau^2 - \partial_\sigma^2)X^\mu = 0$, as well as the constraints $(\partial_+ X)^2 = (\partial_- X)^2 = 0$. Quantum mechanically the state has $n_R = A^2(n-1)^2$ right moving excitations of wave number $k_R = 1$ and $n_L = A^2(n-1)$ left moving excitations with wave number $k_L = n - 1$ (satisfying the level matching condition $n_R k_R = n_L k_L$). All excitations carry one

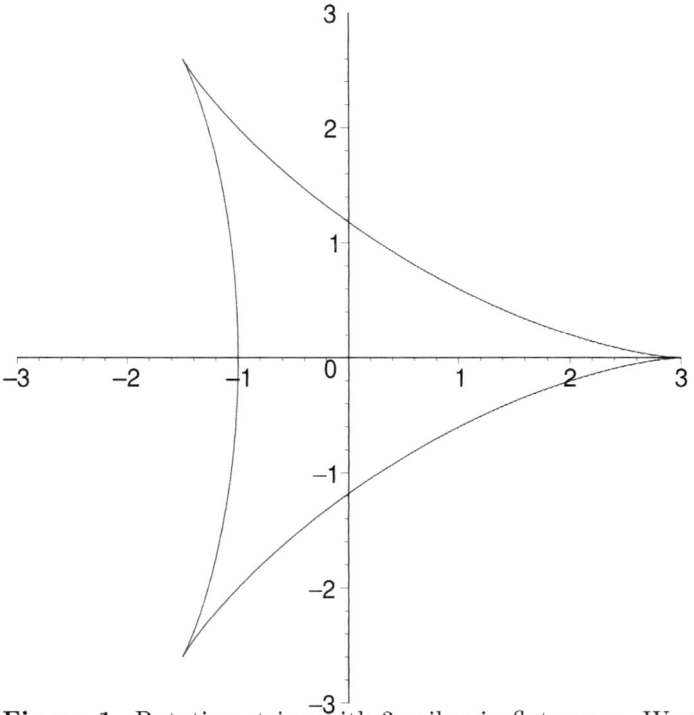

Figure 1. Rotating string with 3 spikes in flat space. We plot (x, y) in eq.(2.3) parametrically as a function of $\sigma = 0 \to 2\pi$ for $n = 3$, $A = 1$ and $\tau = 0$.

unit of angular momentum and therefore the total angular momentum and energy are given by

$$S = n_L + n_R = A^2 n(n-1), \quad E = \sqrt{2(n_L k_L + n_R k_R)} = 2A(n-1), \quad E = 2\sqrt{\frac{n-1}{n}S} \qquad (2.4)$$

which agrees with a classical computation. For $n = 2$ we recover the standard Regge trajectory $E = \sqrt{2S}$ and for $n > 2$ we get a Regge trajectory of modified slope. At fixed time, the shape of the string for $n = 3$ is shown in fig. 1.

3. Spiky strings in the one-loop Bethe Ansatz: 1-cut solution

To obtain the $\ln S$ dependence from the Bethe ansatz at large S we follow the same prescription used in [22]. In the SL(2) sector the Bethe roots are on the real axis and, for large S, they accumulate on various segments or cuts. For the folded string with $J = 2$ and $S \gg 1$, there is only one cut and the root distribution is symmetric. In this case, namely for the lowest twist, there is only one state whereas for $J > 2$ there is more than one state. We consider excited energy states which describe the n-spike solution in the large S limit for a fixed finite J.

We first consider the one loop SL(2) spin chain. The one loop Bethe ansatz equations corresponding to a $XXX_{-\frac{1}{2}}$ nearest neighbor spin chain are

$$\left(\frac{u_k + \frac{i}{2}}{u_k - \frac{i}{2}}\right)^J = \prod_{j \neq k}^{S} \frac{u_k - u_j - i}{u_k - u_j + i}, \qquad k = 1, 2, ..., S \qquad (3.1)$$

Journal of Physics: Conference Series **462** (2013) 012027 doi:10.1088/1742-6596/462/1/012027

Taking the logarithm of (3.1), and further expanding in $\frac{1}{u}$ as appropriate for large S gives [22]

$$\frac{J}{u_k} = 2\pi n_k - 2\sum_{j=1,j\neq k}^{S} \frac{1}{u_k - u_j} \tag{3.2}$$

where u_k are the Bethe roots and J is the length of the corresponding spin chain. The total momentum should vanish due to the cyclicity of the trace which implies

$$\prod_{j=1}^{S} \frac{u_j + \frac{i}{2}}{u_j - \frac{i}{2}} = 1 \tag{3.3}$$

The one loop energy is given by

$$\frac{E - S - J}{2g^2} = \sum_{k=1}^{S} \frac{1}{u_k^2 + \frac{1}{4}} \tag{3.4}$$

At this point one should solve eq.(3.2) for the real numbers u_k subject to the condition (3.3). When the number of roots S is very large, the roots accumulate in cuts on the real axis and one can conveniently approximate the problem by defining a density of roots. Thus we introduce the root distribution function $\rho_0(u) = \sum_k \delta(u - u_k)$. Furthermore, for large S, the cut grows as S and we can make the approximation that $u \gg 1$. Therefore, at leading order in the large S expansion (3.2), we need to solve the equations

$$\fint du \frac{\rho_0(u)}{u' - u} = \pi n_w(u') \tag{3.5}$$

$$\int du \rho_0(u) = S \tag{3.6}$$

$$\fint du \frac{\rho_0(u)}{u} = 0, \tag{3.7}$$

which are, the Bethe equation, the normalization of the density, and the zero momentum condition. The function $\rho_0(u)$ is non-zero on a finite interval that needs to be determined as part of the solution and that we take to extend from $d < x < a$ with $d < 0$ and $a > 0$.

To solve (3.5) we also need the bosonic numbers n_w. Following the intuition from the folded string at all-loops here we identify n_w with the left and right wave numbers of the corresponding flat space solution, i.e. $n_w = -1$ and $n_w = n - 1$ where n is the number of spikes. More precisely we take the function $n_w(u)$ as

$$n_w(u) = \begin{cases} -1, & d < u < 0 \\ n - 1, & 0 < u < a \end{cases} \tag{3.8}$$

To solve (3.5) we introduce a function $F(w) = \sqrt{w - a}\sqrt{w - d}$ which has a cut on the real axis extending precisely from d to a. Furthermore, right above (below) the cut it is purely imaginary with positive (negative) imaginary part. Using this function we define the resolvent as (see [23] for details)

$$G(w) = \frac{1}{\pi}\int_d^a \frac{F(w)}{F(u + i0^+)} \frac{n_w(u)}{u - w} du \tag{3.9}$$

On the cut we get

$$G(x \pm i\epsilon) = \pm\rho(x) + i n_w(x), \quad x \in C, \quad \epsilon \to 0^+. \tag{3.10}$$

with $\rho_0(u)$ determined to be

$$\rho_0(u) = \frac{1}{\pi} \fint_d^a \left| \frac{F(u)}{F(u')} \right| \frac{n_w(u')}{u' - u} du' \tag{3.11}$$

It turns out that for (3.11) to be a solution of (3.5) we need to satisfy the consistency condition

$$\text{Res}\left[\frac{G(w)}{w-u}, \infty\right] = -G_0 = 0 \tag{3.12}$$

where G_0 is the leading coefficient of the expansion of $G(w)$ at large w

$$G(w) \simeq G_0 + G_1 \frac{1}{w} + \ldots, \qquad w \to \infty \tag{3.13}$$

Explicitly for 1-cut we obtain

$$G_0 = -\frac{1}{\pi} \int_d^a \frac{n_w(u)}{F(u + i0^+)} du = i\frac{n-2}{2} + \frac{in}{\pi} \arcsin\left(\frac{a+d}{a-d}\right) \tag{3.14}$$

$$G_1 = -\frac{a+d}{2} G_0 - \frac{1}{\pi} \int_d^a \frac{u n_w(u)}{F(u + i0^+)} du \tag{3.15}$$

On the other hand, equation (3.6) gives

$$S = \int_d^a \rho_0(u) du = \frac{1}{2} \oint_C G(w) dw = i\pi \text{Res}\left[G(w), \infty\right] = -i\pi G_1 \tag{3.16}$$

and eq.(3.7) is

$$0 = \fint_d^a \frac{\rho_0(u)}{u} du = \frac{1}{2} \oint_C \frac{G(w)}{w} dw = i\pi \text{Res}\left[\frac{G(w)}{w}, \infty\right] = i\pi G_0 \tag{3.17}$$

which is satisfied if (3.12) is. In the previous equations we used a contour C that encircles the cut in a clockwise direction. Using (3.14) we then find

$$-\frac{a}{d} = \tan^2\left(\frac{\pi}{2n}\right) \tag{3.18}$$

Using (3.15) and (3.16) gives

$$S = n\sqrt{-ad} \tag{3.19}$$

We now need to compute the energy which, within the Bethe ansatz is given by

$$\frac{E-S-J}{2g^2} = \int_d^a \frac{\rho_0(u)}{u^2 + \frac{1}{4}} du = \pi\left(G(\tfrac{i}{2}) - G(-\tfrac{i}{2})\right) = n \ln \left| \frac{\left(\sqrt{a(\frac{i}{2} - d)} + \sqrt{(a - \frac{i}{2})(-d)}\right)^2}{-\frac{i}{2}(a-d)} \right| \tag{3.20}$$

For fixed n and large S we see from eqs.(3.18) and (3.19) that $a \sim d \sim S \gg 1$ so we can approximate the last equation as

$$\frac{E-S-J}{2g^2} \cong n \ln\left(-\frac{8ad}{a-d}\right), \qquad S \gg 1 \tag{3.21}$$

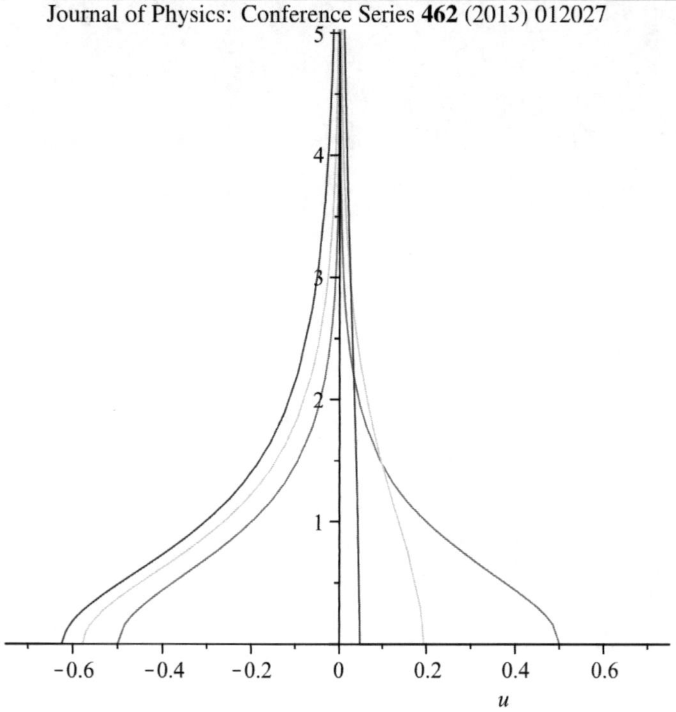

Figure 2. Root density $\rho_0(\bar{u}) = \rho_0(\frac{u}{S})$ for $n = 2$, $n = 3$ and $n = 6$. The larger the n the more skewed the distribution becomes.

Therefore, through eqs.(3.19), (3.21), we have solved the problem by writing E and S in terms of two parameters a and d related by the condition (3.18). Explicitly we can write

$$a = \frac{S}{n}\tan\frac{\pi}{2n}, \qquad d = -\frac{S}{n}\cot\frac{\pi}{2n}, \qquad \frac{E-S-J}{2g^2} = 2n\ln\left(\frac{4S}{n}\sin\frac{\pi}{n}\right) \tag{3.22}$$

The root distribution is asymmetric. Using eq.(3.11) it can be written as

$$\rho_0(u) \;=\; \frac{n}{\pi}\ln\left|\frac{\sqrt{a(u-d)}+\sqrt{-d(a-u)}}{\sqrt{a(u-d)}-\sqrt{-d(a-u)}}\right| \tag{3.23}$$

$$\tag{3.24}$$

Typical plots of the root distribution are shown in Figure 2.

4. Spiky strings in the all loop Bethe ansatz: 1-cut solution

We proceed to find a 1-cut solution to the all-loop Bethe equations. We will follow the procedure used in [22]. The asymptotic all loop Bethe equations in the SL(2) sector read

$$\begin{aligned} 2J\arctan(2u_k) + iJ\ln\frac{1+g^2/(x_k^-)^2}{1+g^2/(x_k^+)^2} &= 2\pi\tilde{n}_k - 2\sum_{\substack{j=-S/2\\ j\neq 0}}^{S/2}\arctan(u_k-u_j) \\ &+ 2i\sum_{\substack{j=-S/2\\ j\neq 0}}^{S/2}\ln\frac{1-g^2/x_k^+x_j^-}{1-g^2/x_k^-x_j^+} - 2\sum_{\substack{j=-S/2\\ j\neq 0}}^{S/2}\theta(u_k,u_j), \end{aligned} \tag{4.1}$$

where $\theta(u,v)$ denotes the dressing phase and $x^{\pm}(u)$ are defined through $u \pm \frac{i}{2} = x^{\pm}(u) + \frac{g^2}{x^{\pm}(u)}$.

As in the one-loop Bethe ansatz, in the large S limit we introduce a root density $\rho(u)$. The all loop equations can then be written as

$$\begin{aligned}
&\frac{J}{u^2+1/4} + iJ\frac{d}{du}\ln\frac{1+g^2/(x^-(u))^2}{1+g^2/(x^+(u))^2} = 2\pi\rho(u) - 2\int_a^d \frac{\rho(u')}{(u-u')^2+1} \\
&+ \int_a^d du'\rho(u')\frac{d}{du}\left(2i\ln\frac{1-g^2/x^+(u)x^-(u')}{1-g^2/x^+(u')x^-(u)} - 2\theta(u,u')\right)
\end{aligned} \tag{4.2}$$

For the highest excited state J is equal to the number of cusps, n.

To simplify the presentation of the all-loop computation we will omit the dressing phase and only restore its contribution in the end. We can now proceed with the all loop equation by splitting off the one-loop piece of the density, $\rho(u) = \rho_0(u) + 2g^2\tilde{\sigma}(u)$,

$$\begin{aligned}
0 &= 2\pi\tilde{\sigma}(u) - 2\int_{-\infty}^{\infty}\frac{\tilde{\sigma}(u')}{(u-u')^2+1} + \frac{1}{2g^2}\int_d^a du'\rho_0(u')\frac{d}{du}\left(2i\ln\frac{1-g^2/x^+(u)x^-(u')}{1-g^2/x^+(u')x^-(u)}\right) \\
&+ \int_{-\infty}^{\infty} du'\tilde{\sigma}(u')\frac{d}{du}\left(2i\ln\frac{1-g^2/x^+(u)x^-(u')}{1-g^2/x^+(u')x^-(u)}\right).
\end{aligned} \tag{4.3}$$

In the above we have extended the integral boundaries to $\pm\infty$ in the terms containing the higher loop density, $\tilde{\sigma}(u)$.

Using the one-loop resolvent at large S, $G\left(\pm\frac{i}{2}\right) = \pm\frac{n}{\pi}\ln S + \ldots$, we find

$$\begin{aligned}
0 &= 2\pi\tilde{\sigma}(u) - 2\int_{-\infty}^{\infty}\frac{\tilde{\sigma}(u')}{(u-u')^2+1} + n\ln S\frac{d}{du}\left(\frac{1}{x^+(u)} + \frac{1}{x^-(u)}\right) \\
&+ \int_{-\infty}^{\infty} du'\tilde{\sigma}(u')\frac{d}{du}\left(2i\ln\frac{1-g^2/x^+(u)x^-(u')}{1-g^2/x^+(u')x^-(u)}\right).
\end{aligned} \tag{4.4}$$

After Fourier transformation[1] and a redefinition of the density,

$$\hat{\sigma}(t) = -2ne^{t/2}\hat{\tilde{\sigma}}(t)\ln S, \tag{4.5}$$

the result is

$$\hat{\sigma}(t) = \frac{t}{e^t-1}\left(K(2gt,0) - 4g^2\int_0^{\infty} dt' K(2gt,2gt')\hat{\sigma}(t')\right) \tag{4.6}$$

Interestingly, the contribution from the dressing phase can be included in the kernel K

$$K(t,t') = K_0(t,t') + K_1(t,t') + K_d(t,t') \tag{4.7}$$

where

$$K_0(t,t') = \frac{2}{tt'}\sum_{n=1}^{\infty}(2n-1)J_{2n-1}(t)J_{2n-1}(t'), \qquad K_1(t,t') = \frac{2}{tt'}\sum_{n=1}^{\infty}2nJ_{2n}(t)J_{2n}(t') \tag{4.8}$$

$$K_d(t,t') = 8g^2\int_0^{\infty} dt'' K_1(t,2gt'')\frac{t''}{e^{t''}-1}K_0(2gt'',t'). \tag{4.9}$$

[1] We use the convention $\hat{f}(t) = \int_{-\infty}^{\infty} dw\, e^{-iwt}f(w)$. Details on the Fourier transforms used here can be found in [22].

Further we find that the energy is given by

$$E - S = 8g^2\, n\hat{\sigma}(0) \ln S + \mathcal{O}(S^0). \tag{4.10}$$

which gives

$$E - S = \frac{n}{2} f(g) \ln S + \mathcal{O}(S^0) \tag{4.11}$$

which matches the string theory result at leading order in large S [19].

The results discussed in section 3 and 4 can be extended to the next order in the large S expansion. We obtain the following all loop result

$$E - S = n + f(g)\frac{n}{2}\left(\ln S + \gamma_E + \ln\left(\frac{2}{n}\sin\frac{\pi}{n}\right)\right) + \frac{n}{2}B_2(g) + \ldots \tag{4.12}$$

where $B_2(g)$ denotes the virtual scaling function of twist 2 operators. Using the first orders in the expansions at strong coupling [12],

$$f(g) = 4g - \frac{3\ln 2}{\pi} + \mathcal{O}(1/g) \tag{4.13}$$

$$B_2(g) = (-\gamma_E - \ln g)\, f(g) - 4g(1 - \ln 2) - \left(1 - \frac{6\ln 2}{\pi} + \frac{3(\ln 2)^2}{\pi}\right) + \mathcal{O}(1/g) \tag{4.14}$$

we find

$$\begin{aligned} E - S &= 2ng\left(\ln\frac{S}{g} + \ln\left(\frac{4}{n}\sin\frac{\pi}{n}\right) - 1\right) \\ &+ \frac{n}{2}\left(1 + \frac{6\ln 2}{\pi} - \frac{3\ln 2}{\pi}\ln\left(\frac{4}{n}\sin\frac{\pi}{n}\right) - \frac{3\ln 2}{\pi}\ln\frac{S}{g}\right) + \mathcal{O}(1/g) \end{aligned} \tag{4.15}$$

We see that the leading strong coupling result is in agreement with the known string theory result [14]. For $n = 2$ we recover the result for the folded string [12], this result is in full agreement with the one-loop computation from string theory [14]. For arbitrary n (4.15) is expected to be in agreement with the one-loop result computed from the sigma model [24].

5. Spiky strings in the one-loop Bethe Ansatz: 2-cuts solution

We start again with equation (3.2) but we no longer drop the term proportional to J

$$⨍ du' \frac{\rho_0(u')}{u - u'} = \pi n_w(u) - \frac{J}{2u} \equiv \pi N_w(u) \tag{6.1}$$

The condition that the total momentum should vanish, and the normalization of the root density give

$$\int du \rho_0(u) \ln\frac{u + \frac{i}{2}}{u - \frac{i}{2}} = 0, \qquad \int du \rho_0(u) = S \tag{6.2}$$

The 1-loop anomalous dimension is

$$\frac{E - S - J}{2g^2} = \int du \frac{\rho_0(u)}{u^2 + \frac{1}{4}} \tag{6.3}$$

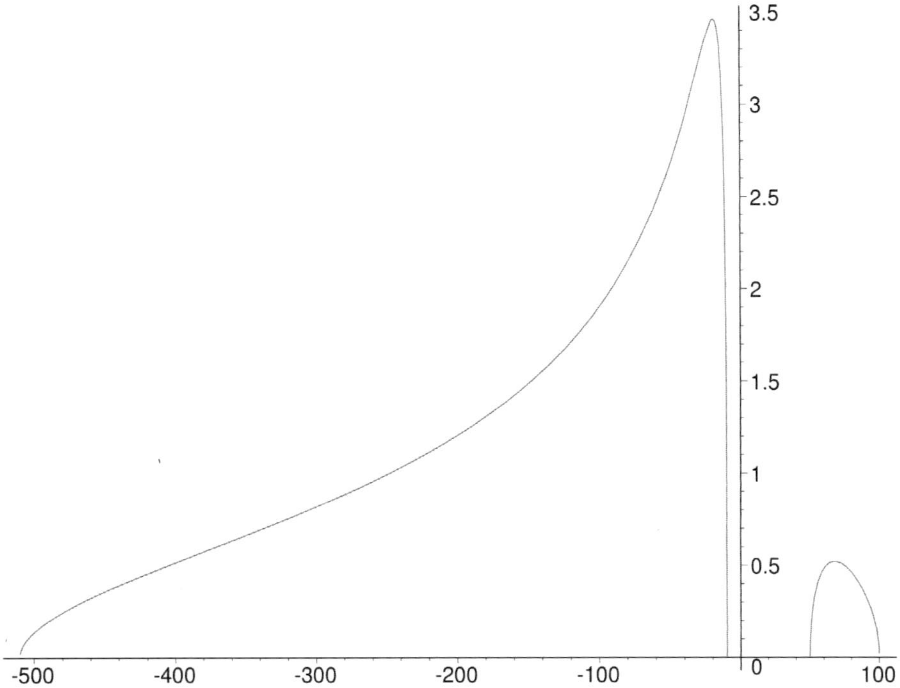

Figure 3. Root density $\rho_0(\bar{u}) = \rho_0(\frac{u}{S})$ for $n = 3$ and $a = 100, b = 50, c = -9.771958775, d = -510.0091708$.

Let us consider two cuts $d \leq c \leq b \leq a$. Extending the bosonic wave number distribution from the previous section we take

$$n_w(u) = \begin{cases} -1, & d < u < c \\ n-1, & b < u < a \end{cases} \tag{6.4}$$

Then we find the solution for the root distribution function is ($r = \frac{(a-b)(c-d)}{(a-c)(b-d)}$)

$$\begin{aligned} \rho_{0L}(u') &= -\frac{2n}{\pi}\sqrt{\frac{(c-u')(b-u')}{(u'-d)(a-u')}}\frac{1}{\sqrt{(a-c)(b-d)}}\left((a-d)\Pi[\frac{(b-a)(u'-d)}{(b-d)(u'-a)},r] + (u'-a)K[r]\right) \\ &+ \frac{J}{2\pi^2 u'}\sqrt{|F(u')|}(A_4 - A_5) \end{aligned} \tag{6.5}$$

for the left interval $u' \in [d, c]$, and

$$\begin{aligned} \rho_{0R}(u') &= \frac{2n}{\pi}\sqrt{\frac{(u'-c)(u'-b)}{(u'-d)(a-u')}}\frac{1}{\sqrt{(a-c)(b-d)}}\left((d-a)\Pi[\frac{(d-c)(u'-a)}{(a-c)(u'-d)},r] + (u'-d)K[r]\right) \\ &+ \frac{J}{2\pi^2 u'}\sqrt{|F(u')|}(-A_4 + A_5) \end{aligned} \tag{6.6}$$

for the right one $u' \in [b, a]$. We show an example of a 2-cut root distribution in figure 3. The existence of this solution implies two relationships among the arbitrary parameters a, b, c, d

$$nA_1 + \frac{J}{2\pi}(A_4 - A_5) = 0, \qquad A_2 + (n-1)A_3 = 0 \tag{6.7}$$

Normalization and momentum conditions give two additional equations

$$\frac{J}{2\pi}(A_2 - A_3) + A_6 + (n-1)A_7 = S, \quad \sqrt{abcd}[A_4 + (n-1)A_5 + \frac{J}{2\pi}(A_8 - A_9)] = 0 \tag{6.8}$$

A_i are a, b, c, d -dependent elliptic integrals defined in [23]. The four equations can be solved in principle for the four parameters a, b, c, d in terms of S, J, n and thus the one-loop dimension can be obtained as $E = E(n, S, J)$. The 1-cut solution of section 3 can be recovered by taking the limit $b = \epsilon$, $c = -\eta\epsilon$ with $\epsilon \to 0_+$.

Acknowledgments
We are grateful to J. Minahan for useful discussions and to the organizers of the QTS-6 conference for allowing us to present this work. M. K. and A.T. were supported in part by NSF under grant PHY-0805948, DOE under grant DE-FG02-91ER40681 and the A.P. Sloan foundation.

References

[1] S. S. Gubser, I. R. Klebanov and A. M. Polyakov, "A semi-classical limit of the gauge/string correspondence," Nucl. Phys. B **636**, 99 (2002) [hep-th/0204051].
[2] M. Kruczenski, "A note on twist two operators in N = 4 SYM and Wilson loops in Minkowski signature," JHEP **0212**, 024 (2002) [arXiv:hep-th/0210115],
Y. Makeenko, "Light-cone Wilson loops and the string / gauge correspondence," JHEP **0301**, 007 (2003) [arXiv:hep-th/0210256].
[3] N. Beisert, B. Eden and M. Staudacher, "Transcendentality and crossing," J. Stat. Mech. **0701**, P021 (2007) [arXiv:hep-th/0610251].
[4] S. Frolov and A. A. Tseytlin, "Semiclassical quantization of rotating superstring in AdS(5) x S(5)," JHEP **0206**, 007 (2002) [hep-th/0204226].
[5] S. Frolov, A. Tirziu and A. A. Tseytlin, "Logarithmic corrections to higher twist scaling at strong coupling from AdS/CFT," Nucl. Phys. B **766**, 232 (2007) [arXiv:hep-th/0611269].
[6] R. Roiban, A. Tirziu and A. A. Tseytlin, "Two-loop world-sheet corrections in AdS5 x S5 superstring," JHEP **0707**, 056 (2007) [arXiv:0704.3638].
R. Roiban and A. A. Tseytlin, "Strong-coupling expansion of cusp anomaly from quantum superstring," JHEP **0711**, 016 (2007) [arXiv:0709.0681].
[7] R. Roiban and A. A. Tseytlin, "Spinning superstrings at two loops: strong-coupling corrections to dimensions of large-twist SYM operators," Phys. Rev. D **77**, 066006 (2008) [arXiv:0712.2479].
[8] M. K. Benna, S. Benvenuti, I. R. Klebanov and A. Scardicchio, "A test of the AdS/CFT correspondence using high-spin operators," Phys. Rev. Lett. **98**, 131603 (2007) [arXiv:hep-th/0611135].
[9] B. Basso, G. P. Korchemsky and J. Kotanski, "Cusp anomalous dimension in maximally supersymmetric Yang-Mills theory at strong coupling," Phys. Rev. Lett. **100**, 091601 (2008) [arXiv:0708.3933].
[10] P. Y. Casteill and C. Kristjansen, "The Strong Coupling Limit of the Scaling Function from the Quantum String Bethe Ansatz," Nucl. Phys. B **785**, 1 (2007) [arXiv:0705.0890].
L. F. Alday and J. M. Maldacena, "Comments on operators with large spin," JHEP **0711**, 019 (2007) [arXiv:0708.0672].
B. Basso and G. P. Korchemsky, "Embedding nonlinear O(6) sigma model into N=4 super-Yang-Mills theory," Nucl. Phys. B **807**, 397 (2009) [arXiv:0805.4194].
[11] L. Freyhult, A. Rej and M. Staudacher, "A Generalized Scaling Function for AdS/CFT," J. Stat. Mech. **0807**, P07015 (2008) [arXiv:0712.2743].
[12] L. Freyhult and S. Zieme, arXiv:0901.2749 [hep-th].
[13] D. Fioravanti, P. Grinza and M. Rossi, "Beyond cusp anomalous dimension from integrability," arXiv:0901.3161 [hep-th].
[14] M. Beccaria, V. Forini, A. Tirziu and A. A. Tseytlin, "Structure of large spin expansion of anomalous dimensions at strong coupling," Nucl. Phys. B **812** (2009) 144 [arXiv:0809.5234 [hep-th]].

[15] M. Kruczenski, "Spiky strings and single trace operators in gauge theories," JHEP **0508**, 014 (2005) [arXiv:hep-th/0410226]

[16] A. V. Belitsky, A. S. Gorsky and G. P. Korchemsky, "Gauge / string duality for QCD conformal operators," Nucl. Phys. B **667**, 3 (2003) [arXiv:hep-th/0304028].
A. V. Belitsky, A. S. Gorsky and G. P. Korchemsky, "Logarithmic scaling in gauge / string correspondence," Nucl. Phys. B **748**, 24 (2006) [arXiv:hep-th/0601112].
A. V. Belitsky, G. P. Korchemsky and R. S. Pasechnik, "Fine structure of anomalous dimensions in N=4 super Yang-Mills theory," Nucl. Phys. B **809**, 244 (2009) [arXiv:0806.3657].

[17] V. A. Kazakov and K. Zarembo, "Classical / quantum integrability in non-compact sector of AdS/CFT," JHEP **0410**, 060 (2004) [arXiv:hep-th/0410105].

[18] N. Dorey, "A Spin Chain from String Theory," arXiv:0805.4387.
N. Dorey and M. Losi, "Spiky Strings and Spin Chains," arXiv:0812.1704 [hep-th].

[19] M. Kruczenski and A. A. Tseytlin, "Spiky strings, light-like Wilson loops and pp-wave anomaly," Phys. Rev. D **77**, 126005 (2008) [arXiv:0802.2039 [hep-th]].

[20] A. Tirziu and A. A. Tseytlin, "Quantum corrections to energy of short spinning string in AdS5," Phys. Rev. D **78**, 066002 (2008) [arXiv:0806.4758 [hep-th]].

[21] N. Beisert, S. Frolov, M. Staudacher and A. A. Tseytlin, "Precision spectroscopy of AdS/CFT," JHEP **0310**, 037 (2003) [arXiv:hep-th/0308117].

[22] B. Eden and M. Staudacher, "Integrability and transcendentality," J. Stat. Mech. **0611**, P014 (2006) [arXiv:hep-th/0603157].

[23] L. Freyhult, M. Kruczenski and A. Tirziu, "Spiky strings in the SL(2) Bethe Ansatz," JHEP **0907**, 038 (2009) [arXiv:0905.3536 [hep-th]].

[24] N. Gromov and P. Vieira, "Complete 1-loop test of AdS/CFT," JHEP **0804** (2008) 046 [arXiv:0709.3487 [hep-th]].

Journal of Physics: Conference Series **462** (2013) 012028 doi:10.1088/1742-6596/462/1/012028

Gravity solutions duals to non-relativistic field theories

Arnab Kundu

Department of Physics and Astronomy. University of Southern California. Los Angeles, CA 90089-0484, USA.

E-mail: akundu@usc.edu

Abstract. In this note we discuss solutions of type IIB supergravity with an isometry group of non-relativistic scaling symmetry and in a special case, the well-known Schrödinger symmetry. In particular we discuss supersymmetric backgrounds with Schrödinger isometry and renormalization group flow geometries with Schrödinger symmetry at the fixed points. We also discuss scale-invariant gravity backgrounds with a general dynamical exponent. This is a summary based on the following works: N. Bobev and A. Kundu, JHEP 0907:098,2009 [arXiv:0904.2873 [hep-th]]; N. Bobev, A. Kundu and K. Pilch, JHEP 0907:107,2009 [arXiv:0905.0673 [hep-th]].

1. Introduction

In recent years gauge-gravity duality has emerged both as a remarkable concept and a novel tool. For a recent and application oriented review of this duality we refer the reader to [1]. The techniques of gauge-gravity duality have been extensively used in understanding aspects of strong coupling dynamics in relativistic systems ranging from quark-gluon plasma to several condensed matter systems [2]. There are several strongly coupled systems which are inherently non-relativistic. Over the past couple of years there has been active research towards constructing holographic duals of non-relativistic field theories.

The primary focus has been to consider non-relativistic field theories which exhibit scale invariance. This scale invariance is characterized by an anisotropic scaling between the spatial and the temporal directions and this anisotropy is parametrized by a real number known as the dynamical exponent. For a special value of this dynamical exponent the non-relativistic scaling symmetry can be extended to include a special conformal generator. The corresponding symmetry group of this extended algebra is known as the Schrödinger group. Physical systems such as free fermions or the cold atoms at unitarity have been observed to possess this symmetry [3]. More details on non-relativistic conformal field theories can be found in *e.g.* [4].

One of the guiding principles behind any holographic construction is to match the global symmetry group of the gauge theory with the isometry group of the proposed dual gravitational background. This has been the basic philosophy behind the pioneering work in [5, 6] where a five-dimensional Schrödinger symmetric background was proposed and the subsequent work in [7, 8, 9] where this background was embedded in ten dimensional supergravity. In this note, we will discuss several other gravitational backgrounds obtained as solutions of type IIB supergravity which possess such non-relativistic scaling symmetry (and for a special value of the

Journal of Physics: Conference Series **462** (2013) 012028 doi:10.1088/1742-6596/462/1/012028

dynamical exponent the Schrödinger symmetry) as the isometry group. These backgrounds are generally obtained by deforming well-known AdS-backgrounds in a specific way.

It is an interesting question to consider the possible values of dynamical exponents realizable within a ten-dimensional type IIB supergravity set-up. In general, given an $AdS_5 \times X_5$ solution it is possible to find a deformed background which has a non-relativistic scaling symmetry characterized by the corresponding dynamical exponent. In this case the dynamical exponents are related to the spectrum of scalar or vector harmonics on the internal manifold X_5 [10, 11].

Note however that it is unclear whether the dual field theories of such supergravity backgrounds share any common feature with real systems (such as the cold atoms at unitarity). Therefore it is not clear what lessons we might learn about strongly coupled non-relativistic field theories using these gravity backgrounds. Hence we stand on the following ground: the supergravity backgrounds we discuss are interesting enough to consider in their own rights and should further be treated as toy models perhaps capable of providing insights into strong coupling dynamics.

The article is organized as follows: in Section 2 we discuss some details of the Schrödinger group and the five-dimensional background which possess this isometry. We move on to Section 3 to describe the embedding of this low-dimensional background into ten dimensional supergravity *via* a solution generating technique. Guided by these results we discuss new gravity solutions in Section 4. This includes supersymmetric backgrounds with Schrödinger symmetry, RG-flow backgrounds with Schrödinger symmetry at the fixed points and gravity backgrounds having scaling symmetry with a general dynamical exponent. Finally we conclude with open questions and future directions in Section 5.

2. The Schrödinger group and the gravity background

We begin with a brief review of the so called Schrödinger group, denoted by $\mathrm{Schr}(d)$. It is the symmetry group of the Schrödinger equation in free space and is the non-relativistic analogue of the relativistic conformal group. The generators of the Schrödinger group are: temporal translation, spatial translations, rotations, Galilean boosts, dilatation and a special conformal transformation.

The corresponding Schrödinger algebra in d-spatial dimensions can be obtained by the so called light cone reduction of the relativistic conformal algebra in $(d+1)$-spatial dimensions. We unpack briefly the operational meaning of the light cone reduction as follows[5]. Let us start with the relativistic conformal group in $(d+1)$-spatial dimensions, denoted by $\mathrm{SO}(d+2,2)$. First we identify the light cone momentum of the $\mathrm{SO}(d+2,2)$ group with the Galilean mass operator. Now we choose all operators in the parent $\mathrm{SO}(d+2,2)$ group that commute with this light cone momentum. This set of operators close under the Schrödinger algebra. Therefore by construction, $\mathrm{Schr}(d) \subset \mathrm{SO}(d+2,2)$.

In this algebra the dilatation and the Galilean mass operator (or the number operator) are simultaneously diagonalizable. So any representation of the Schrödinger algebra is labeled by two numbers. Note that for physical systems the eigenvalues of the number operator should be discrete.

It is simple to understand why this construction is inherently non-relativistic. The Schrödinger algebra thus constructed consists of a dilatation operator under which space and time scale differently: $t \to \lambda^2 t$, $\vec{x} \to \lambda \vec{x}$ where λ is a real number, breaking the Lorentz invariance.

Following our ideology discussed in the introduction, we may ask ourselves whether it is possible to construct a gravitational background with the Schrödinger isometry group. It turns out that such a background can be uniquely realized by deforming the standard AdS metric (we set the radius to unity) in the Poincaré patch in the following manner

$$ds^2 = -\frac{1}{z^4}du^2 + \frac{1}{z^2}\left(-2du dv + d\vec{x}^2 + dz^2\right) , \tag{1}$$

where the term proportional to du^2 is the deformation, u and v are the light-cone coordinates, $\vec{x}$ is the spatial 2-directions and z is the radial coordinate (in other words the renormalization group scale for the dual field theory). The infrared (IR) of the geometry is obtained by setting $z \to \infty$ and the ultraviolet (UV) is located at $z \to 0$. Note that the background in (1) degenerates to a one-dimensional line segment in the strict UV limit and therefore does not have a meaningful UV boundary. In practice this problem can be circumvented by introducing a small but non-vanishing cut-off $z_{\text{cut-off}}$. However, this remains an obstacle in the efforts for a systematic analysis of holographic renormalization in this spacetime. On the other hand, the IR limit of this geometry is well-defined: it simply is an AdS-space in the light cone coordinates.

The coordinates u and $\vec{x}$ are respectively identified with the time and the spatial directions of the dual field theory. The v-coordinate is an additional coordinate present purely in the bulk geometry. Recall that the representations of the Schrödinger group are labeled by two conserved numbers: the conformal dimension and the particle number. Therefore it would be natural to compactify the v-direction and identify the corresponding discrete mass spectrum with the discrete particle number. Note that this is a special requirement imposed by hand and *a priori* there is nothing that prevents the v-direction to be an extended one.

The spacetime in (1) has very interesting causal strucuture. It is conformal to a pp-wave spacetime known to be non-distinguishing [12, 13]. This means that although there are no closed causal curves in the above background, there are distinct points in the spacetime with identical past and future sets. This is the causal structure of a Galilean field theory.

3. Embedding the gravity background in supergravity

We will now briefly discuss the embedding of such backgrounds in supergravity. Shortly after the spacetime in (1) was proposed in [5, 6], an algorithmic approach of constructing similar solutions from known asymptotically AdS solutions of type IIB supergravity was laid out in [7, 8, 9]. This algorithm makes use of a solution generating technique known as the null Melvin twist [14, 15] or the TsT transformation [8]. The basic idea behind the null Melvin twist is simple: it generates a new solution of supergravity equations of motion upon applying a well-defined set of algebraic symmetry operations (involving boosts, T-duality and shift-symmetry) on a given supergravity background. In order to apply these symmetry operations we need to pick one compact and one non-compact U(1)-isometry in the given ten dimensional background.

Let us start with a Freund-Rubin type solution of type IIB supergravity of the form $\text{AdS}_5 \times X_5$

$$ds^2 = ds^2_{\text{AdS}_5} + ds^2_{X_5} \, , \quad F_{(5)} = (1+\star)\, \text{vol}_{\text{AdS}_5} \, , \tag{2}$$

where X_5 is an Einstein manifold. We further assume that X_5 has at least one U(1)-isometry and $\mathcal{K}$ denotes the one-form dual to the corresponding Killing vector. Now we apply the null Melvin twist involving $\mathcal{K}$ and one spatial direction of the AdS-part of the background. This operation yields the following Schrödinger symmetric background [11]

$$ds^2 = ds^2_{\text{Schr}_5} + ds^2_{X_5} \, , \quad F_{(5)} = (1+\star)\, \text{vol}_{\text{Schr}_5} \, , \tag{3}$$

where

$$ds^2_{\text{Schr}_5} = -\frac{\Omega}{z^4} du^2 + \frac{1}{z^2}\left(-2du dv + d\vec{x}^2 + dz^2\right) \, , \quad \text{with} \quad \Omega = ||\mathcal{K}||^2 \, , \tag{4}$$

$$\text{and} \quad B_{(2)} = \frac{1}{z^2} \mathcal{K} \wedge du \, . \tag{5}$$

Note the presence of the function $\Omega(X_5)$ in front of the deformation term in the metric. The general form of the metric generated using the null Melvin twist technique is therefore of the warped product form $\text{Schr}_5 \times_w X_5$.

Journal of Physics: Conference Series **462** (2013) 012028 doi:10.1088/1742-6596/462/1/012028

In [7, 8, 9], X_5 was chosen to be S^5 and the Hopf fiber was chosen to be the compact U(1) along which the twist was performed. The Killing vector associated with the Hopf fiber has a constant norm, and therefore in this case we get $\Omega = 1$ upon proper normalization. The five dimensional piece of this background identically matches with the one in (1). Upon Kaluza-Klein reduction on the compact S^5 we obtain the background given in (1) along with a massive vector field (coming from dimensionally reducing the $B_{(2)}$-field on S^5). Therefore we obtain an Einstein-Proca effective action in five dimensions which contains the background in (1) as a solution.

Following the discussions in [9] we will now briefly comment on the meaning of the null Melvin twist process in the dual gauge theory. Let us recall that the gauge theory dual to $\mathrm{AdS}_5 \times S^5$ is the $\mathcal{N} = 4$ super Yang-Mills (SYM) and the Hopf fiber direction in S^5 corresponds to an $\mathrm{U}(1)_{\mathrm{R}}$ symmetry of the gauge theory. Without the Melvin deformation, compactifying the light like direction v amounts to the Discrete Light Cone Quantization (DLCQ) of the original $\mathcal{N} = 4$ SYM. The Melvin deformation adds a twist to the momentum generator along the light like direction by the momentum generator along the U(1)-direction used in the Melvin twist process. Thus in this case, we end up with the R-symmetry deformed DLCQ version of the $\mathcal{N} = 4$ SYM, which possess the Schrödinger symmetry[1].

Note that the original $\mathcal{N} = 4$ SYM is a maximally supersymmetric theory. However, as shown in [7, 8] this non-relativistic deformation coming from the twist by the R-symmetry direction breaks it completely. Therefore the resulting background in [7, 8, 9] is not supersymmetric. A question naturally arises: can we construct Schrödinger symmetric gravity backgrounds possessing some amount of supersymmetry. We will address this question in the next section.

4. New gravity solutions

4.1. Twisting with a general U(1) *and supersymmetry*

Let us take the specific example when $X_5 = S^5$. Choosing the Hopf fiber as the twisting direction is merely a convenience. Note that the S^5 has an SO(6) isometry group and we can apply a more general three-parameter twist on the Cartan $\mathrm{U}(1)^3$ subgroup [17]. However, in this general construction Ω is indeed a function of the S^5-coordinates and therefore it is not *a priori* known how to obtain an effective five-dimensional description.

It is possible that the function Ω vanishes on some locus in X_5, therefore locally the deformation term may disappear. However, such phenomenon have been analyzed carefully in [18, 19] and it has been concluded that as long as Ω is non-zero on an open set in X_5, the asymptotic and the causal structure of the background are consistent with the dual non-relativistic field theory. This is indeed the case in all examples we consider.

Since we have a more general twist at our disposal, we can revisit the question related to supersymmetry asked at the end of the previous section. Namely we can ask whether the backgrounds in (4) preserve any supersymmetry for some choice of the one-form $\mathcal{K}$. One approach to address this issue is to analyze which Killing spinors of the undeformed $\mathrm{AdS}_5 \times X_5$ background survive the non-relativistic deformation.

Interestingly, we can address this issue without referring to any particular choice of the internal manifold X_5. The only constraint we impose is that X_5 is a Sasaki-Einstein manifold with at least one U(1)-isometry (implying the existence of a non-trivial $\mathcal{K}$). Recall that X_5 needs to be Einstein in order to satisfy the supergravity equations of motion. The condition that X_5 is also a Sasaki manifold comes from requiring that the undeformed $\mathrm{AdS}_5 \times X_5$ background allows for at least one Killing spinor.

Under these assumptions, by analyzing the supersymmetry variation equations it is possible to obtain a necessary and sufficient condition under which a Killing spinor in the undeformed

[1] Alternatively, the Melvin twist can be thought of as modifying the ordinary products of fields to the so called star product. See *e.g.* [16] for more details on such constructions.

$AdS_5 \times X_5$ background will also be a Killing spinor of the $\mathrm{Schr}_5 \times_w X_5$ background and *vice versa*. This condition consists of one algebraic and one differential constraint of the following form [11]:

$$\left(1+\Gamma^{14}\right)\epsilon = 0 \quad \text{and} \quad \mathcal{L}_{\mathcal{K}}\,\epsilon = 0\ , \tag{6}$$

where ϵ is the Killing spinor, Γ denotes the purely real ten-dimensional Dirac gamma matrices, the indices $1, 4$ refers to the flat coordinates corresponding to the u and v-directions respectively and $\mathcal{L}_{\mathcal{K}}$ denotes the Lie derivative along the isometry direction associated with the Killing vector $\mathcal{K}$. Observe that none of the superconformal Killing spinors of the undeformed AdS-background survives the Γ^{14} projection condition. Note however, that this does not impose any condition on the additional Killing spinors (with opposite Γ^{14}-chirality) which may arise only at the Schrödinger background [20].

Let us now go back to the example when $X_5 = S^5$. The Lie derivative acting on the Killing spinors of the undeformed background along the Hopf fiber direction gives the non-zero R-charge of the corresponding Killing spinor. This is in direct violation of the differential constraint imposed in (6) and thus breaks all supersymmetry. This is what was observed in [7, 8]. For the sphere, choosing an appropriate U(1) preserves two real supercharges in general, which can be enhaced to four real supercharges for special choices of the U(1) [11].

Before concluding this section, let us offer some more comments based on other examples of Sasaki-Einstein manifolds. The next simplest example is the $T^{1,1}$ which has an $\mathrm{SU}(2)\times\mathrm{SU}(2)\times\mathrm{U}(1)_{\mathrm{R}}$ symmetry [21]. There are also two infinite families of five-dimensionsal Sasaki-Einstein manifolds with explicitly known metrics. These are the $Y^{p,q}$ family [22] with $\mathrm{SU}(2)\times\mathrm{U}(1)\times\mathrm{U}(1)_{\mathrm{R}}$ symmetry and the $L^{p,q,r}$ family [23, 24] with $\mathrm{U}(1)^2\times\mathrm{U}(1)_{\mathrm{R}}$ symmetry. Note that for a general Sasaki-Einstein manifold there exists a Killing vector with constant norm associated with the $\mathrm{U}(1)_{\mathrm{R}}$-symmetry. This is known as the Reeb vector. Choosing the Reeb vector in the twist will violate the differential constraint in (6) and break supersymmetry completely. However for all Sasaki-Einstein manifolds with explicitly known metrics, we can preserve two real supercharges by choosing an appropriate U(1) other than the R-symmetry. In these cases, the undeformed dual CFT is a quiver gauge theory. Therefore the non-relativistic deformation of such theories is to be understood as the deformed DLCQ of the corresponding quiver gauge theory.

4.2. RG flow geometry with Schrödinger symmetry

Let us now switch gears to discuss RG flow gravity backgrounds dual to non-relativistic field theories. Our approach here is solely based on the solution generating technique discussed earlier. We start with the well-known supergravity solutions dual to relevant deformations of $\mathcal{N}=4$ SYM and apply the null Melvin twist on them and generate the non-relativistic analogue of the flow.

First let us review the relevant deformation of $\mathcal{N}=4$ SYM in some details. This theory has three adjoint chiral superfields, which we denote by Φ_I, $I=1,2,3$. If we give mass to one of these superfields by adding a mass perturbation of the form $m_1\Phi_1^2$ to the superpotential, then the theory flows to an IR $\mathcal{N}=1$ fixed point with $\mathrm{SU}(2)\times\mathrm{U}(1)_{\mathrm{R}}$ symmetry [25]. The dual gravity background of this flow was constructed in [26] and is known as the Pilch-Warner background.

More generally we can also start with a $\mathbb{Z}_2$ quiver gauge theory with an $\mathrm{SU}(N)\times\mathrm{SU}(N)$ gauge group, two hypermultiplets and a pair of adjoint chiral superfields, denoted by $\{\Phi_1,\Phi_2\}$. This defines a CFT with $\mathcal{N}=2$ supersymmetry. Now we can add the following mass perturbation

$$\Delta W \sim m_1\Phi_1^2 + m_2\Phi_2^2\ . \tag{7}$$

Under this mass perturbation, this theory flows to a family of $\mathcal{N}=1$ IR fixed points parametrized by the ratio of the two mass parameters. When $m_1 = m_2$, we get the Pilch-Warner fixed point and for $m_1 = -m_2$ we get the Klebanov-Witten fixed point [27]. The gravity solution interpolating between these two fixed points have been constructed in [28].

Since these gravity solutions are known[2], we can use the machinery of the null Melvin twist to generate the non-relativistic analogue of these flow and interpolating backgrounds. Interestingly for all these backgrounds the internal manifold allows for one U(1)-isometry direction to be used in the Melvin twist procedure. This exercise has been carried out in detail in [17] and it has been shown explicitly that this fixed point structure persists in the Melvin twisted backgrounds which are now dual to the Schrödinger symmetric fixed points. The non-relativistic field theories are again understood as the deformed DLCQ of the corresponding relativistic field theory.

There are some interesting features of these backgrounds. For example, the Schrödinger symmetric Pilch-Warner fixed point is a perfectly regular warped product space with rotation-like terms present in the metric [17]. Furthermore, this background has internal NS and RR-fluxes. To the best of our knowledge, this is the only known Schrödinger symmetric background with all the above-mentioned features.

4.3. Gravity solutions with general dynamical exponents

Note that so far we have looked for gravity solutions having a scaling symmetry of the form: $t \to \lambda^2 t$, $\vec{x} \to \lambda \vec{x}$, which is the scaling symmetry of the Schrödinger group. However we can attempt for a more general scaling symmetry of the form: $t \to \lambda^n t$, $\vec{x} \to \lambda \vec{x}$, where n is some real number known as the dynamical exponent. It is known that for $n \neq 2$ the symmetry algebra does not have a special conformal generator and therefore the corresponding symmetry group is just the Galilean group with a dilatation operator. We call this the Galilean scaling algebra[3]. Unlike the Schrödinger algebra, the representations of Galilean scaling algebra are labeled by a single number: a dimension.

Based on the known results in the Schrödinger (*i.e.* $n = 2$) case in (4) we generalize our ansatz as follows:

$$ds^2 = -\frac{\Omega}{z^{2n}} du^2 + \frac{1}{z^2}\left(-2du dv + d\vec{x}^2 + dz^2\right) + ds^2_{X_5} \,, \tag{8}$$

$$B_{(2)} = \frac{1}{z^n} \mathcal{A} \wedge du \,, \quad F_{(5)} = (1 + \star)\,\mathrm{vol}_{X_5} \,, \tag{9}$$

where X_5 is still an Einstein manifold, Ω is an unknown function on X_5 and $\mathcal{A}$ is an unknown one-form both of which we need to fix from the equations of motion.

The equations of motion separate in two parts: the Maxwell equation and the Einstein equation. The Maxwell equation gives a massive Proca equation along with a transversality condition for the one-form $\mathcal{A}$. Taken together, this results in a Laplace equation for $\mathcal{A}$. On the other hand the Einstein equation gives an inhomogeneous Laplace equation for the scalar function Ω, where the inhomogeneous part consists of the energy-momentum contribution coming from the one-form $\mathcal{A}$.

In order to solve the Maxwell equation, we need to know the spectrum of transverse vector harmonics on X_5. Upon choosing an appropriate vector harmonic on X_5 the Maxwell equation reduces to an algebraic relation between the dynamical exponent n and the eigenvalue of the Laplacian on the corresponding vector harmonic. This algebraic relation determines the allowed values for the dynamical exponents. Then the inhomogeneous Laplace equation for Ω can in principle be solved by expanding the source term into scalar harmonics on X_5. Note that as a special case we can set $\mathcal{A} = 0$ and reproduce the results of [10].

[2] These backgrounds are not always known analytically, but for the twist procedure we do not need an exact analytic form of the background.

[3] We thank M. Rangamani for suggesting this name.

For example, when $X_5 = S^5$ the allowed dynamical exponents are all positive integers greater than one. In this case, we have constructed some explicit solutions in [11]. For a general internal manifold the allowed dynamical exponents can be irrational real numbers as is the case for $T^{1,1}$ [11].

5. Conclusion

We have discussed several interesting gravity backgrounds with a general scaling symmetry. This scaling symmetry is characterized by a real number called the dynamical exponent and we have found a plethora of such dynamical exponents allowed within solutions of the supergravity equations of motion.

For the special case of $n = 2$, *i.e.* when the symmetry group is enhanced to the full Schrödinger group, we have found backgrounds with some amount of supersymmetry. It is therefore natural to ask whether the backgrounds with general dynamical exponents preserve any supersymmetry. This appears to be a difficult problem in complete generality, however it would be very interesting to analyze this for specific examples. This would be particularly interesting since the supersymmetric extension of the Galilean scaling algebra has not been studied extensively in the current literature.

Our backgrounds with general dynamical exponents are warped product spaces and therefore it is not known how to obtain a low-dimensional consistent truncation of these backgrounds. However, for special choices of the internal manifold X_5 and for special values of the dynamical exponent the scalar function Ω may become a pure number; thereby facilitating a straightforward Kaluza-Klein compactification. It would be very interesting to explore this issue further.

Interestingly, although there are no curvature singularities in our backgrounds with general dynamical exponent, the Poincaré patch is geodesically incomplete for $n \geq 2$ [29]. For the Schrödinger case (*i.e.* when $n = 2$), it is possible to extend this patch to a global one which is everywhere regular [29]. However, the analogue of such a global patch is not known for backgrounds with other dynamical exponents. It will be an interesting issue to consider in future.

It is always interesting to find black hole solutions. The only known black hole solutions within this context exist in the asymptotically Schrödinger spacetime. Such solutions are obtained by using the null Melvin twist on the known AdS-Schwarzschild solution. However, for a general dynamical exponent there is no such solution generating technique and one needs to solve the equations of motion in its full glory. It would therefore be very interesting to find black hole solutions with general dynamical exponent within ten-dimensional supergravity or perhaps a lower dimensional effective model. By virtue of the gauge-gravity duality such a solution will automatically provide an arena for studying aspects of a scale-invariant strongly coupled non-relativistic field theory at finite temperature.

In this note we have restricted ourselves within type IIB supergravity. One can also look for similar solutions within type IIA and 11-dimensional supergravity. Some such analysis can be found in *e.g.* [30, 31].

Let us conclude by noting that we are able to find several interesting examples of gravity solutions candidates for a dual of scale-invariant non-relativistic field theory. These field theories are obtained by the so called dipole deformation (in Schrödinger symmetric cases) or deformation by higher dimensional Lorentz violating operators (for a general dynamical exponent) of the conventional gauge theories arising from massless string modes living on the worldvolume of the D-branes. To be a truly compelling proposal, one needs to identify and extract the features of strong coupling dynamics within this framework which might be relevant for real physical systems. These are exciting prospects for future research.

Acknowledgments

I am grateful to the organizers at the University of Kentucky for the wonderful atmosphere of the Quantum Theory and Symmetries 6 and for the opportunity to present this work. I would also like to thank my collaborators Nikolay Bobev and Krzysztof Pilch without whom this work would have not been possible. Further thanks are due to Veronika Hubeny, Clifford V. Johnson, Mukund Rangamani and Nicholas P. Warner for numerous patient explanations, discussions, constant encouragement and support throughout the course of this work. Finally this work is partly supported by the Department of Energy grant DE-FG03-84ER-40168.

References

[1] J. McGreevy, arXiv:0909.0518 [hep-th].
[2] S. A. Hartnoll, Class. Quant. Grav. **26**, 224002 (2009) [arXiv:0903.3246 [hep-th]].
[3] Y. Nishida and D. T. Son, Phys. Rev. D **76**, 086004 (2007) [arXiv:0706.3746 [hep-th]].
[4] C. R. Hagen, Phys. Rev. D **5**, 377 (1972).
[5] D. T. Son, Phys. Rev. D **78**, 046003 (2008) [arXiv:0804.3972 [hep-th]].
[6] K. Balasubramanian and J. McGreevy, Phys. Rev. Lett. **101**, 061601 (2008) [arXiv:0804.4053 [hep-th]].
[7] C. P. Herzog, M. Rangamani and S. F. Ross, JHEP **0811**, 080 (2008) [arXiv:0807.1099 [hep-th]].
[8] J. Maldacena, D. Martelli and Y. Tachikawa, JHEP **0810**, 072 (2008) [arXiv:0807.1100 [hep-th]].
[9] A. Adams, K. Balasubramanian and J. McGreevy, JHEP **0811**, 059 (2008) [arXiv:0807.1111 [hep-th]].
[10] S. A. Hartnoll and K. Yoshida, JHEP **0812**, 071 (2008) [arXiv:0810.0298 [hep-th]].
[11] N. Bobev, A. Kundu and K. Pilch, JHEP **0907**, 107 (2009) [arXiv:0905.0673 [hep-th]].
[12] J. L. Flores and M. Sanchez, Class. Quant. Grav. **20**, 2275 (2003) [arXiv:gr-qc/0211086].
[13] V. E. Hubeny, M. Rangamani and S. F. Ross, Phys. Rev. D **69**, 024007 (2004) [arXiv:hep-th/0307257].
[14] M. Alishahiha and O. J. Ganor, JHEP **0303**, 006 (2003) [arXiv:hep-th/0301080].
[15] E. G. Gimon, A. Hashimoto, V. E. Hubeny, O. Lunin and M. Rangamani, JHEP **0308**, 035 (2003) [arXiv:hep-th/0306131].
[16] A. Bergman, K. Dasgupta, O. J. Ganor, J. L. Karczmarek and G. Rajesh, Phys. Rev. D **65**, 066005 (2002) [arXiv:hep-th/0103090].
[17] N. Bobev and A. Kundu, JHEP **0907**, 098 (2009) [arXiv:0904.2873 [hep-th]].
[18] D. Marolf and S. F. Ross, Class. Quant. Grav. **19**, 6289 (2002) [arXiv:hep-th/0208197].
[19] V. E. Hubeny, M. Rangamani and S. F. Ross, JHEP **0507**, 037 (2005) [arXiv:hep-th/0504034].
[20] A. Donos and J. P. Gauntlett, JHEP **0910**, 073 (2009) [arXiv:0907.1761 [hep-th]].
[21] L. J. Romans, Phys. Lett. B **153**, 392 (1985).
[22] J. P. Gauntlett, D. Martelli, J. Sparks and D. Waldram, Adv. Theor. Math. Phys. **8**, 711 (2004) [arXiv:hep-th/0403002].
[23] M. Cvetic, H. Lu, D. N. Page and C. N. Pope, Phys. Rev. Lett. **95**, 071101 (2005) [arXiv:hep-th/0504225].
[24] M. Cvetic, H. Lu, D. N. Page and C. N. Pope, JHEP **0907**, 082 (2009) [arXiv:hep-th/0505223].
[25] R. G. Leigh and M. J. Strassler, Nucl. Phys. B **447**, 95 (1995) [arXiv:hep-th/9503121].
[26] K. Pilch and N. P. Warner, Adv. Theor. Math. Phys. **4**, 627 (2002) [arXiv:hep-th/0006066].
[27] I. R. Klebanov and E. Witten, Nucl. Phys. B **536**, 199 (1998) [arXiv:hep-th/9807080].
[28] N. Halmagyi, K. Pilch, C. Romelsberger and N. P. Warner, JHEP **0608**, 083 (2006) [arXiv:hep-th/0506206].
[29] M. Blau, J. Hartong and B. Rollier, JHEP **0907**, 027 (2009) [arXiv:0904.3304 [hep-th]].
[30] L. Mazzucato, Y. Oz and S. Theisen, JHEP **0904**, 073 (2009) [arXiv:0810.3673 [hep-th]].
[31] A. Donos and J. P. Gauntlett, JHEP **0907**, 042 (2009) [arXiv:0905.1098 [hep-th]].

Journal of Physics: Conference Series **462** (2013) 012029 doi:10.1088/1742-6596/462/1/012029

Magnetic properties of four dimensional fermions[1]

Oren Bergman[a], Gilad Lifschytz[b] and Matthew Lippert[c]

[a] Department of Physics, Technion, Haifa 32000, Israel.
[b]Department of Mathematics-Physics, University of Haifa at Oranim, Qiryat Tivon 36006, Israel.
[c]Crete Center for Theoretical Physics, Department of Physics, University of Crete, 71003 Heraklion, Crete, Greece.

E-mail: [a]bergman@physics.technion.ac.il, [b]giladl@research.haifa.ac.il, [c]mlippert@physics.uoc.gr

Abstract. We investigate the Sakai-Sugimoto model at nonzero baryon chemical potential in a background magnetic field in the chiral symmetric phase . We find that a new form of baryonic matter shows up, and we investigate its properties. We find a generated axial current, a reduction in the amount of charge participating in dissipative interactions and a metamagnetic like phase transition at low temperature.

1. Introduction

Properties of strongly interacting fermions in $3+1$ dimensions is a vastly important subject. Holographic representation of various systems of strongly interacting fermions can teach us about the possible phenomena one can expect. In particular an addition of a magnetic field can reveal all kinds of interesting properties (as it does in lower dimensions). The Sakai-Sugimoto model [1] is a system of strongly interacting fermions which seems to share quite a few phenomena with QCD. We explore here the behavior of this model at finite baryon density and magnetic field. Other works on parts of the subject include [2, 3, 4, 5, 6, 7]. In particular we look at the phase where chiral symmetry is unbroken, and find that baryonic charge can be carried by a new form of matter. Understanding its properties and possible interpretation is the goal of this note.

2. New form of baryonic matter

The Sakai-Sugimoto model consists of a probe D8-brane and a probe anti D8-brane in the background of N D4-branes compactified on a circle. The matter degrees of freedom are fermions in $3+1$ dimensions in the fundamental representation of $SU(N)$, interacting with $SU(N)$ gauge fields propagating in one higher dimension. The properties of the fermions are encoded in the action of the probe branes. The baryon charge is the charge under the $U(1)$ gauge field living on the probe branes, and can be carried by strings ending on the D8 branes. These strings can start at the horizon and then they have the interpretation as being quarks, or can end on a D4-brane wrapping the S^4, in which case it is identified with a baryon. In the chiral symmetric phase, at zero magnetic field, all the baryonic charge is carried by the quarks.

[1] Based on a talk given by G. Lifschytz.

Let us now turn on a magnetic field. The 8-brane DBI and CS actions in the deconfined background are given by [8],

$$\begin{aligned} S_{DBI} &= \mathcal{N}\int_{u_T}^{\infty} du\, u^{5/2}\sqrt{\left(1-(a_0'(u))^2+f(u)(a_1'(u))^2\right)\left(1+\frac{h^2}{u^3}\right)} \qquad (1) \\ S_{CS} &= -\mathcal{N}\int_{u_T}^{\infty}\left(\partial_2 a_3^V a_0^V(u) a_1^{A\prime}(u) - \partial_2 a_3^V a_0^{V\prime}(u) a_1^A(u) - a_3^V\partial_2 a_0^V a_1^{A\prime} + a_3^V\partial_2 a_1^A a_0^{V\prime}\right), \end{aligned}$$

where now $f = 1-(u_T^3/u^3)$, and where we have included both the 8-brane and anti-8-brane parts, with $\bar{a}_0 = a_0$ and $\bar{a}_1 = -a_1$. The boundary value of the axial field is now a parameter, rather than a field, in the gauge theory, which we set to zero, $a_1(\infty) = 0$. We modify the Chern-Simons action by throwing away boundary terms of the form

$$\frac{1}{2}\partial_2\left(a_3 a_{[1}a_{0]}'\right) + \frac{1}{2}\partial_u\left(a_3\partial_2 a_{[1}a_{0]}\right) \qquad (2)$$

to obtain the correct five-dimensional currents, consistent with gauge invariance.

The integrated equations of motion are then given by

$$\frac{\sqrt{u^5+h^2u^2}\,a_0'(u)}{\sqrt{1-(a_0'(u))^2+f(u)(a_1'(u))^2}} = 3ha_1(u)+d \qquad (3)$$

$$\frac{\sqrt{u^5+h^2u^2}\,f(u)a_1'(u)}{\sqrt{1-(a_0'(u))^2+f(u)(a_1'(u))^2}} = 3ha_0(u)+j_A-\frac{3}{2}h\mu\,, \qquad (4)$$

where d is the total baryon charge density and j_A is the axial current density. As can be seen from this equation there are two source of baryonic charge. First there are strings stretching from the horizon to the D8-brane, being represented as a delta function source at the horizon, in the un-integrated equations of motion. Second there is the term $3ha_1(u)$ representing smeared D4 branes. The charge carried by these is $-3ha_1(u_T)$.

We will now explore the properties of these objects.

2.1. *Axial current*

Looking at equation (4), there is an additional condition imposed by regularity at the horizon, $a_0(u_T) = 0$. Since $f(u_T) = 0$ as well, the consistency of the a_1 equation of motion (4) requires turning on a specific axial current density [8]

$$j_A = \frac{3}{2}h\mu\,. \qquad (5)$$

In terms of physical quantities, the axial current density is

$$J_A = \frac{N_c}{4\pi^2}H\mu_B\,, \qquad (6)$$

which agrees precisely with the result of [9], once we account for the different normalizations (the relative factor of $N_c/2$). What carries this axial current. from the equation of motion it is clear that it is the term $3ha_0(u)$. This term represents the amount of axial current below some u however since $a_0(u_T) = 0$ it is clear that all the axial current comes from regions above the horizon. Thus it is associate with the smeared D4-branes living outside the horizon.

3. Transport properties

We will first consider orthogonal electric and magnetic fields and calculate the ohmic and Hall conductivities [10] employing the same method used by Karch and O'Bannon in the D3-D7 model [11, 12]. We consider the deconfined, chiral-symmetric phase with background magnetic field $f_{23} = \partial_2 a_3 = h$ and electric field $f_{02} = \dot{a}_2 = e$. In order to include spacetime baryonic currents j_μ (though we will call the baryon density $d \equiv j_0$), the spacetime components of the gauge fields will have u-dependent parts $a_\mu(u)$ as well. Primes denote derivatives with respect to u and dots are derivatives with respect to t.

We combine the actions of the 8-brane and the $\bar{8}$-brane by considering the vector part of a_0 and the axial part of a_1, i.e. $a_0^{\overline{D8}} = a_0^{D8}$ and $a_1^{\overline{D8}} = -a_1^{D8}$. The DBI action with these background fields is

$$S_{\rm DBI} = \mathcal{N} \int du\, u^{5/2} \sqrt{X} \tag{7}$$

where

$$X = 1 + \frac{h^2}{u^3} - \frac{e^2}{f(u)u^3} - {a_0'}^2\left(1 + \frac{h^2}{u^3}\right) + f(u){a_1'}^2\left(1 + \frac{h^2}{u^3} - \frac{e^2}{f(u)u^3}\right) + f(u){a_2'}^2 + f(u){a_3'}^2\left(1 - \frac{e^2}{f(u)u^3}\right) - 2\frac{eh}{u^3}\, a_0' a_3' \,. \tag{8}$$

The normalization constant is given by

$$\mathcal{N} = 2\Omega_4 T_{D8} R^5 = \frac{N_c}{6\pi^2}\frac{R^2}{(2\pi\alpha')^3} \tag{9}$$

where the factor of 2 corresponds to the two halves of the embedding. The Chern-Simons action with appropriate addition to make it gauge invaraint is

$$S_{\rm CS} + \Delta S = \frac{3\mathcal{N}}{2}\int \left(h a_{[0}a'_{1]} + e a_{[1}a'_{3]}\right) \,. \tag{10}$$

The charge d and the currents j_i define tilde currents as follows,

$$\tilde{d} = d + 3ha_1 \tag{11}$$

$$\tilde{j}_1 = j_1 + 3h\left(a_0 - \frac{\mu}{2}\right) + 3ea_3 \tag{12}$$

$$\tilde{j}_3 = j_3 - 3ea_1 \,. \tag{13}$$

Solving the equation of motion and plugging back into the action, we find that the DBI part can be written as

$$S_{\rm DBI} = \mathcal{N}\int du\, u^{5/2} \frac{1 + \frac{h^2}{u^3} - \frac{e^2}{f(u)u^3}}{\sqrt{Z}} \,. \tag{14}$$

where

$$Z = \left(1 + \frac{h^2}{u^3} - \frac{e^2}{f(u)u^3}\right)\left(u^5 + \tilde{d}^2 - \frac{\tilde{j}_2^2 + \tilde{j}_3^2}{f(u)}\right) - \left(\frac{hd}{u^3} + \frac{e\tilde{j}_3}{f(u)u^3}\right)^2 - \frac{\tilde{j}_1^2}{f(u)} \,. \tag{15}$$

Journal of Physics: Conference Series **462** (2013) 012029 doi:10.1088/1742-6596/462/1/012029

For the action to be real, Z must clearly be non-negative. The only way this can happen is if all three terms in (15) have double zeros at the same value of u, which we'll call u_*. For the first term, this means each factor must vanish at u_*, so we get a total of four equations. These can be solved and one gets [10]

$$u_*^3 = u_T^3\left(1+\frac{e^2}{h^2+u_T^3}\right) \tag{16}$$

$$j_3 = \left(-\frac{h\tilde{d}_T}{u_T^3+h^2}+3a_{1T}\right)e \tag{17}$$

$$j_2 = \frac{\sqrt{u_T^8+u_T^5h^2+u_T^3\tilde{d}_T^2}}{u_T^3+h^2}e \tag{18}$$

and an axial current

$$j_1 = 3h\left(\frac{\mu}{2}-a_{0*}\right)-3ea_{3*}\ . \tag{19}$$

For small e, $a_3 = \mathcal{O}(e)$ and $a_{0*} = a_{0T} + \mathcal{O}(e^2)$. In addition, regularity at the horizon implies $a_{0T} = 0$, so we find

$$j_1 = \frac{3}{2}h\mu \tag{20}$$

which matches the result found in the previous subsection.

4. Metamagnetic phase transition

We saw that as we turn on a magnetic field D4-branes come up from the horizon and are smeared over the D8-brane. Being above the horizon they do not participate is dissipative phenomena like longitudenal conductivity. It turns out that the process of the D4-brane moving up along the D8-brane is not smooth at low temperature but rather is a first order phase transition [13].

To solve the zero-temperature equations, one notices that dividing the two equations in (3, 4) gives

$$\frac{a_0'}{a_1'} = \frac{3ha_1+d}{3ha_0} \tag{21}$$

which can be integrated, using the boundary conditions $a_1(\infty) = 0$ and $a_0(\infty) = \mu$. One then finds that

$$(3ha_1+d)^2-(3ha_0)^2 = d^2-9h^2\mu^2\ . \tag{22}$$

By defining a new coordinate

$$z = \int_0^u \frac{3hdu}{\sqrt{u^5+h^2u^2+d^2-9h^2\mu^2}} \tag{23}$$

The solution to the equations of motion with the conditions $a_0(z=0) = 0$ and $a_1(z=z_\infty) = 0$ is then

$$a_0 = \frac{d\sinh z}{3h\cosh z_\infty}$$

$$a_1 = \frac{d\cosh z}{3h\cosh z_\infty}-\frac{d}{3h}\ . \tag{24}$$

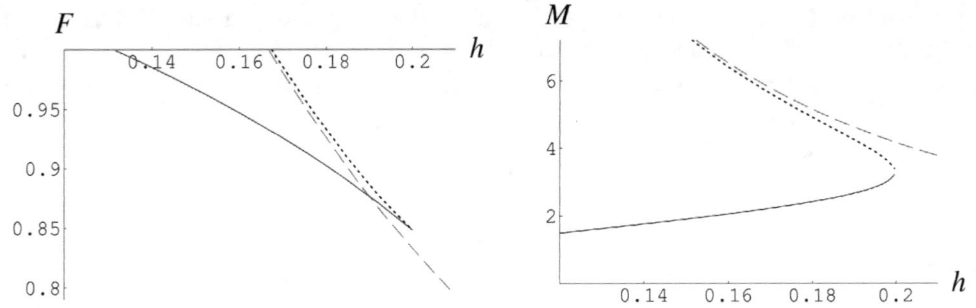

Figure 1. Three solutions for (a) the free energy F and the (b) magnetization M at $T = 0$ with $d = 1$. There is a first-order phase transition from the solid blue $z_\infty \sim h$ solution to the dashed red $z_\infty \to \infty$ solution at $h = 0.19$. The dotted grey solution, which corresponds to the third solution, connects the two stable solutions and is an unstable maximum of the free energy.

From these equations we see that

$$\begin{aligned} \mu &= \frac{d}{3h} \tanh z_\infty \\ a_1(0) &= \frac{d}{3h} \left(\frac{1}{\cosh z_\infty} - 1 \right) , \end{aligned} \tag{25}$$

and thus,

$$z_\infty = \int_0^\infty \frac{3h du}{\sqrt{u^5 + h^2 u^2 + \frac{d^2}{\cosh^2 z_\infty}}} . \tag{26}$$

To find the actual solution for $z_\infty(h, d)$, one needs to solve (26) numerically. We regularize the integral[2] by taking the lower limit of the integral defining z to be ϵ and at the end sending $\epsilon \to 0$. For small d and h there are three solution for z_∞, but for large enough h only one remains which is $z_\infty \sim \log \epsilon \to \infty$. It should be noted that the appearance of three solution and the interesting phase structure which they imply is due to the use of the full DBI action rather than just the Yang-Mills approximation.

The free energy of the three solutions is shown in Fig. 1. At small h the solution $z_\infty \sim h$ is preferred, but at some finite h there is a jump to the branch parametrized by $z_\infty \to \infty$. This first-order phase transition is accompanied by jumps in both the magnetization and the chemical potential.

The preferred solution at large h has

$$\mu = \frac{d}{3h} , \quad F = \frac{d^2}{6h} . \tag{27}$$

We also find that $a_0 = 0$ and $a_1 = -\frac{d}{3h}$ are constant all the way up to $u \to \infty$ where they jump to their asymptotic values $a_0 = \mu$ and $a_1 = 0$.

4.1. Non-zero temperature

We now turn to solving these equations in the more physical case of non-zero temperature. We will see that the properties seen for zero temperature persist. At high temperature there is

[2] Having to insert by hand a regulator is a particular feature of $T = 0$. At non-zero temperature, u_T naturally cuts off integrals over u.

only one solution which means a single phase. As the temperature is lowered, there is a critical temperature $T_c(d)$ below which one finds three solutions to the equations of motion for some range of h. One way to see this in the numerics is to plot $a_1(\infty)$ as a function of $a_1(u_T)$. At high temperature the graph (at all values of h) cross the horizontal axis only once. However, at lower temperature there is a range of h when there are three possible values of $a_1(u_T)$ which gives $a_1(\infty) = 0$. Again, one of the solution is unstable and two are stable. There is a phase transition at some non-zero value of h. At a temperature slightly higher then the critical temperature both $\mu(h)$ and $M(h)$ develop what looks like a wiggle. As the temperature is lowered, we enter a regime where there is a phase transition; the chemical potential and the magnetization become discontinuous. At zero temperature, the jump of the chemical potential is such that the axial current $j_A = \frac{3}{2}h\mu$ attains its maximum value of $d/2$ and does not change any more as h is varied. This implies that in the large h phase the system is fully polarized. At non-zero temperature, j_A jumps close to the maximum value and asymptotes to it at large h. Similarly if we had turned on an axial chemical potential rather then a vector chemical potential, there would be a jump in the vector current in the direction of the magnetic field.

To further describe the phase transition and to try to understand it better, we look at the value of the magnetic field at which the transition occurs as a function of the density and at fixed temperature. The results are plotted in Fig. 2(a) and closely fit

$$h \sim d^{2/3} . \tag{28}$$

We also find that even at non-zero temperature the free energy of the phase at large h is well-approximated by

$$F \sim \frac{d^2}{6h} . \tag{29}$$

As was expected, the behavior here is similar to the zero-temperature case but with finite temperature corrections. One qualitative difference is in the location of the charges in the high h phase. While almost all the charges in the form of smeared D4-branes, they are no longer all near the boundary at $u = \infty$ but are instead spread across all values of u.

In terms of the behavior as a function of the temperature, we find that for a given charge density d above a certain temperature there will be no phase transition for any h; thus, the line of first-order phase transitions ends at a critical point above which there is a smooth cross-over. The critical line in the $T-d$ plane above which there is no phase transition is given in Fig. 2(b) and behaves at large d as

$$T \sim \log d . \tag{30}$$

It appears that no matter how high the temperature is, one can find a large enough density for which a phase transition will occur at some strong magnetic field. Of course, at these parameters, for a given L, the preferred phase maybe the chiral-broken phase. However, as explained before, we can always take L to be large enough that the chiral-symmetric phase is preferred.

5. Interpretation

To try to understand some of the physics of this transition, we explore the phase that is dominant at large h. At least some of its properties are similar to the properties of a free, massless fermion gas in four dimensions in its lowest Landau level.

A massless fermion in four dimensions with charge e in a magnetic field B has the following energy spectrum:

$$E = \sqrt{k_1^2 + 2eBn} \tag{31}$$

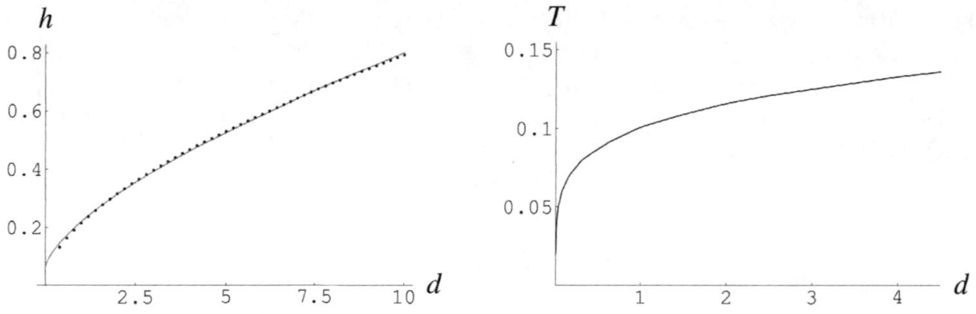

Figure 2. (a) The phase diagram in the d-h plane at $T = 0.07$, and (b) the critical line in the T-d plane.

where k_1 is the momentum in the direction of the magnetic field and n is a natural number labeling the Landau levels in the transverse directions. The degeneracy of each state is roughly B. If all fermions are in the lowest Landau level, the energy of the gas is given by

$$E_{tot} \sim B \int_0^{k_{max}} dk\; k = \frac{1}{2} B k_{max}^2 \;, \tag{32}$$

and the density of fermions is

$$n_f \sim B k_{max} \;, \tag{33}$$

giving a chemical potential and energy

$$\mu \sim \frac{n_f}{B}, \quad E \sim \frac{n_f^2}{B} \;. \tag{34}$$

The lowest Landau level is not spin degenerate, thus when there's a jump to the lowest Landau level, there is an increase in the magnetization. We can estimate when the system crosses from the lowest to the first Landau level; this occurs when it is energetically favorable to increase n rather than add to the parallel momentum, meaning $k_{max}^2 \sim B$, which corresponds to

$$B \sim n_f^{2/3} \;. \tag{35}$$

Many of the properties of the large h phase agree with those expected from free fermions in the lowest Landau level in four dimensions. The behavior of the free energy and chemical potential (27) match the free fermion result (34), and phase transition (28) resembles that of a transition into the lowest Landau level (35).

On the supergravity side, in the large h phase almost all the baryonic charge[3] is represented by smeared D4-branes that crept up the D8-brane. Recently, there have been other indications that smeared D4-branes seem to be related to fermions in the lowest Landau level. For example, in [14] it was argued that in the confined, large h, pion-gradient phase the smeared D4-branes also behave as if they are free fermions and reproduced the thermodynamics of the lowest Landau level.

Furthermore, the smeared D4-branes holographically correspond to bound states of N_c fermions, making up what looks like smeared baryons. This phenomena may be related to

[3] At $T = 0$ this will be all the baryonic charge.

the fact that in the lowest Landau level fermions are confined to move only in one spatial dimension, and thus any attractive potential will make them bind, even one which would not bind them in three spatial dimensions. Such an occurrence has been demonstrated to happen for a different situation related to the magnetically-induced chiral-symmetry breaking [15].

Because the Sakai-Sugimoto model provides a robust holographic model of QCD, it would be interesting to see if the phenomena described here could be found, for example, at RHIC in a strong magnetic field. We see the onset anomaly-driven behavior as soon as a magnetic field is turned on, but this may be an artifact of exactly massless quarks. In real QCD, a strong magnetic field may be required before any of these phenomena become manifest.

There are a few non-relativistic fermionic systems which resemble in some way the relativistic system discussed here. In particular, several condensed matter systems also exhibit first-order quantum phase transitions with discontinuous magnetizations. In the case of metamagnetism [16], the magnetization, which comes from spin ordering, jumps due to a large increase in the density of states near the Fermi surface. This effect sometimes occurs in paramagnetic materials with itinerant electrons in their outer shells. The magnetization, as a function of the external magnetic field, can have a large rise and sometimes a discontinuity due to a rapid transition of many fermion spins from anti-parallel to the magnetic field to parallel.

The phase transition we find in the SS model appears to have some resemblance to metamagnetism but also some differences. We start with a paramagnetic state and see a jump in the magnetization at a particular value of the magnetic field. However, after the jump, the magnetization decreases, while in materials exhibiting metamagnetism, the phase after the transition is ferromagnetic. In addition, while we find evidence the large-magnetic field phase in the SS model is associated with the lowest Landau level, metamagnetism, occur at more modest magnetic fields, and is not related to lowest Landau level physics.

Acknowledgments

I wish to thank Al Shapere and Sumit Das for organizing the QTS6 conference. This work was supported in part by the Israel Science Foundation under grant no. 568/05.

[1] T. Sakai and S. Sugimoto, Prog. Theor. Phys. **113**, 843 (2005) [arXiv:hep-th/0412141]; Prog. Theor. Phys. **114**, 1083 (2006) [arXiv:hep-th/0507073].
[2] O. Bergman, G. Lifschytz and M. Lippert, JHEP **0711**, 056 (2007) [arXiv:0708.0326 [hep-th]].
[3] N. Horigome and Y. Tanii, JHEP **0701**, 072 (2007) [arXiv:hep-th/0608198]; J. L. Davis, M. Gutperle, P. Kraus and I. Sachs, JHEP **0710**, 049 (2007) [arXiv:0708.0589 [hep-th]]; M. Rozali, H. H. Shieh, M. Van Raamsdonk and J. Wu, JHEP **0801**, 053 (2008) [arXiv:0708.1322 [hep-th]]; K. Y. Kim, S. J. Sin and I. Zahed, JHEP **0801**, 002 (2008) [arXiv:0708.1469 [hep-th]].
[4] O. Bergman, G. Lifschytz and M. Lippert, JHEP **0805**, 007 (2008) [arXiv:0802.3720 [hep-th]].
[5] C. V. Johnson and A. Kundu, JHEP **0812**, 053 (2008) [arXiv:0803.0038 [hep-th]].
[6] K. Y. Kim, S. J. Sin and I. Zahed, JHEP **0807**, 096 (2008) [arXiv:0803.0318 [hep-th]].
[7] A. Rebhan, A. Schmitt and S. A. Stricker, arXiv:0811.3533 [hep-th].
[8] O. Bergman, G. Lifschytz and M. Lippert, Phys. Rev. D **79**, 105024 (2009) [arXiv:0806.0366 [hep-th]].
[9] M. A. Metlitski and A. R. Zhitnitsky, Phys. Rev. D **72**, 045011 (2005) [arXiv:hep-ph/0505072]; G. M. Newman and D. T. Son, Phys. Rev. D **73**, 045006 (2006) [arXiv:hep-ph/0510049].
[10] G. Lifschytz and M. Lippert, Phys. Rev. D **80**, 066005 (2009) [arXiv:0904.4772 [hep-th]];
[11] A. Karch and A. O'Bannon, JHEP **0709**, 024 (2007) [arXiv:0705.3870 [hep-th]];
[12] A. O'Bannon, Phys. Rev. D **76**, 086007 (2007) [arXiv:0708.1994 [hep-th]].
[13] G. Lifschytz and M. Lippert, Phys. Rev. D **80**, 066007 (2009) [arXiv:0906.3892 [hep-th]].
[14] E. G. Thompson and D. T. Son, Phys. Rev. D **78**, 066007 (2008) [arXiv:0806.0367 [hep-th]].
[15] D. N. Kabat, K. M. Lee and E. J. Weinberg, Phys. Rev. D **66**, 014004 (2002) [arXiv:hep-ph/0204120].
[16] E. P. Wohlfarth and P. Rhodes., *Philos. Mag.*, **7** (1962) 1817.; M. Shimizu, *J. Phys.* (Paris), **43** (1982) 155.; T. Goto et al., *Solid state Commun.*, **72** (1989) 945; *Physica B*, **300** (2001) 167.

6th International Symposium on Quantum Theory and Symmetries (QTS6) IOP Publishing
Journal of Physics: Conference Series **462** (2013) 012030 doi:10.1088/1742-6596/462/1/012030

Magnetic properties of confined holographic QCD

Oren Bergman[1], **Gilad Lifschytz**[2] **and Matthew Lippert**[123]

[1] Department of Physics, Technion, Haifa 32000, Israel
[2] Department of Mathematics and Physics, University of Haifa, Tivon 36006, Israel
[3] Crete Center for Theoretical Physics, University of Crete, Heraklion 71003, Crete, Greece

E-mail: `mlippert@physics.uoc.gr`

Abstract. We investigate the Sakai-Sugimoto model at nonzero baryon chemical potential in a background magnetic field in the confined phase where chiral symmetry is broken. The D8-brane Chern-Simons term holographically encodes the axial anomaly and generates a gradient of the η' meson, which carries a non-vanishing baryon charge. Above a critical value of the chemical potential, there is a second-order phase transition to a mixed phase which includes also ordinary baryonic matter. However, at fixed baryon charge density, the matter is purely η'-gradient above a critical magnetic field.

1. Introduction

The behavior of QCD under external conditions is an interesting and physically relevant problem. At high temperature the ground state is believed to be a deconfined quark-gluon plasma, and at high density it is believed to be a color-superconductor. The former is relevant for the understanding of the physics at RHIC, and the latter may be relevant for the physics of dense stellar objects such as neutron stars. Background electromagnetic fields provide another kind of external condition. Their effect on the QCD ground state is possibly relevant for magnetars, neutron stars with very large magnetic fields. However, it is often the case that in physically relevant regimes, QCD is strongly coupled, and we cannot reliably use perturbation theory.

An alternative approach to strongly-coupled gauge theories has emerged in recent years from string theory via holographic gauge/gravity duality. This approach is well-poised to address questions related to external conditions since these translate simply to boundary conditions on internal fields in the bulk.

Due to asymptotic freedom, the holographic dual of QCD must go beyond the supergravity approximation. Nevertheless, it may be useful to study string theory in a background where the low-lying excitations resemble those of QCD and in which one can consistently study the supergravity limit. This does not give QCD, but these kinds of models share many of its low-energy properties. The closest so far to QCD is the Sakai-Sugimoto model [1]. This model consists of N_c D4-branes wrapping a circle with anti-periodic boundary conditions for the fermions, N_f D8-branes at a point on the circle, and N_f anti-D8-branes at another point on the circle. The low-lying open string excitations are precisely those of $SU(N_c)$ Yang-Mills theory with N_f flavors of massless quarks. The holographic limit corresponds to $N_c \to \infty$, and N_f is kept finite, so that the D8-branes are treated as probes in the near-horizon background of the D4-branes. In this limit, the background is capped off in the IR, corresponding to confinement, and the D8-branes and anti-D8-branes connect, corresponding to chiral symmetry breaking.

This model has also been studied in various external conditions, including nonzero temperature [2], nonzero baryon density [3, 4], and background electromagnetic fields [5, 6, 7, 8, 9, 10, 11], and it exhibits many properties that are expected of QCD.

Background magnetic fields are particularly interesting in that they may be physically relevant in neutron stars, where they can reach values of about 10^{15} Gauss. On the theory side, background magnetic fields have some interesting effects on the QCD ground state. One effect is the catalysis of chiral symmetry breaking by a strong magnetic field [12]. The effect of a background magnetic field in the Sakai-Sugimoto model was studied in [5, 6], where the catalysis of chiral symmetry breaking was demonstrated explicitly.

Here we will be interested in the effects of a background magnetic field at nonzero baryon chemical potential. This question was recently studied in the the low-energy effective field theory [13, 14, 15], where it was shown that the axial anomaly leads to interesting effects. In the confined phase, the anomaly leads to a non-trivial pion gradient and an associated baryon charge density [15],

$$\nabla\pi^0 = \frac{e}{4\pi^2 f_\pi}\mu_B B \ , \ d = \frac{e}{4\pi^2 f_\pi} B\cdot\nabla\pi^0 \, . \tag{1}$$

We will present here the results of [8] which show that similar effects occur in the one-flavor Sakai-Sugimoto model.[1] In this model, the anomaly is encoded in the Chern-Simons term of the D8-brane action. The model does not include a true electromagnetic field, but we can mimic the effect of a non-dynamical background electromagnetic field using the non-normalizable mode of the D8-brane worldvolume gauge field. This field is actually dual to the baryon current in the gauge theory but is equal in the one-flavor case to the electric current. We will therefore use the same bulk gauge field, but different components, to describe both the baryon chemical potential and the background magnetic field. We will show that these source a third, normalizable component of the gauge field via the Chern-Simons term. In the low-temperature confining background this field has a nonzero boundary value, which is interpreted as the gradient of the $U(1)_A$ pseudo-scalar meson, *i.e.* the η'. This also leads to a baryon number charge density. For small magnetic fields, our result agrees with (1) adapted to the $U(1)_A$ sector. Furthermore, we will show that there is a phase transition at a critical value of the chemical potential to a mixed phase of ordinary baryonic matter and pseudo-scalar gradient matter. In the mixed phase, the relative proportion of ordinary baryonic matter at fixed chemical potential decreases with the magnetic field.

2. Review of finite density HQCD

We first present an abbreviated review of the Sakai-Sugimoto model with nonzero baryon density and with background electromagnetic fields. For more detailed background see, for example, [1, 2] for the model and finite temperature phase structure, [3, 4] for nonzero baryon density, and [5, 6, 7, 8, 11] for electromagnetic fields.

The Euclidean metric dual to the confining phase is given by

$$ds^2 \;=\; u^{3/2}\left((dx_0^E)^2 + d\mathbf{x}^2 + f(u)dx_4^2\right) + u^{-3/2}\left(\frac{du^2}{f(u)} + u^2 d\Omega_4^2\right) \tag{2}$$

where x_4 is a compact coordinate with periodicity $2\pi R_4$, and

$$f(u) = 1 - \frac{u_{KK}^3}{u^3} \ , \ u_{KK} = \frac{4R^2}{9R_4^2} \, . \tag{3}$$

[1] See also [9] and [10], which present some related and overlapping results.

The curvature radius of the space is $R = (\pi g_s N_c)^{1/3}\sqrt{\alpha'}$.

The antipodal embedding of the D8-brane in this background has a U shape that satisfies $x_4'(u) = 0$, with the tip at u_{KK}. The D8-brane worldvolume theory contains a gauge field, which contains both a vector and an axial part depending on the parity under exchanging the two halves of the embedding,

$$a_M(x^\mu, u) = a_M^V(x^\mu, u) + a_M^A(x^\mu, u)\,. \tag{4}$$

The physical gauge field is $A_M = a_M R/(2\pi\alpha')$. There is a discrete spectrum of normalizable radial modes corresponding to various low-spin mesons. In particular, the zero mode of a_u^A is identified with the massless pseudo-scalar corresponding to the Goldstone boson of the broken chiral symmetry. For a single flavor, this is the η'. Working in the gauge $a_u = 0$ [1], the pseudo-scalar reappears in the zero mode of a_μ^A,

$$a_\mu^A(x^\mu, u) = \partial_\mu \varphi(x^\mu)\psi_0(u) + \text{higher modes}\,, \tag{5}$$

where the zero-mode wavefunction $\psi_0(u)$ yields a normalizable field strength. The physical pseudo-scalar field is $\eta'(X) = f_\pi \varphi(x) R^2/(2\pi\alpha')$, where $f_\pi^2 = \frac{N_c u_{KK}^{3/2}}{8\pi^4\alpha'}$.

By contrast, the asymptotic solution for the vector component of the gauge field at large u is

$$a_0^V(u) \approx \mu - \frac{2}{3}\frac{d}{u^{3/2}}\,. \tag{6}$$

where the zero mode μ is u-independent, and therefore non-normalizable, and is identified with the baryon chemical potential. Varying the action with respect to μ allows us to identify the constant d as the baryon charge density. The physical chemical potential is $\mu_B = \mu R/(2\pi\alpha')$, and the physical baryon charge density is $D = d(2\pi\alpha'\mathcal{N}/R)$.

In the absence of sources, the only solution is a constant: $a_0^V(u) = \mu$. A second solution becomes possible when one includes sources at the tip corresponding to D4-branes wrapped on the S^4, which are precisely the baryons of the model. A single baryon carries N_c units of baryon charge and has mass $m_4 = \frac{1}{3}N_c u_{KK}$, which is the mass of a wrapped D4-brane located at $u = u_{KK}$. The physical D4-brane mass and density are given by $M_4 = m_4 R/(2\pi\alpha')$ and $N_4 = n_4(2\pi\alpha'\mathcal{N}/R)$. The baryon charge density is then just $d = N_c n_4$. In the presence of baryon sources, the gauge field at the tip is

$$a_0^V(u_{KK}) = \frac{m_4}{N_c}\,, \tag{7}$$

which implies that the solution exists only for $\mu > m_4/N_c$. This has the obvious phenomenological interpretation that, at low temperature, baryons can only appear when the chemical potential is high enough to produce them. Furthermore, above this point, this "nuclear matter" solution dominates over the vacuum solution. In other words, nuclear matter forms as soon as it can, and there is a phase transition at $\mu_c = m_4/N_c$. This nuclear matter phase transition is marginally second-order, which is different from the expected first-order transition in QCD. However, the result is reasonable since we have ignored baryon interactions.

3. Magnetic properties of the confined phase

To mimic the effect of a background magnetic field, we turn on a background value for the zero mode of a spatial component of the vector gauge field,

$$a_3^V(x_2, u) = h x_2\,. \tag{8}$$

The physical magnetic field is $H = h/(2\pi\alpha')$. Since $a_0^V(u) \neq 0$, the five-dimensional CS term, which comes from the D8-brane CS coupling to F_4, will source the axial field $a_1^A(u)$. As we saw in the previous section, the boundary value of this field corresponds to the (constant) gradient of the pseudo-scalar (in the x^1 direction in this case),

$$a_1^A(\infty) = \nabla\varphi\,. \tag{9}$$

This corresponds to a field, rather than an external parameter, in the gauge theory since the zero mode of the axial field is normalizable. Its value is therefore determined by extremizing the action. Furthermore, since a_1^A is an axial field, it must vanish at the tip,

$$a_1^A(u_{KK}) = 0\,. \tag{10}$$

The D8-brane action with all the relevant fields is $S_{DBI} + S_{CS}$, where

$$\begin{aligned} S_{DBI} &= \mathcal{N}\int_{u_{KK}}^{\infty} du\, u^{5/2}\sqrt{\left(\frac{1}{f(u)} - (a_0^{V\prime}(u))^2 + (a_1^{A\prime}(u))^2\right)\left(1+\frac{h^2}{u^3}\right)} \qquad (11)\\ S_{CS} &= -\mathcal{N}\int_{u_{KK}}^{\infty} du\left(\partial_2 a_3^V a_0^V(u) a_1^{A\prime}(u) - \partial_2 a_3^V a_0^{V\prime}(u) a_1^A(u) - a_3^V \partial_2 a_0^V a_1^{A\prime} + a_3^V \partial_2 a_1^A a_0^{V\prime}\right). \end{aligned}$$

To deal with the last two terms in S_{CS} which give contributions at the boundary, we add to the action the following boundary terms (as in [8]):

$$\Delta S = \frac{\mathcal{N}}{2}\int\left\{\partial_2\left(a_3 a_{[1}a'_{0]}\right) + \partial_0\left(a_2 a_{[1}a'_{3]}\right) + \partial_u\left(a_3\partial_2 a_{[1}a_{0]} + a_2\dot{a}_{[1}a_{3]}\right)\right\}\,. \tag{12}$$

The combined action simplifies to

$$S_{\rm CS} + \Delta S = \frac{3\mathcal{N}}{2}\int\left(h a_{[0}a'_{1]} + e a_{[1}a'_{3]}\right)\,. \tag{13}$$

The corresponding integrated equations of motion are given by

$$\frac{\sqrt{u^5 + h^2u^2}\, a_0^{V\prime}(u)}{\sqrt{\frac{1}{f(u)} - (a_0^{V\prime}(u))^2 + (a_1^{A\prime}(u))^2}} = 3h a_1^A(u) + N_c n_4 \tag{14}$$

$$\frac{\sqrt{u^5 + h^2u^2}\, a_1^{A\prime}(u)}{\sqrt{\frac{1}{f(u)} - (a_0^{V\prime}(u))^2 + (a_1^{A\prime}(u))^2}} = 3h a_0^V(u) + c\,, \tag{15}$$

where the constant of integration in (14) has been identified with the density of D4-brane sources, and the constant of integration in (15) will be determined below.

The baryon charge and currents are defined by

$$J^\mu(x) = \frac{\delta S_{EOM}}{\delta A_\mu(x, u=\infty)} = \lim_{u\to\infty}\left(\frac{\delta\mathcal{L}_{EOM}}{\delta\partial_u A_\mu(x)}\right)\,, \tag{16}$$

which gives the (dimensionless) baryon charge density

$$d = N_c n_4 + \frac{3}{2}h a_1^A(\infty) = N_c n_4 + \frac{3}{2}h\nabla\varphi\,. \tag{17}$$

The origin of the second term can be understood as an additional D4-brane charge inside the D8-brane, which is due to the orthogonal worldvolume field strengths in the (x^2, x^3) and (u, x^1) directions. We can likewise get the (dimensionless) axial current density,

$$j_A = c + \frac{3}{2} h a_0^V(\infty) = c + \frac{3}{2} h\mu \, . \tag{18}$$

We still need to extremize the action with respect to $\nabla\varphi$, which has the effect of setting $j_A = 0$, and therefore $c = -\frac{3}{2}h\mu$.

We can simplify the equations of motion considerably by introducing a new coordinate

$$y = \int_{u_{KK}}^{u} \frac{3hd\tilde{u}}{\sqrt{f(\tilde{u})}\sqrt{\tilde{u}^5\left(1+\frac{h^2}{\tilde{u}^3}\right) + (N_c n_4)^2 - \left(\frac{3}{2}h\mu\right)^2 - 6h\kappa}} \tag{19}$$

where κ is

$$\kappa = \begin{cases} -\frac{3}{2}h(\nabla\varphi)^2 & \text{sourceless case} \\ -\frac{3}{2}h\frac{m_4}{N_c}\left(\mu - \frac{m_4}{N_c}\right) & \text{sourced case} \, . \end{cases} \tag{20}$$

With some algebra, (14) and (15) can be written as

$$a_0^{V\prime}(y) = a_1^A(y) + \frac{N_c n_4}{3h} \tag{21}$$

$$a_1^{A\prime}(y) = a_0^V(y) - \frac{\mu}{2} \, , \tag{22}$$

where the derivative is with respect to y. Let us now analyze the two types of solutions.

3.1. Pseudo-scalar gradient phase

In the absence of sources, $n_4 = 0$, all the baryon charge density comes from the pseudo-scalar gradient:

$$d = \frac{3}{2}h\nabla\varphi \, . \tag{23}$$

The solution to (21) and (22) is given in this case by

$$a_0^V(y) = \frac{\mu}{2}\left(\frac{\cosh y}{\cosh y_\infty} + 1\right) \tag{24}$$

$$a_1^A(y) = \frac{\mu}{2}\frac{\sinh y}{\cosh y_\infty} \, , \tag{25}$$

where $y_\infty \equiv y(u \to \infty)$ can be determined numerically. The pseudo-scalar gradient is then

$$\nabla\varphi = \frac{\mu}{2}\tanh y_\infty \, . \tag{26}$$

The numerical results for $\nabla\varphi$ as a function of μ and h are presented in Figure 1. For $h \ll 1$, *i.e.* sub-string scale magnetic fields, the behavior is linear, and the pseudo-scalar gradient is approximately given by

$$\nabla\varphi \approx \frac{\pi}{2u_{KK}^{3/2}}\mu h \, . \tag{27}$$

In terms of the physical quantities this is

$$\nabla\eta' \approx \frac{N_c}{8\pi^2 f_\pi}\mu_B H \tag{28}$$

for small magnetic fields. This agrees, with appropriate change in normalization,[2] with the one-flavor version of the result (1) from [15].

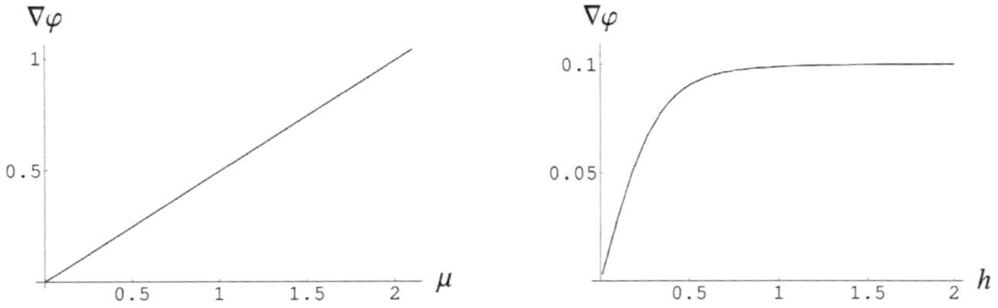

Figure 1. The pseudo-scalar gradient $\nabla\varphi$ as a function of μ for fixed $h = 1$, and as a function of the magnetic field h for fixed $\mu = 0.2$, both with $u_{KK} = 1$.

3.2. Mixed phase

At sufficiently large chemical potential, a solution with D4-brane sources is also possible. In this phase, both the baryons and the pseudo-scalar gradient contribute to the baryon charge density.

Using the boundary conditions at the tip $a_0^V(y=0) = m_4/N_c$ and $a_1^A(y=0) = 0$, the solution to the equations of motion (21) and (22) is now given by

$$a_0^V(y) = \left(\frac{m_4}{N_c} - \frac{\mu}{2}\right)\cosh y + \frac{N_c n_4}{3h}\sinh y + \frac{\mu}{2} \tag{29}$$

$$a_1^A(y) = \left(\frac{m_4}{N_c} - \frac{\mu}{2}\right)\sinh y + \frac{N_c n_4}{3h}(\cosh y - 1)\,. \tag{30}$$

The boundary conditions at infinity then determine the gradient and D4-brane density implicitly in terms of μ and h,

$$\nabla\varphi = \frac{\cosh y_\infty - 1}{\sinh y_\infty}\frac{m_4}{N_c} \tag{31}$$

$$N_c n_4 = \frac{\frac{3}{2}h\mu + \frac{3}{2}h\left(\mu - \frac{2m_4}{N_c}\right)\cosh y_\infty}{\sinh y_\infty}\,. \tag{32}$$

The critical value of the chemical potential corresponds to the point at which the actions of the $\nabla\varphi$ and mixed phases are equal. But, it also coincides, as it did when $h = 0$, with the minimal value of the chemical potential to create a baryon. The phase diagrams in the (μ, h) and (d, h) planes are shown in Figure 2. The critical chemical potential increases from its $h = 0$ value m_4/N_c to $2m_4/N_c$ as $h \to \infty$. For a given h, there is a marginally second-order phase transition from the $\nabla\varphi$ phase to the mixed phase at $\mu_c(h)$, which generalizes the ordinary nuclear matter transition at $h = 0$. We also see that, for a fixed total baryon charge density, the pseudo-scalar gradient phase dominates above a critical magnetic field.

[2] The relative factor of $N_c/2$ is understood as follows. First, the baryon charge of a quark in [15] is $1/N_c$ so $\mu_B^{there} = N_c\mu_B^{here}$. Second, the CS term coupling the baryonic $U(1)_V$ gauge field to the σ_3-component of the $SU(2)$ gauge field has a factor of 2 relative to the purely abelian CS term once the boundary term at spatial infinity is omitted.

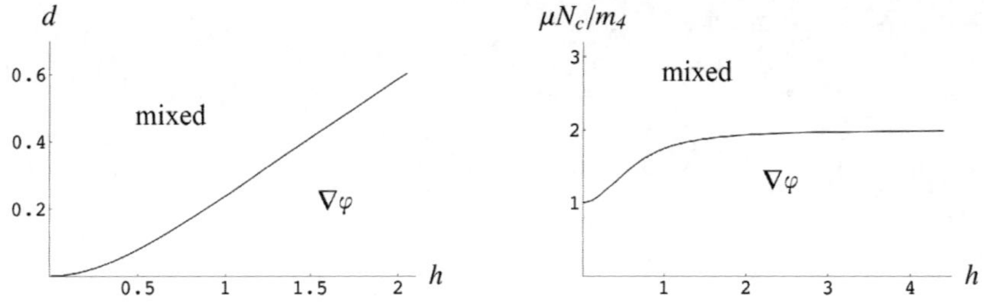

Figure 2. Phase diagram in the (a) canonical and (b) grand canonical ensemble.

The total baryon charge density is given by

$$d = N_c n_4 + \frac{3}{2} h \nabla\varphi = \frac{3h}{2}\left(\mu - \frac{m_4}{N_c}\right)\frac{\cosh y_\infty + 1}{\sinh y_\infty}. \tag{33}$$

Figure 3 shows the fraction, obtained numerically, of that charge carried by baryons. We see that the relative proportion of baryons at fixed h increases with μ. In the limit of large μ, the system is almost entirely baryonic nuclear matter. On the other hand, at fixed μ the proportion of baryons decreases with h. For $m_4/N_c < \mu < 2m_4/N_c$, the proportion of baryons vanishes at the critical magnetic field $h_c(\mu)$ shown in Figure 2, where the $\nabla\varphi$ phase takes over.

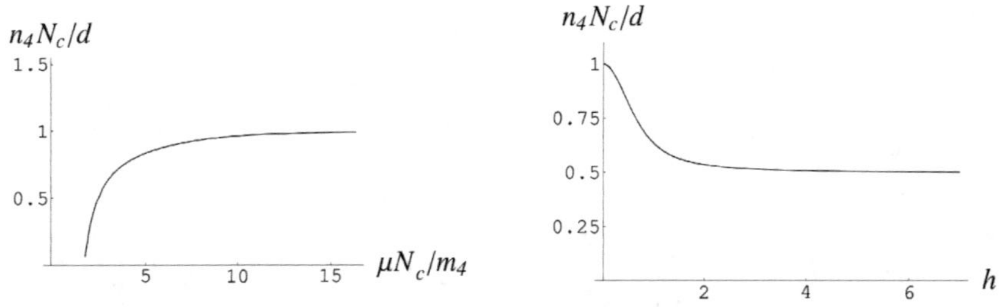

Figure 3. The baryon fraction $n_4 N_c/d$, as a function of μ for fixed $h = 1$ and as a function of h for fixed $\mu = 3m_4/N_c$, both with $u_{KK} = 1$.

3.3. Magnetization

The state described by either the pseudo-scalar gradient phase or the mixed phase responds to the external magnetic field by becoming magnetized. The magnetization M in the grand canonical or canonical ensemble is

$$M(\mu, h) = -\frac{\partial\Omega(\mu, h)}{\partial h}\bigg|_\mu, \qquad M(d, h) = -\frac{\partial F(d, h)}{\partial h}\bigg|_d, \tag{34}$$

where the grand potential is $\Omega(\mu, h) = S[a_0(u), a_1(u)]|_{EOM}$ and the free energy is $F = \Omega + \mu d$.

We focus here on the matter contribution to the magnetization. The magnetic properties of the vacuum were studied in [5, 6]. We will therefore subtract from the quantities above the formally divergent contribution of the vacuum, which gives a finite result that represents the corresponding contribution of just the matter. The numerical results for the magnetizations in the pseudo-scalar gradient and mixed phases are presented in Figure 4. For small h, the response is linear, but for $h \sim O(1)$ the non-linear effect of the DBI action becomes pronounced. We find that the matter is paramagnetic in both phases.

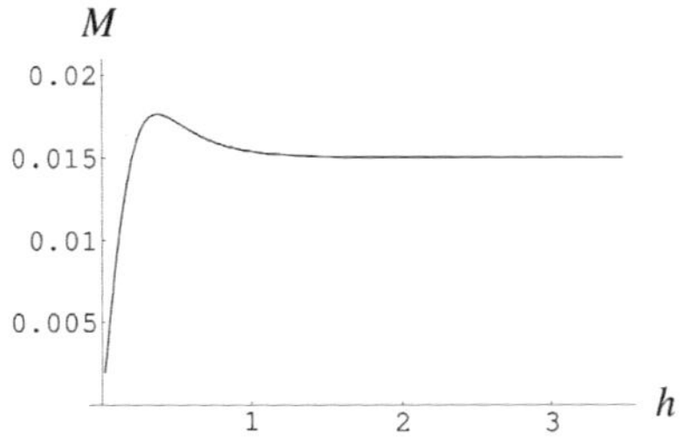

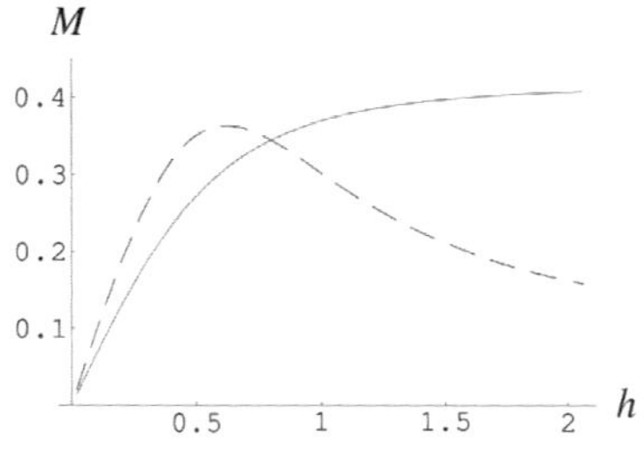

Figure 4. The magnetization M (in units of $\mathcal{N}$) as a function of h in (a) the pseudo-scalar gradient phase for fixed $\mu = 0.2$ and (b) the mixed phase for fixed $\mu = 3m_4/N_c$ (red) and fixed $d = 1$ (dashed blue), all with $u_{KK} = 1$.

4. Conclusions

We have explored the properties of holographic QCD at finite density in a background magnetic field and find the system has a rich phenomenology. In the holographic dual, these axial anomaly-induced properties are induced by the Chern-Simon term on the D8-brane. In the confined phase, turning on a magnetic field induces a gradient for the pseudo-scalar η' field. This gradient carries baryon charge, and at large enough magnetic fields, it is the dominant phase. That is, if we start at zero field with some baryons, as we increase the field those baryons will start being replaced by a gradient of the η' field, eventually disappearing altogether.

Acknowledgments

We would like to thank Al Shapere and Sumit Das for organizing the QTS6 conference and providing the opportunity for ML to present this work. This research was supported in part by the Israel Science Foundation under grant no. 568/05.

[1] T. Sakai and S. Sugimoto, Prog. Theor. Phys. **113**, 843 (2005) [arXiv:hep-th/0412141]; Prog. Theor. Phys. **114**, 1083 (2006) [arXiv:hep-th/0507073].
[2] O. Aharony, J. Sonnenschein and S. Yankielowicz, Annals Phys. **322**, 1420 (2007) [arXiv:hep-th/0604161]; A. Parnachev and D. A. Sahakyan, Phys. Rev. Lett. **97**, 111601 (2006) [arXiv:hep-th/0604173].
[3] O. Bergman, G. Lifschytz and M. Lippert, JHEP **0711**, 056 (2007) [arXiv:0708.0326 [hep-th]].
[4] N. Horigome and Y. Tanii, JHEP **0701**, 072 (2007) [arXiv:hep-th/0608198]; J. L. Davis, M. Gutperle, P. Kraus and I. Sachs, JHEP **0710**, 049 (2007) [arXiv:0708.0589 [hep-th]]; M. Rozali, H. H. Shieh, M. Van Raamsdonk and J. Wu, JHEP **0801**, 053 (2008) [arXiv:0708.1322 [hep-th]]; K. Y. Kim, S. J. Sin and I. Zahed, JHEP **0801**, 002 (2008) [arXiv:0708.1469 [hep-th]].
[5] O. Bergman, G. Lifschytz and M. Lippert, JHEP **0805**, 007 (2008) [arXiv:0802.3720 [hep-th]].
[6] C. V. Johnson and A. Kundu, JHEP **0812**, 053 (2008) [arXiv:0803.0038 [hep-th]].
[7] K. Y. Kim, S. J. Sin and I. Zahed, JHEP **0807**, 096 (2008) [arXiv:0803.0318 [hep-th]].
[8] O. Bergman, G. Lifschytz and M. Lippert, Phys. Rev. D **79**, 105024 (2009) [arXiv:0806.0366 [hep-th]].
[9] E. G. Thompson and D. T. Son, Phys. Rev. D **78**, 066007 (2008) [arXiv:0806.0367 [hep-th]].
[10] A. Rebhan, A. Schmitt and S. A. Stricker, arXiv:0811.3533 [hep-th].
[11] G. Lifschytz and M. Lippert, Phys. Rev. D **80**, 066005 (2009) [arXiv:0904.4772 [hep-th]]; G. Lifschytz and M. Lippert, Phys. Rev. D **80**, 066007 (2009) [arXiv:0906.3892 [hep-th]].
[12] V. A. Miransky and I. A. Shovkovy, Phys. Rev. D **66**, 045006 (2002) [arXiv:hep-ph/0205348].
[13] D. T. Son and A. R. Zhitnitsky, Phys. Rev. D **70**, 074018 (2004) [arXiv:hep-ph/0405216].
[14] M. A. Metlitski and A. R. Zhitnitsky, Phys. Rev. D **72**, 045011 (2005) [arXiv:hep-ph/0505072]; G. M. Newman and D. T. Son, Phys. Rev. D **73**, 045006 (2006) [arXiv:hep-ph/0510049].
[15] D. T. Son and M. A. Stephanov, Phys. Rev. D **77**, 014021 (2008) [arXiv:0710.1084 [hep-ph]].

Journal of Physics: Conference Series **462** (2013) 012031 doi:10.1088/1742-6596/462/1/012031

Spontaneous Broken Local Conformal Symmetry and Dark Energy Candidate

Lu-Xin Liu

Department of Physics, Purdue University, West Lafayette, IN 47907, USA
and National Institute for Theoretical Physics, Department of Physics and Centre for Theoretical Physics, University of the Witwatersrand, Wits, 2050, South Africa

E-mail: `luxin.liu9@gmail.com`

Abstract. The local conformal symmetry is spontaneously broken down to the Local Lorentz invariance symmetry through the approach of nonlinear realization. The resulting effective Lagrangian, in the unitary gauge, describes a cosmological vector field non-minimally coupling to the gravitational field. As a result of the Higgs mechanism, the vector field absorbs the dilaton and becomes massive, but with an independent energy scale. The Proca type vector field can be modelled as dark energy candidate. The possibility that it further triggers Lorentz symmetry violation is also pointed out.

1. Introduction

Astronomical observations have shown strong evidence for the existence of the dark energy, which is assumed to be smoothly distributed through the universe. It accounts for about 70% of the total mass energy of the universe and tends to increase the expanding rate of the universe. In practice, the candidate for the dark energy, besides the cosmological constant, is usually taken to be a dynamical component [1]. Recently, a massive vector field as dark energy candidate has been attracted much attention [2-3]. There, the dark energy is modeled as a massive cosmological vector field, coupling to the gravity non-minimally and permeating the whole universe. Although the vector field can be regarded as an origin for dark energy and drives the inflation of the universe, the origin of the vector field itself may have different physical motivations. For example, as pointed out in Ref.[4], a vector field was introduced as the requirement of local scale invariance. However, one may note that, for a realistic physical model, there is no scale symmetry. Since at low energy the vacuum is invariant under (local) Lorentz transformations, it is conceivable that at higher energy matter might transform under a larger spacetime group. Therefore, if the scale symmetry is realized in nature, if must be a broken one.

Motivated by that, we consider the case that the full symmetry represented by the conformal group, which contains the Lorentz group as a subgroup. Since the breaking mechanism which maintains the preponderance of the dynamical constraints of the symmetry and hence is both theoretically and phenomenologically most attractive is a spontaneous one, in the present paper we consider the spontaneous breaking of conformal symmetry and illustrate an alternative mechanism of introducing a massive vector field into the theory.

Journal of Physics: Conference Series **462** (2013) 012031 doi:10.1088/1742-6596/462/1/012031

As for the spontaneous symmetry breaking, the approach of nonlinear realization has been demonstrated a natural, economical and elegant framework for treating it. This method has been applied to a wide range of physical problems most notably in the form of nonlinear sigma models, supersymmetry and brane theories [5,6]. There, the Lagrangian is invariant with respect to the transformations of some continuous group G, but the ground state is not an invariant of G but only of some subgroup H[7]. In this context, the resulting phenomenological Lagrangian becomes an effective theory at energies far below the scale of spontaneous symmetry breaking. If one takes the Lorentz group as the stability group H, the conformal symmetry can then be globally spontaneously broken down through the approach of nonlinear realization. Consequently, the effective action is expressed in terms of the dynamics of a Nambu-Goldstone field, i.e. the dilaton [8, 9].

Alternatively, the conformal symmetry can also be spontaneously broken down locally. In particle physics, it is well known that the compensating gauge field need be introduced for the local gauge invariance of the theory. It is our purpose of this paper to point out a viable original of the massive cosmological vector field, as well as to give a natural explanation of this vector field at both theoretical and phenomenological levels. It is shown that the vector field behaves as the compensating field resulting from the spontaneous breaking of local conformal invariance. Furthermore, due to the Higgs mechanism, the vector field absorbs the dilaton particle and becomes massive. However, it has an independent mass scale. As a result, the massive vector particle is taken as the source of the dark energy, composing the cosmological gas and driving the universes accelerated expansion. Its mass scale is fixed by the cosmological observation.

The paper is organized as follows. In section 2, we start by giving a brief review of the relevant spontaneous breaking of the global conformal symmetry. The effective action is shown to have a very natural interpretation in terms of Goldstone particles for broken symmetries. Then we generalize the nonlinear realization approach to the spontaneous breaking of the local conformal symmetry. In order to conform to the covariant transformations of the local Maurer-Cartan one-forms, the compensating gauge fields are introduced. Their transformations are then derived accordingly. In section 3, the effective action of the theory is constructed. By using the fact that the inhomogeneous transformations of the Nambu-Goldstone fields are spacetime dependent, we have the degree of freedom of fixing the gauge. It is shown, under the unitary gauge, the NG fields are transformed away and the effective action is reformulated into a vector-tensor theory. It is prominent to note that the resulting massive vector field has an independent energy scale, which is not determined by the conformal symmetry breaking scale. The effective action that describing the general coupling of the vector field to the gravitational field is then obtained. Finally, we further our discussions by pointing out the possibility that the Lorentz invariance violation may be triggered by the spontaneous breaking of local conformal invariance. We hope this work may also shed light on the underlying theories related to Lorentz violation due to the presence of a vector field.

2. Nonlinear Realization of Conformal Symmetry

The global conformal symmetry breaking was discussed in [8, 9]. Here, before considering the case of local conformal symmetry breaking, for completeness we give a brief illustration about the dynamics resulting from its global symmetry breaking. The conformal group G includes Poincare group as its subgroup along with the conformal boosts K_μ and dilatation D symmetries. Choose the Lorentz group as the stability group H. It follows that the conformal symmetry can be spontaneously broken down to the $ISO(1,3)$ symmetry through the nonlinear realization approach. The Coset representative elements $\Omega = G/H$ can be parameterized as

$$\Omega = e^{ix^\mu P_\mu} e^{i\phi D} e^{i\phi^\mu K_\mu} \tag{1}$$

where x^μ parameterize $1+3$ spacetime coordinates, and ϕ,ϕ^μ correspond to the Nambu-Goldstone (NG) fields associated with the broken generators D and K^μ. The effective action describing the dynamics of these NG fields can be constructed from the Maurer-Cartan one-forms $\Omega^{-1}d\Omega$, which can be expanded with respect to the generators of the conformal group as

$$\begin{aligned}\Omega^{-1}d\Omega =& i\omega^\mu P_\mu + i\omega_D D + i\omega_k^\mu K_\mu + i\omega_M^{\mu\nu} M_{\mu\nu} = ie^{-\phi}dx^\mu P_\mu + (id\phi - ie^{-\phi}2\phi^\mu dx_\mu)D+ \\ & ie^{-\phi}2\phi^\nu dx^\mu M_{\mu\nu} + (ie^{-\phi}2\phi_\mu dx^\mu \phi^\nu - ie^{-\phi}\phi^\mu\phi_\mu dx^\nu + id\phi^\nu - id\phi\phi^\nu)K_\nu \end{aligned} \tag{2}$$

The covariant coordinate differentials are found to be $\omega^\mu = e_\nu{}^\mu dx^\nu = e^{-\phi}dx^\mu$, in which $e_\nu{}^\mu = e^{-\phi}\delta_\nu{}^\mu$ is the vierbein. And the covariant derivative of the boost vector field ϕ^μ is given by the one-forms $\omega_k^\mu = \omega^\nu D_\nu \phi^\mu = \omega^\nu(2\phi_\nu\phi^\mu - \phi^{\mu'}\phi_{\mu'}\delta_\nu^\mu + e^\phi\partial_\nu\phi^\mu - e^\phi\partial_\nu\phi\phi^\mu)$. It follows that the kinetic term and the potential term of the dilaton field have the form

$$S_{dilaton} = \int d^4x \det e D_\mu\phi^\mu + \int d^4x \det e = \frac{1}{2}\int d^4x e^{-2\phi}\partial_\mu\phi\partial^\mu\phi + \int d^4x e^{-4\phi} \tag{3}$$

Besides, the role of connection is played by the one-forms $\omega_{M\rho}^{\mu\nu}$ in Eq.(2), i.e.

$$\omega_M^{\mu\nu} = dx^\rho\omega_{M\rho}^{\mu\nu} = e^{-\phi}2\phi^\nu dx^\mu = dx^\rho 2e^{-\phi}\delta_\rho{}^\mu\phi^\nu \tag{4}$$

namely, $\omega_{M\rho}^{\mu\nu} = 2e^{-\phi}\delta_\rho{}^\mu\phi^\nu$. Therefore the effective action of the scalar field Φ coupling to the dilaton can be written as

$$\begin{aligned} S \propto & \int d^4x \det e \eta^{\sigma\lambda} e_\sigma^{-1\mu} e_\lambda^{-1\nu}(\partial_\mu + i\omega_{M\mu}^{\alpha\beta}(M_{\alpha\beta}))\Phi(\partial_\nu + i\omega_{M\nu}^{\alpha\beta}(M_{\alpha\beta}))\Phi \\ = & \int d^4x \det e \eta^{\sigma\lambda} e_\sigma^{-1\mu} e_\lambda^{-1\nu}\partial_\mu\Phi\partial_\nu\Phi = \int d^4x \eta^{\mu\nu}e^{-2\phi}\partial_\mu\Phi\partial_\nu\Phi \end{aligned} \tag{5}$$

For the case of fermion field, the effective action can be derived in a similar approach, with $M_{\alpha\beta}$ replaced by the corresponding representation of the fermion field for the Lorentz group instead.

Consider the global left action of the full group elements g on the Coset elements Eq.(1). The result can be uniquely decomposed as the product of the new Coset elements Ω' and the stability group elements h, i.e.

$$g\Omega = \Omega' h \tag{6}$$

in which the parameters of the full group elements $g = e^{ia^\mu P_\mu}e^{icD}e^{ib^\mu K_\mu}e^{i\theta^{\mu\nu}M_{\mu\nu}}$ are spacetime independent. As a result, the transformations of the Cartan one-forms become

$$\Omega'^{-1}d\Omega' = (h\Omega^{-1}g^{-1})d(g\Omega h^{-1}) = h(\Omega^{-1}d\Omega)h^{-1} + hdh^{-1} \tag{7}$$

or namely $i\omega'(x') = hi\omega(x)h^{-1} + hdh^{-1}$. Therefore, ω^μ, ω_D and ω_k^μ transform covariantly under the action of Eq.(6), whereas $\omega_M^{\mu\nu}$ transforms by a shift hdh^{-1} as required by the connection. On the other hand, the spontaneous breaking of the symmetry group can be spactetime dependent [10,11,12]. In the present context, the transformation of Eq.(6) becomes a local one, i.e.

$$g = e^{ia^\mu(x)P_\mu}e^{ic(x)D}e^{ib^\mu(x)K_\mu}e^{i\theta^{\mu\nu}(x)M_{\mu\nu}} \tag{8}$$

in which the parameters of the group elements g become spacetime dependent. Accordingly, the Cartan one-forms transform as

$$\begin{aligned}\Omega'^{-1}d\Omega' =& (h\Omega^{-1}g^{-1}(x))d(g(x)\Omega h^{-1}) = h\Omega^{-1}(g^{-1}(x)dg(x))\Omega h^{-1} + h(\Omega^{-1}d\Omega)h^{-1} + hdh^{-1} \\ \neq & h(\Omega^{-1}d\Omega)h^{-1} + hdh^{-1}\end{aligned} \tag{9}$$

Introducing the compensating one-forms into Eq.(2), one has

$$\Omega^{-1}d\Omega \to \Omega^{-1}(d+i\hat{E})\Omega = i\omega^m P_m + i\omega_D D + i\omega_k^m K_m + i\omega_M^{mn} M_{mn} \tag{10}$$

Under the action of Eqs.(6,8), it follows that the (local) Cartan one-forms transform covariantly

$$\Omega^{-1}(d+i\hat{E})\Omega \to \Omega'^{-1}(d+i\hat{E}')\Omega' = h(\Omega^{-1}(d+i\hat{E})\Omega)h^{-1} + hdh^{-1} \tag{11}$$

It therefore leads to the infinitesimal transformations of the compensating one-forms

$$\hat{E}' = g(x)\hat{E}g(x)^{-1} - ig(x)dg(x)^{-1} \tag{12}$$

In terms of its expanding $\hat{E} = \hat{E}^m P_m + \hat{A}D + \hat{B}^m K_m + \hat{m}^{mn} M_{mn}$, the transformations of its components are found to be

$$\begin{aligned}
\hat{E}'^m =& \hat{E}^m + 2\hat{E}_{m'}\theta^{m'm}(x) + \hat{E}^m c(x) - \hat{A}a^m(x) + 2a_n(x)\hat{m}^{mn} - da^m(x)\\
\hat{A}' =& \hat{A} + 2\hat{E}^m b_m(x) - 2a^m(x)\hat{B}_m - dc(x)\\
\hat{B}'^m =& \hat{B}^m + b^m(x)\hat{A} + 2\theta^{nm}(x)\hat{B}_n + 2b_n(x)\hat{m}^{mn} - c(x)\hat{B}^m - db^m(x)e^{-c(x)}\\
\hat{m}'^{mn} =& \hat{m}^{mn} - 2\hat{E}^m b^n(x) + 2a^m(x)\hat{B}^n - \theta^{m'm}(x)\hat{m}^{n}{}_{m'} - \theta^{mn'}(x)\hat{m}^{n}_{n'}\\
&+ \theta_{n'}^{m}(x)\hat{m}^{n'n} + \theta^{m}_{n'}(x)\hat{m}^{nn'} - d\theta^{mn}(x)
\end{aligned} \tag{13}$$

Estimating the expansion of local Cartan one-forms of Eq.(10) with the use of the translated form of $\hat{E} = e^{ix^\mu P_\mu} E e^{-ix^\mu P_\mu}$ yields

$$\begin{aligned}
\omega^m =& dx^\mu e^{-\phi}(E_\mu{}^m + \delta_\mu{}^m) = dx^\mu e_\mu{}^m\\
\omega_D =& dx^\mu(-2E_\mu{}^m\phi_m e^{-\phi} + A_\mu + \partial_\mu\phi - 2e^{-\phi}\phi_m\delta_\mu{}^m) = dx^\mu\omega_{D\mu}\\
\omega_k^m =& dx^\mu(2\phi^m\phi_n E_\mu{}^n e^{-\phi} - \phi_n\phi^n E_\mu{}^m e^{-\phi} - \phi^m A_\mu + B_\mu{}^m e^{\phi} + 2\phi_n m_\mu{}^{nm} +\\
&\partial_\mu\phi^m - \partial_\mu\phi\phi^m + 2e^{-\phi}\phi^m\phi_n\delta_\mu{}^n - e^{-\phi}\phi_n\phi^n\delta_\mu{}^m) = dx^\mu\omega_{k\mu}^m\\
\omega_M^{mn} =& dx^\mu(2\phi^n E_\mu{}^m e^{-\phi} + m_\mu{}^{mn} + 2e^{-\phi}\phi^n\delta_\mu{}^m) = dx^\mu\omega_{M\mu}^{mn}
\end{aligned} \tag{14}$$

where the Latin indices $m, n, ...$ represent the local Lorentz indices, and the Greek indices $\mu, \nu, ...$ represent the general coordinate indices. According to Eq.(11), their infinitesimal transformation laws are found to be

$$\omega'^m = \omega^n L_n{}^m; \omega'_D = \omega_D; \omega'^m_k = \omega^n_k L_n{}^m; \omega'^{mn}_M = \omega^{ab}_M L_a{}^m L_b{}^n - d\lambda^{mn}(x) \tag{15}$$

In terms of their components, Eq.(15) can be reexpressed as

$$e'_\mu{}^m = \frac{\partial x^\nu}{\partial x'^\mu} L_n{}^m e_\nu{}^n; \omega'_{D\mu} = \frac{\partial x^\nu}{\partial x'^\mu}\omega_{D\nu}; \omega'^m_{k\mu} = \frac{\partial x^\nu}{\partial x'^\mu} L_n{}^m \omega_k^n \tag{16}$$

along with the inhomogeneous transformations of the spin connection $\omega'^{mn}_{M\mu} = \frac{\partial x^\nu}{\partial x'^\mu}(\omega^{ab}_{M\nu} L_a{}^m L_b{}^n - \partial_\nu\lambda^{mn}(x))$, where L_n^m is the spacetime dependent matrix related to the local Lorentz transformation h, which is parameterized as $h = e^{i\lambda^{\mu\nu}(x)M_{\mu\nu}}$, and $\lambda^{\mu\nu}(x)$ is a function of spacetime determined by the relation of Eq.(6).

3. Effective Actions

The transformation properties of these Cartan one-forms are secured by the transformations (11), and this, in conjunction with the requirement that the phenomenological Lagrangian be invariant under these transformations, describes the underlying theory whose symmetry structure becomes a dynamical type. In fact, the effective actions, which are locally conformal invariant, can be constructed by using the building blocks of the one-forms Eq.(16) as well as their covariant derivatives. Therefore, according to Eq.(16) the rank two metric tensor $g_{\mu\nu} = \eta_{mn} e_\mu^{\ m} e_\nu^{\ n}$ transforms as

$$g'_{\mu\nu} = \eta_{mn} e'^{\ m}_{\mu} e'^{\ n}_{\nu} = \eta_{mn} \frac{\partial x^\gamma}{\partial x'^\mu} L^{\ m}_{a} e_\gamma^{\ a} \frac{\partial x^{\nu'}}{\partial x'^\nu} L^{\ n}_{n'} e_{\nu'}^{\ n'} = \frac{\partial x^\gamma}{\partial x'^\mu} \frac{\partial x^{\nu'}}{\partial x'^\nu} \eta_{an'} e_\gamma^{\ a} e_{\nu'}^{\ n'} = \frac{\partial x^\gamma}{\partial x'^\mu} \frac{\partial x^{\nu'}}{\partial x'^\nu} g_{\gamma\nu'}, \quad (17)$$

and the vierbein has an explicit form $e_\mu^{\ m} = e^{-\phi}(E_\mu^{\ m} + \delta_\mu^{\ m})$.

The metric tensor $g_{\mu\nu}$ and its inverse $g^{\mu\nu}$ are used to raise and lower indices on world tensors. The covariant derivative of the local Lorentz tensor $T^a_{\ b}$ is defined as

$$\nabla_\nu T^a_{\ b} = \partial_\nu T^a_{\ b} - \omega^c_{M\nu b} T^a_{\ c} + \omega^a_{M\nu c} T^c_{\ b} \quad (18)$$

in which the spin connection $\omega^{ab}_{M\nu}$ is given by Eq.(14), whereas the covariant derivative of mixed tensor $T_\mu^{\ m}$ takes the form

$$\nabla_\nu T_\mu^{\ m} = \partial_\nu T_\mu^{\ m} - \Gamma^\lambda_{\nu\mu} T_\lambda^{\ m} + \omega^m_{M\nu n} T_\mu^{\ n} \quad (19)$$

The local Lorentz tensor T^{mn} is transcribed into a world tensor $T^{m\nu}$ by means of the inverse of vierbein in the usual way, i.e. $T^{m\nu} = e^{-1\nu}_{\ \ n} T^{mn}$, and vice versa. And the affine connection $\Gamma^\lambda_{\nu\mu}$ is given by

$$\Gamma^\rho_{\mu\nu} = e^{-1\rho}_{\ \ a} \partial_\mu e_\nu^{\ a} + e^{-1\rho}_{\ \ a} e_\nu^{\ b} \omega^{ac}_{M\mu} \eta_{cb} \quad (20)$$

which can be derived from the requirement that the covariant derivative of the vierbein vanishes

$$\nabla_\nu e_\mu^{\ m} = \partial_\nu e_\mu^{\ m} - \Gamma^\lambda_{\nu\mu} e_\lambda^{\ m} + \omega^m_{M\nu n} e_\mu^{\ n} = 0 \quad (21)$$

Reinterpreting Eq.(21) in terms of the metric tensor $g^{\mu\nu}$, one finds $\nabla_\rho g^{\mu\nu} = 0 = \partial_\rho g^{\mu\nu} + \Gamma^\mu_{\sigma\rho} g^{\sigma\nu} + \Gamma^\nu_{\sigma\rho}$. This, in conjunction with the torsion free condition

$$\Gamma^\rho_{\mu\nu} - \Gamma^\rho_{\nu\mu} = e^{-1\rho}_{\ \ a}(\partial_\mu e_\nu^{\ a} + e_\nu^{\ b} \omega^{ac}_{M\mu} \eta_{cb} - \partial_\nu e_\mu^{\ a} + e^{-1\rho}_{\ \ a} e_\mu^{\ b} \omega^{ac}_{M\nu} \eta_{cb}) = 0 \quad (22)$$

leads to the solution for the affine connection $\Gamma^\rho_{\mu\nu} = \frac{1}{2} g^{\rho\alpha}(\partial_\mu g_{\nu\alpha} + \partial_\nu g_{\mu\alpha} - \partial_\alpha g_{\mu\nu})$. It follows that the rank four Riemann curvature tensor can be written as

$$R^\rho_{\ \sigma\mu\nu} = \partial_\mu \Gamma^\rho_{\nu\sigma} - \partial_\nu \Gamma^\rho_{\mu\sigma} + \Gamma^\alpha_{\nu\sigma} \Gamma^\rho_{\mu\alpha} - \Gamma^\beta_{\mu\sigma} \Gamma^\rho_{\nu\alpha} \quad (23)$$

By means of contracting the Riemann curvature tensor, the associated Ricci tensor is defined as $R_{\mu\nu} = R^\alpha_{\ \nu\mu\alpha}$, whereas the curvature scalar $R = g^{\mu\nu} R_{\mu\nu}$ is defined as the trace of Ricci tensor. By using the affine connection $\Gamma^\lambda_{\nu\mu}$, the covariant derivative of the one-forms $\omega_{D\mu}$ is written as

$$\nabla_\nu \omega_{D\mu} = \partial_\nu \omega_{D\mu} - \Gamma^\alpha_{\nu\mu} \omega_{D\alpha} \quad (24)$$

Considering Eqs.(16,17), the quadratic term for one-forms $\omega_{D\mu}$ is locally conformal invariant

$$\frac{1}{2} g_{\mu\nu} \omega^\mu_D \omega^\nu_D M^2 = \frac{1}{2} \omega_{D\mu} \omega^\mu_D M^2 \quad (25)$$

this will lead to the mass term of the vector field after fixing the gauge of the conformal transformations. Besides, the kinetic terms of $\omega_{D\mu}$ containing no higher than second derivatives have the form

$$-\frac{1}{2}\nabla_\nu\omega_{D\mu}\nabla_\alpha\omega_{D\beta}(g^{\nu\alpha}g^{\mu\beta}-g^{\nu\beta}g^{\mu\alpha}) \tag{26}$$

as well as

$$\nabla_\nu\omega_{D\mu}\nabla^\nu\omega^\mu_D \tag{27}$$

Explicitly, both of them are locally conformal transformation invariant.

Since the vierbein transforms as $e'^{\ m}_{\mu} = \frac{\partial x^\nu}{\partial x'^\mu}L_n^{\ m}e_\nu^{\ n}$, we have

$$d^4x'\det e' = d^4x\det|\frac{\partial x'^\nu}{\partial x^\mu}|\det|\frac{\partial x^\mu}{\partial x'^\nu}|\det e\det L_n^{\ m} = d^4x\det e\det L_n^{\ m} = d^4x\det e \tag{28}$$

This, along with Eqs.(24-27), leads to the general locally conformal invariant effective action, which takes the form

$$I=\int d^4x\det e\Big(-\frac{1}{2}\nabla_\nu\omega_{D\mu}\nabla_\alpha\omega_{D\beta}(g^{\nu\alpha}g^{\mu\beta}-g^{\nu\beta}g^{\mu\alpha})+\varepsilon\nabla_\nu\omega_{D\mu}\nabla^\nu\omega^\mu_D$$
$$+\frac{1}{2}\omega_{D\mu}\omega^\mu_D M^2+\frac{1}{16\pi G}R+\xi\omega_{D\mu}\omega^\mu_D R+\eta\omega_{D\mu}\omega_{D\nu}R^{\mu\nu}\Big) \tag{29}$$

The action (29) is locally conformal transformation invariant, whereas upon making the gauge choice, Eq.(29) can be further simplified and it becomes feasible to identity the remaining of $e_\mu^{\ m}$ as the physical vierbein field, and the remaining of $\omega^{ab}_{M\nu}$ as the physical spin connection. It can be seen that the local conformal gauge invariance of the action prevents the Goldstone field from making an explicit appearance, and thereafter the action will become locally invariant under the linearly realized stability subgroup H. We will work on the unitary gauge. Consider the transformations of the Nambu-Goldstone fields given by Eq.(6). By using Eqs.(1,8), it becomes $e^{ia^\mu(x)P_\mu}e^{ic(x)D}e^{ib^\mu(x)K_\mu}e^{i\theta^{\mu\nu}(x)M_{\mu\nu}}e^{ix^\mu P_\mu}e^{i\phi(x)D}e^{i\phi^\mu(x)K_\mu} = e^{ix'^\mu P_\mu}e^{i\phi'(x')D}e^{i\phi'^\mu(x')K_\mu}e^{i\lambda'^{\mu\nu}(x')M_{\mu\nu}}$. Applying the variation of the full group elements g

$$\delta g = i(a^\mu(x)p_\mu + c(x)D + b^\mu(x)K_\mu + \theta^{\mu\nu}(x)M_{\mu\nu}) \tag{30}$$

to it gives us

$$(1+\delta g)\Omega = (\Omega+\delta\Omega)(1+\delta h) \tag{31}$$

It further leads to $\Omega^{-1}\delta g\Omega - \Omega^{-1}\delta\Omega = \delta h$. Then one can find

$$\Omega^{-1}i(a^\mu(x)p_\mu + c(x)D + b^\mu(x)K_\mu + \theta^{\mu\nu}(x)M_{\mu\nu})\Omega - \Omega^{-1}\delta\Omega = i\lambda^{\mu\nu}(x)M_{\mu\nu} \tag{32}$$

The infinitesimal general coordinate transformation as well as these of the Nambu-Goldstone fields can be derived up to the first order by comparing the coefficients of the generators from each side of Eq.(32). As shown below, the general coordinates and the Nambu-Goldstone fields transform as

$$\delta x^\mu = a^\mu(x) + x^\mu c(x) + 2x\cdot b(x)x^\mu - x^2b^\mu(x) + 2\theta^{\nu\mu}(x)x_\nu$$
$$\delta\phi = c(x) + 2x^\mu b_\mu(x) + ...; \delta\phi^\mu = b^\mu(x)e^\phi - 2\phi^\mu x_\mu b^\mu(x) + 2\phi\cdot b(x)x^\mu + ... \tag{33}$$

Here, $\delta\phi$ etc are the total variation of the fields, i.e. $\delta\phi = \phi'(x') - \phi(x)$, and the intrinsic variation of the these fields is defined as $\delta_{in}\phi = \phi'(x') - \phi(x') = \delta\phi - (\phi(x') - \phi(x)) = \delta\phi - \delta x^\mu\partial_\mu(\phi(x') - \phi(x))$. The broken generators D and K^μ give rise to an inhomogeneous and nonlinear transformation law for the Goldstone field ϕ and ϕ^μ, and these inhomogeneous terms become functions of spacetime and signal the spontaneous breaking of the conformal symmetry. A further choice of the transformations of the fields ϕ and ϕ^μ corresponds to a gauge choice that leaves the Lorentz group as a residual gauge freedom. Choosing the unitary gauge [11,12,13], i.e. $\phi \to \phi' = 0$,$\phi^\mu \to \phi'^\mu = 0$, one then has

$$\Omega \to \Omega' = e^{ix'^\mu P_\mu} \tag{34}$$

Moreover, Eq.(14) becomes

$$\begin{aligned} \omega'^m &= dx'^\mu ({E'}_\mu^{\ m} + {\delta'}_\mu^{\ m}) = dx'^\mu {e'}_\mu^{\ m}; {\omega'}_D = dx'^\mu {A'}_\mu = dx'^\mu {\omega'}_{D\mu} \\ \omega'^m_k &= dx'^\mu {B'}_\mu^{\ m} = dx'^\mu \omega'^m_{k\mu}; {\omega'}^{mn}_M = dx'^\mu {m'}_\mu^{\ mn} = dx'^\mu {\omega'}^{mn}_{M\mu} \end{aligned} \tag{35}$$

Since the NG fields have been transformed away by fixing the gauge, the theory is no longer conformal but rather local Lorentz gauge invariant. For the sake of simplicity, we drop the prime in what follows of this paper. Therefore, the effective action of Eq.(29) is then converted to a vector-tensor theory

$$\begin{aligned} I = &\int d^4x \det e\big(-\frac{1}{2}\nabla_\nu A_\mu \nabla_\alpha A_\beta (g^{\nu\alpha} g^{\mu\beta} - g^{\nu\beta} g^{\mu\alpha}) + \varepsilon \nabla_\nu A_\mu \nabla^\nu A^\mu + \\ &\frac{1}{2} A_\mu A^\mu M^2 + \frac{1}{16\pi G} R + \xi A_\mu A^\mu R + \eta A_\mu A_\nu R^{\mu\nu}\big) \\ = &\int d^4x \det e\big(-\frac{1}{4} F_{\mu\nu} F_{\rho\sigma} g^{\mu\rho} g^{\nu\sigma} + \varepsilon \nabla_\nu A_\mu \nabla^\nu A^\mu + \frac{1}{2} A_\mu A^\mu M^2 + \\ &\frac{1}{16\pi G} R + \xi A_\mu A^\mu R + \eta A_\mu A_\nu R^{\mu\nu}\big) \end{aligned} \tag{36}$$

where the field strength $F_{\mu\nu} = \nabla_\mu A_\nu - \nabla_\nu A_\mu = \partial_\mu A_\nu - \partial_\nu A_\mu$, and the non-minimal coupling of the vector field to the Ricci scalar and to the Ricci tensor are descried by parameters ξ and η respectively. As a result of the Higgs Mechanism [14], the gauge field A_μ, associated with the broken generator D, absorbs the dilaton and acquires a third polarization and becomes massive. However, it is prominent to note that the vector field has an independent energy scale, which is totally different from the case of spontaneous symmetry breaking in the standard model of particle physics. There, the mass of the massive fields is determined by the symmetry breaking scale instead. In the present context, the effective action describes a scenario that the vector field does not decouple at certain energy scale, and it is included as a dynamical degree of freedom in the action accounting for the removal of the Nambu-Goldstone boson from the spectrum.

Modelled as dark energy candidate, the mass scale of the massive vector-like field, which is non-minimally coupled to the gravitational field, can be constrained from observational data of the cosmology. The general equation of motion of the vector field can be derived from Eq.(36) as

$$\nabla_\nu F^{\mu\nu} + 2\varepsilon \nabla_\nu \nabla^\nu A^\mu - A^\mu M^2 - 2\xi A^\mu R - 2\eta A^\nu R^\mu_\nu = 0 \tag{37}$$

Taking different values for the parameters would lead to different physical models [2,15]. Because the dynamics of vectors known to occur in nature are described by a Maxwell term, therefore, for the Maxwell type vector theory, the equation of motion (37) can be converted to the following form after using the relation $[\nabla_\mu, \nabla_\nu] A^\alpha = R^\alpha_{\ \beta\mu\nu} A^\beta$ and setting the parameter ε to zero

$$\nabla_\nu \nabla^\nu A^\mu - \nabla^\mu \nabla_\nu A^\nu + A^\mu M^2 + 2\xi A^\mu R + (1+2\eta) A^\nu R^\mu_\nu = 0 \tag{38}$$

Hence, the Proca type vector field interacts only with the gravitational field and has no standard matter interactions. Accordingly, its mass scale would be intertwined with large distance scale effects of the universe through the Friedmann Equation after one applies the equation of motion of this massive cosmological vector field to cosmology. Specifically, for a homogeneous and isotropic universe, one can assume the smoothly distributed vector field $A^\mu = (A(t), 0, 0, 0)$ to be a function of time with only one non-zero time component. Adopting the Robertson-Walker metric $g_{\mu\nu} = (1, -a(t), -a(t), -a(t))$ gives us $ds^2 = dt^2 - a^2(t)(dx^2 + dy^2 + dz^2)$ for the invariant interval. Therefore, the equation of motion of A^μ(Eq.(38)) leads to the evolution equation for the scale factor $a(t)$

$$(16\xi + 8\eta)\frac{\ddot{a}}{a} + 16\xi\frac{\dot{a}^2}{a^2} = \frac{4M^2}{3} \tag{39}$$

In the limit of large a, it can be shown $a = \exp[H(t - t_0)]$[2], where the Hubber parameter $H = (\frac{M^2}{3(8\xi+2\eta)})^{\frac{1}{2}}$ is related to the mass of the vector particle. Hence, due to the presence of the massive cosmological vector particle, the Friedmann equation gives us a scenario of an inflate universe. The mass of the cosmological massive vector particle which may represent the main component of the universe thus plays the role of cosmological constant.

To summarize, in this paper we demonstrated the existence of a massive vector field, which is the result of the spontaneous broken local conformal symmetry. We illustrate the mechanism of spontaneous symmetry breaking through nonlinear realization along with its correlation with the dark energy. As a result, this paper shows a possible origin of dark energy candidate as a Proca type vector field. As shown above, this vector field has no standard model interactions; it has an independent energy sale, permeating all over the space. Its independent energy scale can be determined by applying to specific cosmological models along with the cosmological observations. In addition, consider an alternative local Lorentz invariant quartic term $\frac{1}{2}\lambda(A_\mu A^\mu)^2$ being added to the effective action (36), where λ is the coupling constant. If the quadratic term has a wrong sign squared mass $-\lambda A_\mu A^\mu M^2$ then the Lorentz symmetry would be spontaneously broken down [16]. Explicitly, the potential $V = \frac{1}{2}\lambda(A_\mu A^\mu)^2 - \lambda A_\mu A^\mu M^2$ is found to have a minimum when the vector field satisfies the condition $A_\mu A^\mu = M^2$. Therefore, the non-zero vacuum expectation value of the vector field A^μ signals the spontaneous breaking of the Lorentz symmetry, and this Lorentz violating vector would play an important role in cosmological models. The impact of its phenomenological consequences on physics should be broad and need further exploration.

The research work was supported in part by a grant of Purdue Research Foundation. The author thanks T.K.Kuo and C.Xiong for support and comments.

References

[1] See for example, P.J.E. Peebles and Bharat Ratra, Rev.Mod.Phys.75:559-606, 2003;Sean M. Carroll, astro-ph/0310342; astro-ph/0107571

[2] C.G. Bohmer and T. Harko, Eur.Phys.J.C50:423-429, 2007

[3] C.Armendariz-Picon, J.Cosmology and Astroparticle Phys, 0407:007, 2004; V.Kiselev, Class.Quant.Grav.21:3323, 2004 ; T. Koivisto and D. Mota, arXiv: 0805.4229

[4] Hao Wei and Rong-Gen Cai J.Cosmology and Astroparticle Phys, 0709:015,2007

[5] M.Gell-Mann and M.Levy, Nuovo Cim.16:705, 1960; S.Weinberg, Phys.Rev. 166,1568, 1968; D.V. Volkov and V.P. Akulov, JETP Lett. 16, 438(1972); Phys. Lett. B46, 109(1973); J. Hughes and J. Polchinski, Nucl. Phys. B278, 147 (1986); P. West,JHEP 0002, 24(2000); T.E. Clark, M. Nitta and T. ter Veldhuis, Phys.Rev. D67: 085026, 2003

[6] S. Bellucci, A. Galajinsky, E. Ivanov, S. Krivonos, Phys.Lett.B555,99,2003; S. Bellucci, A. Galajinsky, E. Ivanov, S. Krivonos, Phys.Rev.D65,104023,2002; S. Bellucci, E. Ivanov, S. Krivonos, Nucl.Phys.B672,123,2003; Phys.Lett.B558,182,2003

[7] S. Coleman, J.Wess, and B. Zumino, Phys. Rev. 177, 2239 (1969); C. Callan, S.Coleman, J.Wess, and B. Zumino, Phys. Rev. 177, 2247 (1969)

[8] A. Salam and J.A. Strathdee, Phys.Rev.184:1760, 1969

[9] S. Bellucci, E. Ivanov and S. Krivonos, Phys.Rev.D66:086001, 2002, Erratum-ibid.D67:049901,2003; E. A. Ivanov, Theor.Math.Phys.139:513-528,2004

[10] G. Grignani and G. Nardelli, Phys.Rev.D45:2719, 1992

[11] K.S. Stelle and Peter C. West, Phys.Rev.D21:1466, 1980

[12] T.E. Clark, S.T. Love, Muneto Nitta and T. ter Veldhuis, Phys.Rev.D72:085014,2005

[13] P. Salgado, F. Izaurieta and Eduardo Rodriguez, Phys.Lett.B574:283, 2003
[14] M. Porrati, JHEP 0204:058, 2002
[15] for example, see J.B.Jimenez and A.L.Maroto, arXiv:0801.1486
[16] V.A.Kostelecky and Stuart Samuel, Phys.Rev.Lett.63:224, 1989; Phys.Rev.D40:1886, 1989

Journal of Physics: Conference Series **462** (2013) 012032 doi:10.1088/1742-6596/462/1/012032

Noninertial symmetry group with invariant Minkowski line element consistent with Heisenberg quantum commutation relations

Stephen G. Low
www.Stephen-Low.net
E-mail: Stephen.Low@alumni.utexas.net

Abstract.
The maximal symmetry of a quantum system with Heisenberg commutation relations is given by the projective representations of the automorphism group of the Weyl-Heisenberg algebra. The automorphism group is the central extension of the inhomogeneous symplectic group with a conformal scaling that acts on extended phase space. We determine the subgroup that also leaves invariant a degenerate Minkowski orthogonal line element. This defines noninertial relativistic symmetry transformations that have the expected classical limit as $c \to \infty$.

1. Introduction

The inhomogeneous Lorentz group defines the relation between inertial states. Clocks locally at rest relative to an inertial state are related to the clocks of other inertial states through the Minkowski proper time line element. Quantum states are rays in a Hilbert space and therefore inertial quantum states are related through the projective representation of the inhomogeneous Lorentz group. Projective representations are equivalence classes of unitary representations of the central extension of the group. The central extension of the inhomogeneous Lorentz group is its cover, the Poincaré group, as this group does not admit an algebraic extension. [1],[2]

The equivalence principle of general relativity enables the noninertial frames of a particle accelerating under gravity to be understood as locally inertial states on a curved manifold. Particles under gravity follow geodesics that are locally inertial trajectories and neighboring locally inertial frames are related by the connection. The clock locally at inertial rest is related to the local clocks of other neighboring locally inertial states in the gravitating system through the Riemannian proper time line element.

Neither general relativity nor special relativity addresses the issue of noninertial states that are not due to gravity, but rather one of the other forces. A special case is a region in which gravity is negligible and the underlying manifold may be considered to be flat. Consider for example an electron in a region that gravity is negligible that encounters an electromagnetic field and therefore has a noninertial trajectory. How are the clocks of such noninertial states related?

We hypothesize that the noninertial relativistic symmetry group relating these states is the most general group consistent with the requirements that

1) the Heisenberg uncertainty principle holds in the noninertial as well as inertial states

2) the proper time given by the Minkowski line element that is invariant for noninertial as well as inertial states

Consider a quantum system in which the position, momentum, energy and time degrees of freedom are represented by the Hermitian representation of the algebra of the Weyl-Heisenberg group $\mathcal{H}(n+1)$ where the number of spacial dimensions is $n=3$. The requirement that the algebra transforms into itself under the action of the relativistic symmetry group means that the symmetry group is a subgroup of the automorphism group of the Weyl-Heisenberg algebra. [3],[4],[5] This automorphism group is

$$\mathcal{A}ut_{\mathcal{H}} \simeq \mathbb{Z}_2 \otimes_s \mathcal{D} \otimes_s \overline{\mathcal{S}p}(2n+2) \otimes_s \mathcal{H}(n+1). \tag{1}$$

$\mathbb{Z}_2$ is the 2 element discrete group, $\mathcal{D}$ is the abelian group isomorphic to the reals under multiplication, $\mathcal{S}p(2n+2)$ is the symplectic group with the over-bar denoting the universal cover and $\mathcal{H}(n+1)$ is the Weyl-Heisenberg group. At this point there is no concept of a relativistic symmetry.

The Minkowski line element is $d\tau^2 = dt^2 - \frac{1}{c^2}dq^2$. This is an invariant for states that are inertially related and the second assertion is that this continues to be true for general noninertial states.

We will show that the homogeneous relativistic symmetry group that is a subgroup of the automorphism group of the Weyl-Heisenberg group that leaves the Minkowski line element invariant is

$$\mathcal{O}a(1,n) \simeq \mathcal{O}(1,n) \otimes_s \mathcal{A}(m), \tag{2}$$

where $m=(n+1)(n+2)/2$ and $\mathcal{A}(m)$ is the abelian group isomorphic to $\mathbb{R}^m$ under addition. The additional generators of the abelian group behave as a power-force stress tensor that is the proper time derivative of the energy-momentum stress tensor. We show that this leads to expected relativistic results in transforming to noninertial states.

This relativistic theory must lead to expected classical results in the limit $c \to \infty$ where the Minkowski line element reduces to the invariant Newtonian time line element dt^2. We have previously studied the most general group that leaves invariant the Newtonian time line element dt^2 that is a subgroup of the automorphisms of the Weyl-Heisenberg group. This results in a group that leads directly to Hamilton's equations and, with the additional requirement of orthonormal position frames, describes the Hamilton symmetry group for noninertial transformations in a classical context. [4]

2. Consistency between a relativistic symmetry group and quantum mechanics

States in quantum mechanics are represented by rays Ψ in a Hilbert space $\mathbf{H}$ that are the equivalence class of states $|\psi\rangle$ in the Hilbert space that are related by a phase

$$\Psi \simeq \left\{ e^{i\omega} \left|\psi\right\rangle | \omega \in \mathbb{R} \right\}, \qquad |\psi\rangle \in H. \tag{3}$$

A relativistic symmetry group $g \in \mathcal{G}$ acts on the states through a projective representation π, $\tilde{\Psi} = \pi(g)\Psi$, with the property that it also has a phase,

$$\pi(\tilde{g}\cdot g) = e^{i\omega(\tilde{g},g)}\pi(\tilde{g})\pi(g), \omega(\tilde{g},g) \in \mathbb{R}. \tag{4}$$

Projective representations are equivalence classes of the unitary representations ϱ of the central extension $\check{\mathcal{G}}$ (denoted by the inverted hat) of the group $\mathcal{G}$ that act on the states, [6],[7]

$$|\tilde{\psi}\rangle = \varrho(g)\left|\psi\right\rangle, \quad g \in \check{\mathcal{G}}, \qquad |\psi\rangle \in \mathbf{H}^{\varrho}. \tag{5}$$

The Hilbert space is determined by the unitary representation ϱ and so we label it as $\mathbf{H}^{\varrho}$. Observables corresponding to a the relativistic symmetry group $\mathcal{G}$ are represented by the Hermitian representations ϱ' of elements Z of the algebra of the centrally extended group$\check{\mathcal{G}}$, $\hat{Z} = \varrho'(Z)$. The action of the projective representation of the group element $g \in \check{\mathcal{G}}$ on these observables is

$$\hat{\tilde{Z}}\,|\tilde{\psi}\rangle = \varrho(g)\hat{Z}\,|\psi\rangle = \varrho(g)\hat{Z}\varrho(g)^{-1}\varrho(g)\,|\psi\rangle = \varrho(g)\hat{Z}\varrho(g)^{-1}\,|\tilde{\psi}\rangle \tag{6}$$

and so $\hat{\tilde{Z}} = \varrho'(\tilde{Z}) = \varrho'(gZg^{-1})$. Therefore, if the representation ϱ is faithful, we have $\tilde{Z} = gZg^{-1}$.

Position, momentum, energy and time observables are the Hermitian representation of the algebra of the Weyl-Heisenberg group $\mathcal{H}(n+1)$ with a general element given by $Z \;= z^{\alpha}Z_{\alpha}$, $\alpha = 1, ..2n{+}2$ where $\{z^{\alpha}\} \in \mathbb{P} \simeq \mathbb{R}^{2n+2}$ and Z_{α} are a dimensionless basis for the Weyl-Heisenberg algebra that satisfy the commutation relations

$$[Z_{\alpha}, Z_{\beta}] = \zeta_{\alpha,\beta} I, \tag{7}$$

where $\zeta_{\alpha,\beta}$ are the components of a symplectic metric. The Hermitian representation of the algebra satisfies $[\hat{Z}_{\alpha}, \hat{Z}_{\beta}] = i\zeta_{\alpha,\beta}\hat{I}$, where $\hat{Z}_{\alpha} = \varrho'(Z_{\alpha})$ and $\hat{I} = \varrho'(I)$ is the unit operator on the Hilbert space. The position-time and momentum-energy Hermitian operators are $\{\hat{Z}_{\alpha}\} = \{\hat{X}_a, \hat{P}_a\}$ with $a = 0, ..n$. Note that there is no relativistic symmetry yet and hence no concept of a mass shell.

The basic physical assumption is that the Heisenberg commutation relations are satisfied by any state related by a relativistic symmetry group $\mathcal{G}$. That is, position, momentum, energy and time observables satisfying the Heisenberg quantum commutation relations will also satisfy the Heisenberg quantum commutation relations for any states related by the projective representations of the symmetry group (6). This implies that if $\{X_a, P_a, I\}$ are a basis of the Weyl-Heisenberg algebra, then $\{\tilde{X}_a, \tilde{P}_a, \tilde{I}\}$ are also a basis of the Weyl-Heisenberg algebra where

$$\tilde{X}_a = gX_ag^{-1}, \tilde{P}_a = gP_ag^{-1}, \tilde{I} = gIg^{-1} = I. \tag{8}$$

and $g \in \check{\mathcal{G}}$ and ϱ is a faithful representation. The maximal group for which this property is true is the automorphism group of the Weyl-Heisenberg group. This results in basic consistency condition that the central extension $\check{\mathcal{G}}$ of the relativistic symmetry group $\mathcal{G}$ must be a subgroup of the automorphism group of the Weyl-Heisenberg group and algebra, $\check{\mathcal{G}} \subseteq \mathcal{A}ut_{\mathcal{H}(n+1)}$.

The automorphism group of the Weyl-Heisenberg group is [3],[4],[5]

$$\mathcal{A}ut_{\mathcal{H}(n+1)} = \overline{\mathcal{DS}p}(2n+2) \otimes_s \mathcal{H}(n+1). \tag{9}$$

The Heisenberg group itself are the inner automorphisms. The outer automorphisms are the cover of the homogeneous group

$$\mathcal{DS}p(2n+2) \simeq \mathbb{Z}_2 \otimes \mathcal{D} \otimes \mathcal{S}p(2n+2). \tag{10}$$

The matrix realization of this group and the group properties are given in [4],[5]. The inhomogeneous group includes translations on extended phase space $\mathbb{P}$,

$$\begin{aligned} \mathcal{IDS}p(2n+2) &= \mathcal{DS}p(2n+2) \otimes_s \mathcal{A}(2n+2) \\ &= \mathbb{Z}_2 \otimes_s \mathcal{D} \otimes_s \mathcal{S}p(2n+2) \otimes_s \mathcal{A}(2n+2) \end{aligned}. \tag{11}$$

The automorphism group is the central extension of this inhomogeneous group, $\mathcal{A}ut_{\mathcal{H}(n+1)} \simeq \mathcal{I}\check{\mathcal{D}}\mathcal{S}p(2n+2)$.

Projective representations of the classical inhomogeneous group are equivalence classes of the unitary representations of the automorphism group that is its central extension. The unitary representation determines the Hilbert space of quantum states. These representations are the largest symmetry of a quantum system that preserves the Weyl-Heisenberg commutation relations.

Journal of Physics: Conference Series **462** (2013) 012032 doi:10.1088/1742-6596/462/1/012032

3. Homogeneous relativistic symmetry group

We determine in this section the homogeneous relativistic symmetry group for noninertial frames that satisfies two conditions

1) It leaves invariant the Minkowski proper time line element.
2) It is a subgroup of the automorphism group of the Weyl-Heisenberg group.

The group $\mathcal{O}a(1,n)$ is dependent on the scale c. We show a homomorphism parameterized by c satisfies the conditions to define a Inönü-Wigner contraction. This contraction results in the Hamilton group that we have previously shown is the relativistic symmetry group for noninertial states in the nonrelativistic ($c\to\infty$) context.

3.1. The group $\mathcal{O}a(1,n)$ and its algebra

The postulates of special relativity requires the invariance of the Minkowski proper time line element

$$d\tau^2 = \eta_{a,b}dx^a dx^b \tag{12}$$

with $a,b.. = 0,..n$ and η is the diagonal matrix $\eta = [\eta_{a,b}] = \mathrm{diag}\{-1,1,...1\}$ with units where $c=1$.

Consider the $2n+2$ dimensional time, position, energy, momentum extended phase space $\mathbb{P}\simeq\mathbb{R}^{2n+2}$ with coordinates $\{z^\alpha\} = \{x^a, p^a\}$ where $\alpha,\beta = 1,...2n+2$, $a,b = 0,1..n$. The Minkowski metric is a degenerate line element on the cotangent space $T_z^*\mathbb{P}$

$$d\tau^2 = \tilde{\eta}_{\alpha,\beta}dz^\alpha dz^\beta \tag{13}$$

where $\tilde{\eta}_{\alpha,\beta}$ are the components of the $(2n+2)\times(2n+2)$ dimensional matrix $\tilde{\eta}$

$$\tilde{\eta} = [\tilde{\eta}_{\alpha,\beta}] = \begin{pmatrix} [\eta_{a,b}] & 0 \\ 0 & 0 \end{pmatrix}.$$

The group $\mathcal{GL}(2n+2,\mathbb{R})$ of nonsingular $(2n+2)\times(2n+2)$ matrices acts naturally on the cotangent space $T_z^*\mathbb{P}$ with basis $\{dz^\alpha|_z\}$. Elements Γ of the subgroup $\mathcal{OGL}a(1,n)\subset\mathcal{GL}(2n+2,\mathbb{R})$ that is defined as the subgroup that leaves invariant the degenerate line element (13) satisfies $\Gamma^{\mathrm{t}}\tilde{\eta}\Gamma = \tilde{\eta}$. Γ may be written in terms of $(n+1)\times(n+1)$ submatrices as $\Gamma = \begin{pmatrix} \Lambda & B \\ \Xi & A \end{pmatrix}$ and therefore

$$\begin{pmatrix} \eta & 0 \\ 0 & 0 \end{pmatrix} = \begin{pmatrix} \Lambda^{\mathrm{t}} & \Xi^{\mathrm{t}} \\ B^{\mathrm{t}} & A^{\mathrm{t}} \end{pmatrix}\begin{pmatrix} \eta & 0 \\ 0 & 0 \end{pmatrix}\begin{pmatrix} \Lambda & B \\ \Xi & A \end{pmatrix} = \begin{pmatrix} \Lambda^{\mathrm{t}}\eta\Lambda & \Lambda^{\mathrm{t}}\eta B \\ B^{\mathrm{t}}\eta\Lambda & B^{\mathrm{t}}\eta B \end{pmatrix}. \tag{14}$$

It follows immediately that $B=0$ and $\Lambda\in\mathcal{O}(1,n)$. As $\det\Gamma = \det\Lambda\det A$, $\det A\neq 0$ and $A\in\mathcal{GL}(n+1,\mathbb{R})$. Therefore,

$$\mathcal{OGL}a(1,n)\simeq(\mathcal{O}(1,n)\otimes\mathcal{GL}(n+1,\mathbb{R}))\otimes_s\mathcal{A}((n+1)^2) \tag{15}$$

that has elements Γ and Γ^{-1} of the form

$$\Gamma = \begin{pmatrix} \Lambda & 0 \\ \Xi & A \end{pmatrix},\quad \Gamma^{-1} = \begin{pmatrix} \Lambda^{-1} & 0 \\ -A^{-1}\Xi\Lambda^{-1} & A^{-1} \end{pmatrix},\quad \begin{matrix} \Lambda\in\mathcal{O}(1,n) \\ A\in\mathcal{GL}(n+1,\mathbb{R}) \end{matrix} \quad \Xi\in\mathcal{A}((n+1)^2) \tag{16}$$

It can be verified that this has the semidirect product structure as claimed following exactly the same steps as given explicitly below for the $\mathcal{O}a(1,n)$ group. The homogeneous relativistic symmetry group, that is called $\mathcal{O}a(1,n)$, is the homogenous subgroup of the automorphism group that leaves invariant the degenerate line element,

$$\mathcal{O}a(1,n) = \mathcal{DS}p(2n+2)\cap\mathcal{OGL}a(1,n). \tag{17}$$

The elements of the $\mathcal{D}\mathcal{S}p(2n+2)$ group are of the form $\Delta\Sigma$ where $\Delta \in \mathbb{Z}_2 \otimes \mathcal{D}$ and $\Sigma \in \mathcal{S}p(2n+2)$. The symplectic matrices satisfy the condition $\Sigma^{\rm t}\zeta\Sigma = \zeta$ and so $\Sigma^{-1} = -\zeta\Sigma^{\rm t}\zeta$ with $\zeta = \begin{pmatrix} 0 & \eta \\ -\eta & 0 \end{pmatrix}$. This may also be written in terms of $(n+1)\times(n+1)$ submatrices $\Sigma_{\mu,\nu}$ $\mu,\nu = 1,2$ with the matrix and inverse having the form

$$\Sigma = \begin{pmatrix} \Sigma_{1,1} & \Sigma_{1,2} \\ \Sigma_{2,1} & \Sigma_{2,2} \end{pmatrix}, \quad \Sigma^{-1} = \begin{pmatrix} \eta\Sigma^{\rm t}_{2,2}\eta & -\eta\Sigma^{\rm t}_{1,2}\eta \\ -\eta\Sigma^{\rm t}_{2,1}\eta & \eta\Sigma^{\rm t}_{1,1}\eta \end{pmatrix}. \tag{18}$$

If Γ in (16) is a subgroup of the outer automorphism group, we have $\Sigma = \Delta^{-1}\Gamma$ or expanding,

$$\begin{pmatrix} \Sigma_{1,1} & \Sigma_{1,2} \\ \Sigma_{2,1} & \Sigma_{2,2} \end{pmatrix} = \Delta^{-1}\begin{pmatrix} \Lambda & 0 \\ \Xi & A \end{pmatrix}. \tag{19}$$

$\Gamma^{-1} = \Delta^{-1}\Sigma^{-1}$ that we can compute from (18) and equate to the inverse calculated in (16)

$$\Gamma^{-1} = \begin{pmatrix} \Lambda^{-1} & 0 \\ -\Lambda^{-1}\Xi A^{-1} & A^{-1} \end{pmatrix} = \Delta^{-2}\begin{pmatrix} \eta A^{\rm t}\eta & 0 \\ -\eta\Xi^{\rm t}\eta & \eta\Lambda^{\rm t}\eta \end{pmatrix} \tag{20}$$

from which it follows that $\Lambda^{-1} = \Delta^{-2}\eta A^{\rm t}\eta$ and $A^{-1} = \Delta^{-2}\eta\Lambda^{\rm t}\eta$. This has a solution if and only if $\Delta = \pm 1_n \in \mathbb{Z}_2 \subset \mathcal{D}$ and $A^{\rm t} = \eta\Lambda^{-1}\eta$. 1_n is the $n\times n$ unit matrix. Noting that $\Lambda^{-1} = \eta\Lambda^{\rm t}\eta$ this gives $A^{\rm t} = \Lambda^{\rm t}$ and therefore $A = \Lambda$. Finally,

$$\Xi^{\rm t} = \eta\Lambda^{-1}\Xi\Lambda^{-1}\eta = \Lambda^{\rm t}\eta\Xi\eta\Lambda^{\rm t}. \tag{21}$$

Thus elements of $\mathcal{O}a(1,n)$ have the form

$$\Gamma(\Lambda,\Xi) = \begin{pmatrix} \Lambda & 0 \\ \Xi & \Lambda \end{pmatrix}, \quad \Lambda \in \mathcal{O}(1,n), \quad \Xi \in \mathcal{A}(m)\ . \tag{22}$$

with $m = (n+1)(n+2)/2$. The group multiplication and inverse of $\mathcal{O}a(1,n)$ are

$$\begin{aligned} \Gamma(\Lambda,\Xi) &= \Gamma(\Lambda',\Xi')\Gamma(\Lambda'',\Xi'') \qquad\qquad \Gamma(\Lambda,\Xi)^{-1} = \Gamma(\Lambda^{-1},-\Lambda^{-1}\Xi\Lambda^{-1}). \\ &= \Gamma(\Lambda'\Lambda'',\Xi'\Lambda''+\Lambda'\Xi''), \end{aligned} \tag{23}$$

The Lorentz group is the subgroup $\Gamma(\Lambda,0)$. The matrix components of the Lorentz matrices may be given as the usual expressions in regular and hyperbolic trigonometry terms of the rotation angles and boost angles.

The elements $\Gamma(1_n,\Xi)$ define an abelian normal subgroup with group multiplication, inverse and automorphisms given by

$$\Gamma(1_n,\Xi')\Gamma(1_n,\Xi'') = \Gamma(1_n,\Xi'+\Xi''), \qquad \Gamma(1_n,\Xi)^{-1} = \Gamma(1_n,-\Xi) \tag{24}$$

$$\Gamma(\Lambda',\Xi')\Gamma(1_n,\Xi)\Gamma(\Lambda',\Xi')^{-1} = \Gamma(1_n,\Lambda'\Xi\Lambda'^{-1})\ . \tag{25}$$

The intersection $\Gamma(\Lambda,0)\cap\Gamma(1_n,\Xi) = 1_n$ and $\Gamma(\Lambda,\Xi) = \Gamma(1_n,\Xi)\Gamma(\Lambda,0)$. Therefore it is the semidirect product

$$\mathcal{O}a(1,n) \simeq \mathcal{O}(1,n) \otimes_s \mathcal{A}((n+1)(n+2)/2). \tag{26}$$

It can be shown that it does not admit an algebraic central extension and therefore the central extension of this group is simply its cover.

A general element of the algebra of $\mathcal{O}a(1,n)$ is $Z = \lambda^{a,b}L_{a,b} + \xi^{a,b}M_{a,b}$. Note that as $\xi^{a,b} = \xi^{b,a}$, that $M_{a,b} = M_{b,a}$. The Lie algebra relations may be directly computed to be

$$\begin{array}{l} [L_{a,b}, L_{c,d}] = -L_{b,d}\eta_{a,c} + L_{b,c}\eta_{a,d} + L_{a,d}\eta_{b,c} - L_{a,c}\eta_{b,d}, \\ [L_{a,b}, M_{c,d}] = -M_{b,d}\eta_{a,c} - M_{b,c}\eta_{a,d} + M_{a,d}\eta_{b,c} + M_{a,c}\eta_{b,d}, \quad [M_{a,b}, M_{c,d}] = 0. \end{array} \tag{27}$$

The $M_{a,b}$ abelian generators transform as a symmetric $(0,2)$ tensor under the Lorentz generators $L_{a,b}$.

Returning to the group, the transformation equations are $d\tilde{z} = \Gamma dz$. Using the definition of Γ in (26) results in

$$d\tilde{x} = \Gamma dx, \qquad d\tilde{p} = \Gamma dp + \Xi dx, \tag{28}$$

that in component form are (with units where $c = 1$)

$$d\tilde{x}^a = \lambda^a_b dx^b, \;\; d\tilde{p}^a = \lambda^a_b dp^b + \xi^a_b dx^b. \tag{29}$$

Then, the proper time line element is invariant as required,

$$d\tau^2 \;\; = \eta_{a,b} d\tilde{x}^a d\tilde{x}^b = \eta_{a,b}\lambda^a_c dx^c \lambda^b_d dx^d = \eta_{a,b} dx^a dx^b. \tag{30}$$

The λ^a_c are the components of the Lorentz transformation that, as usual, depend on the relative rotation angle and hyperbolic boost angle. The mass μ satisfies

$$\begin{array}{rl} c^2 d\tilde{\mu}^2 & = \eta_{a,b} d\tilde{p}^a d\tilde{p}^b = \eta_{a,b}(\lambda^a_c dp^c + \xi^a_c dx^c)\left(\lambda^c_d dp^d + \xi^b_d dx^d\right) \\ & = c^2 d\mu^2 + \eta_{a,b}\xi^a_c \xi^b_d dx^c dx^d + 2\eta_{a,b}\xi^a_c \lambda^b_d dx^c dp^d. \end{array} \tag{31}$$

From basic dimensional analysis, the ξ^a_c have the dimensions of force or power (in units with $c = 1$, these are the same). It is a symmetric tensor satisfying $\xi^a_b = \eta^{a,c}\eta_{b,d}\xi^d_c$ that transforms as an (1,1) tensor under the Lorentz transformation

$$\tilde{\xi}^a_b = \lambda^a_c \lambda^d_b \xi^c_d. \tag{32}$$

These are the properties of a power-force stress tensor that is the proper time derivative of the energy-momentum stress tensor.

The rate of change of the mass squared with respect to the proper time is given by

$$\frac{d\tilde{\mu}^2}{d\tau^2} = \frac{d\mu^2}{d\tau^2} + \frac{1}{c^2}\eta_{a,b}\xi^a_c V^c(\xi^b_d V^d + 2\;\; \lambda^b_d F^d) \tag{33}$$

where $V^a = \frac{dx^a}{d\tau}$ is the *four* velocity and $F^a = \frac{dp^a}{d\tau}$ is the *four* force for the case $n = 3$.

3.2. Three notation

Further insight into the physical meaning of the group may be obtained by converting to $n+1$ notation that for $n = 3$ is the familiar *three* notation $\{x^a\} = \{t, \frac{1}{c}q^i\}$, $\{p^a\} = \{\frac{1}{c}e, p^i\}$, $i, j = 1, ..n$. The Lorentz matrix $\Lambda(\alpha, \beta)$ parameterized by rotation angles $\alpha^{i,j} = -\alpha^{j,i}$ and hyperbolic boost rotations β^i that have the usual form. As usual, we identify velocity as $v^i = c\frac{\beta^i}{|\beta|}\tanh(\beta)$ and define $\gamma(\beta) = \cosh(\beta) = \lambda^0_0$ or equivalently $\gamma(v) = (1 - (\frac{v}{c})^2)^{-\frac{1}{2}}$.

The velocity *four* vectors are given as usual by $\{V^0, V^i\} = \{\gamma, \gamma v^i\} = \gamma\{1, \frac{dx^i}{dt}\}$ where $\gamma = \frac{dt}{d\tau}$. The *four* force likewise is $\{F^0, F^i\} = \{\gamma r, \gamma f^i\}$ where $f^i = \frac{dp^i}{dt}$ and $r = \frac{de}{dt}$ and f^i has the dimensions of force and r has the dimensions of power. The power-force-stress components are

$$\Xi = \begin{pmatrix} \xi^0_0 & \xi^0_i \\ \xi^j_0 & \xi^j_i \end{pmatrix} = \gamma \begin{pmatrix} \frac{1}{c}r & -f_i \\ f^j & \frac{1}{c}m^j_i \end{pmatrix}. \tag{34}$$

Therefore, the transformation equations for the position, time, momentum, energy basis is

$$\begin{array}{ll} d\tilde{t} = \gamma dt + \frac{1}{c}\lambda_i^0 \ dq^i, & d\tilde{p}^i = \lambda_j^i dp^j + \lambda_i^0 \ dt, \\ d\tilde{q}^i = \lambda_j^i dq^j + c\lambda_0^i \ dt, & d\tilde{e} = \gamma de + c\lambda_i^0 \ dp^i - \gamma f_i dq^i + c\gamma r dt \ . \end{array} \quad (35)$$

For $n = 1$ these are simply

$$\begin{array}{ll} d\tilde{t} = \gamma(v)\left(\ dt + \frac{1}{c^2}vdq\right), & d\tilde{p} = \gamma(v)\left(dp^j + \frac{1}{c^2}vde + fdt + \frac{1}{c^2} \ m \ dq\right), \\ d\tilde{q} = \gamma(v)\left(dq + vdt\right), & d\tilde{e} = \gamma(v)\left(de + vdp - fdq + rdt\right). \end{array} \quad (36)$$

and the corresponding group parameter transformations using (23) are

$$\gamma(v)\begin{pmatrix} 1 & v \\ v & 1 \end{pmatrix} = \gamma(v'')\gamma(v')\begin{pmatrix} 1 & v' \\ v' & 1 \end{pmatrix}\begin{pmatrix} 1 & v'' \\ v'' & 1 \end{pmatrix} \quad (37)$$

$$\gamma(v)\begin{pmatrix} r & f \\ f & m \end{pmatrix} = \gamma(v'')\gamma(v')\left(\begin{pmatrix} r' & f' \\ f' & m' \end{pmatrix}\begin{pmatrix} 1 & v'' \\ v'' & 1 \end{pmatrix} + \begin{pmatrix} 1 & v' \\ v' & 1 \end{pmatrix}\begin{pmatrix} r'' & f'' \\ f'' & m'' \end{pmatrix}\right) \quad (38)$$

and so

$$\begin{array}{l} v = (v'' + v')/\left(1 + \frac{v'v''}{c^2}\right), \quad f = \left(f'' + f' + \frac{1}{c^2}\left(r'v'' - v'r''\right)\right)/\left(1 + \frac{v'v''}{c^2}\right), \\ r = \left(r'' + r' - f'v'' + v'f''\right)/\left(1 + \frac{v'v''}{c^2}\right), m = \left(m'' + m' + f'v'' - v'f''\right)/\left(1 + \frac{v'v''}{c^2}\right). \end{array} \quad (39)$$

Consider next the algebra. We have the infinitesimal parameter correspondence

$$\lambda^{0,i} = \frac{1}{c}\beta^i, \lambda^{j,i} = \alpha^{i,j}, \xi^{0,0} = \frac{1}{c}r, \xi^{j,0} = f^j, \xi^{j,i} = \frac{1}{c}m^{j,i} \quad (40)$$

where $\alpha^{i,j} = -\alpha^{i,j}$ and $m^{i,j} = m^{j,i}$ with the corresponding generators

$$L_{0,j} = cK_j, L_{i,j} = J_{i,j}, M_{0,0} = cR, M_{i,0} = N_i, M_{i,j} = cM^{\circ}_{i,j}. \quad (41)$$

A general element of the algebra is $Z = \alpha^{i,j}J_{i,j} + \beta^i K_i + f^i N_i + rR + m^{i,j}M^{\circ}_{i,j}$. The nonzero commutators of the Lie algebra (27) written in terms of these generators.

$$\begin{array}{ll} [J_{i,j}, J_{k,l}] = -J_{j,l}\delta_{i,k} + J_{j,k}\delta_{i,l} + J_{i,l}\delta_{j,k} - J_{i,k}\delta_{j,l}, & [K_i, K_k] = \frac{1}{c^2}J_{i,k}, \\ [J_{i,j}, K_k] = -K_j\delta_{i,k} + K_i\delta_{j,k}, & [K_i, N_k] = -M^{\circ}_{i,k} - R\delta_{i,k}, \\ [J_{i,j}, N_k] = -N_j\delta_{i,k} + N_i\delta_{j,k}, & [K_i, R] = -\frac{2}{c^2}N_i, \\ \left[J_{i,j}, M^{\circ}_{k,l}\right] = M^{\circ}_{j,l}\delta_{i,k} - M^{\circ}_{j,k}\delta_{i,l} + M^{\circ}_{i,l}\delta_{j,k} + M^{\circ}_{i,k}\delta_{j,l} & \\ \left[K_i, M^{\circ}_{k,l}\right] = \frac{-1}{c^2}\left(N_l\delta_{i,k} + N_k\delta_{i,l}\right), & \end{array} \quad (42)$$

3.3. Contraction in the limit $c \to \infty$

The scaling with c given in (42) satisfies the conditions for an Inönü-Wigner [8] contraction $c \to \infty$ for which the nonzero contracted commutators are

$$\begin{array}{ll} [J_{i,j}, J_{k,l}] = -J_{j,l}\delta_{i,k} + J_{j,k}\delta_{i,l} + J_{i,l}\delta_{j,k} - J_{i,k}\delta_{j,l}, & [K_i, N_k] = -M^{\circ}_{i,k} - R\delta_{i,k}, \\ [J_{i,j}, K_k] = -K_j\delta_{i,k} + K_i\delta_{j,k}, & [J_{i,j}, N_k] = -N_j\delta_{i,k} + N_i\delta_{j,k}, \\ \left[J_{i,j}, M^{\circ}_{k,l}\right] = -M^{\circ}_{j,l}\delta_{i,k} - M^{\circ}_{j,k}\delta_{i,l} + M^{\circ}_{i,l}\delta_{j,k} + M^{\circ}_{i,k}\delta_{j,l}. & \end{array} \quad (43)$$

The subgroup spanned by $\{J_{i,j}, K_i, N_i, R\}$ is the algebra of the Hamilton group $\mathcal{H}a(n)$. The full algebra of the group defined by

$$\mathcal{O}(n) \otimes_s \mathcal{A}(n(n+1)/2) \otimes_s \mathcal{H}(n) = \hat{\mathcal{H}a}(n) \otimes_s \mathcal{A}(n(n+1)/2) \quad (44)$$

where $\hat{\mathcal{H}a}(n)$ is the extended Hamilton group

$$\hat{\mathcal{H}a}(n) = \mathbb{Z}_2 \otimes_s \mathcal{H}a(n), \quad \mathcal{H}a(n) = \mathcal{SO}(n) \otimes_s \mathcal{H}(n) \tag{45}$$

where $\{J_{i,j}\}$ are the generators of $\mathcal{O}(n)$, $\{M_{i,j}\}$ are the generators of $\mathcal{A}(n(n+1)/2)$ and $\{K_i, N_i, R\}$ are the generators of the Weyl-Heisenberg group $\mathcal{H}(n)$.

The basis transformation equations (35) contract to the expected transformation equations in the limit [5]

$$\begin{array}{ll} d\tilde{t} = \ dt, & d\tilde{p}^i = \lambda(\alpha)^i_j dp^j + f^i dt, \\ d\tilde{q}^i = \lambda(\alpha)^i_j dq^j + v^i \ \ dt, & d\tilde{e} = de + v_i dp^i - f_i dq^i + r dt. \end{array} \tag{46}$$

where the $\lambda(\alpha)^i_j$ are now the components of a rotation matrix, $\mathcal{O}(n)$.

4. Summary

We started by noting that neither special relativity nor general relativity address the problem of how clocks of noninertial states due to forces other than gravity are related.

The hypothesis that the Minkowski proper time line element is invariant in these states that are not necessarily inertial and requiring that the Heisenberg commutation relations hold in all noninertial states results in the noninertial relativistic symmetry group $\mathcal{O}a(1,n)$. This group gives the expected transformations to noninertial states in terms of a power-force stress tensor that is the proper time derivative of the energy-momentum stress tensor.

The $\mathcal{O}a(1,n)$ group is also the $b \to \infty$ of the $\mathcal{U}(1,n)$ group of reciprocal relativity described in [9]. This gives an understanding of the behavior of reciprocal relativity in the small interaction limit (that is, small forces relative to b) that is analogous to the manner in which the Euclidean group that is the homogeneous group of the Galilei group gives the small velocity limit, relative to c, of the Lorentz group.

Spacetime is an invariant subspace under the actions of the $\mathcal{O}a(1,n)$ group and therefore is observer independent or *absolute*. In this limit, there is an apparent global inertial frame that all observers agree on. Forces appear to be relative to this frame rather than being strictly relative to particle states. Forces and the power-force-stress energy tensor are simply additive and unbounded. Velocities are bounded by c and strictly relative to particle states.

In the $c \to \infty$ limit yields the classical *nonrelativistic* Hamilton theory that describes particles undergoing general noninertial motion. In this case, there is an apparent global inertial rest frame that all observers agree on. Forces and velocities appear to be relative to this frame rather than being strictly relative to particle states. Forces and velocities are simply additive and unbounded.

References

[1] Wigner, E. (1939). On the unitary representations of the inhomogeneous Lorentz group. *Annals of Mathematics*, **40**, 149–204.
[2] Weinberg, S. (1995). *The Quantum Theory of Fields, Volume 1.* Cambridge: Cambridge.
[3] Folland, G. B. (1989). *Harmonic Analysis on Phase Space.* Princeton: Princeton University Press.
[4] Low, S. G. (2007). Relativity group for noninertial frames in Hamilton's mechanics. *J. Math. Phys.*, **48**, 102901.
[5] Low, S. G. (2008). Hamilton relativity group for noninertial states in quantum mechanics. *J. Phys. A: Math Theor.*, **41**, 304034.
[6] Bargmann, V. (1954). On Unitary Ray Representations of Continuous Groups. *Annal. Math.*, **59**, 1–46.
[7] Mackey, G. W. (1958). Unitary Representations of Group Extensions. I. *Acta Math.*, **99**, 265–311.
[8] Gilmore, R. (1974). *Lie Groups, Lie Algebras and Some of their Applications.* New York: John Wiley.
[9] Low, S. G. (2007). Reciprocal relativity of noninertial frames: quantum mechanics. *J. Phys A: Math. Theor.*, **40**, 3999–4016.

Journal of Physics: Conference Series **462** (2013) 012033 doi:10.1088/1742-6596/462/1/012033

Suppressing decoherence by preparing the environment

Olivier Landon-Cardinal[1,2,3] and Richard MacKenzie[2]

[1] Département IRO, Université de Montréal, C.P. 6128, Succ. Centreville, Montréal, QC H3C 3J7 Canada

[2] Département de physique, Université de Montréal, C.P. 6128, Succ. Centreville, Montréal, QC H3C 3J7 Canada

E-mail: `richard.mackenzie@umontreal.ca`

Abstract. To protect a quantum system from decoherence due to interaction with its environment, we investigate the existence of initial states of the environment allowing for decoherence-free evolution of the system. For a class of models in which a two-state system and a dynamical environment interact through a Hamiltonian restricted to be a tensor product, we prove that such states exist if *and only if* the interaction and self-evolution Hamiltonians of the environment share an eigenstate. If decoherence by state preparation is not possible, we show that initial states minimizing decoherence result from a delicate compromise between the environment and interaction dynamics.

1. Introduction

Decoherence provides an elegant framework to explain why an open quantum system coupled to its environment will exhibit a set of preferred states, usually ruling out a coherent superposition of arbitrary states. In [1], Zurek showed that even if the superposition principle treats all quantum states equally, the interaction between the quantum system and its environment would select a restricted number of "pointer" states (einselection) and destroy the phase coherence of superpositions of those pointer states (decoherence). This phenomenon presents a formidable challenge for such applications as quantum computation. In this simplest model of decoherence, it was readily realized that initial states of the environment exist that allow for decoherence-free unitary evolution of the quantum system. These peculiar states were usually neglected on the basis that "in realistic cases, such highly ordered initial states and symmetrical system-environment couplings are unlikely to be relevant" [2] or because the environment self-evolution would preclude such a unitary evolution [3].

In this paper, we investigate the conditions under which such initial states of the environment do exist in a framework where the quantum system interacts with its environment through a tensor product of Hamiltonians and the environment also evolves by itself. The results obtained underline the crucial role of the environment's self-evolution. The ability to identify and prepare such special initial states could be used in order to store quantum states. Indeed, even if the environment dynamics cannot be controlled, it might be possible to prepare it in a specific initial state. The interest of preparing the environment has already been briefly studied from

[3] Present address: Département de physique, Université de Sherbrooke, Sherbrooke, QC J1K 2R1 Canada

a theoretical point of view in [4] for a qubit coupled to a bosonic bath. In quantum dots, longer decoherence times can be obtained by narrowing the distribution of nuclear spin states. Propositions have been made for such a narrowing [5, 6, 7] and it has been demonstrated experimentally [8, 9, 10, 11]. However, our results restrict what can be expected from such a technique.

More precisely, we obtain a mathematical condition for the existence of an initial state allowing decoherence-free evolution in the presence of an interaction between a two-state system and its environment, with in addition a self-evolution of the environment. The interaction is restricted to a tensor product of Hermitian operators acting respectively on the system and on the environment. The result is stated in terms of the spectra of the two Hamiltonians acting on the environment (written $\tilde{H}$ and $H_{\mathcal{E}}$ below). Next, we analyze in detail a particular model, which is an extension of the model introduced by Zurek in [1]. Next, we assess the impact of imperfect state preparation. Finally, we discuss how to choose an initial state that minimizes decoherence when decoherence-free evolution via environment preparation is not in fact possible. Certain mathematical proofs are given in the appendices.

While our original motivation stemmed from the peculiar behavior of models of decoherence for certain initial states of the environment, our work can be related to the substantial body of work on decoherence suppression. Passive strategies to fight against decoherence began with the concept of decoherence-free subspaces introduced in the seminal paper of Lidar, Chuang and Whaley [12], following ideas introduced in [13, 14]. This breakthrough was later generalized to noiseless subsystems where protection arises from symmetry in the noise [15, 16, 17]. Active strategies are generically referred to as quantum error correction [18, 19, 20, 21]. A framework accounting for both noiseless subsystems and quantum error correction called operator quantum error correction (OQEC) was devised in [22, 23]. OQEC led to a large body of work in which conditions are derived for decoherence suppression that are *independent of the state of the environment*. However, recent work [24, 25] deriving the conditions for OQEC in the context of continuous dynamics provides a criterion for arbitrary system-environment Hamiltonians *conditioned on the state of the environment*. This criterion, altough very elegant and general, is difficult to apply operationally. Although it provides a method for testing whether or not a system can be corrected by a given unitary operator, it does not suggest a correction operator that is suitable. On the contrary, our simple, yet less general, model provides a criterion which can be computed efficiently.

2. The model

We consider a quantum system $\mathcal{S}$ and its environment $\mathcal{E}$ with dynamics described by a Hamiltonian of the form

$$H = S \otimes \tilde{H} + \mathbb{1} \otimes H_{\mathcal{E}}. \quad (1)$$

The Hermitian operator S acts in the Hilbert space of $\mathcal{S}$, which we take to be two-dimensional. The quantum system is thus taken to be a quantum bit (qubit) [26]. the Hermitian operators $\tilde{H}$ and $H_{\mathcal{E}}$ act in the Hilbert space of the environment.

The total Hamiltonian (1) induces *pure dephasing* and is typical of a coupling between the system and the environment that commutes with the self-evolution of the system [27].

Without loss of generality, we can assume $S = \sigma^z$. Its eigenstates, of eigenvalues ± 1, are written $|0\rangle$ and $|1\rangle$, respectively.

Suppose that the global system is in a product state at $t = 0$:

$$|\Psi(0)\rangle = |\psi(0)\rangle \otimes |I\rangle$$

where $|\psi(0)\rangle = a\,|0\rangle + b\,|1\rangle$ is an arbitrary normalized pure qubit state. At a later time t, the state evolves to

$$|\Psi(t)\rangle = a\,|0\rangle \otimes |\varepsilon_0(t)\rangle + b\,|1\rangle \otimes |\varepsilon_1(t)\rangle$$

where

$$i\frac{\mathrm{d}}{\mathrm{dt}}\left|\varepsilon_k(t)\right\rangle = H_k\left|\varepsilon_k(t)\right\rangle, \qquad k=0,1, \tag{2}$$

and where we have defined $H_0 \equiv H_\mathcal{E} + \tilde{H}$ and $H_1 \equiv H_\mathcal{E} - \tilde{H}$ with initial condition $|\varepsilon_0(0)\rangle = |\varepsilon_1(0)\rangle = |I\rangle$.

Since $|\Psi(t)\rangle$ is no longer a product state in general, the reduced density matrix $\rho(t)$ of the quantum system $\mathcal{S}$ no longer describes a pure state: the system has decohered. To quantify this, the off-diagonal elements of $\rho(t)$ in the basis $\{|0\rangle, |1\rangle\}$ are reduced by a factor

$$r(t) = \langle\varepsilon_0(t)|\varepsilon_1(t)\rangle.$$

Essentially, $r(t)$ quantifies how distinguishable the states $|\varepsilon_0(t)\rangle$ and $|\varepsilon_1(t)\rangle$ are. If they differ only by a phase, i.e. $|r(t)| = 1$, they are indistinguishable and the system $\mathcal{S}$ remains in (or has returned to) a pure state. If $r(t) = 0$, they can be distinguished with certainty, decoherence is complete, and the state of the quantum system reduces to a statistical mixture $|a|^2|0\rangle\langle 0| + |b|^2|1\rangle\langle 1|$. In general, the system has only undergone partial decoherence and a straightforward calculation shows that its purity is

$$\mathcal{P}(t) = \mathrm{Tr}\,\rho^2(t) = 1 - 2|a|^2|b|^2(1-|r(t)|^2)$$

where $|r(t)|^2$ is the Loschmidt echo [28].

Thus, a qubit prepared in an arbitrary coherent superposition of pointer states will remain pure if and only if the environment can be prepared in a state $|I\rangle$ for which

$$|r(t)| = 1 \text{ for all } t \tag{3}$$

i.e., for which $|\varepsilon_0(t)\rangle$ and $|\varepsilon_1(t)\rangle$ are the same state up to a time-dependent phase at all time.

3. Condition for coherent evolution

Given that the state evolves according to (2), what can be said of the Hamiltonians $\tilde{H}$ and $H_\mathcal{E}$ in order that for some initial state $|I\rangle$, $|\varepsilon_0(t)\rangle$ and $|\varepsilon_1(t)\rangle$ satisfy the coherence criterion (3)? The factor $r(t)$ is given by

$$r(t) = \langle I|\, e^{iH_0t}e^{-iH_1t}\,|I\rangle.$$

Clearly, if $|I\rangle$ is an eigenstate of both H_0 and H_1, then $r(t)$ is a phase and (3) is satisfied, so the existence of a common eigenstate of these Hamiltonians (or, equivalently, of $\tilde{H}$ and $H_\mathcal{E}$) is a sufficient condition. It is also a necessary condition, *although the state $|I\rangle$ itself need not be a common eigenstate.* To understand this, write H_0 in terms of its spectral decomposition, with eigenvalues $\{\lambda_j^{(0)}\}$ and associated projection operators $\{\Pi_j^{(0)}\}$ (so that $H_0 = \sum_j \lambda_j^{(0)}\Pi_j^{(0)}$), and similarly for H_1. Then

$$r(t) = \sum_{j,k} e^{i(\lambda_j^{(0)} - \lambda_k^{(1)})t}\,\langle I|\,\Pi_j^{(0)}\Pi_k^{(1)}\,|I\rangle. \tag{4}$$

In general, $r(t)$ contains terms of different frequencies. However, in order for it to remain of unit magnitude for all t, only one frequency can appear; all others must be associated with vanishing coefficients. The easiest way for this to occur is if $|I\rangle$ is a common eigenstate of H_0 and H_1, in which case the double sum collapses to a single term. A more general possibility is if $|I\rangle$ is a linear combination of common eigenstates of H_0 and H_1 corresponding to the same energy difference. In terms of the original Hamiltonians $\tilde{H}$ and $H_\mathcal{E}$, $|I\rangle$ must be a linear combination of eigenstates of $H_\mathcal{E}$ and each of these must also be a degenerate eigenstate of $\tilde{H}$. Stated otherwise, we must be able to find a basis in which $\tilde{H} = C_M \oplus \tilde{H}'$ and $H_\mathcal{E} = D_M \oplus H'_\mathcal{E}$, where C_M is an

M-dimensional constant matrix (proportional to the identity) and D_M is an M-dimensional diagonal matrix; $|I\rangle$ can be any vector in the first M dimensions.

Thus, preparing an initial state of the environment perfectly maintaining the coherence of the qubit at all times can only be done if both the interaction Hamiltonian and the environment self-evolution Hamiltonian exhibit a specific structure, namely they share an eigenstate. Obviously, this is not likely to occur generically, an observation that can easily be made quantitative.

Notice that if the environment self-evolution Hamiltonian $H_{\mathcal{E}}$ is zero, any eigenstate of the interaction Hamiltonian is a suitable initial state. Thus, it is the dynamics of the environment that usually prevents the existence of such an initial state. In the following section, we will make this situation explicit by computing the loss of coherence induced by adding dynamics to an otherwise-static environment.

4. Environment evolution as a perturbation

In this section, we consider a solvable model due to Zurek [1] to which we add a new term $\mathbb{1} \otimes H_{\mathcal{E}}$ to provide dynamics to an otherwise-static environment. We consider the case in which the environment is initially prepared in an eigenstate of $\tilde{H}$ and show that the self-evolution of the environment will destroy the coherence of the system. In this model, the environment consists of n spin-1/2 particles. The dimension of the environment Hilbert space is $N = 2^n$. Every particle of the environment interacts with the system through a $\sigma^z\sigma^z$ interaction. The eigenstates of the Pauli matrix σ_k^z acting on the kth spin are $\{|0\rangle_k, |1\rangle_k\}$. The strength of the interaction between the system and the kth spin of the environment is measured by a coupling constant $g_k \in \mathbb{R}$. Thus, the Hamiltonian is

$$H = \sigma_{\mathcal{S}}^z \otimes \sum_{k=1}^{n} g_k \sigma_k^z. \tag{5}$$

The eigenstates of $\tilde{H} = \sum_{k=1}^n g_k \sigma_k^z$ are the states in which the kth spin of the environment is in an eigenstate of σ_k^z, i.e., the states $|x\rangle = \bigotimes_{k=1}^n |x_k\rangle_k$ where all $x_k \in \{0,1\}$. A convenient way to order these states is to consider that x is a number between 0 and $N-1$ of which the binary representation is $x = x_1 x_2 \dots x_n$. Thus,

$$\forall x \in \{0,1\}^n \quad \tilde{H}|x\rangle = \omega_x |x\rangle$$

where $\omega_x = \sum_{k=1}^n (-1)^{x_k} g_k$. Throughout this section, we will assume that the coupling constants $\{g_k\}$ are chosen so that the eigenvalues ω_x are distinct.

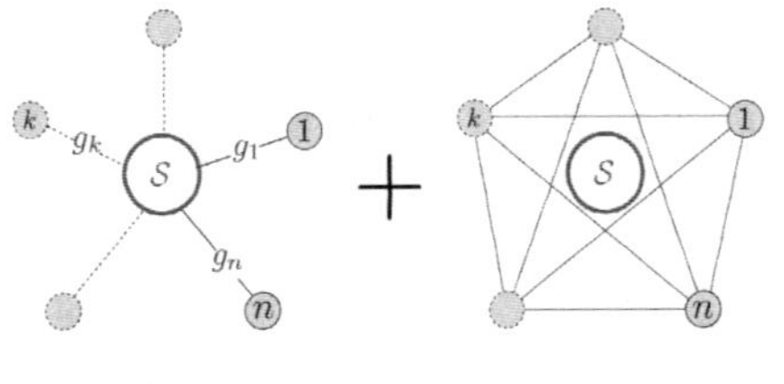

Figure 1. Decoherence model with dynamical environment

So far, the spins of the environment do not interact with each other; thus, preparing the environment in any state $|x\rangle$, being an eigenstate of $\tilde{H}$, will give rise to coherent evolution of the system. We will now add a self-evolution of the environment of the form

$$H_{\mathcal{E}} = \lambda \sum_{x,y \in \{0,1\}^n} |x\rangle\langle y|$$

where $\lambda \in \mathbb{R}$ is a perturbation parameter. This Hamiltonian is proportional to the projector on the state $|\Phi\rangle = \frac{1}{\sqrt{N}} \sum_{x\in\{0,1\}^n} |x\rangle = \bigotimes_{k=1}^{n} |+_k\rangle$, where $|+_k\rangle = \frac{1}{\sqrt{2}}|0_k\rangle + \frac{1}{\sqrt{2}}|1_k\rangle$:

$$H_{\mathcal{E}} = \lambda N |\Phi\rangle \langle\Phi| .$$

Clearly, the interaction and the environment Hamiltonians do not share a common eigenstate because all eigenstates of $\tilde{H}$ have a small overlap with $|\Phi\rangle$.

Suppose that the environment is prepared in the state

$$|0\rangle = \bigotimes_{k=1}^{n} |0\rangle_k \tag{6}$$

which is an eigenstate of $\tilde{H}$ with eigenvalue ω_0.

Let us compute the decoherence factor $r(t)$ for this initial state of the environment. Standard perturbation theory for $\lambda \ll \min_{x,y\in\{0,1\}^n} |\omega_x - \omega_y|$ shows that

$$|r(t)|^2 = 1 - 16\lambda^2 \sum_{x\neq 0} \frac{\sin^4\left(\frac{\omega_x-\omega_0}{2}t\right)}{(\omega_0-\omega_x)^2}.$$

The time average of this quantity is given by

$$\overline{|r(t)|^2} = 1 - 6\lambda^2 \sum_{x\neq 0} \frac{1}{(\omega_0-\omega_x)^2}.$$

Therefore, the time average is reduced by a factor proportional to $\sum_{x\neq 0}\frac{1}{(\omega_0-\omega_x)^2}$, which characterizes the density of energy levels near the unperturbed energy level.

5. Imperfect preparation of the environment

Suppose now that the condition for coherent evolution in the original model (1) is satisfied, so that there is at least one state for which the time evolution preserves the coherence of the system $\mathcal{S}$. We have assumed implicitly our ability to prepare perfectly the initial state of the environment. Notice that this strong assumption is nonetheless weaker than requiring control over the environment dynamics at all times. However, preparing the environment in a given initial state is a very difficult task which might only be achieved partially.

Let us give a simple example in the case of Zurek's model (5). When trying to prepare the state (6), the goal is to prepare all spins in the state $\alpha_k |0\rangle_k + \beta_k |1\rangle_k$ with $\alpha_k = 1$ and $\beta_k = 0$. Suppose that we are only able to ensure that $|\beta_k|^2 \leq \varepsilon \ll 1$ for all k, i.e., all spins of the environment are prepared with a small error. In that case, the average value $\overline{|r(t)|^2}$ is bounded by

$$\overline{|r(t)|^2} \geq \left((1-\varepsilon)^2 + \varepsilon^2\right)^n \overset{\varepsilon\ll 1}{\longrightarrow} 1 - 2n\varepsilon$$

which is attained if $|\beta_k|^2 = \varepsilon$ for all k. Thus, for small independent errors on each of the n spin of the environment, the coherence loss is proportional to n. This particular example indicates that even if an initial state allowing for decoherence-free evolution exists, the difficulty in preparing it will grow with the size of the environment, as one would expect intuitively.

A simple calculation shows the robustness of our method under imperfect preparation, an important feature in view of potential practical implementation. If the environment is prepared in the desired state $|I\rangle$, the system remains pure

$$\left|\Psi^{\mathcal{SE}}(t)\right\rangle = U(t)\left(\left|\psi^{\mathcal{S}}\right\rangle \otimes |I\rangle\right) = R(t)\left|\psi^{\mathcal{S}}\right\rangle \otimes V(t)|I\rangle ,$$

whereas for *imperfect* preparation, say state $|\tilde{I}\rangle$, the evolution leads to $|\tilde{\Psi}^{\mathcal{SE}}(t)\rangle$. The evolution being unitary, the fidelity [26] remains constant, i.e.

$$\mathcal{F}\left(|\Psi^{\mathcal{SE}}(t)\rangle, |\tilde{\Psi}^{\mathcal{SE}}(t)\rangle\right) = |\langle I|\tilde{I}\rangle|.$$

To compare the trace over the environment of both states, we use Uhlmann's characterization of fidelity [29] as the maximal overlap over all purifications, which implies

$$\mathcal{F}\left(\mathrm{Tr}_{\mathcal{E}}\,|\Psi^{\mathcal{SE}}\rangle\langle\Psi^{\mathcal{SE}}|, \mathrm{Tr}_{\mathcal{E}}\,|\tilde{\Psi}^{\mathcal{SE}}\rangle\langle\tilde{\Psi}^{\mathcal{SE}}|\right) \geq |\langle\Psi^{\mathcal{SE}}|\,\tilde{\Psi}^{\mathcal{SE}}\rangle|.$$

Hence, the fidelity between the states of the system evolving under perfect (on the one hand) and imperfect (on the other) initialization of the environment is bounded:

$$\mathcal{F}\left(R(t)\,|\psi^{\mathcal{S}}\rangle, \mathrm{Tr}_{\mathcal{E}}\,|\tilde{\Psi}^{\mathcal{SE}}\rangle\langle\tilde{\Psi}^{\mathcal{SE}}|\right) \geq |\langle I|\tilde{I}\rangle|. \tag{7}$$

This calculation yields the same result as the much more general work of [30, 25] on the robustness of OQEC.

6. Imperfect control of the environment

We now address a situation which is in a sense opposite to the one just considered. Rather than having perfect control over the dynamics (so that a common eigenstate of $\tilde{H}$ and $H_{\mathcal{E}}$ can be made to exist) but an imperfect ability to prepare the initial state, suppose that we can prepare perfectly any state we wish but that no such common eigenstate exists. Given that the coherence of $\mathcal{S}$ cannot be preserved, is there an optimal choice of initial state, that is, one for which the ensuing decoherence is in some sense minimized? To address this question, we must first specify what we mean by "optimal." Do we wish to minimize average decoherence, in which case we would not "see" brief but significant drops in the coherence? Alternatively, do we wish to minimize the maximum decoherence, in which case a significant drop in coherence will make a state appear to be a bad choice even though it may be good on average? In the following we will adopt this latter criterion.

It is somewhat easier to use the combinations $H_{0,1}$ rather than $\tilde{H}$ and $H_{\mathcal{E}}$. We wish to find the state $|I\rangle$ for which the minimum value of $|r(t)|$ is maximal, in the case where several frequencies are present in the sum (4). The general case appears difficult to analyze, but one might expect that the best choice of $|I\rangle$ is an eigenstate of one of the two Hamiltonians and a combination of two eigenstates of the other Hamiltonian. For instance, if $\Pi_1^{(0)}\,|I\rangle = (\Pi_1^{(1)} + \Pi_2^{(1)})\,|I\rangle = |I\rangle$, then (4) becomes

$$r(t) = e^{i\lambda_1^1 t}\left(e^{-i\lambda_1^2 t}\,\langle I|\,\Pi_1^{(1)}\,|I\rangle + e^{-i\lambda_2^2 t}\,\langle I|\,\Pi_2^{(1)}\,|I\rangle\right)$$

and

$$|r_{\min}| = |\,\langle I|\,\Pi_1^{(1)}\,|I\rangle - \langle I|\,\Pi_2^{(1)}\,|I\rangle\,|.$$

This expectation turns out not to be the best choice, in general. To see this, we examine the simplest example in which no common eigenstate exists, namely, an environment consisting of a single qubit with $H_{0,1}$ describing its interaction with non-parallel magnetic fields of the same intensity. Let the precession frequency be ω and let the direction of the magnetic fields corresponding to H_0 and H_1 be $\hat{m}_0 = (\sin\alpha, 0, \cos\alpha)$ and $\hat{m}_1 = (-\sin\alpha, 0, \cos\alpha)$, with $0 < \alpha < \pi/2$, respectively. The initial state can be taken to be a spin aligned along a third direction, $\hat{v} = (\sin\theta\cos\phi, \sin\theta\sin\phi, \cos\theta)$, say. Its evolution according to each of the Hamiltonians is simple: the spin precesses around $\hat{m}_i$ with frequency ω so $|\varepsilon_i(t)\rangle = |\hat{v}_i(t)\rangle$ where $\hat{v}_i(t)$ is $\hat{v}$ rotated about $\hat{m}_i$ by angle ωt. Then $r(t) = \langle\varepsilon_0(t)|\varepsilon_1(t)\rangle$ and $|r(t)| = \cos(\gamma(t)/2)$, where

$\gamma(t)$ is the angle between $\hat{v}_0(t)$ and $\hat{v}_1(t)$. Thus we would like to find the vector $\hat{v}$ for which the maximum angle between $\hat{v}_0(t)$ and $\hat{v}_1(t)$ as they precess is minimized. If, as was conjectured above, we choose $\hat{v} = \hat{m}_0$, then it is easy to see that the maximum angle is the lesser of 4α and $2\pi - 4\alpha$, reached after half a precession. Although it is surprisingly difficult to find this maximum angle for an arbitrary $\hat{v}$, the optimal choice turns out to depend on the angle α as indicated in Figure 2.

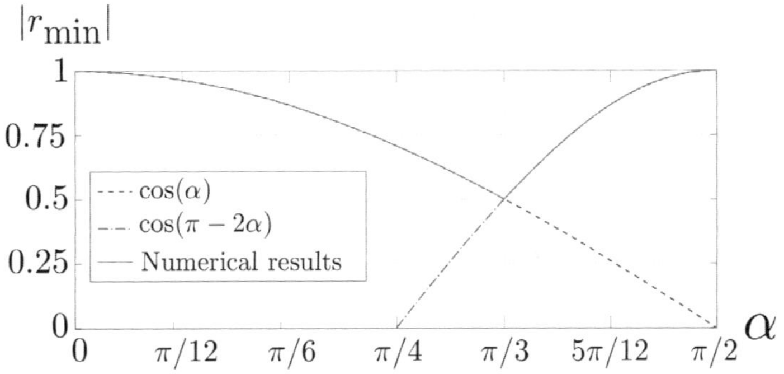

Figure 2. Numerical determination of $|r_{min}|$ as a function of α

If $\alpha \leq \pi/3$, it is best to choose $\hat{v} = \hat{y}$, perpendicular to $\hat{m}_0$ and $\hat{m}_1$, resulting in $\gamma_{\max} = 2\alpha$ (attained after a quarter-rotation) and

$$|r_{\min}| = \cos\alpha.$$

If $\alpha \geq \pi/3$, an optimal choice is $\hat{v}$ collinear with $\hat{m}_0$ (or $\hat{m}_1$), resulting in $\gamma_{\max} = 2\pi - 4\alpha$ (attained after a half-rotation) and

$$|r_{\min}| = \cos(\pi - 2\alpha).$$

Thus, the optimal value of $|r_{\min}|$ is close to unity if the directions of the two magnetic fields are almost parallel. More surprising is the fact that the best choice of initial state is far from what one would naively have guessed. In the case in which $\alpha \leq \pi/3$, the optimal choice is $\hat{v} = \hat{y}$, which does not correspond to an eigenstate of either Hamiltonian.

To sum up, two regimes emerge. On the one hand, for $\alpha > \pi/3$, the system-environment interaction prevails whereas the self-evolution of the environment is only a perturbation. In this regime, the system and the environment play symmetric roles. Reducing decoherence in this case boils down to minimizing the entanglement. The optimal initial states are thus the pointer states of the environment, i.e. the eigenstates of $\tilde{H}$. On the other hand, for $\alpha < \pi/3$, the self-evolution Hamiltonian $H_{\mathcal{E}}$ dominates over the interaction Hamiltonian $\tilde{H}$. However, the evolution *of the quantum system* relies essentially on the interaction. Thus, for the evolution of the system, $\tilde{H}$, no matter how small, *cannot* be considered a perturbation with respect to $H_{\mathcal{E}}$. In that case, the evolution of the environment is dominated by its own dynamics but the impact on the quantum system is mediated by the interaction Hamiltonian. Hence, finding an analytical criterion characterizing an initial state of the environment that optimally limits the subsequent decoherence of the system remains an unresolved challenge.

Acknowledgments
We thank the National Science and Engineering Research Council of Canada for funding, and David Poulin for useful discussions.

Journal of Physics: Conference Series **462** (2013) 012033 doi:10.1088/1742-6596/462/1/012033

References

[1] Zurek W H 1982 *Phys. Rev. D* **26** 1862–1880
[2] Schlosshauer M 2007 *Decoherence and the Quantum to Classical Transition* (Springer-Verlag Berlin)
[3] Paz J P and Zurek W H 2001 *Coherent Matter Waves, Lectures from the 72nd Les Houches Summer School, 1999* (Springer-Verlag, Berlin) pp 533–614
[4] Dajka J, Mierzejewski M, Luczka J and Haenggi P 2009 *Physica E* **In Press** ISSN 1386-9477
[5] Klauser D, Coish W A and Loss D 2006 *Phys. Rev. B* **73** 205302
[6] Ribeiro H and Burkard G 2009 *Phys. Rev. Lett.* **102** 216802
[7] Coish W A 2009 *Nat. Phys.* **5** 710–711
[8] Greilich A, Shabaev A, Yakovlev D R, Efros A L, Yugova I A, Reuter D, Wieck A D and Bayer M 2007 *Science* **317** 1896–1899
[9] Reilly D, Taylor J, Petta J, Marcus C, Hanson M and Gossard A 2008 *Science* **321** 817
[10] Xu X, Yao W, Sun B, Steel D G, Bracker A S, Gammon D and Sham L J 2009 *Nature* **459** 1105–1109
[11] Latta C, Hogele A, Zhao Y, Vamivakas A N, Maletinsky P, Kroner M, Dreiser J, Carusotto I, Badolato A, Schuh D, Wegscheider W, Atature M and Imamoglu A 2009 *Nat Phys* **5** 758–763
[12] Lidar D A, Chuang I L and Whaley K B 1998 *Phys. Rev. Lett.* **81** 2594–2597
[13] Duan L M and Guo G C 1998 *Phys. Rev. A* **57** 737–741
[14] Zanardi P and Rasetti M 1997 *Phys. Rev. Lett.* **79** 3306–3309
[15] Knill E, Laflamme R and Viola L 2000 *Phys. Rev. Lett.* **84** 2525–2528
[16] De Filippo S 2000 *Phys. Rev. A* **62** 052307
[17] Yang C P and Gea-Banacloche J 2001 *Phys. Rev. A* **64** 032309
[18] Bennett C H, DiVincenzo D P, Smolin J A and Wootters W K 1996 *Phys. Rev. A* **54** 3824–3851
[19] Knill E and Laflamme R 1997 *Phys. Rev. A* **55** 900–911
[20] Shor P W 1995 *Phys. Rev. A* **52** R2493–R2496
[21] Steane A M 1996 *Phys. Rev. Lett.* **77** 793–797
[22] Kribs D, Laflamme R and Poulin D 2005 *Phys. Rev. Lett.* **94** 180501
[23] Kribs D, Laflamme R, Poulin D and Lesosky M 2006 *Quantum Information & Computation* **6** 382–399
[24] Oreshkov O, Lidar D A and Brun T A 2008 *Phys. Rev. A* **78** 022333
[25] Oreshkov O 2008 *Topics in quantum information and the theory of open quantum systems* Ph.D. thesis University of Southern California
[26] Nielsen M A and Chuang I L 2000 *Quantum Information and Quantum Information* (Cambridge University Press)
[27] Hornberger K 2009 *Entanglement and Decoherence* (*Lecture Notes in Physics* vol 768) (Springer Berlin / Heidelberg) pp 221–276
[28] Cucchietti F M, Paz J P and Zurek W H 2005 *Phys. Rev. A* **72** 052113
[29] Uhlmann A 1976 *Rep. Math. Phys.* **9** 273–279
[30] Oreshkov O 2008 *Phys. Rev. A* **77** 032333
[31] Katō T 1966 *Perturbation theory for linear operators* (Springer)

6th International Symposium on Quantum Theory and Symmetries (QTS6) IOP Publishing
Journal of Physics: Conference Series **462** (2013) 012034 doi:10.1088/1742-6596/462/1/012034

How fuzzballs resolve the information paradox

Samir D. Mathur
Department of Physics,
The Ohio State University,
Columbus, OH 43210, USA

E-mail: mathur@mps.ohio-state.edu

Abstract. When a shell collapses through its horizon, semiclassical physics suggests that information cannot escape from this horizon. One might hope that nonperturbative quantum gravity effects will change this situation and avoid the 'information paradox'. We note that string theory has provided a set of states over which the wavefunction of the shell can spread, and that the number of these states is large enough that such a spreading would significantly modify the classically expected evolution. In this article we perform a simple estimate of the spreading time, showing that it is much shorter than the Hawking evaporation time for the hole. Thus information can emerge from the hole through the relaxation of the shell state into a linear combination of fuzzballs.

A longstanding puzzle in quantum gravity is the following: is there anything at the black hole horizon?

There are many reasons to take the horizon seriously as a place where 'physics happens'. The Bekenstein entropy S_{bek} of the hole is proportional to the area of the horizon [1]. This suggests that we attach one 'bit' of information to each planck area of the horizon, and perhaps quantize the horizon area to give an integral number of bits [2]. In loop quantum gravity the entropy arises from spin degrees of freedom on links that intersect the horizon [3]. In the Schwarzschild coordinate frame, all matter that ever fell towards the horizon appears to be frozen forever just outside the horizon. This picture was concretized in the 'membrane paradigm' [4], where one puts an artificial cutoff surface just outside the horizon, and finds boundary conditions at this surface which will reproduce correct mechanical and thermal effects in the region outside this surface. These boundary conditions can be summarized by imagining a fictitious membrane placed at this surface, with appropriate mechanical and thermal properties. In particular, there is a very high temperature near the membrane, so one might imagine that an infalling observer would get burnt up and eventually reradiated out to infinity [5]. This would be good, because to avoid the information paradox we need to have some way of returning the information in infalling objects back to infinity.

But there turns out to be a flaw in these attempts to assign genuine dynamics to the horizon. It is well known that the Schwarzschild coordinate system fails at the horizon, and the regular coordinates there – the Kruskal coordinates – show that *nothing* special happens at the horizon of the traditional black hole. Nothing can 'stay' at the horizon; all matter must fall in towards the singularity. The high temperature in the Schwarzschild frame disappears in the Kruskal frame; it is seen to be nothing but the Rindler tempertaure observed in an accelerated system of coordinates. Careful renormalization of the stress tensor shows that there is no energy density at the horizon [6]; the energy of Rindler quanta is cancelled by vacuum polarization energy, since the physical vacuum at the horizon is naturally described in Kruskal rather than Schwarzschild coordinates.

What then, about the fate of information falling into the hole? Over the past decade, many physicists had settled down to the following set of beliefs (i) information is not lost; it emerges in the Hawking radiation from the hole (ii) nothing special happens at the horizon, because the Kruskal description looks like the logical one (iii) the above two beliefs are consistent because we need only small delicate correlations among Hawking quanta to carry the information, and subleading corrections to Hawking's computation [7] should be able to encode the required information in the quanta.

But if these beliefs were correct, one has to wonder why over the years no one produced a model showing how it was possible for information to get encoded in the radiation while preserving a near-semiclassical Hamiltonian at the horizon. The reason for this failure emerged last year, when it was proved that (iii) is in fact incorrect: it is *not* possible to make small corrections to Hawking's leading order state and recover any significant information in the radiation [8]. More precisely, if matrix elements of the Hamiltonian for low energy modes are within ϵ of their semiclassical values, then no more than a fraction 2ϵ of the information can be encoded in the radiation. The proof of this theorem used a well known inequality from quantum information theory called 'strong subadditivity' [9]. But no *elementary* proof of this inequality is known, so it is perhaps not surprising that its consequences were not built into our naive thinking about encoding of information in the radiation.

It would now seem that we are back to square one on the information puzzle. Must we reconcile ourselves to information loss? We will now argue that another set of results over the past few years gives us a picture of how there can be real dynamics at the horizon, which will lead to an escape of information in the radiation.

A black hole has given values of of its conserved quantum numbers: mass, charge and angular momentum. In earlier work, one assumed a simple ansatz for the metric. Thus for the nonrotating hole, one assumed independence of the angular directions θ, ϕ for all metric components. In a higher dimensional gravity theory, the extra directions have to be assumed compact, and the ansatz assumed no dependence of the metric on the compact coordinates z_i either. Einstein's equations then gave a unique solution, possessing a horizon and a singularity – the traditional black hole.

But it has been now recognized that there are $Exp[S_{bek}]$ *different* solutions of Einstein's equations with the *same* quantum numbers [10]. All are regular, with no horizon and no singularity. None of them are spherically symmetric, and all involve the compact directions in a nontrivial way. So it was the simplicity of our initial ansatz which led us to miss these solutions. Further, string theory has allowed us to list all possible states for simple sets of quantum numbers. The regular states mentioned above map in a one-to one and onto way onto the set of all allowed states, while the spherical solution with horizon and singularity is not realized at all.

This situation gives us an appealing way out of the information problem:

(a) The region far from the hole has an infinite dimensional phase space due to its infinite volume. Very close to the horizon, however, there is again a very large phase space [11], from the $Exp[S_{bek}]$ solutions – termed 'microstates' – giving the entropy of the hole (fig.1).

(b) Most of these solutions have nontrivial structure only very close to the horizon. This is because rapid variation of the metric with position costs a lot of energy, and the large redshift near $r = 2M$ is needed to lower this to the given total energy M. The detailed structure of the solutions can be inferred from the explicitly solutions which have been constructed. The geometry has *ergoregions*, with the direction of frame dragging changing rapidly from place to place (fig.2). This rapid space-time time dependence is what keeps the geometrical structure

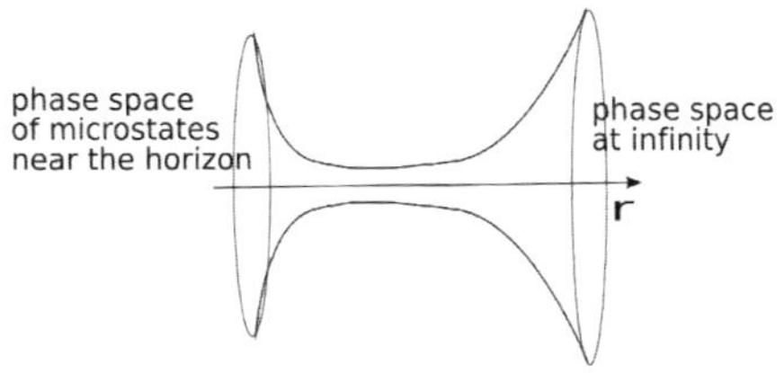

Figure 1. Schematic depiction of phase space in the black hole: there is a large phase space near the horizon from the microstate solutions of Einstein's equations.

from collapsing towards $r = 0$. (For an analogy, see fig.3; a circular elastic band shrinks to a point, but an rapidly vibrating band stays open with a nontrivial average radius.)

Figure 2. Schematic description of a microstate solution of Einstein's equations. There are 'local ergoregions' with rapidly changing direction of frame dragging near the horizon. The geometry closes off without having an interior horizon or singularity due to its peculiar topological structure.

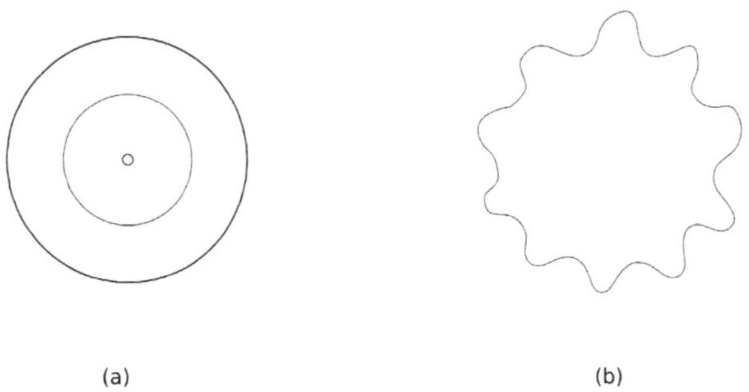

Figure 3. A circular elastic band in flat space will shrink to a point under its tension, but rapid oscillations maintain a nonzero average radius for the loop.

(c) Consider a spherically symmetric shell made of gravitational radiation, collapsing to make a hole. The motion upto $r = 2M + \epsilon$ is described by traditional dynamics in the Schwarzschild frame ($\epsilon \ll 2M$). But at $r < 2M + \epsilon$ the wavefunction of the shell can spread over the $Exp[S_{bek}]$ alternative solutions of Einstein's equations. The amplitude for this spread can be estimated by computing the *tunneling* amplitude between the shell and one of the microstate solutions. This amplitude $Exp[-S_{cl}]$ is very *small*, since the involved objects are macroscopic and the classical action of tunneling S_{cl} is big. But the *number* of states that we can tunnel to very *large*, given by $Exp[S_{bek}]$. It turns out that $S_{cl} \sim S_{bek}$ [13], so instead of making a long lived black hole, the shell state becomes a linear superposition of microstate solutions in a time which is much shorter than the Hawking evaporation time [14].

(d) It has been checked that ergoregion emission from the microstate solutions reproduces exactly the Hawking radiation rate expected from those solutions [12]. But this emission does not involve information loss, since there is no horizon.

(e) The further evolution of the collapsing shell is given by the evolution of the linear superposition of microstates into which it has tunneled. It is plausible that the short time evolution of the collective modes of these complicated microstates can be mimicked by evolution in the interior of the traditional black hole geometry. This effective evolution would then corresponds to the interior of the black hole in the Penrose diagram. The crucial fact here is that there are a large number of degrees of freedom in the fuzzball solutions. Further, there is a scaling symmetry in the fuzzball solutions, describing how close to the horizon the fuzzball structure is. An infalling object will interact with these fuzzball degrees of freedom and be absorbed into them with an amplitude given by the fermi golden rule. The large entropy (and fast evolution [15]) of the fuzzballs implies that different infalling degrees of freedom enter different parts of the fuzzball phase space, and thus do not interfere with each other to leading order in their further evolution. Thus the infalling Hilbert space is injectively mapped to a subset of the black hole degrees of freedom, and the subsequent evolution can be given a dual description, which mimics infall through the horizon. This is similar to the basic AdS/CFT duality map, where the infalling object is absorbed into a large number of open string degrees of freedom, and the injective map then allows us to map the infalling gravitational objects into an effective evolution in an AdS geometry.

With this picture we seem to have an explicit realization of the membrane paradigm. But the membrane now has real degrees of freedom on it, given by the rapidly oscillating solutions of Einstein's equations that correspond to the black hole microstates. It would be exciting to explore if the collective dynamics of the microstates reproduces the very precise properties of the fictitious membrane of the membrane paradigm.

Acknowledgments
This work was supported in part by DOE grant DE-FG02-91ER-40690.

References

[1] J. D. Bekenstein, Phys. Rev. D **7**, 2333 (1973).
[2] J. D. Bekenstein, Lett. Nuovo Cim. **11**, 467 (1974).
[3] C. Rovelli and L. Smolin, Nucl. Phys. B **442**, 593 (1995) [Erratum-ibid. B **456**, 753 (1995)] [arXiv:gr-qc/9411005].
[4] K. S. . Thorne, R. H. . Price and D. A. . Macdonald, *NEW HAVEN, USA: YALE UNIV. PR. (1986) 367p*
[5] L. Susskind, L. Thorlacius and J. Uglum, Phys. Rev. D **48**, 3743 (1993) [arXiv:hep-th/9306069].
[6] D. W. Sciama, P. Candelas and D. Deutsch, Adv. Phys. **30**, 327 (1981).
[7] S. W. Hawking, Commun. Math. Phys. **43**, 199 (1975) [Erratum-ibid. **46**, 206 (1976)]; S. W. Hawking, Phys. Rev. D **14**, 2460 (1976).
[8] S. D. Mathur, Class. Quant. Grav. **26**, 224001 (2009) [arXiv:0909.1038 [hep-th]].
[9] E. H. Lieb and M. B. Ruskai, J. Math. Phys. **14**, 1938 (1973).
[10] O. Lunin and S. D. Mathur, Nucl. Phys. B **623**, 342 (2002) [arXiv:hep-th/0109154]; O. Lunin and S. D. Mathur, Phys. Rev. Lett. **88**, 211303 (2002) [arXiv:hep-th/0202072]; O. Lunin, J. M. Maldacena and L. Maoz, arXiv:hep-th/0212210; I. Kanitscheider, K. Skenderis and M. Taylor, JHEP **0706**, 056 (2007) [arXiv:0704.0690 [hep-th]]. S. Giusto, S. D. Mathur and A. Saxena, Nucl. Phys. B **710**, 425 (2005)

[arXiv:hep-th/0406103]; I. Bena and N. P. Warner, Adv. Theor. Math. Phys. **9**, 667 (2005) [arXiv:hep-th/0408106]; V. Jejjala, O. Madden, S. F. Ross and G. Titchener, Phys. Rev. D **71**, 124030 (2005) [arXiv:hep-th/0504181]. V. Balasubramanian, E. G. Gimon and T. S. Levi, arXiv:hep-th/0606118; J. de Boer, F. Denef, S. El-Showk, I. Messamah and D. Van den Bleeken, JHEP **0811**, 050 (2008) [arXiv:0802.2257 [hep-th]]; S. D. Mathur, arXiv:0810.4525 [hep-th].

[11] S. D. Mathur, arXiv:0706.3884 [hep-th].

[12] V. Cardoso, O. J. C. Dias, J. L. Hovdebo and R. C. Myers, Phys. Rev. D **73**, 064031 (2006) [arXiv:hep-th/0512277]; B. D. Chowdhury and S. D. Mathur, Class. Quant. Grav. **25**, 135005 (2008) [arXiv:0711.4817 [hep-th]].

[13] S. D. Mathur, arXiv:0805.3716 [hep-th].

[14] S. D. Mathur, Int. J. Mod. Phys. D **18**, 2215 (2009) [arXiv:0905.4483 [hep-th]].

[15] P. Hayden and J. Preskill, JHEP **0709**, 120 (2007) [arXiv:0708.4025 [hep-th]]; Y. Sekino and L. Susskind, JHEP **0810**, 065 (2008) [arXiv:0808.2096 [hep-th]].

A limit on the ABJ model

J. A. Minahan[1], O. Ohlsson Sax[1] and C. Sieg[2]

[1] Department of Physics and Astronomy, Uppsala University
SE-751 08 Uppsala, Sweden and [2]The Niels Bohr International Academy
The Niels Bohr Institute
Blegdamsvej 17, DK-2100, Copenhagen Ø, Denmark

E-mail: joseph.minahan@fysast.uu.se, olof.ohlsson-sax@physics.uu.se, csieg@nbi.dk

Abstract.
There is a large amount of evidence that the ABJM model is integrable in the planar limit. Less clear is whether or not the ABJ model is integrable. Here we investigate a limit of the ABJ model in the weak coupling limit where one 't Hooft parameter is much greater than the other. At the 4 loop level in the $SU(2) \times SU(2)$ sector the anomalous dimensions of single trace operators map to two Heisenberg spin chains with nearest neighbor interactions with an overall coefficient that is a function of one of the 't Hooft parameters. We conjecture the form of this function and show that is consistent with observations about the ABJ model concerning unitarity and parity, including strong coupling statements.

The ABJM model [1] is a $U(N)_k \times U(N)_{-k}$ superconformal Chern-Simons theory with $\mathcal{N} = 6$ supersymmetry. In the planar limit the two point functions of single trace operators map to a spin chain of alternating type that is believed to be integrable [2–5]. The integrability is remarkably similar to that found for $\mathcal{N} = 4$ super Yang-Mills. However, one important difference is the presence of an unknown function $h^2(\lambda)$, where λ is the 't Hooft coupling $\lambda = N/k$. This function appears in the Bethe equations as well as the spin chain magnon dispersion relation

$$E(p) = \sqrt{Q^2 + 4h^2(\lambda)\sin^2\tfrac{p}{2}} - Q\,, \tag{1}$$

where Q is an R-charge. In $\mathcal{N} = 4$ SYM, $Q = 1$ and $h^2(\lambda) = \frac{\lambda}{4\pi}$. For the ABJM model $Q = 1/2$, but $h^2(\lambda)$ is only known at its asymptotic limits, where for weak coupling $h^2(\lambda) = \lambda^2 + \mathrm{O}(\lambda^4)$, while for strong coupling $h^2(\lambda) = \frac{1}{2}\lambda - \frac{\ln 2}{\sqrt{2}\pi}\sqrt{\lambda} + \mathrm{O}(1)$.

The ABJ model [6] is an interesting extension of ABJM, where now the gauge group is $U(M)_k \times U(N)_{-k}$ but the amount of supersymmetry is unchanged. In the planar limit there are now two 't Hooft constants, $\lambda = M/k$, $\hat{\lambda} = N/k$, but whether or not ABJ is integrable is not clear. It turns out that at the two-loop level, the lowest order of perturbation theory, the anomalous dimension mixing matrix still maps to an integrable spin-chain [7,8]. However, the corresponding string dual for this theory is type IIA on $AdS_4 \times CP^3$ with a b-flux through a CP^1 on the CP^3. This b-flux corresponds to a θ-angle for the world-sheet theory, where $\theta = 2\pi(N - M)/k + \pi$ [9], which generically breaks parity. The θ-angle cannot affect the classical integrability [10, 11] since it does not change the equations of motion. However, at the quantum level it is expected to have an effect. This is what happens for the $O(N)$ model, where the two parity preserving values, $\theta = 0, \pi$ are known to be integrable. However, the IR behavior for these two values of θ are completely different, for example the $\theta = 0$ point has a gap while the $\theta = \pi$ point is gapless. It is not clear at all how to smoothly interpolate between these two IR points by varying θ, leaving the integrability in doubt.

For the purpose of this note let us ask what the behavior would be if the planar ABJ model *were* integrable for any value of λ and $\hat{\lambda}$. In this case it seems that the only possible modification to the ABJM results is to replace the function $h^2(\lambda)$ with $h^2(\lambda, \hat{\lambda})$ in the dispersion relation and the Bethe equations. At the two loop level it has been shown that it simply replaces λ^2 with $\lambda\hat{\lambda}$ [7,8]. This two loop behavior motivates us to define two new variables $\bar{\lambda}$ and σ where

$$\bar{\lambda} = \sqrt{\lambda\hat{\lambda}}\,, \qquad \sigma = \frac{\lambda - \hat{\lambda}}{\bar{\lambda}}\,, \tag{2}$$

and then write $h^2(\bar{\lambda}, \sigma)$ in terms of these two variables.

Since parity transformations in the ABJM/ABJ theories include the map $A_\mu \leftrightarrow \hat{A}_\mu$, the ABJ theories are not in general parity invariant. One might expect this would translate into parity violation in the spin chain, manifested by different dispersion relations for odd and even site magnons, or equivalently with $h^2(\bar{\lambda}, \sigma) \neq h^2(\bar{\lambda}, -\sigma)$. However, at least up to and including the four-loop level there turns out to be no parity violation.

In [12,13] we reported on a four loop calculation to compute the first corrections to $h^2(\lambda, \hat{\lambda})$. Since $h^2(\lambda, \hat{\lambda})$ can be extracted from the magnon dispersion relation, it is not necessary to consider the most general spin chains. Instead we can restrict ourselves to an $SU(2) \times SU(2)$ subsector which is closed. The magnons on the odd sites are the excitations in one $SU(2)$ while the magnons on the even sites are the excitations for the other $SU(2)$. Furthermore, the odd-site magnons do not interact with the even-site magnons until the eight loop level so at four loops they can be treated as noninteracting.

The function $h^2(\bar{\lambda}, \sigma)$ is assumed to have the following form consistent with parity conservation,

$$h^2(\bar{\lambda}, \sigma) = \bar{\lambda}^2 + \bar{\lambda}^4 h_4(\sigma)\,, \qquad h_4(\sigma) = h_4 + \sigma^2 h_{4,\sigma}\,, \tag{3}$$

A σ^4 term would also be consistent with parity, but a quick inspection of possible Feynman diagrams shows that such a term does not appear.

The 4-loop correction $h_4(\sigma)$ is extracted from 4-loop diagrams by considering the expansion of the dispersion relation

$$E(p) = 2\bar{\lambda}^2(1 - \cos p) + 2\bar{\lambda}^4(h_4(\sigma) - 3 + (4 - h_4(\sigma))\cos p - \cos 2p) + \mathcal{O}(\bar{\lambda}^6)\,. \tag{4}$$

These terms arise from the permutation structures that sit in the spin chain Hamiltonian, or equivalently the dilatation operator D. From equation (4) we see that D has the form

$$D = L + \bar{\lambda}^2 D_2 + \bar{\lambda}^4 D_4(\sigma) + \mathcal{O}(\bar{\lambda}^6)\,, \tag{5}$$

with

$$\begin{aligned} D_{2,\text{even}} &= \{\} - \{1\}\,, \\ D_{2,\text{odd}} &= \{\} - \{2\}\,, \\ D_{4,\text{odd}}(\sigma) &= (h_4(\sigma) - 4)\{\} + (6 - h_4(\sigma))\{1\} - \{1,3\} - \{3,1\}\,, \\ D_{4,\text{even}}(\sigma) &= (h_4(\sigma) - 4)\{\} + (6 - h_4(\sigma))\{2\} - \{2,4\} - \{4,2\}\,. \end{aligned} \tag{6}$$

The notation $\{\dots\}$ describes the permutation structures, where

$$\{a_1, a_2, \dots, a_m\} = \sum_{i=1}^{L} \mathrm{P}_{2i+a_1\,2i+a_1+2}\,\mathrm{P}_{2i+a_2\,2i+a_2+2} \cdots \mathrm{P}_{2i+a_m\,2i+a_m+2}\,, \tag{7}$$

Inspection of (6) shows that $h_4(\sigma)$ can be extracted from D by considering those contributions that involve a single exchange. There are still dozens of Feynman diagrams to consider that are listed in [12,13]. Here we just report the result,

$$h_4 = -16 + 4\zeta(2) \approx -9.42\,, \qquad h_{4,\sigma} = -\zeta(2)\,. \tag{8}$$

Note that the result for h_4 is consistent with a dampening of the quadratic behavior at weak coupling to the linear behavior at strong coupling. Also note that h_4 has mixed transcendentality while $h_{4,\sigma}$ has highest transcendentality.

In [12, 13] it was also shown that one could take $\lambda \gg \hat{\lambda}$ and consider the rescaled dilatation operator $\mathcal{D} = \hat{\lambda}^{-1}(D - 2L)$, which at the 4-loop level reduces to

$$\mathcal{D} = \lambda(1 - \zeta(2)\lambda^2)(2\{\} - \{1\} - \{2\}) \tag{9}$$

for the $SU(2) \times SU(2)$ sector. But this is simply the Hamiltonian for two decoupled Heisenberg spin chains. We further believe that the form of $\mathcal{D}$ persists to higher loop level, that is the Hamiltonian will still be two decoupled Heisenberg chains with an overall coefficient that is some function of λ, $f(\lambda)$.

The form of the 4-loop expansion is very suggestive and we conjecture that the function is[1]

$$f(\lambda) = \frac{1}{\pi}\,\sin(\pi\,\lambda)\,. \tag{10}$$

Let us now motivate why $f(\lambda)$ in (10) has the right behavior.

First, as was argued in [6], unitary ABJ theories with gauge group $U(N+\ell)_k \times U(N)_{-k}$ do not exist if $\ell > k$. In terms of the 't Hooft parameters this corresponds to $\lambda > \hat{\lambda} + 1$. In the limit where $\lambda \gg \hat{\lambda}$ this translates into $\lambda > 1$ violating the unitary bound. If we now consider the function in (10) we see that it changes sign as λ passes through 1. But if the overall coefficient is negative, then so are the anomalous dimensions. Included in these is the single trace-operator $\mathrm{Tr}[(Y^1Y_3^\dagger Y^2Y_4^\dagger - Y^1Y_4^\dagger Y^2Y_3^\dagger)]$ which is a superpartner of $\mathrm{Tr}[Y^AY_A^\dagger]$ that too must have the same negative anomalous dimension. But then $\mathrm{Tr}[Y^AY_A^\dagger]$ would have a total dimension that is slightly less than 1, violating the unitary bound for superconformal field theories in 3 dimensions [14]. Hence, the spin-chain also indicates that there is no such unitary theory.

A second test of (10) is related to the first and relies on parity. Based on the brane picture there are reasons to believe that the $U(N+\ell)_k \times U(N)_{-k}$ theory is equivalent to $U(N)_k \times U(N+k-\ell)_{-k}$ [6]. This corresponds to the 't Hooft pair equivalence $(\lambda, \hat{\lambda}) \equiv (\hat{\lambda}, 2\hat{\lambda} + 1 - \lambda)$. However, if parity is preserved at least for the spin chain calculations then one should find the same anomalous dimensions for $(2\hat{\lambda} + 1 - \lambda, \hat{\lambda})$. In the limit where $\lambda \gg \hat{\lambda}$ this indicates an equivalence between the theory with λ and the theory with $1 - \lambda$, which is consistent with the $f(\lambda)$ in (10).

Of course more evidence is needed to confirm the form of (10). A six-loop calculation to compute the next term in the expansion is a daunting proposition but may be feasible, since the number of Feynman graphs that must be computed is quite restricted. We are presently investigating such a calculation.

Acknowledgments

We would like to thank N. Nekrasov for helpful discussions. The research of J. A. M. is supported in part by the Swedish research council and the STINT foundation. J. A. M. thanks the CTP at MIT for kind hospitality during the course of this work.

References

[1] O. Aharony, O. Bergman, D. L. Jafferis, and J. Maldacena, *$\mathcal{N}$= 6 Superconformal Chern-Simons-Matter Theories, M2-Branes and Their Gravity Duals*, *JHEP* **10** (2008) 091, [arXiv:0806.1218].

[2] J. A. Minahan and K. Zarembo, *The Bethe Ansatz for Superconformal Chern-Simons*, *JHEP* **09** (2008) 040, [arXiv:0806.3951].

[3] D. Gaiotto, S. Giombi, and X. Yin, *Spin Chains in $\mathcal{N}$= 6 Superconformal Chern-Simons-Matter Theory*, *JHEP* **04** (2009) 066, [arXiv:0806.4589].

[4] N. Gromov and P. Vieira, *The All Loop AdS4/CFT3 Bethe Ansatz*, *JHEP* **01** (2009) 016, [arXiv:0807.0777].

[5] D. Bak and S.-J. Rey, *Integrable Spin Chain in Superconformal Chern-Simons Theory*, *JHEP* **10** (2008) 053, [arXiv:0807.2063].

[6] O. Aharony, O. Bergman, and D. L. Jafferis, *Fractional M2-Branes*, *JHEP* **11** (2008) 043, [arXiv:0807.4924].

[1] We thank N. Nekrasov for suggesting this possibility to us.

[7] D. Bak, D. Gang, and S.-J. Rey, *Integrable Spin Chain of Superconformal U(M)Xu(N) Chern- Simons Theory*, *JHEP* **10** (2008) 038, [arXiv:0808.0170].

[8] J. A. Minahan, W. Schulgin, and K. Zarembo, *Two Loop Integrability for Chern-Simons Theories with* $\mathcal{N} = 6$ *Supersymmetry*, *JHEP* **03** (2009) 057, [arXiv:0901.1142].

[9] O. Aharony, A. Hashimoto, S. Hirano, and P. Ouyang, *D-Brane Charges in Gravitational Duals of 2+1 Dimensional Gauge Theories and Duality Cascades*, arXiv:0906.2390.

[10] G. Arutyunov and S. Frolov, *Superstrings on* $AdS_4 \times Cp^3$ *as a Coset Sigma-Model*, *JHEP* **09** (2008) 129, [arXiv:0806.4940].

[11] B. Stefanski, jr., *Green-Schwarz Action for Type IIA Strings on* $AdS_4 \times Cp^3$, *Nucl. Phys.* **B808** (2009) 80–87, [arXiv:0806.4948].

[12] J. A. Minahan, O. Ohlsson Sax, and C. Sieg, *Magnon Dispersion to Four Loops in the Abjm and Abj Models*, arXiv:0908.2463.

[13] J. A. Minahan, O. Ohlsson Sax, and C. Sieg, *Anomalous Dimensions at Four Loops in* $\mathcal{N} = 6$ *Superconformal Chern-Simons Theories*, arXiv:0912.3460.

[14] S. Minwalla, *Restrictions Imposed by Superconformal Invariance on Quantum Field Theories*, *Adv. Theor. Math. Phys.* **2** (1998) 781–846, [hep-th/9712074].

On the Physics of the Riemann Zeros

Yang-Hui He[1], Vishnu Jejjala[2] and Djordje Minic[3]

[1]Rudolf Peierls Centre for Theoretical Physics, Oxford University, 1 Keble Road, OX1 3NP, U.K., Merton College, Oxford, OX1 4JD, U.K., Department of Mathematics, City University, Northampton Square, London EC1V 0HB, U.K., and School of Physics, NanKai University, Tianjin, 300071, P.R. China , [2]Centre for Research in String Theory, Department of Physics, Queen Mary, University of London, Mile End Road, London E1 4NS, U.K., [3]IPNAS, Department of Physics, Virginia Tech, Blacksburg, VA 24061, U.S.A.

E-mail: [1]yang-hui.he@merton.ox.ac.uk,[2]v.jejjala@qmul.ac.uk,[3]dminic@vt.edu

Abstract. We discuss a formal derivation of an integral expression for the Li coefficients associated with the Riemann ξ-function which, in particular, indicates that their positivity criterion is obeyed, whereby entailing the criticality of the non-trivial zeros. We conjecture the validity of this and related expressions without the need for the Riemann Hypothesis and discuss a physical interpretation of this result within the Hilbert–Pólya approach. In this context we also outline a relation between string theory and the Riemann Hypothesis.

1. Introduction: Riemann Zeros and the Li Criterion

This talk is devoted to the physics of the non-trivial zeros of the Riemann ξ function, the so-called Riemann zeros [1]. First, we review Li's Criterion, with an emphasis on requisite properties of the Riemann ξ-function. In the central section of the talk, we show how using the exact formula for the number of non-trivial zeros $N(T)$ up to a given height in the critical strip, and its derivative as eigenvalue distributions in the critical strip, would imply the positivity of Li's coefficients. Finally, we interpret this result from a physicist's perspective emphasizing a possibly exciting connection to string theory [2].

Let us, perhaps more for the sake of notation, first remind the reader of the statement of the Riemann Hypothesis [3–5]: The analytic continuation, from $z \in \mathbb{R}_{>1}$ to the whole complex plane $z \in \mathbb{C}$ of $\zeta(z) := \sum\limits_{n=1}^{\infty} n^{-z}$ has all its zeros in the critical strip $\mathcal{S} := \{z \mid 0 < \mathrm{Re}(z) < 1\}$ lying on the critical line $\mathcal{L} := \{z \mid \mathrm{Re}(z) = \frac{1}{2}\}$. Now, $\zeta(z)$ obeys a remarkable functional equation $\zeta(z) = 2^z \pi^{z-1} \sin(\frac{1}{2}\pi z)\Gamma(1-z)\zeta(1-z)$, which inspired Riemann to define the ξ-function, which will be central to our discussions:

$$\xi(z) := \frac{1}{2} z(z-1)\pi^{-\frac{z}{2}}\Gamma(\frac{z}{2})\zeta(z) \ . \tag{1.1}$$

There are many advantages to considering $\xi(z)$ instead of $\zeta(z)$, which we now summarize.

Journal of Physics: Conference Series **462** (2013) 012036 doi:10.1088/1742-6596/462/1/012036

1.1. The Riemann ξ-Function

First, note that $\zeta(z)$ has trivial zeros at all negative even integers, which are conveniently cancelled by the corresponding simple poles of the Γ-function. Hence, $\xi(z)$ has only non-trivial zeros, all located within the critical strip, and affords an elegant Weierstraß product expansion, known as the *Hadamard product*:

$$\xi(z) = \xi(0) \prod_{\rho \in \mathcal{S}} (1 - \frac{z}{\rho}) \ , \tag{1.2}$$

where $\xi(0) = \frac{1}{2}$, and ρ are the non-trivial zeros of $\zeta(z)$ in $\mathcal{S}$ since none of the other factors in the definition of $\xi(z)$ vanishes therein. It is an obvious but important fact that $\mathcal{S}$ extends to both the upper and the lower half-planes and that all the ρ occur in *conjugate pairs* above and below the real line. We shall denote $\mathcal{S}_+$ as the critical strip above the real line and correspondingly $\mathcal{L}_+$ as the upper critical line.

Second, the functional equation becomes particularly simple:

$$\xi(z) = \xi(1 - z) \ . \tag{1.3}$$

Indeed, the symmetry about the critical line of $\mathrm{Re}(z) = \frac{1}{2}$ becomes manifest. Hence, not only are the zeros symmetric about the real axis, they are also symmetric about the critical line. The Riemann Hypothesis postulates that all the zeros in fact lie on $\mathcal{L}$. It is known that the first 10^{13} zeros lie on the critical line [6].

Third, consider the conformal mapping $z \mapsto \frac{z}{z-1}$, which maps $\mathcal{L}_+$ to the boundary circle $|z| = 1$ of the unit disk and the entirety of the left half of $\mathcal{S}_+$, *i.e.*, $\{z \mid 0 < \mathrm{Re}(z) < \frac{1}{2}\}$, to the interior of the open unit disk. Consider, therefore, the function

$$\phi(z) := \xi(\frac{z}{z-1}) = \xi(\frac{1}{1-z}) \ , \tag{1.4}$$

where we have used (1.3) in the second equality. Since all zeros are symmetric around the critical line, to consider the zeros of ξ within $\mathcal{S}$ therefore amounts to considering the zeros of $\phi(z)$ within the open unit disk. Indeed, the Riemann Hypothesis would require that there be no such zeros and indeed that all the critical zeros are on the boundary circle. Since the number of zeros of an analytic function $f(z)$ in a region R is equal to $\int_R \mathrm{d}z \, \frac{f'(z)}{f(z)}$, we have that: The Riemann Hypothesis is equivalent to $\frac{\phi'(z)}{\phi(z)}$ being analytic (*i.e.*, holomorphic and without poles) within the open unit disk $|z| < 1$.

Fourth, a beautiful exact result relates the number of zeros up to height T within $\mathcal{S}$, and the ξ-function:

$$N(T) := \#\{\rho \in \mathcal{S} \ , 0 < \mathrm{Im}(\rho) < T | \ \xi(\rho) = 0\} = \frac{1}{\pi} \mathrm{Im} \log \xi(\frac{1}{2} + iT) \ . \tag{1.5}$$

Extensive work has been done in estimating (1.5). We have, using the definition (1.1) for the first equality, that

$$N(T) = \frac{1}{\pi}\vartheta(T) + 1 + \frac{1}{\pi} \mathrm{Im} \log \zeta(\frac{1}{2} + iT) \sim \frac{T}{2\pi} \log \frac{T}{2\pi} - \frac{T}{2\pi} - \frac{7}{8} + \mathcal{O}(\log T) \ , \tag{1.6}$$

where historically $\vartheta(T) = \mathrm{Im} \log \Gamma(\frac{i}{2}T + \frac{1}{4}) - \frac{T}{2} \log \pi$ is known as the average part and $\frac{1}{\pi} \mathrm{Im} \log \zeta(\frac{1}{2} + iT) \sim \mathcal{O}(\log T)$ is the fluctuating part around the essentially $T \log T$ growth of the former. This is the inspiration behind classical (the average) and quantum mechanical (the fluctuation) interpretations of the critical zeros [7, 8].

We note that $N(T)$ is a real step function, increasing by unity each time a new critical zero is encountered:

$$N(T) = \sum_{\rho \in \mathcal{S}_+} \theta(T - \operatorname{Im} \rho) = \sum_{j=1}^{\infty} \alpha_j \, \theta(T - \mu_j) \; . \tag{1.7}$$

It is vital to explain the above in detail. We can explicitly write the upper critical zeros as $\rho_{j,k} = \sigma_{j,k} + i\mu_j$, indexed by $j, k \in \mathbb{Z}_{>0}$, where $\sigma_{j,k} \in (0,1)$ and $\mu_j \in \mathbb{R}_{>0}$. The zeros are ordered so that $\mu_{j+1} > \mu_j$. Crucially, as we do not assume the Riemann Hypothesis, the real part of ρ need not be $\frac{1}{2}$. Furthermore, we do not assume simplicity of the zeros, which is as yet also unknown [9]. If, for example, we have a double root, we explicitly count this twice. By the functional identity, zeros of the ξ-function not on the critical line are paired within the upper critical strip: if $\sigma + i\mu$ is a zero, then so is $(1 - \sigma) + i\mu$. The index $k = 1, \ldots, \alpha_j$ enumerates the zeros with the same imaginary part. We order the zeros so that $\sigma_{j,k+1} \geq \sigma_{j,k}$. The α_j then counts the number of zeros with imaginary part μ_j including the multiplicities of the roots. If, for example, there are a pair of simple roots with imaginary part μ_j, then in an epsilon interval around μ_j, the counting function $N(T)$ jumps by two. Contrariwise, if $\rho \in \mathcal{L}_+$ is the only simple root with imaginary part μ_j, then $N(T)$ jumps by one. We know that the number of roots with imaginary part in the interval $(0, T)$ is finite. Indeed, the asymptotics of the expression are given in (1.6). In summary, $N(T)$ is a strictly increasing step function as we move up in height T regardless of the Riemann Hypothesis. The Riemann Hypothesis is, of course, the statement that $\sigma_j = \frac{1}{2}$ for all zeros.

The rewriting of $N(T)$ lends itself to a wonderful interpretation: $N(T)$ is a cumulative density function defined over the critical strip. In other words, its derivative, $\widetilde{\rho}(T) := N'(T)$ is a density function of distributions. That is, one could conceive of a physical system whose energy levels (eigenvalues of the Hamiltonian) are thus distributed; this is along the school of thought of Hilbert–Pólya [7]. Of course, the resulting eigenvalue density is highly non-smooth, but is, rather, a sum of delta-functions:

$$\widetilde{\rho}(T) := N'(T) = \sum_{j=1}^{\infty} \alpha_j \, \delta(T - \mu_j) \; . \tag{1.8}$$

1.2. Li's Criterion

Another striking property of the ξ-function was noted by Li [10] not too long ago. Let $\{k_n\}$, for positive integers n, be defined by

$$k_n := \frac{1}{(n-1)!} \frac{\mathrm{d}^n}{\mathrm{d}z^n} \left[z^{n-1} \log \xi(z) \right]_{z=1} \; ; \tag{1.9}$$

then we have that: The Riemann Hypothesis is equivalent to the condition that $k_n \geq 0$ for all $n \in \mathbb{Z}_{>0}$. This is Li's Criterion. Li also showed two equivalent ways of writing these numbers, namely

$$\frac{\phi'(z)}{\phi(z)} = \sum_{n=0}^{\infty} k_{n+1} z^n \; ; \tag{1.10}$$

$$k_n = \sum_{\rho \in \mathcal{S}} \left[1 - \left(1 - \frac{1}{\rho} \right)^n \right] \; , \tag{1.11}$$

where we recall that $\phi(z) := \xi(\frac{1}{1-z})$.

2. Density Function for the Distribution of Critical Zeros

Thus armed with all necessary ingredients, let us proceed to rewrite the Li coefficients in a suggestive form. We will do so *formally* and point out the subtleties involved at the end of the manipulations [1].

First, integrating (1.10) and using the definition as well as the functional equation (1.4), we have that, for z in the unit disk,

$$\log \xi(\frac{z}{z-1}) = \log \xi(\frac{1}{1-z}) = \sum_{n=1}^{\infty} k_n \int \mathrm{d}z\ z^{n-1} = -\log 2 + \sum_{n=1}^{\infty} \frac{k_n}{n} z^n \ , \tag{2.1}$$

where $-\log 2$ is easily checked to be the constant of integration.

We note, however, that the above expansion has radius of convergence strictly less than unity, whereby excluding the unit circle, to which, crucially, $\mathcal{L}$ is mapped under our conformal transformation $z \mapsto \frac{z}{z-1}$. Therefore, it is imperative to analytically continue. Let us start with (2.1) and rewrite the expansion about the point $z = -1$. Note that the region of convergence for this is an open of circle of radius two centered about $z = -1$, which, in particular, encloses the entirety of the closed unit disk, except the point $z = 1$, where there is a pole for $\phi(z) = \xi(\frac{1}{1-z})$. Therefore, we have that

$$\log \xi(\frac{1}{1-z}) = \log \frac{1}{2} + \sum_{n=1}^{\infty} \frac{k_n}{n} z^n = b_0 + \sum_{n=1}^{\infty} b_n (z+1)^n \ , \tag{2.2}$$

where we have written the two expansions together for comparison.

Expanding $(z+1)^n$, we readily obtain an expression for the Li coefficients in terms of the new expansion coefficients b_n (studied further in [11]), $k_n = n \sum_{j=n}^{\infty} \binom{j}{n} b_j$. We now wish to solve for the k_n explicitly and demonstrate positivity.

We first see, using the counting formula (1.5) and the expansion (2.2), that

$$N(\mu) = \frac{1}{\pi} \operatorname{Im} \log \xi(\frac{1}{2} + i\mu) = \sum_{n=1}^{\infty} \frac{b_n}{\pi} \operatorname{Im} \left(\frac{2\mu + i}{2\mu - i} + 1 \right)^n \ , \tag{2.3}$$

where we have used the substitution $\frac{1}{2} + i\mu = \frac{1}{1-z}$, or $z = \frac{2\mu+i}{2\mu-i}$. Then, since

$$\operatorname{Im} \left(\frac{2\mu + i}{2\mu - i} + 1 \right)^n = \frac{(4\mu)^n}{(4\mu^2+1)^{\frac{n}{2}}} \sin(n \tan^{-1} \frac{1}{2\mu}) = 2^n \cos^n \theta \ \sin(n\theta) \ , \qquad \cos\theta := \frac{2\mu}{\sqrt{4\mu^2+1}} \ , \tag{2.4}$$

and due to the definition of the Chebyshev polynomial of the second kind

$$U_{m-1}(\cos\phi) := \frac{\sin(m\phi)}{\sin\phi} \ , \tag{2.5}$$

we arrive at the conclusion that for all $m \in \mathbb{Z}_{>0}$:

$$k_m = 32m \int_0^{\infty} \mathrm{d}\mu \ \frac{\mu}{(4\mu^2+1)^2} \ N(\mu) \ U_{m-1}(\frac{4\mu^2-1}{4\mu^2+1}) \ . \tag{2.6}$$

This is a beautiful and remarkable integral expression for the Li coefficients which works wonderfully numerically [1].

Let us now proceed to simplify the integral (2.6). First, recall that Chebyshev polynomials of the second kind are related to the Chebyshev polynomials of the first kind by the equation

$$\int \mathrm{d}x \; U_n(x) = \frac{1}{n+1} T_{n+1}(x) \; . \tag{2.7}$$

Then using the integral expression for k_m we have that [1]

$$k_m = 2\sum_{j=1}^{\infty} \alpha_j \left(1 - T_m\left(\frac{4\mu_j^2 - 1}{4\mu_j^2 + 1}\right)\right) , \quad m \in \mathbb{Z}_{>0} \; . \tag{2.8}$$

Recalling the definition of the Chebyshev polynomial of the first kind, that $T_m(\cos\theta) := \cos(m\theta)$ and remembering that all α_j, regardless of actual value, are strictly positive integers, each summand in (2.8) is non-negative. In fact, because $\frac{4\mu_j^2-1}{4\mu_j^2+1} < 1$ strictly for any imaginary part as we know the first μ_j is at least 14,

$$T_m\left(\frac{4\mu_j^2 - 1}{4\mu_j^2 + 1}\right) < 1 \qquad \text{for any given } m \in \mathbb{Z}_{>0} \text{ and for all } j \in \mathbb{Z}_{>0} \; , \tag{2.9}$$

so each summand is actually strictly positive. In conclusion, this would imply that all Li's coefficients $k_m > 0$. Indeed, we have numerically evaluated [1] the first few k_m coefficients given in (2.8), using the first 10^4 critical zeros, and see that they are very close to the results obtained from (1.10) and (2.6), and moreover constitute an increasing sequence of strictly positive numbers. In particular, the above discussions would force all critical zeros to lie entirely on $\mathcal{L}$.

We seem to have arrived the positivity of the Li's coefficients without the assumption of the Riemann Hypothesis! This, of course, is not quite true and the criticality of the zeros of the ξ-function had in fact been inherently invoked in the choice of a branch cut for the logarithm. Nevertheless, we have *formally* arrived the integral formula (2.6) which does not seem to depend on the precise location of the zeros and the subsequent choice of branch-cut. Note that Li's criterion can be applied to other zeta functions [10] and thus it is natural to conjecture that our integral and summation formulas for the Li coefficients should exist in those cases as well.

3. The Physics of the Critical Zeros and String Theory

Attempts to use a physical argument to explain the Riemann Hypothesis date back to Hilbert and Pólya [7], [8]. Were one to find a matrix

$$H = \frac{1}{2} \cdot \mathbb{I} + i \cdot \mathcal{T} \; , \tag{3.1}$$

whose eigenvalues are all the zeros of $\xi(z)$ and if in addition the matrix $\mathcal{T}$ were Hermitian, then all the zeros would lie on the critical line $\mathcal{L}$. Therefore, one would wish to find the appropriate operator H and its corresponding eigenfunctions, or, in physical terms, the Hamiltonian and its wave functions.

What can one say about the quantum mechanical picture [1]? Let us investigate the secular equation of the operator in (3.1). It should by construction have roots corresponding to the zeros of $\xi(z)$, that is,

$$\xi(z) = \det(z \cdot \mathbb{I} - H) \ . \tag{3.2}$$

To arrive at the appropriate matrix description, we start by considering the expansion (2.1), which, combined with (2.8), gives

$$\log \xi(\frac{1}{1-s}) = -\log 2 + \sum_{n=1}^{\infty}\sum_{j=1}^{\infty} \frac{2\alpha_j}{n}\left[1 - T_n(\frac{4\mu_j^2-1}{4\mu_j^2+1})\right] s^n \ . \tag{3.3}$$

Put $\cos\theta_j := \frac{4\mu_j^2-1}{4\mu_j^2+1}$. Using the definition of the Chebyshev polynomial and reversing the sums, we obtain

$$\log \xi(\frac{1}{1-s}) = -\log 2 + \sum_{j=1}^{\infty} 2\alpha_j \sum_{n=1}^{\infty}\left(\frac{1}{n} - \frac{1}{n}\cos(n\theta_j)\right) s^n. \tag{3.4}$$

Substituting $z = \frac{1}{1-s}$ and using that $e^{\pm i\theta_j} = 1 - (\frac{1}{2} \pm i\mu_j)^{-1}$, we find

$$\xi(z) = \frac{1}{2}\prod_{j=1}^{\infty}\left(\frac{(z-(\frac{1}{2}+i\mu_j))(z-(\frac{1}{2}-i\mu_j))}{(\frac{1}{2}+i\mu_j)(\frac{1}{2}-i\mu_j)}\right)^{\alpha_j} , \tag{3.5}$$

which is simply a refined version of the Hadamard product, a reassuring check. Note that each factor in the product obeys the functional identity.

Equation (3.5) is to be compared with (3.2). We see that there exists a matrix such that its eigenvalues coincide with μ_j, which is consistent with the Hilbert–Pólya picture. As the expression vanishes for zeros on the critical line, the refined Hadamard product is already in the form of the determinant of an infinite matrix, and the eigenvalues of $\mathcal{T}$ are real.

3.1. Topological string theory, Gromov-Witten invariants and the critical zeros

The profound proofs of the Riemann Hypothesis for the case of finite fields by Weil, Deligne and others are essentially geometric, and thus one might expect a deeply geometric resolution of the Riemann Hypothesis [12]. Motivated by this expectation in this concluding section we present a possibly exciting connection between Gromov-Witten (GW) invariants, topological string theory and the Riemann zeros [2].

First we recall a reformulation of the Riemann Hypothesis in terms of an elementary problem due to Lagarias [13]: Let $H_n = \sum_{j=1}^{n}\frac{1}{j}$ be the n-the harmonic number and let $\sigma_1 = \sum_{d|n} d$ is the sum of divisors (Ramanujan) function, then the Riemann Hypothesis is true if and only if

$$\sigma_1(n) < H_n + e^{H_n}\log H_n \tag{3.6}$$

for $n > 1$. In the following we notice the appearance of $\sigma_1(n)$ in the genus-one free energy of a particular topological string theory [2].

We start with the following statement from [14]: Take the type II string on the local Calabi-Yau threefold defined by

$$uv = H(y,z) = y^2 - P_n(z)^2 + 1 \ , \tag{3.7}$$

where $H(y,z)=0$ defines a Riemann surface and $P_n(z)$ is a degree n polynomial. The genus one amplitude can be given exactly (due to Seiberg-Witten (SW) theory) by

$$F_1(\tau) = -\log \eta(\tau) \; ; \tag{3.8}$$

this is also the genus-one free energy of the topological string (i.e. topologically twisted from the $\mathcal{N}=2$ SW theory). From the definition of the Dedekind eta-function, we can expand, setting $q=\exp(2\pi i\tau)$,

$$\begin{aligned} F_1(\tau) &= -\log\eta(\tau) = -\frac{1}{24}\log q - \sum_{i=1}^{\infty}\log(1-q^i) = -\frac{\pi i\tau}{12} - \sum_{i=1}^{\infty}\sum_{j=1}^{\infty}\frac{q^{ij}}{j} \\ &= -\frac{\pi i\tau}{12} - \sum_{i=1}^{\infty}\sum_{j=1}^{\infty}\frac{q^{ij}}{j} = -\frac{\pi i\tau}{12} - \sum_{n=1}^{\infty}(\sum_{d|n} d)q^n = -\frac{\pi i\tau}{12} - \sum_{n=1}^{\infty}\sigma_1(n)q^n \; . \end{aligned} \tag{3.9}$$

Thus, the Gromov-Witten invariants at genus $g=1$ and number of marked points $m=0$, for the above Calabi-Yau X is the Ramanujan σ_1 function! That is:

$$N_{g=1,m=0}(n;X) = \sigma_1(n) \; . \tag{3.10}$$

Then, the statement, due to Lagarias' theorem [13], is that the asymptotics of the free energy of this particular topological string at genus-one world-sheet embedding should be $\sigma_1(n) < H_n + e^{H_n}\log H_n$, if and only if Riemann Hypothesis holds. Therefore: (1), we know that the Lagarias bound on $\sigma_1(n)$ in terms of the harmonic numbers is equivalent to Riemann Hypothesis and (2), we know of examples of Gromov-Witten invariants which are equal to $\sigma_1(n)$. Hence the question: does there exist an independent geometric bound on such invariants? (At this point one might wonder whether the Gromov-Witten invariants could encode the critical zeros. This seems unnatural because GW invariants are integers indexed by integers whilst the critical zeros are not, whence it is hard to imagine, though no entirely impossible, that a combinatorial problem should be associated thereto.)

Also, by considering rational curves of degree d on the K3 surface we can deduce that [2] $\sum_{n=0}^{k}\tau(n)N_1(k-n) \leq k\sigma_1(n)$, for each $k=0,1,2,\ldots$, and thus restate the Riemann Hypothesis in the following form [2]:

$$\sum_{n=0}^{k}\tau(n)N_1(k-n) \leq k(H_k + e^{H_k}\log H_k), \quad k=0,1,2,\ldots \tag{3.11}$$

where $\tau(n)$ is the famous Ramanujan multiplicative tau-function (for which, $\tau(0)=0$), and $N_1(j)$ is the number of elliptic curves with j nodes and passing through a given point on a generic, smooth K3 surface. Note that according to the reference [15] there exists a universal constant a such that

$$Rat(X) \leq \chi(X) + a \; , \tag{3.12}$$

for every generic smooth surface X, with $\chi(X)$ being the Euler number. We need a version of this statement for elliptic curves.

The papers on the so-called Calabi-Yau differential equations have studied the integer coefficients of mirror maps, which we know to be an interesting phenomenon in and of itself.

(In particular, the mirror symmetry might be related to the Birch and Swinnerton-Dyer conjecture [16].) The reference [17] is particularly interesting. We know that, after judicious transformation, the coefficients of the mirror map are precisely the GW invariants. In [17], an expression of the coefficients in terms of harmonic numbers is given. The strategy is clear. There appear to exist two directions: one giving the GW invariants in terms of $\sigma_1(n)$ and the other, the GW invariants via the mirror map, in terms of the harmonic numbers. One thus must relate the two and the question is whether the relation is strong enough to give the Lagarias bound.

Acknowledgments

We are indebted to Michael Berry, Hung Bui, Philip Candelas, Mark Coffey, Roger Heath-Brown, Jon Keating, Eric Sharpe, Al Shapere and Andrew Wiles for interesting conversations and questions regarding the material presented in this talk. YHH is supported in part by an Advanced Fellowship from the STFC, UK in association with the Rudolf Peierls Centre for Theoretical Physics, University of Oxford, a Supernumerary Fellowship of Merton College, Oxford, and impending funding from City University, London and a Chang-Jiang Chair Professorship from the Chinese Ministry of Education, at Nan-Kai University, Tian-Jin. VJ is supported by STFC. DM is supported in part by the U.S. Department of Energy under contract DE-FG05-92ER40677. DM thanks the organizers of the QTS6 conference for the opportunity to present this work. VJ and DM also wish to thank the magnanimity of the Warden, Fellows, and Scholars of Merton College, Oxford University.

References

[1] Y.-H. He, V. Jejjala and D. Minic, "Eigenvalue Density, Li's Positivity, and the Critical Strip," arXiv:0903.4321[math-ph] and references therein.
[2] Y.-H. He, V. Jejjala and D. Minic, work in progress.
[3] B. Riemann, "Ueber die Anzahl der Primzahlen unter einer gegebenen Größe," Monat. der Königl. Preuss. Akad. der Wissen. zu Berlin aus der Jahre **1859**, 671 (1860).
[4] H. M. Edwards, *Riemann's zeta function*, New York: Academic Press (1974).
[5] E. C. Titchmarsh, *The theory of the Riemann zeta function*, Oxford: Clarendon Press (1986).
[6] X. Gourdon, "The 10^{13} first zeros of the Riemann zeta function, and zeros computation at very large height" (2004), `http://numbers.computation.free.fr/Constants/Miscellaneous/zetazeros1e13-1e24.pdf`.
[7] D. Hilbert, unpublished (c. 1914); G. Pólya, unpublished (c. 1914).
[8] M. V. Berry and J. Keating, "H = xp and the Riemann zeros" in *Supersymmetry and trace formulae: chaos and disorder*, I. V. Lerner and J. P. Keating, eds., New York: Plenum (1999). Also, "The Riemann zeros and eigenvalue asymptotics," SIAM Review **41**, 236 (1999) and references therein.
[9] J. B. Conrey, A. Ghosh, and S. M. Gonek, "Simple zeros of the Riemann zeta function," Proc. London Math. Soc. **76**, 497 (1998).
[10] X.-J. Li, "The positivity of a sequence of numbers and the Riemann hypothesis," J. Numb. Th. **65**, 325 (1997).
[11] M. W. Coffey, "On certain sums over the nontrivial zeta zeros," arXiv:0912.2390.
[12] E. Bombieri, "Problems of the millennium: The Riemann hypothesis," Clay Mathematics Institute (2000). `http://www.claymath.org/millennium/Riemann_Hypothesis/Official_Problem_Description.pdf`.
[13] J. C. Lagarias, ""An Elementary Problem Equivalent to the Riemann Hypothesis," Amer. Math. Monthly **109**, 534 (2002) and references therein.
[14] M. Aganagic, V. Bouchard and A. Klemm, "Topological Strings and (Almost) Modular Forms," Commun. Math. Phys. **277**, 771 (2008) and references therein.
[15] R. Blache, "Explicit bounds for rational curves on Kähler surfaces of non-negative Kodaira dimension," Proc. London Math. Soc. **74**, 81 (1997)., Conjecture A.
[16] Y.-H. He, "On Fields over Fields", arXiv:1003.2986 and references therein.
[17] C. Krattenthaler and T. Rivoal, "On the integrality of the Taylor coefficients of mirror maps, II," Commun. Number Theory Phys. **3**, 555 (2009) and references therein.

Invariant Cones in Lie Algebras and Positive Energy Representations and Contractions of Conformal Algebras

Patrick Moylan

Department of Physics, Pennsylvania State University, The Abington College, Abington, PA 19001 USA

E-mail: `pjm11@psu.edu`

Abstract.

We recall some important results, due to Kostant and others, about invariant convex cones in Lie algebras and positive energy representations. We apply these results to a study of positive energy representation of the conformal groups in n dimensions, and we present a proof of the converse of a theorem attributed to I.E. Segal, which relates positive energy representations to positivity of the action of the generator of time translations for representations of the n-dimensional conformal group.

We also discuss related notions of deformation and contractions of Lie algebras and describe a deformation of the Poincaré subalgebra of the conformal algebra which generalizes the usual treatment. We consider the positive energy representations of the anti-deSitter subalgebras in the physically important four dimensional case, and apply this generalization to argue that the singelton representations cannot have nontrivial contractions to representations of the Poincaré algebra. We believe that our results represent a sharpening of the meaning of "kinematical confinement", introduced by Flato, Fronsdal and their coworkers.

1. Invariant Convex Cones in Simple Lie Algebras

Let V be a finite dimensional vector space over $\mathbb{R}$. A *cone* in V is a closed convex subset stable under scalar multiplication by nonnegative real numbers i.e. C is a cone if C is closed and convex and such that

$$tC \subset C \;\forall\; t \geq 0\,.$$

By convexity we mean [1]

$$\lambda\,\mathbf{v} + \mu\,\mathbf{w} \in C \text{ if } \mathbf{v}, \mathbf{w} \in C$$

where $\lambda \geq 0$, $\mu \geq 0, \lambda + \mu = 1$. A cone C in a Lie algebra $\mathcal{G}$ is called *invariant* if it is invariant relative to the adjoint group i.e. if

$$\exp(\mathrm{ad}X)C \subset C\,\forall X \in \mathcal{G}\,.$$

A causal cone is one for which $C \cap -C = 0$.

The classification of invariant cones in Lie algebras was initiated by Kostant and Vinberg, and the classification of invariant cones in simple Lie algebras was accomplished by S.M. Paneitz in 1980 as part of his Ph.D. thesis and then independently by G. I. Olshanskii [3]. Classification schemes for arbitrary Lie algebras are reported in [4]. A main theorem in the subject is the following theorem of Kostant [2].

Theorem 1.1. (Kostant)

Let G be connected semisimple Lie group acting in a real finite dimensional vector space V. Let K be a maximal compact subgroup of G. Then there exists a a closed G-invariant convex cone C in V satisfying $C \cap -C = 0$ *if and only if V has a nonzero K-invariant vector which lies in* $\mathscr{K}$, *the Lie algebra of K.*

For simple Lie algebras there is a corollary of this theorem, namely: *there exists a non-trivial casual cone in* $\mathscr{G} \iff$ *G/K is Hermitian symmetric* [2]. A list of hermitian symmetric algebras may be found in [5]. It suffices for us to note that they include $so(2,n)$ $(n > 3)$, $sp(n,\mathbb{R})$ and $su(p,q)$ $(p \geq q \geq 1)$. Let $\mathscr{C}$ be the (one-dimensional) center of $\mathscr{K}$. Choose $Z \in \mathscr{C}$ suitably nomalized. Note that for $so(2,n)$, $Z = X_0$ is the Einstein energy [6].

We let C_{Min} be the closed cone generated by Ad(G)Z. It is a theorem that any (invariant) cone in $\mathscr{G}$ contains C_{Min}. Self-dual cones are those for which $C_{Min} = C_{Max} = (C_{Min})^*$. If $\mathscr{G} \cong sl(n,\mathbb{R})$, then C_{Min} is self-dual [2]. Since $so(2,3) \cong sp(2,\mathbb{R})$ [5], we have for $\mathscr{G} = so(2,3)$ the existence of a minimal, invariant self-dual cone in $\mathscr{G}$.

2. Lie Algebra Deformations and Contractions

Given V, a finite dimensional vector space of dimension n over $\mathbb{R}$, let $\mathscr{M}_n$ be the space of all Lie structures on V i.e. the space of all equivalence classes of bilinear mappings

$$\mu : V \times V \to V \quad \Big([x,\, y] = \mu(x,\, y)\Big)$$

such that $\mu(x,\, y) = -\mu(y,\, x)$ and

$$\sum_{\mathrm{cycl}\{x,y,z\}} \mu(x,\, \mu(y,\, z)) = 0 \text{ (Jacobi identity)}$$

where two such maps are equivalent if they give isomorphic Lie algebras.

Let $\Lambda \in \mathbb{R}^+$, then a deformation of a given Lie algebra $\mathscr{G}$ over $\mathbb{R}$ is a continuous mapping $\psi : [0,\Lambda] \longrightarrow \mathscr{M}_n$ with $\mathscr{G} = (V, \psi(0))$ i.e. $\psi(0)$ is the Lie structure of $\mathscr{G}$, where continuous means continuous in the topology $\mathscr{M}_n$ inherits from the inclusion of the space of structure constants in $\mathbb{R}^{n^3}$ (i.e. in the *Segal* topology)[7], [8]. ψ is a trivial deformation if $\psi(t)$ for all $t \in [0,\, \Lambda]$ is equivalent to $\psi(0)$.

The process of deformation of a Lie algebra may be viewed, at least in special circumstances, as the inverse of contraction of a Lie algebra, which idea goes back to Segal [9] and Inönü and Wigner [10] and formalized by Saletan [11]. We now turn to this notion of contraction of a Lie algebra. Let $\mathscr{G} = (V, [\cdot,\cdot])$ be an n-dimensional Lie algebra over $\mathbb{R}$ with an underlying n-dimensional vector space V over $\mathbb{R}$ and a Lie bracket $[\cdot,\cdot]$. Consider a continuous family $(\phi_\lambda)_{\lambda\in\mathbb{R}^+}$ of surjective mappings $\phi_\lambda \in Hom(V,W)$ where W is another vector space of the same dimension as V. If the ϕ_λ are injective, we may define a new bracket on V as follows:

$$[x,y]_\lambda = \phi_\lambda^{-1}[\phi_\lambda(x), \phi_\lambda(y)] \;\; \forall\, \lambda \in \mathbb{R}^+ ,\;\; \forall\; x,\, y \in V. \tag{1}$$

Set $\mathscr{G}_\lambda = (V, [\cdot,\cdot]_\lambda)$ and assume that $\mathscr{G}_1 = \mathscr{G}$, so that ϕ_1 defines an isomorphism of $\mathscr{G}$ onto its image. If

$$\lim_{\lambda\to 0} [x,\, y]_\lambda := [x,\, y]_0 \tag{2}$$

exists for all $x,\, y \in V$, then it defines a (possibly) new Lie algebra and we call this new Lie algebra, $\mathscr{G}_0 = (V,\, [\cdot,\cdot]_0)$, a contraction of the Lie algebra $\mathscr{G}$ or (simply) the *contracted of* $\mathscr{G}$. For the case when

$V = W$ and so $\phi_\lambda \in GL(V)$ our definition reduces to the usual definition given in the literature c.f. [8], [12], [13].

We now make use of the family of mappings $(\phi_\lambda)_{\lambda\in\mathbb{R}^+}$ to define contractions of representations of Lie algebras [14], [15]. We start with a given Lie algebra $\mathscr{G} = (W, [\cdot,\cdot])$ and an infinitesmally unitarizable representation $(d\pi(\mathscr{G}),\mathscr{H})$ of $\mathscr{G}$ defined on a fixed Hilbert space, $\mathscr{H}$. Given a continuous family $(\Pi_\lambda)_{\lambda\in\mathbb{R}^+}$ of closed invertible linear transformations of $\mathscr{H}$ with $\Pi_1 = \mathrm{Id}$ (Id = the identity on $\mathscr{H}$.) it is easy to see that the map

$$\mathscr{G}_\lambda = (V, [\cdot,\cdot]_\lambda) \ni X \to d\pi_\lambda(X) = \Pi_\lambda^{-1}\, d\pi(\phi_\lambda\,(X))\Pi_\lambda \tag{3}$$

defines a representation of $\mathscr{G}_\lambda$ on $\mathscr{H}$. In order to assure infinitesimal unitarizability of the contracted representation, the family $(\Pi_\lambda)_{\lambda\in\mathbb{R}^+}$ should be chosen so that $d\pi_\lambda(X)$ is skew-symmetric. To check the representation condition we have [15]:

$$[d\pi_\lambda(X), d\pi_\lambda(Y)] = \Pi_\lambda^{-1}[d\pi(\phi_\lambda(X)), d\pi_\lambda(\phi(Y))]\Pi_\lambda =$$

$$= \Pi_\lambda^{-1} d\pi([\phi_\lambda(X),\phi_\lambda(Y)])\Pi_\lambda = \Pi_\lambda^{-1} d\pi\,(\phi_\lambda([X,Y]_\lambda))\,\Pi_\lambda = d\pi_\lambda([X,Y]_\lambda) \tag{4}$$

where in the last line we have made use of eqn. (1). We define the representation of the contracted Lie algebra, $\mathscr{G}_0$, as $d\pi_0(X) = \lim_{\lambda\to 0} d\pi_\lambda(X)$ provided this limit exists, and we call it the *contracted of the representation* $(d\pi(\mathscr{G}),\mathscr{H})$.

3. Positive Energy Representations of the Conformal Group in *n* dimensions

Now we want to appy the above to obtain some physically useful results about representations of the conformal group in n dimensions and some of its subgroups. Consider the quadratic form $Q(x)$ defined on $\mathbb{R}^{n+2}$ by

$$Q(x) = x_{-1}^2 + x_0^2 - x_1^2 - x_2^2 - \ldots - x_n^2 \tag{5}$$

where $x = (x_{-1},x_0,x_1,\ldots,x_n) \in \mathbb{R}^{n+2}$. Thus $\mathbb{R}^{n+2}$ equipped with the metric defined by $Q(x)$ is $n+2$ dimensional Minkowski space. We denote n dimensional Minkowski space by M_0. Let $G = SO_0(2,n)$ denote the connected component of the group of linear transformations of $\mathbb{R}^{n+2}$ preserving the symmetric bilinear form which is associated to $Q(x)$ by polarization. We shall call G the n-dimensional conformal group, and we denote the universal cover of G by $G^\sim = SO_0(2,n)^\sim$. Let $\mathscr{G}$ be the Lie algebra of G. $\mathscr{G}$ is identified with the set of all matrices (a_{ij}) $(-1 \le i,j \le n)$ such that $a_{ii} = 0$ $(0 \le i \le n)$, $a_{ij} = -a_{ji}$ $(1 \le i \le j \le n)$, $a_{0,j} = a_{j,0}$ $(1 \le j \le n)$, $a_{-1,j} = a_{j,-1}$ $(1 \le j \le n)$ and $a_{-1,0} = -a_{0,-1}$. We define subalgebras $\mathscr{K}$, $\mathscr{G}'$, $\mathscr{A}$, $\mathscr{N}_+$ and $\mathscr{N}_-$ as follows. Let E_{ij} be the matrix such that the (i,j) component is equal to 1 and the other components are all equal to 0. Let $\mathbf{L}_{ij} = E_{ij} - E_{ji}$ $(1 \le i \le j \le n)$, $\mathbf{L}_{0i} = E_{i,o} + E_{o,i}$ $(1 \le i \le n)$, $\mathbf{L}_{-1,i} = E_{i,-1} + E_{-1,i}$ $(1 \le i \le n)$ and $\mathbf{L}_{-10} = E_{-1,0} - E_{0,-1}$. Let $\mathscr{K}$ be the subalgebra spanned by: $\mathbf{L}_{ij}$ $(1 \le i,j \le n)$ and $\mathbf{L}_{-1,0}$; $\mathscr{G}'$ be the subalgebra spanned by: $\mathbf{L}_{ij}$ $(1 \le i,j \le n-1)$, $\mathbf{L}_{0i}$ $(1 \le i \le n-1)$, $\mathbf{L}_{-1,i}$ $(1 \le i \le n-1)$ and $\mathbf{L}_{-1,0}$; $\mathscr{A}$ be the subalgebra spanned by $\mathbf{L}_{-1,n}$; and $\mathscr{N}_+$ ($\mathscr{N}_-$) be the subalgebra spanned by $\mathbf{P}_i = \frac{1}{2}(\mathbf{L}_{n,i} + \mathbf{L}_{-1,i})(0 \le i \le n-1)$ ($\tilde{\mathbf{P}}_i = \frac{1}{2}(\mathbf{L}_{n,i} - \mathbf{L}_{-1,i})(0 \le i \le n-1)$). Denote the analytic subgroups of G corresponding to $\mathscr{K}$, $\mathscr{G}'$, $\mathscr{A}$, $\mathscr{N}_+$ and $\mathscr{N}_-$ by K, H, A, N_+ and N_-, respectively. $H = SO_0(2,n-1)$ is the anti-de Sitter subgroup, and we have an Iwasawa like decomposition of G i.e. the map $H' \times A \times N_+ \longrightarrow G$ is an injective diffeomorphism onto an open, dense subset of G, where H' is $SO(2,n-1)$ [16].

Consider the $n+1$ dimensional isotropic cone in $\mathbb{R}^{n+2}$ defined by

$$C = \{x \in \mathbb{R}^{n+2} | Q(x) = 0\}. \tag{6}$$

Let $\mathbb{R}^{n+2^*}$ and C^* be the sets of nonzero elements in $\mathbb{R}^{n+2}$ and C, respectively. Let $P \subset G$ be the stabilizer subgroup of $e = e_{-1} + e_n$ where $e_i = (0,0,\,,\,,\,0,\,1,\,0,\,,\,,\,,0) \in \mathbb{R}^{n+2}$ (i.e. e_i is the vector with 1 in the ith slot and zeros elsewhere). $P \cong SO_0(1,n-1) \times_s N_+$, where $\times_s$ denotes semi-direct product, i.e. P is the n dimensional Poincar´e group. The orbit of e under G is $\mathcal{C}$. Hence $C^* \cong G/P$.

Now we consider a representation π of $SO(2,n)^\sim$ and $d\pi$ the associated representation of $so(2,n)$ which lifts to a representation of the enveloping algebra. For brevity, write X for $d\pi(X)$ with $X \in \mathscr{E}(\mathscr{G}_{\mathbb{C}})$, the enveloping algebra of the complexification $\mathscr{G}_{\mathbb{C}}$of $\mathscr{G}$. We let $\mathscr{H}$ be a Cartan subalgebra of $\mathscr{G}_{\mathbb{C}}$ and let Δ be the root system of $(\mathscr{G},\mathscr{H})$. Define $\mathscr{G}^\lambda = \{X \in \mathscr{G} | [h,X] = \lambda(h)X \;\; \forall\; h \in \mathscr{H} \;\text{ and }\; \lambda \in \Delta\}$. Set $\mathscr{N} = \Sigma_{\lambda>0}\mathscr{G}^\lambda$ and $\tilde{\mathscr{N}} = \Sigma_{\lambda<0}\mathscr{G}^\lambda$. Let V be a $\mathscr{E}(\mathscr{G}_{\mathbb{C}})$ module with action $d\pi$ of $\mathscr{E}(\mathscr{G}_{\mathbb{C}})$ on V. Let: a) $V = \Sigma_\lambda^\oplus V_\lambda$ with $V_\lambda = \{v \in V | H_i v = \lambda(H_i)v\}$; b) $\exists\, v_0 \in V \ni$ (i)$X^+(X^-)\, v_0 = 0$ with $X^+ \in \mathscr{N}$ $(X^- \in \tilde{\mathscr{N}})$; (ii)$H_i\, v_0 = \Lambda_0(H_i)\, v_0$; (iii)$\mathscr{E}(\mathscr{G}_{\mathbb{C}})\, v_0 = V$. If $d\pi$ comes from a unitary representation of $G^\sim$, then it is an infinitesmally unitarizable *lowest (highest) weight representation* with lowest (highest) weight Λ_0.

An irreducible, infinitesmally unitarizable representation $d\pi$ of $\mathscr{G}$ on a Hilbert space $\mathscr{H}$ is a representation with positive energy if

$$C_\pi = \{X \in \mathscr{G} \mid id\pi(X) \quad \text{is a nonnegative self adjoint operator}\}$$

is a non-zero and proper cone in $\mathscr{G}$.

Theorem 3.1.

Let $d\pi$ be an irreducible, infinitesmally unitarizable representation of $\mathscr{G}$ on $\mathscr{H}$. Then $d\pi$ is positive energy if and only if $d\pi$ is a highest (lowest) weight representation.

To prove the theorem we refer the reader to [17] where it is shown that if $id\pi(X_0)$ a positive, self-adjoint operator then the representation $(d\pi(\mathscr{G}),\mathscr{H})$ is a lowest (highest) weight representation. The theorem then follows from the observation that any cone in $\mathscr{G} = so(2,n)$ must contain C_{Min}.

Next consider the 3 dimensional (simple) subalgebra (TDS) spanned by:

$$\{L_{-10} = -X_0\,,\; L_{0\,n}\,,\; L_{-1\,n} = -S\} \tag{7}$$

X_0 is the Einstein energy, and S is the generator of scale.

Recall

$$P_0 = i\,T_0 = \frac{i}{2}(L_{-10} + L_{0\,n}) \tag{8}$$

We have

$$[S,\, X_0] = L_{0n} \;\;,\;\; [S,\, L_{0\,n}] = X_0 \;\;,\;\; [X_0,\, L_{0\,n}] = S\,. \tag{9}$$

Additionally

$$L_{-10}(\alpha) = e^{\alpha S}(L_{-10})e^{-\alpha S}\,,\; L_{-1i}(\alpha) = [L_{i0},L_{-10}(\alpha)]\,,\; L_{\mu\nu}(\alpha) = L_{\mu\nu}(\mu,\nu = 0,1,2,...n-1) \tag{10}$$

These $L_{ab}(\alpha)$'s are a basis for a conjugate copy of $so(2,n-1)$, conjugate under $so(2,n)$,

and

$$P_\mu = \frac{i}{2}\lim_{\alpha\to\infty} e^{-\alpha}\, L_{-1\mu}(\alpha)\,,\; L_{\mu\nu} \tag{11}$$

where P_μ and $L_{\mu\nu}$ are the generators of the n-dimensional Poincar´e Lie algebra, $\mathscr{P}$.

Theorem 3.2. *$d\pi$ is a positive energy representation of $so(2,n)$ $\Longleftrightarrow$ $d\pi(P_0)$ $(-d\pi(P_0))$ is a positive, self-adjoint operator.*

The proof in the $\Leftarrow$ direction is due to Segal and is given in [17]. It uses the conformal inversion. Although the other direction (the converse of Segal's theorem) seems to be known to experts in the field, to our knowledge an explicit proof of it, at least in the physics literature, seems to be lacking, and so we record it here: we observe that positivity of the Einstein energy implies

$$(\psi, d\pi(iX_0(\alpha)\psi) = (\psi, id\pi(e^{\alpha S}X_0e^{-\alpha S})\psi) = (\psi', d\pi(iX_0)\psi') > 0 \;\forall\; \alpha > 0 \tag{12}$$

where $\psi' = e^{-\alpha S}\psi$. Now eqn. (9) implies

$$-X_0(\alpha) = e^{\alpha S}(-X_0)e^{-\alpha S} = \{\cosh(\alpha)(-X_0) + \sinh(\alpha)L_{on}\} \to -2iP_0\, e^{\alpha} \text{ as } \alpha \to \infty\,. \tag{13}$$

The desired result is easily seen to be a consequence of these two equations.

4. A Deformation of the Poincaré Subalgebra and the Contracted of the Di and Rac

Denote by $\mathscr{E}(\mathscr{G}')$ and $\mathscr{E}(\mathscr{P})$ be the universal enveloping algebras of $\mathscr{G}' = so(2,n-1)$ and $\mathscr{P}$, respectively. We introduce the following element of $\mathscr{E}(\mathscr{G}')$: $\mathbf{Q}_2 = \sum_{i=1}^{n-1}\mathbf{L}_{0i}^2 - \frac{1}{2}\sum_{i,\,j=1}^{n-1}\mathbf{L}_{ij}^2$. Let $\mathscr{K}(\mathscr{P})$ be the skew field of $\mathscr{P}$ and $\mathscr{K}(\mathscr{G}'_\lambda)$ be the skew field of $\mathscr{G}'_\lambda$ (see below for definition of $\mathscr{G}'_\lambda$). Define a commutative algebraic extensions of $\mathscr{K}(\mathscr{P})$ by $\mathscr{K}(\mathscr{P})^{\rm ext} = \left\{a + bY \,\middle|\, a,b \in \mathscr{K}(\mathscr{P}), [Y,a] = 0 \;\forall\, a \in \mathscr{K}(\mathscr{P}), Y^2 = \mathbf{P}^2\right\}$ where Y commutes with all elements of $\mathscr{K}(\mathscr{P})$ and $\mathbf{P}^2 = \Sigma_{k=0}^{n-1}\mathbf{P}_k\,\mathbf{P}^k$. Now define a mapping τ_λ from $\mathscr{G}'_\lambda$ to $\mathscr{K}(\mathscr{P})$ by

$$\tau_\lambda(\tilde{\mathbf{L}}_{\mu\nu}) = \mathbf{L}_{\mu\nu}\;,\;\; \tau_\lambda(\tilde{\mathbf{L}}_{-1\mu}) = \frac{i\lambda}{2Y}\Big[\mathbf{Q}_2\,,\,\mathbf{P}_\mu\Big] + \mathbf{P}_\mu\,. \tag{14}$$

The $\lambda^{-1}\tau_\lambda(\tilde{\mathbf{L}}_{-1\mu})$ and $\tau_\lambda(\tilde{\mathbf{L}}_{\mu\nu})$ satisfy the commutation relations of the generators of $\mathscr{G}'$. The $\tau_\lambda(\tilde{\mathbf{L}}_{-1\mu})$ and $\tau_\lambda(\tilde{\mathbf{L}}_{\mu\nu})$ are a basis for an isomorphic copy $\mathscr{G}'_\lambda$ of $\mathscr{G}'$.

For the remainder of paper we take $n = 4$, but it seems clear that much of the following holds for the higher dimensions [18] and even for other groups like $SL(n,\mathbb{R})$ [20], especially regarding contractions of the positive energy representations of the anti-deSitter group [15]. Let $\tau_{\lambda=1}(\tilde{Y}) = Y$, then $\tau = \tau_{\lambda=1}$ and define a commutative algebraic extension of $\mathscr{K}(\mathscr{G}'_\lambda)$ by

$$\mathscr{K}(\mathscr{G}'_\lambda)^{\rm ext} = \left\{a + b\,\tilde{Y} + c\,\tilde{Y}^2 + d\,\tilde{Y}^3 \;\middle|\; a,\,b,\,c,\,d \in \mathscr{K}(\mathscr{G}_\lambda)\right\}$$

where $\tilde{Y}$ commutes with all elements of $\mathscr{K}(\mathscr{G}'_\lambda)$. Then τ_λ can be extended to a homomorphism of $\mathscr{K}(\mathscr{G}')^{\rm ext}$ into $\mathscr{K}(\mathscr{P})^{\rm ext}$ in an obvious way, which, because of Lemma 4.2, is actually surjective in the case $n = 4$ which we now consider. Denote this extension also by $\tau = \tau_{\lambda=1}$. Elements of $\mathscr{K}(\mathscr{G}')^{\rm ext}$ have a tilde to keep them distinct from elements of $\mathscr{K}(\mathscr{P})^{\rm ext}$, and we introduce $*$ structures on $\mathscr{K}(\mathscr{P})^{\rm ext}$ and $\mathscr{K}(\mathscr{G}')^{\rm ext}$.

Lemma 4.1.

Let $\tau(\mathscr{G}'_\lambda)$ be the isomorphic copy of $\mathscr{G}'$ having basis elements $\mathbf{L}_{ij} \in \mathscr{G}$ and $\mathbf{L}_{-1\mu} \in \mathscr{K}(\mathscr{P})^{\rm ext}$ defined by eqns. (14). Then (for $\lambda = 1$) the following holds:

$$\mathbf{C}_2 = -Y^2 - \left[\frac{\mathbf{W}}{Y^2} + \frac{9}{4}\mathbf{I}\right]\;,\;\; \mathbf{C}_4 = \left[Y^2 + \frac{1}{4}\right]\frac{\mathbf{W}}{Y^2}\,. \tag{15}$$

($\mathbf{I}$ is the identity in $\mathscr{K}(\mathscr{G}'_\lambda)$ and $\mathbf{C}_2$ and and $\mathbf{C}_4$ are the 2nd and 4th order Casimir operators of $\mathscr{G}'$ as defined in [18], [19].)

Lemma 4.2.
Solutions $\mathbf{P}_\mu(\lambda)$ to eqns. (14) are given by $\mathbf{P}_\mu(\lambda) = \mathbf{D}^{-1}\,\mathbf{A}_\mu{}^\nu \mathbf{L}_{\nu\,-1}$ *with*

$$\mathbf{A}_\mu{}^\nu = -\mathbf{C}_4^{\lambda'}\,\delta_\mu^\nu + \frac{i}{2}\left[\left(\mathbf{Q}_2 + \frac{1}{4}\right)\delta_\mu^\nu - \frac{3}{2}\mathbf{L}_\mu^\nu - \mathbf{L}_{\mu\rho}\mathbf{L}^{\rho\nu} - \mathbf{Q}_4\varepsilon^\nu_{\mu\rho\tau}\mathbf{L}^{\rho\tau}\right]\frac{Y}{\lambda} -$$

$$- \left[(\mathbf{Q}_2 + \frac{1}{4} - \mathbf{C}_2^{\lambda'})\delta_\mu^\nu - \mathbf{L}_\mu^\nu - \mathbf{L}_{\mu\rho}\mathbf{L}^{\rho\nu}\right]\frac{Y^2}{\lambda^2} + (\frac{1}{2}\delta_\mu^\nu - \mathbf{L}_\mu^\nu)\frac{Y^3}{\lambda^3}\,, \tag{16}$$

and

$$\mathbf{D} = \mathbf{Q}_4 + \frac{1}{4}\mathbf{Q}_2 - \mathbf{C}_4^{\lambda'} + \frac{3}{16}\mathbf{I} + i\left(\mathbf{Q}_2 + \frac{1}{2}\right)\frac{Y}{\lambda} - \left(\mathbf{Q}_2 - \mathbf{C}_2^{\lambda'} - \frac{1}{2}\right)\frac{Y^2}{\lambda^2} + 2i\frac{Y^3}{\lambda^3}\,. \tag{17}$$

Also Y^2 satisfies $Y^4 + \lambda^2\,\mathbf{C}_2^{\lambda'}Y^2 + \lambda^4\,\mathbf{C}_4^{\lambda'} = 0$, *with* $\mathbf{C}_2' = \left(\mathbf{C}_2 + \frac{5}{2}\mathbf{I}\right)$ *and* $\mathbf{C}_4' = \left(\mathbf{C}_4 + \frac{1}{4}\mathbf{C}_2 + \frac{9}{16}\mathbf{I}\right)$
and

$$\mathbf{W} = \sum_{\mu=0}^{3}\sum_{\nu=0}^{3}\sum_{\rho=0}^{3}\left(\mathbf{P}_\mu\,\mathbf{P}^\nu\,\mathbf{L}_{\nu\rho}\,\mathbf{L}^{\rho\mu} - \frac{1}{2}\mathbf{P}_\rho\,\mathbf{P}^\rho\,\mathbf{L}_{\mu\nu}\,\mathbf{L}^{\nu\mu}\right)\,. \tag{18}$$

Theorem 4.1.
Let † be such that $\tilde{\mathbf{L}}^\dagger_{\mu\nu} = -\tilde{\mathbf{L}}_{\mu\nu}$, $\tilde{\mathbf{L}}^\dagger_{4\mu} = -\tilde{\mathbf{L}}_{4\mu}$ *and if* $\tilde{Y}^\dagger = \tilde{Y}$, *then* $\tilde{\mathbf{P}}_\mu = \left(\tilde{\mathbf{D}}^{-1}\tilde{\mathbf{A}}_\mu^\rho\tilde{\mathbf{L}}_{-1\rho}\right)$ *and*

$$\tilde{\mathbf{P}}^\dagger_\mu = \left(\tilde{\mathbf{D}}^{-1}\tilde{\mathbf{A}}_\mu^\rho\tilde{\mathbf{L}}_{-1\rho}\right)^\dagger = \tilde{\mathbf{L}}^\dagger_{-1\rho}\tilde{\mathbf{A}}^\dagger_{\mu\rho}(\tilde{\mathbf{D}}^\dagger)^{-1} = -\tilde{\mathbf{P}}_\mu\,. \tag{19}$$

Furthermore $\left[\tilde{\mathbf{P}}_\mu\,,\,\tilde{\mathbf{P}}_\nu\right] = 0$ *and* $ker(\tau)\Big|_{\mathscr{E}(\mathscr{G}')} = 0$.

Now let the vector space V be the underlying vector space of the Lie algebra of $\mathscr{P} = (V, [\cdot\,,\,\cdot]_0)$ and define a map $\phi_\lambda : \mathscr{P} \to \mathscr{K}(\mathscr{P})^{\text{ext}}$ by

$$\phi_\lambda(\mathbf{L}_{\mu\nu}) = \tau_\lambda(\tilde{\mathbf{L}}_{\mu\nu})\,;\; \phi_\lambda(\mathbf{P}_\mu) = \tau_\lambda(\tilde{\mathbf{L}}_{-1\mu})\,. \tag{20}$$

Due to $ker(\tau)\Big|_{\mathscr{E}(\mathscr{G}')} = 0$ in the just stated theorem, ϕ_λ must be a vector space isomorphism onto its image and thus we can define a new Lie bracket on $\mathscr{P}$ by:

$$[a\,,\,b]_\lambda = \phi_\lambda^{-1}([\phi_\lambda(a)\,,\,\phi_\lambda(b)])\;\forall\; a,\, b \text{ in } \mathscr{P}\,. \tag{21}$$

Thus, if we let $\psi : I \subset I\!R \to \mathscr{M}_n$ be defined as $\psi(\lambda) = [\,\cdot,\,\cdot\,]_\lambda$ with $\lambda \in I$, then ψ is a (nontrivial) deformation of $\mathscr{P}$, with $\psi(1)$ being the Lie structure of $\mathscr{G}'$. Since τ_λ is an isomorphism, we can apply it in the other direction to obtain contractions of representations of $\mathscr{G}'$.

A classification of the infinitesmally unitarizable, irreducible lowest weight representations of $\mathscr{G} = so(2,3)$ using a modern representation theoretic approach involving Verma modules and singular vectors is given in [21]. The results there are: let $D(E_0, s_0)$ be a given such lowest weight representation with lowest weight $\Lambda \ni \Lambda(H_1) = s_0$ and $\Lambda(H_2) = E_0$. Then we have: i) $D(E_0, s_0) = D(\frac{1}{2}, 0)$ (Rac); ii) $D(E_0, s_0) = D(1, \frac{1}{2})$ (Di); iii) $D(E_0 > \frac{1}{2}, s_0 = 0)$; iv) $D(E_0 > 1, s_0 = \frac{1}{2})$; v) $D(E_0 \geq s_0 + 1, s_0 \geq 1)$. The massless representations are: $D(E_0 = s_0 + 1, s_0 \geq 1)$.

All of the representations given there have nontrivial contractions to representations of $\mathscr{P}$, except for the case of the singelton representations (cf. [15] for details on the contractions of the massless representations in n dimensions). For the singletons, i.e. the Di and the Rac (which is the minimal representation [22] of $SO(2,3)$), what goes wrong is that it is necessary that both $d\pi(\tilde{\mathbf{D}})^{-1}$ and $d\pi(\tilde{\mathbf{D}})^{\dagger^{-1}}$ of Theorem 4.1 must exist on a suitable, dense domains in $\mathscr{H}$, in order that there exists a representation $d\tilde{\pi}^{+}_{\frac{1}{2}}$ of $\mathscr{P}$ on $D(1/2,0) \subset \mathscr{H}$ (e.g. for the Rac). For both the *Di* and *Rac* is $\tilde{Y}^2 = -1, -\frac{1}{4}$ which implies not only that $d\pi(\tilde{\mathbf{P}}_\mu)$ are not skew-symmetric operators, but, e.g. for the *Rac*: $d\pi(\tilde{\mathbf{D}}) = 0$ (with $\tilde{Y} = -i$)! Thus the $d\tilde{\pi}(\mathbf{P}_\mu)$ cannot exist in any sense as operators on the Hilbert space of the representation with $d\pi(\tau_\lambda(\tilde{\mathbf{L}}_{-1\mu}))$ being the action of $\tilde{\mathbf{L}}_{-1\mu}$ in the Rac representation. Using this fact it is straighforward to see that the contracted of $d\pi(\mathscr{G}')$ where $d\pi$ is the *Di* or *Rac* representation, which is induced by the isomorphism τ_λ cannot exist except trivially i.e. as the identity representation of $\mathscr{P}$.

5. References

[1] Eggleston 1958 *Convexity* (Cambridge: Cambridge Univ. Press)
[2] Paneitz S M 1981 *Jour. Funct. Anal.* **43** 313-359
[3] Olshanskii G I 1981 *Funct. Anal. and Its Applications* **15**, 4, 53-66
[4] Hilgert J, Hoffman K H 1988 *Semigroup Forum* **37** 241-252
[5] Helgason S 1978 *Differential Geometry, Lie Groups and Symmetric Spaces'* (New York: Academic Press)
[6] Paneitz S M, Segal I E 1982 *Jour. Funct. Anal.*, **47**, 78-142; Paneitz S M 1983 *Jour. Funct. Anal.*, **54**, 18-112.
[7] Ritter W G 2003 Quantum Field Theory and the Space of All Lie Algebras Preprint: Harvard Univ., (arXiv: math-ph/0304031)
[8] Levy-Nahas M 1967 *J. Math. Phys.* **8** 1211-1222 (This paper contains explicit information about the Segal topology i.e. the topology in the space $\mathscr{M}_n$ of structure constants.)
[9] Segal I E 1951 *Duke Math. J.* **18** 221-265.
[10] In¨o n¨u E, Wigner E P 1953 *Proc. Nat. Acad. Sci. U S A* **39** 510-524.
[11] Saletan E 1961 *J. Math. Phys.* **2** 1-21.
[12] L¨o hmus Ya H 1967 *Contractions of Lie Groups*, Volume 4 of Proceedings of the 1967 Summer School on Elementary Particle Physics held at Otep¨a¨a, Estonian SSR, August 1-13.
[13] Weimar-Woods E 2000 *Reviews in Math. Phys.* **12**, 11, 1505-1529.
[14] Dooley A H, Rice J W 1985 *Trans. Amer. Math. Soc.* **289**, 1, 185-202.
[15] Angelopoulos E, Laoues M 1998 *Reviews in Math. Phys.* **10**, 3, 271-299.
[16] Sekiguchi J 1980 *Nagoya Math. Jour.* **79** 211-230.
[17] Mack G 1977 *Comm. Math. Phys.*, **55** 1-21.
[18] Moylan P 2004 *Studies in Mathematical Physics Research* Ch. 2, (New Jersey: Nova Science)
[19] Havliček M, Moylan P 1993 *Jour. Math. Phys.* **34**, 11, 5320-5332.
[20] Salom I, Šijacčki D 2009 arXiv.0906.2106v1 [math-ph] 11Jun 2009
[21] V.K. Dobrev, E. Sezgin, *Lecture Notes in Physics*, **379**, (Springer, 1990), 227.
[22] Kostant B 1990 *The vanishing of scalar curvature and the minimal representation of $SO(4,4)$* in Progr. Math. **92** (Boston: Birkh¨a user), 85-124; Kobayashi T, Mano G 2008 *The Schr¨o dinger model for the minimal representaion of the indefi nite orthogonal group $O(p,q)$*, arXiv: 0712.1769v2 [math.RT] 19 Jul 2008; Kobayashi, T. Orsted B 2003 *Adv. Math.* **180** 486-595;
[23] Flato M, Fronsdal C 1980 *Phys. Lett. B* **97**, 2, 236-240; Flato M, Fronsdal C, Sternheimer D 1999 *Singleton Physics*, arXiv:hep-th/9901043 v1 12 Jan 1999;

Journal of Physics: Conference Series **462** (2013) 012038 doi:10.1088/1742-6596/462/1/012038

String worldsheet theory in hamiltonian framework and background independence

Partha Mukhopadhyay

The Institute of Mathematical Sciences, C.I.T. Campus, Taramani, Chennai 600113, India

E-mail: parthamu@imsc.res.in

Abstract. We analyze exact conformal invariance of string worldsheet theory in non-trivial backgrounds using hamiltonian framework. In the first part of this talk we consider the example of type IIB superstrings in Ramond-Ramond pp-wave background. In particular, we discuss the quantum definition of energy-momentum (EM) tensor and two methods of computing Virasoro algebra. One of the methods uses dynamical supersymmetries and indirectly establishes (partially) conformal invariance when the background is on-shell. We discuss the problem of operator ordering involved in the other method which attempts to compute the algebra directly. This method is supposed to work for off-shell backgrounds and therefore is more useful. In order to understand this method better we attempt a background independent formulation of the problem which is discussed in the second half of the talk. For a bosonic string moving in an arbitrary metric-background such a framework is obtained by following DeWitt's work (Phys.Rev.85:653-661,1952) in the context of particle quantum mechanics. In particular, we construct certain background independent analogue of quantum Virasoro generators and show that in spin-zero representation they satisfy the Witt algebra with additional anomalous terms that vanish for Ricci-flat backgrounds. We also report on a new result which states that the same algebra holds true in arbitrary tensor representations as well.

1. Introduction and summary

This talk will be based on the papers [1, 2, 3, 4] and certain new results in [5]. In [1, 2, 3] we try to understand exact conformal invariance of string worldsheet theory in certain pp-wave backgrounds using hamiltonian framework[1]. Usually backgrounds that are nearly flat are considered for such analysis. The reason is that for flat background the theory is exactly solvable and the relevant vacuum is known. One can then define the quantum EM tensor in the weak background field approximation by normal ordering the interaction terms with respect to the flat-space-vacuum. The pp-wave background on the other hand is not close to flat and therefore the corresponding worldsheet theory is strongly interacting. We will discuss this problem in section 2.

The end result of the analysis in section 2 will show that there exists an operator ordering ambiguity in the computation of the Virasoro anomaly if one tries to calculate them directly using the quantum EM tensor. In order to have a better understanding of the situation we seek a framework where the computations can be done in a background independent way. Such a framework was developed by DeWitt in [10] for particle quantum mechanics. As will be

[1] Two-dimensional non-linear sigma model is usually studied using lagrangian framework [6]. Hamiltonian framework has also been used to some extent earlier in [7, 8] (see also [9]).

discussed in section 3, the classical worldsheet theory can be interpreted to be a theory of a particle moving in an infinite-dimensional curved background subject to certain potential. It was shown in [4] that the background independent analogue of the quantum Virasoro generators constructed in the sense of DeWitt satisfy, in spin-zero representation, the Witt algebra with additional anomalous terms that vanish for Ricci-flat backgrounds. We also discuss how this might possibly suggest a resolution of the ordering problem for the pp-wave.

In order to understand the implications of the Ricci-flatness condition mentioned above various other questions regarding this construction need to be answered. One such question, namely extending the analysis to higher rank tensor representations, has been considered in [5]. The results show that the same Ricci-flatness condition holds true in general. We will discuss this in section 3. We end with some final remarks in section 4.

2. Hamiltonian framework for worldsheet theory in pp-wave

Here we will consider a pp-wave background, with or without R-R flux, such that the only non-trivial component of the metric is[2],

$$G_{++} = K(\vec{x}) \ , \tag{2.1}$$

where the vector sign refers to the transverse directions x^I, $I = 1 \cdots D-2$, D being 26 or 10 depending on bosonic/superstring theory. A plane-wave is a pp-wave for which $K(\vec{x})$ is quadratic in x^I.

Proving conformal invariance for the type IIB R-R plane-wave background [13] using hamiltonian framework was first attempted by Kazama and Yokoi in [14]. In order to formulate the worldsheet theory as a CFT the authors considered Green-Schwarz superstring in semi-light-cone gauge following [15]. The Virasoro algebra was also calculated using a direct method in local form. In this method one directly calculates the commutators $[\mathcal{T}(\sigma), \mathcal{T}(\sigma')]$, $[\tilde{\mathcal{T}}(\sigma), \tilde{\mathcal{T}}(\sigma')]$ and $[\mathcal{T}(\sigma), \tilde{\mathcal{T}}(\sigma')]$, where $\mathcal{T}(\sigma)$ and $\tilde{\mathcal{T}}(\sigma)$ are the right and left moving EM tensor components and σ is the worldsheet space coordinate. The main result of [14] that is relevant to us may be summarized as follows[3],

> *If the EM tensor is ordered according to MNO, then the Virasoro algebra is not satisfied due to non-zero anomalous terms. However, this anomaly vanishes if the the EM tensor is ordered with a different prescription called phase-space normal ordering (PNO).* (2.2)

This result poses a puzzle as defining the EM tensor according to PNO would imply that the theory does not have a smooth flat-space limit. This, as we will explain below, is contradictory with the universality property discussed in [16, 17].

Let us consider an on-shell pp-wave background with flat transverse space in string theory. The following theorem can be proved (see [17]) for the corresponding worldsheet theory,

> *There exists a universal sector of worldsheet operators whose correlation functions are evaluated to be same as that in flat background.* (2.3)

[2] In absence of R-R flux, the pp-wave should correspond to an exact CFT whenever it is Ricci-flat (i.e. vanishing transverse laplacian of K) [12]. We will consider a particular case of constant R-R flux in type IIB string theory, in which case conformal invariance should require the Ricci tensor to be proportional to the square of the flux strength [13].

[3] MNO refers to *massless normal ordering* [14], the one that is relevant to flat background. The reason for this nomenclature is that the worldsheet theory for flat background may be understood as the massless limit of that corresponding to R-R plane-wave in light-cone gauge. PNO was also discussed in the literature earlier (for example [8]), but we will not go into the technical details of its definition.

This implies that the worldsheet theory corresponding to any pp-wave should admit a smooth flat space limit. In fact it requires that the spectrum of conformal dimensions within the universal sector of Hilbert space (i.e. the states created by universal operators) be universal. This argument implies that PNO can not be the right way to order the EM tensor.

2.1. *Testing universality argument in bosonic string theory*

In [1] we considered the same problem in the simpler setting of bosonic string theory and showed that the results are consistent with the above universality argument. It is easy to check that the PNO ordered EM tensor gives rise to a spectrum of negative conformal dimensions in flat background. The physical spectrum was computed for the plane-wave in [1] by explicitly constructing the string field theory (SFT) quadratic term and diagonalizing it. It indeed shows the existence of negative conformal dimensions for the PNO ordering. On the other hand the MNO-ordered EM tensor reproduces the result that is expected from light-cone gauge analysis.

The Virasoro algebra was also calculated for the off-shell pp-wave using the same method as in [14]. In addition to the standard central charge terms, this produces certain additional operator anomaly terms. The calculations have been done using both PNO and MNO ordered EM tensor and in both the cases the operator anomaly terms, ordered accordingly, were shown to be proportional to the equation of motion for the pp-wave.

2.2. *A closer look at R-R pp-wave*

Although the results of [1] support the universality argument, the case of R-R pp-wave seems to be more subtle. In particular, the above results do not explain the observation in (2.2). This was explained in [2] which we discuss below.

We first write,

$$\mathcal{T} = \mathcal{T}^{(0)} + \delta\mathcal{T} \ , \quad \tilde{\mathcal{T}} = \tilde{\mathcal{T}}^{(0)} + \delta\mathcal{T} \ , \tag{2.4}$$

where the superscript (0) refers to the flat background, i.e. the free part and $\delta\mathcal{T}$ is the interaction part containing non-trivial background fields. The correct way of defining the quantum EM tensor was shown to be as follows,

> *The free part of the EM tensor is ordered according to MNO, while the interaction part needs to be ordered according to PNO.* (2.5)

The reason for the PNO ordering for the interaction term is as follows. Given the supergravity background characterized by the function $K(\vec{x}) = \int d\vec{k}\ \tilde{K}_{sugra}(\vec{k})e^{i\vec{k}.\vec{x}}$, the relevant couplings $\tilde{K}_{ws}(\vec{k})$ in the worldsheet sigma model are in general related to $\tilde{K}_{sugra}(\vec{k})$ by a field redefinition. The effect of such field redefinition can be absorbed into the normal ordering prescription of the corresponding interaction term. It turns out that if the term is ordered according to PNO then the field redefinition becomes identity. By constructing the SFT quadratic term and diagonalizing it one finds that the definition in (2.5) reproduces the correct physical spectrum as found in light-cone analysis [18][4].

The explanation of (2.2) goes as follows. The operator anomaly terms corresponding to the three commutators mentioned above (2.2) have the following structure,

$$\mathcal{A}^R(\sigma,\sigma') \;=\; A^R(\sigma,\sigma') + A_F(\sigma,\sigma') \ ,$$

[4] The field redefinition discussed above involves the transverse laplacian of K which is proportional to the R-R flux strength when the background is on-shell. Therefore for the bosonic example considered in [1], where there is no R-R flux, the field redefinition becomes trivial on-shell. This is why in the bosonic case the EM tensor, with both the free and interaction parts ordered according to MNO, gives the right spectrum, as mentioned toward the end of subsection 2.1.

$$\begin{aligned}\mathcal{A}^L(\sigma,\sigma') &= A^L(\sigma,\sigma') + A_F(\sigma,\sigma') \ , \\ \mathcal{A}(\sigma,\sigma') &= A(\sigma,\sigma') + A_F(\sigma,\sigma') \ , \end{aligned} \tag{2.6}$$

respectively, where A^R, A^L and A are the bosonic and A_F is the fermionic contribution. The calculations are done by introducing a UV-regulator ϵ on the worldsheet. All such terms turn out to have a special structure: $\mathcal{O}_\epsilon(\sigma,\sigma')\delta_\epsilon(\sigma-\sigma')$, where $\mathcal{O}_\epsilon(\sigma,\sigma')$ is a composite operator anti-symmetric in σ and σ' and $\delta_\epsilon(\sigma-\sigma')$ is the Poisson kernel representation of Dirac delta function which approaches the latter in the limit $\epsilon \to 0$. Notice that at the classical level, where $\mathcal{O}_\epsilon(\sigma,\sigma')$ remains the same under reordering of operators, such a term must vanish as expected. In the quantum theory this term is well-defined at a finite ϵ. However, first reordering the term in a different way and then taking $\epsilon \to 0$ produces a different answer.

The above result indicates that the direct method of calculating the Virasoro anomaly suffers from operator ordering ambiguity. It turns out that if the anomaly terms are ordered according to PNO then the correct supergravity equation of motion (EOM) is reproduced. This is the reason for the observation in (2.2). However, there is no good understanding why such an ordering prescription should work.

We will come back to this question in section 3 where a background independent construction will be described. In this construction covariant Ricci-anomaly terms will arise whose technical origin is quite different. This indicates that the observation that *PNO-prescription* described above gives the right supergravity EOM may be misleading and does not generalize to arbitrary backgrounds.

2.3. Dynamical supersymmetry argument for conformal invariance

The ordering ambiguity described above may be attributed to the fact that the relevant vacuum is non-perturbative and it is not generically known how to define quantum operators in a controlled manner. However, in this particular case the background is supersymmetric (in fact, as much as flat space) which may be used to gain such control. It turns out that in semi-light-cone gauge the following relations hold true in both flat and R-R plane-wave backgrounds [3],

$$\{Q_{\dot a}, Q_{\dot b}\} = 2\delta_{\dot a\dot b}\oint \frac{d\sigma}{2\pi}\,\mathcal{T}_\perp(\sigma)\ , \quad \{\tilde Q_{\dot a}, \tilde Q_{\dot b}\} = 2\delta_{\dot a\dot b}\oint \frac{d\sigma}{2\pi}\,\tilde{\mathcal{T}}_\perp(\sigma)\ , \tag{2.7}$$

where $Q_{\dot a}$ and $\tilde Q_{\dot a}$ are properly scaled dynamical supercharges and $\mathcal{T}_\perp$ and $\tilde{\mathcal{T}}_\perp$ are the transverse parts of the EM tensor components. As shown in [3], the above relations can be used in certain Jacobi identities to relate the second order susy variations of the transverse EM tensor components and certain integrated forms of the anomaly terms,

$$\begin{aligned}\{[\mathcal{T}_\perp(\sigma), Q_{\dot a}], Q_{\dot b}\} + \{[\mathcal{T}_\perp(\sigma), Q_{\dot b}], Q_{\dot a}\} &= 2\delta_{\dot a\dot b}\left[\oint \frac{d\sigma'}{2\pi}\,\mathcal{A}^R(\sigma,\sigma') + \cdots\right] \ , \\ \left\{[\tilde{\mathcal{T}}_\perp(\sigma), \tilde Q_{\dot a}], \tilde Q_{\dot b}\right\} + \left\{[\tilde{\mathcal{T}}_\perp(\sigma), \tilde Q_{\dot b}], \tilde Q_{\dot a}\right\} &= 2\delta_{\dot a\dot b}\left[\oint \frac{d\sigma'}{2\pi}\,\mathcal{A}^L(\sigma,\sigma') + \cdots\right] \ , \\ \left\{[\tilde{\mathcal{T}}_\perp(\sigma), Q_{\dot a}], Q_{\dot b}\right\} + \left\{[\tilde{\mathcal{T}}_\perp(\sigma), Q_{\dot b}], Q_{\dot a}\right\} &= 2\delta_{\dot a\dot b}\left[-\oint \frac{d\sigma'}{2\pi}\mathcal{A}(\sigma',\sigma) + \cdots\right] \ , \end{aligned} \tag{2.8}$$

where the ellipses refer to certain known terms. The idea is to compute the anomalies indirectly by calculating the second order susy variations on the left hand side independently. Although such a computation does not suffer from ordering ambiguity in the sense discussed earlier, the problem arise in the form of divergences as non-(anti)commuting operators appear at the same point. However, this subtlety can be taken care of by considering a point-split definition of the EM tensor. The final results show that all the integrated anomaly terms appearing on the right hand sides of (2.8) are zero.

The above argument establishes conformal invariance only partially as the anomaly terms have been computed with one coordinate integrated over. The complete proof would require one to extend this argument by incorporating local susy currents.

3. A background independent hamiltonian framework

The susy argument works only for on-shell backgrounds which are supersymmetric. In order to understand background EOM from Virasoro anomaly one needs to start with an off-shell background and compute the Virasoro algebra using the direct method discussed in subsections 2.1 and 2.2. However, as discussed earlier this method leads to ambiguous results for the anomaly terms. In [4] we explore a background independent framework, hereafter called DeWitt-Virasoro (DWV) construction, where we attempt to formulate the problem in a vacuum independent way. Such a framework, if it exists, should define the worldsheet theory, at least formally, over the space of all possible geometrical backgrounds. Starting from this framework it should also be possible to derive all the know properties of the 2-d CFT for a given exact background by specializing to it.

3.1. DeWitt-Virasoro construction

The approach considered in [4] is to describe the string quantum mechanics in a coordinate independent way following DeWitt's argument [10] in the context of particle quantum mechanics. We consider the simplest case of a bosonic string moving in an arbitrary metric-background and re-write the worldsheet theory in a language where it describes a single particle moving in an infinite-dimensional curved background. The relevant map is given by,

$$x^i = \oint \frac{d\sigma}{2\pi}\, X^\mu(\sigma)e^{-im\sigma}, \quad g_{ij}(x) = \oint \frac{d\sigma}{2\pi}\, G_{\mu\nu}(X(\sigma))e^{i(m+n)\sigma}, \quad a^i(x) = \oint \frac{d\sigma}{2\pi}\, \partial X^\mu(\sigma)e^{-im\sigma} \tag{3.9}$$

where the index identifications are as follows: $i = \{\mu, m\}$, $j = \{\nu, n\}$ etc. The general coordinate transformation (GCT) in the physical spacetime: $X^\mu \to X'^\mu(X)$ induces the same in the infinite-dimensional spacetime: $x^i \to x'^i(x)$ under which $g_{ij}(x)$ transforms as metric tensor and $a^i(x)$ as a vector field. The classical worldsheet lagrangian takes the following form in terms of the new set of variables,

$$L(x,\dot{x}) \;=\; \frac{1}{2}g_{ij}(x)\left[\dot{x}^i\dot{x}^j - a^i(x)a^j(x)\right] . \tag{3.10}$$

GCT is a point canonical transformation which can be described by a unitary transformation in the quantum theory. DeWitt's analysis in [10] shows how to construct the quantum mechanics such that general covariance is manifest in its position space representation. As a result it gives generally covariant expressions for all the matrix elements.

Following DeWitt's method we formally define a background independent notion of the quantum Virasoro generators and algebra, hereafter called DWV generators and algebra. We first write down the classical Virasoro generators using the particle language which then leads to the following natural definition (consistent with the hermiticity property) of the quantum DWV generators,

$$\begin{aligned} L_{(i)} &= \frac{1}{4}\left[\hat{\pi}^\star_k g^{kl+i}(\hat{x})\hat{\pi}_l - (\hat{\pi}^\star_k a^{k+i}(\hat{x}) + a^{k+i}(\hat{x})\hat{\pi}_k) + g_{kl}(\hat{x})a^k(\hat{x})a^{l+i}(\hat{x})\right] , \\ \tilde{L}_{(i)} &= \frac{1}{4}\left[\hat{\pi}^\star_k g^{kl+\bar{i}}(\hat{x})\hat{\pi}_l + (\hat{\pi}^\star_k a^{k+\bar{i}}(\hat{x}) + a^{k+\bar{i}}(\hat{x})\hat{\pi}_k) + g_{kl}(\hat{x})a^k(\hat{x})a^{l+\bar{i}}(\hat{x})\right] , \end{aligned} \tag{3.11}$$

where given the index identification below eqs.(3.9), we have defined: $\bar{i} = \{\mu, -m\}$, $(i) = m$ and $i + j = \{\mu, m+n\}$. The $\hat{\pi}$-operators are given by (in $\hbar = \alpha' = 1$ unit) [19],

$$\hat{\pi}_j = \hat{p}_j + \frac{i}{2}\gamma_j(\hat{x}) , \quad \hat{\pi}^\star_j = \hat{p}_j - \frac{i}{2}\gamma_j(\hat{x}) , \tag{3.12}$$

Journal of Physics: Conference Series **462** (2013) 012038 doi:10.1088/1742-6596/462/1/012038

where $\gamma_j = \gamma^k_{jk}$ are the contracted Christoffel symbols. DeWitt's technique in [10] allows us to calculate the matrix elements in spin-zero representation. We find that the DWV algebra in spin-zero representation is given by the Witt algebra with additional anomaly terms that vanish for Ricci-flat backgrounds,

$$\langle\chi| \left\{ \begin{array}{l} [\hat{L}_{(i)}, \hat{L}_{(j)}] = (i-j)\hat{L}_{(i+j)} \\ [\hat{\tilde{L}}_{(i)}, \hat{\tilde{L}}_{(j)}] = (i-j)\hat{\tilde{L}}_{(i+j)} \\ [\hat{L}_{(i)}, \hat{\tilde{L}}_{(j)}] = \frac{1}{8}\left(\hat{\pi}^{\star k+i} r_{kl}(\hat{x}) a^{l+\bar{j}}(\hat{x}) - a^{k+i}(\hat{x}) r_{kl}(\hat{x}) \hat{\pi}^{l+\bar{j}}\right) \end{array} \right\} |\psi\rangle \ , \tag{3.13}$$

where $|\chi\rangle$ and $|\psi\rangle$ are two arbitrary spin-zero states and r_{ij} is the Ricci tensor.

The DWV generators constructed above are background independent and are not normal ordered with respect to any particular vacuum. However, as shown in [4], the quantum Virasoro generators in flat and pp-wave backgrounds can be obtained by normal ordering the DWV generators in a suitable way. The resulting quantum Virasoro algebra contains the central charge terms in addition to the operator anomaly terms appearing in (3.13). This enables us to compare the (ambiguous) operator anomaly terms in the bosonic sector in (2.6) with the one in (3.13), at least in the spin-zero representation. This comparison tells us that all the operator anomaly terms A^R, A^L and A in (2.6) must vanish. Furthermore it also implies that the Ricci-term in (3.13) was missing in the previous computation. This apparent discrepancy can be resolved by the following observation. By going to the position space representation it can be argued that the matrix element of the Ricci-term in (3.13) vanishes, though the Ricci tensor itself does not, because of certain restrictions on the index contraction for pp-wave and that the Ricci tensor depends only on the transverse coordinates.

3.2. Higher rank tensor generalization

An important question for the construction described above is how to generalize this to higher spin representations which are very relevant in string theory. In particular, one would like to know if the DWV algebra in (3.13) holds true in arbitrary spin representations. We studied this question in [5] and found the answer in affirmative.

In flat background one usually obtains the higher rank tensor states by applying suitable creation operators on the ground state. It turns out that such state-operator mapping is not suitable for the background independent framework. What turn out to be more relevant in this case are the Schödinger wavefunctions. Such wavefunctions can be obtained either by multiplying a tensor field with a scalar wavefunction or by applying covariant derivatives on it.

In [5] we consider a new framework where an arbitrary tensor state of rank n, which is by itself coordinate invariant, is expanded as follows,

$$|\psi_{(n)}\rangle = \int dw\ \psi^{i_1 i_2 \cdots i_n}(x) |_{i_1 i_2 \cdots i_n}; x\rangle \ , \tag{3.14}$$

where $\psi^{i_1 i_2 \cdots i_n}(x)$ is the tensor wavefunction under consideration and $|_{i_1 i_2 \cdots i_n}; x\rangle$ are the rank-n position eigenstates which are newly introduced. The problem then reduces to finding the representation of the momentum operator in such basis states. This is done by suitably generalizing DeWitt's argument in [10]. The result is given by,

$$\langle_{i_1 \cdots i_n}; x | \hat{\pi}_k |_{j_1 \cdots j_{\tilde{n}}}; \tilde{x}\rangle \ = \ -i\delta_{n,\tilde{n}} \left[\Delta_{i_1 \cdots i_n j_1 \cdots j_n}(x,\tilde{x}) \partial_k \delta(x,\tilde{x}) + F_{i_1 \cdots i_n j_1 \cdots j_n k}(x,\tilde{x}) \delta(x,\tilde{x})\right] \ , \tag{3.15}$$

where $\partial_k \equiv \frac{\partial}{\partial x^k}$,

$$\delta(x,\tilde{x}) = \frac{\delta(x-\tilde{x})}{(g(x)g(\tilde{x}))^{\frac{1}{4}}} \ , \tag{3.16}$$

Journal of Physics: Conference Series **462** (2013) 012038 doi:10.1088/1742-6596/462/1/012038

$\delta(x-\tilde{x})$ being the Dirac delta function, $g = |\det g_{ij}|$,

$$F_{i_1\cdots i_n j_1\cdots j_n k}(x,\tilde{x}) = -\nabla_k \Delta_{i_1\cdots i_n j_1\cdots j_n}(x,\tilde{x}) , \tag{3.17}$$

where ∇_k is the covariant derivative with respect to x^k and

$$\Delta_{i_1\cdots i_n j_1\cdots j_n}(x,\tilde{x}) = \Delta_{i_1 j_1}(x,\tilde{x})\cdots\Delta_{i_n j_n}(x,\tilde{x}) . \tag{3.18}$$

To define $\Delta_{ij}(x,\tilde{x})$ we proceed as follows. Let us first consider two points X^μ and $\tilde{X}^\mu$ in the physical spacetime sufficiently close to each other so that a unique geodesic passes through them. Then following [20] we define the bi-vector of geodetic parallel transport $D_{\mu\nu}(X,\tilde{X})$. The two vector-indices μ and ν are associated to the two points X and $\tilde{X}$ respectively such that under a GCT the bi-vector transform as a vector at each point separately. Moreover, $D_{\mu\nu}(X,\tilde{X})$, contracted with another vector at one of the points, gives the same vector parallel transported to the other point along the geodesic. One generalizes this to a string by considering two string embedding $X^\mu(\sigma)$ and $\tilde{X}^\mu(\sigma)$ such that two points belonging to the two strings at the same value of σ are connected by a unique geodesic. This gives us the parallel transport bi-vector $D_{\mu\nu}(X(\sigma),\tilde{X}(\sigma))$ on the worldsheet whose infinite-dimensional counterpart is given by $\Delta_{ij}(x,\tilde{x})$,

$$\Delta_{ij}(x,\tilde{x}) = \oint \frac{d\sigma}{2\pi} D_{\mu\nu}(X(\sigma),\tilde{X}(\sigma))e^{i(m+n)\sigma} . \tag{3.19}$$

The above definition can be generalized to any other well-behaved paths which are not necessarily geodesics. However, the expression in (3.15), being sensitive to the expansion of $\Delta_{ij}(x,\tilde{x})$ only up to first order in separation, is insensitive to the choice of path.

Although the expression in (3.15) makes the tensorial properties of the relevant matrix element manifest[5], the actual calculations are done using an explicit representation of $\Delta_{ij}(x,\tilde{x})$ given in terms of the vielbeins. Given this basic ingredient, one can compute the DWV generators in tensor representations. These expressions contain all the terms that are present in the scalar representation. Moreover, they receive additional contributions involving spin connection.

Finally, following [4] we proceed to compute the DWV algebra. As expected, the computation gets more complicated than the previous one because of the additional contributions. However, we show that all the unwanted contributions cancel in various ways to reproduce the same algebra in (3.13). There arise three kinds of additional terms which depend on Riemann tensor, covariant derivative of spin connection and higher power of spin connection. It turns out that because of the particular definition in (3.11) these three kinds of terms are organized in such a way that they cancel due to various identities in general relativity.

4. Final remarks

The fact that the same Ricci-flatness condition is obtained as the condition of vanishing anomaly in the DWV algebra in arbitrary tensor representations is interesting. An important question that arises at this point is how to understand the conventional CFT's as special cases of the DWV construction. This is understood for flat and pp-wave backgrounds in spin-zero representation. However, because of the new way of describing the tensor states as in (3.14) it is interesting to explore how exactly the conventional CFT computations should be realized in our background independent description. It would also be interesting to understand the central charge terms in the Virasoro algebra in a background independent way. We hope to come back to these questions in future.

[5] Under a GCT, $\hat{x}^i \to \hat{x}'^i(\hat{x})$: $\hat{\pi}_k \to \hat{\pi}'_k = \lambda_k^{\ l}(\hat{x})\hat{\pi}_l$, where $\lambda_k^{\ l}(\hat{x})$ is the inverse of the Jacobian matrix for the transformation. This implies that the matrix element on the left hand side of (3.15) has a covariant vector index k at x. This is obviously true for the second term on the right hand side. It is also true for the first term as $\delta(x,\tilde{x})$ is a bi-scalar.

Acknowledgments
I thank Sumit R. Das and Alfred Shapere, the organizers of the symposium Quantum Theory and Symmetries 6, for such an exciting event.

References
[1] Mukhopadhyay P 2008 *JHEP* **0811**, 034 (2008) (*Preprint* arXiv:0807.0923 [hep-th]).
[2] Mukhopadhyay P 2009 *JHEP* **0905**, 059 (2009) (*Preprint* arXiv:0902.3750 [hep-th]).
[3] Mukhopadhyay P 2009 *Phys. Rev.* D **80**, 126007 (2009) (*Preprint* [arXiv:0907.2155 [hep-th]).
[4] Mukhopadhyay P 2009 (*Preprint* arXiv:0912.3987 [hep-th]).
[5] Mukhopadhyay P 2010 DeWitt-Virasoro construction in tensor representations (*to appear*).
[6] Friedan D 1980 *Phys. Rev. Lett.* **45**, 1057
Friedan D 1985 *Annals Phys.* **163**, 318
Alvarez-Gaume L, Freedman D Z and Mukhi S 1981 *Annals Phys.* **134**, 85
Fradkin E S and Tseytlin A A, 1985 *Phys. Lett.* B **160**, 69
Fradkin E S and Tseytlin A A 1985 *Nucl. Phys.* B **261** 1
Sen A 1985 *Phys. Rev.* D **32**, 2102
Sen A 1985 *Phys. Rev. Lett.* **55**, 1846
Callan C G, Martinec E J, Perry M J and Friedan D 1985 *Nucl. Phys.* B **262**, 593
Callan C G, Klebanov I R and Perry M J 1986 *Nucl. Phys.* B **278**, 78
Fridling B E and Jevicki A 1986 *Phys. Lett.* B **174**, 75
[7] Maharana J and Veneziano G 1987 *Nucl. Phys.* B **283**, 126
Akhoury R and Okada Y 1987 *Phys. Rev.* D **35**, 1917
Akhoury R and Okada Y 1989 *Nucl. Phys.* B **318**, 176
Das A K and Roy S 1987 *Z. Phys.* C **36**, 317
Das A K, Maharana J and Roy S 1989 *Phys. Rev.* D **40**, 4037
Das A K, Maharana J and Roy S 1990 *Nucl. Phys.* B **331**, 573
Fubini S, Maharana J, Roncadelli M and Veneziano G 1989 *Nucl. Phys.* B **316**, 36
Alam S 1989 *Phys. Rev.* D **40**, 4047
[8] Diakonou M, Farakos K, Koutsoumbas G and Papantonopoulos E 1990 *Phys. Lett.* B **247**, 273
Diakonou M, Farakos K, Koutsoumbas G and Papantonopoulos E 1990 *Phys. Lett.* B **240** (1990) 351;
[9] Jain S, Mandal G and Wadia S R 1987 *Phys. Rev.* D **35**, 778
Jain S, Mandal G and Wadia S R 1987 *Phys. Rev.* D **35**, 3116
[10] DeWitt B S 1952 *Phys. Rev.* **85**, 653
DeWitt B S 1957 *Rev. Mod. Phys.* **29**, 377
[11] Lawrence A E and Martinec E J 1996 *Class. Quant. Grav.* **13**, 63 (*Preprint* hep-th/9509149).
[12] Amati D and Klimcik C 1988 *Phys. Lett.* B **210**, 92
Amati D and Klimcik C 1989 *Phys. Lett.* B **219**, 443
Horowitz G T and Steif A R 1990 *Phys. Rev. Lett.* **64**, 260
[13] Blau M, Figueroa-O'Farrill J, Hull C and Papadopoulos G 2002 *JHEP* **0201**, 047 (*Preprint* [arXiv:hep-th/0110242])
[14] Kazama Y and Yokoi N 2008 *JHEP* **0803**, 057 (*Preprint* arXiv:0801.1561 [hep-th])
[15] Berkovits N and Marchioro D Z 2005 *JHEP* **0501**, 018 (*Preprint* arXiv:hep-th/0412198)
[16] Mukhopadhyay P 2006 (*Preprint* arXiv:hep-th/0611138)
[17] Mukhopadhyay P 2007 *JHEP* **0706**, 061 (*Preprint* arXiv:0704.0085 [hep-th])
[18] Metsaev R R 2002 *Nucl. Phys.* B **625**, 70 (*Preprint* hep-th/0112044)
Metsaev R R and Tseytlin A A 2002 *Phys. Rev.* D **65**, 126004 (*Preprint* hep-th/0202109)
[19] Omote M and Sato H 1972 *Prog. Theor. Phys.* **47**, 1367
[20] DeWitt B S and Brehme R W 1960 *Annals Phys.* **9**, 220

Journal of Physics: Conference Series **462** (2013) 012039 doi:10.1088/1742-6596/462/1/012039

The Hamiltonian Approach to Yang-Mills (2+1): An Update and Corrections to String Tension

V. P. Nair

Physics Department, City College of the CUNY, New York, NY 10031, USA

E-mail: vpn@sci.ccny.cuny.edu

Abstract. Yang-Mills theories in 2+1 (or 3) dimensions are interesting as nontrivial gauge theories in their own right and as effective theories of QCD at high temperatures. I shall review the basics of our Hamiltonian approach to this theory, emphasizing symmetries with a short update on its status. We will show that the calculation of the vacuum wave function for Yang-Mills theory in 2+1 dimensions is in the lowest order of a systematic expansion. Expectation values of observables can be calculated using an effective interacting chiral boson theory, which also leads to a natural expansion as a double series in the coupling constant (to be interpreted within a resummed perturbation series) and a particular kinematical factor. The calculation of the first set of corrections in this expansion shows that the string tension is modified by about -0.3% to -2.8% compared to the lowest order value. This is in good agreement with lattice estimates.

1. Introduction

As many of you know, Yang-Mills theory in $2+1$ dimensions has been the focus of research for my collaborators and myself for many years. In the first part of this talk, I shall give an update on the status of our attempts to understand this theory. In the second part, I shall explain our recent calculations of corrections to the string tension. This latter part is more than just a calculation of specific results; it also formulates a systematic calculational framework which may prove useful for further calculations, such as glueball masses.

Let me begin by recalling why this theory is of interest to us. Yang-Mills theories have, of course, been a great puzzle for many decades. Since the realistic case of $3+1$ dimensions is highly nontrivial and difficult, physicists have often turned to the time-honored path of seeking guidance from lower dimensions. Yang-Mills theories in $1+1$ dimensions are exactly solvable, but they have no propagating degrees of freedom and, as a consequence, are a little too trivial for us. In $2+1$ dimensions, there are propagating degrees of freedom, the theory has nontrivial dynamics, yet seems to be within reach of some level of analytical investigation. This is aided by the fact that the theory has a dimensional coupling constant and, as a related fact, because it is super-renormalizable.

2. A short review and status report

Our approach to this problem has been to use Hamiltonian analysis and to solve the Schrödinger equation for the vacuum wave function [1, 2]. Generally, Hamiltonian methods are not easy to implement in a field theory with its infinite number of degrees of freedom and regularization and renormalization issues. But for this particular problem, it has some advantages, because, at constant time, we are dealing with two-dimensional gauge fields and, for these, our experience with conformal field theory is a useful guide.

As is standard in a Hamiltonian analysis, we then start with the $A_0 = 0$ gauge. The spatial components of the gauge potential can be combined as $A = A_z = \frac{1}{2}(A_1 + iA_2)$, $\bar{A} = A_{\bar{z}} = \frac{1}{2}(A_1 - iA_2)$. The crucial first step for us is the parametrization of these potentials as $A = -\partial M\, M^{-1}$, $\bar{A} = M^{\dagger -1}\bar{\partial} M^{\dagger}$. Here M is an $SL(N, \mathbb{C})$-matrix for an $SU(N)$-gauge theory. More generally, M is a complex matrix which is an element of $G^{\mathbb{C}}$, the complexification of G which is the Lie group in which the gauge transformations take values.

Time-independent gauge transformations act on M via $M(\vec{x}) \longrightarrow g(\vec{x})M(\vec{x})$, $g(\vec{x}) \in SU(N)$. The hermitian matrix $H = M^{\dagger}M$ is thus gauge-invariant. The fact that the gauge transformations are made homogeneous is why the matrix parametrization for A, $\bar{A}$ is so useful.

Wave functions $\Psi(H)$ are gauge-invariant and depend only on $H = M^{\dagger}M$; this is basically the Gauss law. Further, the volume element for the space of gauge-invariant configurations can be explicitly evaluated and the inner product for the wave functions obtained as

$$\langle 1|2\rangle = \int d\mu(H)\ \exp\left(2c_A S_{wzw}(H)\right)\ \Psi_1^*\Psi_2 \tag{1}$$

where $S_{wzw}(H)$ is the Wess-Zumino-Witten action for the hermitian field H and $d\mu(H)$ is the Haar measure for H viewed as an element of $SL(N, \mathbb{C})/SU(N)$. The number c_A is the quadratic Casimir value for the adjoint representation; it is defined in terms of the structure constants f^{abc} (of the Lie algebra) as $c_A\delta^{ab} = f^{amn}f^{bmn}$ and is equal to N for $SU(N)$. The WZW action is given by

$$S_{wzw}(H) = \frac{1}{2\pi}\int \mathrm{Tr}(\partial H \bar{\partial} H^{-1}) + \frac{i}{12\pi}\int \epsilon^{\mu\nu\alpha}\mathrm{Tr}(H^{-1}\partial_\mu H\ H^{-1}\partial_\nu H\ H^{-1}\partial_\alpha H) \tag{2}$$

Actually we can strengthen the statements given above a little further and argue that the wave functions Ψ and the Hamiltonian can be taken to be functions of a scaled version of the current of the WZW action, namely, of $J = (2/e)\partial H\, H^{-1}$. One way to see this is to consider the Wilson loop variable,

$$W(C) = \mathrm{Tr}\ \mathcal{P}e^{-\oint_C A} = \mathrm{Tr}\ \mathcal{P}\exp\left(\frac{e}{2}\oint_C J\right) \tag{3}$$

where, in the second step, we have simply used the matrix parametrization for A, $\bar{A}$. Since all gauge-invariant quantities can be constructed from the Wilson loop variable (by suitable choices of the contour of integration C), we may restrict attention to functions of J.

Taking the wave functions to be functions of the current J, functional derivatives with respect to $A, \bar{A}$ may be obtained via the chain rule of differentiation. This leads to the expression for the Hamiltonian operator in terms of the J as $\mathcal{H} = \mathcal{H}_0 + \mathcal{H}_1$ with

$$\begin{aligned} \mathcal{H}_0 &= m\int J_a(\vec{z})\frac{\delta}{\delta J_a(\vec{z})} + \frac{2}{\pi}\int_{w,z}\frac{1}{(z-w)^2}\frac{\delta}{\delta J_a(\vec{w})}\frac{\delta}{\delta J_a(\vec{z})} + \frac{1}{2}\int_z :\bar{\partial}J^a(z)\bar{\partial}J^a(z): \\ \mathcal{H}_1 &= ie\int_{w,z} f_{abc}\frac{J^c(w)}{\pi(z-w)}\frac{\delta}{\delta J_a(\vec{w})}\frac{\delta}{\delta J_b(\vec{z})} \end{aligned} \tag{4}$$

where $m = e^2 c_A/2\pi = e^2 N/2\pi$. Notice that the first term of the Hamiltonian is counting powers of J in the wave function on which it acts, giving an energy contribution m for each J. This will turn out to be the essence of the mass gap in the theory.

The Hamiltonian operator involves products of J's, $\delta/\delta J$'s and nonlocal functions such as $(z-w)^{-1}$. As in any field theory, all calculations have to be done with proper regularization.

Journal of Physics: Conference Series **462** (2013) 012039 doi:10.1088/1742-6596/462/1/012039

Table 1. Comparison of magnetic mass calculations

m/e^2	Method
0.25	Resummation of perturbation theory [4]
0.35	Lattice, common factor for glueball masses [5]
0.51	Lattice, maximal abelian gauge [6]
0.52	Lattice, Landau gauge [6]
0.44	Lattice, $\lambda_3 = 2$ gauge [6]
0.38	Resummation of perturbation theory [7]
0.28	Resummation of perturbation theory [8, 9]
0.37	Gauge-invariant lattice definition [10]
0.32	Calculation via our Hamiltonian method

We will not go into this matter here, except to say that all the results presented here have been checked using regularized expressions, with a single regulator from beginning to end, to avoid possible conflicts between regularizations. For more details, see [1, 2, 3].

One interesting result can be read off from what we have done so far. By comparison with resummed perturbation theory, we can see that m serves as the propagator mass for gluons (which are generated by J). This can then be identified with the magnetic screening mass if we interpret the theory as the high temperature limit of QCD. For the gauge group $SU(2)$, $m = e^2/\pi \approx 0.32e^2$. There are a number of other calculations for the magnetic screening mass, so it is interesting to compare these values. Table 1 shows that while there is some variation, the values are not outrageously different from e^2/π.

In our earlier work [1, 2], we solved the Schrödinger equation $\mathcal{H}\Psi_0 = 0$ for the vacuum wave function Ψ_0 to the leading order in a strong-coupling expansion to obtain $\Psi_0 = \exp(-\frac{1}{2}S)$, where

$$\begin{aligned} S(H) &= \int \bar{\partial} J^a \left[\frac{1}{(m + \sqrt{m^2 - \nabla^2})} \right] \bar{\partial} J^a + \mathcal{O}(J^3) \\ &\approx \frac{1}{2m} \int \bar{\partial} J^a \bar{\partial} J^a = \frac{1}{4g^2} \int d^2x \; F^a_{ij} F^a_{ij} \end{aligned} \tag{5}$$

where the second line is the approximation to modes of J with momentum $\ll e^2$. Thus the computation of expectation values reduces in this limit to a calculation in a Euclidean two-dimensional Yang-Mills theory with a coupling $g^2 = me^2$. The expectation value of the Wilson loop variable for the representation R then takes the form

$$\langle W(C) \rangle \sim \exp\left(-\sigma_R \mathcal{A}_C\right), \qquad \sqrt{\sigma_R} \;=\; e^2 \sqrt{\frac{c_A c_R}{4\pi}} \tag{6}$$

where $\mathcal{A}_C$ is the area enclosed by the curve C and σ_R is the string tension. This compares favorably, to within $1 - 3\%$, with the lattice calculations [11, 12], see table 2.

I have not described in detail the process of solving the Schrödinger equation; this is because there is another argument which can be used to obtain the results to this order [2]. For this purpose, absorb the factor $\exp(2c_A S_{wzw}(H))$ from the inner product into the wave function by $\Psi = e^{-c_A S_{wzw}(H)}\Phi$. Then the Hamiltonian acting on Φ is given by $\mathcal{H} \longrightarrow \mathcal{H}_\Phi = e^{c_A S_{wzw}(H)}\, \mathcal{H}\, e^{-c_A S_{wzw}(H)}$. We shall now consider this expression when H can be approximated as $H = e^{t^a \varphi^a} \approx 1 + t^a \varphi^a + \cdots$; this small φ limit is appropriate for perturbation theory. The new Hamiltonian is then

$$\mathcal{H}_\Phi \approx \frac{1}{2} \int \left[-\frac{\delta}{\delta\phi^a \delta\phi^a} + \phi^a (m^2 - \nabla^2) \phi^a \right] + \cdots \tag{7}$$

Journal of Physics: Conference Series **462** (2013) 012039 doi:10.1088/1742-6596/462/1/012039

Table 2. Comparison of predictions from (6) with lattice calcuations

Group	Representations					
	k=1 Fund.	k=2 antisym	k=3 antisym	k=2 sym	k=3 sym	k=3 mixed
$SU(2)$	0.345 0.335					
$SU(3)$	0.564 0.553					
$SU(4)$	0.772 0.759	0.891 0.883		1.196 1.110		
$SU(5)$	0.977 0.966					
$SU(6)$	1.180 1.167	1.493 1.484	1.583 1.569	1.784 1.727	2.318 2.251	1.985 1.921
$SU(N)$ $N\to\infty$	0.1995 N 0.1976 N					

The upper entries follow from (6), the lower entries are the lattice values. k is the rank of the representation.

where $\phi^a = \sqrt{c_A(-\nabla^2)/8\pi m}\ \varphi^a$. We see that the leading term in $\mathcal{H}_\Phi$ corresponds to a free field of mass m (actually *dim G* fields, counting the multiplicity due to the index a.) The field ϕ may be taken as representing the gluon (actually a gauge-invariant version with some resummations involved) and this is consistent with what we said about the gluon mass being m. The vacuum wave function in this approximation may be trivially calculated as $\Phi_0 = \exp(-\frac{1}{2}S)$, with $S = \int \phi^a\ \sqrt{m^2 - \nabla^2}\ \phi^a$. If this is transformed back to the Ψ-version, we find

$$\Psi_0 \approx \exp\left[-\frac{c_A}{\pi m}\int (\bar\partial\partial\varphi^a)\left[\frac{1}{m+\sqrt{m^2-\nabla^2}}\right](\bar\partial\partial\varphi^a) + \cdots\right] \tag{8}$$

This was obtained in an approximation of small φ. However, we do know that the full wave function must be a function of the current J. So we can ask the question: What function of J can we write such that it reduces to (8) in the small φ approximation when $J \approx (2/e)\partial\varphi$? The answer, to quadratic order in J, is clearly the solution (5). Notice that this argument only depends on the form of integration measure which, in turn, is determined by a two-dimenional anomaly calculation. So it is rather robust; for more details, see [2].

Another interesting line of development has been about glueball masses. The authors of [13] have followed the general line of reasoning we have outlined, but used a somewhat different wave function given by

$$\Psi_0 = \exp\left[-\frac{1}{4m}\int \bar\partial J\ K[L]\ \bar\partial J\right], \qquad K[L] = \frac{J_2(4\sqrt{L})}{\sqrt{L}\ J_1(4\sqrt{L})} \tag{9}$$

where $L = \mathcal{D}\bar\partial/m^2$ and $J_1,\ J_2$ are Bessel functions of orders 1 and 2 respectively. The kernel K is, despite appearances, very close to the kernel in (5), as is clear from the graphical comparison in figure 1 (I thank A. Yelnikov for this picture). Glueball masses were obtained by calculating the two-point function for different color-singlet composite operators, characterized by spin, parity and charge conjugation properties (j^{PC}-notation). We will not display the results here for lack of space, but they are given in [13] and are in reasonable agreement with the lattice data of reference [11, 14].

There have been a number of other developments as well. The formalism has been extended to the Yang-Mills-Chern-Simons theory and issues related to dynamical mass generation and screening of Wilson loops have been analyzed [15].

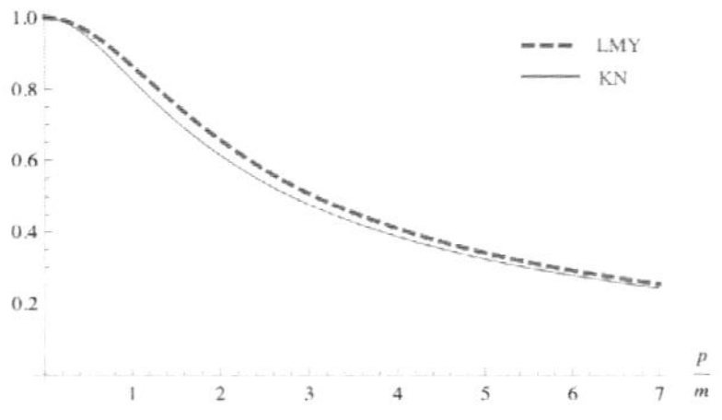

Figure 1. Comparison of KKN and LMY kernels for the Gaussian term of the vacuum wave function as a function of the momentum p

Another interesting question has been about the screening of $\mathbb{Z}_N$-invariant representations. The energy of a glue-lump state made of a heavy scalar field and the gluons can be estimated in the limit of large e^2 and large N [3]. This gives the energy at which an adjoint string breaks as $E_* \approx 7.92\, m$, for $SU(2)$, to be compared with the lattice value of $E_* \approx 8.68\, m$ [16]; we are off by about 8.8%. Considering the difficulties in calculating both in our method and the lattice, this may be taken as not too bad. The calculation is more a qualitative demonstration of feasibility: it is possible to incorporate screening in our formalism.

We have also worked out the extension of the formalism to the case of space being S^2, i.e., for $\mathbb{R} \times S^2$ [17], as a prelude to calculations on the torus which may be used to analyze finite-temperature effects. In this regard, I also mention the recent paper by Abe, where some of the torus formalism is developed and arguments on deconfinement are presented [18].

3. Corrections to string tension

We now turn to the string tension formula (6) [19]. This is important for several reasons. First, we want to see the vacuum wave function (5) and the corresponding string tension (6) as the lowest order in a systematic expansion scheme. Secondly, notice that, in the large N limit, our formula differs from the lattice value by about 1%. A different lattice calculation shows the difference to be about 1.55% [20]. In both cases, the lattice calculations are considered accurate enough that the deviations are statistically significant. So we need a systematic method which has the potential to calculate corrections.

There are two types of corrections possible. For the first set, we can calculate the corrections to the propagators (or the coefficient of $\bar{\partial} J\, \bar{\partial} J$ in the wave function) and then use the result in the calculation of the Wilson loop. This set of corrections will be independent of the representation R of the Wilson line. The second set would involve vertex type corrections and can be sensitive to the representation of the Wilson line. We will leave out these for now, focusing on the first set. The corrections are expected to be purely numerical, so, at first glance, seem to be impossible to control in any way. However, we shall see that a systematic expansion is possible along the following lines.

We rewrite the derivation of the vacuum wave function as a recursive procedure for the solution of the Schrödinger equation from which it will be clear that (5) is the lowest order result in a systematic expansion. For this purpose, we will treat m and e as independent parameters, setting $m = e^2 c_A/2\pi$ only at the end of all calculations. For $\Psi_0^* \Psi_0 = e^F$, this gives F as a power series in e, of the form

$$\begin{aligned} F \;=\; \int f^{(2)}_{a_1 a_2}(x_1,x_2)\; J^{a_1}(x_1) J^{a_2}(x_2) \;+\; \frac{e}{2}\, f^{(3)}_{a_1 a_2 a_3}(x_1,x_2,x_3)\; J^{a_1}(x_1) J^{a_2}(x_2) J^{a_3}(x_3) \\ +\; \frac{e^2}{4}\, f^{(4)}_{a_1 a_2 a_3 a_4}(x_1,x_2,x_3,x_4)\; J^{a_1}(x_1) J^{a_2}(x_2) J^{a_3}(x_3) J^{a_4}(x_4) \;+\; \cdots \end{aligned} \qquad (10)$$

$f^{(2)}$, $f^{(3)}$, etc., are determined recursively in powers of e, with each function having an expansion in powers of e^2; e.g., $f^{(2)} = f_0^{(2)} + e^2 f_1^{(2)} + \cdots$, etc. To the lowest order, with $q, \bar{q}$ denoting the momenta,

$$f_{0\ a_1 a_2}^{(2)}(x_1, x_2) = \delta_{a_1 a_2}\left[-\bar{q}^2/(m+E_q)\right]_{x_1,x_2} \approx \delta_{a_1 a_2}\ \bar{q}^2/2m \tag{11}$$

which leads to (5). To calculate corrections to order e^2, we need the lowest order results for $f^{(3)}$ and $f^{(4)}$. These are given in [19].

We are interested in corrections to $f^{(2)}$ since that determines the string tension. From the recursive solution of the Schrödinger equation, we find

$$e^2\ f_2^{(2)}(q) \approx \frac{\bar{q}^2}{2m}(1.1308) + \ldots \tag{12}$$

Seemingly, this is a 113% correction, but, as we shall see, other "loop" corrections are important.

Since the measure of integration has the WZW action, we can transform the functional integration over $\Psi_0^*\Psi_0 = e^F$ into the integration over a two-dimensional chiral boson field φ, $\bar{\varphi}$ (not the φ we used in parametrizing H as $e^{t^a\varphi^a}$). In other words, for an observable $\mathcal{O}$,

$$\begin{aligned} \langle\mathcal{O}\rangle &= \int d\mu(H)\ e^{2c_A S_{wzw}(H)}\ e^{F(J)}\ \mathcal{O}(J) = \int [d\varphi d\bar{\varphi}]\ e^{-S(\varphi)}\ \mathcal{O}(\sqrt{2\pi/mc_A}\ \bar{\varphi}t^a\varphi) \\ S(\varphi) &= \int (Z_2\bar{\varphi}\bar{\partial}\varphi + Z_1\bar{\varphi}\bar{C}\varphi)\ -\ F(Z_1\sqrt{2\pi/mc_A}\ \bar{\varphi}t^a\varphi) \end{aligned} \tag{13}$$

$(Z_1, Z_2$ are renormalization constants, we will not discuss these further here; see [19].)[1] Notice that, effectively, the current J^a is replaced by $Z_1\sqrt{2\pi/mc_A}\ \bar{\varphi}t^a\varphi$. The function $F(Z_1\sqrt{2\pi/mc_A}\ \bar{\varphi}t^a\varphi)$ contains vertices, $F^{(2)}$ with two currents (quartic in φ, $\bar{\varphi}$), $F^{(3)}$ with three currents, etc. For example, we may diagrammatically represent $F^{(2)}$, with two φ's and two $\bar{\varphi}$'s, as

$$F^{(2)} = \tfrac{2\pi}{mc_A}\int(\bar{\varphi}t^a\varphi)_x\ f^{(2)}(x,y)(\bar{\varphi}t^a\varphi)_y\ =$$ 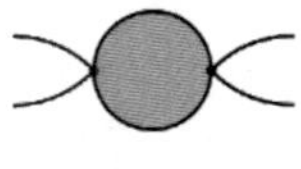

The corrections to $F^{(2)}$, which is what we are interested in, may be viewed as Wilsonian renormalization corrections to the quartic vertex in this two-dimensional field theory.

In computing the corrections to $F^{(2)}$, we can treat the vertices $F^{(3)}$, $F^{(4)}$, etc., perturbatively since they carry powers of e. However, the lowest term in the vertex $F^{(2)}$, corresponding to $f_{0a_1a_2}^{(2)}(x_1, x_2)$, has no powers of e and hence its contributions must be summed up. This means that the current-current correlator becomes

$$\langle\bar{\varphi}t^a\varphi(x)\ \bar{\varphi}t^b\varphi(y)\rangle = \delta^{ab}\frac{c_A}{\pi}\int\frac{d^2k}{(2\pi)^2}e^{ik(x-y)}\ \frac{k}{\bar{k}}\left(\frac{m}{E_k}\right) \tag{14}$$

Here $E_k = \sqrt{k^2+m^2}$; the m/E_k factor arises from the summation of corrections due to $F_0^{(2)}$, shown diagrammatically in figure 2. Our computational strategy is then the following: We will construct loop diagrams generated by $F^{(3)}$ (3 factors of $\bar{\varphi}t^a\varphi$) and $F^{(4)}$ (4 factors of $\bar{\varphi}t^a\varphi$), taking account of the arbitrary insertions of $F^{(2)}$'s, and the corresponding factors of m/E_k, as in figure 3. There are also renormalizations (due to $F^{(2)}$) we have to take into account. We then sum up $F^{(2)}$ insertions in all diagrams (of order e^2) generated by $F^{(3)}$ and $F^{(4)}$, using powers of m/E_k to group them appropriately. We will compute corrections to order e^2 and up to 4 powers of m/E_k. Denoting the factors of m/E_k by shaded circles at the vertices, the corrections to the

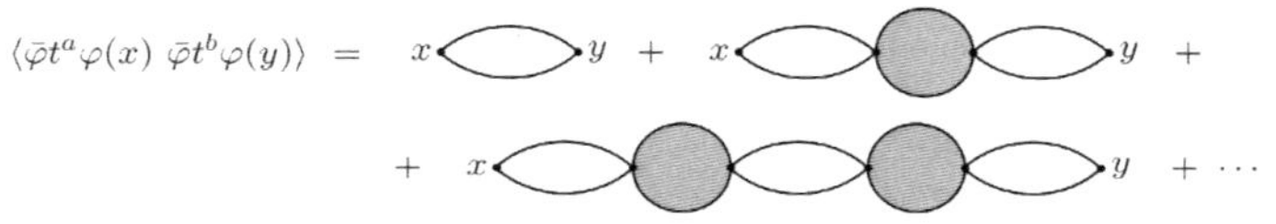

Figure 2. The current-current correlator including all contributions from $F^{(2)}$

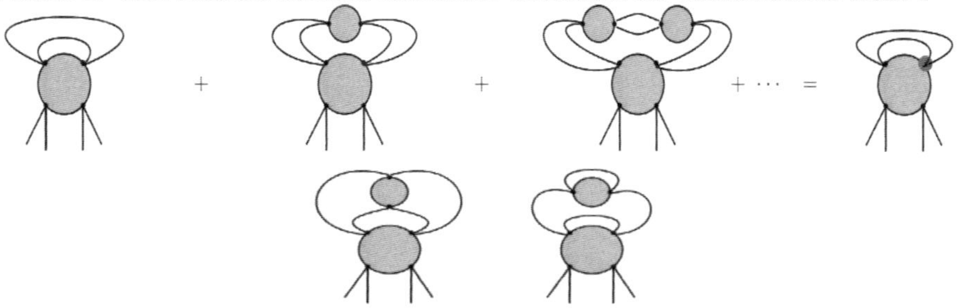

Figure 3. Corrections from $F^{(2)}$ summed up as a factor of m/E_k (shaded circle at vertex) and sample renormalization diagrams

low momentum limit of $f^{(2)}$ may summarized as in figure 4. We show the coefficients of $\bar{q}^2/2m$, for small q, $\bar{q}$.

Let C_n denote the partial sum of corrections up to terms with $(m/E_k)^n$, starting with $C_0 = 1.1308$ from the recursive procedure (12). We then find $C_1 = 0.5496$, $C_2 = 0.2730$, $C_3 = 0.0373$. There is a small ambiguity in an integral for the last diagram in figure 5 [19], so that C_4 is in the range -0.05843 to -0.00583. The partial sums are systematically decreasing in value, showing that the ordering of diagrams by powers of m/E_k does constitute a sensible expansion. Also, the corrections to the order we have calculated are small. For the string tension, we find

$$\sqrt{\sigma_R} = e^2 \sqrt{\frac{c_A c_R}{4\pi}} \begin{cases} (1 - 0.02799 + \cdots) \\ (1 - 0.0029 + \cdots) \end{cases} \tag{15}$$

This correction, of the order of -2.8% to -0.03%, is entirely consistent with lattice calculations.

[1] $\bar{C}$ may be set to zero at this point, it was included to show how Z_1, Z_2 appear in $S(\varphi)$. Also we may note that the transformation to φ, $\bar{\varphi}$ is analogous to the fermionization of the WZW model.

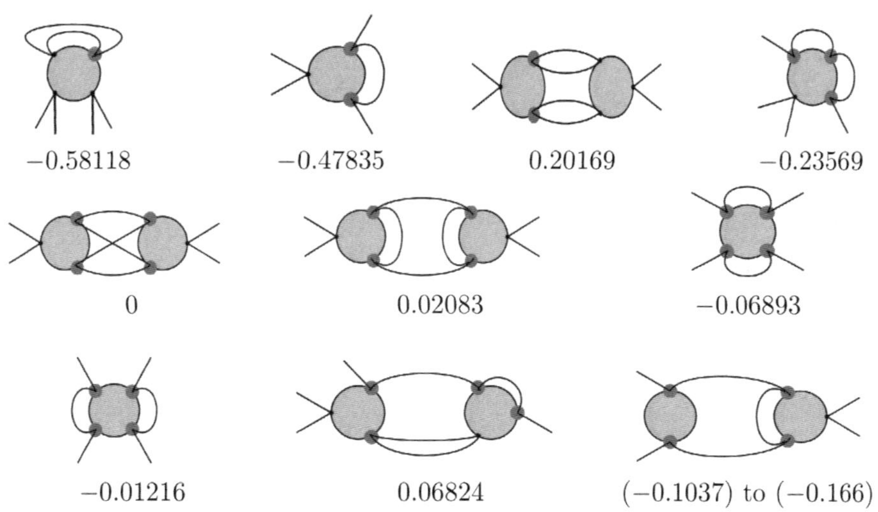

Figure 4. Corrections to the low momentum limit of the $F^{(2)}$ vertex

Terms of order $(m/E_k)^5$ are expected to contribute at the level of a fraction of 1%, and likewise for diagrams with two or more current loops.

4. Discussion

Perhaps the most important point to be mentioned is that one has a systematic expansion and a scheme for calculations. At first glance, the possibility of any expansion scheme in 2+1 dimensional Yang-Mills theory seems very remote, to say the least. The coupling constant e^2 does not constitute an expansion parameter. It simply tells us that modes of momenta $\ll e^2$ should be treated nonperturbatively, while modes of momenta $\gg e^2$ can be treated perturbatively; it is at best a marker for this separation. Nevertheless, from what we have discussed, a systematic calculational procedure is possible. This is done by keeping m and e^2 as independent until the end, giving a parametric way to classify various contributions to any calculation. Further, there is a kinematic factor, of the form $(m/E_k)^n$, which helps in the further classification of corrections. The corrections to the lowest order results for the string tension then come out to be rather small.

In the language of ordinary perturbation theory, we may note that our analysis involves at least three different resummations. First, the definition of H in terms of A, $\bar{A}$ is an infinite series with nonlocal terms. Second, in including m in $\mathcal{H}_0$, there is another resummation, since m is really of order e^2. Thirdly, we also carry out the summation of contributions due to $F^{(2)}$ to get factors of m/E_k. An *a priori* attempt at these resummations, even with some guidance from the Hamiltonian analysis, is clearly very difficult.

Where do we go from here? Higher order corrections to the string tension, a better computation of glueball masses, extension to the torus and finite-temperature effects as well as the inclusion of matter fields are some of the questions we could now explore systematically.

References

[1] D. Karabali and V.P. Nair, Nucl. Phys. **B464**, 135 (1996); Phys. Lett. **B379**, 141 (1996); D. Karabali, Chanju Kim and V.P. Nair, Nucl. Phys. **B524**, 661 (1998).
[2] D. Karabali, Chanju Kim and V.P. Nair, Phys. Lett. **B434**, 103 (1998); D. Karabali and V.P. Nair, Phys. Rev. **D77**, 025014 (2008).
[3] A. Agarwal, D. Karabali and V.P. Nair, Nucl. Phys. **B790**, 216 (2008).
[4] The possibility of calculating the magnetic mass by summing Feynman diagrams was addressed in J.M. Cornwall, Phys. Rev. **D26**, 1453 (1982). The paper J.M. Cornwall, W-S. Hou, J.E. King, Phys. Lett. **B153**, 173 (1985) gave the bound $m/e^2 \geq 0.58$; the later paper J.M. Cornwall, Phys. Rev. **D57**, 3694 (1998) gave a modified procedure with the estimate $0.248 \approx 0.25$, as quoted.
[5] W. Buchmüller and O. Philipsen, Nucl.Phys. B443 (1995) 47.
[6] F. Karsch *et al*, Nucl. Phys. **B474**, 217 (1996); F. Karsch, M. Oevers and P. Petreczky, Phys. Lett. **B442**, 291 (1998).
[7] G. Alexanian and V.P. Nair, Phys. Lett. **B352**, 435 (1995).
[8] W. Buchmüller and O. Philipsen, Phys. Lett. **B397**, 112 (1997).
[9] R. Jackiw and S.Y. Pi, Phys. Lett. **B368**, 131 (1996); *ibid* **B403**, 297 (1997).
[10] O. Philipsen, Phys. Lett. **B521**, 273 (2001); talk at *Lattice 2001*, hep-lat/0110114.
[11] M. Teper, Phys. Rev. **D59**, 014512 (1999) and references therein.
[12] B. Lucini and M. Teper, Phys. Rev. **D66**, 097502 (2002); B. Bringoltz and M. Teper, Phys. Lett. B **645**, 383 (2007).
[13] R. G. Leigh, D. Minic and A. Yelnikov, Phys. Rev. Lett. **96**, 222001 (2006); Phys. Rev. **D76**, 065018 (2007).
[14] H.B. Meyer and M.J. Teper, Nucl. Phys. **B668**, 111 (2003).
[15] D. Karabali, Chanju Kim and V.P. Nair, Nucl. Phys. **B566**, 331 (2000).
[16] S. Kratochvila and P. de Forcrand, Nucl. Phys. **B671**, 103 (2003); see also O. Philipsen and H. Wittig, Phys. Lett. **B451**, 146 (1999); P.W. Stephenson, Nucl. Phys. **B550**, 427 (1999).
[17] A. Agarwal and V. P. Nair, Nucl. Phys. **B816**, 117 (2009) [arXiv:0807.2131].
[18] Yasuhiro Abe, arXiv:0804.3125 [hep-th].
[19] D. Karabali, V.P. Nair and A. Yelnikov, arXiv:0906.0783 [hep-th], to be published in Nucl. Phys. B.
[20] J. Kiskis and R. Narayanan, JHEP 0809:080 (2008) [arXiv:0807.1315].

Journal of Physics: Conference Series **462** (2013) 012040 doi:10.1088/1742-6596/462/1/012040

On Quasinormal modes and Quantization of Area of Black Holes

Yongjoon Kwon and Soonkeon Nam

Department of Physics, Kyung Hee University, Seoul, 130-701, Korea

E-mail: emwave@khu.ac.kr, nam@khu.ac.kr

Abstract. Perturbations of black holes give very important tools of understanding physical properties of black holes. The relaxation of the perturbation is described by quasinormal modes with complex frequency. Using this, there has been some proposal of Hod and later by Maggiore that one can obtain quantization of the area of black holes. These ideas, when applied to rotating black holes were less than controversial. In this work we address how the are can be quantized with equal spacing. We also comment on application of this to other black holes in three dimensional spacetime.

1. Introduction

Black holes are one of the most fascinating objects in physics which have been confirmed to exist in our Universe. The term black hole was first used in public by J.A. Wheeler at the Sigma Xi - Phi Beta Kappa lecture in 1967 held at New York Hilton hotel [1]. Classical behavior of a black hole is rather simple, and depend only on a few parameters, due to the well known 'No Hair Theorem'. Again Wheeler was involved coining this [2]. Whether these black hole solutions are stable under linear perturbation was the question raised by Regge and Wheeler in 1957, when they wrote down the wave equation of linear fluctuation around a spherically symmetric Schwarzschild black hole of mass M [3]. The 'Regge-Wheeler' equation has a form of Schrodinger equation, with an effective potential

$$V_{\text{eff}} = \left(1 - \frac{2GM}{r}\right)\left[\frac{l(l+1)}{r^2} + \frac{2GM}{r^3}\right], \tag{1}$$

where l is the angular momentum mode number, and G is the Newton constant. This seemingly easy Schrödinger equation still requires numerical solution. It was Chandrasekhar and Detweiler [4] who have done some numerical calculations and obtained the first few modes. The wave should vanish at the black hole horizon and there is no incoming wave from the asymptotic infinity. These boundary conditions allow only specific values of the frequency, like normal modes. Due to the perfect absorption at the horizon, the wave decays away, so there is an imaginary term in the frequency. So we call these modes quasinormal modes.

In particular the quasinormal mode becomes, for large imaginary value, as follows:

$$\omega_n M \sim 0.0437 - \frac{i}{4}\left(n + \frac{1}{2}\right) + O\left(\frac{1}{\sqrt{n+1}}\right). \tag{2}$$

This behavior is independent of the value of the angular momentum to leading order. The real part of the quasinormal mode become a universal number.

When people performed some analytic calculations [5], it was shown that the real part behaves as

$$0.0437 \approx \frac{\log 3}{8\pi}. \tag{3}$$

The imaginary part can be easily computed for the large damping case. One can use the Born approximation. When perform the Fourier transformation of the Regge-Wheeler potential for the approximation, we obtain some Gamma functions. The poles of the Gamma functions occur when we impose the boundary conditions. These lead to the discrete values of the imaginary part of the quasinormal frequency.

For the Reissner-Nordstrom black hole, the linear wave equation around it was studied by Zerilli [6]. The Kerr black hole case was much more involved and was first written down by Teukolsky [7,8], who used the Newman-Penrose formalism.

These quasinormal modes are important probes of black holes. Just like we can get some idea of the property of wine inside a casket by tapping on it, quasinormal modes capture some of the properties of classical black holes.

2. Black hole area quantization

It is believed that the quantum behavior of black holes could play a significant role as a testing ground for the quantum theory of gravity. Sometime ago Bekenstein [9] proposed that since the black hole horizon area is an adiabatic invariant, it should be quantized. The argument was through Ehrenfest principle [10] which says that only an adiabatically invariant quantities are quantizable. When we consider the minimum change of the horizon area by an absorption of a test particle into the black hole, it is reasonable that this minimum area should be proportional to the Planck area $l_P^2 = \frac{\hbar G}{c^3}$. Therefore intuitively we have

$$A_n = \gamma n l_P^2 = \gamma n \hbar, \quad n = 0, 1, 2, \cdots, \tag{4}$$

where γ is an undetermined dimensionless constant. (We are using the units $c = G = 1$). There has been a lot of work in relation to this observation [11].

An important development on this line of though was done by Hod [12] who suggested that the dimensionless parameter γ can be determined by the quasinormal mode of a black hole. He argued that the real part of the asymptotic quasinormal mode ω_R, can be regarded as the transition frequency in the semiclassical limit and therefore the when there is a quantum emitted there will be change in the black hole mass, $\Delta M = \hbar\omega_R$. Since $A = 16\pi M^2$, we have

$$\Delta A = 32\pi M \Delta M = 32\pi M \frac{\hbar \log 3}{8\pi M} = 4 \log \hbar. \tag{5}$$

Therefore he could determine the dimensionless parameter γ of Bekenstein to be $4 \log 3$. Some people in loop quantum gravity found some significance in this number. Furthermore we have an equally spaced area spectrum.

A different view on this problem was provided by Kunstatter [13]. By viewing as the quasinormal mode as a mechanical system with frequency ω_R, then noticed that the quantity

$$I = \int \frac{dE}{\omega} \tag{6}$$

is an adiabatic invariant, at least for a vibrational system with frequency ω. Once we have an adiabatic invariant we can perform Bohr-Sommerfeld quantization. For example when we plug in the ADM mass for the energy of Schwarzschild black hole, and use ω_R for the frequency in the denominator of the integrand, we get $M^2 = \frac{\log 3}{4\pi} n\hbar$, in agreement with Hod's result.

More recently, Maggiore [14] argued that we should view the quasinomal mode as a damped harmonic oscillator, especially for the highly excited modes. Actually it is these highly excited modes that we can safely use the Bohr-Sommerfeld quantization. Furthermore the transition frequency should be the characteristic classical frequency ω_c where

$$\omega_c \approx (|\omega_I|)_k - (|\omega_I|)_{k-1}. \tag{7}$$

With this new identification, the change in the mass is $\Delta M \hbar \omega_c$, and for highly damped quasinormal modes, it is not dependent on the real part but rather the imaginary part of the quasinormal mode spectrum. The result is that we now have $A_n = 8\pi n\hbar$.

When these methods were applied to rotating Kerr black hole, the result was that the area spectrum is no longer equally spaced [15-18]. This lead to some confusion. In these works, the area was taken to be the adiabatic invariant. By using the area law for the entropy, $S = \frac{A}{4}$, we then have $I \sim \int \frac{T_H dS}{T_H}$, where we have multiplied the Hawking temperature T_H. The reason for this is that quasinormal modes is proportional to the Hawking temperature in the asymptotic region. If we use the first law of black hole thermodynamics we the get

$$I \sim \int \frac{dM - \Omega dJ}{\omega} = n\hbar \tag{8}$$

where Ω is the angular velocity at the horizon and j is the angular momentum of the black hole. It turns out that thus obtained area of black hole has a very complicated area spectrum, which is not equally spaced.

To us this is not satisfactory and we would like to have an equally spaced area spectrum, because Bekenstein's argument was rather general and also intuitively it is correct. Why was the adiabatic invariant was quantized? As mentioned before, Ehrenfest principle says that any classical adiabatic invariant corresponds to a quantum object with discrete spectrum. However this spectrum need not be equally spaced. There is a particular subclass of adiabatic invariants are quantized with equally spaced spectrum. It is the action variable

$$\mathcal{I} = \frac{1}{2\pi} \oint pdq. \tag{9}$$

When we do Bohr-Sommerfeld quantization, we have $\mathcal{I} = n\hbar$. I.e. an action variable is an adiabatic invariant but the reverse is not necessarily true. So instead of considering the area itself as the adiabatic invariant, we need to identify the action variable. It is the integration of the energy divided by the transition frequency! The change in energy is the change in the mass M.

3. Area Spectrum of a Rotating BTZ Black Hole

To demonstrate our suggestion that we have to quantize the action variable, and that we do indeed get an equally spaced area spectrum even for a rotating case, we would like to consider the rotating BTZ black hole solution in 2+1 dimensional spacetime [19].

$$ds^2 = -\left(-M + \frac{r^2}{l^2} + \frac{J^2}{4\,r^2}\right) dt^2 + \frac{dr^2}{\left(-M + \frac{r^2}{l^2} + \frac{J^2}{4\,r^2}\right)} + r^2 \left(d\phi - \frac{J}{2\,r^2}\,dt\right)^2 , \tag{10}$$

where the cosmological constant is given by $\Lambda = -\frac{1}{l^2}$. The mass M and angular momentum J of the black hole can be expressed in terms of the outer and inner horizons, $r_\pm$, as follows:

$$M = \frac{r_+^2 + r_-^2}{l^2} \quad , \quad J = \frac{2\,r_+\,r_-}{l} . \tag{11}$$

The two families of the quasinormal modes of the rotating BTZ black hole for a massive scalar field are given by [20, 21]

$$\omega_R = -\frac{m}{l} - i\frac{(r_+ + r_-)\,(2k+1+\sqrt{1+\mu})}{l^2}\ , \quad \omega_L = \frac{m}{l} - i\frac{(r_+ - r_-)\,(2k+1+\sqrt{1+\mu})}{l^2}\ , \quad (12)$$

where $m \in Z$ and $k = 0, 1, 2, ...$ The m and k are the angular quantum number and the overtone quantum number respectively, and μ is the mass parameter, $\mu \equiv u^2\, l^2/\hbar^2$, where u is the mass of the scalar field. At large k for a fixed $|m|$, in particular for $k \gg |m|$, the two families of quasinormal modes gives two possible transition frequencies, ω_{Rc} and ω_{Lc}. We find the two transition frequencies corresponding to each quasinormal mode;

$$\omega_{Rc} = \frac{2\,(r_+ + r_-)}{l^2} = \frac{2\sqrt{M + J/l}}{l}\ , \quad \omega_{Lc} = \frac{2\,(r_+ - r_-)}{l^2} = \frac{2\sqrt{M - J/l}}{l}\ . \quad (13)$$

We consider the two action variables corresponding to each possible transition frequency. We obtain the two quantization conditions;

$$\mathcal{I}_{\mathcal{R}} = \int \frac{dM}{\omega_{Rc}} = l\sqrt{M + J/l} = n_R\,\hbar\ , \ \ \mathcal{I}_{\mathcal{L}} = \int \frac{dM}{\omega_{Lc}} = l\sqrt{M - J/l} = n_L\,\hbar\ , \quad (14)$$

where n_R, $n_L = 0, 1, 2, ..$. Notice that we have $n_R \geq n_L$.

Because the total horizon area is quantized and equally spaced;

$$A_{tot} = A_{out} + A_{in} = 2\,\pi\, n_R\,\hbar\ , \qquad n_R = 0, 1, 2, \ldots \quad (15)$$

Interestingly, the difference of the two horizon areas is also quantized

$$A_{sub} \equiv A_{out} - A_{in} = 2\,\pi\, l\,\sqrt{M - J/l} = 2\,\pi\, n_L\,\hbar\ , \qquad n_L = 0, 1, 2, \ldots \quad (16)$$

We therefore find the quantizations of the outer and inner horizon areas;

$$A_{out} = \pi\,(n_R + n_L)\,\hbar\ , \ \ A_{in} = \pi\,(n_R - n_L)\,\hbar\ , \quad (17)$$

where $n_R \geq n_L$. Therefore we find that both the outer and inner horizon areas are equally spaced with the same spacing; $\Delta A_{out} = \pi\,\hbar$, $\Delta A_{in} = \pi\,\hbar$ .By Bekenstein-Hawking area law, the entropy spectrum is given by

$$S = \frac{4\,\pi\, r_+}{\hbar} = \frac{A_{tot} + A_{sub}}{\hbar} = 2\,\pi\,(n_R + n_L)\ . \quad (18)$$

Hence it also has equal spacing, $\Delta S = 2\,\pi$ consistent with Bekenstein's proposal; Notice that the spacing of the entropy spectrum is half of the non-rotating BTZ black hole case and same as for the Schwarzschild black hole [18]. So this simple exercise shows that even for rotating black holes, albeit in three dimensions, has equal spacing [23].

4. The BTZ black hole with the gravitational Chern-Simons term

Let us try to do a slightly more There is a small debate in the gravity community arguing which is more fundamental? Geometry or information? Or as Wheeler has put, "It or Bit?"'. This issue cannot be resolved in cases where the entropy is proportional to the horizon area. This is the case for black hole solutions in Einstein gravity, where the action is the Einstein-Hilbert action. However, for the cases where there are additional terms in the action, entropy is no longer proportional to the horizon area of the black hole. For these cases comparison of the area and the entropy spectra becomes very interesting as we will see below. The question is, which of these has equal spacing, if any? For this purpose we will consider black holes which arises when there is gravitational Chern-Simons term in the action:

$$\mathcal{S} = \frac{1}{16\pi G}\int_M d^3x\sqrt{-g}\left(R+\frac{2}{l^2}\right) - \frac{l}{96\pi G\nu}\int_M d^3x\sqrt{-g}\,\varepsilon^{\lambda\mu\nu}\Gamma^r_{\lambda\sigma}\left(\partial_\mu\Gamma^\sigma_{r\nu}+\frac{2}{3}\Gamma^\sigma_{\mu\tau}\Gamma^\tau_{\nu r}\right). \quad (19)$$

It again has the rotating BTZ black hole as a solution. The metric of the rotating BTZ black hole is exactly the metric given in the previous section. Also, the quasinormal modes are same as for the rotating BTZ black hole without the Chern-Simons term. However we have to use ADT mass rather than ADM mass. ADT mass of a BTZ black hole is given as follows:

$$\mathcal{M} = M + \frac{J}{3\,\nu\,l}\,, \quad (20)$$

where M and J are the mass and the angular momentum for the BTZ black hole in the absence of the Chern-Simons term. So, the change of the energy of the black hole should be considered as the change of the ADT mass $\mathcal{M}$, i.e. $dE = d\mathcal{M}$. The action variable $\mathcal{I}$ is given by

$$\mathcal{I} = \int\frac{dE}{\omega_c} = \int\frac{d\mathcal{M}}{\omega_c} = \int\frac{dM}{\omega_c} + \frac{1}{3\,\nu\,l}\int\frac{dJ}{\omega_c}\,. \quad (21)$$

The two action variables, $\mathcal{I}_+$ and $\mathcal{I}_-$, are obtained and quantized as follows:

$$\mathcal{I}_\pm = \frac{(3\,\nu\pm1)\,l}{24\,G\,\nu}\sqrt{8\,G\,M\pm8\,G\,J/l} = n_R\,\hbar\,. \quad (22)$$

We find that the total horizon area spectrum is given by

$$A_{tot} \equiv A_{out}+A_{in} = 2\,\pi\,l\sqrt{8\,G\,M+8\,G\,J/l} = \frac{48\,\pi\,G\,\nu}{(3\,\nu+1)}\,n_R\,\hbar\,, \quad (23)$$

where $n_R = 0, 1, 2, \ldots$ This spectrum is equally spaced but the spacing is dependent on the coupling constant ν. The spectra of the outer and inner horizon areas are obtained as follows:

$$A_\pm = \frac{24\,G\,\pi\,\nu}{9\,\nu^2-1}\Big[(3\,\nu-1)\,n_R \pm (3\,\nu+1)\,n_L\Big]\,\hbar\,, \quad (24)$$

where $(3\,\nu-1)\,n_R \geq (3\,\nu+1)\,n_L$. $A_+ = A_{out}$ and $A_- = A_{in}$. These area spectra are not equally spaced and dependent on the coupling constant ν, so that they are affected by the Chern-Simons term.

Now let us find the entropy spectrum. There is ν dependence in the relation between horizon areas and the entropy as follows:

$$S = \frac{1}{4\,G\,\hbar}\left(A_{out}+\frac{1}{3\,\nu}\,A_{in}\right) \quad (25)$$

However, in the entropy spectrum ν dependence drops out completely and we have and equally spaced spectrum:

$$S = 2\,\pi\,(n_R+n_L)\,. \quad (26)$$

The entropy spectrum (26) is independent of the coupling constant ν and exactly same as for the rotating BTZ black hole without the Chern-Simons term obtained in the previous section. Therefore the Chern-Simons term does not affect the entropy spectrum of the BTZ black hole, even though it affects the area spectra. This suggests that the entropy is more fundamental than the horizon area. We will consider another black hole solution in the next section which will enforce this point.

5. Area spectrum vs Entropy Spectrum

Again the action we will consider will have the gravitational Chern-Simons term together with the usual Einstein Hilbert action. There is a new solution called the warped AdS_3 black holes with a deformation parameter $\nu \geq 1$. The case $\nu = 1$ is related to the BTZ black hole considered in the previous section. The metric is given by [24, 25]

$$ds^2 = l^2 \Bigg[dt^2 + \frac{dr^2}{(\nu^2+3)(r-r_+)(r-r_-)} + \left(2\nu r - \sqrt{(\nu^2+3)\,r_+ r_-}\right) dt\, d\theta$$
$$+\frac{r}{4}\left(3(\nu^2-1)\,r + (\nu^2+3)(r_+ + r_-) - 4\nu\sqrt{(\nu^2+3)\,r_+ r_-}\right) d\theta^2\Bigg] , \tag{27}$$

where r_+ and r_- are the outer and inner horizons, respectively. The coordinates t and r are dimensionless. The entropy of the warped AdS_3 black hole is given by [25, 26]

$$S = \frac{\pi l}{24 G \nu \hbar}\left[\left(9\nu^2+3\right) r_+ - \left(\nu^2+3\right) r_- - 4\nu\sqrt{(\nu^2+3)\,r_+ r_-}\right] . \tag{28}$$

As we can see it has rather complicated ν dependence. The conserved charges such as the ADT mass $\mathcal{M}$ and angular momentum $\mathcal{J}$ are given by [25, 26]

$$\mathcal{M} = \frac{(\nu^2+3)}{24 G}\left(r_+ + r_- - \frac{1}{\nu}\sqrt{(\nu^2+3)\,r_+ r_-}\right) , \tag{29}$$

$$\mathcal{J} = \frac{(\nu^2+3)\,\nu\, l}{96 G}\left[\left(r_+ + r_- - \frac{1}{\nu}\sqrt{(\nu^2+3)\,r_+ r_-}\right)^2 - \frac{(5\nu^2+3)}{4\nu^2}(r_+ - r_-)^2\right], \tag{30}$$

which satisfy the first law of the black hole thermodynamics, $d\mathcal{M} = \tilde{T}\, dS + \tilde{\Omega}\, d\mathcal{J}$, with the Hawking temperature $\tilde{T}$ and the angular velocity of the horizon $\tilde{\Omega}$ given by [25]

$$\tilde{T} \equiv \frac{T}{l} = \frac{(\nu^2+3)}{4\pi l}\frac{(r_+ - r_-)}{(2\nu r_+ - \sqrt{(\nu^2+3)\,r_+ r_-})} , \tag{31}$$

$$\tilde{\Omega} \equiv \frac{\Omega}{l} = -\frac{2}{(2\nu r_+ - \sqrt{(\nu^2+3)\,r_+ r_-})\, l} . \tag{32}$$

From the metric we can read off the inner and the outer horizon areas,

$$A_{out} = \pi l \left(2\nu r_+ - \sqrt{(\nu^2+3)\,r_+ r_-}\right) , \tag{33}$$

$$A_{in} = \pi l \left|2\nu r_- - \sqrt{(\nu^2+3)\,r_+ r_-}\right| . \tag{34}$$

When we calculate the spectra of the outer and inner horizon areas we get [namkwon2]:

$$A_{out} = \frac{48\pi G \nu^2}{5\nu^2+3}\left(n_R + \frac{5\nu^2+3}{4\nu^2} n_L\right)\hbar , \tag{35}$$

$$A_{in} = \frac{48\pi G \nu^2}{5\nu^2+3}\left|n_R - \frac{5\nu^2+3}{4\nu^2} n_L\right|\hbar . \tag{36}$$

These area spectra are not equally spaced and dependent on the coupling constant ν. The entropy spectrum is obtained by rewriting the entropy expression (28) in terms of the horizon areas as follows:

$$S = \frac{(9\nu^2+3)\,A_{out} \pm (\nu^2+3)\,A_{in}}{48 G \nu^2 \hbar} . \tag{37}$$

Because of the Chern-Simons term, the entropy is proportional to not only the outer horizon area but also the inner horizon area, and again there is ν dependence. Nevertheless, when we plug in the area spectra, miraculous canceling of ν happens and the entropy spectrum is as follows:

$$S = 2\,\pi\,(n_R + n_L)\;. \tag{38}$$

This entropy spectrum is independent of the coupling constant ν and the spacing is given by $\triangle S = 2\,\pi$ [27]. Furthermore this is the exactly same as for the rotating BTZ black holes with and without the Chern-Simons term. It implies that the entropy spectra of the black holes have a universal behavior regardless of the presence of the gravitational Chern-Simons term. Therefore we would like to claim that the entropy spectrum rather than the area spectrum of a black hole should be equally spaced. Furthermore it strongly suggest the entropy which has a universal behavior is more fundamental than the area. Wheeler towards the end of his career was leaning towards information rather than the geometry as the fundamental entity of gravitation. We pleasantly find some examples which agrees with his deep insights.

Acknowledgments
This work is supported by the National Research Foundation of Korea grant funded by the Korea government (MEST) (No. 2009-0063068).

References
[1] Frolov V P and Novikov I D 1998 Black Hole Physics, (Kluwer Academic Pub., Boston, 1998). See Fig. 1.1 where the handwritten note of Wheeler confirms this fact.
[2] Ruffini R and Wheeler J A 1971 *Physics Today* **24** 30
[3] T. Regge T and J.A. Wheeler J A 1957 *Phys. Rev.* **108** 1063
[4] Chandrasekhar S and Detweiler S 1975 *Proc. R. Soc. London* **A344** 441
[5] See for example Motl L and Neitzke A 2003 *Adv. Theor. Math. Phys.* **7** 307
[6] Zerilli F J 1974 *Phys. Rev.* **D9** 860
[7] Teukolsky S A 1972 *Phys. Rev. Lett.* **29** 1114
[8] Teukolsky S A 1973 *Astrophys. J.* **185** 635
[9] Bekenstein J D 1974 *Lett. Nuovo Cim.* **11** 467.
[10] See, for example, Born M 1969 *Atomic physics* (8th edition) (London: Blackie & Son Ltd.) p 113.
[11] Bekenstein J D and Mukhanov V F 1995 *Phys Lett.* B **360** 7 (*Preprint* gr-qc/9505012);
Louko J and Makela J 1996 *Phys. Rev.* D **54** 4982 (*Preprint* gr-qc/9605058);
Louko J and Makela J 1998 *Phys. Rev.* D **57** 4899 (*Preprint* gr-qc/9708029);
Vaz C and Witten L 2001 *Phys. Rev.* D **63** 024008 (*Preprint* gr-qc/0006039);
Makela J, Repo P, Luomajoki M and Piilonen J 2001 *Phys. Rev.* D **64** 024018 (*Preprint* gr-qc/0012055);
Barvinsky A, Das S and Kunstatter G 2001*Class. Quant. Grav.* **18** 4845 (*Preprint* gr-qc/0012066);
Barvinsky A and Das S 2002 *Found. Phys.* **32** 1851 (*Preprint* hep-th/0209039);
Gour G 2002 *Phys. Rev.* D **66** 104022 (*Preprint* gr-qc/0210024);
Gour G and Medved A J M 2003 *Class. Quant. Grav.* **20** 2261 (*Preprint* gr-qc/0211089);
Gour G and Medved A J M 2003 *Class. Quant. Grav.* **20** 1661 (*Preprint* gr-qc/0212021);
Padmanabhan T and Patel A 2003 (*Preprint* gr-qc/0309053);
Tamaki T and Nomura H 2004 *Phys. Rev.* D **70** 044041 (*Preprint* hep-th/0405191);
Khriplovich I B 2005 *Int. J. Mod. Phys.* D **14** 181 (*Preprint* gr-qc/0407111);
Das S, Mukhopadhyay H and Ramadevi P 2005 *Class. Quant. Grav.* **22** 453 (*Preprint* hep-th/0407151);
Khriplovich I B 2008 *Phys. Atom. Nucl.* **71** 671 (*Preprint* gr-qc/0506082);
Kiselev V V 2005 (*Preprint* gr-qc/0509038);
Kothawala D, Padmanabhan T and Sarkar S 2008 *Phys. Rev.* D **78** 104018 (*Preprint* arXiv:0807.1481);
Ropotenko K 2009 *Phys. Rev.* D **80** 044022 (*Preprint* arXiv:0906.1949);
Medved A J M 2009 *Mod. Phys. Lett.* A **24** 2601 (*Preprint* arXiv:0906.2641)
[12] Hod S 1998 *Phys. Rev. Lett.* **81** 4293 (*Preprint* gr-qc/9812002)
[13] Kunstatter G 2003 *Phys. Rev. Lett.* **90** 161301 (*Preprint* gr-qc/0212014)
[14] Maggiore M 2008 *Phys. Rev. Lett.* **100** 141301 (*Preprint* arXiv:0711.3145)

[15] Setare M R and Vagenas E C 2005 *Mod. Phys. Lett.* A **20** 1923 (*Preprint* hep-th/0401187); Fernando S 2009
[16] Setare M R and Vagenas E C 2005 *Mod. Phys. Lett.* A **20** 1923 (*Preprint* hep-th/0401187)
[17] Vagenas E C 2008 *J. High Energy Phys.* JHEP **11** 073 (*Preprint* arXiv:0804.3264)
[18] Medved A J M 2008 *Class. Quant. Grav.* **25** 205014 (*Preprint* arXiv:0804.4346)
[19] Banados M, Teitelboim C and Zanelli J 1992 *Phys. Rev. Lett.* **69** 1849 (*Preprint* hep-th/9204099)
[20] Cardoso V and Lemos J P S 2001 *Phys. Rev.* D **63** 124015 (*Preprint* gr-qc/0101052)
[21] Birmingham D 2001 *Phys. Rev.* D **64** 064024 (*Preprint* hep-th/0101194)
[22] Setare M R 2004 *Class. Quant. Grav.* **21** 1453 (*Preprint* hep-th/0311221)
[23] Kwon Y and Nam S 2010 arXiv:1001.5106, to appear in *Class. Quant. Grav.*
[24] Moussa K A, Clement G and Leygnac C 2003, *Class. Quant. Grav.* **20** L277 (*Preprint* gr-qc/0303042)
[25] Anninos D, Li W, Padi M, Song W and Strominger A 2009, *JHEP* **03** 130 (*Preprint* arXiv:0807.3040)
[26] Bouchareb A and Clement G 2007, *Class. Quant. Grav.* **24** 5581 (*Preprint* arXiv:0706.0263)
[27] Kwon Y and Nam S 2010 arXiv:1002.0911.

Journal of Physics: Conference Series **462** (2013) 012041 doi:10.1088/1742-6596/462/1/012041

Phase transitions at strong coupling in the 2+1-d abelian Higgs model

R. B. MacKenzie

Groupe de physique des particules, Université de Montréal, C.P. 6128, succ. centre-ville, Montréal, Québec, Canada, H3C 3J7

E-mail: rbmack@lps.umontreal.ca

Faïza Nebia-Rahal

Groupe de physique des particules, Université de Montréal, C.P. 6128, succ. centre-ville, Montréal, Québec, Canada, H3C 3J7

E-mail: faiza@lps.umontreal.ca

M. B. Paranjape

Groupe de physique des particules, Université de Montréal, C.P. 6128, succ. centre-ville, Montréal, Québec, Canada, H3C 3J7

E-mail: paranj@lps.umontreal.ca

Abstract. We study, using numerical Monte-Carlo simulations, an effective description of the 2+1 dimensional Abelian Higgs model which is valid at strong coupling, in the broken symmetry sector. In this limit, the massive gauge boson and the massive neutral Higgs decouple leaving only the massive vortices. The vortices have no long range interactions. We find a phase transition as the mass of the vortices is made lighter and lighter. At the transition, the contributions to the functional integral come from a so-called infinite vortex anti-vortex loop. Adding the Chern-Simons term simply counts the linking number between the vortices. We find that the Wilson loop exhibits perimeter law behaviour in both phases, although the polarization cloud increases by an order of magnitude at the transition. We also study the 't Hooft loop. We find the 't Hooft loop exhibits perimeter law behaviour in the presence of the Chern-Simons term but is trivial in its absence. Thus we have a theory with perimeter law for both the Wilson loop and the 't Hooft loop, but contains no massless particles.

1. Abelian Higgs model at strong coupling

The abelian Higgs model corresponds to the Lagrangian:

$$\mathcal{L} = -\frac{1}{4}F_{\mu\nu}F^{\mu\nu} + \frac{1}{2}D_\mu\phi(D^\mu\phi)^* - \frac{\lambda}{4}(|\phi|^2 - \eta^2)^2. \qquad (1)$$

At strong coupling, $\lambda, e \to \infty$, keeping λ/e^2 fixed, the neutral scalar mass $\sqrt{2\lambda}\eta$ and the gauge boson mass $e\eta$ diverge but the vortex mass $\mu = \eta^2 f(2\lambda/e^2)$ can be held finite and variable, see [1] for details.

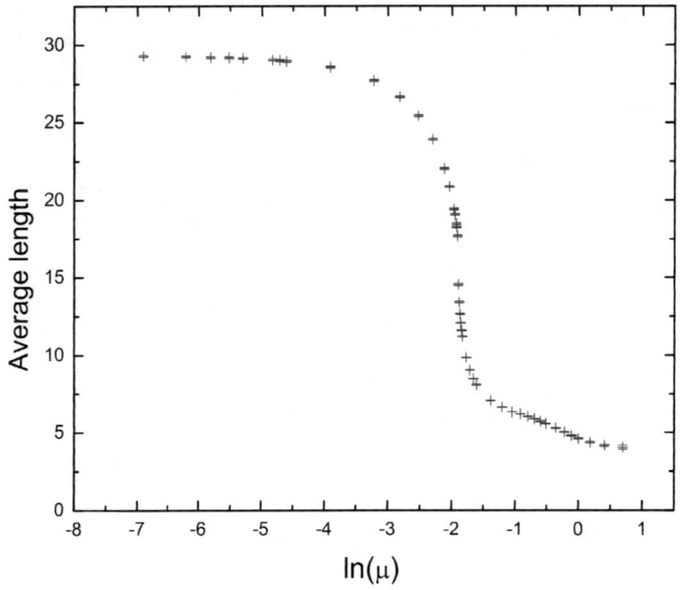

Figure 1. (color online) Average length of the loops as a function of $\ln(\mu)$.

1.1. Evidence for a phase change

We calculate the Euclidian vacuum persistence amplitude using the functional integral and lattice Monte Carlo simulations. The field configurations that contribute to this amplitude correspond to closed vortex loops, which describe the virtual process of the creation of a vortex anti-vortex pair and its subsequent annihilation. The Euclidian action for such a configuration is, to first approximation, simply the vortex mass multiplied by the length of the vortex loop [2]. Thus we generate configurations of closed loops on a 3 dimensional lattice, and weigh each such configuration using the Boltzmann factor corresponding to the total length of the loops multiplied by μ. For large μ only small minimal loops enter the calculation, however as μ is lowered, larger and larger loops become important. After a specific value, $\mu = .152$, there is a sudden transition where about half of the loops reorganize and form a so-called infinite loops. for smaller values of μ, the total length of the vortex loops increases, however this occurs simply by appending to the length of the infinite loop, the number of finite loops nor their total length do not appreciably change.

An infinite loop corresponds to a loop that has a length substantially larger than what a loop should have if it were of a linear dimension equal to the size of the lattice. The loops are closed non-self intersecting random walks. On a cubic lattice their linear dimension should be approximately $N^{(3/5)}$ [3]. Our lattice is actually a much more intricate lattice, where it is not known what the size of an infinite loop should be. We consider any loop longer than 10000 steps to be infinite.

Clear evidence for a change of phase is seen in Figure (1) of the average length of the loops versus the logarithm of the mass of the vortices, $\ln\mu$. We see that as μ is diminished, the average length of the loops increases dramatically about six-fold.

2. Wilson loop

We compute the expectation value of the Wilson loop [4], using our Monte Carlo simulation. We place the Wilson loop along a prescribed rectangular contour. The Wilson loop operator simply measures the total linking number of the Wilson loop contour with the dynamical vortex loops.

$$W(L,T) = \left\langle e^{-i\frac{q}{e}\oint A_\mu dx_\mu} \right\rangle, \tag{2}$$

We then evaluate this for various different spatial sizes, we find for example: which show a perfect

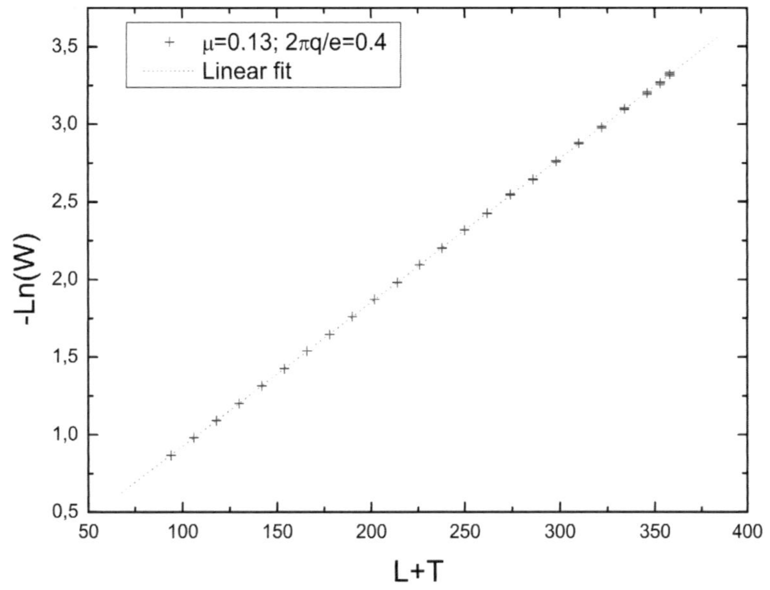

Figure 2. (color online) Plot of $-\ln(W)$ for $\mu = .13$. The dashed lines are a linear fit, we find $-\ln(W) = .009280(\pm 4\times 10^{-6})\cdot(L+T) + .00350(\pm 5\times 10^{-4})$

perimeter law. The perimeter law exists on both sides of the transition, only the slope changes by about 10 fold [1].

3. Chern-Simons term

It is possible to add the Chern-Simons term to the action:

$$\mathcal{L}_{CS} = 16\pi^3\kappa \int d^3x\, \epsilon^{\mu\nu\lambda} A_\mu F_{\nu\lambda} \tag{3}$$

The Chern-simons term converts charged particles or magnetic flux tubes into anyons. Hence the vortices behave as fractional statistics particles. With this normalization, the term is just equal to $2\kappa \times L_T$ where L_T is the total linking number of all the vortex loops with each other. The important distinction is that in Euclidean space this term, being t-odd remains imaginary [5] in Euclidean space.. Hence it cannot be used to define the Boltzmann factor in the Monte Carlo simulation. However the real part of the action provides a perfectly good measure on the space of vortex loops, which respects all the symmetries of the theory with the Chern-Simons term added. Hence we can use the ensemble obtained by the Monte Carlo procedure to determine the expectation values with the Chern-Simons term added, then:

$$\langle \mathcal{O} \rangle \approx \langle \mathcal{O} \rangle_N = \frac{\sum_{i=1}^{N} \mathcal{O}(C_i) e^{iS_{CS}}}{\sum_i e^{iS_{CS}}} \tag{4}$$

In principle, the set of Green's functions obtained from this measure will satisfy the Wightman axioms [6], and can be used to reconstruct [7] the full Chern-Simons added quantum field theory.

4. Wilson loop and 't Hooft loop in the Chern-Simons theory

The calculation of these two order parameters follows in a straightforward fashion in the Chern-Simons theory, for details see [8]. The Wilson loop operator is unchanged since the Chern-simons term has no import on the distribution of the linking number of the Wilson loop. Hence, at fixed total linking number, the average of the Wilson loop is equal to its average without the restriction of fixed total linking number and the effect of the Chern-Simons term cancels in Eqn.(4). The Wilson lop continues to display perimeter law.

The 't Hooft loop [9] on the other hand is completely dependent on the Chern-Simons term. The 't Hooft loop corresponds to the insertion of a singular flux tube into the functional integral, along a fixed rectangular path. The Chern-Simons term simply measures the linking number of this flux tube with the rest of the dynamical vortex loops. It is evident that the 't Hooft loop is trivial in the absence of the Chern-Simons term. However, for fixed value of the Chern-Simons coefficient, the 't Hooft loop behaves exactly as the Wilson loop for changes in the dimensions of rectangular path. since it is 2κ times the linking number of the rectangular loop, it behaves exactly as the Wilson loop as a function of the parameters of the rectangular path. Therefore it displays the perimeter law.

5. Conclusions

We have computed using Monte Carlo simulations in an effective description of the 2+1 dimensional Abelian Higgs model, properties of the vacuum. Although we find a phase change, the expectation value of the Wilson loop and the 't Hooft loop does not change. The 't Hooft loop is trivial without the Chern-simons term, the Wilson loop displays perimeter law behaviour in both phases. The two order parameters continue to display the perimeter law in both phases, even in the presence of the Chern-Simons term. This is a counter example to the understanding that such behaviour necessitates the existence of massless particles, [10]. Here we have no massless particles, however the Chern-Simons term exercises a subtle long range, statistics changing interaction, which is able to permit the perimeter law behaviour for both order parameters.

6. Acknowledgments

We thank Al Shapere and Sumit Das for a wonderful, well organized QTS6 conference and Jacques Richer of the Réseau Québecois de Calcul de Haute Performance for many discussions and help with the calculations. We also thank NSERC of Canada for financial support.

References

[1] R. MacKenzie, F. Nebia-Rahal and M. B. Paranjape, "Phase transitions in a 3-d lattice loop gas," arXiv:0710.3236 [hep-lat].
[2] H.B. Nielsen and P.Olesen, Nucl. Phys. **B61**, 45 (1973).
[3] Madras, N. and Slade, G. 'The Self-Avoiding Walk', Birkhäuser, Boston, (1993).
[4] K. G. Wilson, "Confinement of quarks," Phys. Rev. D **10**, 2445 (1974).

[5] G. Alexanian, R. MacKenzie, M. B. Paranjape and J. Ruel, "Path integration and perturbation theory with complex Euclidean actions," Phys. Rev. D **77**, 105014 (2008) [arXiv:0802.0354 [hep-th]].

[6] A. S. Wightman, "Quantum Field Theory in Terms of Vacuum Expectation Values," Phys. Rev. **101**, 860 (1956).

[7] K. Osterwalder and R. Schrader, "Axioms For Euclidean Green's Functions" Commun. Math. Phys. **31**, 83 (1973); "Axioms For Euclidean Green's Functions 2," Commun. Math. Phys. **42**, 281 (1975).

[8] R. MacKenzie, F. Nebia-Rahal, M. B. Paranjape and J. Richer, "Phase transitions in a theory of anyons", in preparation.

[9] G. t'Hooft, Nucl. Phys. B138, 1 (1978).

[10] A. Ukawa, P. Windey and A. H. Guth, "Dual Variables For Lattice Gauge Theories And The Phase Structure Of Z(N)Systems," Phys. Rev. D **21**, 1013 (1980).

Journal of Physics: Conference Series **462** (2013) 012042 doi:10.1088/1742-6596/462/1/012042

Dimer models and quiver gauge theories

Ramadevi Pichai

Department of Physics, Indian Institute of Technology Bombay, Mumbai-400 076

E-mail: ramdevi@phy.iitb.ac.in

Abstract. $\mathcal{N} = 1$ quiver gauge theories on coincident $D3$ branes placed at a tip of a Calabi-Yau singularity C are dual to string theories on $AdS_5 \times X_5$ where X_5 are Sasaki-Einstein spaces. We present a neat combinatorial approach called dimer model to understand interrelations between toric quiver gauge theories and toric data representing the Calabi-Yau singularities.

1. Introduction

$\mathcal{N} = 1$ quiver gauge theories arise on D-branes placed at the tip of the singular Calabi-Yau C and they are dual to string theory on $AdS_5 \times X_5$ [1] where X_5 is called the Sasaki-Einstein manifolds having a topology of $S^2 \times S^3$ topology. Mathematically, C is the cone over X_5:

$$ds^2(C) = dr^2 + r^2 dX_5^2 \,. \tag{1}$$

Quiver gauge theory on a D-brane placed transverse to orbifold C^3/Γ is obtained from $U(r)$ gauge theory subjected to orbifolding action (r= rank of the orbifold group Γ). Clearly, orbifolding results in breaking the gauge group $U(r) \rightarrow U(1)^r$. There are possibilities of appearance of bi-fundamental fields and the adjoint fields. The matter content and their interactions of these quiver gauge theories have information about the Calabi-Yau singularities. As indicated in Fig.1, there are systematic algorithms[2] to determine toric data for the Calabi-Yau space from the matter content and gauge groups of quiver gauge theories. Conversely, one could obtain quiver gauge theory data from the toric data. Eventhough, these algorithms are straightforward, the method is tedious. Few years ago, Hanany et al[3] showed that there exists an elegant approach called dimer tiling approach which directly gives both the gauge theory data as well as the toric data.

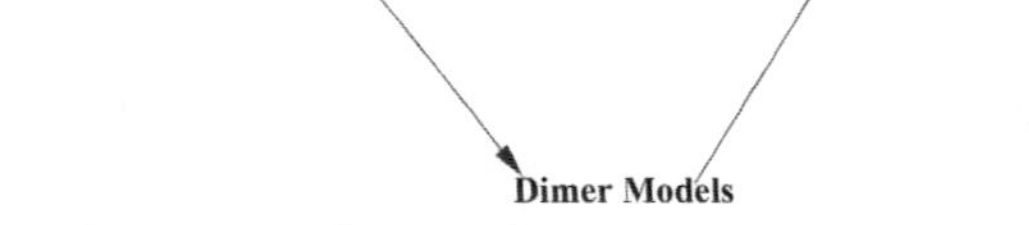

Figure 1. Flow chart

The plan of the article is as follows:In section 2, we first review quiver gauge theories and the toric diagrams representing toric Calabi-Yau three-folds. Then, we present the salient features of the algorithms. In section 3, we briefly discuss statistical mechanical models called dimer models and their powerfulness in directly determining both the quiver gauge theories and the toric diagram. In the concluding section, we summarise our results.

2. Quiver gauge theories and toric Calabi-Yau spaces

2.1. Quiver gauge theories

The matter content of quiver gauge theories are usually represented by a quiver diagram as shown in Fig.2. The number of nodes r in the quiver diagram gives the number of gauge groups and the number of edges m determines the number of adjoint & bi-fundamental fields. The

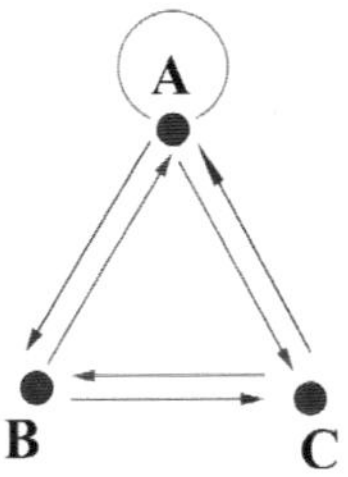

Figure 2. Quiver diagram

line from node A to itself denotes the adjoint field. The directed edge from node A to node B implies that a bifundamental field X_{AB} charged -1 with respect to gauge group G_A and $+1$ with respect to gauge group G_B . For the example shown in Fig.2, we see that there are three gauge groups $r = 3$, one adjoint and six bi-fundamental fields ($m = 7$). The diagram directly gives charge matrix d_{ai} where $a = (1, 2 \ldots r)$ is the gauge group index and $i = (1, 2 \ldots m)$ is the matter field index. These d_{ai}'s appear in the D-term equations: $D_a = \sum_i d_{ai}|X_i|^2 - \xi_a$. One of the rows of the charge matrix can always be eliminated to give a $(r-1) \times m$ matrix Δ. The interaction amongst the fields is given by a superpotential $W[\{X\}]$. The quiver gauge theories are called *toric quiver gauge theories* if all the gauge groups have the same rank and W contains each bifundamental field in two terms with relative signs. For example, the following superpotential

$$\begin{aligned} W &= X_{BA}X_{AB}X_{BC}X_{CB} - X_{CB}X_{BC}X_{CA}X_{AC} \\ &\quad + X_{AC}X_{CA}X_{AA} - X_{AB}X_{BA}X_{AA} \,, \end{aligned} \quad (2)$$

plus the quiver diagram in Fig.2 gives the toric quiver gauge theory information. It will be interesting to determine the toric Calabi-Yau corresponding to these quiver gauge theories. We will now briefly present toric diagrams representing toric Calabi-Yau spaces and then discuss the algorithms relating the Calabi-Yau spaces to the quiver gauge theories.

2.2. Non-compact toric Calabi–Yau spaces

Non-compact toric Calabi-Yau spaces are spaces described as T^3 fibration over R^3. These spaces can also be obtained from partial resolutions of orbifolds of C^3. There is an equivalent description of these toric Calabi-Yau spaces in Witten's gauged linear sigma model. Every toric Calabi-Yau spaces can be represented by a toric diagram. The toric diagram corresponding to C^3 and orbifolds of C^3 are drawn in Fig.3. We call these orbifolds as parent spaces. The numbers against every node i in the toric diagram indicate multiplicity of the sigma model fields. The

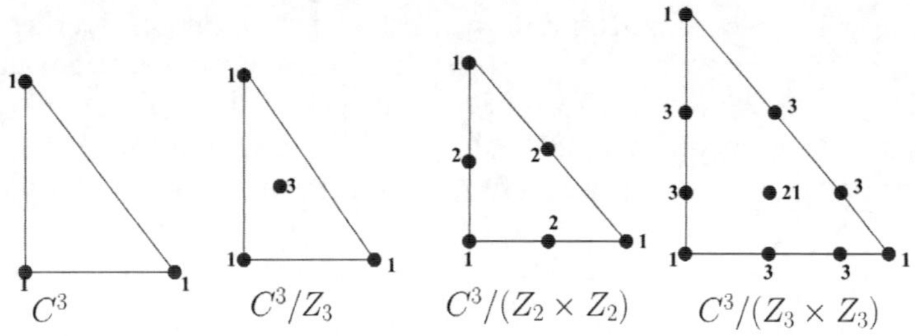

Figure 3. Toric diagrams for C^3 and their orbifolds

partial resolutions of these parent spaces gives non-orbifold toric Calabi-Yau spaces (daughter spaces) as shown in Fig.4. There is a systematic algorithm of relating a quiver gauge theories

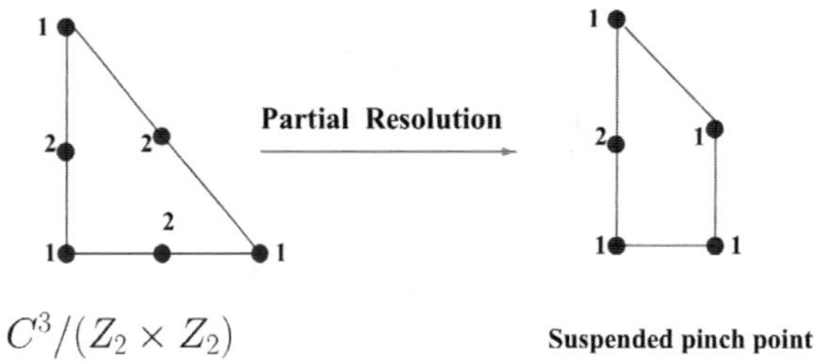

Figure 4. Partial resolution of a parent space into a daughter space

to the corresponding Calabi-Yau space which we call forward algorithm [2]. In the following subsection, we elaborate the essential details of the algorithm.

2.3. Forward Algorithm
(Quiver gauge theory $\rightarrow$ toric diagrams)
Given a gauge theory data (quiver diagram+W)[2], the F-term equations

$$\frac{\partial W}{\partial X_i} = 0 \ ,$$

are all not independent. These can be solved by introducing $r+2$ new fields v_j as:

$$X_i = \prod_j v_j^{K_{ij}}$$

where $i \in (1, m)$ and $j \in (1, r+2)$. For W (2) we can determine the K matrix (analogue of charge matrix Δ) in the following way: First we choose the v_i's to be

$$v_1 = X_{BA}, v_2 = X_{CA}, v_3 = X_{CB}, v_4 = X_{BC} \ ,$$

Journal of Physics: Conference Series **462** (2013) 012042 doi:10.1088/1742-6596/462/1/012042

and then determine the other matter fields X_i's using the F-term equations. We find $X_{AA} = v_2 v_4$, $X_{AC} = v_1 v_5$ and $X_{AB} = v_2 v_5$. Hence, the K matrix for this example is

$$K = \begin{pmatrix} & v_1 & v_2 & v_3 & v_4 & v_5 \\ X_{BA} & 1 & 0 & 0 & 0 & 0 \\ X_{CA} & 0 & 1 & 0 & 0 & 0 \\ X_{AA} & 0 & 0 & 1 & 1 & 0 \\ X_{AB} & 0 & 1 & 0 & 0 & 1 \\ X_{BC} & 0 & 0 & 0 & 1 & 0 \\ X_{AC} & 1 & 0 & 0 & 0 & 1 \\ X_{CB} & 0 & 0 & 1 & 0 & 0 \end{pmatrix}$$

Using this K matrix, we determine another rectangular matrix $(r+2) \times c$ matrix T satisfying $K.T \geq 0$. Here c is known only from the evaluation of the T matrix. Similar to K matrix relating X_i's to v_j's, the T matrix relates v_j fields to c new fields p_α's in the following way:

$$v_j = \prod_\alpha p_\alpha^{T_{j\alpha}} .$$

Clearly, the extra $c-(r-2)$ fields are redundant fields which are eliminiated using a $(c-r-2) \times c$ matrix Q_F such that $TQ_F^t = 0$. We say such a Q_F as the co-kernel of T. Similarly we could write a Q_D matrix related to the charge matrix Δ. Rewriting X_i's in terms of p_α's

$$X_i = \prod_\alpha p_\alpha^{\sum_j K_{ij} T_{j\alpha}} ,$$

we can see that $(K.T)Q_D^t = \Delta^t$. Concatenating Q_D and Q_F gives a matrix Q. This Q plays a crucial role to obtain the toric data. For the toric quiver gauge theories, we can find a rectangular matrix G such that $Q.G^t = 0$.Each column of G is a vector which can be plotted to give a toric diagram.

For the quiver theory example, G is

$$G = \begin{pmatrix} p_1 & p_2 & p_3 & p_4 & p_5 & p_6 \\ 0 & 1 & 0 & 0 & -1 & 1 \\ 1 & 1 & 1 & 0 & 1 & 0 \\ 1 & 1 & 1 & 1 & 1 & 1 \end{pmatrix}$$

which represents the suspended pinch point toric diagram in Fig.4. So far, we briefly presented the methodology of obtaining the toric data corresponding to a quiver gauge theory. It appears that each step is invertible and hence we should be able to tackle the converse problem of obtaining a quiver gauge theory from the toric data matrix G. Unlike the forward algorithm which gives a unique toric data, the inverse algorithm results in more than one quiver gauge theory corresponding to a toric diagram[2]. The inverse algorithm does not give the adjoint matter fields. Also, the algorithm is not practical to implement for large toric diagrams. Interestingly, an elegant combinatorial approach involving a statistical mechanical model called dimer models determines both the gauge-theory data as well as the Calabi-Yau geometry toric data[3]. We describe dimer models in the following section.

3. Dimer models

Dimer model refers to a bipartite graph with vertices colored either black or white such that no two adjacent vertices have the same color. Each edge is called a dimer. These graphs are on a torus and hence periodic as indicated. For such a graph, we can draw a *fundamental domain* as

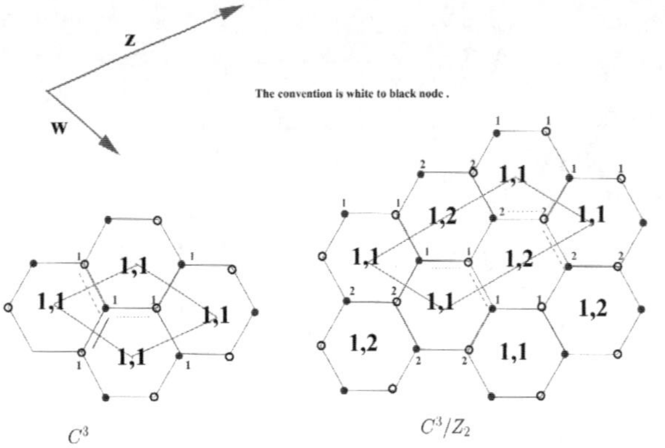

Figure 5. Dimer tiling for C^3 and C^3/Z_2

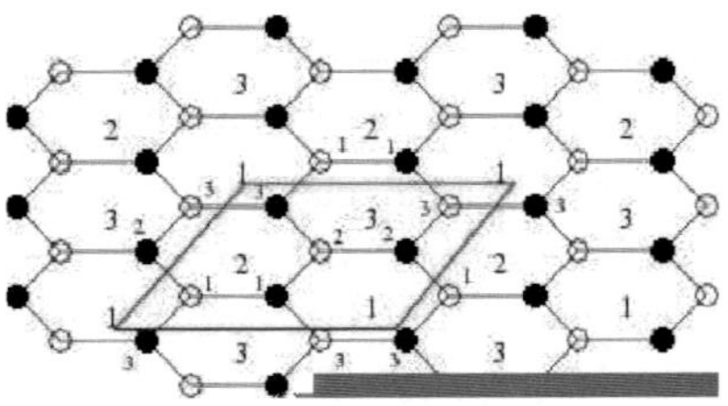

Figure 6. Dimer tiling for C^3/Z_3

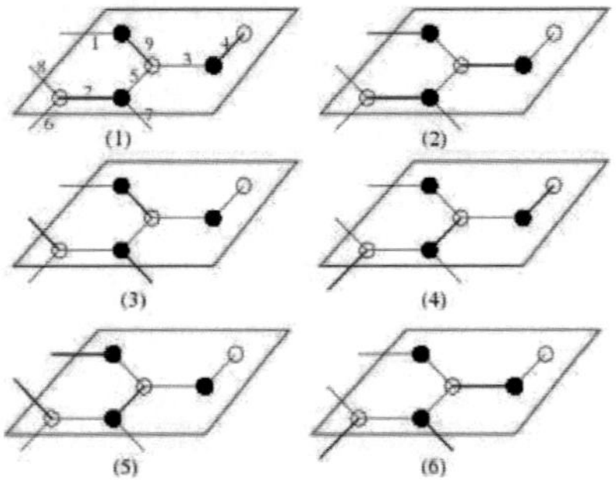

Figure 7. Perfect matchings for C^3/Z_3

shown by a parallellogram in Fig.5. The dimer tilings for C^3/Z_3 with the fundamental domain is shown in Fig.6. Clearly, within the fundamental domain, we can count the number of vertices and the edges connecting them. Every dimer model tiling is characterised by a quantity called *perfect matching*. Perfect matching is defined as a subset of edges such that every vertex in the fundamental domain is an endpoint of only one edge. For instance, in Fig.5, the dotted line(s) gives one perfect matching and dashed line(s) gives another perfect matching and so on. Similarly, there are six perfect matchings for the C^3/Z_3 case as indicated by blue edges in Fig.7. We can draw the edges representing difference between any two perfect matchings and see that they give *closed curves* enclosing a face of the dimer diagram. The total number of perfect matchings p_α's enumerated for a dimer diagram will determine c automatically without working out the T matrix. In order to determine the toric diagram associated with a a dimer diagram, we introduce *height function* $h = (h_W, h_Z)$ to each perfect matching. These height functions plotted on a two dimensional plane gives a toric diagram. In Fig.7, the height function for matchings (1),(5) and (6) are (0,0), and the matchings (2), (3) and (4) will have heights $(-1, 0)$, $(1, -1)$ and $(0, 1)$ respectively. Plotting the six height functions, we can see that the internal point in the toric diagram is $(0, 0)$ and its multiplicity is 3. There is another method of determining R-charge for every perfect matching in closed string theory[4]. Using basis vectors $(1/3, 0, 0)$, $(0, 1/3, 0)$,$(0, 0, 1/3)$[4], the perfect matchings $1, 5, 6$ will have R-charge as $(1/3, 1/3, 1/3)$ which corresponds to a marginal twisted sector. It can seen that the internal point in the toric diagram

Table 1. Quiver gauge theory and dimer diagram correspondence

Quiver theory	dimer diagram
Gauge groups	Faces
bi-fundamental fields	edges
superpotential W	vertices

corresponds to a marginal twisted sector state. We have seen that every perfect matching can be given a height function and represents a point in the toric diagram. Interestingly, all the faces of the dimer diagram can be obtained from a subset of perfect matchings[5] which motivated us to conjecture the following:
All the face symmetries of the dimer diagram can be written in terms of perfect matchings corresponding to an internal point in the toric diagram whenever present.
We have verified the conjecture for many non-orbifold toric Calabi-Yau spaces(daughter spaces).

For the dimer diagram and perfect matchings as shown in Fig.6 and Fig.7, we see that the three faces F_i's can be obtained using perfect matchings p_1,p_5 and p_6 corresponding to the internal point as follows: $F_1 = p_5 - p6$, $F_2 = p_1 - p_5$ and $F_3 = p_6 - p_1$. For a face/closed curve, we require the R-charge to be zero which is another proof for the conjecture. For a general dimer diagram, it may not be practical to draw all the perfect matchings. There is an algebraic method of enumerating the perfect matchings by looking at the terms in the determinant of a matrix called *Kastelyne matrix* $K(Z, W)$. This algebraic approach is really useful for large toric diagrams. Let a_{ij}^k represent an edge-weight for edge k between white vertex i and black vertex j. Then $K_{ij}(Z, W) = \sum_k a_{ij}^k Z^{\langle k,\gamma_Z\rangle} W^{\langle k,\gamma_W\rangle}$ where $\langle k, \gamma_Z\rangle$ gives the intersection number of edge k with the oriented γ_Z cycle of the torus which will be $\pm 1, 0$. Taking the row index as white-vertex numbers and column index as black-vertex index, the matrix K will be:

$$K(Z,W) = \begin{pmatrix} a_{11} & a_{12}W & a_{13}Z \\ a_{21} & a_{22} & a_{23} \\ \frac{a_{31}}{Z} & a_{32} & \frac{a_{33}}{W} \end{pmatrix}$$

The determinant of $K(Z, W)$ will give six terms corresponding to the six perfect matchings in Fig.7:

$$det\ K = -a_{13}a_{22}a_{31} + a_{12}a_{23}a_{31}\frac{W}{Z} - a_{11}a_{23}a_{32} - a_{12}a_{21}a_{33} + a_{13}a_{21}a_{32}Z + a_{11}a_{22}a_{33}\frac{1}{W}\ .$$

The last term, for example, is an algebraic representation of perfect matching numbered (2) in Fig.7. So far, we have described perfect matchings and the Kastelyne matrix for a dimer model which gives us the toric data directly. The next step is to obtain quiver gauge theory data from the dimer diagram. The exact correspondence between the faces, edges and vertices in the dimer diagram to the gauge theory data is given in the Table.1. The terms in the superpotential is given by the bifundamental fields intersecting the vertices with a sign convention as indicated in Fig.8. Now, we could write the charge matrix from the dimer diagram:

$$d_{ai} = \langle F_a, e_i\rangle\ ,\ \ (Q_D)_{a\alpha} = \langle F_a, p_\alpha\rangle\ ,$$

where F_a's are the faces and e_i's are the edges in the dimer diagram. It is convenient to introduce matching matrix

$$\mathcal{M}_{i\alpha} = \langle e_i, p_\alpha\rangle$$

Journal of Physics: Conference Series **462** (2013) 012042 doi:10.1088/1742-6596/462/1/012042

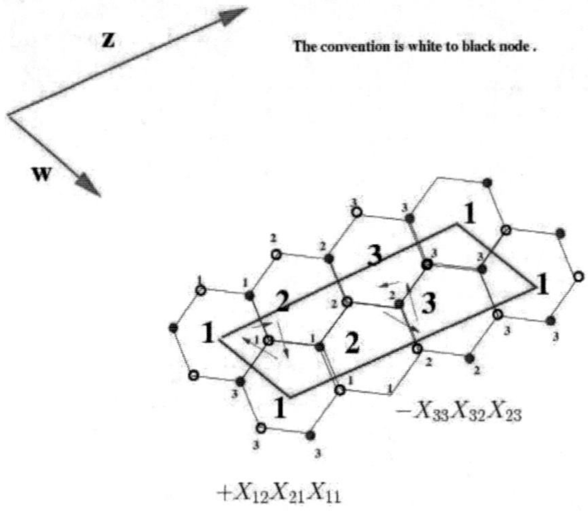

Figure 8. Superpotential terms

which is 1 if the edge e_i is in the perfect matching p_α. Q_F is given by

$$Q_F \mathcal{M}^t = 0 \ .$$

In terms of matching matrix, the matter fields X_i's and the charge matrix can be written as

$$X_i = \prod_\alpha p_\alpha^{\mathcal{M}_{i\alpha}} \text{ and } d = Q_D \mathcal{M}^t \ . \tag{3}$$

Having presented the essential ingredients relating dimer model to the toric and gauge theory data, we will now present the inverse algorithm and indicate when adjoint fields must be introduced [5].

3.1. *Our results*

Starting from the dimer diagram for the parent space $C^3/(Z_m \times Z_n)$, we can remove one or more edges (this is called **Higgsing**), we get daughter theories and exotic theories. For these theories, we have verified the conjecture that the face symmetries are obtainable from the perfect matchings corresponding to any internal point in the toric diagram[5].

In order to see whether the theory requires adjoint fields, we first check whether the matching matrix is consistent: For instance, the number of edges participating in any perfect matching are same (see Fig.7). This means that every column (representing perfect matching)in the matching matrix must have equal number of 1's.

Matching matrix for $C^3/(Z_2 \times Z_3)$ is

$$\mathcal{M} = \begin{pmatrix}
 & p_1 & p_2 & p_3 & p_4 & p_5 & p_6 & p_7 & p_8 & p_9 & p_{10} & p_{11} & p_{12} & p_{13} & p_{14} & p_{15} & p_{16} & p_{17} \\
X_1 & 1 & 0 & 0 & 1 & 0 & 1 & 0 & 1 & 0 & 1 & 0 & 0 & 0 & 0 & 0 & 1 & 0 \\
X_2 & 1 & 0 & 1 & 0 & 0 & 1 & 1 & 0 & 0 & 0 & 1 & 0 & 0 & 0 & 1 & 0 & 0 \\
X_3 & 1 & 0 & 0 & 0 & 1 & 0 & 1 & 1 & 0 & 1 & 0 & 0 & 0 & 1 & 0 & 0 & 0 \\
X_4 & 1 & 0 & 0 & 1 & 0 & 1 & 0 & 1 & 0 & 0 & 1 & 0 & 1 & 0 & 0 & 0 & 0 \\
X_5 & 1 & 0 & 1 & 0 & 0 & 1 & 1 & 0 & 0 & 1 & 0 & 1 & 0 & 0 & 0 & 0 & 0 \\
X_6 & 1 & 0 & 0 & 0 & 1 & 0 & 1 & 1 & 0 & 0 & 1 & 0 & 0 & 0 & 0 & 0 & 1 \\
Y_1 & 0 & 1 & 1 & 1 & 0 & 1 & 0 & 0 & 0 & 0 & 0 & 0 & 0 & 0 & 1 & 1 & 0 \\
Y_2 & 0 & 1 & 1 & 0 & 1 & 0 & 1 & 0 & 0 & 0 & 0 & 0 & 0 & 1 & 1 & 0 & 0 \\
Y_3 & 0 & 1 & 0 & 1 & 1 & 0 & 0 & 1 & 0 & 0 & 0 & 0 & 1 & 1 & 0 & 0 & 0 \\
Y_4 & 0 & 1 & 1 & 1 & 0 & 1 & 0 & 0 & 0 & 0 & 0 & 1 & 1 & 0 & 0 & 0 & 0 \\
Y_5 & 0 & 1 & 1 & 0 & 1 & 0 & 1 & 0 & 0 & 0 & 0 & 1 & 0 & 0 & 0 & 0 & 1 \\
Y_6 & 0 & 1 & 0 & 1 & 1 & 0 & 0 & 1 & 0 & 0 & 0 & 0 & 0 & 0 & 0 & 1 & 1 \\
Z_1 & 0 & 0 & 0 & 0 & 0 & 0 & 0 & 0 & 1 & 1 & 0 & 0 & 0 & 1 & 1 & 1 & 0 \\
Z_2 & 0 & 0 & 0 & 0 & 0 & 0 & 0 & 0 & 1 & 0 & 1 & 0 & 1 & 1 & 1 & 0 & 0 \\
Z_3 & 0 & 0 & 0 & 0 & 0 & 0 & 0 & 0 & 1 & 1 & 0 & 1 & 1 & 1 & 0 & 0 & 0 \\
Z_4 & 0 & 0 & 0 & 0 & 0 & 0 & 0 & 0 & 1 & 0 & 1 & 1 & 1 & 0 & 0 & 0 & 1 \\
Z_5 & 0 & 0 & 0 & 0 & 0 & 0 & 0 & 0 & 1 & 1 & 0 & 1 & 0 & 0 & 0 & 1 & 1 \\
Z_6 & 0 & 0 & 0 & 0 & 0 & 0 & 0 & 0 & 1 & 0 & 1 & 0 & 0 & 0 & 1 & 1 & 1
\end{pmatrix}$$

We will briefly outline our approach of obtaining quiver gauge theory corresponding to non-orbifold Calabi-Yau spaces. Starting from the matching matrix $\mathcal{M}$ of a parent space, we would like to obtain the matching matrix for a daughter theory. This corresponds to deleting the row corresponding to an edge representing a bifundamental field which is given a vev. Also,delete the columns representing those perfect matchings which contains the specific edge to obtain the reduced matching matrix $\mathcal{M}_r$. Then, we determine the kernel $Q_{F(r)}$ for the reduced matching matrix $\mathcal{M}_r$. For $Q_{F(r)}$, we obtain the kernel T_r and hence the dual of T_r giving K_r.

If the new matching matrix K_rT_r is consistent, then we can obtain the superpotential from K_r and draw the dimer diagram for the daughter space. In case K_rT_r is not consistent, we add one or more rows to K_r with entries 0 or 1 so that the number of 1's in every column is equal. The number of additional rows will indicate the number of adjoint fields which will give a consistent $\mathcal{M}$ and W.

4. Conclusions

Dimer models gives an elegant method of obtaining quiver gauge theories on daughter spaces (non-orbifold toric CY_3). Perfect matchings of the dimer diagram will determine the toric data including the multiplicity of the sigma model fields. Also, the charge matrix, matter fields and the superpotential of the quiver gauge theories can be directly obtained from the dimer diagram. Imposing a consistency requirement on the matching matrix gives a methodical way of incorporating adjoint fields in the quiver gauge theory.

We hope to look at counting of gauge invariant operators from the dimer model approach. There are interesting works on dimer tiling approaches to understand Chern-Simons gauge theories on coincident $M2$ branes. We are presently looking at some of these problems.

Acknowledgments

The author would like to thank IRCC, IIT Bombay for the grant.

References

[1] Benvenuti S, Franco S, Hanany A, Martelli D and Sparks J 2005 An Infinite family of superconformal quiver gauge theories with Sasaki-Einstein duals, *JHEP* **0506** 064, (*Preprint* hep-th/0411264)
[2] Feng B, Hanany A and He Y-H 2001 D-brane gauge theories from toric singularities and toric duality Nucl. Phys. B **595** 165 (*Preprint* hep-th/0003085)
[3] Hanany A and Kennaway K D Dimer models and toric diagrams *Preprint* hep-th/0503149
[4] Sarkar T 2007 On Dimer Models and Closed String Theories *JHEP* **0710** 010, *Preprint* hep-th 0705.3575
[5] Agarwal P, Ramadevi P and Sarkar T 2008 A note on dimer models and D-brane gauge theories *JHEP* **06** 054

Journal of Physics: Conference Series **462** (2013) 012043 doi:10.1088/1742-6596/462/1/012043

The Geometry of Non-Ideal Fluids

S. G. Rajeev

Department of Physics and Astronomy, Department of Mathematics, University of Rochester, Rochester, NY 14618

E-mail: s.g.rajeev@rochester.edu

Abstract.
Arnold showed that the Euler equations of an ideal fluid describe geodesics on the Lie algebra of incompressible vector fields. We generalize this to fluids with dissipation and Gaussian random forcing. The dynamics is determined by the structure constants of a Lie algebra, along with inner products defining kinetic energy, Ohmic dissipation and the covariance of the forces. This allows us to construct tractable toy models for fluid mechanics with a finite number of degrees of freedom. We solve one of them to show how symmetries can be broken spontaneously.In another direction, we derive a deterministic equation that describes the most likely path connecting two points in the phase space of a randomly forced system: this is a WKB approximation to the Fokker-Plank-Kramer equation, analogous to the instantons of quantum theory. Applied to hydrodynamics, we derive a PDE system for Navier-Stokes instantons.

1. The Euler Equations

The velocity of an ideal fluid satisfies

$$\frac{\partial v_i}{\partial t} = -v^k \partial_k v_i - \nabla_i p \tag{1}$$

along with the auxiliary condition of incompressibility

$$\partial_k v_k = 0. \tag{2}$$

Arnold [1] advocated an elegant geometric interpretation for these equations: they describe geodesics on the Lie algebra of incompressible vector fields.

Recall that the commutator

$$[u, v]^k = u^j \partial_j v^k - v^j \partial_j u^k. \tag{3}$$

of two incompressible (i.e., zero divergence) vector fields is still incompressible: such vector fields form a Lie algebra. Now, any inner product on a Lie algebra defines a left-invariant Riemannian metric on the corresponding Lie group [2]. If the Lie algebra is non-abelian, this is a curved metric. But, the evolution of the tangent vectors (momenta) along a geodesic decouple from the co-ordinate variables on the group, because of the left invariance under the group action.

1.1. Hamiltonian Formalism

A more physical point of view is to think of the Lie algebras as defining Poisson brackets on the (angular) velocities

$$\{v_a, v_b\} = f_{ab}^c v_c \tag{4}$$

The inner product defines the kinetic energy

$$E = \frac{1}{2} G^{ab} v_a v_b \tag{5}$$

The resulting Hamilton's equations are the geodesic equations of the underlying left-invariant metric on the Lie group:

$$\frac{dv_a}{dt} = \{v_a, E\} = f_{ab}^c G^{bd} v_c v_d \tag{6}$$

More precisely, this is how the tangent vector evolves along a geodesic.

For an ideal fluid, the kinetic energy provides a natural inner product:

$$E = \frac{1}{2} \int v^2 dx \tag{7}$$

Arnold's observation is that the Euler equations (1) are precisely the geodesic equations defined by this metric (7) on the Lie algebra (3) above. A more detailed derivation can be found in [3, 4].

More generally, the kinetic energy is a quadratic function of the velocities, so defines a Riemannian metric on the configuration space. In the absence of potential energy, the system moves along the geodesics of this metric: the time variable is proportional to the arc-length. Arnold's point is that an ideal fluid is a conservative (hamiltonian) system without any potential energy; so it moves along the geodesics of the metric defined by its kinetic energy. A compressible fluid has potential energy, stored in the deformation of density from its equilibrium value; so its time evolution is not along geodesics. Also, a viscous fluid is not conservative, so the Navier-Stokes equations are not geodesic equations either.

The curvature of a Riemannian manifold describes the stability of geodesic flows [5, 2]. If the sectional curvature $R(u, v, u, v) < 0$ two geodesics that start at the same point along tangent vectors u, v will diverge from each other. Arnold showed that in all except a finite number of directions, the sectional curvature is negative for the Euler flow, perhaps explaining the instability of fluid flow.

1.2. Example: Euler Equations of A Rigid Body

The most well known system to which we can apply the hamiltonian formalism above is the rigid body. The Lie algebra is that of rotations

$$\{L_i, L_j\} = \epsilon_{ijk} L_k \tag{8}$$

where L_i are the components of angular momentum as measured in the comoving frame of the body. Because this is not an inertial frame, the angular momentum may not be conserved even in the absence of external torque. The kinetic energy is (in the frame where the moment of inertia tensor is diagonal)

$$\frac{1}{2}\left[G_1 L_1^2 + G_2 L_2^2 + G_3 L_3^2\right] \tag{9}$$

leading to the Euler equations of a rigid body. These are equations for geodesics on a triaxial ellipsoid: the group $SO(3)$ with a left-invariant metric.

The curvatures in the principal planes are

$$K_{23} = \frac{(G_2 - G_3)^2 + 2G_1(G_2 + G_3) - 3G_1^2}{4G_1 G_2 G_3} \tag{10}$$

and cyclic permutations thereof.

1.2.1. An instability of coin flip For a cylinder with

$$h \leq \sqrt{\frac{3}{2}} r \tag{11}$$

(e.g., coin) rotations with axis in the plane of symmetry are unstable: curvature is negative. To see this instability we would have to mark a point of the circumference of a coin. The rotation of a coin around an axis that lies on the plane of the coin is unstable (i.e., the marked point will deviate) with respect to a perturbation of the axis to another one which also lies in the plane of the coin. An experiment to see this phenomenon is not too hard to set up with modern equipment (video cameras) but appears not to have been performed ever.

1.3. An Anomaly

Recall that in hamiltonian mechanics there is always a volume form in phase space that is invariant under time evolution: this is Liouville's theorem. In our case we can ignore the position variables as they decouple from the evolution of velocities. If we simply take the divergence,

$$\frac{\partial V_a}{\partial v_a} = f^a_{ab} \frac{\partial E}{\partial v_b}. \tag{12}$$

If the Lie algebra is unimodular,

$$f^a_{ab} = 0, \tag{13}$$

this is zero

$$\frac{\partial V_a}{\partial v_a} = 0. \tag{14}$$

Thus the constant measure is invariant. Any semi-simple Lie algebra is unimodular. But some nilpotent and solvable ones are not.

Finite dimensional non-unimodular algebras have some other measure μ on momentum space that is preserved by the geodesic flow:

$$\frac{\partial \left[\mu V_a\right]}{\partial v_a} = 0. \tag{15}$$

We will find it explicitly in some special cases below. This subtle point, that the constant measure in velocity space may not be invariant under time evolution, is reminiscent of an 'anomaly' in quantum field theory. The anomalies of quantum field theory are determined by the cohomology of the Lie algebra twisted by its action on some space of fields; the invariant measure μ also determines an element of the (untwisted9 Lie algebra cohomology; when this element is not non-trivial we have an anomaly as well: μ is not constant. We will see that, as in field theory, such an anomaly can lead to spontaneous symmetry breaking.

1.4. Example: The Poincare Half Plane

The only non-abelian Lie algebra in two dimensions is the affine Lie Algebra

$$\{v_0, v_1\} = v_1. \tag{16}$$

This algebra is *not* unimodular. The invariant measure on velocity space is

$$\mu dv = \frac{dv_0 dv_1}{v_1}. \tag{17}$$

The point is that v_0 generates scale transformations in v_1, so to be invariant we must divide the naive measure $dv_0 dv_1$ by v_1. This example therefore is a toy model where we can investigate the consequences of the anomaly we mentioned earlier.

Any quadratic form on this Lie algebra can be reduced to

$$E = \frac{1}{2}\left[v_0^2 + v_1^2\right]. \tag{18}$$

The corresponding group manifold is the upper half-plane, with the Poincaremetric. The geodesic equations are, using the above hamiltonian formalism

$$\frac{dv_0}{dt} = -v_1^2,\ \frac{dv_1}{dt} = v_0 v_1 \tag{19}$$

Solutions are semi-circles in momentum space:

$$v_0 = -\rho \tanh \rho t,\ v_1 = \rho \text{sech}\ \rho t. \tag{20}$$

We will see later how dissipation and random forcing affect this dynamics.

2. Geometry of Dissipative Motion

A conservative dynamical system is described by a Poisson bracket and hamiltonian (energy). Dissipation is described by (a possibly degenerate, but always non-negative) tensor Γ^{ab}, which determines the gradient of energy:

$$\frac{d\xi_a}{dt} = \{E, \xi_a\} - \Gamma_{ab}\frac{\partial E}{\partial \xi_b} \tag{21}$$

It is straightforward to check that the Energy decreases:

$$\frac{dE}{dt} = -\Gamma_{ab}\frac{\partial E}{\partial \xi_a}\frac{\partial E}{\partial \xi_b} \leq 0 \tag{22}$$

In the cases we study in this paper, the Poisson bracket is linear (a Lie algebra), the energy is a quadratic function and Γ_{ab} is constant.

$$\frac{dv_a}{dt} = -\Gamma_{ab}G^{bc}v_c + f_{ab}^c G^{bd} v_c v_d \tag{23}$$

Even this special case is quite complex: it includes Navier-Stokes, with the choices

$$E = \frac{1}{2}(v, Gv) = \frac{1}{2}\int v^2 dx,\ (v, \Gamma v) = \gamma \int (\partial_i v_j)^2 dx. \tag{24}$$

Thus the quadratic form of kinetic energy is the L^2 norm while that of dissipation is the H^1 Sobolev norm.

3. The Langevin And Fokker-Planck-Kramers Equations

A standard model of randomness [6] is a Gaussian white noise: a force whose correlations vanish fast compared to the dynamical and dissipative time scales. If the random force is a sum of a large number of more or less independent forces,it will be a Gaussian. The covariance of the Gaussian introduces a third positive tensor into the game.

$$\frac{dv_a}{dt} = -\Gamma_{ab}G^{bc}v_c + f_{ab}^c G^{bd} v_c v_d + \eta_a,\ < \eta_a(t)\eta_b(t') >= D_{ab}\delta(t-t'). \tag{25}$$

Journal of Physics: Conference Series **462** (2013) 012043 doi:10.1088/1742-6596/462/1/012043

This stochastic differential equation (Langevin equation) implies the Fokker-Planck-Kramers Equation for the evolution of the probability density of velocity:

$$\mu \frac{\partial P}{\partial t} = \frac{\partial}{\partial v_a}\left[\mu\left\{D_{ab}\frac{\partial P}{\partial v_b} - \left(-\Gamma_{ab}G^{bc}v_c + f^c_{ab}G^{bd}v_c v_d\right)P\right\}\right] \tag{26}$$

Here, μ is the invariant density of the ideal dynamics.

The traditional derivation[7, 8] of this FPK equation does not include the density μ : it is implicitly assumed that the invariant (Liouville) density is the standard one. In order that the total probability be conserved:

$$\frac{\partial}{\partial t}\int \mu P dv = 0. \tag{27}$$

we must order the differentiations and insert factors of μ as above. This is one of our innovations in this paper.

3.1. Maxwell-Boltzmann is Equilibrium Solution for Unimodular Lie Algebras

If the Einstein relation

$$\beta D_{ab} = \Gamma_{ab} \tag{28}$$

holds, $P = e^{-\beta E}$ is a static solution of the FPK equations. But it may not be normalizable, because $\int \mu e^{-\beta E} dv$ might diverge. For unimodular Lie algebras, $\mu = 1$ and the Maxwell-Boltzmann distribution $P = e^{-\frac{1}{2}\beta(v,Gv)}$ is an equilibrium solution. An example is the random motion of a rigid body. This case has been studied in detail by astronomers interested in the motion of dust grains in the interstellar medium[9]. We now turn to an example with $\mu \neq 1$.

3.2. The Langevin Equation on the Half-Plane

The geodesic equations still follow for zero dissipation if we choose as hamiltonian any function $H(\rho)$ of $\rho = \sqrt{v_0^2 + v_1^2}$. We can choose $H(\rho)$ such that the FKP equation is easier to solve. But physically, the kinetic energy would still be $E = \frac{1}{2}\rho^2$. The Fokker-Planck-Kramer reduces to

$$\frac{\partial Q}{\partial t} = D\frac{\partial^2 Q}{\partial \rho^2} - \gamma\rho\frac{\partial Q}{\partial \rho} + \frac{D}{\rho^2}\frac{\partial}{\partial \theta}\left[\cosh^2\theta\frac{\partial Q}{\partial \theta}\right] - H'(\rho)\frac{\partial Q}{\partial \theta}, \quad P = e^{-\beta\frac{v^2}{2}}Q \tag{29}$$

This equation is separable for $H = -\frac{k}{\rho}$. In fact the solution can be found in closed form in terms of spheroidal functions.

Numerical solution of toy model shows that the details of the dynamics are not important: the vestige of the dynamics that is important to the long term behavior is the invariant measure μ. We can even set $H = 0$ and get a qualitatively similar equilibrium distribution. In this case the solution is quite simple[10]:

$$P(v_0, v_1)\frac{dv_0 dv_1}{v_1} = \frac{v_1}{\rho^2}e^{-\frac{\beta}{2}\rho^2}\left[1 - \frac{1}{\sqrt{\beta}\rho}e^{-\frac{\beta\rho^2}{4}}\operatorname{erf}\left(\frac{\sqrt{\beta}\rho}{2}\right)\right]dv_0 dv_1 \tag{30}$$

The main point is that this solution does not have a peak at minimum energy $v_0 = v_1 = 0$; instead the peak is at some finite value of v_1. The anomaly drives a spontaneous breaking of the rotation invariance in the energy $E = \frac{1}{2}[v_0^2 + v_1^2]$. Even if the two components of velocity appear symmetrically in energy, their commutation relations are different. These commutation relation determine the measure of integration which is singular at $v_1 = 0$. It is this singularity that pushes the most likely equilibrium value of v_1 away from zero.

Thus we see that anomaly we discovered can drive a spontaneous symmetry breakdown: a steady state flow along the x_1 axis is generated in the equilibrium. Perhaps this is a toy model for the formation of flows such as hurricanes and ocean currents which persist for long times despite the instability of the underlying fluid system.

4. The WKB Approximation to the FPK Equation

It is often important to know the probability that a certain initial condition will evolve to a given final condition under the influence of random forces. It might be important to compute this even when this probability is small: there could be final states that are catastrophic so that even a small chance of them happening cannot be ignored. If the probability is small, we can use the WKB approximation as for the Schrodinger equation; put $P = e^{-\Phi}$ and ignore second derivatives of Φ:

$$\frac{\partial\Phi}{\partial t} + D_{ab}\frac{\partial\Phi}{\partial v_a}\frac{\partial\Phi}{\partial v_b} + \left(-\Gamma_{ab}G^{bc}v_c + f_{ab}^{c}G^{bd}v_c v_d\right)\frac{\partial\Phi}{\partial v_a} = 0 \tag{31}$$

These are the Hamilton-Jacobi equations of the hamiltonian

$$H = D_{ab}w^a w^b + \left(-\Gamma_{ab}G^{bc}v_c + f_{ab}^{c}G^{bd}v_c v_d\right)w^a, \tag{32}$$

where w_a are the canonical conjugates of the variables v_a:

$$\{v_a, v_b\}_1 = 0 = \left\{w^a, w^b\right\}, \ \left\{v_a w^b\right\}_1 = \delta_a^b \tag{33}$$

Kramer used this WKB approximation to the FPK in chemical kinetics; we extend them to a more general framework. In quantum mechanics, the WKB approximation works when the quantum tunneling amplitudes are small; the path with the greatest probability is called the instanton. The analogue in our case is the solution to the Hamilton's equations following from the above efectie hamiltonian.

$$\frac{dv_a}{dt} = 2D_{ab}w^b - \Gamma_{ab}G^{bc}v_c + f_{ab}^{c}G^{bd}v_c v_d, \quad \frac{dw^c}{dt} = \left(\Gamma_{ab}G^{bc} - 2f_{ab}^{c}G^{bd}v_d\right)w^a \tag{34}$$

Notice that this effective hamiltonian mechanics of rare transitions has twice the number of degrees of freedom as the original conservative dynamics we started with: we can specify initial and final conditions independently. The action of the path that solves this equation with given initial and final states gives the log of the transition probability:

$$\Phi(v', v, T) = \int_0^T [w^a \dot{v_a} - H(v, w)]dt \tag{35}$$

4.1. Instantons of Random Motion on the Half-Plane

With an isotropic dissipation tensor, these instanton equations are quite simple for the half-plane:

$$\frac{dv_0}{dt} = w_0 - \gamma v_0 - v_1^2, \quad \frac{dv_1}{dt} = w_1 - \gamma v_1 + v_0 v_1 \tag{36}$$

$$\frac{dw_0}{dt} = \gamma w_0 - v_1 w_1, \quad \frac{dw_1}{dt} = \gamma w_1 + 2v_1 w_0 - v_0 w_1 \tag{37}$$

We could not find the general solution to this system. But notice the special solution with $w_a = 2\gamma v_a$. This is just the time reverse of the dissipative solution! This solution leads to the Maxwell-Boltzmann distribution: agrees with the exact solution for large energies. WKB is not good for small energies, where the exact answer departs from MB distribution. This is the phenomenon of spontaneous symmetry breakdown we discovered in the last section.

4.2. Navier-Stokes Instantons

We can now derive the PDE describing the most likely time evolution of a randomly forced dissipative fluid:

$$\frac{\partial v_i}{\partial t} = F_i + \gamma \partial^2 v_i - v_j \partial_j v_i - \partial_i p, \quad \partial_k v_k = 0. \tag{38}$$

$$\frac{\partial w_i}{\partial t} = -\gamma \partial^2 w_i + w_j \partial_j v_i + v_j \partial_j w_i - \partial_i \phi, \; \partial_k w_k = 0. \tag{39}$$

The covariance of the fluctuations is likely to be an integral operator[6]

$$D_{ij}(x, x') = \int e^{2\pi i k \cdot (x - x')} \left[\delta_{ij} - \frac{k_i k_j}{k^2} \right] \tilde{D}(k) dk. \tag{40}$$

If we choose

$$\tilde{D}(k) = |k|^{-2} \tag{41}$$

we get an auxiliary differential equation to complete the above system

$$\partial^2 F_i = 2 w_i. \tag{42}$$

With this choice, the random forcing grows with spatial scale, while dissipation is more important at short scales: what is expected in fluid mechanics. The above equation should be a good approximation to calculate probabilities and paths of large deviations from Navier-Stokes. Such rare events are important in many situations: the path of a hurricane, or the chances of a tsunami or a rogue wave. Similar instanton equations can be derived for Rotating Shallow Water equations, the Stochastic Loewner Evolution[11], two dimensional incompressible flow etc. We hope to return to these questions in later publications. Numerical solution of our system is not much harder than the solution of Navier-Stokes itself: only twice as many variables; moreover, the new variables appear linearly.

Acknowledgments

I thank M. Gordina, G. Krishnaswamy, V.V. Sreedhar and C. Mueller for discussions. This work was supported in part by a grant from the US Department of Energy under contract DE-FG02-91ER40685.

References

[1] Arnold V. I. 1966 *Ann. Inst. Poly. Genoble* 16 319.
[2] Milnor J. 1976 *Adv. Math.* **21** 293.
[3] Rajeev S. G. 2009 *Geometry of the Motion of Ideal Fluids and Rigid Bodies*, Lectures at the Chennai Mathematical Institute Symposium on Fluid Mechanics ; arXiv:0906.0184[math-phys]
[4] Rajeev S. G. 2009 *Ann. Phys.* **324** 2586; arxiv:0907.2401[math-phys]
[5] Chavel I. 2006 *Riemannian Geometry: A Modern Introduction*, Cambridge University Press
[6] Kraichnan R. H. 1965 *Proc. Symp. Applied Mat.* **13** 199.
[7] Kramers H. A. 1940 *Physica* **7** 284.
[8] Chandrasekhar S. 1943 *Rev. Mod. Phys.* **15** 1.
[9] G. W. Ford, J. T. Lewis and J., McConnel, Phys. rev. A19,(1979) 907; M. Efroimsky, J.Math.Phys. 41 (2000) 1854; arxiv:astro-ph/9909220
[10] Abramowitz M. and Stegun I. A. 1972 *Handbook of Mathematical Functions*, p. 753, Dover .
[11] Friedrich R. M. 2009 *The Global Geometry of Stochastic Loewner Evolutions* arXiv:0906.5328v1 [math-ph].

Journal of Physics: Conference Series **462** (2013) 012044 doi:10.1088/1742-6596/462/1/012044

Probes of universality in AdS/CFT

Adam Ritz

Department of Physics and Astronomy, University of Victoria
Victoria BC, V8P 5C2, Canada

Abstract. We discuss some universal relations that hold for conformal field theories which admit a dual classical gravitational description in anti-de Sitter space. These relations therefore apply in a suitable large N limit, and imply that various hydrodynamic transport coefficients for conserved currents and thermodynamic state functions are entirely fixed by central charges.

1. Introduction

From the Wilsonian viewpoint [1], the primary building blocks of the space of quantum field theories are those special points which exhibit conformal invariance. Indeed, the quantum field theories which play a role in nature are all understood to be either conformal field theories (CFT) themselves, or relevant deformations of a CFT. Among other things, such field theories describe the interactions of all known elementary particles, and the universal scaling behaviour near critical points in statistical mechanics.

In attempting to study CFTs at the quantitative level, there are several avenues available beyond $d = 1 + 1$ where symmetry alone is sufficiently powerful to constrain the allowed space of theories. One approach to arrive at an interacting CFT is by taking a critical limit of a statistical-mechanical model (or, more generally, by following the renormalization group flow to the endpoint) [1]. Another is to start with a Lagrangian formulation of a classically conformal theory, and use extra symmetry (such as supersymmetry) to argue that the quantum theory must be conformally invariant as well [2]. A prime example of such a CFT is the $\mathcal{N} = 4$ supersymmetric Yang-Mills theory in 3+1 dimensions. However, the last decade has exhibited another remarkable approach to arrive at a CFT, namely through a higher-dimensional theory of (quantum) gravity in Anti-de Sitter (AdS) space, and the AdS/CFT correspondence [3, 4, 5].

In the limit where classical gravity is a reliable approximation, which involves taking both the number of CFT degrees of freedom and their interactions to be parametrically large, this latter approach is particularly tractable and provides a new class of tools. It is therefore natural to ask which class of CFTs, and thus which universality classes, can be described in this tractable classical gravity regime. In exploring this question, particularly in the condensed matter context, it is important to distinguish two notions of universality:

- *Static universality*: This reflects the universal scaling behaviour of thermodynamic functions, such as the specific heat, as one approaches the critical point. Static universality classes are generally determined by the spatial dimension and the symmetry of the order parameter distinguishing the phases. A classic example is the fact that the liquid-gas critical point and the ferromagnetic critical point are both in the same static universality class, that of the $d = 3$ Ising CFT.

Journal of Physics: Conference Series **462** (2013) 012044 doi:10.1088/1742-6596/462/1/012044

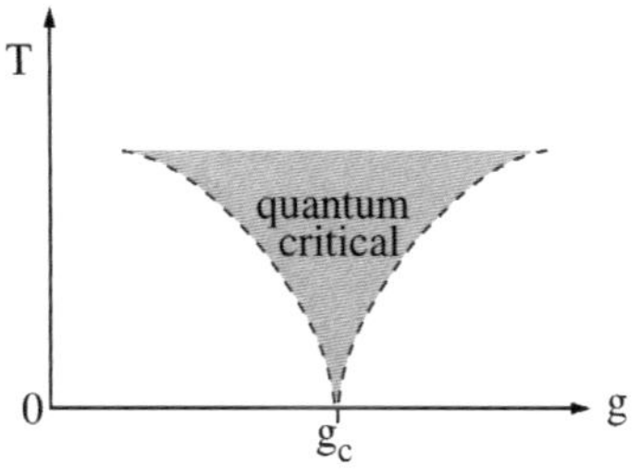

Figure 1. A schematic illustration of the phase diagram near a quantum critical point at $T = 0$ and $g = g_c$, where g is some coupling or control parameter for the system. The shaded region is controlled by the CFT associated with the critical point.

- *Dynamic universality*: This goes beyond static universality and describes classes of systems in which dynamical transport coefficients also exhibit universal scaling laws [6]. Two fixed points in the same static universality class may exhibit quite distinct dynamical behaviour. Indeed, the liquid-gas and ferromagnetic critical points are in different dynamical universality classes [6]. The additional characteristic which plays a crucial role in defining these classes is the presence or otherwise of conserved currents, which provide additional hydrodynamic degrees of freedom.

Given this dichotomy, it is natural to ask whether there are special fixed points in which these two notions of universality are the same, so that all static and dynamical behaviour is described near the fixed point by a Lorentzian CFT, formulated in spacetime (rather than just space). Physical examples are indeed provided by quantum critical phenomena [7, 8], where the phase transition occurs at zero temperature, driven by quantum fluctuations (see Fig. 1). Such CFTs are relativistic, although their speed of "light" v need not equal 3×10^8m/s. The CFT then dictates static and dynamical quantities in any nearby regime dominated by a single dynamical scale, i.e. temperature, which breaks conformal invariance.

It has become clear that, for these reasons, the quantum critical regime is a primary target to be modeled using the AdS/CFT correspondence. Understanding CFTs in the language of classical gravity has already proved useful in studies of quantum critical transport [9], and this motivates further exploration of the relation of the AdS/CFT correspondence to quantum criticality. In this contribution, we will discuss a number of constraints that naturally apply to any CFT dual to classical gravity in AdS (see [10, 11, 12] for more details). In broad terms, these constraints relate the thermodynamic and hydrodynamic peroperties of the CFT at finite temperature to each other, and furthermore to the central charges which determine the zero-temperature correlators of conserved currents.

Recall that at very short distances, the effects of temperature are irrelevant, and the natural physical questions involve the leading short-distance singularities of the correlation functions. On the other hand, at long distances the effects of temperature become important, and the natural questions are related to thermodynamics and hydrodynamic transport phenomena. In CFTs, however, short and long distances are related by a scaling symmetry, and one may anticipate a universal relation between the long-distance transport coefficients and the parameters which describe the short-distance singularities. Unfortunately, in general this expectation seems to be quantitatively true only in 1+1 dimensions. Nonetheless, this expectation does indeed hold true in CFTs dual to classical gravity in any dimension.

At zero temperature, the (Euclidean) vacuum correlation functions of the energy-momentum tensor $T_{\mu\nu}$ and a $U(1)$ conserved current J_μ in a CFT are fixed to be

$$\langle J_\mu(x) J_\nu(0)\rangle = \frac{k}{x^{2(d-1)}}\frac{1}{\omega_{d-1}^2} I_{\mu\nu}\,, \tag{1.1}$$

$$\langle T_{\mu\nu}(x)T_{\alpha\beta}(0)\rangle = \frac{c}{x^{2d}}\frac{1}{\omega_{d-1}^2}\left(I_{\mu\alpha}I_{\nu\beta} + I_{\mu\beta}I_{\nu\alpha} - \frac{2}{d}\delta_{\mu\nu}\delta_{\alpha\beta}\right), \tag{1.2}$$

where $I_{\mu\nu} \equiv \delta_{\mu\nu} - 2x_\mu x_\nu/x^2$, and k and c are central charges, which are dimensionless constants. [We use units in which $\hbar$=v=1 where v is the speed of "light" in the CFT.] The factors of $\omega_{d-1} \equiv 2\pi^{d/2}/\Gamma(d/2)$ are inserted for notational convenience. At non-zero temperature T, the equilibrium state is characterized by pressure P, as well as by the charge susceptibility $\chi = \langle Q^2\rangle/(VT)$, where Q is the conserved charge associated with the current J_μ, and we take the thermodynamic limit $V\to\infty$. The susceptibility can be evaluated by introducing a small chemical potential μ, so that $\chi(T) = \partial\rho/\partial\mu|_{\mu=0}$, where $\rho(T,\mu) = \langle Q\rangle/V$ is the charge density. In a CFT, temperature remains the only scale which dictates,

$$P(T) = c'T^d, \qquad \chi(T) = k'T^{d-2}, \tag{1.3}$$

where c' and k' are dimensionless constants. Physically, c and c' provide a measure of the total number of degrees of freedom in the system, while k and k' measure the number of charged degrees of freedom. In two dimensions, c is uniquely related to c' [13, 14], while k is uniquely related to k':

$$c' = \frac{\pi}{6}c, \qquad k' = \frac{1}{2\pi}k, \tag{1.4}$$

which means that thermodynamics is uniquely fixed by the central charges. The reason is that in two dimensional CFTs, the vacuum state is related to the thermal state by a symmetry transformation. In $d > 2$, the conformal symmetry group is not large enough to enforce a relation similar to Eq. (1.4), and therefore thermodynamics is not determined by the central charges. However, there does exist a large class of CFTs in $d > 2$, whose pressure is determined by the central charge c, and whose susceptibility is determined by the central charge k, resembling the two-dimensional case [10, 11]. The crucial property of these models is that they admit a dual description in terms of classical gravity on a $(d+1)$ dimensional anti-de Sitter space (AdS). Namely, we have the following relations [10, 11]:

$$\frac{c'}{c} = \frac{1}{4\pi^{d/2}}\left(\frac{4\pi}{d}\right)^d \frac{\Gamma(d/2)^3}{\Gamma(d)}\frac{(d-1)}{d(d+1)}, \qquad \frac{k'}{k} = \frac{1}{2\pi^{d/2}}\left(\frac{4\pi}{d}\right)^{d-2}\frac{\Gamma(d/2)^3}{\Gamma(d)}. \tag{1.5}$$

In $d = 2$ they reproduce the universal relations (1.4).

The CFTs which admit a dual gravitational description have many more universal properties beyond the above relation between thermodynamics and the central charges. A surprising feature of these CFTs (and of their relevant deformations) is that momentum transport in these models is completely determined by thermodynamics. In particular, their viscosity is given by $\eta = s/4\pi$ in any dimension [15, 16], where $s = (\partial P/\partial T)$ is the entropy density. This is surprising because transport coefficients are typically determined by the mean-free path even in CFTs [17], and are not fixed by thermodynamics. Moreover, charge transport in these models is also completely determined by thermodynamics. In particular, the dc electrical conductivity σ obeys a similar relation, so we have

$$\frac{\eta}{s} = \frac{1}{4\pi}, \qquad \frac{\sigma}{\chi} = \frac{1}{4\pi T}\frac{d}{d-2}. \tag{1.6}$$

Even though the ratio σ/χ was derived for integer $d \geqslant 3$, it can be analytically continued to any real positive $d > 2$. The singularity at $d = 2$ is precisely what one expects in 1+1 dimensional CFTs.

In what follows, we describe the derivation of these relations between central charges and thermodynamics, and between thermodynamics and hydrodynamics. In the final section, we will discuss some comparisons to known systems, and deviations from these universal relations beyond the classical gravity limit.

2. Thermodynamics from central charges

We will focus on quantum field theories which admit a dual description in terms of classical gravity in Anti-de Sitter (AdS) space within the AdS/CFT correspondence [3, 4, 5]. For such field theories, a large-volume thermal state in a d-dimensional CFT is described by a $(d+1)$-dimensional black hole in AdS. The black hole solution follows from the Einstein-Maxwell action,

$$S = \frac{1}{16\pi G}\int d^{d+1}x\,\sqrt{-g}\left[R + \frac{d(d-1)}{L^2}\right] - \frac{1}{4g_{d+1}^2}\int d^{d+1}x\,\sqrt{-g}\,F^2\,, \tag{2.7}$$

where L^2 sets the value of the cosmological constant, and g_{d+1}^2 is the $(d+1)$-dimensional gauge coupling constant, which has dimension of $(\text{length})^{d-3}$. An equilibrium state at finite temperature and density is described by the Reissner-Nordstrom black hole in AdS. The thermodynamics of these black holes has been studied extensively, see for example [18]. The metric in the thermodynamic limit is given by

$$ds^2 = \frac{r^2}{L^2}\left(-V(r)dt^2 + d\mathbf{x}^2\right) + \frac{L^2}{r^2}\frac{dr^2}{V(r)}, \tag{2.8}$$

where $V(r) = 1 - m/r^d + m_q^2/r^{2d-2}$, and the boundary is at $r \to \infty$. The parameter m determines the mass of the black hole, and m_q determines its charge. The background gauge field is,

$$A_t = \mu - \frac{g_{d+1}^2 L^{d-1}\rho}{(d-2)r^{d-2}}, \tag{2.9}$$

where the charge density ρ is defined by the variation of the action with respect to the boundary value of the gauge field $A_t^{(b)} = A_t(r \to \infty)$, and is given by $\rho = \mu(d-2)r_0^{d-2}/(g_{d+1}^2 L^{d-1})$ in terms of the chemical potential μ. This relation is fixed by requiring that A_t vanishes at the horizon $r = r_0$.

We will actually work in the limit $\mu \to 0$, so that temperature is the dominant scale breaking conformal invariance, given by $T = r_0 d/(4\pi L^2)$. The charge density then vanishes, but the results above allow us to compute the susceptibility.

Central charges:- To find the central charge c in dimension $d \geqslant 3$, we consider the vacuum geometry $(T = 0)$ and a convenient momentum space representation of the correlator (1.2),

$$G_{\mu\nu,\alpha\beta}(k) = \left(P_{\mu\alpha}P_{\nu\beta} + P_{\mu\beta}P_{\nu\alpha} - \frac{2}{d-1}P_{\mu\nu}P_{\alpha\beta}\right)G(k^2), \tag{2.10}$$

where $P_{\mu\nu} = \delta_{\mu\nu} - k_\mu k_\nu/k^2$. The central charge is related to $G(k^2)$ by

$$\frac{c}{(x^2)^d} = \frac{d+1}{d-1}\omega_{d-1}^2\int \frac{d^dk}{(2\pi)^d}\,e^{ikx}G(k^2)\,. \tag{2.11}$$

Choosing k along x_i with $i \neq 1$ or 2, one has $G_{12,12}(k) = G(k^2)$, and therefore it suffices to evaluate $G_{12,12}$ to find the central charge. According to the AdS/CFT prescription, $G_{12,12}$ is given by the second variation of the gravitational action with respect to the boundary value of the metric perturbation h_{12}. The h_{12} perturbation decouples from all other perturbations, and obeys the equation of motion coming from the action of a massless scalar in the AdS_{d+1} background. The two-point correlation function for the massless scalar can be evaluated using the standard AdS/CFT prescription, and one finds for the central charge,

$$c = \frac{d+1}{d-1}\frac{L^{d-1}}{4\pi G_N^{d+1}}\frac{\Gamma(d+1)\pi^{d/2}}{\Gamma(d/2)^3}\,. \tag{2.12}$$

The value of the central charge k can be found in a similar manner using the Maxwell action for the gauge field dual to the current J^μ. However, we will skip the details here, as it was obtained previously in the work of Freedman et al. [19]:

$$k = \frac{L^{d-3}}{g^2_{d+1}} \frac{\Gamma(d)(d-2)}{2\pi^{d/2}\Gamma(d/2)} \omega^2_{d-1} \,. \tag{2.13}$$

Thermodynamics:- The entropy is proportional to the $d-1$ dimensional area of the horizon, $S = A_{d-1}/4G_N$ where G_N is the $d+1$-dimensional Newton's constant. Dividing by the (infinite) $(d-1)$–volume V, and working with $T \gg \mu$ as described above, one finds for the entropy density,

$$s = \frac{1}{4G_N} \left(\frac{4\pi L}{d} \right)^{d-1} T^{d-1}. \tag{2.14}$$

To find the susceptibility, we need the relation between ρ and μ to linear order in μ. Expressing r_0 in terms of temperature when $\mu \to 0$, $T = r_0 d/(4\pi L^2)$, we find $\rho(T, \mu) = \chi(T)\,\mu$, where the susceptibility is

$$\chi = \frac{(d-2)L^{d-3}}{g^2_{d+1}} \left(\frac{4\pi}{d} \right)^{d-2} T^{d-2}. \tag{2.15}$$

Comparing the results above for s and χ to those for c and k, we deduce the universal relations for c'/c and k'/k given in Eq. (1.5).

3. Hydrodynamics from central charges

Momentum transport:- The well-known universal result for the shear viscosity to entropy ratio, $\eta/s = 1/(4\pi)$ can now be translated one step further to a universal relation of η to the central charge c. Given [20],

$$\eta = -\lim_{\omega \to 0} \mathrm{Im}\, \frac{1}{\omega} \langle T_{xy} T_{xy} \rangle (\omega, \mathbf{q} = \mathbf{0})_{\mathrm{ret}}, \tag{3.16}$$

one obtains

$$\eta = \frac{s}{4\pi} = \left[\frac{1}{16\pi^{d/2+1}} \frac{d-1}{d+1} \left(\frac{4\pi}{d} \right)^d \frac{\Gamma(d/2)^3}{\Gamma(d)} \right] c T^{d-1}. \tag{3.17}$$

This exhibits the universal relation between hydrodynamics, thermodynamics, and the vacuum in the shear channel associated with the conserved energy momentum tensor.

Charge transport:- An analogous procedure allows us to compute the electrical conductivity of CFTs with a dual gravity description. To define conductivity, we imagine gauging a global $U(1)$ symmetry of the theory with a small coupling e. To leading order in e, the effects of the gauge field can be ignored, and the electromagnetic response can be determined from the original theory. This essentially amounts to sending $J_\mu \to eJ_\mu$, and a factor of e^2 will appear in both the conductivity and the susceptibility. The dc conductivity is determined from the real-time retarded correlation function of U(1) currents in thermal equilibrium,

$$\sigma = e^2 \lim_{\omega \to 0} \mathrm{Im}\, \frac{1}{\omega} \langle J_x J_x \rangle (\omega, \mathbf{q}{=}0)_{\mathrm{ret}} \,. \tag{3.18}$$

To evaluate the retarded correlator, we use the standard AdS/CFT recipe of [21, 20], and consider Maxwell fields propagating on the $(d{+}1)$ dimensional background (2.8) at $m_q{=}0$. Skipping the details [11], the Kubo formula (3.18) then gives the conductivity,

$$\sigma = \frac{e^2}{g^2_{d+1}} \left(\frac{L}{z_0} \right)^{d-3} . \tag{3.19}$$

Journal of Physics: Conference Series **462** (2013) 012044 doi:10.1088/1742-6596/462/1/012044

On the other hand, using the susceptibility (2.15), we arrive at the simple result (1.6) for the conductivity to susceptibility ratio. For systems in which charge transport proceeds by diffusion, conductivity is related to the diffusion constant D by the Einstein relation $\sigma = \chi D$. Therefore, our result can be interpreted as a remarkably simple diffusion constant in d spacetime dimensions,

$$D = \frac{\sigma}{\chi} = \frac{1}{4\pi T}\frac{d}{d{-}2}\,. \tag{3.20}$$

which agrees with the known results in $d{=}4$ [20], and $d{=}3,6$ [22].The electrical conductivity takes a particularly simple form in 2+1 dimensional CFTs where it is *frequency-independent* $\sigma(\omega) = e^2/g_4^2$ [9], and shows no crossover regime at $\hbar\omega \sim k_{\rm B}T$.

Putting the pieces together, we have

$$\sigma = \frac{\chi}{4\pi T}\frac{d}{d-2} = \left[\frac{1}{8\pi^{d/2+1}}\frac{d}{d-2}\left(\frac{4\pi}{d}\right)^{d-2}\frac{\Gamma(d/2)^3}{\Gamma(d)}\right]kT^{d-3}, \tag{3.21}$$

again exhibiting the relation between hydrodynamics, thermodynamics, and the vacuum, this time in relation to the conserved U(1) current.

4. Discussion

In this concluding section, we will comment on some corollaries of these results, some comparisons with known systems, and deviations from universality away from the classical gravity (i.e. large N) limit.

Transport coefficient ratios:- The general relations, $\eta \sim c\,T^{d-1}$, and $\sigma \sim k\,T^{d-3}$ also imply that the ratio of viscosity to conductivity is proportional to the ratio of the central charges,

$$\frac{\eta\, e^2}{\sigma T^2} = \frac{8\pi^2(d{-}1)(d{-}2)}{d^3(d{+}1)}\,\frac{c}{k} \propto \frac{g_{d+1}^2}{G_N}\,. \tag{4.22}$$

If c and k indeed provide a suitable measure of the number of degrees of freedom in the system, it is natural to conjecture a bound of the form $\sigma T^2 \leqslant \lambda_d\,\eta\, e^2$, with some order one constant λ_d, given that k is only sensitive to charged degrees of freedom.[1] Another suggestive viewpoint is that the ratio in (4.22) is proportional to the ratio of bulk couplings, g_{d+1}^2/G_N, and from this dual gravitational point of view, such an inequality between c and k is related to a version of the "weak gravity conjecture" [24] in AdS space. However, one should keep in mind that the definition of σ (or η) involves an arbitrary choice of normalization for the corresponding current and therefore any bound on conductivity will more naturally involve a quantity which is independent of the normalization, such as σ/χ [11].

Nonetheless, it is interesting to note that the ratio c/k also determines when the U(1) symmetry is unstable to condensation forming a superfluid (or superconducting) phase. In $d = 3$, the instability occurs when [25]

$$q^2 \geqslant (3 + 2\Delta(\Delta - 3))\frac{k}{c} \tag{4.23}$$

with q (Δ) the charge (dimension) of the operator breaking U(1).

Free-field comparisons:- It is natural to ask if there are "conventional" CFTs which obey the AdS/CFT relations for c'/c and k'/k. This cannot happen for free (or weakly-interacting)

[1] There are, however, known counter-examples to the decrease of c and k along renormalization group trajectories in supersymmetric theories [23], but in these examples k corresponds to the R-current, and therefore $k \propto c$ by supersymmetry.

theories when d is odd, because c' is proportional to $\zeta(d)$, while c contains no such factors. In even dimensions, this is no longer the case and in $d=4$, for a free theory with an arbitrary number of scalar, fermionic and vector degrees of freedom one finds that $3/8 \leqslant \frac{c/c'|_{\text{free}}}{c/c'|_{\text{AdS}}} \leqslant 9/4$. Importantly, we observe that free theories with $2n_s + n_f = 8n_v$ (including e.g. $\mathcal{N}=4$ SYM) do satisfy the AdS relation. Thus we conclude that, at least in $d=4$, the AdS relation c'/c is necessary but not sufficient for a given CFT to possess a gravity dual.

In $d=3$, as noted above we do not expect free theories to exhibit the AdS relation, but we can consider an interacting example, namely the 3-dimensional $O(N)$ model at large N with fields ϕ^α, $\alpha = 1,..,N$ subject to a constraint $\phi^\alpha\phi^\alpha = 1$. In this system, the ratio of $c'/c|_{O(N)} \approx 0.2041$ was computed in the large-N limit by Sachdev [26], which differs by only a few percent from the holographic answer, $c'/c = \pi^3/162 \approx 0.1914$ [10]. With regard to the proposal of Klebanov and Polyakov [27] – that the large-N dual is a higher-spin gauge theory in AdS – this result amounts to a prediction for the bulk spin-two sector and implies a (small) quantitative difference with pure Einstein gravity. A similar comparison is possible for the ratio k'/k, where a natural variant of the vector current studied above lies in the adjoint, $J^{\alpha\beta}_\mu = \phi^{[\alpha}\partial_\mu\phi^{\beta]}$. Computation of the central charge at large N leads to $k=2$ [28], while the charge susceptibility was obtained in [29] leading to $k'/k|_{O(N)} \approx 0.1713$, which differs by about 24% from the result $k'/k = \pi/24 \approx 0.1309$ for models with gravitational duals in AdS. On the other hand, the conductivity in the $O(N)$ model is large, σ/χ is $\mathcal{O}(N)$ [8], reflecting the fact that the model becomes weakly coupled at large N. Therefore, the comparison with the $O(N)$ model in $d=3$ provides an example of a situation where two systems have very similar static thermodynamic properties, but vastly different transport properties.

Deviations from universality:- We have argued that in a large class of CFTs in $d>2$, there are universal relations between the thermodynamic and transport properties, and the central charges which dictate the short distance behaviour of current-current correlators. One way of defining this class of theories is that they possess dual descriptions within AdS at the level of classical gravity and Maxwell electrodynamics.

It is natural to ask about the regime of validity, or alternatively the constraints on the CFTs which may enter such universality classes. Indeed, the analysis we have performed using the AdS/CFT duality required the validity of a classical gravity approximation, and thus some kind of large-N limit. On the gravity side of the duality, universality follows from the uniqueness of the lowest dimension operator which determines the dynamics of the metric and/or the gauge field dual to the current in question, i.e. the uniqueness of the Einstein-Hilbert and Maxwell actions respectively. From this point of view, once we move to finite N, it appears that a large number of higher derivative corrections will also be required, thus limiting the possibility for universal behaviour.

Examples of higher curvature corrections to the Einstein-Hilbert action have been shown to correct the universal relation $\eta/s = 1/(4\pi)$ [30]. Here we will discuss an analogous correction to σ, arising from a bulk higher-derivative coupling to the Weyl tensor [12],

$$\mathcal{L} = \frac{1}{4g^2_{d+1}}\left(F^{\mu\nu}F_{\mu\nu} - 4\gamma C^{\mu\nu\rho\sigma}F_{\mu\nu}F_{\rho\sigma} + \cdots\right), \tag{4.24}$$

This setting provides a very clean test of the universality relation (3.21) at higher order, because the geometry dual to the zero temperature state, namely pure AdS, is Weyl flat. It follows immediately that the 2-point function $\langle J_\mu(x)J_\nu(0)\rangle$ at $T=0$, and thus the central charge k, are uncorrected by turning on the perturbation γ,

$$k(\gamma) = k(\gamma = 0). \tag{4.25}$$

However, the Weyl tensor does not vanish for the AdS black hole and a straightforward calculation shows that [12]

$$\sigma = \frac{\pi LT}{g_5^2}\left(1+\frac{8\gamma}{L^2}\right), \qquad D = \frac{\sigma}{\chi} = \frac{1}{2\pi T}\left(1+\frac{16\gamma}{L^2}+\cdots\right), \tag{4.26}$$

and the universal relation is violated beyond the classical gravity limit. This result is not unexpected, but it is interesting that causality [30] provides limits on γ ensuring that deviations cannot be large [12].

It would be interesting to see if there are other symmetries which might preserve these universality relations beyond the classical gravity limit, i.e. beyond leading order in N. It would also be interesting to explore other regimes, analogous to quantum critical points in which these techniques may be useful. Recall that the primary link was the presence of just one dimensionful scale breaking conformal invariance. One can easily imagine scenarios where this scale is something other than temperature. For example, it would be interesting to consider cold systems dominated instead by the chemical potential, which should have a dual description in terms of extremal black holes in AdS.

Acknowledgements: I'd like to thank the organizers for their invitation to this stimulating meeting, and Pavel Kovtun and John Ward for enjoyable collaboration on the work described here, which was supported in part by NSERC, Canada.

References

[1] Wilson K G and Kogut J B 1974 *Phys. Rept.* **12** 75
[2] See *e.g.* Leigh R G and Strassler M J 1995 *Nucl. Phys. B* **447** 95
[3] Maldacena J M 1998 *Adv. Theor. Math. Phys.* **2** 231
[4] Gubser S S et al. 1998 *Phys. Lett. B* **428** 105
[5] Witten E 1998 *Adv. Theor. Math. Phys.* **2** 253
[6] Hohenberg P C and Halperin B I 1977 *Rev. Mod. Phys.* **49** 435
[7] For a review, see Sachdev S 2008 *Nature Physics* **4** 173
[8] Sachdev S 1999 *Quantum phase transitions*, (Cambridge University Press, UK)
[9] Herzog C P et al. 2007 *Phys. Rev. D* **75** 085020
[10] Kovtun P and Ritz A 2008 *Phys. Rev. Lett.* **100** 171606
[11] Kovtun P and Ritz A 2008 Phys. Rev. D **78** 066009
[12] Ritz A and Ward J 1009 *Phys. Rev. D* **79** 066003
[13] Bloete H W J et al. 1986 *Phys. Rev. Lett.* **56** 742
[14] Affleck I 1986 *Phys. Rev. Lett.* **56** 746
[15] Kovtun P et al. 2005 *Phys. Rev. Lett.* **94** 111601
[16] Buchel A 2005 *Phys. Lett. B* **609** 392
[17] Damle K and Sachdev S 1997 *Phys. Rev. B* **56** 8714
[18] Chamblin A et al. 1999 *Phys. Rev. D* **60** 064018
[19] Freedman D Z et al. 1999 *Nucl. Phys. B* **546** 96
[20] Policastro G et al. 2002 *JHEP* **0209** 043
[21] Son D T and Starinets A O 2002 *JHEP* **0209** 042
[22] Herzog C P 2002 *JHEP* **0212** 026
[23] Anselmi D et al. 1998 *Phys. Rev. D* **57** 7570
[24] Arkani-Hamed N et al. 2007 *JHEP* **0706** 060
[25] Denef F and Hartnoll S A 2009 *Phys. Rev. D* **79** 126008
[26] Sachdev S 1993 *Phys. Lett. B* **309** 285
[27] Klebanov I R and Polyakov A M 2002 *Phys. Lett. B* **550** 213
[28] Petkou A 1996 *Annals Phys.* **249** 180
[29] Chubukov A V et al. 1994 *Phys. Rev. B* **49** 11919
[30] Buchel A et al. 2005 *Nucl. Phys. B* **707** 56; Brigante M et al. 2008 *Phys. Rev. D* **77** 126006; Kats Y and Petrov P *JHEP* **0901** 044

Hot Holographic Giant Loop

G. Grignani[1], J. L. Karczmarek[2] and G. W. Semenoff[2]

[1]Department of Physics, University of Perugia
via A. Pascoli, Perugia, Italy
[2]Department of Physics, University of British Columbia
Vancouver, British Columbia, Canada V6T 1Z1

Abstract. We consider the Polyakov loop operator in finite temperature planar N = 4 supersymmetric Yang-Mills theory defined on a spatial S^3 and in representations where the number of boxes in the Young Tableau k is large and scales so that kN remains finite in the large N limit. We review the argument that, in the de-confined phase of the gauge theory, and for symmetric representations with row Young tableau, there is a quantum phase transition in the expectation value of the Polyakov loop operator which occurs as the size of the representation is increased beyond a critical value of kN. We also argue that for completely antisymmetric representations with column tableau, there is no such phase transition. The AdS/CFT dual of such large representation loops are thought to be probe D-branes with k units of fundamental string charge dissolved in their world-volumes. Our results for both symmetric and antisymmetric representations are consistent with what is known about these branes on the thermal AdS black hole background.

1. Introduction

A beautiful picture of the quark de-confinement phase transition of finite temperature Yang-Mills theory has emerged in the context of AdS/CFT duality [1]. In the most conservative version of this duality, the gauge theory is $\mathcal{N} = 4$ supersymmetric Yang-Mills theory, which has conformal symmetry and is not normally viewed as a confining gauge theory. In fact, conformal invariance dictates that its non-Abelian gauge field mediated interactions are Coulomb-like at all distance scales. However, in context of AdS/CFT holography, the gauge theory must be defined on a spatial three-dimensional sphere, S^3. Then "confinement" is simply due the impossibility of solving the gauge theory Gauss' law for a color charged state on a compact space. Furthermore the $\mathcal{N} = 4$ theory has only adjoint representation degrees of freedom, which are not capable of combining with a fundamental representation external source to make a color singlet. For this reason, the question of confinement can then be addressed by testing the response of the gauge theory to insertion of a fundamental representation external charge, which must necessarily create a color-charged state. This response is measured by the expectation value of the Polyakov loop operator, and the question of confinement or de-confinement can be couched in terms of the realization of the center symmetry of the adjoint gauge theory and under which the Polyakov loop transforms.

Being in finite volume, de-confinement can only be a sharp phase transition in the large N limit. The idea is that, as the temperature gets larger, the thermally excited degrees of freedom which tend to screen the color charge are more numerous. When N is infinite, this screening can be so efficient that above a certain critical temperature, charged states are actually allowed.

We will generally work with the large N planar limit of Yang-Mills theory, also called the 't Hooft limit where we take N to infinity while holding the 't Hooft coupling $\lambda \equiv g_{YM}^2 N$ finite. Here, g_{YM} is the coupling constant of the gauge theory. In planar Yang-Mills theory defined on a spatial S^3, the de-confinement phase transition can be seen even in the weakly coupled gauge theory [2]-[3].

$\mathcal{N}=4$ supersymmetric Yang-Mills theory is thought to be exactly dual to type IIB superstring theory on an $AdS_5 \times S^5$ background space-time with N units of Ramond-Ramond 4-form flux. The 't Hooft coupling of the gauge theory is related to the fundamental string tension, $\frac{1}{4\pi\alpha'}$, and the radius of curvature of the AdS_5 and S^5 spaces, R, by $\lambda = R^4/(\alpha')^2$. The planar limit of the Yang-Mills theory is equivalent to the classical (tree level) limit of the string theory. The further limit of strong 't Hooft coupling in the planar Yang-Mills theory is dual to the limit of string theory as the radius of curvature of the background space is much larger than the fundamental string scale. This limit produces classical type IIB supergravity in an $AdS_5 \times S^5$ background.

At large 't Hooft coupling, the planar Yang-Mills theory is therefore efficiently described by its dual, classical IIB supergravity. There, the gravitational dual of de-confinement is the collapse of hot Anti-de Sitter space to an anti-de Sitter-Schwarzschild black hole. Unlike asymptotically flat space, where a thermodynamic state of a gravitational system is unstable to black hole collapse no matter what the temperature, at sufficiently low temperatures, anti-de Sitter space is thought to be stable. It contains a hot gas of super-gravitons. Moreover, when it is heated to a certain critical temperature, comparable to the inverse radius of curvature of the AdS space, the hot gas eventually undergoes gravitational collapse resulting in a space-time which is an anti-de Sitter black hole. The temperature and entropy of the resulting space-time are identified with the Hawking temperature and Beckenstein-Hawking entropy of the black hole. The phase transition between the two phases is first order and is called the Hawking-Page transition [4]. It is thought to be the gravity dual of the de-confinement transition of finite temperature Yang-Mills theory, extrapolated from the weak coupling to the strong coupling regime. It is conjectured that the strong and weak coupling limits are connected by a line of de-confining phase transitions which stretches across the intermediate λ regime.

In this Article, we shall review some of our recent work on the properties of the behavior of some particular Polyakov loop expectation values in de-confined phase of the gauge theory [5]. This work was motivated by an observation in string theory concerning the gravity dual of the Polyakov loops [6]. We will elaborate on this motivation in Section 2.

2. Giant Wilson loops

In the duality between gauge field theory and string theory, the expectation value of the Wilson loop is normally thought to correspond to an open fundamental string amplitude in string theory. This has been made precise for the Maldacena-Wilson loop [7] of the $\mathcal{N}=4$ supersymmetric Yang-Mills theory,

$$W_{\rm M}[C] = {\rm Tr}\mathcal{P}e^{\oint_C d\tau\left(iA_\mu(x)\dot{x}^\mu(\tau)+\phi^I(x)\theta^I|\dot{x}(\tau)|\right)} \quad (1)$$

where $x^\mu(\tau)$ parameterizes a closed curve C. $\phi^I(x)$, $I = 1, ..., 6$ are the scalar quark fields of $\mathcal{N}=4$ super Yang-Mills theory and θ^I is a (not necessarily τ-independent) unit 6-vector.

When λ is large, the loop (1) is computed by finding the minimum area of a fundamental string worldsheet embedded in $AdS_5 \times S^5$ and which has boundary located on the contour C that is placed at the boundary of AdS_5 and at a point θ^I on S^5.

If the contour C in (1) links periodic Euclidean time in the finite temperature Yang-Mills theory, $W_M[C]$ carries center charge and it can be nonzero only when the center symmetry is spontaneously broken, that is, in the de-confined phase. Then, the boundary of the disc in the disc amplitude of the string dual must also link periodic Euclidean time, but in the Euclidean AdS background. Whether such a disc exists depends on whether the time circle is contractible.

Journal of Physics: Conference Series **462** (2013) 012045 doi:10.1088/1742-6596/462/1/012045

It is not contractible in the hot AdS background, and it is contractible on the AdS black hole background. There is therefore no tree level disc amplitude on the hot AdS background. This is interpreted as implying that the expectation value of (1) vanishes there. The black hole background, on the other hand can have a classical disc amplitude, so the expectation value of (1) can be non-zero. This difference was pointed out by Witten as further evidence for the identification of the black hole formation with de-confinement [1].

In the gauge theory, if C wraps the time circle, $W_{\rm M}[C]$ carries center charge and its expectation value is governed by the realization of the center symmetry. Its expectation value is related to the holonomy of a heavy W-boson which would be created by Higgsing $SU(N+1)$ gauge symmetry to $SU(N)\times U(1)$ and where θ^I gives the orientation of the scalar condensate. One could imagine larger representation objects made from W-bosons, for example a bound state of a large number of the W-bosons which transforms in a higher representation of the gauge group.

In the zero temperature Yang-Mills theory defined on a spatial R^3, an interesting phenomenon occurs for loops in representations where the number of boxes k in the Young tableau is large so that $\frac{k}{N}$ is finite in the large N limit. The dual fundamental string worldsheet is replaced by a D-brane with world-volume electric flux [8].This object is called a "giant Wilson loop".

This was found by studying highly supersymmetric $\frac{1}{2}$-BPS loops, where some results are known for all values of the coupling constant [9]. For the anti-symmetric representation, the dual is a D5-brane whose world volume is a direct product $AdS_2\times S^4$ where $AdS_2\subset AdS_5$ and $S^4\subset S^5$. For a symmetric representation, it is a D3-brane with world volume $AdS_2\times S^2\subset AdS_5$.

It is interesting to ask whether these D-branes exist in the finite AdS black hole geometry where they would be dual to a gauge theory loop linking periodic Euclidean time and in the de-confined phase. This question was studied by Hartnoll and Kumar [6] who searched for solutions of the appropriate Born-Infeld actions on the black hole background.

For the D5-brane wrapped on $S^4\subset S^5$ which corresponds to a totally antisymmetric representation on the gauge theory side, there seem to be solutions for any $\frac{k}{N}$ with the usual cutoff at $k=N$ dictated by the maximum size of an antisymmetric representation on the gauge theory side and a maximum radius for embedding S^4 in S^5 on the supergravity side. However, in the case of the D3-brane, which should correspond to a totally symmetric representation, Hartnoll and Kumar could not find any solutions at all. It is interesting to ask whether this difference between the two cases is visible in the gauge theory. This question is what motivated our work on the gauge theory which is summarized in Ref. [5] and which we shall review in the following Sections.

3. Confinement-de-confinement and the Polyakov loop

Let us begin by reviewing the role of the Polyakov loop operator and the origin of center symmetry in an adjoint gauge field theory [10]-[11]. The Polyakov loop,

$$\left\langle {\rm Tr}\mathcal{P}e^{i\int_0^\beta d\tau A_0(\tau,\vec x)}\right\rangle = \frac{\int [dA_\mu ...] e^{-\int \frac{1}{2g_{YM}^2}{\rm Tr}F_{\mu\nu}^2+...}\,{\rm Tr}\mathcal{P}e^{i\oint_0^\beta d\tau A_0(\tau,\vec x)}}{\int [dA_\mu ...] e^{-\int \frac{1}{2g_{YM}^2}{\rm Tr}F_{\mu\nu}^2+...}} \tag{2}$$

measures the trace of the holonomy of the gauge group on the Euclidean time circle. In pure Yang-Mills theory, or a theory such as $\mathcal{N}=4$ super Yang-Mills theory where all matter fields transform in the adjoint representation of the gauge group, the Euclidean path integral has a center symmetry. It arises from gauge transformations which, in such a theory need not be periodic in Euclidean time, they need only preserve the periodicity of the local fields. As such, they can obey a boundary condition

$$g(\tau+\beta,\vec x)=cg(\tau,\vec x)$$

where c where $c = e^{2\pi i/N}$ is the generator of the Z_N center of the $SU(N)$ gauge group. Because c commutes with everything, all local operators which transform in the adjoint representation, for example $F_{\mu\nu}(\tau,\vec{x}) \to g(\tau,\vec{x})F_{\mu\nu}(\tau,\vec{x})g^\dagger(\tau,\vec{x})$, remain periodic after the gauge transform. However, the Polyakov loop transforms as

$$\mathrm{Tr}\mathcal{P}e^{i\int_0^\beta d\tau A_0(\tau,\vec{x})} \to c\ \mathrm{Tr}\mathcal{P}e^{i\int_0^\beta d\tau A_0(\tau,\vec{x})} \tag{3}$$

The center transformation appears as a global symmetry of the Euclidean path integral and it acts on Polyakov loops. When it is a good symmetry, any combination of Polyakov loops with non-vanishing center-charge must have vanishing expectation value. This is interpreted as confinement. When the center symmetry is spontaneously broken, the expectation value can be non-zero. This is interpreted as de-confinement. The representation of center symmetry has been used to study the de-confining phase transition of Yang-Mills theory [12]-[15].

The expectation value of the Polyakov loop operator is related to the free energy of Yang-Mills theory with a classical color source inserted. If one quantizes Yang-Mills theory with the constraint that a non-dynamical color charge is located at point $\vec{x}$, the free energy of the system Γ is obtained from the Euclidean path integral in (2).

$$e^{-\beta\Gamma} = \left\langle \mathrm{Tr}\mathcal{P}e^{i\int_0^\beta d\tau A_0(\tau,\vec{x})}\right\rangle \tag{4}$$

Γ is the energy that is needed to insert the external charge. When the expectation value is zero, as in the confining phase, the energy is infinite. When the expectation value is non-zero, the energy is finite.

4. Effective field theory

Our goal is to study the realization of the center symmetry at weak coupling. In this Section, we will review how this can be done using an effective unitary matrix model. In recent work, the weak coupling limit of both Yang-Mills theory and $\mathcal{N}=4$ super Yang-Mills theory have been studied for the case where the space-time is $S^3\times S^1$ [2]-[3]. This is an interesting case in that, at weak coupling, the spectrum of the vector gluon fields is completely gapped. If they are conformally coupled to the curvature of the S^3, the scalar fields of $\mathcal{N}=4$ theory are also gapped, as are the fermionic quarks. Then, the only low energy degree of freedom is the temporal component A_0 and, in the regime where the temperature is much less than the gap, all of the gapped fields can be integrated out to find an effective action for the zero mode on S^3 of A_0, or alternatively the unitary matrix variable $U \equiv \mathrm{Tr}\mathcal{P}e^{\oint A_0}$, and in the effective theory, an order parameter for the center symmetry is

$$\langle \mathrm{Tr}\ U\rangle = \frac{\int dU e^{-S_{\rm eff}[U]}\mathrm{Tr}\ U}{\int dU e^{-S_{\rm eff}[U]}} \tag{5}$$

This is a unitary one-matrix model. At one-loop order, the effective action is

$$S_{\rm eff}[U] = -\sum_{n=1}^{\infty}\left[z_B(x^n) + (-1)^{n+1}z_F(x^n)\right]\frac{|\mathrm{Tr}U^n|^2}{n} \tag{6}$$

where

$$x = e^{-\frac{\beta}{R}}\ ,\ z_B(x) = \frac{6x+12x^2-x^3}{(1-x)^3}\ ,\ z_F(x) = \frac{16x^{\frac{3}{2}}}{(1-x)^3} \tag{7}$$

Here, we have presented the action for $\mathcal{N}=4$ Yang-Mills theory. The one for pure Yang-Mills theory is similar and is presented in Ref. [3]. $S_{\rm eff}$ in (6) does not depend on the 't Hooft coupling

λ. This is due to the fact that it is the one-loop approximation. Dependence on λ starts at two loops. For pure Yang-Mills theory, it has been partially found up to order two loops [16] where the first order nature of the phase transition is confirmed.

The effective action S_{eff} inherits gauge invariance from its parent theory:

$$S_{\text{eff}}[U] = S_{\text{eff}}[VUV^{-1}] \tag{8}$$

as well as center symmetry

$$S_{\text{eff}}[U] = S_{\text{eff}}[cU] \tag{9}$$

Further, the action is of order N^2, $S_{\text{eff}}[U = 1] \sim N^2$.

The analysis of unitary matrix models and the existence of a phase transition of the type that we shall discuss has a long history dating back to the seminal work of Gross and Witten [17]. Gauge invariance (8) allows one to diagonalize the unitary matrices to form a model of the eigenvalues. The large N limit can then be solved by a saddle-point method and it is found that it has a phase transition from a center symmetric to a center symmetry broken phase at a critical temperature. The phase transition occurs at $T_C \simeq 0.38$ which is marginal to the regime $T << 1$. We will assume that it is within the range of validity of the effective field theory technique. In the following we will explore the de-confined phase. We will assume that we are at temperatures just above the critical one and we will assume that the effective matrix model gives an accurate description of the physics there.

5. Higher representations

We could also consider order parameters for center symmetry breaking in any representation of the gauge group that carries non-zero center charge,

$$\langle \text{Tr}_R U(x) \rangle = \frac{\int [dU] e^{-S_{\text{eff}}[U]} \text{Tr}_R\, U}{\int [dU] e^{-S_{\text{eff}}[U]}} \ . \tag{10}$$

where $\text{Tr}_R\, U$ is called the character. This expectation value should also exhibit the de-confining phase transition and it is interesting to ask whether it is at the same temperature for all representations. For small representations, it is easy to argue that the phase transition comes from the matrix ensemble and the transition temperature is always the same, independent of which representation we choose. On the other hand, we shall show in the following that (10) can have interesting behavior which depends on the size and nature of the representation when the representation is large. Two types of representation are easy to analyze: the completely symmetric representations $\mathcal{S}_k$ whose Young tableaux are a single row with k boxes and completely antisymmetric representations $\mathcal{A}_k$ whose Young tableaux are a single column with k boxes. We shall consider large values of k so that the ratio $\frac{k}{N}$ remains finite as $N \to \infty$ and we will study the behavior of these representations as $\frac{k}{N}$ is varied.

Note that we have not normalized the trace in (10), as one would normally do by dividing by a factor of the dimension of the representation. Our reason for not doing so is to be able to compare our results directly with holographic duality where the appropriate operator is the un-normalized one. This immediately introduces the interesting possibility that the expectation value is bigger than one – for example if the averaging over U were concentrated at the unit matrix – the expectation value would simply be equal to the dimension of the representation. The free energy would be negative, indicating that the system would "attract" the heavy quark. This apparent attraction is not dynamical, it is statistical, simply due to the increase in entropy from the multiplicity of states of the quark.

The center charge of a representation is equal to the number of boxes in the Young Tableau corresponding to that representation, modulo N. Thus, both representations $\mathcal{S}_k$ and $\mathcal{A}_k$ have center charge k mod N. The expectation value (10) is therefore expected to vanish in the confined phase when this charge is non-zero. On the other hand, the expectation value can be non-zero in the de-confined phase.

If the matrix $U = \text{diag}[e^{i\phi_1}, ..., e^{i\phi_N}]$ were diagonal, the permutation symmetry can be used to order the eigenvalues in a completely symmetric or completely antisymmetric representation so that they occur in order of non-decreasing index:

$$\text{Tr}_{\mathcal{S}_k} U = \sum_{a_1 \leq a_2 \leq ... \leq a_k} e^{i\phi_{a_1}} e^{i\phi_{a_2}} ... e^{i\phi_{a_k}} \quad , \quad \text{Tr}_{\mathcal{A}_k} U = \sum_{a_1 < a_2 < ... < a_k} e^{i\phi_{a_1}} e^{i\phi_{a_2}} ... e^{i\phi_{a_k}} \tag{11}$$

It is convenient to obtain these expressions from generating functions

$$\text{Tr}_{\mathcal{S}_k} U = \oint \frac{dt}{2\pi i t^{k+1}} \prod_{a=1}^{N} \frac{1}{1 - te^{i\phi_a}} \quad , \quad \text{Tr}_{\mathcal{A}_k} U = \oint \frac{dt}{2\pi i t^{k+1}} \prod_{a=1}^{N} \left(1 + te^{i\phi_a}\right) \tag{12}$$

where the contour integral projects onto the term in a Taylor expansion of the integrand which contains k eigenvalues. The contour in the integral over t encircles the origin. It can be moved away from the origin if it does not cross singularities of the integrand. In the case of the anti-symmetric representation (12) when N is finite, the integrand is a polynomial and the contour can be moved anywhere. For the symmetric representation (12) it should remain within the unit circle.

(12) can be written as the covariant expressions for the free energies,

$$\beta\Gamma_{\mathcal{S}_k} \equiv -\frac{1}{N} \ln \langle \text{Tr}_{\mathcal{S}_k} U \rangle = -\frac{1}{N} \ln \frac{1}{2\pi i} \oint dt \frac{1}{t^{k+1}} \langle \exp\left[-\text{Tr} \ln(1 - tU)\right] \rangle \,, \tag{13}$$

$$\beta\Gamma_{\mathcal{A}_k} \equiv -\frac{1}{N} \ln \langle \text{Tr}_{\mathcal{A}_k} U \rangle = -\frac{1}{N} \ln \frac{1}{2\pi i} \oint dt \frac{1}{t^{k+1}} \langle \exp\left[\text{Tr} \ln(1 + tU)\right] \rangle \tag{14}$$

The characters in (13) and (14) have center charge k and therefore they must vanish in the center symmetric confining phase. They can be non-zero in the de-confined phase.

In the large N limit, the quantities in (13) and (14) can be computed using two saddle point approximations. The first occurs while integrating over unitary matrices in the expectation value bracket in (13) and (14) and which is defined in (10). Because of the gauge symmetry, this is an eigenvalue model. The gauge symmetry can be used to diagonalize U. At large N, the eigenvalues become classical variables and their distribution is found by minimizing S_{eff} plus the logarithm of a Jacobian from the unitary integral measure. As long as $k \ll N^2$, the loop operators in (13) do not modify the eigenvalue distribution in the leading order at large N. It is given by a density $\rho(\phi)$ which is $\frac{1}{N}$ times the number of eigenvalues between ϕ and $\phi + d\phi$ and is normalized, $\int_{-\pi}^{\pi} d\phi \rho(\phi) = 1$. In the large N limit the expectation values in Eq. (13) are computed using the eigenvalue density,[1]

$$\beta\Gamma_{\mathcal{S}_k} = -\frac{1}{N} \ln \frac{1}{2\pi i} \oint dt \frac{1}{t} \exp\left(-N \int_{-\pi}^{\pi} d\phi \rho(\phi) \ln(1 - te^{i\phi}) - k \ln t\right) \,. \tag{15}$$

$$\beta\Gamma_{\mathcal{A}_k} = -\frac{1}{N} \ln \frac{1}{2\pi i} \oint dt \frac{1}{t} \exp\left(N \int_{-\pi}^{\pi} d\phi \rho(\phi) \ln(1 + te^{i\phi}) - k \ln t\right) \,. \tag{16}$$

The second use of a saddle-point approximation is to evaluate the integral over t in (15) and (16). Let $\hat{t}$ satisfy the saddle-point equation

$$R_{\mathcal{S}_k}(\hat{t}) \equiv \int_{-\pi}^{\pi} d\phi \rho(\phi) \frac{\hat{t}e^{i\phi}}{1 - \hat{t}e^{i\phi}} = \frac{k}{N} \quad , \quad R_{\mathcal{A}_k}(\hat{t}) \equiv \int_{-\pi}^{\pi} d\phi \rho(\phi) \frac{\hat{t}e^{i\phi}}{1 + \hat{t}e^{i\phi}} = \frac{k}{N} \,. \tag{17}$$

[1] We will argue that using the leading order N^0 density $\rho(\phi)$ in (15) is sufficient to obtain $\beta\Gamma_{\mathcal{S}_k/\mathcal{A}_k}$ to leading order N^0 accuracy.

The functions $R_{\mathcal{S}_k/\mathcal{A}_k}(t)$ in (17) are related to the resolvent of the matrix model and are holomorphic functions of t with cut singularities on the unit circle determined by the support of $\rho(\phi)$. Once the solution $\hat{t}$ of the saddle point is determined, the free energy is given by

$$\beta\Gamma_{\mathcal{S}_k} = \int_{-\pi}^{\pi} d\phi\rho(\phi)\ln(1-\hat{t}e^{i\phi}) + \frac{k}{N}\ln\hat{t}\ . \tag{18}$$

$$\beta\Gamma_{\mathcal{A}_k} = -\int_{-\pi}^{\pi} d\phi\rho(\phi)\ln(1+\hat{t}e^{i\phi}) + \frac{k}{N}\ln\hat{t}\ . \tag{19}$$

The generating function technique that we have used in the above is well known. See Ref. [18] for a recent application in a different context.

The reader might have the concern that the presence of the loop variable in the path integral, though it does not alter the eigenvalue distribution to the leading order N^0, it will have an effect at order $1/N$ and a $1/N$ correction in the order N^2 part of the action would contribute a term of order N which competes with the free energy which we are computing. To see why this is not a problem, consider the free energy in the large N limit is given by

$$N\beta\Gamma = \inf_{(\rho,t)}\left[N^2S[\rho] + \lambda\int\rho - \lambda + N\int\rho\ln\left(1-te^{i\phi}\right) + k\ln t\right] - \inf_{\rho}\left[N^2S[\rho] + \lambda\int\rho - \lambda\right] \tag{20}$$

where $S[\rho]$ is the effective action consisting of S_{eff} plus a contribution from the integration measure. The saddle-point equations are

$$\frac{\delta S}{\delta\rho} + \frac{1}{N}\ln\left(1-te^{i\phi}\right) + \frac{\lambda}{N^2} = 0\ ,\quad \int\rho = 1\ ,\quad \int\rho\frac{te^{i\phi}}{1-te^{i\phi}} = \frac{k}{N}\frac{1}{t} \tag{21}$$

for the first infimum and

$$\frac{\delta S}{\delta\rho} + \frac{\lambda}{N^2} = 0\ ,\quad \int\rho = 1 \tag{22}$$

for the second infimum. The eigenvalue density which satisfies (22) is $\hat{\rho}_0$. Then the density which satisfies (21) differs from it by a correction of order $\frac{1}{N}$, $\hat{\rho}_0 + \frac{1}{N}\hat{\rho}_1$. However, since $\hat{\rho}_0$ satisfies (22), it is easy to see, that if we are interested in $N\beta\Gamma$ only to accuracy of order N, we can get simply use $\hat{\rho}_0$ in the equation which determines $\hat{t}$ and, to the same accuracy (where we trust the order N but not the order N^0 contribution), in the expression for $N\beta\Gamma$ in (20). This justifies our use of the "probe approximation" where we use the eigenvalue distribution of the effective unitary matrix model to compute the generating functions in (15) to (19). We note that a similar probe approximation is made when analyzing the dual objects on the string theory side of the AdS/CFT correspondence.

Before we proceed further, let us consider a simple example, the confined phase. Center symmetry is an invariance under a simultaneous translation of all eigenvalues $\phi_a \to \phi_a + 2\pi/N$. In the center-symmetric confined phase, the distribution is translation invariant, eigenvalues are uniformly distributed on the unit circle and

$$\rho_{\text{conf}} = \frac{1}{2\pi} \tag{23}$$

In the de-confined phase, on the other hand, the eigenvalues would have a non-constant distribution. Let us put off discussing the de-confined phase until later.

With the confining eigenvalue distribution (23) we can integrate over ϕ in the saddle-point equations (17),

$$R_{\mathcal{S}_k}(\hat{t}) = \begin{cases} 0 & |\hat{t}|<1 \\ -1 & |\hat{t}|>1 \end{cases} = \frac{k}{N}\ ,\quad R_{\mathcal{A}_k}(\hat{t}) = \begin{cases} 0 & |\hat{t}|<1 \\ 1 & |\hat{t}|>1 \end{cases} = \frac{k}{N} \tag{24}$$

Similarly, and consistent with this, we can integrate the free energies in (18) and (19),

$$\beta\Gamma_{\mathcal{S}_k} = \begin{cases} \frac{k}{N}\ln\hat{t} & |t|<1 \\ i\pi + \left(1+\frac{k}{N}\right)\ln\hat{t} & |t|>1 \end{cases}, \quad \beta\Gamma_{\mathcal{A}_k} = \begin{cases} \frac{k}{N}\ln\hat{t} & |t|<1 \\ \left(\frac{k}{N}-1\right)\ln\hat{t} & |t|>1 \end{cases} \tag{25}$$

In the case of the symmetric representation, the saddle-point equation (24) has a solution only when $\frac{k}{N}=0$. We interpret the absence of a solution when $\frac{k}{N}\neq 0$ as meaning that the expectation value of any representation with non-zero center charge vanishes. Certainly, if there is no saddle-point of a periodic function of a variable ϕ, the integration is not dominated by any particular value of ϕ and ϕ must be integrated over its entire range. This would average the expectation value of any operator with non-zero center charge to zero. It is in the other case, when there is a saddle point, where the large N limit forces one to evaluate the integrand at the saddle point and the expectation value is generically non-zero. Further, we see that the free energy has an imaginary part when $|\hat{t}|>1$ which indicates an instability. This apparent pathology is consistent with the observation after (12) that the integration contour should remain inside the unit circle. We shall henceforth ignore the region $|t|>1$ for symmetric representations.[2]

Similarly, for the anti-symmetric representation, the saddle-point equation (24) has a solution only when either $\frac{k}{N}=0$ or $\frac{k}{N}=1$, the two cases where the antisymmetric representation is center neutral. This is also interpreted as confinement, the expectation value vanishes in all other cases. Note that it has an expected $k\to N-k$ duality, though it comes from interchanging two saddle points, one with $|\hat{t}|<1$ and one with $|\hat{t}|>1$. Neither of these saddle-points alone exhibit this duality.

In the next Section, we will review the computation of the higher representation Polyakov loops with more complicated eigenvalue distributions.

6. Giant Polyakov loop

Let us begin with the symmetric representation $\mathcal{S}_k$. We consider three examples of eigenvalue distributions. First, the confining phase has $\rho_{\rm conf}=\frac{1}{2\pi}$ considered above. As a second example consider a simple prototype of a de-confined distribution (while it is not realistic for $\mathcal{N}=4$ Yang-Mills theory, this is actually the solution of the strong coupling phase of the Gross-Witten model [17]) $\rho(\phi)=\frac{1}{2\pi}(1+2p\cos\phi)$. there the parameter $p=\frac{1}{N}\langle{\rm Tr}\ U\rangle=\int d\phi\rho(\phi)e^{i\phi}$ is the fundamental representation loop. Positivity of the density requires $0\leq p\leq\frac{1}{2}$.

There is one solution of $R_{\mathcal{S}_k}(\hat{t})=\frac{k}{N}$ in the region $|\hat{t}|<1$ at $\hat{t}=\frac{k}{N}/p$. (If $\frac{k}{N}$ and p are such that $|\hat{t}|>1$, both $R_{\mathcal{S}_k}$ and $\Gamma_{\mathcal{S}_k}$ should be extended there by analytic continuation.) The free energy is

$$\Gamma_{\mathcal{S}_k}=\frac{k}{N}\ln\left[\frac{k/N}{ep}\right], \tag{26}$$

where $e=2.718\ldots$. $\Gamma_{\mathcal{S}_k}$ has the interesting feature that, as $\frac{k}{N}$ is increased, it changes sign from negative to positive. This results in a phase transition which occurs when $\frac{k}{N}=\left(\frac{k}{N}\right)_{\rm crit}=ep$. When $\frac{k}{N}<\left(\frac{k}{N}\right)_{\rm crit}$, $\Gamma_{\mathcal{S}_k}$ is negative and the loop expectation value, $e^{-N\Gamma}$, is exponentially large. When $\frac{k}{N}>\left(\frac{k}{N}\right)_{\rm crit}$, $\Gamma_{\mathcal{S}_k}$ is positive and the loop vanishes for $N\to\infty$. This phase transition implies that, even in the de-confined phase, sufficiently large symmetric representations are still confined.

As a check of the saddle-point approximation to the t-integral in this simple example, observe that, if for the moment we assume that k and N are finite, we can integrate (15) explicitly to get $e^{-N\Gamma_{\mathcal{S}_k}}=\frac{N^k}{k!}p^k$. Using the Stirling formula and taking $k\sim N\to\infty$ reproduces (26).

[2] We do later consider the analytic continuation of the free energy and the solution of the saddle point equation from the region $|t|<1$ to the entire complex plane.

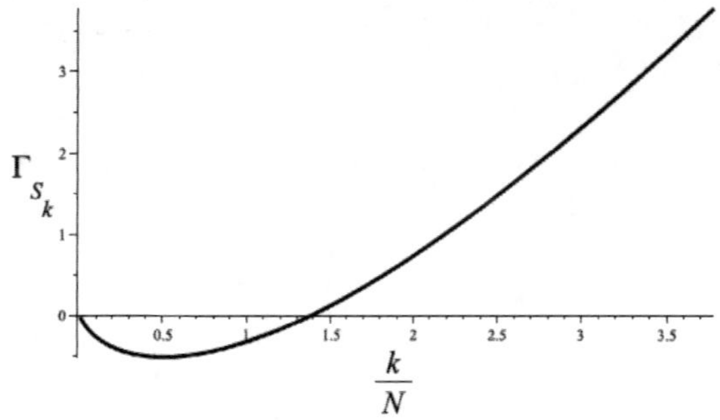

Figure 1. The free energy $\Gamma_{\mathcal{S}_k}(\theta)$ as a function of $\frac{k}{N}$ in the semi-circle distribution with $p = 0.51$.

To see this behavior in another example, consider the semi-circle distribution which, for $|\phi| < 2\arcsin\sqrt{2-2p}$, is

$$\rho(\phi) = \frac{\cos\frac{\phi}{2}}{\pi(2-2p)}\sqrt{2-2p-\sin^2\frac{\phi}{2}} \tag{27}$$

and which vanishes in the gap $2\arcsin\sqrt{2-2p} \le |\phi| \le \pi$. We still use the fundamental loop, p, as a parameter and now $\frac{1}{2} \le p \le 1$. This is the distribution in the weak coupling phase of 2-dimensional lattice Yang-Mills theory [17]. It is also an approximation to the de-confined distribution for weakly coupled $\mathcal{N} = 4$ Yang-Mills theory [3][19]. For sufficiently weak coupling, it could be accurate near the phase transition where $p = \frac{1}{2}$. The saddle point computation can be done explicitly near $t = 0$ and analytically continued. The free energy is

$$\begin{aligned}\Gamma_{\mathcal{S}_k} &= (2\theta\cosh\theta - \sinh\theta)\frac{\sinh\theta + \sqrt{\sinh^2\theta + 2 - 2p}}{2-2p} \\ &- \frac{1}{2} - \ln\left[\frac{\sinh\theta + \sqrt{\sinh^2\theta + 2 - 2p}}{2-2p}\right] ,\end{aligned} \tag{28}$$

where θ is defined by $\hat{t} = e^{2\theta}$ and is determined by the saddle-point equation

$$\frac{k}{N} + \frac{1}{2} = \cosh\theta\left[\frac{\sinh\theta + \sqrt{\sinh^2\theta + 2 - 2p}}{2-2p}\right] , \tag{29}$$

which can be solved for $\sinh(\theta)$. The free energy is zero when $k = 0$, negative for small k, goes to zero at a critical $\frac{k}{N}$ and is positive thereafter. This is so for any value of p in the allowed range. A graph of $\Gamma_{\mathcal{S}_k}$ versus $\frac{k}{N}$ for $p = 0.51$ is plotted in Fig. 1. With this value of p, the free energy becomes positive at $\theta \simeq 0.50$ which corresponds to $\frac{k}{N}_{\text{crit.}} \simeq 1.4$.

Now, we consider the antisymmetric representation. For a large class of distributions gapped around $\phi = \pi$ and with $\frac{dR_{\mathcal{A}_k}(\hat{t})}{d\hat{t}} > 0$, which includes the semi-circle distribution (27), we can argue that $\Gamma_{\mathcal{A}_k}$ is always negative and the phase transition that we are discussing does not occur. To begin, by changing variables in (18), we observe that $\Gamma_{\mathcal{A}_k} = \Gamma_{\mathcal{A}_{N-k}}$. This symmetry is reflected in the saddle-point equation (17) which, using our assumption that $\rho(\phi) = \rho(-\phi)$, can be re-written as

$$\frac{1}{2}\int_{-\pi}^{\pi} d\phi\rho(\phi)\frac{\hat{t}^{\frac{1}{2}}e^{i\frac{\phi}{2}} - \hat{t}^{-\frac{1}{2}}e^{-i\frac{\phi}{2}}}{\hat{t}^{\frac{1}{2}}e^{i\frac{\phi}{2}} + \hat{t}^{-\frac{1}{2}}e^{-i\frac{\phi}{2}}} = \frac{k}{N} - \frac{1}{2} \tag{30}$$

and implies $\hat{t}(k/N) = 1/\hat{t}(1-k/N)$. The free energy,

$$\Gamma_{\mathcal{A}_k} = -\int_{-\pi}^{\pi} d\phi\rho(\phi)\ln\left(\hat{t}^{\frac{1}{2}}e^{i\frac{\phi}{2}} + \hat{t}^{-\frac{1}{2}}e^{-i\frac{\phi}{2}}\right) + \left(\frac{k}{N} - \frac{1}{2}\right)\ln\hat{t} \tag{31}$$

is symmetric under $\frac{k}{N} \to 1 - \frac{k}{N}$. Moreover, with a gapped distribution, $\rho(\pi) = 0$, and the integral in (30) is continuous at $\hat{t} = 1$. From $\frac{dR_{\mathcal{A}_k}(\hat{t})}{d\hat{t}} > 0$, $\hat{t}(k)$ is monotone, and one can see in (30) that $\hat{t} = 0$ corresponds to $\frac{k}{N} = 0$, $\hat{t} = \infty$ to $\frac{k}{N} = 1$ and $\hat{t} = 1$ to $\frac{k}{N} = \frac{1}{2}$. Furthermore, since $\frac{d\Gamma_{\mathcal{A}_k}}{d(k/N)} = \ln\hat{t}(k)$, $\frac{d^2\Gamma_{\mathcal{A}_k}}{d(k/N)^2} > 0$, thus $\Gamma_{\mathcal{A}_k}$ is a convex function which decreases from 0 to a negative minimum as $\frac{k}{N}$ goes from 0 to $\frac{1}{2}$ and then increases back to zero when $\frac{k}{N}$ goes from $\frac{1}{2}$ to 1. $\Gamma_{\mathcal{A}_k}$ does not become positive and there is no phase transition of the kind that we found for symmetric representations. When the distribution is ungapped, or when $\frac{dR_{\mathcal{A}_k}(\hat{t})}{d\hat{t}}$ becomes negative (for example when $p < 0$), interesting behavior can occur, discussion of which we put off to a later time.

7. Conclusion

We have found a difference between the symmetric and antisymmetric representation Polyakov loops in the gauge theory which is qualitatively similar to the one found by Hartnoll and Kumar [6] for the dual objects in supergravity: the antisymmetric loop is non-zero in the de-confined phase for all allowed $\frac{k}{N}$ and the dual D5-brane exists whereas the gauge theory symmetric representation loop has a phase transition at a critical value of $\frac{k}{N}$. The numerical search for the dual D3-brane in [6] combined with analytic arguments at large values of the parameter $\kappa = \frac{\sqrt{\lambda}}{4}\frac{k}{N}$ found no solution at all. We have also carried out a further numerical search for the D3-brane, which was partially reported in [5]. We explored the region of small κ and found no evidence for a solution except for extremely small values of κ which we cannot rule out. If we assume that there is no solution and that this means that the expectation value of the gauge theory quantity vanishes, it indicates that, even in the de-confined phase, the gauge theory is still confining for large charge symmetric representation sources. Of course, coming from the string theory, this is accurate only for large λ. The conclusion is that the critical value of $\frac{k}{N}$ becomes coupling constant dependent and the function $\left[\frac{k}{N}\right]_{\text{crit.}}(T,\lambda)$ goes to zero faster than $\frac{4}{\sqrt{\lambda}}$ as $\lambda \to \infty$.

We note that a phase transition of similar nature, but in a somewhat different context of branes with large angular momentum has recently been discussed [20].

The authors acknowledge hospitality of the Galileo Galilei Institute, Aspen Center for Physics and Perimeter Institute. This work is supported in part by NSERC of Canada and the INFN of Italy.

References

[1] E. Witten, "Anti-de Sitter space, thermal phase transition, and confinement in gauge theories," Adv. Theor. Math. Phys. **2**, 505 (1998) [arXiv:hep-th/9803131].

[2] B. Sundborg, "The Hagedorn transition, de-confinement and N = 4 SYM theory," Nucl. Phys. B **573**, 349 (2000) [arXiv:hep-th/9908001].

[3] O. Aharony, J. Marsano, S. Minwalla, K. Papadodimas and M. Van Raamsdonk, "The Hagedorn / deconfinement phase transition in weakly coupled large N gauge theories," Adv. Theor. Math. Phys. **8**, 603 (2004) [arXiv:hep-th/0310285].

[4] S. W. Hawking and D. N. Page, "Thermodynamics Of Black Holes In Anti-De Sitter Space," Commun. Math. Phys. **87**, 577 (1983).
[5] G. Grignani, J. L. Karczmarek and G. W. Semenoff, "Hot Giant Loop Holography," arXiv:0904.3750 [hep-th].
[6] S. A. Hartnoll and S. Prem Kumar, "Multiply wound Polyakov loops at strong coupling," Phys. Rev. D **74**, 026001 (2006) [arXiv:hep-th/0603190].
[7] J. M. Maldacena, "Wilson loops in large N field theories," Phys. Rev. Lett. **80**, 4859 (1998) [arXiv:hep-th/9803002].
[8] N. Drukker and B. Fiol, "All-genus calculation of Wilson loops using D-branes," JHEP **0502**, 010 (2005) [arXiv:hep-th/0501109].
S. Yamaguchi, "Bubbling geometries for half BPS Wilson lines," Int. J. Mod. Phys. A **22**, 1353 (2007) [arXiv:hep-th/0601089];
S. Yamaguchi, "Wilson loops of anti-symmetric representation and D5-branes," JHEP **0605**, 037 (2006) [arXiv:hep-th/0603208].
J. Gomis and F. Passerini, "Holographic Wilson loops," JHEP **0608**, 074 (2006) [arXiv:hep-th/0604007];
J. Gomis and F. Passerini, "Wilson loops as D3-branes," JHEP **0701**, 097 (2007) [arXiv:hep-th/0612022];
J. Gomis, S. Matsuura, T. Okuda and D. Trancanelli, "Wilson loop correlators at strong coupling: from matrices to bubbling geometries," JHEP **0808**, 068 (2008) [arXiv:0807.3330 [hep-th]].
K. Okuyama and G. W. Semenoff, "Wilson loops in N = 4 SYM and fermion droplets," JHEP **0606**, 057 (2006) [arXiv:hep-th/0604209];
[9] J. K. Erickson, G. W. Semenoff and K. Zarembo, "Wilson loops in N = 4 supersymmetric Yang-Mills theory," Nucl. Phys. B **582**, 155 (2000) [arXiv:hep-th/0003055]; V. Pestun, "Localization of gauge theory on a four-sphere and supersymmetric Wilson loops," arXiv:0712.2824 [hep-th].
[10] A. M. Polyakov, "Thermal Properties Of Gauge Fields And Quark Liberation," Phys. Lett. B **72**, 477 (1978).
[11] L. Susskind, "Lattice Models Of Quark Confinement At High Temperature," Phys. Rev. D **20**, 2610 (1979).
[12] L. D. McLerran and B. Svetitsky, "A Monte Carlo Study Of SU(2) Yang-Mills Theory At Finite Temperature," Phys. Lett. B **98**, 195 (1981);
[13] B. Svetitsky and L. G. Yaffe, "Critical Behavior At Finite Temperature Confinement Transitions," Nucl. Phys. B **210**, 423 (1982);
[14] L. G. Yaffe and B. Svetitsky, "First Order Phase Transition In The SU(3) Gauge Theory At Finite Temperature," Phys. Rev. D **26**, 963 (1982);
[15] A. Dumitru, Y. Hatta, J. Lenaghan, K. Orginos and R. D. Pisarski, "Deconfining phase transition as a matrix model of renormalized Polyakov loops," Phys. Rev. D **70**, 034511 (2004) [arXiv:hep-th/0311223].
[16] O. Aharony, J. Marsano and M. Van Raamsdonk, "Two loop partition function for large N pure Yang-Mills theory on a small S**3," Phys. Rev. D **74**, 105012 (2006) [arXiv:hep-th/0608156].
[17] D. J. Gross and E. Witten, "Possible Third Order Phase Transition In The Large N Lattice Gauge Theory," Phys. Rev. D **21** (1980) 446.
[18] S. A. Hartnoll and S. P. Kumar, "Higher rank Wilson loops from a matrix model," JHEP **0608**, 026 (2006) [arXiv:hep-th/0605027].
[19] J. Jurkiewicz and K. Zalewski, "Vacuum Structure Of The U(N $\to$ Infinity) Gauge Theory On A Two-Dimensional Lattice For A Broad Class Of Variant Actions," Nucl. Phys. B **220**, 167 (1983).
[20] D. Gang, J. S. Park and S. Yamaguchi, "Operator with large spin and spinning D3-brane," JHEP **0911**, 024 (2009) [arXiv:0908.3938 [hep-th]].

Journal of Physics: Conference Series **462** (2013) 012046 doi:10.1088/1742-6596/462/1/012046

Probing individual topological tunneling events of a quantum field: Switching statistics of a superconducting nanowire

Nayana Shah

Department of Physics, University of Cincinnati, Cincinnati, OH 45221-0011

E-mail: nayana.shah@uc.edu

Abstract. Phase slips are topological fluctuation events that carry the superconducting order-parameter field between distinct current carrying states and impart a non-zero resistance to superconducting nanowires. They play a fundamental role in determining the fate of superconductivity in nanowires. Conversely, superconducting nanowires provide an ideal setting for accessing non-trivial fluctuations driven by thermal activation and—at low temperatures—by quantum tunneling of a one-dimensional field. However, this potential has not been fully realized because resistance measurements, on the one hand, are capable of capturing only the averaged phase-slip behavior, and on the other hand, are incapable of pinning down the low temperature phase-slip behavior, as the measured resistance values drop below the noise floor. On going beyond the linear-response regime, the I-V characteristics show a hysteretic behavior. As the current is ramped up repeatedly, the state switches from a superconductive to a resistive one, doing so at somewhat random current values below the depairing critical current. The distribution of these switching currents was studied recently. I will report on the rather counter-intuitive temperature dependence of the distribution and its theoretical understanding via a stochastic model. It will be shown that although, in general, several phase-slip events are necessary to induce switching, there is an experimentally accessible temperature- and current-range for which a single phase-slip event is sufficient to switch the wire to the normal (resistive) state. I will conclude by arguing that switching-current statistics provide an effective probe to resolve individual phase-slip events and in addition offer unprecedented access to quantum phase-slip tunneling events.

1. Introduction

Quantum tunneling [1, 2, 3] is one of the most spectacular signatures of quantum physics. It clearly demarcates the quantum from the classical since it involves tunneling through classically forbidden regions. It lies at the heart of whole range of observed or proposed phenomena and encompasses all branches of modern physics that emanated from or metamorphized by the quantum revolution. Examples range from alpha decay of nuclei which is perhaps the most classic manifestation of quantum tunneling, to instanton solutions [4, 5] corresponding to the tunneling between different vacuua in quantum chromodynamics as well as the cosmological implications of the fate of a false vacuum [6, 7]. Examples in condensed matter systems abound [8] and quantum tunneling is essential also for experimental methods such as scanning tunneling microscopy and for technological applications.

The question of whether one can observe Macroscopic Quantum Tunneling (MQT) [9, 10] *i.e.* quantum tunneling on a macroscopic scale is intriguing and has spurred a whole area of research at the intersection of fields such as foundations of quantum mechanics, quantum measurement, quantum computation and information [11, 12, 13], condensed matter physics [8] and ultra-cold atomic physics [14]. Indeed, macroscopic systems exist for which the temperature can be lowered much below the frequency of small oscillations (about the metastable equilibrium) and are expected to furnish tunneling probabilities that are not unobservable. The coupling to the environment and dissipation has been argued to play a crucial role in dictating whether the observation of MQT is in fact possible [15, 16, 17] or not.

In accordance with theoretical expectations [18, 16], Josephson junction circuitry has proven to be an ideal setting for the observation of MQT, wherein the macroscopic degree of freedom corresponds to the phase difference of the superconducting order parameter across a Josephson junction or the flux in a Superconducting Quantum Interference Device (SQuID) [19, 20, 21, 22, 8]. A current-biased Josephson junction can be effectively represented by a mechanical analog of a 'particle' moving in a tilted cosine (washboard) potential and the MQT corresponds to the particle escaping from a metastable minimum of the potential by tunneling through the barrier [23]. In an underdamped Josephson junction, a single tunneling event can take the system from a zero-voltage state (corresponding to the phase point or the particle confined to a local minimum) to a finite voltage state (corresponding to a delocalized phase or the particle running through the washboard potential). It is important to note that it is not a single electron but the phase variable describing the collective (superconducting) state of a macroscopic number of electrons that is tunneling.

Encouraged by the observation of MQT of a quantum variable, one cannot but wonder about the next intriguing question: is it possible to experimentally observe MQT of a quantum field? If a Josephson junction is the 'zero-dimensional' case then one would like to consider the 'one-dimensional' case in which the phase is a function of position. Lo and behold, a superconducting wire (with length greater and diameter much less than the coherence length) precisely furnishes this desirable configuration: the superconducting order parameter is a complex field with a fixed amplitude and the phase winding along the length of the wire to form a helix. Furthermore, in a so-called phase slip event [24], the complex-valued order parameter field goes from one metastable minimum to another in which the number of phase-windings is reduced by one; *i.e.* the phase slips by 2π and the phase-difference across the wire is correspondingly lowered by 2π. In going from one topologically distinct super-current carrying state to another, the order-parameter field essentially passes through a configuration in which its amplitude drops to zero somewhere along the wire (typically over a length-scale of the order of coherence length) where the phase slips by 2π. Therefore, during the course of a phase-slip process the superconducting order loses its coherence and the wire acquires a non-zero resistance. In fact, such resistive phase-slip fluctuations are responsible for determining the very essence of superconductivity in quasi-one-dimensional wires [25, 23].

Phase-slip fluctuations in which the order-parameter field jumps over the potential barrier via thermal activation are well-understood in the vicinity of the transition temperature. The LAMH theory [26, 27, 28] for thermally-activated phase-slips (TAPS) derived within the (time-dependent) Ginzburg-Landau formalism provides good fits for the broad resistive transition (as against a sharp transition to the zero resistance state in bulk superconductors) in quasi-one-dimensional wires. The search for quantum phase slips (QPS) [29, 30] in which the order-parameter field tunnels through the potential barrier has been mostly elusive. In this proceeding I will report on a recent experiment [31] and its theoretical interpretation [32, 33] which we believe provides evidence for the detection of a single quantum phase slip process and hence for an observation of the MQT of a quantum field.

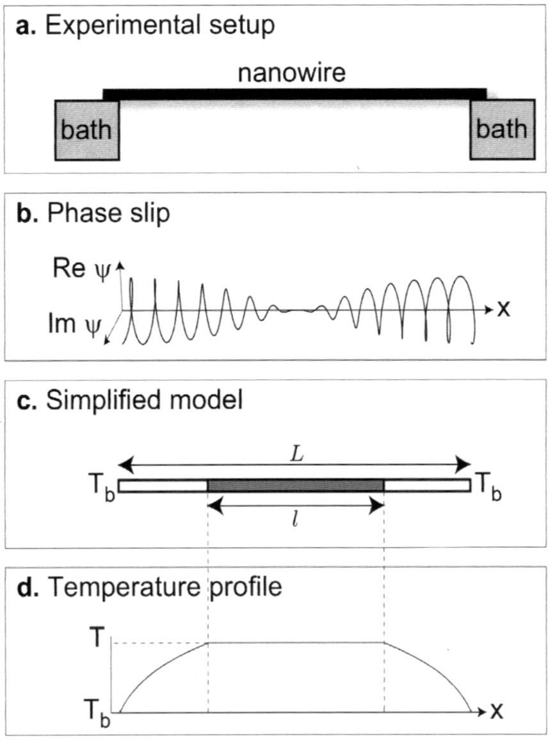

Figure 1. Illustration of the system and the model: a. Schematic of an experimental set-up in which a superconducting nanowire is suspended between two thermal baths. b. Sketch showing the attenuation of the order parameter in the core of a phase-slip. c. Schematic of the simplified model. All phase slips are taken to occur in the central (*i.e.*, shaded) segment of length l, which is assumed to be at a uniform temperature T; heat is carried away through the end segments, which are assumed to have no heat capacity. The temperature at the ends of the wire is fixed to be T_{b}. d. Sketch of a typical temperature profile.

2. System and Model

Recent advances in the fabrication of long and homogeneous superconducting nanowires—using carbon nanotube [34] or DNA templating [35]—has made it possible to access regimes in which the probability for QPS at low enough temperatures is in the potentially observable range. However, in the low temperature window in which QPS are expected to be significant enough, the measured resistance values drop below the noise floor, thereby making it difficult and challenging to unambiguously observe QPS. A recent experiment [31] seeks to address this problem by going beyond the linear-response regime and modulating the bias current as a periodic (triangular or sinusoidal) function of time. As the current is ramped up repeatedly, the state switches from a nominally superconductive (*i.e.*, barring the exponentially suppressed yet non-zero resistance imparted by the phase slip fluctuations) to a nominally normal (*i.e.*, resistive) one; doing so at somewhat random current values below the depairing critical current [36, 37, 23]. Since one expects phase-slip fluctuations to be responsible for the premature stochastic switching [30, 38, 39], the distribution of these switching currents and its temperature evolution promises to provide a useful and less ambiguous probe of phase-slip fluctuations.

To be able to theoretically understand the switching current distribution or alternatively the mean switching time τ_{s} of the bistable current-biased nanowire that can be extracted from the switching-current statistics [40], one needs to first identify the physical mechanism that causes the wire to switch. Given that the experimental set-up consists of a free-standing wire connected to thermal baths at its two ends (see Figure 1a), it has been argued [38, 32] that if the Joule heating caused by resistive phase slip fluctuations is not overcome sufficiently rapidly by conductive cooling, it effectively reduces the depairing current, ultimately to below the applied current, thus causing switching to the highly resistive state.

To study the dynamics of switching, it is necessary to explicitly take into consideration the fact that the resistive fluctuations of the superconducting nanowire consist of discrete phase-slip events that take place at random moments of time and are centered at random spatial locations. Instead of solving a full 1+1 dimensional stochastic differential equation, it is reasonable to consider a simplified model (as shown in panels c and d of Figure 1) in which all the phase slips and hence the heating takes place within a central segment of length l to which a uniform temperature T is assigned and the heat is conducted away through the end segments (within which the heat capacity can be ignored). The dynamics of the nanowire is then given by a stochastic ordinary differential equation for the time-evolution of the temperature T and is dictated by the competition between discrete heating (since each phase slip heats the wire by a "quantum" of energy $hI/2e$, for a given value of the applied current I) and continuous cooling.

3. Results and Conclusion

The recent experimental measurement (introduced in the previous section) of switching current statistics of a superconducting nanowire found that the width of the distribution increases as the temperature of the bath T_b is lowered [31]. Given that the stochasticity in the switching process stems from the inherent stochasticity of the phase slip fluctuations, one might naively assume that the mean switching time τ_s for a given value of T_b and current I would correspond to the inverse of the phase-slip rate $\Gamma(T_b, I)$. Under this assumption the observed temperature dependence is counter-intuitive since the phase-slip rate is expected to go down as the temperature is lowered (see the dashed contour lines in Figure 2a). Even after modifying Γ to incorporate quantum phase slips, one is not able to fit the τ_s curves extracted from the observed switching-current distribution with those of Γ.

The model discussed in the previous section indeed yield τ_s that drastically deviate from Γ except in a certain range of T_b and I as shown in panels (a) and (c) of Figure 2. One can understand this as follows. 'Thermal runaway'—heating by rare sequences of closely-spaced phase slips—constitutes the mechanism of superconductive-to-resistive switching within our model. The number of phase-slips in such a phase-slip sequence can be easily estimated by evaluating the number $N(T_b, I)$ of phase-slips needed to cause switching in the absence of cooling. In the range of T_b and I in which N is greater than one, the value of τ_s is of course expected to be larger than Γ^{-1}. The larger the number of phase-slips in the sequence inducing the superconductive-to-resistive thermal runaway, the smaller the stochasticity in the switching process and, hence, the sharper the distribution of switching currents (see Figure 2b). Upon entering the regime in which a single plase slip is sufficient to switch the wire to the normal (resistive) state, τ_s coincides with Γ^{-1} and the distribution width decreases upon lowering the temperature in accordance with the naive expectation for thermally activated phase slips.

The experimentally observed increase in the distribution width is thus explained by the model even if $\Gamma(T_b, I)$, which is an input function in the model, is chosen to be determined solely by thermally activated phase slips (TAPS). One can, in fact, successfully fit the experimental curves for a wide range of T_b [31, 33]. However, contrary to the non-monotonic behavior of the distribution width expected for this choice of Γ (Figure 2), the observed distribution width does not reverse the trend. By choosing a form of Γ that phenomenologically incorporates quantum phase slips (QPS), it becomes possible to fit the experimental curves down to the lowest temperatures. Moreover, based on the comparison with theory, one can identify a temperature

Journal of Physics: Conference Series **462** (2013) 012046 doi:10.1088/1742-6596/462/1/012046

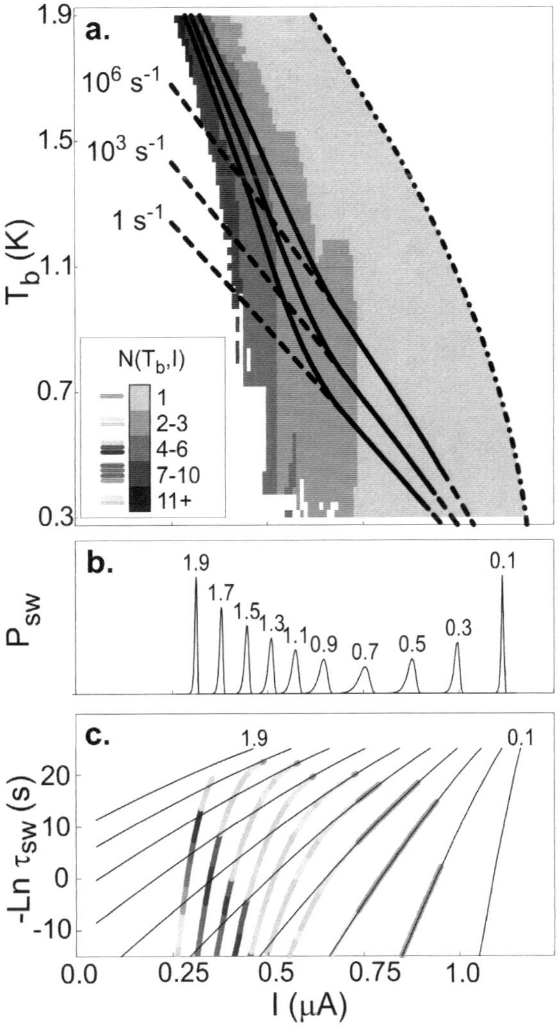

Figure 2. Switching statistics a. Map showing $N(T_b, I)$ (see text) and the contour lines (solid lines) for the inverse of mean switching time, $\tau_s^{-1} = 1$, 10^3, 10^6 s^{-1}; the contour lines (dashed lines) for the phase-slip rate Γ are also shown. The thermodynamic (depairing) critical current (dashed-dotted line) is plotted for reference. b. Switching-current distributions P_{SW} obtained at various values of T_b and for $r = 58\,\mu$A/s. c. The logarithms of τ_s^{-1} (colored lines) and of Γ (thinner black lines) as a function of I, obtained for the same set of $T_b$values as in panel (a). The colors of τ_s^{-1} plots correspond to different values of $N(T_b, I)$ [as indicated in the legend of panel (a)].

window in which the experiments are probing a single quantum phase slip event via its drastic consequence of switching the entire nanowire from the superconductive to the resistive state.

Fluctuations play an important role in determining the behavior of a superconducting nanowire, which provides an ideal laboratory not only for observation of MQT of a quantum field, but also for studying quantum phase transitions [34, 41, 42, 43, 44]. Experimental characterization of QPS is bound to enhance the understanding of various mechanisms for quantum phase transitions and their relevance for the experiments. A better understanding of low-temperature fluctuations has also implications for varied technological applications such as integration of superconducting wires in electronic circuitry (e.g., as controllable switching elements), implementation of nanowire-based devices [35, 45, 46], and the potential use of nanowires as qubits in quantum computers [47].

Acknowledgments

I would like to gratefully acknowledge my collaborators, D. Pekker, P. M. Goldbart (theory) and M. Sahu, A. Bezryadin (experiment). Many thanks also to the conference organizers, S. Das and A. Shapere for inviting me to present in the conference on Quantum Theory and Symmetries – it was a rare opportunity to interact and reconnect with theorists from different branches of physics that share the history of the quantum revolution.

References

[1] Tomsovic S (ed) 1998 *Tunneling in Complex Systems* (World Scientific Publishing)
[2] Razavy M 2003 *Quantum Theory of Tunneling* (World Scientific Publishing)
[3] Ankerhold J 2007 *Quantum Tunneling in Complex Systems: The Semiclassical Approach* (*Springer Tracts in Modern Physics* vol 224) (Springer-Verlag)
[4] Coleman S 1985 *Aspects of Symmetry: Selected Erice Lectures* (Cambridge University Press)
[5] Rajaraman R 1987 *Solitons and Instantons: An Introduction to Solitons and Instantons in Quantum Field Theory* (North-Holland Personal Library, Elsvier Science B. V.)
[6] Coleman S 1977 *Phys. Rev. D* **15** 2929–2936
[7] Callan C G and Coleman S 1977 *Phys. Rev. D* **16** 1762–1768
[8] Kagan Yu and Leggett A J (eds) 1992 *Quantum Tunnelling In Condensed Media* (*Modern Problems in Condensed Matter Sciences* vol 34) (North-Holland)
[9] Leggett A J 1978 *J. Phys. (Paris), Colloq. 39,* C6–1264
[10] Takagi S 2002 *Macroscopic quantum tunneling* (Cambridge University Press)
[11] Makhlin Y, Schön G and Shnirman A 2001 *Rev. Mod. Phys.* **73** 357–400
[12] Chiorescu I, Nakamura Y, Harmans C J P M and Mooij J E 2003 *Science* **299** 1869–1871
[13] Manucharyan V E, Koch J, Glazman L I and Devoret M H 2009 *Science* **326** 113–116
[14] Ueda M and Leggett A J 1998 *Phys. Rev. Lett.* **80** 1576–1579
[15] Caldeira A O and Leggett A J 1981 *Phys. Rev. Lett.* **46** 211–214
[16] Caldeira A O and Leggett A J 1983 *Annals of Physics* **149** 374 – 456
[17] Weiss U *Quantum Dissipative Systems* (*Series in Modern Condensed Matter Physics* vol 10) (World Scientific Publishing)
[18] Ivanchenko Y M and Zilberman L A 1969 *Sov. Phys. JETP* **28** 1272
[19] Voss R F and Webb R A 1981 *Phys. Rev. Lett.* **47** 265–268
[20] Devoret M H, Martinis J M and Clarke J 1985 *Phys. Rev. Lett.* **55** 1908–1911
[21] Martinis J M, Devoret M H and Clarke J 1987 *Phys. Rev. B* **35** 4682–4698
[22] Rouse R, Han S and Lukens J E 1995 *Phys. Rev. Lett.* **75** 1614–1617
[23] Tinkham M 1996 *Introduction to Superconductivity* (McGraw-Hill, New York)
[24] Little W A 1967 *Phys. Rev.* **156** 396–403
[25] Skocpol W J and Tinkham M 1975 *Reports on Progress in Physics* **38** 1049–1097
[26] Langer J S and Ambegaokar V 1967 *Phys. Rev.* **164** 498–510
[27] McCumber D E 1968 *Phys. Rev.* **172** 427–429
[28] McCumber D E and Halperin B I 1970 *Phys. Rev. B* **1** 1054–1070
[29] Grabert H and Weiss U 1984 *Phys. Rev. Lett.* **53** 1787–1790
[30] Giordano N 1990 *Phys. Rev. B* **41** 6350–6365

[31] Sahu M, Bae M H, Rogachev A, Pekker D, Wei T C, Shah N, Goldbart P M and Bezryadin A 2009 *Nature Physics* **5** 503–508
[32] Shah N, Pekker D and Goldbart P M 2008 *Phys. Rev. Lett.* **101** 207001
[33] Pekker D, Shah N, Sahu M, Bezryadin A and Goldbart P M 2009 *Phys. Rev. B* **80** 214525
[34] Bezryadin A, Lau C N and Tinkham M 2000 *Nature* **404** 971
[35] Hopkins D S, Pekker D, Goldbart P M and Bezryadin A 2005 *Science* **308** 1762–1765
[36] Bardeen J 1962 *Rev. Mod. Phys.* **34** 667–681
[37] Romijn J, Klapwijk T M, Renne M J and Mooij J E 1982 *Phys. Rev. B* **26** 3648–3655
[38] Tinkham M, Free J U, Lau C N and Markovic N 2003 *Phys. Rev. B* **68** 134515
[39] Altomare F, Chang A M, Melloch M R, Hong Y and Tu C W 2006 *Phys. Rev. Lett.* **97** 017001
[40] Fulton T A and Dunkleberger L N 1974 *Phys. Rev. B* **9** 4760–4768
[41] Lopatin A V, Shah N and Vinokur V M 2005 *Phys. Rev. Lett.* **94** 037003
[42] Rogachev A, Bollinger A T and Bezryadin A 2005 *Phy. Rev. Lett.* **94** 017004
[43] Shah N and Lopatin A 2007 *Phys. Rev. B* **76** 094511
[44] Del Maestro A, Rosenow B, Shah N and Sachdev S 2008 *Phys. Rev. B* **77** 180501
[45] Pekker D, Bezryadin A, Hopkins D S and Goldbart P M 2005 *Phys. Rev. B* **72** 104517
[46] Johansson A, Sambandamurthy G, Shahar D, Jacobson N and Tenne R 2005 *Phys. Rev. Lett.* **95** 116805
[47] Mooij J E and Nazarov Y V 2006 *Nature Physics* **2** 169

Journal of Physics: Conference Series **462** (2013) 012047 doi:10.1088/1742-6596/462/1/012047

GLSM's, gerbes, and Kuznetsov's homological projective duality

Eric Sharpe

Physics Department, Robeson Hall (0435), Virginia Tech, Blacksburg, VA 24061

E-mail: ersharpe@vt.edu

Abstract. In this short note we give an overview of recent work on string propagation on stacks and applications to gauged linear sigma models. We begin by outlining noneffective orbifolds (orbifolds in which a subgroup acts trivially) and related phenomena in two-dimensional gauge theories, which realize string propagation on gerbes. We then discuss the 'decomposition conjecture,' equating conformal field theories of strings on gerbes and strings on disjoint unions of spaces. Finally, we apply these ideas to gauged linear sigma models for complete intersections of quadrics, and use the decomposition conjecture to show that the Landau-Ginzburg points of those models have a geometric interpretation in terms of a (sometimes noncommutative resolution of) a branched double cover, realized via nonperturbative effects rather than as the vanishing locus of a superpotential. These examples violate old unproven lore on GLSM's (namely, that geometric phases must be related by birational transformations), and we conclude by observing that in these examples (and conjecturing more generally in GLSM's), the phases are instead related by Kuznetsov's 'homological projective duality.'

1. Introduction

In this short note we will outline a few of the results in [1, 2, 3, 4, 5]. In general terms, these papers outline the definition and some applications of string compactifications on (smooth, Deligne-Mumford) stacks and, in particular, gerbes.

The original motivation for understanding string compactifications on stacks was to investigate what new string vacua, new conformal field theories, arise from the near-geometries provided by stacks, and to better understand some existing theories that were already special cases of string compactifications on stacks.

This note will focus on the special case of stacks that are gerbes. A string on a gerbe is described by a noneffective orbifold (an orbifold in which a subgroup acts trivially on the space) or noneffective gauge theory. Despite the fact that the group action is trivial, physics nevertheless knows about that (trivially-acting) subgroup via nonperturbative effects, as we shall outline in section 2.

Massless spectra resulting from such noneffective group actions appear to violate cluster decomposition, a foundational property of most quantum field theories, but these theories are nevertheless nontrivial because they are equivalent to sigma models on disjoint unions of ordinary spaces, a result known as the 'decomposition conjecture,' and which we review in section 3.

The decomposition conjecture has several applications, including examples in Gromov-Witten theory and the geometric Langlands program. In section 4, we focus on a different application of the decomposition conjecture, namely to understand the Landau-Ginzburg points of certain puzzling gauged linear sigma models. We point out that the Landau-Ginzburg points in question have, for the most part, a $\mathbf{Z}_2$ gerbe structure to which the decomposition conjecture can be applied, and find a mathematically sensible description of the Landau-Ginzburg points in terms of branched double covers (the double cover arising from the application of the decomposition conjecture to a $\mathbf{Z}_2$ gerbe). The resulting theories realize geometry from GLSM's in a novel fashion, and also violate the lore that all geometric phases in GLSM's are birational to one another. In examples, we observe that geometric phases are related by Kuznetsov's 'homological projective duality' (instead of 'birationality'), and conjecture that this is true more generally.

2. Noneffective orbifolds and gauge theories

This note will concern applications of noneffective orbifolds to physics and geometry, so let us beginning by defining a noneffective orbifold. It is a quotient $[X/G]$, where a subgroup of G, call it K, acts trivially on X. Such quotients are also examples of K-gerbes.

To understand why such a quotient is physically different from the corresponding effective orbifold $[X/(G/K)]$, let us work through an example. Consider the orbifold $[X/D_4]$, where the $\mathbf{Z}_2$ center of D_4 acts trivially on X. Now, D_4 can be described as the central extension

$$1 \longrightarrow \mathbf{Z}_2 \longrightarrow D_4 \longrightarrow \mathbf{Z}_2 \times \mathbf{Z}_2 \longrightarrow 1$$

so that only the $\mathbf{Z}_2 \times \mathbf{Z}_2$ acts nontrivially. We shall see that the one-loop partition function of $[X/D_4]$ is very different from the one-loop partition function of $[X/\mathbf{Z}_2 \times \mathbf{Z}_2]$, thus making it clear that there is a difference.

We can enumerate the elements of D_4 as

$$D_4 \;=\; \{1, z, a, b, az, bz, ab, ba = abz\}$$

where z generates the $\mathbf{Z}_2$ center, and the elements of $D_4/\mathbf{Z}_2 = \mathbf{Z}_2 \times \mathbf{Z}_2$ are

$$\mathbf{Z}_2 \times \mathbf{Z}_2 \;=\; \{1, \overline{a}, \overline{b}, \overline{ab}\}$$

where the overline indicates the coset formed through multiplication by z. Now, the one-loop partition function of $[X/D_4]$ has the form

$$Z(D_4) \;=\; \frac{1}{|D_4|} \sum_{g,h \in D_4; gh=hg} Z_{g,h}$$

Journal of Physics: Conference Series **462** (2013) 012047 doi:10.1088/1742-6596/462/1/012047

Each of the $Z_{g,h}$ twisted sectors that appears, is the same as a $\mathbf{Z}_2 \times \mathbf{Z}_2$ twisted sector (since the $\mathbf{Z}_2$ center of D_4 acts trivially), appearing with multiplicity $|\mathbf{Z}_2|^2 = 4$, except for the sectors

$$\bar{a}\underset{\bar{b}}{\square}, \quad \bar{a}\underset{\overline{ab}}{\square}, \quad \bar{b}\underset{\overline{ab}}{\square}$$

which are omitted, because the lifts of the group elements to D_4 do not commute in D_4.

Thus, we see that

$$\begin{aligned} Z(D_4) &= \frac{|\mathbf{Z}_2 \times \mathbf{Z}_2|}{|D_4|} |\mathbf{Z}_2|^2 \left(Z(\mathbf{Z}_2 \times \mathbf{Z}_2) \; - \; \text{some twisted sectors} \right) \\ &= 2 \left(Z(\mathbf{Z}_2 \times \mathbf{Z}_2) \; - \; \text{some twisted sectors} \right) \end{aligned}$$

In particular, the D_4 orbifold is significantly different from the $\mathbf{Z}_2 \times \mathbf{Z}_2$ orbifold, so we see here explicitly that string theory does indeed know about even trivially-acting groups.

With a little more work, we can find a cleaner interpretation of this result. Discrete torsion acts as a sign on precisely the omitted twisted sectors, so we see that

$$Z([X/D_4]) \; = \; Z\left([X/\mathbf{Z}_2 \times \mathbf{Z}_2] \coprod [X/\mathbf{Z}_2 \times \mathbf{Z}_2] \right)$$

with discrete torsion in one component.

Although we have focused so far on orbifolds, the same issue exists in two-dimensional gauge theories, where it manifests as a question of whether *e.g.* an abelian gauge theory with matter of charge 2 is the same as if matter is charge 1. Briefly, the perturbative physics is identical, but nonperturbatively these two theories can differ.

For example, consider the supersymmetric $\mathbf{P}^{N-1}$ model, in which the chiral superfields are taken to have charge k rather than charge 1. In the ordinary $\mathbf{P}^{N-1}$ model, the axial $U(1)_A$ is broken to $\mathbf{Z}_{2N}$ by instantons, whereas here it is broken to $\mathbf{Z}_{2kn}$. Similarly, A model correlation functions are different. In the ordinary $\mathbf{P}^{N-1}$ model, it is straightforward to show that

$$\langle X^{N(d+1)-1} \rangle_d \; = \; q^d$$

where X corresponds to the generator of the classical cohomology ring. Here, by contrast,

$$\langle X^{N(kd+1)-1} \rangle_d \; = \; q^d$$

As a result, quantum cohomology rings differ. In the ordinary $\mathbf{P}^{N-1}$ model, the quantum cohomology ring is

$$\mathbf{C}[x]/(x^N - q)$$

whereas in the present case it is

$$\mathbf{C}[x]/(x^{kN} - q)$$

In each case, we see meaningfully different physics.

On a compact worldsheet, the distinction above follows from the fact [6] that to uniquely specify a Higgs field, one must specify to which bundle it couples. For example, if a $U(1)$ gauge field couples to a line bundle L, then to say a scalar ϕ has charge Q means $\phi \in \Gamma(L^{\otimes Q})$. Different bundles implies different zero modes, which implies different anomalies, and hence different physics. For noncompact worldsheets, there is an analogous argument using periodicity of the theta angle [6].

This phenomenon is not specific to two dimensions, but also has analogues in four dimensions. For example, the same effect is at the heart of the (nonperturbative) distinction between $SU(n)$ and $SU(n)/\mathbf{Z}_n$ gauge theories, and the distinction between Spin(n) and $SO(n)$ gauge theories. For $\mathcal{N}=1$ supersymmetry in four dimensions, there are results [7, 8] describing Seiberg duality between Spin(n) gauge theory with massive spinors and $SO(n)$ gauge theories with $\mathbf{Z}_2$ monopoles. In $\mathcal{N}=4$ supersymmetry in four dimensions, this phenomenon is crucial for the physical understanding of the geometric Langlands program [9].

3. Decomposition conjecture

Consider a quotient $[X/H]$, where

$$1 \longrightarrow G \longrightarrow H \longrightarrow K \longrightarrow 1$$

and G acts trivially, and is a finite group. We claim that

$$\mathrm{CFT}\left([X/H]\right) = \mathrm{CFT}\left(\left[\left(X \times \hat{G}\right)/K\right]\right)$$

(together with some B field, as specified in [4]), where $\hat{G}$ is the set of irreducible representations of G. As the right-hand side is a disjoint union, this relates strings on gerbes to strings on disjoint unions of spaces (or non-gerbe stacks). We call this the *decomposition conjecture.*

When K acts trivially on $\hat{G}$, the decomposition conjecture above reduces to the statement that

$$\mathrm{CFT}\left([X/H]\right) = \mathrm{CFT}\left(\coprod_{\hat{G}}(X, B)\right)$$

where the B field is determined by the image of the characteristic class of the gerbe under the map defined by an element of $\hat{G}$:

$$H^2(X, Z(G)) \stackrel{Z(G)\rightarrow U(1)}{\longrightarrow} H^2(X, U(1))$$

The decomposition conjecture satisfies a number of checks, for example:

- For global quotients by finite groups, one can check explicitly that partition functions match at arbitrary worldsheet genus.
- This implies (by virtue of D-branes) a statement about K-theory, namely that

  $$K_H(X) = \text{twisted } K_K(X \times \hat{G})$$

  which can be checked independently.

- It is consistent with standard results on sheaves on gerbes, namely that sheaves on gerbes decompose (in the same way), and that Ext group elements between sheaves on different components vanish.
- It implies results in Gromov-Witten theory, which are being checked in papers including [10, 11, 12, 13, 14, 15].
- One can compute the Toda mirrors to Fano toric stacks, and and compare quantum cohomology computations on either side of the duality.

The example of the previous section, $[X/D_4]$, fits into this framework: the decomposition conjecture predicts that the CFT of $[X/D_4]$ should look like two copies of $[X/\mathbf{Z}_2 \times \mathbf{Z}_2]$, one copy with a flat B field corresponding to nonzero discrete torsion, which is what we found explicitly.

The decomposition conjecture also appears implicitly in the work of [9] on the physical realization of geometric Langlands; there, after dimensionally-reducing along a Riemann surface to two dimensions, one often finds a sigma model on a disjoint union of moduli spaces. This can be understood from the fact that the moduli space of the four-dimensional theory in such cases is a gerbe, as described in [4].

4. Application to GLSM's

Let us now apply these ideas to study gauged linear sigma models (GLSM's) in two dimensions [16]. In particular, let us consider the GLSM describing the complete intersection of four quadric (degree two) hypersurfaces in $\mathbf{P}^7$. This GLSM has

- Eight chiral superfields ϕ_i, each of charge 1, corresponding to homogeneous coordinates on $\mathbf{P}^7$
- Four chiral superfields p_a, each of charge -2, one for each hypersurface in the complete intersection

and a superpotential

$$W = \sum_a p_a G_a(\phi)$$

where the G_a's are degree two homogeneous polynomials.

Let us analyze the space of supersymmetric vacua in this theory, in semiclassical regimes. The D-term constraint is

$$\sum_i |\phi_i|^2 - 2\sum_a |p_a|^2 - r = 0$$

When $r \gg 0$, the ϕ_i cannot all vanish (from the D-terms above), hence vanishing of F terms requires that $p_a = G_a = 0$, and the theory appears to flow in the IR to a nonlinear sigma model on the complete intersection $\mathbf{P}^7[2,2,2,2]$.

The other limit of r is more interesting. It is helpful to rewrite the superpotential as

$$W = \sum_a p_a G_a(\phi) = \sum_{ij} \phi_i A^{ij}(p) \phi_j$$

where A^{ij} is a symmetric matrix with entries linear in p's, that encodes the G_a's. Written in this form, it is clear that the superpotential is really a mass matrix for the ϕ_i. In the limit that $r \ll 0$, the D terms specify that the p_a cannot all be zero, and the superpotential above appears to imply that the ϕ_i are all massive, suggesting that this limit flows to a nonlinear sigma model on $\mathbf{P}^3$. However, since the $r \gg 0$ limit is Calabi-Yau, this limit should also be Calabi-Yau, a contradiction.

The correct analysis of this limit relies on two subtleties. The first subtlety is that the ϕ_i are not massive everywhere, some of their masses vanish along the locus $\{\det A = 0\} \subset \mathbf{P}^3$. The second subtlety is the fact that the p's have nonminimal charge, so over the part of $\mathbf{P}^3$ where the ϕ_i are all massive, we have a nonminimally-charged abelian gauge theory, which (as outlined previously) describes a local noneffective $\mathbf{Z}_2$ orbifold, and hence a $\mathbf{Z}_2$ gerbe.

As a result of these subtleties and the decomposition conjecture, physics sees the $\mathbf{Z}_2$ gerbe as a double cover of $\mathbf{P}^3$ away from the locus $\{\det A = 0\}$. Moreover, it can be shown that there is a Berry phase around the locus $\{\det A = 0\}$, which has the effect of interchangeing the two sheets of the cover. Thus, physics seems to be seeing a branched double cover of $\mathbf{P}^3$, namely Clemens' octic double solid [17]. As a consistency check, the locus $\{\det A = 0\}$ has the right degree for this branched double cover to be Calabi-Yau, as one would expect of a GLSM describing a Calabi-Yau at another limit in Kähler moduli space.

The result we have presented so far is noteworthy for at least two reasons:

- This is a novel realization of geometry in a GLSM, considering that GLSM's ordinarily build Calabi-Yau's as vanishing loci of potentials.
- This branched double cover is *not* birational to the geometry appearing at the other limit in Kähler moduli space, namely $\mathbf{P}^7[2,2,2,2]$. This result contradicts standard lore in the GLSM community, namely that geometries appearing at limit points of the same GLSM are all birational to one another.

Analogous results for nonabelian GLSM's have also been obtained in [18].

This example, the GLSM for $\mathbf{P}^7[2,2,2,2]$, is just one of a number of examples which can be analyzed in the same form. In a subset of those cases, including $\mathbf{P}^7[2,2,2,2]$, there is additional structure to uncover, deriving from the fact that the branched double cover is singular, but the GLSM physics behaves as if it is smooth at those singularities. In such cases, when the branched double cover is singular (a subset of all examples of the form above), we believe the Landau-Ginzburg model is actually describing a 'noncommutative resolution' of the branched double cover worked out by Kuznetsov.

This particular notion of 'noncommutative space' is one described by *e.g.* [19, 20, 21, 22, 23, 24, 25, 26, 27], and is distinct from other notions of noncommutative space appearing previously in the physics literature such as [28, 29]. Specifically, a 'noncommutative space' in this sense is defined via its sheaves.

In the present case, the noncommutative resolution in question is defined by sheaves of B-modules over $\mathbf{P}^3$, where B is the sheaf of even parts of Clifford algebras associated with the universal quadric over $\mathbf{P}^3$ defined by the GLSM superpotential.

Physically, those sheaves of B-modules are precisely [30] the same as matrix factorizations at the Landau-Ginzburg point. Intuitively, we can understand this result as follows. First, let us recall matrix factorizations for a quadratic superpotential: even though the bulk theory is massive, one still has $D0$-branes with a Clifford algebra structure [31]. In the present case, we have a Landau-Ginzburg model (obtained via a finite amount of renormalization group flow) fibered over $\mathbf{P}^3$, so a Born-Oppenheimer analysis gives sheaves of Clifford algebras (determined by the superpotential) and modules thereof. This is ultimately why the D-branes precisely duplicate the definition of the noncommutative resolution.

Thus, in addition to a novel realization of geometry, and an example in which the two geometric limits of a GLSM are not birational to one another, we see in addition that the GLSM is physically realizing a noncommutative resolution, in the sense discussed earlier.

As an aside, one way to study this noncommutative resolution is via D-brane probes. It can be shown [32, 33] that the moduli space of D-branes propagating on this noncommutative resolution is a necessarily non-Kähler small resolution of the singular space. The non-Kähler structure makes it essentially impossible for the D-brane moduli space to be the target of the closed string theory, as it would break worldsheet supersymmetry. It is allowed here because it is the open string moduli space, not where the closed strings propagate. (Another example where the closed string target space is different from the D-brane moduli space is orbifolds [2][section 8.2]. There, closed strings see an orbifold – a quotient stack – whereas D-branes see a resolution [34].)

Although the phases of the GLSM above are not birational, they nevertheless do have a precise mathematical relationship: they are related by Kuznetsov's "homological projective duality" [19, 20, 21]. We conjecture that phases in all GLSM's are always related by homological projective duality.

5. Summary

In this note we have outlined some recent results and applications of string compactifications on near-geometries provided by gerbes.

- We began by describing how physics sees noneffective group actions, which is the physics at the heart of string propagation on gerbes.
- We outlined the decomposition conjecture for strings propagating on gerbes, which states that the CFT of a nonlinear sigma model on a gerbe matches the CFT of a nonlinear sigma model on a disjoint union of spaces or non-gerbe stacks.
- We described the application of the decomposition conjecture to GLSM's, resulting in novel realizations of geometry, some violations of old lore on GLSM's, a physical realization of Kuznetsov's homological projective duality, and a concrete realization of strings on noncommutative resolutions.

6. Acknowledgements

We would like to thank our collaborators on the papers reviewed here, namely M. Ando, A. Caldararu, J. Distler, S. Hellerman, A. Henriques, and T. Pantev, for many useful discussions.

E.S. was partially supported by NSF grants PHY-0755614, DMS-0705381.

[1] T. Pantev, E. Sharpe, "Notes on gauging noneffective group actions," hep-th/0502027.
[2] T. Pantev, E. Sharpe, "String compactifications on Calabi-Yau stacks," Nucl. Phys. **B733** (2006) 233-296, hep-th/0502044.
[3] T. Pantev, E. Sharpe, "GLSM's for gerbes (and other toric stacks)," Adv. Theor. Math. Phys. **10** (2006) 77-121, hep-th/0502053.
[4] S. Hellerman, A. Henriques, T. Pantev, E. Sharpe, M. Ando, "Cluster decomposition, T-duality, and gerby CFTs," Adv. Theor. Math. Phys. **11** (2007) 751-818, hep-th/0606034.
[5] A. Caldararu, J. Distler, S. Hellerman, T. Pantev, E. Sharpe, "Non-birational twisted derived equivalences in abelian GLSMs," arXiv: 0709.3855.
[6] J. Distler, R. Plesser, private communication.
[7] P. Pouliot, "Chiral duals of non-chiral susy gauge theories," Phys. Lett. **B359** (1995) 108-113, hep-th/9507018.
[8] M. Strassler, "Duality, phases, spinors and monopoles in $SO(n)$ and Spin(n) gauge theories," JHEP **9809** (1998) 017, hep-th/9709081.
[9] A. Kapustin, E. Witten, "Electric-magnetic duality and the geometric Langlands program," Comm. Number Theory Phys. **1** (2007) 1-236, hep-th/0604151.
[10] E. Andreini, Y. Jiang, H.-H. Tseng, "On Gromov-Witten theory of root gerbes," arXiv: 0812.4477.
[11] E. Andreini, Y. Jiang, H.-H. Tseng, "Gromov-Witten theory of product stacks," arXiv: 0905.2258.
[12] E. Andreini, Y. Jiang, H.-H. Tseng, "Gromov-Witten theory of etale gerbes, i: root gerbes," arXiv: 0907.2087.
[13] H.-H. Tseng, "On degree zero elliptic orbifold Gromov-Witten invariants," arXiv: 0912.3580.
[14] A. Gholampour, H.-H. Tseng, "On Donaldson-Thomas invariants of threefold stacks and gerbes," arXiv: 1001.0435.
[15] X. Tang, H.-H. Tseng, "Duality theorems of étale gerbes on orbifolds," arXiv: 1004.1376.
[16] E. Witten, "Phases of $N = 2$ theories in two dimensions," Nucl. Phys. **B403** (1993) 159-222, hep-th/9301042.
[17] H. Clemens, "Double solids," Adv. Math. **47** (1983) 107-230.
[18] K. Hori, D. Tong, "Aspects of non-Abelian gauge dynamics in two-dimensional $N = (2,2)$ theories," JHEP **0705** (2007) 079, hep-th/0609032.
[19] A. Kuznetsov, "Homological projective duality," Publ. Math. Inst. Hautes Études Sci. **105** (2007) 157-220, math.AG/0507292.
[20] A. Kuznetsov, "Derived categories of quadric fibrations and intersections of quadrics," Adv. Math. **218** (2008) 1340-1369, math.AG/0510670.
[21] A. Kuznetsov, "Homological projective duality for Grassmannians of lines," math.AG/0610957.
[22] M. Kontsevich, "Course on non-commutative geometry," ENS, 1998, lecture notes at http://www.math.uchicago.edu/; ,arinkin/langlands/kontsevich.ps.
[23] Y. Soibelman, "Lectures on deformation theory and mirror symmetry, IPAM, 2003, http://www.math.ksu.edu/; ,soibel/ipam-fina.l.ps.
[24] K. Costello, "Topological conformal field theories and Calabi-Yau categories," Adv. Math. **210** (2007) 165-214, math.QA/0412149.
[25] M. van den Bergh, "Three-dimensional flops and noncommutative rings," Duke Math.J. **122** (2004) 423-455.
[26] M. van den Bergh, "Non-commutative crepant resolutions," pp. 749-770 in *The legacy of Niels Henrik Abel: the Abel bicentennial, Oslo, 2002*, Springer, 2004.
[27] M. van den Bergh, "Non-commutative quadrics," arXiv: 0807.3753.
[28] N. Seiberg, E. Witten, "String theory and noncommutative geometry," JHEP **9909** (1999) 032, hep-th/9908142.
[29] D. Roggenkamp, K. Wendland, "Limits and degenerations of unitary conformal field theories," Comm. Math. Phys. **251** (2004) 589-643, hep-th/0308143.
[30] A. Kuznetsov, private communication.
[31] A. Kapustin, Y. Li, "D-branes in Landau-Ginzburg models and algebraic geometry," JHEP **0312** (2003) 005, hep-th/0210296.
[32] N. Addington, "The derived category of the intersection of four quadrics," arXiv: 0904.1764.
[33] N. Addington, "Spinor sheaves on singular quadrics," arXiv: 0904.1766.
[34] M. Douglas, B. Greene, D. Morrison, "Orbifold resolution by D-branes," Nucl. Phys. **B506** (1997) 84-106, hep-th/9704151.

Journal of Physics: Conference Series **462** (2013) 012048 doi:10.1088/1742-6596/462/1/012048

Time-dependent flow from an AdS Schwarzschild black hole

George Siopsis

Department of Physics and Astronomy, The University of Tennessee, Knoxville, TN 37996-1200, USA

E-mail: siopsis@tennessee.edu

Abstract. I discuss two examples of time-dependent flow which can be described in terms of an AdS Schwarzschild black hole via holography. The first example involves Bjorken hydrodynamics which should be applicable to the formation of the quark gluon plasma in heavy ion collisions. The second example is the cosmological evolution of our Universe.

1. Introduction

The AdS/CFT correspondence affords us a means of studying the behavior of gauge theory fluids at strong coupling in terms of gravity duals in a space of one extra dimension. This is done by mapping properties of solutions of the Einstein field equations with a negative cosmological constant onto the conformal boundary via holographic renormalization [1, 2]. Understanding the transport properties of strongly coupled time-dependent fluids is a challenge due to the lack of exact time-dependent supergravity solutions. Here I discuss how this problem can be circumvented by applying holographic renormalization to static AdS Schwarzschild black hole solutions. I discuss two examples: Bjorken (boost-invariant) hydrodynamics which should be related to the quark-gluon plasma produced in heavy ion colliders (RHIC and the LHC) and the cosmological evolution of the Universe.

2. AdS$_d$ Schwarzschild black holes

AdS Schwarzschild black holes are exact solutions of the Einstein field equations in the presence of a negative cosmological constant. The metric can be written as

$$ds^2_{\text{b.h.}} = \frac{R^2}{z^2}\left(-f(z)dt^2 + d\Sigma^2_{K,d-2} + \frac{dz^2}{f(z)}\right) \quad , \quad f(z) = 1 + K\frac{z^2}{R^2} - 2\mu z^{d-1} \tag{1}$$

I shall choose units so that the AdS radius $R = 1$. The horizon radius satisfies $f(z_+) = 0$ and the Hawking temperature is

$$T_H = \frac{d-1+K(d-3)z_+^2}{4\pi z_+} \tag{2}$$

The mass and entropy of the hole are, respectively,

$$M = \frac{(d-2)(1+Kz_+^2)}{16\pi G z_+^{d-1}} Vol(\Sigma_{K,d-2}) \ , \quad S = \frac{1}{4Gz_+^{d-2}} Vol(\Sigma_{K,d-2}) \tag{3}$$

The parameter K determines the curvature of the horizon and the boundary of AdS space. For $K = 0, +1, -1$ we have, respectively, a flat ($\mathbb{R}^{d-2}$), spherical ($\mathbb{S}^{d-2}$) and hyperbolic ($\mathbb{H}^{d-2}/\Gamma$, topological black hole, where Γ is a discrete group of isometries) horizon (boundary).

For simplicity, I shall concentrate on a flat ($K = 0$) hole which is the large hole limit of any AdS Schwarzschild black hole. It has a flat horizon at $z_+ = (2\mu)^{-\frac{1}{d-1}}$. This black hole is dual to a gauge theory plasma residing on the conformal boundary via holographic renormalization [1, 2]. To see this, write the metric (1) in terms of Fefferman-Graham coordinates

$$ds^2 = \left(g_{\mu\nu}dx^\mu dx^\nu + dz_{FG}^2\right)/z_{FG}^2 \tag{4}$$

Near the boundary ($z_{FG} = 0$), expand

$$g_{\mu\nu} = g_{\mu\nu}^{(0)} + z_{FG}^2 g_{\mu\nu}^{(2)} + \cdots + z_{FG}^{d-1} g_{\mu\nu}^{(d-1)} + h_{\mu\nu}^{(d-1)} z_{FG}^{d-1} \ln z_{FG}^2 + \mathcal{O}(z_{FG}^d) \tag{5}$$

where $g_{\mu\nu}^{(0)} = \eta_{\mu\nu}$. The vacuum expectation value of the stress-energy tensor of the plasma on the boundary is related to the bulk metric through

$$\langle T_{\mu\nu}\rangle = \frac{d-1}{16\pi G} g_{\mu\nu}^{(d-1)} \tag{6}$$

From this, we read off the energy density and pressure, respectively,

$$\epsilon = \langle T^{tt}\rangle = (d-2)\frac{\mu}{8\pi G} \quad , \quad p = \langle T^{ii}\rangle = \frac{\mu}{8\pi G} \tag{7}$$

obeying $p = \frac{1}{d-2}\epsilon$, as expected for a conformal fluid. We deduce the equation of state

$$p = \frac{1}{16\pi G_d}\left(\frac{4\pi T_H}{d-1}\right)^{d-1} \tag{8}$$

and energy and entropy densities, respectively, as functions of temperature

$$\epsilon = \frac{d-2}{16\pi G}\left(\frac{4\pi T_H}{d-1}\right)^{d-1} \quad , \quad s = \frac{dp}{dT} = \frac{1}{4G}\left(\frac{4\pi T_H}{d-1}\right)^{d-2} \tag{9}$$

Thus we obtain a static fluid at temperature $T = T_H$ (eq. (2)).

3. Bjorken flow

Following a suggestion by Bjorken [3], in order to understand the quark gluon plasma produced in heavy ion collisions, one needs to study boost invariant hydrodynamics. This is because there is a "plateau" in particle production in the central rapidity region. The emergent nuclei are highly Lorentz-contracted pancakes receding at speed of light, resulting in a flow which is independent of the Lorentz frame (boost invariant). Given appropriate initial conditions (see table 1 for RHIC and the LHC), the hydrodynamic equations respect their symmetry (boost invariance) and lead to simple solutions. For a conformal fluid, one obtains

$$\epsilon \sim \tau^{-4/3} \quad , \quad T \sim \tau^{-1/3} \quad , \quad s \sim \tau^{-1} \tag{10}$$

where τ is the longitudinal proper time defined in (11).

Suppose that the plasma lives on a $(d-1)$-dimensional Minkowski space spanned by coordinates $\tilde{x}^\mu$ ($\mu = 0, 1, \ldots, d-2$) and the colliding beams are along the $\tilde{x}^1$ direction. Introducing coordinates τ, y (proper time and rapidity in the longitudinal plane, respectively)

$$\tilde{x}^0 = \tau\cosh y \quad , \quad \tilde{x}^1 = \tau\sinh y \tag{11}$$

Table 1. Initial conditions for the quark gluon plasma at RHIC and the LHC

	τ_0 (fm/c)	ϵ_0 (GeV/fm^3)	T (GeV)	$\sqrt{s}$ (GeV)
RHIC	0.2	10	0.5	200
LHC	0.1	10	1	5,500

the Minkowski metric reads

$$d\tilde{s}^2 = d\tilde{x}_\mu d\tilde{x}^\mu = -d\tau^2 + \tau^2 dy^2 + (d\tilde{x}^\perp)^2 \tag{12}$$

where $\tilde{x}^\perp = (\tilde{x}^2, \dots, \tilde{x}^{d-2})$ are the transverse coordinates.

The stress-energy tensor is

$$T^{\mu\nu} = \text{diag}\left[\epsilon(\tau) \quad \frac{p(\tau)}{\tau^2} - 2\frac{d-3}{d-2}\frac{\eta(\tau)}{\tau^3} \quad \dots \quad p(\tau) + \frac{2}{d-2}\frac{\eta(\tau)}{\tau}\right] . \tag{13}$$

The local conservation law yields

$$\partial_\tau \epsilon + \frac{1}{\tau}(\epsilon + p) - 2\frac{d-3}{d-2}\frac{\eta}{\tau^2} = 0 \tag{14}$$

Conformal symmetry implies tracelessness ($T^\mu_\mu = 0$), therefore

$$\epsilon = (d-2)p = \epsilon_0 \tau^{-\frac{d-1}{d-2}} \tag{15}$$

Then conservation of entropy in perfect fluid implies

$$T = \frac{T_0}{\tau^{1/(d-2)}} \ , \quad s = \frac{\dot{p}}{\dot{T}} = \frac{s_0}{\tau} \ , \quad s_0 = \frac{d-1}{d-2}\frac{\epsilon_0}{T_0} \tag{16}$$

Notice that the energy and entropy densities have the same dependence on temperature as in the static case (eq. (9)). If we identify initial data (at $\tau = 1$) with their corresponding values in the static case,

$$T_0 = T_H \ , \quad \epsilon_0 = \frac{(d-2)\mu}{8\pi G} \tag{17}$$

then eq. (9), with T_H replaced by T (16), describes the evolution of the energy and entropy densities in a Bjorken flow.

One may similarly find the time dependence of viscosity corrections. At high temperatures one expects that the ratio η/s asymptotes to a constant. Therefore,

$$\eta = \eta_0 \tau^{-1} + \dots \ , \quad \epsilon = \epsilon_0 \tau^{-\frac{d-1}{d-2}} - 2\eta_0 \tau^{-2} + \dots$$

$$T = T_0\left[\frac{1}{\tau^{\frac{1}{d-2}}} - \frac{2\eta_0}{(d-1)\epsilon_0 \tau} + \dots\right] \ , \quad s = s_0\left[\frac{1}{\tau} - \frac{2(d-2)\eta_0}{(d-1)\epsilon_0}\frac{1}{\tau^{\frac{2d-5}{d-2}}} + \dots\right] \tag{18}$$

For the gravity dual of Bjorken flow, a solution of the Einstein equations was constructed [4, 5, 6] as a series expansion in large longitudinal time keeping the parameter

$$v = \tilde{z}\tau^{-1/(d-2)} \tag{19}$$

fixed. At leading order the metric that leads to boost invariant hydrodynamics reads

$$ds^2_{\text{Bjorken}} = \frac{1}{\tilde{z}^2}\left[-\left(1 - 2\mu v^{d-1}\right)d\tau^2 + \tau^2 dy^2 + (d\tilde{x}^\perp)^2 + \frac{d\tilde{z}^2}{1 - 2\mu v^{d-1}}\right] \tag{20}$$

Higher-order corrections can be systematically calculated and lead to the ratio

$$\frac{\eta}{s} = \frac{1}{4\pi} \tag{21}$$

as with sinusoidal perturbations of black holes [7].

Interestingly, the above asymptotic expression (20) for the metric can also be obtained from a static Schwarzschild black hole as follows [8]. Instead of approaching the conformal boundary with $z =$ const. hypersurfaces (as $z \to 0$), use $\tilde{z} =$ const. hypersurfaces, where

$$t = \frac{d-2}{d-3}\tau^{\frac{d-3}{d-2}} \;, \quad x^1 = \tau^{\frac{d-3}{d-2}} y \;, \quad x^\perp = \frac{\tilde{x}^\perp}{\tau^{1/(d-2)}} \;, \quad z = \frac{\tilde{z}}{\tau^{1/(d-2)}} \tag{22}$$

They coincide initially (at $\tau = 1$) with their static counterparts, but they "flow" because the new coordinates (22) of the black hole metric are τ-dependent.

Applying the transformation (22) to the exact black hole metric (1), or more precisely, to a patch which includes the boundary $z \to 0$, one obtains

$$ds^2_{\text{b.h.}} = \frac{1}{\tilde{z}^2}\left[-\left(1 - 2\mu v^{d-1}\right)d\tau^2 + \tau^2 dy^2 + (d\tilde{x}^\perp)^2 + \frac{d\tilde{z}^2}{1 - 2\mu v^{d-1}}\right] + \dots \tag{23}$$

which matches the bulk metric of Bjorken flow (20) to leading order in $1/\tau$ with fixed v. Thus, one obtains Bjorken hydrodynamics asymptotically (as $\tau \to \infty$) from a static black hole!

Higher-order corrections to Bjorken flow dictated by the black hole may be found by refining the transformation [9]. This entails introducing corrections which are of $o(1/\tau)$ and making sure that the application of the transformation to the black hole metric does not introduce dependence of the metric on the rapidity and the transverse coordinates. It can be done systematically at each order in the $1/\tau$ expansion. However, at higher orders, to maintain boost invariance on the boundary, one must perturb the Schwarzschild metric. This perturbation leads to the ratio (21) [9] in agreement with the time-dependent asymptotic solution of Einstein equations.

4. Cosmology

In a popular scenario for cosmology, our Universe is modeled by a 3-brane embedded in a higher-dimensional bulk. The cosmological evolution of the brane is equivalent to its motion within the bulk space. The bulk may be occupied by a black hole, or a more complicated (or unknown) solution to the Einstein equations (allowing for energy exchange between the brane and the bulk).

My goal is to understand how the cosmological evolution emerges in the context of the AdS/CFT correspondence. I shall show that the equations of cosmological evolution emerge via holographic renormalization, even when one starts from a static AdS Schwarzschild black hole, provided the boundary conditions are chosen appropriately [10]. I shall concentrate on the case of a five-dimensional bulk $\mathcal{M}$ with a physically relevant four-dimensional conformal boundary $\partial\mathcal{M}$.

One usually fixes the geometry of the conformal boundary by adopting Dirichlet boundary conditions. However, for cosmological evolution the boundary geometry must remain dynamical. This may lead to fluctuations of the bulk metric which are not normalizable, but such fears are unfounded: the boundary geometry can be dynamical if one correctly introduces boundary counterterms needed in order to cancel infinities [11].

The Einstein field equations are obtained by varying the bulk action $I_{\mathcal{M}}$ which consists of the five-dimensional Einstein-Hilbert action on $\mathcal{M}$ with a cosmological term, the Gibbons-Hawking boundary term and boundary counterterms needed to render the system finite.

In addition to a black hole solution in the bulk, one thus obtains the boundary term

$$\delta I_{\mathcal{M}} = \frac{1}{2}\int_{\partial\mathcal{M}} d^4x\sqrt{-\det g^{(0)}}T^{(CFT)}_{\mu\nu}\delta g^{(0)\mu\nu} \tag{24}$$

where $g^{(0)}_{\mu\nu}$ $(\mu,\nu = 0,1,2,3)$ is the boundary metric and $T^{(CFT)}_{\mu\nu}$ is the stress-energy tensor of the dual CFT on the conformal boundary $\partial\mathcal{M}$. Dirichlet boundary conditions fix $g^{(0)}_{\mu\nu}$, therefore the additional term (24) vanishes.

To keep $g^{(0)}_{\mu\nu}$ dynamical, I shall adopt *mixed* boundary conditions. To this end, introduce the boundary action

$$I_{\partial\mathcal{M}} = I^{(EH)}_{\partial\mathcal{M}} + I^{(matter)}_{\partial\mathcal{M}} \ . \tag{25}$$

consisting of the four-dimensional Einstein-Hilbert action

$$I^{(EH)}_{\partial\mathcal{M}} = -\frac{1}{16\pi G_4}\int_{\partial\mathcal{M}} d^4x\sqrt{-\det g^{(0)}}(\mathcal{R}-2\Lambda_4) \ , \tag{26}$$

where G_4 (Λ_4) is Newton's (cosmological) constant in the four-dimensional boundary and $\mathcal{R}$ is the four-dimensional Ricci scalar constructed from $g^{(0)}_{\mu\nu}$, and an unspecified action for matter fields,

$$I^{(matter)}_{\partial\mathcal{M}} = \int_{\partial\mathcal{M}} d^4x\sqrt{-\det g^{(0)}}\mathcal{L}^{(matter)} \ . \tag{27}$$

which may reside on the boundary and have no bulk duals.

To the variation of the bulk action one must now add the variation of the new boundary action $\delta I_{\partial\mathcal{M}}$ under a change in the boundary metric.

$$\delta I_{\mathcal{M}} + \delta I_{\partial\mathcal{M}} = 0 \tag{28}$$

This leads to the possibility of *mixed* boundary conditions,

$$\mathcal{R}_{\mu\nu} - \frac{1}{2}g^{(0)}_{\mu\nu}\mathcal{R} - \Lambda_4 g^{(0)}_{\mu\nu} = 8\pi G_4\left(T^{(CFT)}_{\mu\nu} + T^{(matter)}_{\mu\nu}\right) \ . \tag{29}$$

which are simply the four-dimensional Einstein field equations! Moreover, the variation of the boundary matter action under a change in the matter fields yields the standard four-dimensional matter field equations.

For the most straightforward application of the AdS/CFT correspondence, write the metric (1) in Fefferman-Graham coordinates

$$\frac{dz_{FG}}{z_{FG}} = \frac{dz}{z\sqrt{f(z)}} \tag{30}$$

which gives (with an appropriate integration constant)

$$z^4_{FG} = \frac{16}{K^2+4\mu}\,\frac{1+\frac{K}{2}z^2-\sqrt{f(z)}}{1+\frac{K}{2}z^2+\sqrt{f(z)}} \ , \quad z^2 = \frac{z^2_{FG}}{\alpha+\beta z^2_{FG}+\gamma z^4_{FG}} \tag{31}$$

where $\alpha = 1$, $\beta = -\frac{K}{2}$, $\gamma = \frac{K^2+4\mu}{16}$. The metric (1) reads

$$ds^2_{\text{b.h.}} = \frac{1}{z^2_{FG}}\left[dz^2_{FG} - \frac{\left(1-\gamma z^4_{FG}\right)^2}{1+\beta z^2_{FG}+\gamma z^4_{FG}}dt^2 + \left(1+\beta z^2_{FG}+\gamma z^4_{FG}\right)d\Sigma^2_{K,3}\right] \tag{32}$$

Through holographic renormalization [1, 2], we deduce the energy density and pressure, respectively,

$$\langle T_{tt}^{(CFT)}\rangle = 3\langle T_{ii}^{(CFT)}\rangle = \frac{3\gamma}{4\pi G_5} \tag{33}$$

on a static Einstein Universe $\mathbb{R}\times\Sigma_{K,3}$ with metric

$$ds_0^2 = g^{(0)}_{\mu\nu}dx^\mu dx^\nu = -dt^2 + d\Sigma^2_{K,3} \tag{34}$$

It should be noted that the total energy $E = \langle T_{tt}^{(CFT)}\rangle Vol(\Sigma_{K,3})$ is larger than the mass of the black hole by a constant (Casimir energy) in the case of a curved horizon ($K\neq 0$). The two quantities agree for flat horizons ($K=0$). The additional piece is due to a change of the vacuum state from Minkowski to the conformal vacuum (curved metrics can be conformally mapped on Minkowski space).

For cosmology, instead of a static boundary, we need one with the form of a *Robertson-Walker* (RW) spacetime

$$ds_0^2 = g^{(0)}_{\mu\nu}dx^\mu dx^\nu = -d\tau^2 + a^2(\tau)d\Sigma^2_{K,3} \tag{35}$$

To this end, I shall choose a different foliation away from the black hole, consisting of hypersurfaces whose metric is asymptotically RW. To implement this, change coordinates $(t,z)\to(\tau,z_{FG})$ and bring the black-hole metric in the form

$$ds^2_{\text{b.h.}} = \frac{1}{z^2_{FG}}\left[dz^2_{FG} - \mathcal{N}^2(\tau,z_{FG})d\tau^2 + \mathcal{A}^2(\tau,z_{FG})d\Sigma^2_{K,3}\right] , \tag{36}$$

where $\mathcal{N}(\tau,z_{FG})\to 1$ and $\mathcal{A}(\tau,z_{FG})\to a(\tau)$ as we approach the boundary $z=0$. Comparison with the static case suggests the *ansatz*

$$\mathcal{A}^2 = \alpha(\tau) + \beta(\tau)z^2_{FG} + \gamma(\tau)z^4_{FG} , \tag{37}$$

where $\alpha(\tau),\beta(\tau),\gamma(\tau)$ are functions to be determined.

$\mathcal{N}$ is constrained by the τz_{FG} component of the Einstein equations,

$$\mathcal{N} = \frac{\dot{\mathcal{A}}}{\delta(\tau)} . \tag{38}$$

Agreement with the boundary metric fixes

$$\alpha(\tau) = a^2(\tau) \quad , \quad \delta(\tau) = \dot{a}(\tau) . \tag{39}$$

The diagonal components of the Einstein equations collectively yield

$$\beta = -\frac{\dot{a}^2+K}{2} . \tag{40}$$

The rest of the Einstein equations are satisfied provided

$$\beta\dot{\beta} = 2(\dot{\alpha}\gamma + \alpha\dot{\gamma}) , \tag{41}$$

which is integrated to give

$$\gamma = \frac{(\dot{a}^2+K)^2+4\mu}{16a^2} , \tag{42}$$

where we fixed the integration constant by comparing with the static case. It follows that the metric *ansatz* is uniquely specified. It agrees with the black hole metric (1) provided

$$\frac{(z')^2}{f(z)} - f(z)(t')^2 = \frac{z^2}{z_{FG}^2} , \quad \frac{z'\dot{z}}{f(z)} - f(z)t'\dot{t} = 0 , \quad \frac{\dot{z}^2}{f(z)} - f(z)\dot{t}^2 = -\frac{z^2\mathcal{N}^2}{z_{FG}^2} , \quad 1 = \frac{z\mathcal{A}}{z_{FG}} \tag{43}$$

The last equation fixes $z(\tau, z_{FG})$. Two of the other three equations determine the derivatives $\dot{t} = \frac{\dot{A}z'}{f\dot{a}}$ and $t' = -\frac{z^2\dot{a}}{z_{FG}f}$. These expressions actually satisfy all three equations. One can verify consistency of the system by calculating the mixed derivative $\dot{t}'$ using each of the two expressions and showing that they match. Upon integration, we obtain a unique function $t(\tau, r)$, up to an irrelevant constant.

As an example, consider pure AdS space in Poincaré coordinates ($\mu = 0$, $K = 0$). We have

$$t(\tau, z_{FG}) = -\frac{2\dot{a}z_{FG}^2}{4a^2 - \dot{a}^2 z_{FG}^2} + \int^{\tau} \frac{d\tau'}{a(\tau')} \tag{44}$$

At the boundary ($z_{FG} = 0$), t reduces to conformal time, $\int^{\tau} d\tau'/a(\tau')$. It receives corrections as we move into the bulk. This is generally the case.

In particular, for a de Sitter (dS) boundary, $a(\tau) = e^{H\tau}$, the metric reads

$$ds^2 = \frac{1}{z^2}\left[dz^2 + \left(1 - \frac{H^2}{4}z^2\right)\{-d\tau^2 + e^{2H\tau}d\vec{x}^2\}\right] \tag{45}$$

There is an *apparent horizon* at $z_+ = \frac{2}{H}$ and the temperature (from the conical singularity of Euclidean space) is

$$T = \frac{H}{2\pi} \tag{46}$$

which coincides with the temperature of four-dimensional dS space in the conformal vacuum!

General explicit expressions are not needed to extract physical results since we already know the explicit form of the metric in the new coordinates. The energy density and pressure, respectively, are

$$\langle (T^{(CFT)})_{\tau\tau} \rangle = \frac{3}{64\pi G}\frac{(\dot{a}^2+K)^2+4\mu}{a^4} , \quad \langle (T^{(CFT)})_i^i \rangle = \frac{(\dot{a}^2+K)^2+4\mu-4a\ddot{a}(\dot{a}^2+K)}{64\pi G a^4}$$

where there is no summation over i (it can be chosen in any spatial direction due to isotropy).

We deduce the conformal anomaly

$$g^{(0)\mu\nu}\langle T_{\mu\nu}^{(CFT)} \rangle = -\frac{3\ddot{a}(\dot{a}^2+K)}{16\pi G a^3} \tag{47}$$

It should be pointed out that the above results can also be obtained by a direct (four-dimensional) calculation by observing that the RW metric is conformally equivalent to the flat Minkowski metric.

The temperature on the RW boundary is found to be

$$T = \frac{T_H}{a} \tag{48}$$

where T_H is the Hawking temperature (temperature of a static boundary) and a is the conformal factor. The entropy is the Bekenstein-Hawking entropy of the hole (3) obeying the law of thermodynamics $dE = TdS - pdV$.

The boundary conditions yield the equation of cosmological evolution

$$H^2 + \frac{K}{a^2} - \frac{\Lambda_4}{3} = \frac{1}{16\pi G}\left[\left(H^2 + \frac{K}{a^2}\right)^2 + \frac{4\mu}{a^4}\right] + \frac{8\pi G_4}{3}\rho \tag{49}$$

where $\rho = T_{00}^{(matter)}$ is the energy density of four-dimensional "ordinary" matter (without a bulk dual) and $H = \frac{\dot{a}}{a}$ is the Hubble parameter. This equation is of the expected form [12], reflecting the conformal anomaly and the presence of a radiative energy component with energy density $\sim a^{-4}$.

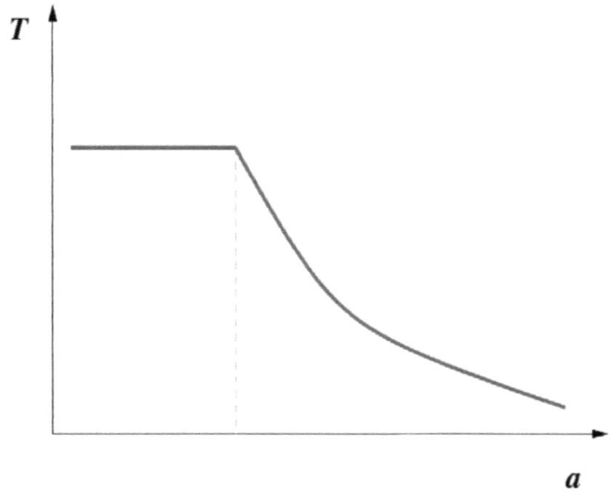

Figure 1. Evolution of the Universe.

To end on an amusing note, consider a phase transition (depicted in fig. 1) between an exponentially expanding (dS) Universe whose bulk dual contains no black hole ($\mu = 0$) at constant temperature (46) and our Universe of decreasing temperature (48) which is dual to a Schwarzschild black hole, as we discussed. The transition occurs when the black hole forms. Did the formation of a black hole in the bulk cause the end of inflation?

5. Conclusion

Black holes and their perturbations are a powerful tool in understanding the hydrodynamic behavior of a gauge theory fluid at strong coupling. Bjorken flow can be described in terms of a dual black hole in the bulk. RHIC and the LHC may probe black holes and provide information on string theory as well as non-perturbative QCD effects. The cosmological evolution may also be understood in terms of an AdS_5 Schwarzschild black hole.

Acknowledgments

Work supported in part by the Department of Energy under grant DE-FG05-91ER40627.

References

[1] de Haro S, Solodukhin S N and Skenderis K 2001 *Commun. Math. Phys.* **217** 595

[2] Skenderis K 2002 *Class. Quant. Grav.* **19** 5849
[3] Bjorken J D 1983 *Phys. Rev.* D **27** 140
[4] Janik R A and Peschanski R 2006 *Phys. Rev.* D **73** 045013
[5] Janik R A 2007 *Phys. Rev. Lett.* **98** 022302
[6] Nakamura S and Sin S-J 2006 *J. High Energy Phys.* JHEP09(2006)020
[7] Policastro G, Son D T and Starinets A O 2001 *Phys. Rev. Lett.* **87** 081601
[8] Alsup J and Siopsis G 2008 *Phys. Rev. Lett.* **101** 031602
[9] Alsup J and Siopsis G 2009 *Phys. Rev.* D **79** 084005
[10] Apostolopoulos P S, Siopsis G and Tetradis N 2009 *Phys. Rev. Lett.* **102** 151301
[11] Compère G and Marolf D 2008 *Class. Quant. Grav.* **25** 195014
[12] Kiritsis E 2005 *J. Cosmol. Astropart. Phys.* JCAP10(2005)014

Journal of Physics: Conference Series **462** (2013) 012049 doi:10.1088/1742-6596/462/1/012049

Graviton Scattering Made Simple(r)[1]

Marcus Spradlin

Department of Physics, Brown University, Box 1843, Providence RI 02912 USA

E-mail: `Marcus_Spradlin@brown.edu`

Abstract. I summarize a few recent developments in unlocking the hidden structure of graviton scattering amplitudes. In particular: (1) I show how the 'bonus relations' for graviton amplitudes may be used to provide simple analytic proofs of the equivalence of various distinct formulas for the n-graviton MHV amplitude which have appeared in the literature (and, in particular, providing a direct proof of the original BGK formula); (2) review the 'tree formula' for MHV graviton scattering, which has several nice properties including the fact that it manifests both S_{n-2} symmetry as well as $1/z^2$ falloff at large z; and (3) comment on the solution of the on-shell recursion relation for arbitrary tree-level graviton amplitudes.

1. Introduction

The past few years have witnessed dramatic progress in our understanding of the mathematical structure of scattering amplitudes. These developments have led to purely theoretical insights and also, as a pleasant spin-off, to powerful new methods for carrying out previously intractable calculations. The overwhelming majority of recent process has been in maximally supersymmetric gauge theory (SYM), where we now have a pretty good understanding of scattering amplitudes at both weak and strong coupling, and in fact at all values of the coupling for $n = 4, 5$ particles. There are reasons to expect graviton scattering amplitudes to have even richer structure [4], but progress on that front has been much slower.

In this talk I discuss some recent progress on even the simplest graviton amplitudes at tree level. In fact I argue that there is still a lot of room for improvement on the simplest of simplest graviton amplitudes—maximally-helicity violating (MHV) amplitudes at tree level.

MHV amplitudes are the simplest non-vanishing amplitudes in both Yang-Mills theory and in gravity. These describe for example the production of any number of positive helicity particles from a pair of scattering negative helicity particles, as depicted in fig. 1. Amplitudes become more and more complicated as the number of additional negative helicity particles in the outgoing state increases.

2. The Parke-Taylor and Berends-Giele-Kuijf Formulas for MHV Amplitudes

The MHV n-gluon scattering amplitude is given by the formula

$$A(1, \ldots, 2) = \frac{1}{\langle 1\,2\rangle\langle 2\,3\rangle \cdots \langle n\,1\rangle} \tag{1}$$

[1] Based on work done in collaboration with J. Drummond, D. Nguyen, A. Volovich, and C. Wen in [1, 2, 3].

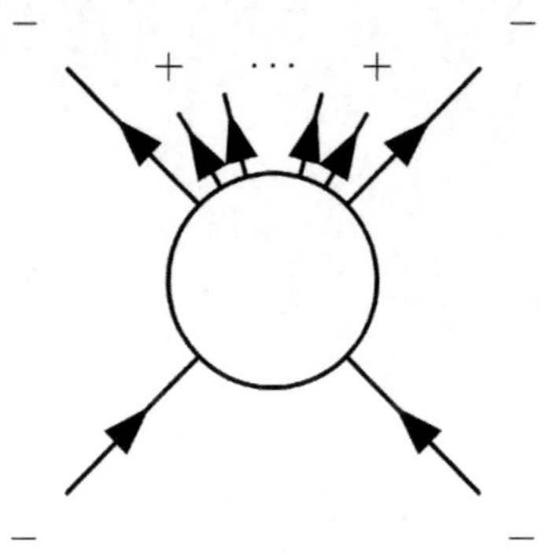

Figure 1. A maximally helicity violating (MHV) amplitude.

conjectured by Parke and Taylor [5] and proven by Berends and Giele in 1988 [6]. Here and in what follows we use the spinor helicity notation:

$$p_{a\dot{a}} = \lambda_a \widetilde{\lambda}_{\dot{a}} \qquad \langle i\, j \rangle = \epsilon_{ab} \lambda_i^a \lambda_i^b \qquad [i\, j] = \epsilon_{\dot{a}\dot{b}} \widetilde{\lambda}_i^{\dot{a}} \widetilde{\lambda}_j^{b} \tag{2}$$

$$(p_i + p_j)^2 = \langle i\, j \rangle [i\, j] \qquad [i|p_j + p_k + \cdots |p_l\rangle = [i\, j]\langle j\, l\rangle + [i\, k]\langle k\, l\rangle + \cdots . \tag{3}$$

The existence of the formula (1) has implications far beyond simply knowing the answer for a certain scattering amplitude, which one may or may not be particularly interested in. Rather, the existence of such a simple formula for something which would otherwise require enormously tedious Feynman diagram calculations is telling us that the theory has a rich hidden structure, and moreover that if one actually is interested in computing some particular amplitude, it behooves one to understand this structure.

For MHV graviton amplitudes, Berends, Giele and Kuijf conjectured in 1987 the formula [7]

$$\mathcal{M}_n = \sum_{\mathcal{P}(2,\ldots,n-2)} \frac{[1\,2][n{-}2\; n{-}1]}{\langle 1\; n{-}1\rangle} \left(\prod_{i=1}^{n-3} \prod_{j=i+2}^{n-1} \langle i\, j\rangle \right) \prod_{k=3}^{n-3} [k|p_{k+1} + \cdots + p_{n-1}|n\rangle . \tag{4}$$

The BGK formula serves perfectly well if one's interest is to know the answer for any particular graviton scattering amplitude, but it serves poorly as a guide towards discovering hidden structure in gravity, for several reasons:

- It contains vestigial references ··· to an ordering of the gravitons (this is usually a consequence of having used KLT relations [8]), even though gravitons carry no color structure so the amplitude should be independent of the ordering;
- it only manifests a subgroup S_{n-3} of the full S_n symmetry that the amplitude should have under permutation of all n gravitons; and finally,
- even that piece of the symmetry is realized in a fairly 'brutal' way (by an explicit sum over the permutations $\mathcal{P}$ of $(2, \ldots, n-1)$).

In essence, the formula is 'not sufficiently gravitational'.

3. A Tale of Two Families of Formulas [1]

Several variants of the BGK formula have appeared in the literature in recent years (see for example [9, 10, 11]), and the complexity of the various formulas is such that in general it is not possible to prove directly that two different formulas are in fact equivalent to each other.

Journal of Physics: Conference Series **462** (2013) 012049 doi:10.1088/1742-6596/462/1/012049

Rather one must to resort to numerical experimentation to convince oneself that they are. In fact the original 1987 BGK formula itself was only very recently proven to be correct by Mason and Skinner [12].

Some forms, like BGK, have manifest S_{n-3} symmetry, while others have a slightly larger manifest S_{n-2} symmetry, which however comes at a price I will describe shortly.

3.1. *A High-z Limit*

Four-dimensional amplitudes of gluons or gravitons are functions of the spinor helicity variables $(\lambda_i, \widetilde{\lambda}_i)$. All gravity amplitudes have exceptionally soft behavior, falling off as $1/z^2$ (see [9, 13, 14, 15, 10, 16, 17] and [4] for the most complete treatment) as two particles are taken to infinity in a particular complex direction

$$\lambda_1 \to \lambda_1 + z\lambda_2, \qquad \widetilde{\lambda}_2 \to \widetilde{\lambda}_2 - z\widetilde{\lambda}_1, \qquad z \to \infty \tag{5}$$

This is a property of amplitudes as a whole—individual Feynman diagrams can diverge in this limit, but miraculously their sum magically conspires to enable $1/z^2$ falloff. In contrast, gluon amplitudes vanish only as $1/z$ in the same limit [18].

3.2. *On-Shell Recursion from $1/z$ Falloff*

Falloff as $1/z$ for large z, a feature of both Yang-Mills theory and gravity, is already sufficiently strong to imply the on-shell recursion relation [18]. To see this, consider the contour integral

$$0 = \oint dz \frac{A(z)}{z} \tag{6}$$

with a contour circling infinity. Then deform the contour inward to write it as a sum over residues,

$$0 = A(0) + \sum_k \frac{1}{z_k} \mathrm{Res}_{z=z_k} A(z)\,. \tag{7}$$

Now the only poles in a tree-level amplitude are those which come from an internal propagator going on-shell. Moreover the residue at any such pole is a product of two tree amplitudes. Therefore this formula gives us a relation schematically of the form

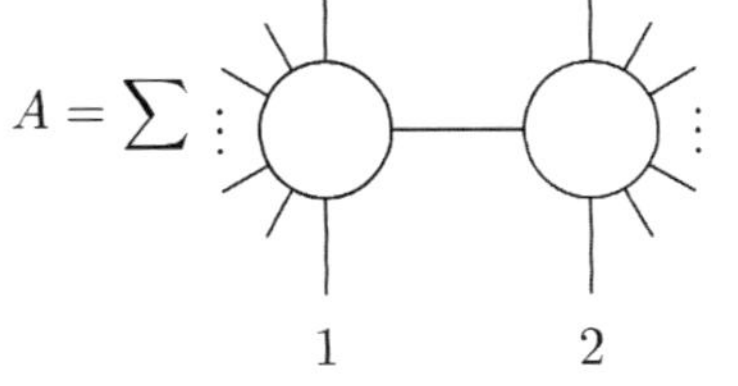

In words: any amplitude may be expressed as a sum over all on-shell factorizations in which the two 'special' legs 1 and 2 are separated from each other. Note that this remarkable on-shell recursion only works for theories with spin. The mantra is that scalar field theories may have simple Lagrangians, but their amplitudes are quite messy. In contrast gauge and gravity theories have horrendous Lagrangians, but very nice amplitudes.

3.3. Solving the Recursion for MHV Amplitudes

For MHV amplitudes only a single type of factorization can appear, which means that the on-shell recursion takes the extremely simple form

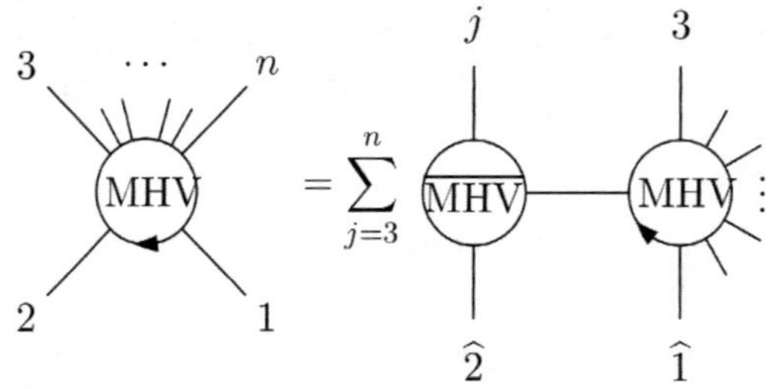

which can easily be solved explicitly. Depending on how one chooses to organize the calculation one can put the result into slightly different forms (see for example [9, 11]). Of particular importance to us will be a particular formula of Elvang and Freedman [11] which can be written as

$$M_n^{\text{MHV}} = \sum_{\mathcal{P}(3,\ldots,n)} [A^{\text{MHV}}(1,2,\ldots,n)]^2 G^{\text{MHV}}(1,2,\ldots,n) \tag{8}$$

where $A(1,2,\ldots,n)$ is nothing but the MHV gluon scattering amplitude shown above in eq. (1), and $G(1,2,\ldots,n)$ is a particular 'gravity dressing factor.' The sum is taken over the $(n-2)!$ permutations of the external lines which preserve the two 'special' legs. It is a generic feature of MHV formulas derived from the on-shell recursion that they have $(n-2)!$ terms. In contrast, recall that the BGK formula has $(n-3)!$ terms.

3.4. Bonus Relations from $1/z^2$ Falloff

The vanishing of amplitudes like $1/z^2$, which is a special feature of spin-2 (i.e., gravity), implies the existence of nontrivial hidden relations between amplitudes. To see this, consider the contour integral

$$0 = \oint dz\, M(z)\,,$$

again with a contour circling infinity. Deforming the contour to write it as a sum over residues gives the relation

$$0 = \sum_k \text{Res}_{z=z_k} M(z) \tag{9}$$

which can be expressed graphically as

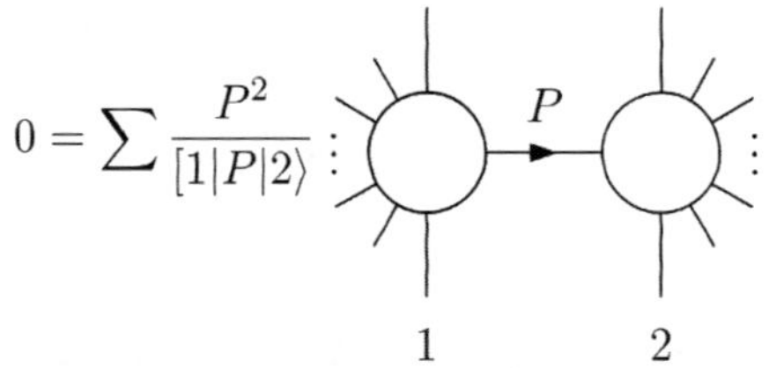

For MHV amplitudes in particular we have simply

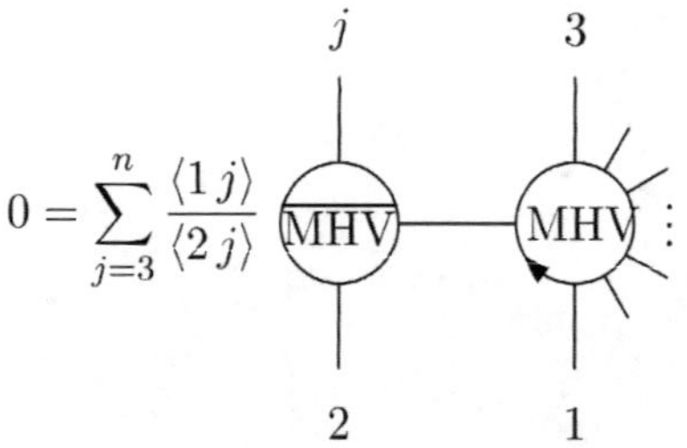

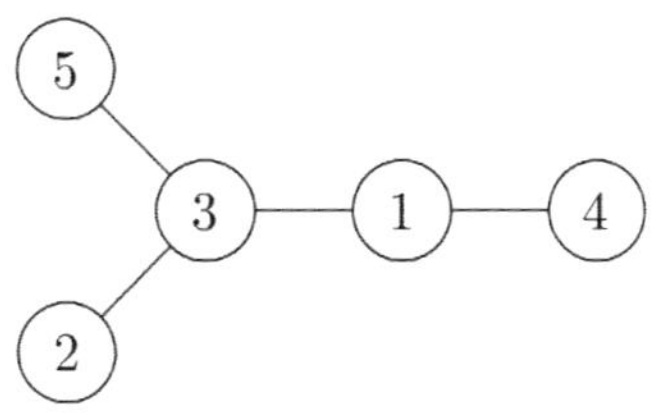

Figure 2. An example of a labeled tree diagram for $n = 7$ graviton scattering.

This linear relation can be used to eliminate any one of the $n-2$ terms in the on-shell recursion for the MHV graviton amplitude. When applied recursively, this can be used to systematically reduce $(n-2)!$-term formulas to $(n-3)!$-term formulas.

3.5. Summary and Wishlist

All known formulas in the literature are of two types: they have either $(n-2)!$ or $(n-3)!$ terms. Until recently, all of the former had been proven by using the on-shell recursion, while all of the latter (like the original BGK formula) were just conjectures (like BGK) that could only be numerically compared to the former. Using the 'bonus relation' for gravity amplitudes we have demonstrated the relation between these two classes of formulas and provided the first direct proof that the 1987 BGK formula satisfies the on-shell recursion [1]. The $(n-2)!$-term formulas have a larger manifest symmetry, but they do not manifest the $1/z^2$ falloff: the sum over $(n-2)!$ permutations contains redundant information which can be 'squeezed' out, using the bonus relation, leaving a sum over "only" $(n-3)!$ terms.

Ideally, we might hope to expect that a truly gravitational formula should have:

- manifest $1/z^2$ falloff, term by term,
- no vestigial reference to $\cdots$,
- manifest symmetry that is naturally S_{n-2} or larger!), and finally,
- "simplicity and beauty."

4. Twistor Space

In search of such a formula we now turn our attention to twistor space. Arkani-Hamed et. al. have recently shown [19] that amplitudes have some nice properties when expressed in 'mixed' twistor space representations. This means that for some particles we choose to Fourier transform $\lambda_i \to \mu_i$ while or others we choose to Fourier transform $\widetilde{\lambda}_J \to \widetilde{\mu}_J$. For a judicious choice of variables, the on-shell recursion becomes a simple integral equation roughly of the form (see [20] for a closely related formulation)

$$M_n = \int \sum_{k=3}^{n-1} M_k M_{n-k+2} \, .$$

In [3] our goal was to solve this 'simple' integral equation, already for the simplest example of MHV amplitudes. One very nice feature of twistor space is that cancellations which are far from obvious in physical space seem to happen automatically in twistor space.

5. The 'Tree Formula' for MHV Graviton Amplitudes [3]

By working in the 'link representation' introduced in [19], one can write down and prove a formula for the n-particle MHV graviton amplitude as a sum over all labeled tree diagrams with vertices labeled $\{1, \ldots, n-2\}$ (note one must choose two of the particles to treat as special; a sample diagram for $n = 7$ is shown in fig. 2).[2] The value of a diagram is simply

$$\left[\frac{1}{\langle n{-}1\, n\rangle^2} \prod_{a=1}^{n-2} \frac{1}{(\langle a\, n{-}1\rangle \langle a\, n\rangle)^2}\right] \prod_{\text{edges } ab} \frac{[a\, b]}{\langle a\, b\rangle} \langle a\, n{-}1\rangle \langle b\, n{-}1\rangle \langle a\, n\rangle \langle b\, n\rangle \tag{10}$$

Features of this formula include

- No vestigial reference to any cyclic ordering of the gravitons,
- manifest S_{n-2} symmetry
- without the need for an explicit sum over permutations, and
- manifest $1/z^2$ falloff, term by term.

Finally we note that the formula only has $(n-2)^{n-4}$ terms involving the basic structures $\langle i\, j\rangle \cdots [k\, l] \cdots$, an enormous savings over BGK-like formulas which have of order $n!^2$ terms.

The tree formula has a particularly nice property of making the soft limits manifest. In 1965 Weinberg showed that graviton amplitudes have a universal soft limit [22] which does not receive quantum corrections to any loop order. In the limit as the momentum $p_1 \to 0$, the amplitude must behave as

$$\lim_{p_1 \to 0} \frac{M(1^+, 2^+, \cdots, n^-)}{M(2^+, \ldots, n^-)} = \sum_{i=2}^{n-2} g(i^+), \qquad g(i^+) = \frac{\langle i\, n-1\rangle}{\langle 1\, n-1\rangle} \frac{\langle i\, n\rangle}{\langle 1\, n\rangle} \frac{[1\, i]}{\langle 1\, i\rangle}. \tag{11}$$

The tree formula beautifully manifests this fact: the only diagrams which survive this limit are those in which vertex "1" is attached to the rest of the diagram via a single line. Taking the limit amputates all these appendages, leaving precisely the diagrams contributing to the $n-1$ graviton amplitude, times the indicated factor.

6. All Tree-Level Graviton Amplitudes [2]

It is actually possible to solve the on-shell recursion for all tree-level graviton amplitudes [2], using two ingredients. The first is the solution for all tree-level SYM amplitudes due to Drummond and Henn [23] which schematically takes the form

$$A^{\text{all}} = A^{\text{MHV}} \sum_\alpha R_\alpha(\lambda_i, \widetilde{\lambda}_i, \eta_i) \tag{12}$$

where the R_α are certain specific dual-conformally invariant objects. The second ingredient is the Elvang-Freedman formula for MHV graviton amplitudes, which can be cast into the form

$$M^{\text{MHV}} = \sum_{\mathcal{P}(3,\ldots,n)} [A^{\text{MHV}}(1, 2, \ldots, n)]^2 G^{\text{MHV}}(\lambda_i, \widetilde{\lambda}_i), \tag{13}$$

where G^{MHV} is a particular quantity which we call the 'gravity dressing factor'.

Inspired by these two results we were able to solve the on-shell recursion for all gravity amplitudes, obtaining a formula which naturally merges eqs. (12) and (13) into the form

$$M^{\text{all}} = \sum_{\mathcal{P}(3,\ldots,n)} [A^{\text{MHV}}(1, 2, \ldots, n)]^2 \sum_\alpha [R_\alpha(\lambda_i, \widetilde{\lambda}_i, \eta_i)]^2 G(\lambda_i, \widetilde{\lambda}_i)\,. \tag{14}$$

[2] The tree formula can be shown to be equivalent (in a re-arranged form) to a formula presented by Bern, Dixon, Perelstein and Rozowsky using 'half-soft' factors [21].

Explicit formulas for some gravity dressing factor G may be found in [2]; they generalize the quantity $G^{\rm MHV}$ appearing in the MHV amplitude eq. (13).

A notable feature of eq. (14) is that the R's are precisely the same ones that appear in gluon scattering, except they are now literally squared—this can be understood simply as a result of moving up from $\mathcal{N} = 4$ to $\mathcal{N} = 8$ on-shell superspace using the rule

$$\delta^4(\theta) \to [\delta^4(\theta)]^2 \equiv \delta^8(\theta)\,. \tag{15}$$

Other than this minor modification, all of the super-structure of graviton scattering amplitudes is identical to that of gluon amplitudes; the gravity dressing factors G depend only on the spinor helicity variables $(\lambda_i, \tilde{\lambda}_i)$ and not on the supercoordinates η_i.

The above formulas for non-MHV amplitudes, while explicit, are almost certainly not the end of the story since they do not possess any of the nice features of the MHV tree formula. So I am sure that we are still quite far from fully unlocking the structure of general graviton amplitudes, even at tree level!

Acknowledgements
This work was supported in part by the US Department of Energy under contract DE-FG02-91ER40688 and the US National Science Foundation under grants PHY-0638520 and PHY-0643150.

References
[1] M. Spradlin, A. Volovich and C. Wen, Phys. Lett. B **674** (2009) 69 [arXiv:0812.4767 [hep-th]].
[2] J. M. Drummond, M. Spradlin, A. Volovich and C. Wen, Phys. Rev. D **79** (2009) 105018 [arXiv:0901.2363 [hep-th]].
[3] D. Nguyen, M. Spradlin, A. Volovich and C. Wen, arXiv:0907.2276 [hep-th].
[4] N. Arkani-Hamed, F. Cachazo and J. Kaplan, arXiv:0808.1446 [hep-th].
[5] S. J. Parke and T. R. Taylor, Phys. Rev. Lett. **56**, 2459 (1986).
[6] F. A. Berends and W. T. Giele, Nucl. Phys. B **306**, 759 (1988).
[7] F. A. Berends, W. T. Giele and H. Kuijf, Phys. Lett. B **211**, 91 (1988).
[8] H. Kawai, D. C. Lewellen and S. H. H. Tye, Nucl. Phys. B **269**, 1 (1986).
[9] J. Bedford, A. Brandhuber, B. J. Spence and G. Travaglini, Nucl. Phys. B **721**, 98 (2005) [arXiv:hep-th/0502146].
[10] Z. Bern, J. J. Carrasco, D. Forde, H. Ita and H. Johansson, Phys. Rev. D **77**, 025010 (2008) [arXiv:0707.1035 [hep-th]].
[11] H. Elvang and D. Z. Freedman, JHEP **0805**, 096 (2008) [arXiv:0710.1270 [hep-th]].
[12] L. Mason, D. Skinner, arXiv:0808.3907.
[13] F. Cachazo and P. Svrcek, arXiv:hep-th/0502160.
[14] N. E. J. Bjerrum-Bohr, D. C. Dunbar, H. Ita, W. B. Perkins and K. Risager, JHEP **0601**, 009 (2006) [arXiv:hep-th/0509016].
[15] P. Benincasa, C. Boucher-Veronneau and F. Cachazo, JHEP **0711**, 057 (2007) [arXiv:hep-th/0702032].
[16] N. Arkani-Hamed and J. Kaplan, JHEP **0804**, 076 (2008) [arXiv:0801.2385].
[17] M. Bianchi, H. Elvang and D. Z. Freedman, JHEP **0809**, 063 (2008) [arXiv:0805.0757].
[18] R. Britto, F. Cachazo, B. Feng and E. Witten, Phys. Rev. Lett. **94**, 181602 (2005) [arXiv:hep-th/0501052].
[19] N. Arkani-Hamed, F. Cachazo, C. Cheung and J. Kaplan, arXiv:0903.2110.
[20] L. Mason and D. Skinner, arXiv:0903.2083.
[21] Z. Bern, L. J. Dixon, M. Perelstein and J. S. Rozowsky, Nucl. Phys. B **546**, 423 (1999) [arXiv:hep-th/9811140].
[22] S. Weinberg, Phys. Rev. **140**, B516 (1965).
[23] J. M. Drummond and J. M. Henn, JHEP **0904**, 018 (2009) [arXiv:0808.2475].

Journal of Physics: Conference Series **462** (2013) 012050 doi:10.1088/1742-6596/462/1/012050

The Xerographic Distribution: Scientific Reasoning in a Large Universe

Mark Srednicki and James Hartle

Department of Physics, University of California, Santa Barbara, CA 93109 USA

E-mail: mark@physics.ucsb.edu hartle@physics.ucsb.edu

Abstract. As observers of the universe we are physical systems within it. If the universe is very large in space and/or time, the probability becomes significant that the data on which we base predictions is replicated at other locations in spacetime. Predictions of our future observations therefore require an assumed probability distribution—the *xerographic distribution*—for our location among the possible ones. It is the combination of basic theory plus the xerographic distribution that can be predictive and testable by further observations. This is illustrated by examining a toy model of a classical deterministic universe with a fixed flat metric.

Many contemporary cosmological models predict that the universe is much larger than our observable Hubble volume, and is possibly infinite. In such a large (or infinite) universe, the probability that there are multiple copies of 'our' data, D_0, approaches (or equals) 100%.

The data D_0 consists of every scrap of information that humanity possesses, including all memories and all records of every observation, individual and collective, that has ever been made about anything. 'Humanity' refers to a particular physical subsystem of the universe, consisting of human beings and all associated technology. Terms such as 'our', 'we', etc., refer to this subsystem.

The possible existence of multiple copies of D_0 presents a challenge to the conventional use of scientific reasoning [1]. To see why, consider a hypothetical situation in which our data D_0 happens to be consistent with a completely classical dynamical theory (of particles and/or fields), and a fixed, flat spacetime metric. In such a situation, a complete physical theory T consists of the dynamical equations (for the particles and/or fields) and an initial condition.

A theory T is compatible with our data D_0 if it predicts the occurrence of *at least one* instance of D_0 throughout spacetime [2]. In order to avoid issues of infinities, we suppose that D_0 is compatible with at least some theories T with finite spatial volume and finite temporal extent (in the coordinate system that makes the dynamical equations as simple as possible).

If a theory T predicts *exactly one* instance of D_0 anywhere in the finite spacetime, then we can use T to make predictions about our future observations, and hence test T.

If, however, a theory T predicts *more than one* instance of D_0 (at various locations in spacetime), and furthermore predicts that *different sets of future observations* will occur at each of these locations, then we cannot use T to make definite predictions about our own future observations. This is because we do not know which copy of D_0 we correspond to; we could be any one of them.

Note that this is not due to any incompleteness of the theory T; it makes definite predictions for what will happen at each of the instances of D_0. These predictions can be called *third-person*

Journal of Physics: Conference Series **462** (2013) 012050 doi:10.1088/1742-6596/462/1/012050

predictions. Third-person predictions could be completely verified only by a hypothetical 'third-person' observer outside the system described by T. We, however, are observers within the system. Therefore, what we would like to have is a *first-person* prediction of what 'we' will see next. The theory T, though fully complete and deterministic in the usual sense, cannot provide such a prediction by itself.

To make a prediction of what 'we' will see next, we need to decide which copy corresponds to 'us', or, more generally, posit a probability distribution—the *xerographic distribution*—on the set of copies. More specifically, if the theory T predicts that there are N copies of D_0 throughout the finite spacetime, then we assign probability ξ_i that the ith copy (in some numbering scheme) corresponds to us. Given a *theoretical framework* (T, ξ) consisting of *both* a theory T *and* a xerographic distribution ξ, probabilistic predictions of what 'we' will see next can be made.

Now the problem is how to choose the xerographic distribution ξ. The most important point to remember is that ξ is *not* determined by the theory T, which predicts the future of each instance of D_0, but does not tell us which instance is ours. Therefore, the xerographic distribution is an *additional ingredient* that must be posited, along with a conventional theory T, in order to have a framework (T, ξ) that is capable of generating first-person predictions for (probabilities of) what 'we' will see.

The simplest and most obvious choice of ξ is uniform on the possibilities: $\xi_i = 1/N$. This can be considered as an application of the 'principle of indifference': we have no reason to choose one copy over another, and therefore we should assign them all equal probability.

There are, however, situations in which we may indeed have a good reason for choosing one copy over another.

Suppose that some of the copies are 'Boltzmann brains' (BBs for short) [3], in which the data D_0 is not a faithful record of actual events and circumstances, but simply a chance fluctuation. Other copies are 'ordinary observers' (OOs for short), where the data does faithfully record actual events and circumstances. BBs and OOs can be considered as endpoints of a continuum of some measure of the faithfulness of records and memories to external reality; the degree of faithfulness of a given copy could in principle be determined by a third-person observer. For simplicity of exposition we will divide all copies of D_0 into OOs (which have a high degree of faithfulness) and BBs (which have a low degree of faithfulness).

We may wish to discount the possibility that we are BBs. This is because we believe that we currently have a coherent picture of the universe and at least some of the laws that govern it; if we are BBs, this picture is fraudulent. Therefore, whatever evidence led us to consider the theory T (which predicts the existence of BBs) cannot be trusted, and there is no reason to favor T over any other theory that allows for BBs with our data D_0, including theories that have a completely different external reality. The belief that we are BBs is therefore 'self-undermining' [4].

We can avoid the problem of possibly holding self-undermining beliefs by choosing $\xi_i = 0$ for any copy i which is a BB. This choice has the nature of a 'working assumption'. We choose to *assume* that our memories and records faithfully reflect external reality. Such an assumption implicitly underlies all of science.

To see the consequences of these different possible choices for ξ, consider a theory T that predicts the existence of one OO consistent with D_0, and also a large number of BBs. Furthermore, while the future of the OO is a straightforward continuation of its past, the futures of the BBs are all wildly different, and inconsistent with their (faulty) records.

If we take a uniform ξ, then we predict that it is overwhelmingly likely that 'our' future is very unlike our past. Let's now wait a few seconds ... [5]; we note that our present has turned out to be quite like our past, and so this framework—the theory T that predicts one OO and lots of BBs, and the uniform xerographic distribution—is ruled out with a high degree of confidence.

If, on the other hand, we take $\xi_{i=\text{OO}} = 1$ and $\xi_{i=\text{BB}} = 0$, then we predict that 'our' future is very much like our past. Let's now wait a few seconds .... We note that the present is quite like the past, and so this framework—the theory T that predicts one OO and lots of BBs, and the xerographic distribution that is peaked on the OO—is fully supported by the subsequent data.

More generally, different choices of ξ can be treated in the same manner as different choices of T. Once a framework (T, ξ) is specified, we can see if it makes any high-probability predictions of what 'we' will see, and, if so, test these predictions. If the predictions are successful, the framework (T, ξ) remains a viable one. If the predictions are unsuccessful, a new framework (with either a different theory T or a different xerographic distribution ξ or both) must be postulated, and the process of prediction and testing repeated. The details of a standard Bayesian analysis of this scientific process are given in [1].

Some scientists believe that the uniform ξ must always be chosen (see e.g. [6]), so that if a framework $(T, \xi_{\text{uniform}})$ makes an incorrect prediction, then the theory T must be discarded, and ξ_{uniform} retained. The present authors see no fundamental justification for this rigid philosophical stance. *Any* of our scientific assumptions might be wrong, and so differing assumptions should be competed and compared on various criteria, including how predictive they are.

Modern cosmological theories must also include the complicating features of both quantum mechanics and general relativity. Since quantum mechanics is inherently probabilistic, and provides a scheme for computing probabilities, it might seem that the xerographic distribution is computable from the underlying theory. As our purely deterministic example shows, however, this is not possible. Quantum mechanics is no more capable than classical mechanics of telling us which copy is 'ours', provided that there is a nonzero third-person probability of the existence of more than one copy. General relativity complicates the issue by making the counting of copies dependent on the choice of a reference volume. This is especially problematic in the case of an infinite universe; some sort of cutoff procedure must be adopted, numbers of copies computed, and then appropriate limits taken.

How to deal with the complications of quantum mechanics and general relativity (in, say, computing the ratio of the number of OOs to BBs) comprises the 'measure problem' of modern cosmology [7]. Some scientists hope that further theoretical developments in our understanding of quantum gravity will solve this problem—effectively, tell us which reference volumes to use and how to regulate infinities. Whether or not this is the case, we will still be left with the issue of comparing and competing different xerographic distributions on the (now suitably regulated and well-defined) set of copies of our data.

Thus, in a universe containing more than one copy of our data, the set of physical theories under consideration must be supplemented by a set of xerographic distributions under consideration. Best scientific practice dictates that a rigid dogmatic choice of either should be avoided.

We thank Raphael Bousso, Brandon Carter, Sean Carroll, Ben Freivogel, Steve Giddings, Alan Guth, Thomas Hertog, Matthew Kleban, Don Page, Steve Shenker, and Alex Vilenkin for numerous helpful conversations. This work was supported in part by the National Science Foundation under grants PHY05-55669 and PHY07-57035.

References

[1] M. Srednicki and J. Hartle, *Science in a Very Large Universe*, arXiv:0906.0042.

[2] J. B. Hartle and M. Srednicki, *Are we typical?*. Phys. Rev. D **75**, 123523 (2007) [arXiv:0704:2530].

[3] A. Albrecht and L. Sorbo, *Can the universe afford inflation?*, Phys. Rev. D **70**, 063528 (2004) [arXiv:hep-th/0405270]. There is an extended set of references in D. N. Page, *Cosmological measures without volume weighting*, JCAP **0810**, 025 (2008) [arXiv:0808:0351],

[4] S. Carroll, *What we know, and don't, and why*, http://preposterousuniverse.blogspot.com/2004/12/what-we-know-and-dont-and-why.html

[5] J. R. Gott, *Boltzmann brains—I'd rather see than be one*, arXiv:0802.0233.

[6] J. Garriga and A. Vilenkin, *Prediction and explanation in the multiverse*, Phys. Rev. D **77**, 043526 (2008) [arXiv:0711.2559]; R. Bousso, B. Freivogel, and I.-S. Yang, *Boltzmann babies and the proper time measure*, Phys. Rev. D **77**, 103514 (2008) [arXiv:0712.3324].

[7] R. Bousso, B. Freivogel, and I-S. Yang, *Properties of the scale factor measure*, *Phys. Rev. D* **79**, 063513 (2009) [arXiv:0808.3770]; A. De Simone, A. H. Guth, A. Linde, M. Noorbala, M. P. Salem, and A. Vilenkin, *Boltzmann brains and the scale-factor cutoff measure of the multiverse*, arXiv:0808.3778; and references therein.

Lattice formulation of 2D $\mathcal{N} = (2,2)$ SQCD and twisted supercharges [1]

Fumihiko Sugino

Okayama Institute for Quantum Physics, Kyoyama 1-9-1, Kita-ku, Okayama 700-0015, Japan

E-mail: fumihiko_sugino@pref.okayama.lg.jp

Abstract. We discuss on two kinds of lattice formulation of two-dimensional $\mathcal{N} = (2,2)$ supersymmetric QCD preserving one of its supercharges. The preserved supercharge is of the same form obtained by topological twist. One formulation is based on A model twist, and the other is based on B model twist. For the A model twist case, by employing the Ginsparg-Wilson operator, the lattice model can be constructed for general number of flavors of fundamental and anti-fundamental matter multiplets under the gauge group $G = U(N)$ or $SU(N)$. It preserves both of one supercharge (Q) and chiral flavor symmetry of matter multiplets. Furthermore, superpotential can be introduced with exact holomorphic or anti-holomorphic structure on lattice. For the B model twist case, the lattice action takes a simpler form compared with the A model twist case, because the admissibility condition is not required to the formulation. It can be applied to a restricted cases of $G = U(N)$ and of fundamental matters or anti-fundamental matters only.

1. Introduction

Lattice formulation of quantum field theory has been the most solid and conventional method to give its constructive definition as well as to explore its nonperturbative aspects, since Wilson's formulation of lattice gauge theory. Lattice formulation of supersymmetric (gauge) theory has notorious difficulty in realizing supersymmetry (SUSY) on lattice. Namely, SUSY induces infinitesimal translations, although they are not symmetries on the lattice which is a grid with a finite spacing a. Such a mismatch vanishes in the continuum limit $a \to 0$ and SUSY is restored *at the classical level.* However, loop effects *at the quantum level* cause UV divergent radiative correction terms having inverse powers of a, which possibly prevent from restoring the SUSY in the continuum limit. If so, it will be necessary to introduce counter terms and fine-tune their parameters to be cancelled with the harmful radiative corrections.

Recently, for supersymmetric gauge theory with extended SUSY, various lattice models in which a part of supercharges is preserved on the lattice, have been constructed. The exact SUSY on the lattice helps reducing the number of parameters needed to be fine-tuned in the continuum limit. In dimensions lower than four it effectively works, and in particular for two dimensions there are lattice models where the target continuum SUSY theory can be obtained with no need of fine-tuning.

The supercharges preserved on the lattice are nilpotent up to internal symmetry, not inducing translations. It reminds us of scalar supercharges obtained from topological twist [6, 7].

[1] This talk is based on the work [1, 2, 3, 4, 5].

Journal of Physics: Conference Series **462** (2013) 012051 doi:10.1088/1742-6596/462/1/012051

For example, two-dimensional $\mathcal{N} = (2,2)$ supersymmetric Yang-Mills (SYM) theory or supersymmetric QCD (SQCD), there are two kinds of lattice formulations. One preserves the supercharge $Q = -\frac{1}{\sqrt{2}}(Q_L + \bar{Q}_R)$ which is the same as a scalar supercharge obtained by A-model twist, the other preserves $Q' = -\frac{1}{\sqrt{2}}(\bar{Q}_L + \bar{Q}_R)$ of the same form as the one obtained by B-model twist. The two-dimensional $\mathcal{N} = (2,2)$ theory has two $U(1)_R$-symmetries, $U(1)_A$ and $U(1)_V$ at the classical level. The Q invariant formulation respects $U(1)_A$ on the lattice, while the Q' invariant one $U(1)_V$.

Let us note that we are considering the theory on flat space-time, where topological twists just amount to renaming fields variables and the continuum theory does not change by the twists. However, lattice models can become different depending on which of Q and Q' is exactly preserved.

In the following sections, we discuss two kinds of lattice models for two-dimensional $\mathcal{N} = (2,2)$ SQCD, which is SYM theory coupled to n_+ fundamental and n_- anti-fundamental matter multiplets, with $G = U(N)$ or $SU(N)$ gauge group. Lattice is two-dimensional square lattice with the spacing a, and compact link variables $U_\mu = e^{iaA_\mu}$ are used as gauge fields. Remaining field contents are defined on lattice sites. Each of the lattice models preserves either Q or Q'.

2. Continuum $\mathcal{N} = (2,2)$ SQCD in two dimensions

The continuum theory of two-dimensional $\mathcal{N} = (2,2)$ SQCD is obtained from four-dimensional $\mathcal{N} = 1$ SQCD by dimensional reduction. The multiplets $V = (A_\mu, \phi, \bar{\phi}; \lambda; D)$ are from four-dimensional $\mathcal{N} = 1$ vector multiplets. A_μ is a two-dimensional vector field, ϕ, $\bar{\phi}$ are complex scalars, λ is a gaugino, and D is an auxiliary field. The fundamental matter multiplets $\Phi_{+I} = (\phi_{+I}; \psi_{+IR}, \psi_{+IL}; F_{+I})$ and the anti-fundamental matter multiplets $\Phi_{-I'} = (\phi_{-I'}; \psi_{-I'R}, \psi_{-I'L}; F_{-I'})$ are from four-dimensional $\mathcal{N} = 1$ chiral multiplets. $I = 1, \cdots, n_+$ and $I' = 1, \cdots, n_-$ are flavor indices of fundamentals and anti-fundamentals, respectively. ϕ_{+I} is a scalar, ψ_{+IR}, ψ_{+IL} are fermions, and F_{+I} is an auxiliary field. Similar notations are used for the anti-fundamentals. There are also anti-chiral multiplets belonging to the fundamental representation $\Phi^\dagger_{+I} = (\phi^\dagger_{+I}; \bar{\psi}_{+IR}, \bar{\psi}_{+IL}; F^\dagger_{+I})$ and to the anti-fundamental representation $\Phi^\dagger_{-I'} = (\phi^\dagger_{-I'}; \bar{\psi}_{-I'R}, \bar{\psi}_{-I'L}; F^\dagger_{-I'})$. The classical theory has two kinds of R-symmetries. $U(1)_V$ originates from the chiral symmetry of four-dimensional $\mathcal{N} = 1$ theory, and $U(1)_A$ is from a rotational symmetry on a dimensionally reduced two-dimensional plane.

The continuum action is expressed in terms of dimensionally reduced four-dimensional $\mathcal{N} = 1$ superfields as

$$
\begin{aligned}
S &= S_{\rm SYM} + S_{\rm mat} + S_{\rm pot} + S_{{\rm FI},\vartheta}, && (1)\\
S_{\rm SYM} &= \frac{1}{8g^2}\int d^2x\ {\rm tr}\left(W^\alpha W_\alpha|_{\theta\theta} + \bar{W}_{\dot\alpha}\bar{W}^{\dot\alpha}\Big|_{\bar\theta\bar\theta}\right), && (2)\\
S_{\rm mat} &= \int d^2x \left[\sum_{I=1}^{n_+}\Phi^\dagger_{+I}e^V\Phi_{+I} + \sum_{I'=1}^{n_-}\Phi_{-I'}e^{-V}\Phi^\dagger_{-I'}\right]\Bigg|_{\theta\theta\bar\theta\bar\theta}, && (3)\\
S_{\rm pot} &= \int d^2x\ W(\Phi_+, \Phi_-)|_{\theta\theta} + \int d^2x\ \overline{W}(\Phi^\dagger_+, \Phi^\dagger_-)\Big|_{\bar\theta\bar\theta}, && (4)\\
S_{{\rm FI},\vartheta} &= \int d^2x\ {\rm tr}\left(-\kappa D + \frac{\vartheta}{2\pi}F_{01}\right), && (5)
\end{aligned}
$$

$S_{\rm pot}$ gives superpotential terms, and $S_{{\rm FI},\vartheta}$ represents the Fayet-Iliopoulos (FI) and ϑ-terms.

Journal of Physics: Conference Series **462** (2013) 012051 doi:10.1088/1742-6596/462/1/012051

3. SYM part (A model twist)

After taking the Wess-Zumino gauge, the Euclidean action of the continuum SYM theory becomes

$$\begin{aligned} S^{(E)}_{\rm SYM} &= \frac{1}{g^2}\int d^2x \ {\rm tr}\left(\frac{1}{2}F_{\mu\nu}F_{\mu\nu} + D_\mu\phi D_\mu\bar\phi + \frac{1}{4}[\phi,\bar\phi]^2 - D^2 \right. \\ &\left. +4\bar\lambda_R D_z\lambda_R + 4\bar\lambda_L D_{\bar z}\lambda_L + 2\bar\lambda_R[\bar\phi,\lambda_L] + 2\bar\lambda_L[\phi,\lambda_R]\right). \end{aligned} \tag{6}$$

Q-SUSY transformation acts component fields as

$$\begin{aligned} &QA_\mu = \psi_\mu, \qquad Q\psi_\mu = iD_\mu\phi, \\ &Q\phi = 0, \\ &Q\bar\phi = \eta, \qquad Q\eta = [\phi,\bar\phi], \\ &Q\chi = iD + iF_{01}, \qquad QD = -QF_{01} - i[\phi,\chi], \end{aligned} \tag{7}$$

where Q is nilpotent in the sense that

$$Q^2 = (\text{infinitesimal gauge transformation with the parameter } \phi). \tag{8}$$

Here, the gaugino fields were renamed as

$$\begin{aligned} &\psi_0 \equiv \frac{1}{\sqrt2}(\lambda_L + \bar\lambda_R), \qquad \psi_1 \equiv \frac{i}{\sqrt2}(\lambda_L - \bar\lambda_R), \\ &\chi \equiv \frac{1}{\sqrt2}(\lambda_R - \bar\lambda_L), \qquad \eta \equiv -i\sqrt2(\lambda_R + \bar\lambda_L). \end{aligned} \tag{9}$$

Under the $U(1)_A$ rotation, the fields change as

$$\begin{aligned} &\phi \to e^{i2\alpha}\,\phi, \qquad \bar\phi \to e^{-i2\alpha}\,\bar\phi, \\ &\psi_\mu \to e^{i\alpha}\,\psi_\mu, \qquad \chi \to e^{-i\alpha}\,\chi, \qquad \eta \to e^{-i\alpha}\,\eta. \end{aligned} \tag{10}$$

Note that the action (6) is expressed as Q-exact form:

$$S^{(E)}_{\rm SYM} = Q\frac{1}{g^2}\int d^2x \ {\rm tr}\left[-i\chi(F_{01} - D) + \frac{1}{4}\eta[\phi,\bar\phi] - i\psi_\mu \mathcal{D}_\mu\bar\phi\right]. \tag{11}$$

The action is manifestly symmetric under $\lambda_R \leftrightarrow \bar\lambda_L$ and $\lambda_L \leftrightarrow \bar\lambda_R$, but not $\lambda_R \leftrightarrow \lambda_L$ and $\bar\lambda_R \leftrightarrow \bar\lambda_L$.

3.1. Q-SUSY on lattice (A model twist)

As gauge fields on lattice, we consider compact link variables $U_\mu(x) = e^{iaA_\mu(x)}$. x represents a lattice site, and all the other fields are put on sites. Then, the Q-SUSY satisfying (8) can be realized on the lattice as

$$\begin{aligned} &QU_\mu(x) = i\psi_\mu(x)U_\mu(x) \\ &Q\psi_\mu(x) = i\psi_\mu(x)\psi_\mu(x) + ia\nabla_\mu\phi(x) \\ &Q\phi(x) = 0 \\ &Q\bar\phi(x) = \eta(x), \quad Q\eta(x) = [\phi(x),\bar\phi(x)] \\ &Q\chi(x) = iD(x) + \frac{i}{2}\widehat\Phi(x), \quad QD(x) = -\frac{1}{2}Q\widehat\Phi(x) - i[\phi(x),\chi(x)], \end{aligned} \tag{12}$$

where $a\nabla_\mu\phi(x) \equiv U_\mu(x)\phi(x+\hat{\mu})U_\mu(x)^{-1} - \phi(x)$ and

$$\widehat{\Phi}(x) \equiv \frac{-i(U_{01}(x) - U_{10}(x))}{1 - \frac{1}{\epsilon^2}||1 - U_{01}(x)||^2} \tag{13}$$

corresponds to $2F_{01}$ in the naive continuum limit. $U_{\mu\nu}(x)$ are plaquette variables defined by

$$U_{\mu\nu}(x) \equiv U_\mu(x)U_\nu(x+\hat{\mu})U_\mu(x+\hat{\nu})^{-1}U_\nu(x)^{-1}, \tag{14}$$

and ϵ is a positive number chosen as $0 < \epsilon < \frac{2}{N}$ for $G = U(N)$, $0 < \epsilon < \frac{2\sqrt{2}}{\sqrt{N}}$ for $G = SU(N)$ with $N = 2, 3, 4$, and $0 < \epsilon < 2\sin\left(\frac{\pi}{N}\right)$ for $G = SU(N)$ with $N \geq 5$, which is introduced to impose the admissibility condition for the plaquette variable

$$||1 - U_{01}(x)|| < \epsilon \tag{15}$$

at each site x. Here, the norm of an arbitrary $M \times M$ complex matrix A is defined as $||A|| \equiv \sqrt{\frac{1}{M}\,\mathrm{tr}(AA^\dagger)}$. Note that, since ϵ is independent of the lattice spacing a, gauge fields become almost unconstrained as approaching the continuum limit. As is seen below, the admissibility condition plays a crucial role to construct a Q invariant lattice action exhibiting correct weak coupling behavior.

3.2. *Q invariant lattice action (A model twist)*

The lattice action can be constructed as Q-exact form:

$$S_{\mathrm{SYM}}^{(\mathrm{lat\ A})} = Q\frac{1}{g_0^2}\sum_x \mathrm{tr}\left[\chi(x)\left\{-\frac{i}{2}\widehat{\Phi}(x) + iD(x)\right\} + \frac{1}{4}\eta(x)[\phi(x), \bar{\phi}(x)] - i\sum_\mu \psi_\mu(x)a\nabla_\mu\bar{\phi}(x)\right] \tag{16}$$

for admissible gauge fields satisfying (15) for $\forall x$, and

$$S_{\mathrm{SYM}}^{(\mathrm{lat\ A})} = \infty \tag{17}$$

(meaning that the Boltzmann weight vanishes) otherwise. Since the action (16) has the form Q acting to gauge invariant terms, it can be seen to be invariant under Q transformations from the nilpotency (8).

After the Q action in RHS of (16), we have

$$\begin{aligned} S_{\mathrm{SYM}}^{(\mathrm{lat\ A})} = \frac{1}{g_0^2}\sum_x \mathrm{tr}\Bigg[& \frac{1}{4}\widehat{\Phi}(x)^2 + a^2\sum_\mu \nabla_\mu\phi(x)\nabla_\mu\bar{\phi}(x) + i\chi(x)Q\widehat{\Phi}(x) \\ & + i\sum_\mu \psi_\mu(x)a\nabla_\mu\eta(x) - D(x)^2 \\ & + \frac{1}{4}[\phi(x), \bar{\phi}(x)]^2 - \chi(x)[\phi(x), \chi(x)] - \frac{1}{4}\eta(x)[\phi(x), \eta(x)] \\ & - \sum_\mu \psi_\mu(x)\psi_\mu(x)\left(\bar{\phi}(x) + U_\mu(x)\bar{\phi}(x+\hat{\mu})U_\mu(x)^{-1}\right)\Bigg]. \end{aligned} \tag{18}$$

Without the admissibility (15) and the denominator of $\widehat{\Phi}$, gauge kinetic terms would be proportional to

$$-\,\mathrm{tr}\,(U_{01}(x) - U_{10}(x))^2 = \mathrm{tr}\,(2 - U_{01}(x)^2 - U_{10}(x)^2), \tag{19}$$

leading to the classical minima

$$U_{01}(x) = \begin{pmatrix} \pm 1 & & \\ & \ddots & \\ & & \pm 1 \end{pmatrix} \qquad \text{(up to gauge transformations)} \tag{20}$$

for $\forall x$. The minima have a huge degeneracy, and we should take into account fluctuations around all the minima, although the weakly coupled continuum theory is reproduced as fluctuations around a single minimum $U_{01}(x) = 1$. Thus, the connection of the lattice model to the continuum theory would become unclear. To avoid such situation and to single out the vacuum $U_{01}(x) = 1$, the admissibility and $\widehat{\Phi}$ are employed here [2].

In the lattice perturbation theory, we can show that the lattice model does not need any fine-tuning to obtain the target continuum theory with the full $\mathcal{N} = (2, 2)$ SUSY. Moreover, by computer simulation, ref. [8] performed a nonperturbative check that the full $\mathcal{N} = (2, 2)$ SUSY is restored in the continuum limit of the lattice theory with $G = SU(2)$.

The FI and ϑ-terms (5) can be latticized maintaining the Q invariance [3, 4].

4. SYM part (B model twist)

In B-model twist case, we consider $G = U(N)$ gauge group. It is convenient to take linear combinations of field variables as

$$\begin{aligned} \psi'_0 &= \frac{1}{\sqrt{2}}(\lambda_L - \lambda_R), \quad \psi'_1 = \frac{i}{\sqrt{2}}(\lambda_L + \lambda_R), \\ \chi' &= \frac{1}{\sqrt{2}}(\bar{\lambda}_L + \bar{\lambda}_R), \quad \eta' = \frac{i}{\sqrt{2}}(\bar{\lambda}_L - \bar{\lambda}_R), \\ X_0 &= -\frac{1}{2}(\phi + \bar{\phi}), \quad X_1 = -\frac{i}{2}(\phi - \bar{\phi}), \\ D' &= D - D_\mu X_\mu, \quad \mathcal{A}_\mu = A_\mu - iX_\mu, \quad \mathcal{A}_\mu^\dagger = A_\mu + iX_\mu. \end{aligned} \tag{21}$$

Then, Q'-SUSY transforms the fields as

$$\begin{aligned} &Q'\mathcal{A}_\mu = 0, \\ &Q'\mathcal{A}_\mu^\dagger = 2\psi'_\mu, \qquad Q'\psi'_\mu = 0, \\ &Q'\chi = -i\mathcal{F}_{01}, \\ &Q'\eta' = D', \qquad Q'D' = 0, \end{aligned} \tag{22}$$

where $\mathcal{F}_{01}$ is the field strength of the complexified gauge potentials $\mathcal{A}_\mu$. Note that Q' is exactly nilpotent: $(Q')^2 = 0$.

The action (6) can be recast into Q'-exact form:

$$S^{(E)}_{\text{SYM}} = Q' \frac{1}{g^2} \int d^2x \ \text{tr} \left[-\eta'(D' + 2D_\mu X_\mu) + i\chi' \mathcal{F}^\dagger_{01} \right]. \tag{23}$$

It is manifestly symmetric under $\lambda_R \leftrightarrow \lambda_L$ and $\bar{\lambda}_R \leftrightarrow \bar{\lambda}_L$, but not $\lambda_R \leftrightarrow \bar{\lambda}_L$ and $\lambda_L \leftrightarrow \bar{\lambda}_R$.

4.1. Q'-SUSY on lattice (B model twist)

In the lattice formulation based on B twist, we introduce positive definite hermitian variables $V_\mu(x) = e^{-aX_\mu(x)}$ defined on sites for scalar fields. Then, $V_\mu(x)U_\mu(x) \sim e^{ia\mathcal{A}_\mu(x)}$ near the continuum limit.

The nilpotent Q'-SUSY can be realized on the lattice as

$$\begin{aligned} &Q'U_\mu(x) = i\psi'_\mu(x)U_\mu(x), \qquad Q'\psi_\mu(x) = i\psi_\mu(x)\psi_\mu(x), \\ &Q'V_\mu(x) = i\psi'_\mu(x)V_\mu(x), \\ &Q'\chi'(x) = 1 - U_{01}(V^{-1}U)(x), \\ &Q'\eta'(x) = D'(x), \qquad Q'D'(x) = 0, \end{aligned} \tag{24}$$

where $U_{01}(V^{-1}U)(x)$ means a plaquette variable constructed by the links $V_\mu^{-1}U_\mu \in GL(N,\mathbf{C})$.

4.2. Q′ invariant lattice action

On the analogy of the continuum case (23), we define the lattice action by

$$S_{\rm SYM}^{(\rm lat\,B)} = Q'\frac{1}{g_0^2}\sum_x \mathrm{tr}\left[-\eta'(x)(D'(x)+W(x)) + \chi'(x)\left(1-U_{01}(V^{-1}U)(x)\right)^\dagger\right], \tag{25}$$

where $W(x)$ is a lattice counterpart of $2D_\mu X_\mu$ chosen as

$$W(x) \equiv 2\sum_x \left[V_\mu(x)^{-1} + U_\mu(x-\hat\mu)^{-1}V_\mu(x-\hat\mu)U_\mu(x-\hat\mu) - 2\right]. \tag{26}$$

As a result of the Q' action in RHS of (25), we have

$$\begin{aligned} S_{\rm SYM}^{(\rm lat\,B)} &= \frac{1}{g_0^2}\sum_x \mathrm{tr}\Big[-D'(x)^2 - D'(x)W(x) + \left|1-U_{01}(V^{-1}U)(x)\right|^2 \\ &\qquad +\chi'(x)Q'U_{10}(VU)(x) + \eta'(x)Q'W(x)\Big]. \end{aligned} \tag{27}$$

The minimum of the bosonic part should satisfy

$$W(x) = 0, \qquad U_{01}(V^{-1}U)(x) = 1, \tag{28}$$

from which we obtain [2]

$$0 = \sum_x \mathrm{tr}\,W(x) = \sum_x \mathrm{tr}\Big[\left(V_\mu(x)^{-1/2} - V_\mu(x)^{1/2}\right)^2\Big], \tag{29}$$

leading to $V_\mu(x) = 1$, and thus $U_{01}(x) = 1$. In this lattice formulation, a single desirable minimum $U_{01}(x) = 1$ is obtained for gauge fields, and flat directions for scalar fields are lifted.

5. Matter part

As seen in the SYM part, Q-SUSY respects the symmetry under $\lambda_R \leftrightarrow \bar\lambda_L$ and $\lambda_L \leftrightarrow \bar\lambda_R$, while Q'-SUSY respects the symmetry under $\lambda_R \leftrightarrow \lambda_L$ and $\bar\lambda_R \leftrightarrow \bar\lambda_L$. Correspondingly, for the matter part, Q-SUSY respects the symmetry under

$$\psi_{\pm IR} \leftrightarrow \bar\psi_{\mp IL}, \qquad \psi_{\pm IL} \leftrightarrow \bar\psi_{\mp IR}, \tag{30}$$

while Q'-SUSY respects the symmetry under

$$\psi_{\pm IR} \leftrightarrow \psi_{\pm IL}, \qquad \bar\psi_{\pm IR} \leftrightarrow \bar\psi_{\pm IL}. \tag{31}$$

[2] Note that $V_\mu(x)^{1/2}$ is unambiguously defined, since $V_\mu(x)$ is positive definite.

Journal of Physics: Conference Series **462** (2013) 012051 doi:10.1088/1742-6596/462/1/012051

5.1. *Q invariant lattice formulation (A model twist)*

In ref. [3], the Wilson terms were introduced to suppress the bosonic and fermionic doublers. Note that the Wilson terms play a role of $\mathcal{O}(1/a)$ mass terms coupling Φ_{+I} with Φ_{-I} and coupling $\Phi^\dagger_{+I}$ with $\Phi^\dagger_{-I}$. So, it is possible to maintain the Q invariance only when $n_+ = n_-(\equiv n)$. The Wilson terms breaks chiral flavor symmetry $U(n_+) \times U(n_-)$ to its diagonal subgroup $U(n)$.

Ref. [4] applied Ginsparg-Wilson formulation [9] to realize both of Q-SUSY and the chiral flavor symmetry for general $n_\pm$. As an interesting feature of this construction, holomorphic structure of superpotential is not ruined by lattice artifacts.

When $n_+ \neq n_-$, $U(1)_A$ R-symmetry is anomalous. The Ginsparg-Wilson formulation reproduces the correct anomaly in the form of the trace of the Ginsparg-Wilson Dirac operator $\hat{D}$:

$$\mathrm{Tr}\left(\gamma_3 a\hat{D}\right) \simeq \frac{1}{\pi}\int d^2x \ \mathrm{tr}\ F_{01}, \tag{32}$$

which arises from the $U(1)_A$ transformation to path-integral measure of the lattice theory.

When using the Ginsparg-Wilson formulation, ϵ appearing in the admissibility condition is chosen as follows. For $G = U(N)$ without ϑ-term, $0 < \epsilon < \frac{1}{5}$ if $N = 1, \cdots, 100$, and $0 < \epsilon < \frac{2}{\sqrt{N}}$ if $N \geq 101$. For $G = U(N)$ with ϑ-term, $0 < \epsilon < \frac{1}{5}$ if $N = 1, \cdots, 25$, and $0 < \epsilon < \frac{1}{\sqrt{N}}$ if $N \geq 26$. For $G = SU(N)$, $0 < \epsilon < \frac{1}{5}$ if $N = 2, 3, \cdots, 31$, and $0 < \epsilon < 2\sin\left(\frac{\pi}{N}\right)$ if $N \geq 32$.

5.2. *Q' invariant lattice formulation (B model twist)*

From the symmetry (31) for fermions, Q' invariant formulation seems more natural to preserve the chiral flavor symmetry.

As discussed in ref. [5], in the case $n_- = 0$ (only fundamental matters) or $n_+ = 0$ (only anti-fundamental matters), the Q' invariant lattice action can be shown to have a single minimum at $V_\mu(x) = 1, U_{01}(x) = 1$.

Since the Q' invariant lattice action is not required to impose the admissibility condition, it takes simpler form compared with the Q invariant action. However, it can be applied to the gauge group $G = U(N)$, not $G = SU(N)$ at present. It is difficult to introduce superpotential with preserving Q'-SUSY. It is because the anti-holomorphic part of superpotential $\overline{W}$ is written as Q'-exact while the holomorphic part W is not as Q'-exact.

6. Summary and discussion

We have presented the two kinds of lattice formulations of two-dimensional $\mathcal{N} = (2, 2)$ SQCD preserving one supercharge on the lattice.

One is the Q invariant formulation, which is based on A model twist. The lattice model can be constructed for $G = U(N)$ or $SU(N)$. Moreover, by employing the Ginsparg-Wilson operator which realizes exact chiral flavor symmetry on the lattice, it is possible to formulate the model for general $n_\pm$ and for general superpotential.

The other is the Q' invariant formulation, based on B model twist. It overcomes difficulty of degenerate minima and flat directions without using the admissibility condition, and thus the lattice action has a simpler form compared with the case of the Q invariant formulation. However, its applicability is restricted to $G = U(N)$ and to $(n_+, n_-) = (n_+, 0)$ or $(0, n_-)$ at present.

There are different lattice formulations of SYM theories based on orbifolding-deconstruction approach from matrix models [10, 11, 12, 13]. Ref. [14] pointed out that the supercharge preserved in such formulations corresponds to Q' from B model twist. Since bifundamental fields play a key role in the formulation, it is applicable to $G = U(N)$, but difficult to $G = SU(N)$, because it is impossible to impose a traceless condition on fundamental fields in a gauge invariant way.

In ref. [15], two-dimensional $\mathcal{N} = (2,2)$ SQCD has been constructed in the orbifolding-deconstruction approach. Let us start from matrix models with eight supercharges, corresponding to $\mathcal{N} = (4,4)$ in two dimensions. By orbifolding and deconstruction, $U(N_c) \times U(N_f)$ quiver gauge theory in two dimensions is obtained. If $U(N_f)$ adjoint fields are killed by hand, $U(N_f)$ gauge symmetry reduces to global flavor symmetry with Q'-SUSY intact. Then, SQCD with $n_+ = n_-(= N_f)$ and without superpotential can be constructed on two-dimensional lattice. The lattice theory has somewhat different structure from our Q' invariant formulation.

For future work, a number of interesting subjects are waiting for us. Firstly, the Ginsparg-Wilson formulation to two-dimensional $\mathcal{N} = (2,2)$ SQCD [4] can be applied to the lattice study of gauged linear sigma models. Due to the exact chiral structure, we can separately couple matter multiplets belonging to different representations of G to different chiral sectors of the model. Two-dimensional $\mathcal{N} = (2,2)$ SQCD models with various superpotentials have been analytically investigated based on the effective twisted superpotentials [16, 17, 18]. The number of the vacua or the Witten index of the models has been computed for various $N, n_\pm$, and an analog of the Seiberg duality in four dimensions has been discussed based on the symmetry of Grassmannians. It will be worth confirming those properties and exploring new aspects, which are not yet investigated there, from the first principle computation using this lattice formulation.

Secondly, concerning nonperturbative aspects of superstring theory, 2d $\mathcal{N} = (8,8)$ SYM with $G = U(N)$ is known as matrix string theory [19], which is conjectured to give type IIA superstring theory in low energy regime at large N. It will be important to construct its lattice formulation with no fine-tuning required.

Finally, we hope that the above lattice formulation will give some insights to explore cases of higher-dimensional theories.

Acknowledgments

The author thanks Tetsuo Horigane, Daisuke Kadoh, Issaku Kanamori, Hiroki Kawai, Yoshio Kikukawa and Hiroshi Suzuki for lots of stimulating and enjoyable discussion and for collaboration. The work of the author is partly supported by Grant-in-Aid for Scientific Research (C), 21540290.

References

[1] Sugino F 2004 *J. High Energy Phys.* JHEP **0401**(2004)015 (Preprint hep-lat/0311021)
[2] Sugino F 2004 *J. High Energy Phys.* JHEP **0403**(2004)067 (Preprint hep-lat/040101)
[3] Sugino F 2009 *Nucl. Phys.* B **808** 292 (Preprint 0807.2683 [hep-lat])
[4] Kikukawa Y and Sugino F 2009 *Nucl. Phys.* B **819** 76 (Preprint 0811.0916 [hep-lat])
[5] Kadoh D, Sugino F and Suzuki H 2009 *Nucl. Phys.* B **820** 99 (Preprint 0903.5398 [hep-lat])
[6] Witten E 1988 *Commun. Math. Phys.* **117** 353
[7] Witten E 1991 *Int. J. Mod. Phys.* A **6** 2775
[8] Kanamori I and Suzuki H 2009 *Nucl. Phys.* B **811** 420 (Preprint 0809.2856 [hep-lat])
[9] Ginsparg P H and Wilson K G 1982 *Phys. Rev.* D **25** 2649
[10] Kaplan D B, Katz E and Ünsal M 2003 *J. High Energy Phys.* JHEP **0305**(2003)037 (Preprint hep-lat/0206019)
[11] Cohen A G, Kaplan D B, Katz E and Ünsal M 2003 *J. High Energy Phys.* JHEP **0308**(2003)024 (Preprint hep-lat/0302017)
[12] Cohen A G, Kaplan D B, Katz E and Ünsal M 2003 *J. High Energy Phys.* JHEP **0312**(2003)031 (Preprint hep-lat/0307012)
[13] Kaplan D B and Ünsal M 2005 *J. High Energy Phys.* JHEP **0509**(2005)042 (Preprint hep-lat/0503039)
[14] Catterall S 2008 *J. High Energy Phys.* JHEP **0801**(2008)048 (Preprint 0712.2532 [hep-th])
[15] Matsuura S 2008 *J. High Energy Phys.* JHEP **0807**(2008)127 (Preprint 0805.4491 [hep-th])
[16] Witten E 1993 *Nucl. Phys.* B **403** 159 (Preprint hep-th/9301042)
[17] Hori K and Tong D 2007 *J. High Energy Phys.* JHEP **0705**(2007)079 (Preprint hep-th/0609032)
[18] Hanany A and Hori K 1998 *Nucl. Phys.* B **513** 119 (Preprint hep-th/9707192)
[19] Dijkgraaf R, Verlinde E P and Verlinde H L 1997 *Nucl. Phys.* B **500** 43 (Preprint hep-th/9703030)

Some illustrative examples of Argyres-Seiberg-Gaiotto duality

Yuji Tachikawa
Institute for Advanced Study, 1 Einstein Drive, Princeton, NJ 08540 USA
E-mail: yujitach@ias.edu

Abstract. In the past few years we saw significant advancement in understanding the strong-weak duality of $\mathcal{N}=2$ supersymmetric theories in four dimensions, initiated by the seminal works by Argyres and Seiberg [1], and then greatly generalized by Gaiotto [2]. A most surprising feature of this new duality is that the dual of a normal supersymmetric gauge theory can involve a strange mixture of weakly-coupled gauge multiplets and non-Lagrangian superconformal field theories. Here we illustrate these new dualities by reviewing known examples and then by demonstrating new dual pairs. This work is based on the author's collaboration with Benini and Benvenuti [3].

1. Introduction

The maximally supersymmetric Yang-Mills theory with SU(N) gauge group, which is with $\mathcal{N}=4$ supersymmetry, is known to be invariant under the S-duality. Let us define the complexified gauge coupling τ to be $\tau=\theta/2\pi+4\pi i/g^2$ where θ is the theta angle and g the coupling constant. Then the theory is invariant under the transformation $S:\tau\to-1/\tau$, accompanied by the exchange of W-bosons and monopoles. The theory has more mundane invariance $T:\tau\to\tau+1$ which shifts the theta angle by 2π without affecting anything. As is well-known, S and T generate the group SL$(2,\mathbb{Z})$ and any point on the upper half plane on which τ lives can be mapped by its action into the fundamental region shown with shaded gray in Fig. 1. The fundamental region does not touch the real axis, which means that you can always go to a duality frame where the theory is not infinitely strongly coupled.

The situation is the same with $\mathcal{N}=2$ supersymmetric SU(2) gauge theory with four flavors in the fundamental representation. Here it is customary to redefine the coupling constant by a factor of two, and to take $\tau=\theta/\pi+8\pi i/g^2$. This theory is known to be invariant under the two transformations S and T above.

In general, $\mathcal{N}=2$ supersymmetric SU(N) gauge theory with $N_f=2N$ hypermultiplets in the fundamental representation has zero beta function. It is believed to have the strong-weak duality under the transformation S, judging from the behavior of the Seiberg-Witten curve, for example. However, the theory does not have the invariance under T, because it shifts the theta angle by π. The SU(2) gauge theory is special in this sense, because the doublet and the anti-doublet representation are the same in this special case. Thus one could achieve the invariance under T by combining the shift of the theta angle by π and the charge conjugation. But this is not available for $N>2$, and the duality group is generated by S and T^2. The fundamental

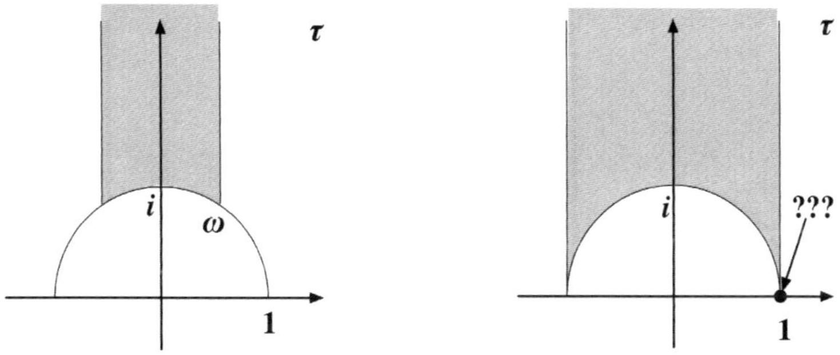

Figure 1. Fundamental regions. Left: $\mathcal{N} = 4$ SU(N), and $\mathcal{N} = 2$ SU(2) with $N_f = 4$. Right: $\mathcal{N} = 2$ SU(3) with $N_f = 6$.

region then touches the real axis, see Fig. 1. Therefore there is an infinitely strongly coupled point in the space of the coupling constant.

Argyres and Seiberg [1] proposed the dual description of SU(3) with six flavors which solves this conundrum. They argued, using various field theoretical analyses, that the SU(3) theory with coupling τ is equivalent to an SU(2) theory with coupling constant $\tau' = 1/(1-\tau)$ coupled to one flavor of doublet and also to Minahan and Nemeschansky's superconformal theory with flavor symmetry E_6. Then the infinitely-strongly-coupled limit $\tau = 1$ of the original theory becomes a weakly-coupled limit of this dual description. Their proposal was ground-breaking: no one had ever thought of coupling a dynamical gauge field to those mysterious superconformal theories before their work.

At that time, however, their observation seemed rather esoteric and not very relevant to our general understanding of the dynamics of $\mathcal{N} = 2$ theories. It was Gaiotto [2] who showed that, on the contrary, this S-duality involving non-Lagrangian superconformal theories coupled to dynamical gauge fields is a common phenomenon which naturally follows from the analysis of M5-branes.

2. Review of the construction via M5-branes

2.1. $\mathcal{N} = 4$

Let us first consider the realization of $\mathcal{N} = 4$ SU(N) theory in terms of M5-branes. This is surprisingly simple: one only needs to compactify N M5-branes on T^2. First, M5-branes on S^1 is almost by definition give D4-branes, on which lives maximally supersymmetric SU(N) gauge theory in 5 dimensions. Then, a further compactification on S^1 gives maximally supersymmetric SU(N) theory in 4 dimensions when one takes the low energy limit and makes all the Kaluza-Klein modes decouple. One finds that the complexified gauge coupling constant of the resulting four-dimensional theory is given by the modulus of the torus T^2. Now, the way one assigns the modulus to a given torus is not unique: the torus with the modulus τ is equivalent to the torus with the modulus $-1/\tau$. This statement then turns into the equivalence of the $\mathcal{N} = 4$ SU(N) theory at coupling τ and at coupling $-1/\tau$, see Fig. 2.

2.2. $\mathcal{N} = 2$ SU(2) *with four flavors*

Then let us discuss how the S-duality of $\mathcal{N} = 2$ SU(2) theory with four flavors can be geometrically understood using M5-branes. Gaiotto showed in [2] that this theory can be constructed by wrapping two M5-branes on a sphere with four punctures, see Fig 3. The exponential q of the coupling constant is given by the cross ratio of the four points. One can also think of the torus which is a double cover of the same sphere with branch points at these

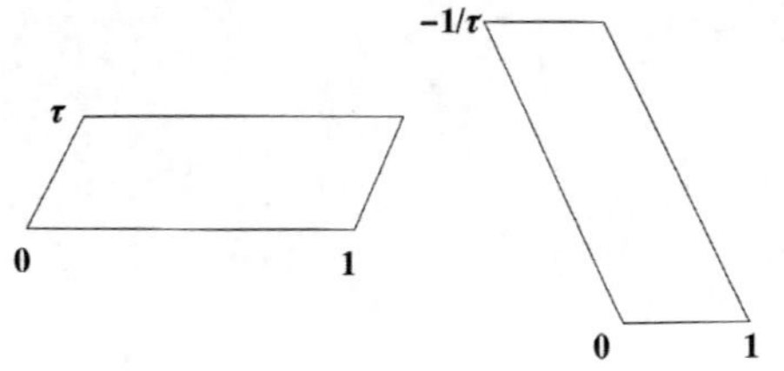

Figure 2. Two ways to label the same torus.

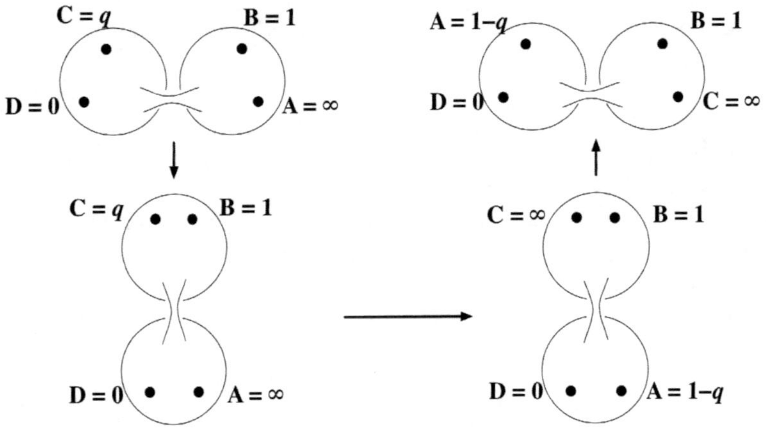

Figure 3. Relabeling of a sphere with four punctures. Step 1: split the sphere vertically. Step 2: perform the coordinate change. Step 3: rotate the sphere by 90 degrees.

four punctures. Then the low-energy coupling constant is given by the modulus of the torus. In either case, there is no unique way to associate a coupling constant to a given four-punctured sphere. This translates to the S-duality of the SU(2) theory with four flavors.

In fact one can read off more about the S-duality from this construction. Four hypermultiplets in the doublet representation transform under the vector representation of SO(8) flavor symmetry. Each of four punctures can be associated to a particular SU(2) subgroup of this flavor symmetry. First let us split four hypermultiplets into two pairs of two hypermultiplets, with flavor symmetry $\mathrm{SO}(4) \times \mathrm{SO}(4)$. Then we decompose each SO(4) into $\mathrm{SU}(2) \times \mathrm{SU}(2)$. Call the punctures at $z = \infty$, 1, q and 0 respectively A, B, C and D, and call the corresponding flavor symmetry $\mathrm{SU}(2)_{A,B,C,D}$. Then the vector representation of SO(8) transforms as

$$2_A \otimes 2_B \oplus 2_C \otimes 2_D. \tag{1}$$

Under the S-duality which sends $q \to 1 - q$, the positions of the punctures are reshuffled, see Fig 3. Now the hypermultiplets transform as

$$2_A \otimes 2_D \oplus 2_B \otimes 2_C, \tag{2}$$

which is the decomposition of a chiral spinor representation of SO(8). This clearly exemplifies that the S-duality of this theory involves the triality of the flavor symmetry SO(8). It is surprising that this simple explanation was not known until 2009.

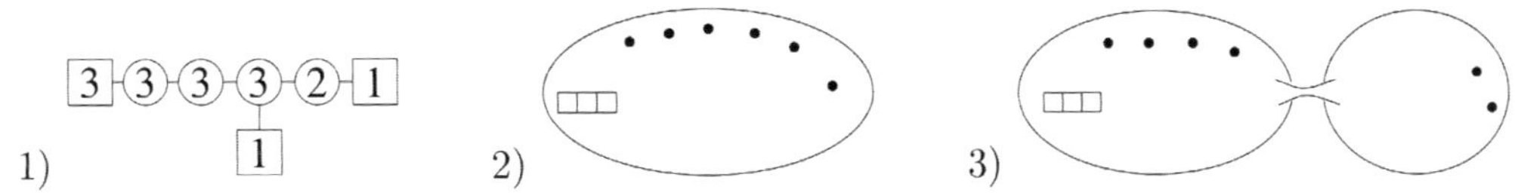

Figure 4. Example of a linear quiver theory and its realization by M5 branes. 1) The gauge theory. 2) M5-branes on a sphere with punctures. 3) The limit where SU(2) becomes weak.

2.3. General procedure

In general, one can consider a superconformal gauge theory with the gauge group

$$\mathrm{SU}(d_1) \times \mathrm{SU}(d_2) \times \cdots \times \mathrm{SU}(d_{n-1}) \times \mathrm{SU}(d_n) \,, \tag{3}$$

with a bifundamental hypermultiplet between each pair of consecutive gauge groups $\mathrm{SU}(d_a) \times \mathrm{SU}(d_{a+1})$. We put $k_a = 2d_a - d_{a-1} - d_{a+1}$ extra fundamental hypermultiplets for $\mathrm{SU}(d_a)$ to make it superconformal. A theory which falls within this construction is called a superconformal linear quiver theory.

Since k_a is non-negative, we have

$$d_1 < d_2 < \cdots < d_l = \cdots = d_r > d_{r+1} > \cdots > d_n \,. \tag{4}$$

We denote $N = d_l = \cdots = d_r$; we refer to the parts to the right of d_r and to the left of d_l as two tails of this superconformal quiver. k_a is non-negative, which means that $d_a - d_{a+1}$ is monotonically non-decreasing for $a > r$. Thus we can associate naturally a Young diagram to the tail by requiring that it has a row of width $d_a - d_{a+1}$ for each $a \geq r$. In the following, a Young diagram with the columns of height $3, 3, 2$ will be denoted as $\{3^2, 2\}$, etc.

Gaiotto showed that this gauge theory can be obtained by wrapping N M5-branes on a sphere with $n+3$ punctures. The punctures are at

$$z = \infty,\ 1,\ q_1,\ q_1 q_2,\ \ldots,\ q_1 \cdots q_n,\ 0. \tag{5}$$

The punctures at $z \neq 0, \infty$ are all of the same type, and called the simple punctures. The punctures at $z = \infty, 0$ encode the structure of the left and the right tails, respectively, and thus labeled by the corresponding Young diagrams. q_i is the exponentiated complexified gauge coupling of the i-th SU gauge group. A puncture at $z = z_i$ controls how the worldvolume fields on the coincident M5-branes diverge at $z = z_i$, much as a 't Hoot loop in a gauge theory controls how the gauge field diverges close to the loop. The type of the puncture specifies exactly how they diverge. For more details, see [2].

As an example, consider the quiver in Fig. 4, 1). A circle with a number d stands for $\mathrm{SU}(d)$ gauge group, and a box with a number k stands for k extra fundamental hypermultiplets associated to the gauge group connected by a line to the box in question. The corresponding configuration of punctures on 3 M5-branes is given in Fig. 4, 2). There, the puncture at $z = 0$ for the tail SU(3)–SU(2)–U(1) is denoted by the same symbol as the one for the simple punctures, because the worldvolume fields around a simple puncture and the puncture associated to this tail behave in a same way. This is as it should be, because the S-duality of the rightmostSU(2) can exchange these two punctures.

Now let us make the coupling of the SU(2) gauge group very weak. Equivalently, we take q of this gauge group to be very small. This creates a long neck between the second and the third simple punctures from the right, see Fig. 4, 3). In other words, if we split off two simple punctures on a sphere from the rest of the Riemann surface on which three M5 branes wrap,

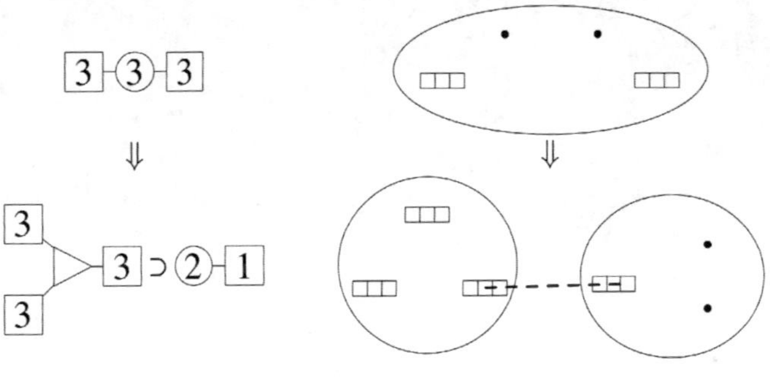

Figure 5. S-duality involving Minahan-Nemeschansky's E_6 SCFT. The configuration of punctures on M5-branes is shown on the right.

there is a weakly coupled SU(2) gauge group with one extra fundamental hypermultiplet coupled to it. In general, if we start by wrapping N M5-branes and split off a sphere with $N-1$ simple punctures from the rest of the Riemann surface, we generate a tail of the form

$$\mathrm{SU}(N-1) \times \mathrm{SU}(N-2) \times \cdots \times \mathrm{SU}(3) \times \mathrm{SU}(2) \tag{6}$$

with one extra fundamental hypermultiplet for the rightmost SU(2), and the coupling constant of the leftmost $\mathrm{SU}(N-1)$ becomes very weak.

3. Application

3.1. E_6

Using the procedure reviewed in the previous section, we can now readily find that the S-dual of a linear quiver often involves non-trivial superconformal field theory. The first example is the original duality of Argyres and Seiberg. SU(3) theory with six flavors can be thought of as a linear quiver of the form shown in the upper row of Fig. 5. The infinite coupling limit involves making two simple punctures collide. In a different coordinate system, this implies that we split off a sphere with two simple punctures. As we saw, this generates a weakly-coupled SU(2) gauge group with one extra hypermultiplet, see the bottom row of Fig. 5.

Let us study what the rest of the Riemann surface stands for. We have three punctures of the same type $\{1^3\}$, which makes $\mathrm{SU}(3)^3$ flavor symmetry manifest. Now, two out of the three punctures came from the punctures which we originally had, and they represented two pairs of three fundamental hypermultiplets. Therefore $\mathrm{SU}(3)^2$ flavor symmetry is a subgroup of SU(6) flavor symmetry of the six fundamental hypermultiplets. From the standpoint of wrapped M5-branes, however, there is no difference between two punctures originally present and the final puncture generated by the decoupling of the tail. Therefore, any two out of three SU(3) flavor symmetry should enhance to SU(6) symmetry. This is possible only when the $\mathrm{SU}(3)^3$ flavor symmetry is in fact inside E_6 flavor symmetry, see Fig. 6. This was discussed in [2].

3.2. E_7

Next, let us consider the quiver shown in the first line of Fig. 7. The gauge group is $\mathrm{SU}(4)\times\mathrm{SU}(2)$ with the bifundamental hypermultiplets charged under the two SU factors, and there are in addition six fundamental hypermultiplets for the node SU(4). In term of four M5-branes wrapping a sphere, we have three simple punctures, one puncture of type $\{1^4\}$ and one of type $\{2^2\}$. We can go to a limit where a sphere with three simple punctures splits. A dual superconformal tail with gauge groups $\mathrm{SU}(3) \times \mathrm{SU}(2)$ appears. After the neck is pinched off, we have a sphere with one puncture of type $\{2^2\}$ and two punctures of type $\{1^4\}$. This

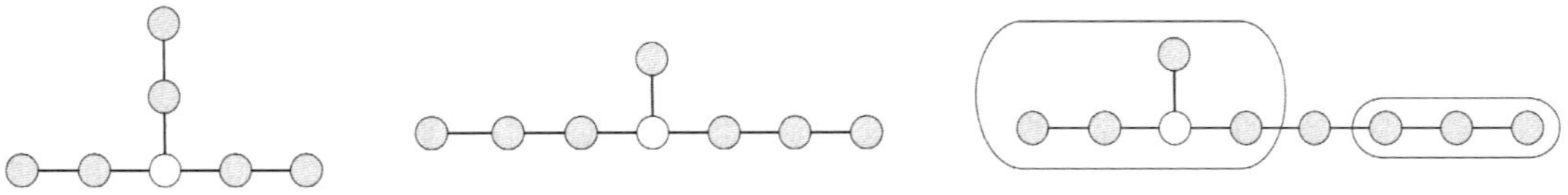

Figure 6. Dynkin diagram of $E_{6,7,8}$. The E_6 diagram for E_6 shows three SU(3) subgroups, any two out of which enhance to SU(6) via the node at the center. The E_7 diagram shows the subgroup $\mathrm{SU}(2) \times \mathrm{SU}(4)^2$. The E_8 diagram shows the subgroup $\mathrm{SO}(10) \times \mathrm{SU}(4)$ and the subgroup $\mathrm{SU}(2) \times \mathrm{SU}(3) \times \mathrm{SU}(6)$.

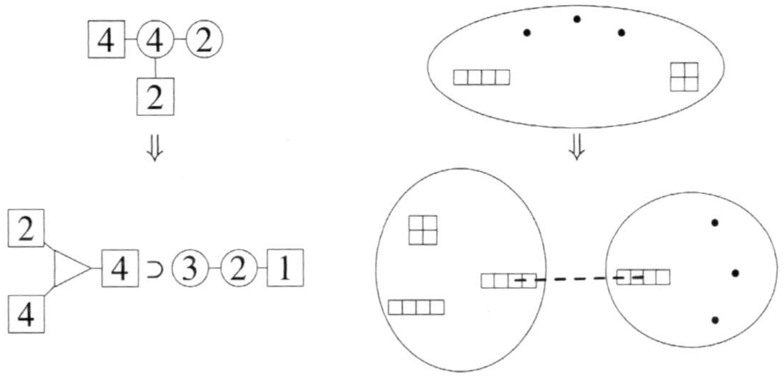

Figure 7. S-duality involving Minahan-Nemeschansky's E_7 SCFT.

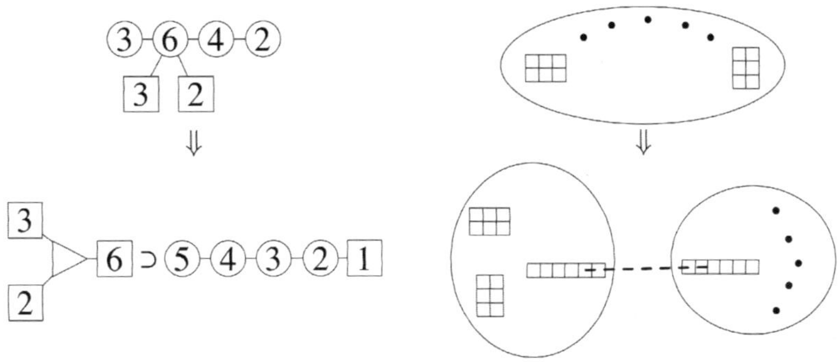

Figure 8. First example of S-duality involving Minahan-Nemeschansky's E_8 SCFT.

description shows the flavor symmetry $\mathrm{SU}(2) \times \mathrm{SU}(4)^2$. In the original description, it is clear that $\mathrm{SU}(2) \times \mathrm{SU}(4)$ enhances to SU(6). From the point of view of the M5-branes, the two SU(4) cannot be distinguished. Therefore, the other combination of $\mathrm{SU}(2) \times \mathrm{SU}(4)$ should also enhance to SU(6). This is only possible when the total flavor symmetry enhances to E_7, see Fig. 6. This was discussed in [3].

3.3. E_8

To discuss S-dualities involving superconformal theory with E_8 flavor symmetry, let us consider the quiver with the gauge group $\mathrm{SU}(3) \times \mathrm{SU}(6) \times \mathrm{SU}(4) \times \mathrm{SU}(2)$ with bifundamental hypermultiplets between the consecutive SU factors; one has in addition five fundamental hypermultiplets for SU(6), see the first line of Fig. 8. We can go to a limit where a

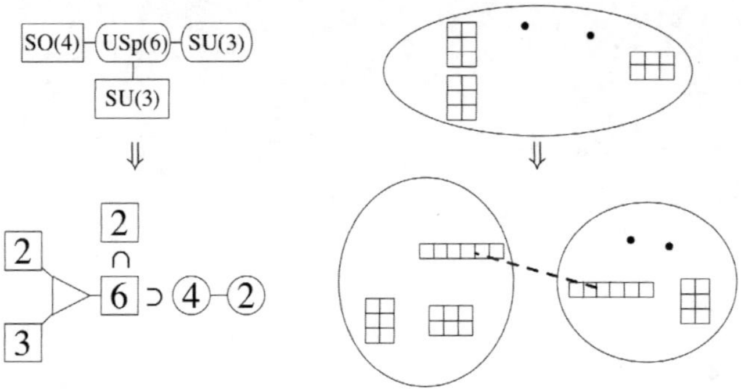

Figure 9. Second example of S-duality involving Minahan-Nemeschansky's E_8 SCFT.

sphere with five simple punctures splits off. A dual superconformal tail with gauge groups $SU(5) \times SU(4) \times SU(3) \times SU(2)$ appears. We tune the gauge coupling of the SU(5) group to zero, leaving a theory which is given by wrapping six M5-branes on a sphere with one puncture of type $\{1^6\}$, one of type $\{2^3\}$ and another of type $\{3^2\}$. This description shows the flavor symmetry $SU(2) \times SU(3) \times SU(6)$. In the original description, it is clear that $SU(3) \times SU(2)$ enhances to SU(5). With a small effort, one can show that the flavor symmetry further enhances to E_8. This was discussed in [3].

Finally, let us consider a new example, whose quiver is shown in the first line of Fig. 9. This theory has the gauge group $USp(6) \times SU(3)$, with five fundamental hypermultiplet for the USp(6), and a hypermultiplet in the bifundamental of $USp(6) \times SU(3)$. As briefly discussed in [2] and detailed in [4], we get a tail of the form $SU(2N)$–$USp(2N)$ with two extra fundamental hypermultiplet for $USp(2N)$ when we put two punctures of type $\{N^2\}$ on a sphere. From this perspective it is natural to split five hypermultiplets into two and three, and this quiver has the realization as six M5-branes wrapped on a sphere with two simple punctures, two punctures of type $\{3^2\}$, and one of type $\{2^3\}$. By splitting off a sphere with two simple punctures and one of type $\{2^3\}$, a tail with the gauge group $SU(4) \times SU(2)$ is generated, with the gauge coupling of SU(4) becoming weak. We are again left with a theory corresponding to a sphere with one puncture of type $\{1^6\}$, one of type $\{2^3\}$ and another of type $\{3^2\}$.

The enhancement of the flavor symmetry to E_8 is more transparent in this description. In the original description we have SO(10) flavor symmetry acting on five fundamental hypermultiplets, of which the subgroup $SU(2) \times SU(2) \times SU(3)$ is manifest in the dual description. Here, one SU(2) is directly carried by the puncture $\{3^2\}$, but the other SU(2) toghether with the SU(4) to which the gauge fields couple is a subgroup of SU(6) carried by the puncture $\{1^6\}$. Therefore, the $SU(2) \times SU(3) \times SU(6)$ flavor symmetry of the theory corresponding to this three-punctured sphere should be such that the subgroup $SU(2) \times SU(3) \times SU(2)$ enhances to SO(10). This is only possible when the total flavor symmetry is in fact E_8, see Fig. 6.

Now, let us determine the effective numbers n_v and n_h of the hyper- and vector multiplets of the E_8 theory. In the first realization, the original quiver gauge theory with gauge group $SU(3) \times SU(6) \times SU(4) \times SU(2)$ had

$$n_v = 8 + 35 + 15 + 3 = 61, \qquad n_h = 18 + 30 + 24 + 8 = 80. \tag{7}$$

The S-dual description has

$$n_v = n_v[E_8] + 24 + 16 + 8 + 3, \qquad n_h = n_h[E_8] + 20 + 12 + 6 + 2. \tag{8}$$

Equating (7) and (8), we conclude

$$n_v[E_8] = 11, \qquad n_h[E_8] = 40. \tag{9}$$

In the second realization, the original theory with gauge group USp(6) $\times$ SU(3) had

$$n_v = 21 + 8 = 29, \qquad n_h = 12 + 18 + 18 = 48. \tag{10}$$

The S-dual description has

$$n_v = n_v[E_8] + 15 + 3, \qquad n_h = n_h[E_8] + 8. \tag{11}$$

Equating (10) and (11), we again find (9), which shows the consistency of our construction.

Another S-duality involving E_8 theory had been discovered in [5]. The original theory was USp(6) gauge theory with 11 half-hypers in the fundamental representation and with a half-hyper in the three-index antisymmetric traceless tensor. The dual is the E_8 theory coupled to an SO(5) gauge multiplet, via the subgroup SO(5) $\times$ SO(11) $\subset$ SO(16) $\subset E_8$. Here, the original had

$$n_v = 21, \qquad n_h = 33 + 7 = 40. \tag{12}$$

The dual had

$$n_v = n_v[E_8] + 10, \qquad n_h = n_h[E_8]. \tag{13}$$

Again, the consistency of (12) and (13) requires (9). The result (9) is also consistent with [6].

4. Conclusions

In this talk, we saw how the construction of $\mathcal{N} = 2$ theories using wrapped M5-branes helps us understand the S-duality involving strange non-Lagrangian superconformal theories such as the duality of Argyres and Seiberg. As explicit examples, we saw how Minahan and Nemeschansky's superconformal theories with exceptional flavor symmetry arise naturally in S-dual descriptions of standard linear quiver gauge theories.

The general procedure reviewed above makes it clear that M5-branes wrapped on a sphere with three punctures give basic building blocks of $\mathcal{N} = 2$ theories of this class. Such a building block is naturally labeled by three Young diagrams which specify the types of punctures put on the stack of M5-branes. The results presented in Sec. 3 show that some of them are theories which have been long known, but the rest are completely new $\mathcal{N} = 2$ superconformal field theories in four dimensions. They are in some sense as elementary as free hypermultiplets, and warrant further study.

Acknowledgements

The author thanks the hospitality of University of Kentucky during the conference "Quantum Theory and Symmetry 6". He also thanks S. Benvenuti and F. Benini for a nice collaboration, and D. Gaiotto for helpful discussions. He is supported in part by the NSF grant PHY-0503584, and by the Marvin L. Goldberger membership at the Institute for Advanced Study.

References

[1] P. C. Argyres and N. Seiberg, "S-Duality in $\mathcal{N}$= 2 Supersymmetric Gauge Theories," JHEP **0712** (2007) 088 [arXiv:0711.0054 [hep-th]].
[2] D. Gaiotto, "$\mathcal{N}$= 2 Dualities," arXiv:0904.2715 [hep-th].
[3] F. Benini, S. Benvenuti and Y. Tachikawa, "Webs of Five-Branes and $\mathcal{N}$= 2 Superconformal Field Theories," JHEP **0909** (2009) 052 [arXiv:0906.0359 [hep-th]].
[4] D. Nanopoulos and D. Xie, "$\mathcal{N}$ = 2 SU Quiver with USp Ends or SU Ends with Antisymmetric Matter," JHEP **0908** (2009) 108 [arXiv:0907.1651 [hep-th]].
[5] P. C. Argyres and J. R. Wittig, "Infinite Coupling Duals of $\mathcal{N}$ = 2 Gauge Theories and New Rank 1 Superconformal Field Theories," JHEP **0801** (2008) 074 [arXiv:0712.2028 [hep-th]].
[6] O. Aharony and Y. Tachikawa, "A Holographic Computation of the Central Charges of d = 4, $\mathcal{N}$= 2 SCFTs," JHEP **0801** (2008) 037 [arXiv:0711.4532 [hep-th]].

Journal of Physics: Conference Series **462** (2013) 012053 doi:10.1088/1742-6596/462/1/012053

Holographic Entanglement Entropy, Fractional Quantum Hall Effect and Lifshitz-like Fixed Point

Tadashi Takayanagi

Institute for the Physics and Mathematics of the Universe (IPMU), The University of Tokyo, 5-1-5 Kashiwanoha, Kashiwa, 277-8583, Japan

E-mail: tadashi.takayanagi@ipmu.jp

Abstract. The entanglement entropy has been very important in various subjects such as quantum information theory, condensed matter physics, and quantum gravity. In the first half of this talk, I will give an overview of recent progresses in the holographic calculation of entanglement entropy. Next, I will apply it to string theory duals of anisotropic scale invariant theories (Lifshitz-like points), which are obtained from a string theory realization of fractional quantum Hall effect.

1. Introduction

The idea of holography has been crucially important in string theory. Holography claims that the degrees of freedom in $(d+2)$-dimensional quantum gravity are much more reduced than we naively think, and will be comparable to those of quantum many body systems in $d+1$ dimensions [1, 2]. This was essentially found by remembering that the entropy of a black hole is not proportional to its volume, but to its area of the event horizon Σ (the Bekenstein-Hawking formula [3]):

$$S_{BH} = \frac{\text{Area}(\Sigma)}{4G_N} , \tag{1}$$

where G_N is the Newton constant. Owing to the discovery of the AdS/CFT correspondence [4], we know explicit examples where the holography is manifestly realized. The AdS/CFT argues that the quantum gravity on $(d+2)$-dimensional anti-de Sitter spacetime (AdS_{d+2}) is equivalent to a certain conformal field theory in $d+1$ dimensions (CFT_{d+1}) [4].

Even after quite active researches of AdS/CFT for these ten years, fundamental mechanism of the AdS/CFT correspondence still remains a mystery, in spite of so many of evidences in various examples. In particular, we cannot answer which region of AdS is responsible to particular information in the dual CFT. To make any progresses for this long standing problem, we believe that it is important to understand and formulate the holography in terms of a universal observable, rather than quantities which depend on the details of theories such as specific operators or Wilson loops etc. We only expect that a quantum gravity in some spacetime is dual to (i.e. equivalent to) a certain theory which is governed by the law of quantum mechanics. We would like to propose that an appropriate quantity which can be useful in this universal viewpoint is the entanglement entropy. Indeed, we can always define the entanglement entropy in any quantum mechanical system.

The entanglement entropy S_A in quantum field theories or quantum many body systems is a non-local quantity as opposed to correlation functions. It is defined as the von Neumann entropy S_A of the reduced density matrix when we 'trace out' (or smear out) degrees of freedom inside a d-dimensional space-like submanifold B in a given $(d+1)$-dimensional QFT, which is a complement of A. S_A measures how the subsystems A and B are correlated with each other. Intuitively we can also say that this is the entropy for an observer in A who is not accessible to B as the information is lost by the smearing out in region B. This origin of entropy looks analogous to the black hole entropy. Indeed, this was the historical motivation of considering the entanglement entropy in quantum field theories [5, 6]. Interestingly, the leading divergence of S_A is proportional to the area of the subsystem A, called the area law [5, 6] (refer also to the review articles [7, 8]).

Now we come back to our original question where in AdS given information in CFT is saved. Since the information included in a subsystem B is evaluated by the entanglement entropy S_A, we can formulate this question more concretely as follows: "Which part of AdS space is responsible for the calculation of S_A in the dual gravity side ?" We propose that a holographic formula of the entanglement entropy in [9, 10]:

$$S_A = \frac{\text{Area}(\gamma_A)}{4G_N^{(d+2)}} , \tag{2}$$

where γ_A is the d-dimensional minimal surface γ_A whose boundary is given by the $(d-1)$-dimensional manifold $\partial\gamma_A = \partial A$ (see Fig. 1); the constant $G_N^{(d+2)}$ is the Newton constant of the general gravity in AdS_{d+2}. This formula can be applied equally well to asymptotically AdS static spacetimes.

The purpose of this talk is to review this holographic description and discuss its applications. In particular, the examples we study will include confining gauge theories, fractional quantum Hall systems and Lifshitz-like fixed points, where the entanglement entropy plays the role of quantum order parameters.

2. Holographic Entanglement Entropy

We start with a review of the definition and properties of the entanglement entropy.

2.1. Definition of Entanglement Entropy

Consider a quantum mechanical system with many degrees of freedom such as spin chains. More generally, we can consider arbitrary lattice models or quantum field theories (QFTs). We put such a system at zero temperature and then the total quantum system is described by the pure ground state $|\Psi\rangle$. We assume no degeneracy of the ground state. Then, the density matrix is that of the pure state

$$\rho_{tot} = |\Psi\rangle\langle\Psi| . \tag{3}$$

The von Neumann entropy of the total system is clearly zero $S_{tot} = -\text{tr}\,\rho_{tot}\log\rho_{tot} = 0$.

Next we divide the total system into two subsystems A and B. In the spin chain example, we just artificially cut off the chain at some point and divide the lattice points into two groups. Notice that physically we do not do anything to the system and the cutting procedure is an imaginary process. Accordingly the total Hilbert space can be written as a direct product of two spaces $\mathcal{H}_{tot} = \mathcal{H}_A \otimes \mathcal{H}_B$ corresponding to those of subsystems A and B. The observer who is only accessible to the subsystem A will feel as if the total system is described by the reduced density matrix ρ_A

$$\rho_A = \text{tr}_B\ \rho_{tot} , \tag{4}$$

where the trace is taken only over the Hilbert space $\mathcal{H}_B$.

Journal of Physics: Conference Series **462** (2013) 012053 doi:10.1088/1742-6596/462/1/012053

Now we define the entanglement entropy of the subsystem A as the von Neumann entropy of the reduced density matrix ρ_A

$$S_A = -\mathrm{tr}_A\, \rho_A \log \rho_A \ . \tag{5}$$

This quantity provides us with a convenient way to measure how closely entangled (or how "quantum") a given wave function $|\Psi\rangle$ is.

It is also possible to define the entanglement entropy $S_A(\beta)$ at finite temperature $T = \beta^{-1}$. This can be done just by replacing (3) with the thermal one $\rho_{thermal} = e^{-\beta H}$, where H is the total Hamiltonian. When A is the total system, $S_A(\beta)$ is clearly the same as the thermal entropy.

2.2. Properties

There are several useful properties which the entanglement entropy enjoys generally. We summarize some of them as follows (the derivations and other properties of the entanglement entropy can be found in e.g. the textbook [11]):

- If the density matrix ρ_{tot} is pure such as in the zero temperature system, then we find the following relation assuming B is the complement of A:

$$S_A = S_B \ . \tag{6}$$

  This manifestly shows that the entanglement entropy is not an extensive quantity. This equality (6) is violated at finite temperature.
- For any three subsystems A, B and C that do not intersect each other, the following inequalities hold :

$$S_{A+B+C} + S_B \leq S_{A+B} + S_{B+C} \ , \tag{7}$$
$$S_A + S_C \leq S_{A+B} + S_{B+C} \ , \tag{8}$$

  These inequalities are called the strong subadditivity [12], which is the most powerful inequality obtained so far with respect to the entanglement entropy.
- By setting B empty in (7), we can find the subadditivity relation

$$S_{A+B} \leq S_A + S_B \ . \tag{9}$$

  The subadditivity (9) allows us to define an interesting quantity called mutual information $I(A, B)$ by

$$I(A, B) = S_A + S_B - S_{A+B} \geq 0 \ . \tag{10}$$

2.3. Entanglement Entropy in QFTs

Consider a QFT on a $(d+1)$-dimensional manifold $\mathbf{R} \times \mathbf{N}$, where $\mathbf{R}$ and N denote the time direction and the d-dimensional space-like manifold, respectively. We define the subsystem by a d-dimensional submanifold $A \subset N$ at fixed time $t = t_0$. We call its complement the submanifold B. The boundary of A, which is denoted by ∂A, divides the manifold N into two submanifolds A and B. Then we can define the entanglement entropy S_A by the previous formula (5). Sometimes this kind of entropy is called geometric entropy as it depends on the geometry of the submanifold A. Since the entanglement entropy is always divergent in a continuum theory, we introduce an ultraviolet cut off a (or a lattice spacing). Then the coefficient in front of the divergence turns out to be proportional to the area of the boundary ∂A of the subsystem A as first pointed out in [5, 6],

$$S_A = \gamma \cdot \frac{\mathrm{Area}(\partial A)}{a^{d-1}} + \text{subleading terms} \ , \tag{11}$$

where γ is a constant which depends on the system. This behavior can be intuitively understood since the entanglement between A and B occurs at the boundary ∂A most strongly. This result (11) was originally found from numerical computations [6, 5] and checked in many later arguments.

The simple area law (11), however, does not always describe the scaling of the entanglement entropy in generic situations. Indeed the entanglement entropy of 2D CFT scales logarithmically with respect to the length l of A [13, 14]. If we assume the total system is infinitely long, it is given by the simple formula [13, 14]

$$S_A = \frac{c}{3}\log\frac{l}{a}\ , \tag{12}$$

where c is the central charge of the CFT. We will later see that the scaling behavior (12) is consistent with the generic structure (22) expected from AdS/CFT.

The result for a finite size system at finite temperature has been obtained in [15] for a free Dirac fermion (i.e. $c = 1$) in two dimensions. In the high temperature expansion, the result becomes (we set $L = 1$)

$$\begin{aligned} S_A(\beta, l) &= \frac{1}{3}\log\left[\frac{\beta}{\pi a}\sinh\left(\frac{\pi l}{\beta}\right)\right] + \frac{1}{3}\sum_{m=1}^{\infty}\log\left[\frac{(1-e^{2\pi\frac{l}{\beta}}e^{-2\pi\frac{m}{\beta}})(1-e^{-2\pi\frac{l}{\beta}}e^{-2\pi\frac{m}{\beta}})}{(1-e^{-2\pi\frac{m}{\beta}})^2}\right] \\ &+2\sum_{k=1}^{\infty}\frac{(-1)^k}{k}\cdot\frac{\frac{\pi k l}{\beta}\coth\left(\frac{\pi k l}{\beta}\right)-1}{\sinh\left(\pi\frac{k}{\beta}\right)}\ . \end{aligned} \tag{13}$$

Using this expression, we can find the relation between the thermal entropy $S^{thermal}(\beta)$ and the entanglement entropy

$$S^{thermal}(\beta) = \lim_{\epsilon\to 0}\Big(S(\beta, 1-\epsilon) - S(\beta, \epsilon)\Big)\ . \tag{14}$$

2.4. Holographic Entanglement Entropy Proposal

Here we would like to explain the holographic calculation of the entanglement entropy. For a detailed review article, refer to [8]. In order to simplify the notations and reduce ambiguities, we consider the setup of the AdS/CFT correspondence, though it will be rather straightforward to extend our results to general holographic setups.

The AdS/CFT correspondence argues that (quantum) gravity in the $(d+2)$-dimensional anti de-Sitter space AdS_{d+2} is equivalent to a $(d+1)$-dimensional conformal field theory CFT_{d+1} [4]. Below we mainly employ the Poincare metric of AdS_{d+2} with radius R:

$$ds^2 = R^2\,\frac{dz^2 - dx_0^2 + \sum_{i=1}^{d-1}dx_i^2}{z^2}\ . \tag{15}$$

The dual CFT_{d+1} is supposed to live on the boundary of AdS_{d+2} which is $R^{1,d}$ at $z \to 0$ spanned by the coordinates (x^0, x^i). The extra coordinate z in AdS_{d+2} is interpreted as the length scale of the dual CFT_{d+1} in the RG sense. Since the metric diverges in the limit $z \to 0$, we put a cut off by imposing $z \geq a$. Then the boundary is situated at $z = a$ and this cut off a is identified with the ultraviolet cut off in the dual CFT. Under this interpretation, a fundamental principle of AdS/CFT, known as the bulk to boundary relation, is simply expressed by the equivalence of the partition functions in both theories

$$Z_{CFT} = Z_{AdS\ Gravity}\ . \tag{16}$$

Since (non-normalizable) perturbations in the AdS background by exciting fields in the AdS side are dual to the shift of background in the CFT side, we can compute the correlation functions in the CFT by taking the derivatives with respect to the perturbations. In generic parameter regions, the gravity should be treated in string theory to take the quantum corrections into account. Nevertheless, in particular interesting limit, typically strong coupling limit of CFT, the quantum corrections become negligible and we can employ supergravity to describe AdS spaces. Moreover, most of our examples shown in this article are simple enough that we can apply general relativity. In this situation, the right-hand side of (16) is reduced to the exponential $e^{-S_{EH}}$ of the on-shell Einstein-Hilbert action. So far we applied the AdS/CFT to the pure AdS spacetime (15). However, the AdS/CFT can be applied to any asymptotically AdS spacetimes including the AdS black holes.

Now we are in a position to present how to calculate the entanglement entropy in CFT_{d+1} from the gravity on AdS_{d+2}. This argument here can be straightforwardly generalized to any static backgrounds.

To define the entanglement entropy in the CFT_{d+1}, we divide the (boundary) time slice N into A and B as we explained before (see Fig. 1). In the Poincare coordinate (15), we are setting $N = R^d$ and the CFT_{d+1} is supposed to live on the boundary $z = a \to 0$ of AdS_{d+2}. To have its dual gravity picture, we need to extend this division $N = A \cup B$ to the time slice M of the bulk spacetime. In the setup (15), M is the $(d+1)$-dimensional hyperbolic spacetime H_{d+1}. Thus we extend ∂A to a surface γ_A in the entire M such that $\partial\gamma_A = \partial A$. Notice that this is a surface in the time slice M, which is a Euclidean manifold. Of course, there are infinitely many different choices of γ_A. We claim that we have to choose the minimal area surface among them. This means that we require that the variation of the area functional vanishes; if there are multiple solutions, we choose the one whose area takes the minimum value. This procedure singles out a unique minimal surface and we call this γ_A again (see Fig. 1).

In this setup we propose that the entanglement entropy S_A in CFT_{d+1} can be computed from the following formula [9, 10]

$$S_A = \frac{\mathrm{Area}(\gamma_A)}{4G_N^{(d+2)}} \,. \tag{17}$$

We stress again that the manifold γ_A is the d-dimensional minimal area surface in AdS_{d+2} whose boundary is given by ∂A. Its area is denoted by $\mathrm{Area}(\gamma_A)$. Also $G_N^{(d+2)}$ is the $(d+2)$-dimensional Newton constant of the AdS gravity. We can easily show that the leading divergence $\sim a^{-(d-1)}$ in (17) is proportional to the area of the boundary ∂A and this immediately reproduces the area law property (11).

The appearance of the formula (17) looks very similar to the area law of the Bekenstein-Hawking formula (1) of black hole entropy. Indeed, we can regard our formula (17) as a generalization of (1) because in the presence of event horizon such as the AdS Schwarzschild black hole solutions, the minimal surface tends to wrap the horizon.

2.5. Entanglement Entropy from AdS_3/CFT_2

Consider $\mathrm{AdS}_3/\mathrm{CFT}_2$ as one of the simplest setups of AdS/CFT. Since the entanglement entropy in two-dimensional CFT can be analytically obtained as we mentioned, we can test our holographic formula explicitly. The central charge of CFT is related to radius of AdS_3 via

$$c = \frac{3R}{2G_N^{(3)}} \,. \tag{18}$$

We are interested in the entanglement entropy S_A in an infinitely long system when A is an interval of length l. To compute this via the holographic formula (17), we need to find a geodesics

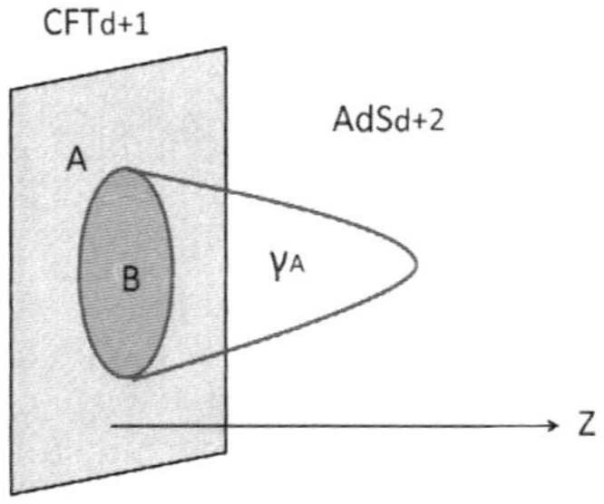

Figure 1. The holographic calculation of entanglement entropy via AdS/CFT.

between the two points $(x^1, z) = (-l/2, a)$ and $(x^1, z) = (l/2, a)$ in the Poincare coordinate (15). It is actually given by the half circle

$$(x, z) = \frac{l}{2}(\cos s, \sin s) \ , \qquad (\epsilon \leq s \leq \pi - \epsilon) \ , \tag{19}$$

where $\epsilon = \frac{2a}{l}$. The length of γ_A can be found as

$$\text{Length}(\gamma_A) = 2R \int_{\epsilon}^{\pi/2} \frac{ds}{\sin s} = -2R \log(\epsilon/2) = 2R \log \frac{l}{a} \ . \tag{20}$$

Finally the entropy can be obtained as follows

$$S_A = \frac{\text{Length}(\gamma_A)}{4G_N^{(3)}} = \frac{c}{3} \log \frac{l}{a} \ . \tag{21}$$

This perfectly agrees with the result (12) in the CFT side.

2.6. General Predictions from Holographic Calculations

For conformal field theories in higher dimensions ($d > 1$), our holographic method predicts the following general form of S_A for relativistic quantum field theories, assuming that ∂A is a smooth and compact manifold

$$\begin{aligned} S_A \ &= \ p_1 \, (l/a)^{d-1} + p_3 \, (l/a)^{d-3} + \cdots \\ &\cdots + \begin{cases} p_{d-1} \, (l/a) + p_d \ , & d\text{: even} \\ p_{d-2} \, (l/a)^2 + \tilde{c} \log{(l/a)} \ , & d\text{: odd} \end{cases} \end{aligned} \tag{22}$$

where l is the typical length scale of ∂A. This result includes the known result for $d = 1$ (12) with $\tilde{c} = c/3$. Also, in the case of (3+1)-dimensional conformal field theories ($d = 3$), the scaling law (22) has been confirmed by direct field theoretical calculations based on Weyl anomaly [10], where again the coefficient of the logarithmic term $\tilde{c}$ is given in terms of central charges of (3+1) CFTs. For $d =$ even, Eq. (22) has been the only known analytical result for (interacting) conformal field theories. Even though we assumed conformal field theories in the above, the

Journal of Physics: Conference Series **462** (2013) 012053 doi:10.1088/1742-6596/462/1/012053

same scaling formula (22) should be true for a quantum field theory with a UV fixed point, i.e., at a relativistic quantum critical point.

When the boundary ∂A is not a smooth manifold such as the one with cusp singularities, we will have other terms $(l/a)^{d-2}, (l/a)^{d-4}, \cdots$ which do not obey the scaling law in (22). For example, for a three-dimensional CFT ($d = 2$), if ∂A has a cusp with the angle Ω, then S_A includes a logarithmic term $\sim -f(\Omega)\log l/a$, for a certain function f.

Also, if we consider a gapped system in three dimensions ($d = 2$) which is described (at low energies) by a topological field theory (called topologically ordered phase), the scaling of the entanglement entropy is the same as (22) with $d = 2$. The constant $p_d = p_2$ in a topologically ordered phase is, however, invariant under a smooth deformation of the boundary ∂A. In this case, $S_{top} = p_2$ is called the topological entanglement entropy [17, 16].

2.7. Holographic Proof of Strong Subadditivity

One of the most important properties of the entanglement entropy is the strong subadditivity [12] given by the inequalities (7) and (8). This represents the concavity of the entropy and is somehow analogous to the second law of thermodynamics. Actually, it is possible to check that our holographic formula (17) satisfies this property in a rather simple argument as shown in [18] (see also [19] for the explicit numerical studies).

Let us start with three regions A, B and C on a time slice of a given CFT so that there are no overlaps between them. We extend this boundary setup toward the bulk AdS. Consider the entanglement entropy S_{A+B} and S_{B+C}. In the holographic description (17), they are given by the areas of minimal area surfaces γ_{A+B} and γ_{B+C} which satisfy $\partial\gamma_{A+B} = \partial(A+B)$ and $\partial\gamma_{B+C} = \partial(B+C)$ as before. Then it is easy to see that we can divide these two minimal surfaces into four pieces and recombine into (i) two surfaces γ'_B and γ'_{A+B+C} or (ii) two surfaces γ'_A and γ'_C, corresponding to two different ways of the recombination. Here we again meant γ'_X is a surface which satisfies $\partial\gamma'_X = \partial X$. Since in general γ'_Xs are not minimal area surface, we have Area$(\gamma'_X) \geq$Area(γ_X). Therefore, as we can easily find that this argument immediately leads to

$$\begin{aligned} \text{Area}(\gamma_{A+B}) + \text{Area}(\gamma_{B+C}) &= \text{Area}(\gamma'_B) + \text{Area}(\gamma'_{A+B+C}) \geq \text{Area}(\gamma_B) + \text{Area}(\gamma_{A+B+C})\ , \\ \text{Area}(\gamma_{A+B}) + \text{Area}(\gamma_{B+C}) &= \text{Area}(\gamma'_A) + \text{Area}(\gamma'_C) \geq \text{Area}(\gamma_A) + \text{Area}(\gamma_C)\ . \end{aligned} \tag{23}$$

In this way, we are able to check the strong subadditivity (7) and (8).

3. Entanglement Entropy and Confinement/deconfinement Transition

In recent discussions in condensed matter physics, the entanglement entropy is expected to play a role of an appropriate order parameter describing quantum phases and phase transitions. For example, it can be particularly useful for a system which realizes a topological order, such as fractional quantum Hall systems. At low energies, such systems can be described by a topological field theory, and the correlation functions are not useful order parameter as they are trivial. However, the entanglement entropy can capture important information of the topological ground state [17, 16].

The main purpose of this section is to apply the entanglement entropy to the confinement/deconfinement transition of gauge theories [20, 21]. It is much easier to employ our holographic calculation as we need to deal with strongly coupled gauge theories.

3.1. Confinement/Deconfinement Transition

One of the most interesting applications of the entanglement entropy is that it can be used as an order parameter for the confinement/deconfinement phase transition in the confining gauge theory. When we divide one of the spatial direction into a line segment with length l and its complement, the entanglement entropy between the two regions measures the effective degrees of freedom at the energy scale $\Lambda \sim 1/l$. Then, in the confining gauge theory, the behavior entanglement entropy should become trivial (i.e. S_A approaches to a constant) as l becomes large, i.e. the infrared limit $\Lambda \to 0$.

Such a transition can be captured by the holographic entanglement entropy if there are confining backgrounds dual to the confining gauge theories [20, 21]. We find that there are two candidates for the minimal surface with the same endpoints at the boundary in the confining background. One is the two disconnected straight lines extending from the endpoints of the line segment to inside the bulk, and the other is the curved line connecting these two points. The connected curve correspond to the deconfinement phase in dual gauge theory because the entanglement entropy depends on the length l, while the disconnected lines independent of the length l correspond to the confinement phase. In general, there is a critical length l_c above which the disconnected lines are favored, while below which the connected curve is favored, as we will see below explicitly.

For example, we consider the AdS soliton solution [22]

$$ds^2 = R^2\frac{dr^2}{r^2 f(r)} + \frac{r^2}{R^2}(-dt^2 + f(r)d\theta^2 + dx_1^2 + dx_2^2) \ , \tag{1}$$

where $f(r) = 1 - r_0^4/r^4$ and the θ direction is compactified with the radius $L = \pi R^2/r_0$ to avoid the conical singularity at $r = r_0$. This can be obtained from the double Wick rotation of the AdS Schwarzschild solution. The dual gauge theory is $N = 4$ super Yang-Mills on $R^{1,2} \times S^1$, but the supersymmetry is broken due to the anti-periodic boundary condition for fermions along the θ direction. Then the scalar fields acquire non-zero masses from radiative corrections, and the theory becomes almost the same as the $(2+1)$-dimensional pure Yang-Mills, which shows the confinement behavior [22].

To define the entanglement entropy, let us divide the boundary region into two parts A and B: A is defined by $-l/2 \le x_1 \le l/2$, $0 \le x_2 \le V(\to \infty)$ and $0 \le \theta \le L$, and B is the complement of A. The minimal surface γ_A, whose boundary coincides with the endpoint ∂A, can be obtained by minimizing the area

$$\text{Area} = LV\int_{-l/2}^{l/2} dx_1 \frac{r}{R}\sqrt{\left(\frac{dr}{dx_1}\right)^2 + \frac{r^4 f(r)}{R^4}} \ . \tag{2}$$

Regarding x_1 as a time, then the energy conservation leads to

$$\frac{dr}{dx_1} = \frac{r^2}{R^2}\sqrt{f(r)\left(\frac{r^6 f(r)}{r_*^6 f(r_*)} - 1\right)} \ , \tag{3}$$

where r_* is the minimal value of r. When integrating this relation, we should also take the boundary condition into account

$$\frac{l}{2} = \int_{r_*}^{r_\infty} dr \frac{R^2}{r^2\sqrt{f(r)\left(\frac{r^6 f(r)}{r_*^6 f(r_*)} - 1\right)}} \ , \tag{4}$$

which relates r_* with l. Here we introduced the UV cutoff at $r = r_\infty$. After eliminating l in (2) and (4), we find the entanglement entropy as

$$S_A^{(con)} = \frac{LV}{2RG_N^{(5)}}\int_{r_*}^{r_\infty} \frac{r^4\sqrt{f(r)}}{r^6 f(r) - r_*^6 f(r_*)} \ . \tag{5}$$

It is important that l is bounded from above due to the relation (4)

$$l \leq l_{max} \simeq 0.22L \ . \tag{6}$$

Then, when l becomes large, there is no minimal surface that connects the two boundaries of ∂A. Instead, the disconnected straight lines actually dominate before l becomes greater than l_{max}. The entanglement entropy is easy to be found

$$S_A^{(discon)} = \frac{VL}{2G_N^{(5)}} \int_{r_0}^{r_\infty} dr \frac{r}{R} = \frac{VL}{4G_N^{(5)} R} (r_\infty^2 - r_0^2) \ . \tag{7}$$

When ΔS_A becomes positive at the critical length $l_c (< l_{max})$, the disconnected surface dominates, i.e., becomes minimal. Then there happens a phase transition at $l = l_c$, which corresponds to the confinement/deconfinement phase transition in dual gauge theory [20, 21].

A similar analysis has been done in [21] in the more general backgrounds including the Klebanov-Strassler solution [23]. These results indicate that the entanglement entropy can be a good order parameter for a phase transition. A benefit of the holographic entanglement entropy is that in order to detect the confinement/deconfinement, we do not need finite temperature black brane solutions, which are often difficult to get analytically.

In summary, our holographic analysis predicts the following behavior of the finite part of the entanglement entropy in $(d+1)$-dimensional confining large N gauge theories at vanishing temperature (we subtracted the area law divergence $\sim a^{-(d-1)}$):

$$\begin{aligned} &S_A(l)|_{finite} = -VF(l) \ , \\ \text{where} \quad &F(l) \simeq c_1 N^2 l^{-(d-1)} \quad (l \to 0) \ , \\ &F(l) = c_2 N^2 \quad (l > l_c) \ , \end{aligned} \tag{8}$$

where V is the volume of the non-compact $d-2$ directions transverse to the separation. The numerical coefficients c_1 and c_2 depend on each theory.

Remarkably, the numerical computation of the entanglement entropy in the lattice gauge theory has been done in [24], and the non-analytic behavior (8) has been confirmed. These results would also support the validity of the holographic formula (17) of the entanglement entropy in the AdS/CFT correspondence.

4. Fractional Quantum Hall Effect and Entanglement Entropy from String Theory

Another example where the entanglement entropy plays the role of quantum order parameter is the fractional quantum Hall effect (FQHE). Thus we would like to embed FQHE in string theory using D-branes and discuss its entanglement entropy via AdS/CFT.

The simplest series of the fractional quantum Hall states is known as the Laughlin state which has $\nu = \frac{1}{k}$ with k being an odd integer. At low energies, the Laughlin states are described by a (2+1)-dimensional $U(1)$ Chern-Simons theory coupled to an external electromagnetic field $\tilde{A}$ [25]:

$$S_{QHE} = \frac{k}{4\pi} \int A \wedge dA + \frac{e}{2\pi} \int \tilde{A} \wedge dA, \tag{9}$$

where A is the $U(1)$ gauge field that describes the internal degrees of freedom and e is the charge of the electrons. It follows straightforwardly from (9) that the Hall conductivity (defined by $j_x = \sigma_{xy} E_y$) is fractionally quantized: $\sigma_{xy} = \frac{\nu e^2}{h}$. Since the Chern-Simons gauge theory is topological [26] and has no propagating degree of freedom in contrast to the Maxwell theory, the FQHE provides an example for the gapped many-body system whose ground state can be described by a topological field theory. Therefore we would like to realize this topological theory in a holographic setup.

We start with the $\mathcal{N} = 4$ super Yang-Mills in four dimensions and compactify one of its three spatial directions (denoted by θ). Then we impose the anti-periodic boundary condition on θ, which makes the fermions massive. The quantum corrections give a mass to scalar fields and in the end we obtain a pure Yang-Mills theory in $(2+1)$ dimensions in the IR limit. This theory is dual to the AdS_5 soliton (or double Wick-rotated AdS_5 black brane) in IIB string theory as we explained in (1). Since the cycle ∂_θ shrinks at $r = r_0$, the two-dimensional space spanned by (r, θ) is topologically a two-dimensional disk D_2. The boundary of the disk is at $r = \infty$ and this is the boundary of the AdS. The dual gauge theory lives in $R^{1,2}$ whose coordinate is (t, x, y).

Now we would like to deform this theory so that it includes the Chern-Simons term. Remember the WZ-term of a D3-brane,

$$\frac{1}{4\pi}\int_{D3} \chi F \wedge F = -\frac{1}{4\pi}\int_{D3} d\chi \wedge A \wedge F, \tag{10}$$

where χ is the axion field in IIB theory. If we assume the background axion field $\chi = \frac{k}{L}\theta$, this leads to the CS coupling

$$\frac{k}{4\pi}\int_{R^{1,2}} A \wedge F. \tag{11}$$

In this way, we get a Maxwell-Chern-Simons theory which flows into the level k pure Chern-Simons theory in the IR. This argument can be easily generalized to N D3-branes and we get the non-abelian Chern-Simons term $\frac{k}{4\pi}\int_{R^{1,2}} \mathrm{Tr}\left(A \wedge dA + \frac{2}{3}A^3\right)$.

The axion field $\chi = \frac{k}{L}\theta$ can be regarded as k D7-branes located at $r = r_0$. Therefore we find that the $U(N)_k$ Yang-Mills-Chern-Simons theory is holographically dual to this AdS_5 soliton background with the k D7-branes. The IR limit of the CFT side is dual to the small r region of the dual background and this is given by the D7-brane theory. Due to the Chern-Simons term $\frac{N}{4\pi}\int_{R^3} A \wedge F$ from the RR 5-form flux, we get $U(k)_N$ pure Chern-Simons theory. Thus this holography in the low energy limit is equivalent to the level-rank duality of the Chern-Simons theory. Indeed, the level-rank duality becomes more direct for $U(N)_k$ than for $SU(N)_k$ as shown in [27]. Indeed, we can prove the following identity for the partition functions on S^3 [27]:

$$Z(S^3, U(N)_k) = Z(S^3, U(k)_N), \qquad Z(S^3, SU(N)_k) = \sqrt{\frac{k}{N}} Z(S^3, SU(k)_N). \tag{12}$$

Now we would like to examine the gauge theory dual to the gravity background (1) with k D7-branes. To interpret our model as one for the quantum Hall effect, we need to couple it to an external gauge field. To this end, we assume that the background RR 2-from field has the form

$$B_{RR} = A_{RR} \wedge d\tilde{\theta}, \tag{13}$$

where A_{RR} is a 1-form in $R^{1,2}$ and will serve as the external gauge field. We define $\tilde{\theta} \equiv \frac{2\pi}{L}\theta$ so that the period of $\tilde{\theta}$ is 2π.

The relevant RR-coupling on D3-branes reads

$$\frac{1}{4\pi^2}\int_{D3} B_{RR} \wedge \mathrm{Tr}F = \frac{1}{2\pi}\int_{R^{1,2}} A_{RR} \wedge \mathrm{Tr}F. \tag{14}$$

Therefore in the IR limit, the QFT side becomes the Chern-Simons theory coupled to an external gauge field

$$S_{D3} = \frac{k}{4\pi}\int \mathrm{Tr}\left(A \wedge dA + \frac{2}{3}A \wedge A \wedge A\right) + \frac{1}{2\pi}\int A_{RR} \wedge \mathrm{Tr}F, \tag{15}$$

This is a non-abelian version of the standard Chern-Simons description of the quantum Hall effect.

It is also important to identify the anyons in this system. A D5-brane wrapped on S^5 is the baryon vertex in $AdS_5 \times S^5$. Due to the string creation, it comes with N F-strings stretching between the D5 and N D3-branes. The F-strings induce a unit charge for each $U(1)$ gauge theory in $U(1)^N \subset U(N)$. Thus when two baryons are interchanged, the full wave-function acquires a phase factor $e^{\frac{\pi i N}{k}}$, which means that the baryons are anyons for generic values of N. We would like to ask the reader to refer to the original paper [28] for the holographic calculations of the Hall conductivity.

The topological entanglement entropy S_{top} [16, 17] is defined as the finite part of the entanglement entropy

$$S_A = \gamma \frac{l}{a} + S_{top} \,, \tag{16}$$

where a is the UV cutoff and γ a certain numerical factor proportional to the number of UV degrees of freedom. The entanglement entropy S_A is defined as the von-Neumann entropy when we trace out a subsystem A on a time-slice. We assume A is a two-dimensional disk in a time-slice. l is the length of the boundary ∂A of A. When the theory has a mass gap, the quantity S_{top} can be shown to be invariant under a continuous deformation of the region A [16, 17].

In the holographic calculation [9], we introduce the (negative) deficit angle $\delta = 2\pi(1-n)$ on the ∂A at $r=\infty$. Then we extend this deficit angle surface toward the bulk AdS, called γ. Since the entanglement entropy is defined by $S_A = -\frac{\partial}{\partial n}\log Z_n$, where Z_n is the partition function on the manifold with the deficit angle $\delta = 2\pi(1-n)$, in our setup we obtain

$$S_A = \frac{\text{Area}(\gamma)}{4G_N} + S_{D7}, \tag{17}$$

where the part S_{D7} is the contribution of k D7-branes. The action principle tells us that γ is the minimal area surface [9] whose boundary coincides with ∂A. It is clear from this holographic expression (17) that the topological entanglement entropy is given by S_{D7} as the Chern-Simons term appears due to the D7-branes. Since we are interested in the (2+1)-dimensional Yang-Mills-Chern-Simons theory, the Kaluza-Klein modes need to be negligible. This requires that the size of region A be much larger than the compactified radius L. When there is no D7-brane, the surface γ ends at the bubble wall $r = r_0$ as shown in [20, 21, 29]. In our case with D7-branes, and with their backreactions neglected, the deficit angle surface γ is extended into the (2+1)-dimensional theory on the k D7-branes. Since the large N limit corresponds to the large level limit, the D7-brane theory becomes the classical Chern-Simons theory. Then, the AdS/CFT tells us that S_{top} is given by S_{D7}, i.e. the topological entanglement entropy of $U(k)_N$ Chern-Simons theory. Using the level-rank duality, this reproduces the S_{top} of the original CS theory. In this way, we find how to reproduce the correct topological entanglement entropy via AdS/CFT. Its more direct calculation from the bulk gravity itself will be an important future problem.

5. Lifshitz-like Fixed Points and Entanglement Entropy

5.1. Holographic Dual of Lifshitz-like Fixed Points

It is natural to try to extend the AdS/CFT correspondence to a holography for the following anisotropic spacetime

$$ds^2 = r^{2z}\left(-dt^2 + \sum_{i=1}^{p} dx_i^2\right) + r^2 \sum_{j=p+1}^{d} dy_j^2 + \frac{dr^2}{r^2}, \tag{18}$$

where $0 \leq p \leq d-1$, and the parameter $z(\neq 1)$ measures the degree of Lorentz symmetry violation and anisotropy. Since the metric (18) is invariant under the scaling $(t, x_i, y_j, r) \to (\lambda^z t, \lambda^z x_i, \lambda y_j, \frac{r}{\lambda})$, we expect that on the field theory side it is dual to a fixed point which is invariant under the scaling transformation

$$(t, x_i, y_j) \to (\lambda^z t, \lambda^z x_i, \lambda y_j). \tag{19}$$

In general, fixed points with the anisotropic scaling property (19) are called Lifshitz(-like) fixed points. This generalization of AdS/CFT correspondence to Lifshitz-like fixed points (18) was first proposed and analyzed in [30] in the particular case of $p = 0$. The simplest case with $p = 0$ represents non-relativistic fixed points with dynamical critical exponent z, which appear in many examples of quantum criticality in condensed matter physics (see the references in [30]). The other cases where $1 \leq p \leq d-1$ are not only generalizations of $p = 0$ case but can also be interpreted as space-like anisotropic fixed points. Lifshitz fixed points with space-like anisotropic scale invariance appear in realistic magnets such as MnP and the axial next-nearest-neighbor Ising model.

To understand holographic duals of such gravity backgrounds, it is the best to embed them into string theory, where microscopic interpretations are often possible by using D-branes. One such example can be obtained by considering the near horizon of the backreacted geometry for a D3-D7 system which we studied in the previous section as found in [31]. The metric in Einstein frame is given by

$$ds_E^2 = \tilde{R}^2 \left[r^2(-dt^2 + dx^2 + dy^2) + r^{\frac{4}{3}} dw^2 + \frac{dr^2}{r^2} \right] + R^2 ds_{X_5}^2, \tag{20}$$

where $R^2 = \frac{12}{11}\tilde{R}^2$ and R is the standard radius of $AdS_5 \times S^5$. The dilaton now depends on the radial coordinate as $e^\phi = e^{\phi_0} r^{2/3}$. The metric (20) is invariant under the scaling

$$(t, x, y, w, r) \to \left(\lambda t, \lambda x, \lambda y, \lambda^{\frac{2}{3}} w, \frac{r}{\lambda} \right), \tag{21}$$

and therefore is expected to be holographically dual to Lifshitz-like fixed points with space-like anisotropic scale invariance. Note that the metric corresponds to $z = 3/2$, $p = 2$ and $d = 3$.

We can also straightforwardly generalize them to black brane solutions which have regular event horizons. The metric in the Einstein frame is

$$ds_E^2 = \tilde{R}^2 \left[r^2(-F(r)dt^2 + dx^2 + dy^2) + r^{\frac{4}{3}} dw^2 + \frac{dr^2}{r^2 F(r)} \right] + R^2 ds_{X_5}^2, \tag{22}$$

where

$$F(r) = 1 - \frac{\mu}{r^{\frac{11}{3}}}. \tag{23}$$

The constant μ represents the mass parameter of the black brane. The dilaton and RR fields remain the same.

Requiring the smoothness of the Euclidean geometry of (22) gives the Hawking temperature

$$T_H = \frac{11}{12\pi} \mu^{\frac{3}{11}}. \tag{24}$$

The Bekenstein-Hawing entropy is then

$$S_{BH} = \gamma \cdot \left(\frac{\pi^3}{\mathrm{Vol}(X_5)} \right) \cdot N^2 \cdot T_H^{\frac{8}{3}} \cdot V_2 \cdot L, \tag{25}$$

where $\gamma \simeq 3.729$, and V_2 represents the area in the (x, y) direction. The entropy (25) is proportional to N^2 and thus is consistent with the planar limit of a certain gauge theory.

Journal of Physics: Conference Series **462** (2013) 012053 doi:10.1088/1742-6596/462/1/012053

5.2. Holographic Entanglement Entropy

For an anisotropic system, an interesting question is "how does the scaling behavior of the entanglement entropy depend on the direction along which the subsystems are delineated?" Here we will study the entanglement entropy of various subsystems A of the (x, y, w) space at the boundary $(r \to \infty)$ of the 5D part of the D3-D7 scaling solution (20). The field theoretical computation of the entanglement entropy is expected to be difficult as the system will be strongly coupled. We will instead compute its holographic dual on the gravity side using the formula (17).

Let's first consider the subsystem A cut out along the x-direction:

$$x \in [0, \ell_x < L_x], \qquad y \in [0, L_y], \qquad w \in [0, L_w]. \tag{26}$$

We finally obtain the holographic entanglement entropy for the subsystem divided out along the x-direction:

$$\begin{aligned} S_{EE-x} &= \left(\frac{11}{12}\right)^3 \frac{\pi^2}{\mathrm{Vol}(X_5)} \cdot N^2 L_y L_w \frac{1}{d-1}\left[\frac{1}{a^{d-1}} - \left(\frac{2}{\ell_x}\right)^{d-1}\left(\frac{\sqrt{\pi}\,\Gamma(\frac{1+d}{2d})}{\Gamma(\frac{1}{2d})}\right)^d\right] \\ &= N^2 L_y L_w \left[\frac{\gamma_1}{a^{5/3}} - \frac{\gamma_2}{(\ell_x)^{5/3}}\right], \end{aligned} \tag{27}$$

with $d = \frac{8}{3}$. γ_1 and γ_2 are numerical constants.

The next subsystem we can consider is to divide along the w-direction:

$$w \in [0, \ell_w < L_w]. \tag{28}$$

This leads to the entanglement entropy of subsystem along the w-direction

$$\begin{aligned} S_{EE-w} &= \left(\frac{11}{12}\right)^3 \frac{\pi^2}{\mathrm{Vol}(X_5)} \cdot N^2 L_x L_y \frac{1}{D-1}\left[\frac{1}{a^{D-1}} - \left(\frac{3}{2}\right)^{d_w-1}\left(\frac{2}{\ell_w}\right)^{d_w-1}\left(\frac{\sqrt{\pi}\,\Gamma(\frac{1}{2d_w}+\frac{1}{2})}{\Gamma(\frac{1}{2d_w})}\right)^{d_w}\right] \\ &= N^2 L_x L_y \left[\frac{\gamma_1'}{a^2} - \frac{\gamma_2'}{(\ell_w)^3}\right], \end{aligned} \tag{29}$$

where $D - 1 \equiv \frac{2(d_w-1)}{3} = 2$. The negative finite part has the same form as the result for the subsystem divided along the x-direction with $d = \frac{8}{3}$ replaced by $d_w = 4$. We can confirm that both results are consistent with the anisotropic scale invariance of our system.

6. Conclusions

In this talk, we reviewed a holographic calculation of entanglement entropy and its application to various systems. In all examples we analyzed we find complete agreement between the holographic analysis and the field theoretic results. Moreover, our holographic method predicts in strongly coupled higher dimensional field theories, where the field theoretic analysis is very complicated. The most important future problem is to rigorously prove our holographic entanglement entropy formula from the first principle of AdS/CFT correspondence. Another interesting problem to understand the holography in de Sitter spacetime as almost nothing has been known on this in spite of many efforts. Since the entanglement entropy is quite general quantity and needs only quantum mechanics for its definition, we can always study this quantity for any kinds of holography.

Acknowledgments
I am very grateful to Sumit Das and Al Shapere for organizing this wonderful conference and for their kind hospitality. I would like to thank Tatsuo Azeyanagi, Mitsutoshi Fujita, Matt Headrick, Tomoyoshi Hirata, Wei Li, Tatsuma Nishioka and Shinsei Ryu for wonderful collaborations relevant to this talk.

References

[1] G. 't Hooft, [arXiv:gr-qc/9310026].
[2] L. Susskind, J. Math. Phys. **36**, 6377 (1995) [arXiv:hep-th/9409089]; D. Bigatti and L. Susskind, [arXiv:hep-th/0002044].
[3] J. D. Bekenstein, Phys. Rev. D **7**, 2333 (1973); S. W. Hawking, Commun. Math. Phys. **43**, 199 (1975) [Erratum-ibid. **46**, 206 (1976)].
[4] J. M. Maldacena, Adv. Theor. Math. Phys. **2**, 231 (1998) [Int. J. Theor. Phys. **38**, 1113 (1999)] [arXiv:hep-th/9711200].
[5] L. Bombelli, R. K. Koul, J. H. Lee and R. D. Sorkin, Phys. Rev. D **34**, 373 (1986).
[6] M. Srednicki, Phys. Rev. Lett. **71**, 666 (1993) [arXiv:hep-th/9303048].
[7] A. Riera and J. I. Latorre, Phys. Rev. A **74** (2006) 052326 [arXiv:quant-ph/0605112];
J. I. Latorre, J. Phys. A **40** (2007) 6689 [arXiv:quant-ph/0611271];
L. Amico, R. Fazio, A. Osterloh, V. Vedral, [arXiv:quant-ph/0703044];
J. Eisert, M. Cramer, M.B. Plenio. [arXiv:0808.3773 [quant-ph]];
P. Calabrese and J. Cardy, J. Phys. A **42** (2009) 504005 [arXiv:0905.4013 [cond-mat.stat-mech]];
H. Casini and M. Huerta, J. Phys. A **42** (2009) 504007 [arXiv:0905.2562 [hep-th]];
[8] T. Nishioka, S. Ryu and T. Takayanagi, J. Phys. A **42** (2009) 504008 [arXiv:0905.0932 [hep-th]].
[9] S. Ryu and T. Takayanagi, Phys. Rev. Lett. **96**, 181602 (2006) [arXiv:hep-th/0603001].
[10] S. Ryu and T. Takayanagi, JHEP **0608** (2006) 045 [arXiv:hep-th/0605073].
[11] See for example, M. Nielsen, and I. L. Chuang, "Quantum Computation and Quantum Information", Cambridge university press, (2000).
[12] E. H. Lieb and M. B. Ruskai, " A fundamental property of quantum-mechanical entropy," Phys. Rev. Lett. **30** (1973) 434–436; quantum-mechanical entropy," J. Math. Phys. **14** (1973) 1938–1941. With an appendix by B. Simon.
[13] C. Holzhey, F. Larsen and F. Wilczek, Nucl. Phys. B **424**, 443 (1994) [arXiv:hep-th/9403108].
[14] P. Calabrese and J. Cardy, J. Stat. Mech. **0406**, P002 (2004) [arXiv:hep-th/0405152].
[15] T. Azeyanagi, T. Nishioka and T. Takayanagi, Phys. Rev. D **77** (2008) 064005 [arXiv:0710.2956 [hep-th]].
[16] A. Kitaev and J. Preskill, Phys. Rev. Lett. **96**, 110404 (2006) [arXiv:hep-th/0510092].
[17] M. Levin and X. G. Wen, Phys. Rev. Lett. **96**, 110405 (2006) [arXiv:cond-mat/0510613].
[18] M. Headrick and T. Takayanagi, Phys. Rev. D **76** (2007) 106013 [arXiv:0704.3719 [hep-th]].
[19] T. Hirata and T. Takayanagi, JHEP **0702** (2007) 042 [arXiv:hep-th/0608213].
[20] T. Nishioka and T. Takayanagi, JHEP **0701** (2007) 090 [arXiv:hep-th/0611035].
[21] I. R. Klebanov, D. Kutasov and A. Murugan, Nucl. Phys. B **796** (2008) 274 [arXiv:0709.2140 [hep-th]].
[22] E. Witten, Adv. Theor. Math. Phys. **2**, 505 (1998) [arXiv:hep-th/9803131].
[23] I. R. Klebanov and M. J. Strassler, JHEP **0008**, 052 (2000) [arXiv:hep-th/0007191].
[24] A. Velytsky, Phys. Rev. D **77** (2008) 085021 [arXiv:0801.4111 [hep-th]];
P. V. Buividovich and M. I. Polikarpov, Nucl. Phys. B **802** (2008) 458 [arXiv:0802.4247 [hep-lat]];
P. V. Buividovich and M. I. Polikarpov, Phys. Lett. B **670** (2008) 141 [arXiv:0806.3376 [hep-th]];
P. V. Buividovich and M. I. Polikarpov, PoS CONFINEMENT **8**, 039 (2008), [arXiv:0811.3824 [hep-lat]];
Y. Nakagawa, A. Nakamura, S. Motoki and V. I. Zakharov, arXiv:0911.2596 [hep-lat].
[25] X-G. Wen, "Quantum Field Theory of Many-Body Systems," *Oxford University Press (2004).*
[26] E. Witten, Commun. Math. Phys. **121** (1989) 351.
[27] S. G. Naculich and H. J. Schnitzer, theory," JHEP **0706** (2007) 023 [arXiv:hep-th/0703089].
[28] M. Fujita, W. Li, S. Ryu and T. Takayanagi, JHEP **0906**, 066 (2009) [arXiv:0901.0924 [hep-th]].
[29] A. Pakman and A. Parnachev, JHEP **0807** (2008) 097 [arXiv:0805.1891 [hep-th]].
[30] S. Kachru, X. Liu and M. Mulligan, Phys. Rev. D **78** (2008) 106005 [arXiv:0808.1725 [hep-th]].
[31] T. Azeyanagi, W. Li and T. Takayanagi, JHEP **0906**, 084 (2009) [arXiv:0905.0688 [hep-th]].

Journal of Physics: Conference Series **462** (2013) 012054 doi:10.1088/1742-6596/462/1/012054

Quantum canonical transformations in star-product formalism

T Dereli[1], T Hakioğlu[2] and A Teğmen[3] [1]

[1] Physics Department, Koç University, 80910 Sarıyer-Istanbul, TURKEY
[2] Physics Department, Bilkent University, 06533 Ankara, TURKEY
[3] Feza Gürsey Institute, 34684 Çengelköy-Istanbul, TURKEY

E-mail: tdereli@ku.edu.tr, hakioglu@fen.bilkent.edu.tr and tegmen@science.ankara.edu.tr

Abstract. We study construction of the star-product version of three basic quantum canonical transformations which are known as the generators of the full canonical algebra. By considering the fact that star-product of c-number phase-space functions is in complete isomorphism to Hilbert-space operator algebra, it is shown that while the constructions of gauge and point transformations are immediate, generator of the interchanging transformation deforms this isomorphism. As an alternative approach, we study all of them within the deformed form. How to transform any c-number function under linear-nonlinear transformations and the intertwining method are shown within this argument as the complementary subjects of the text.

1. Introduction

In this work, we construct quantum canonical transformations (QCTs) in phase-space formalism developed by H. Weyl, E. Wigner, H. Groenewold and J. Moyal [1]. We will say Moyal or star-product formalism in short for this phase-space formalism. In doing so, we consider a conjecture stating that any QCT can be written as a composition of three basic CTs. These are gauge, point and interchange CTs [2, 3]. In fact a similar conjecture in classical mechanics has been stated by using only linear and point transformations [4]. The same triplet in quantum mechanics may also be used to generate the classical CTs but this point is out of the scope of this work.

First we give the three basic QCTs in this section. In section 2, we see that if we employ the isomorphism between the Hilbert-space operator algebra and star-product of phase-space functions to get the star-product version of the triplet, the constructions of gauge and point transformations are immediate but the generator of interchange transformation deforms this isomorphism. But still as an alternative way, an ordinary exponential version of it can be constructed. In section 3, as the new approach, all the basic CTs are derived in the ordinary exponential form. Section 4 shows that intertwining transformation is in fact a CT. Finally, section 5 contains a short summary and conclusions including some unanswered matters related with the topic.

So it is natural to start with the definition of a QCT: Given arbitrary complex function (unitary or non-unitary) $\hat{F}(\hat{q}, \hat{p})$ which is so called generating function (GF) on quantum phase

[1] On sabbatical from Department of Physics, University of Ankara, 06100 Ankara TURKEY

space, QCTs are given by the equations

$$\begin{aligned} \hat{F}(\hat{q},\hat{p})\,\hat{q}\,\hat{F}^{-1}(\hat{q},\hat{p}) &= \hat{Q}(\hat{q},\hat{p}), \\ \hat{F}(\hat{q},\hat{p})\,\hat{p}\,\hat{F}^{-1}(\hat{q},\hat{p}) &= \hat{P}(\hat{q},\hat{p}), \end{aligned} \tag{1}$$

transforming q-numbers into the other variables $\hat{Q}$ and $\hat{P}$, where the new variables must, of course, satisfy the canonicity condition $[\hat{Q},\hat{P}] = \hat{Q}\hat{P} - \hat{P}\hat{Q} = \mathrm{i}\hbar$ and $\hat{F}^{-1}$ is the algebraic inverse of $\hat{F}$.

Now we list the explicit forms of the three basic QCTs: Th gauge transformation is generated by the function $f(\hat{q})$ such as

$$\begin{aligned} \exp[\lambda f(\hat{q})]\,\hat{q}\,\exp[-\lambda f(\hat{q})] &= \hat{q}, \\ \exp[\lambda f(\hat{q})]\,\hat{p}\,\exp[-\lambda f(\hat{q})] &= \hat{p} + \mathrm{i}\hbar\lambda\,\partial_q f, \end{aligned} \tag{2}$$

where λ is a pure imaginary number with a continuous parameter. On the other hand, the explicit form of the point transformation is given by

$$\begin{aligned} \exp[\lambda f(\hat{q})\,\hat{p}]\,\hat{q}\,\exp[-\lambda f(\hat{q})\,\hat{p}] &= A(\hat{q}), \\ \exp[\lambda f(\hat{q})\,\hat{p}]\,\hat{p}\,\exp[-\lambda f(\hat{q})\,\hat{p}] &= (\partial_q A)^{-1}\,\hat{p}, \end{aligned} \tag{3}$$

where the function $A(\hat{q})$ is given by the operation

$$A(\hat{q}) = \exp[-\mathrm{i}\hbar\lambda f(\hat{q})\partial_q]\,\hat{q}. \tag{4}$$

Finally, the interchange transformation

$$\hat{I}\,\hat{q}\,\hat{I}^{-1} = \hat{p}, \qquad \hat{I}\,\hat{p}\,\hat{I}^{-1} = -\hat{q} \tag{5}$$

is achieved by a Fourier transform operator whose definition is given by the action

$$\hat{I}^{\pm} = \frac{1}{(2\pi)^{1/2}} \int_{-\infty}^{\infty} \cdot \exp(\pm\mathrm{i}\hat{q}\hat{q}'/\hbar)\,\mathrm{d}q' \tag{6}$$

satisfying

$$\hat{I}\,\hat{q}' = -\mathrm{i}\hbar\partial_q\hat{I}, \qquad \hat{I}\,\hat{p}' = -\hat{q}\,\hat{I}. \tag{7}$$

The inverse of the transformation means that $\hat{q}$ goes to minus $\hat{p}$ and $\hat{p}$ goes to $\hat{q}$.

Now let us give an immediate example related with the conjecture mentioned above. Our example is the well-known linear CTs:

$$\begin{aligned} \hat{F}\,\hat{q}\,\hat{F}^{-1} &= \hat{Q} = a\,\hat{q} + b\,\hat{p}, \\ \hat{F}\,\hat{p}\,\hat{F}^{-1} &= \hat{P} = c\,\hat{q} + d\,\hat{p}, \end{aligned} \tag{8}$$

where a, b, c, d are real constants satisfying $a\,d - b\,c = 1$ and $a + d + 2 \neq 0$. Such a transformation can be decomposed into a five-step sequence:

$$\hat{F}_L = \exp[\,\gamma\hat{q}\,\hat{p}\,]\exp[\,\beta\hat{q}^2\,]\,\hat{I}\exp[\,\alpha\hat{q}^2\,]\,\hat{I}^{-1} = \hat{\mathcal{P}}\,\hat{G}_2\,\hat{I}\,\hat{G}_1\,\hat{I}^{-1} \tag{9}$$

where $\mathcal{G}$ and $\mathcal{P}$ stand for gauge and point respectively and α, β, γ are the pure imaginary numbers compatible with a linear CT [3]. At this stage, it should be note that such a decomposition is not unique, that means of course, one can find many decompositions giving the same transformation. The total transformation after the five steps appears as

$$\begin{aligned} \hat{q} &\to (1 + 4\hbar^2\alpha\beta)\,\hat{q} - 2i\hbar\alpha\,\hat{p}, \\ \hat{p} &\to 2i\hbar\beta k\,\hat{q} + \tilde{p}/k \end{aligned} \tag{10}$$

2. Canonical transformations in Moyal formalism

In this section, as our main purpose, we will try to construct the three basic CTs in Moyal formalism. Moyal formalism is based on an interesting but very powerful product given by

$$\star = \exp\left[\frac{i\hbar}{2}(\overleftarrow{\partial}_q\overrightarrow{\partial}_p - \overleftarrow{\partial}_p\overrightarrow{\partial}_q)\right] \tag{11}$$

which is so called star-product or Moyal product, where the arrows indicate the direction that the derivatives act. Its explicit action can be seen easily by the general product rule

$$\begin{aligned} F(q,p)\star G(q,p) &= F(q+i\hbar\partial_p/2, p-i\hbar\partial_q/2)G(q,p), \\ &= G\left(q-i\hbar\partial_p/2, p+i\hbar\partial_q/2\right)F(q,p). \end{aligned} \tag{12}$$

Note that, as an advantage, it can be applied to the both sides. On the other hand, by the Groenewold isomorphism given by the integral transform

$$\begin{aligned} \hat{F}(\hat{q},\hat{p})\,\hat{G}(\hat{q},\hat{p}) = \frac{1}{(2\pi)^2}\int_{-\infty}^{\infty} & \mathrm{d}\sigma\,\mathrm{d}\tau\,\mathrm{d}q\,\mathrm{d}p\,[F(q,p)\star G(q,p)] \\ & \times\exp\{\mathrm{i}\,[\sigma(\hat{q}-q)+\tau(\hat{p}-p)]/\hbar\}, \end{aligned} \tag{13}$$

one can make a correspondence between the operator product in quantum phase-space and the Moyal product in c-number phase-space:

$$\hat{F}(\hat{q},\hat{p})\,\hat{G}(\hat{q},\hat{p}) \leftrightarrow F(q,p)\star G(q,p). \tag{14}$$

We should note that after the transformation, the resulting functions $\hat{F}$ and $\hat{G}$ in quantum phase-space become Weyl-ordered or symmetrically ordered. Therefore such a transform provides us a quantization method. Conversely, the inverse transform gives us a de-quantization method. Therefore, by considering this isomorphism, CTs in Moyal formalism can be defined easily by the equations

$$\begin{aligned} F(q,p)\star q\star F^{-1}(q,p) = Q(q,p), \\ F(q,p)\star p\star F^{-1}(q,p) = P(q,p) \end{aligned} \tag{15}$$

satisfying the Moyal bracket

$$\{Q\,,P\}^M = Q\star P - P\star Q = i\hbar \tag{16}$$

as the canonicity condition. Now we are ready to construct the triplet in terms of star-product. First one is the gauge transformation. Since, by (2) and the correspondence (14), the GF $G=\exp_\star[-\lambda f(q)]$ where the star-exponential is given by the expansion [5]

$$\exp_\star[\lambda\, f(q,p)] = 1+\lambda\, f(q,p)+\frac{\lambda^2}{2!}\, f(q,p)\star f(q,p)+\cdots\,, \tag{17}$$

contains only q as the variable, the quantum exponential and the star-exponential are equivalent and the new GF generates the same transformation

$$G\star q\star G^{-1} = q, \qquad G\star p\star G^{-1} = p+\mathrm{i}\hbar\lambda\,\partial_q f, \tag{18}$$

just like the previous one (2). When we consider the point transformation

$$\mathcal{P}\star q\star\mathcal{P}^{-1} = A(q), \qquad \mathcal{P}\star p\star\mathcal{P}^{-1} = (\partial_q A)^{-1}\star p \tag{19}$$

generated by the GF $\mathcal{P}=\exp_\star[-\lambda f(q)\star p]$, we see similarly that all hats are gone, and the operator product is replaced by the star-product as expected.

But the matter is the last one, i.e., the interchange transformation. The matter arises from the fact that c-number phase-space functions are not operators taking integral as in (6). But we can find a c-number phase-space function resulting an interchange transformation.

3. Canonical transformations in Moyal formalism revisited

According to the definition of a CT in the star-product formalism, the interchange transformation must be in the form

$$\mathcal{I} \star q \star \mathcal{I}^{-1} = p, \qquad \mathcal{I} \star p \star \mathcal{I}^{-1} = -q. \tag{20}$$

If we rewrite the transformation in the form

$$\mathcal{I} \star q = p \star \mathcal{I}, \qquad \mathcal{I} \star p = -q \star \mathcal{I}, \tag{21}$$

we get a new definition of the transformation without needing the inverse of the GF $\mathcal{I}$ [6]. This point is important, because getting the inverse of the generator becomes usually a tedious matter. Expansion of the star-products on both sides of (21) gives us the system of partial differential equations

$$(q-p)\mathcal{I} = -\frac{i\hbar}{2}(\partial_q - \partial_p)\mathcal{I}, \qquad (q+p)\mathcal{I} = -\frac{i\hbar}{2}(\partial_q + \partial_p)\mathcal{I}. \tag{22}$$

The solution as the generator of the interchange transformation appears in an ordinary exponential form with no $\star$ sign:

$$\mathcal{I} = \exp[\mathrm{i}(q^2 + p^2)/\hbar]. \tag{23}$$

Therefore it is natural to think of getting the other generators in the ordinary exponential form, by converting every transformation into a system of partial differential equations as in the following.

The new GF $\mathcal{G} = \exp[-\lambda f(q)]$ of the gauge transformation is immediate without solving any differential equation due to the single variable dependence of G. Therefore they generate the same transformation as before:

$$\mathcal{G} \star q \star \mathcal{G}^{-1} = q, \qquad \mathcal{G} \star p \star \mathcal{G}^{-1} = p + \mathrm{i}\hbar\lambda\, \partial_q f. \tag{24}$$

On the other hand, getting the generator of the point transformation is not so easy. But similar to the previous one, the converted definition can be given in the form

$$\wp \star q = Q(q) \star \wp, \qquad \wp \star p = \tilde{Q}(q)\, p \star \wp, \qquad \tilde{Q}(q) = (\partial_q Q)^{-1}. \tag{25}$$

This system can be solved by looking for the solutions of type $\wp = \exp\{\lambda[f(q)p + g(q)]\}$. Such a trial function gives the function $f(q)$ as the solution to the algebraic equation

$$Q\,(v) = q - \frac{\mathrm{i}\hbar\lambda}{2} f, \qquad v = q + \frac{\mathrm{i}\hbar\lambda}{2} f. \tag{26}$$

The function $g(q)$ is determined by the integral

$$g(q) = -\frac{i\hbar}{\lambda} \int \frac{(\partial_q \tilde{Q})(v)}{1 + \tilde{Q}(v)} \frac{\partial f}{\partial q}\, \mathrm{dq} \tag{27}$$

resulting from the differential equation (25).

It may be remarkable to point out that this approach, i.e., converting the transformation into a differential equation system, can be used to determine both the generating function and the transformation itself. For example, given transformation $Q = 1/q$, $P = -q^2\, p$, the GF

$$f(q) = \pm\frac{2}{\hbar\lambda}(1-q^2)^{1/2}, \qquad g(q) = -\frac{1}{2\lambda}\ln(q^2-1) \tag{28}$$

is obtained by solving the equations in (26) and (27). Conversely, given generating function $f(q) = q$, $g(q) = \text{const.}$, for instance, the result is the scaling transformation

$$Q = k\,q, \quad P = p/k, \quad k = \frac{2 - \mathrm{i}\hbar\lambda}{2 + \mathrm{i}\hbar\lambda}, \quad \lambda \neq 2i/\hbar. \tag{29}$$

After having the construction of the basic transformations in the ordinary exponential form, we are ready to decompose the transformations. For this aim, we consider again the linear CTs

$$Q = a\,q + b\,p, \qquad P = c\,q + d\,p. \tag{30}$$

For the linear CTs we have the system

$$F_L \star q = Q \star F_L, \quad F_L \star p = P \star F_L. \tag{31}$$

This time the solution is given by this ordinary exponential form:

$$F_L(q,p) = \exp\{2\mathrm{i}A[\,bp^2 - cq^2 + (a-d)qp\,]/\hbar\}, \quad A = 1/(a+d+2). \tag{32}$$

On the other hand, if we remember its quantum correspondent (9), we can construct the decomposed form of the compact one in the five-step sequence

$$\begin{aligned} F_L(q,p) &= \exp(\lambda qp) \star \exp(\beta q^2) \star \exp[\mathrm{i}(q^2+p^2)/\hbar] \star \exp(\alpha q^2) \star \Big[\exp[\mathrm{i}(q^2+p^2)/\hbar]\Big]^{-1} \\ &= \wp \star \mathcal{G}_2 \star \mathcal{I} \star \mathcal{G}_1 \star \mathcal{I}^{-1}. \end{aligned} \tag{33}$$

The second example differs from the others, here we consider the transformation of a function. In our example linear potential system $H_0 = p^2 + q$ transforms to the free particle system $H_1 = p^2$ via the transformation rule

$$F \star H_0 \star F^{-1} = H_1. \tag{34}$$

It can be shown that transformations of functions can also be decomposed similarly. The decomposition

$$\begin{aligned} F &= \exp[\mathrm{i}(q^2+p^2)/\hbar] \star \exp[\lambda(fp+q)] \\ &\quad \star \Big[\exp\mathrm{i}(q^2+p^2)/\hbar\Big]^{-1} \star \exp(\mathrm{i}q^3/3\hbar) \star \exp[\mathrm{i}(q^2+p^2)/\hbar] \\ &= \mathcal{I} \star \wp \star \mathcal{I}^{-1} \star \mathcal{G} \star \mathcal{I} \end{aligned} \tag{35}$$

consists of gauge, point and interchange again, where the GF $\wp$ of the point transformation in the forth step is given by the equations

$$\begin{aligned} f(q) &= \frac{\mathrm{i}}{\hbar\lambda}\left[2q + 1 - (1+8q)^{1/2}\right], \\ g(q) &= \frac{1}{2\lambda}\ln\left[\frac{1+(1+8q)^{1/2}}{1+8q}\right]. \end{aligned} \tag{36}$$

4. Intertwining as a canonical transformation

In this section, we will show that the intertwining method can be treated within this argument. Now, suppose that there exists a c-number phase-space function $L(q,p)$ making a link between two Hamilton functions:

$$L(q,p) \star H_0(q,p) \star L^{-1}(q,p) = H_1(q,p), \tag{37}$$

where the Hamilton functions are in the potential form

$$H_0 = p^2 + V_0(q), \qquad H_1 = p^2 + V_1(q). \tag{38}$$

It is clear that the transformation equation (37) can be rewritten in the form

$$L(q,p) \star H_0(q,p) = H_1(q,p) \star L(q,p). \tag{39}$$

Expansion of the star products on both sides of (39) gives us the differential equation

$$V_1\left(q + \frac{\mathrm{i}\hbar}{2}\partial_p\right) L(q,p) = V_0\left(q - \frac{\mathrm{i}\hbar}{2}\partial_p\right) L(q,p) + 2\mathrm{i}\hbar\, p\, \partial_q L(q,p). \tag{40}$$

relating the two potentials V_0 and V_1. For the solution if we choose the ansatz

$$L(q,p) = p - \mathrm{i}\varphi(q), \tag{41}$$

we get the well known consistency condition

$$V_1(q) + V_0(q) = 2\varphi^2(q), \tag{42}$$

and the Riccati equation

$$V_1(q) = V_0(q) + 2\hbar\partial_q\varphi(q) \tag{43}$$

with the solution $\varphi(q)$. If we look at the ansatz (41) again, we see that the function $L(q,p)$ is in fact a gauge transformation given by

$$L(q,p) = \exp\left[-\int \varphi(q)\mathrm{d}q/\hbar\right] \star p \star \exp\left[\int \varphi(q)\mathrm{d}q/\hbar\right]. \tag{44}$$

As the second step, if we replace the term p in the middle by the decomposition

$$p = \exp\left[\mathrm{i}(q^2+p^2)/\hbar\right] \star \exp(\ln q) \star \exp\left[\mathrm{i}(q^2+p^2)/\hbar\right]^{-1}, \tag{45}$$

we conclude that the function $L(q,p)$ is the sequence

$$L = \mathcal{G}_1^{-1} \star \mathcal{I} \star \mathcal{G}_2 \star \mathcal{I}^{-1} \star \mathcal{G}_1 \tag{46}$$

of the fundamental transformations and this says obviously that the intertwining method is a CT.

5. Conclusions and summary

Three fundamental CTs, especially the interchange transformation, has been constructed in terms of the ordinary exponential form. It has been shown that the ordinary form may present a powerful tool in determining both the GF and the CT. Within this formalism, transform of functions under the linear (gauge) transformations runs just like in quantum mechanics. The behavior under the nonlinear CTs has been shown and justified by a well-known example.

It should be noted that the GFs in the $\star$-exponential form do not correspond to the quantized versions of those given in the ordinary form. For example, quantized version of the interchange operator $\exp[\mathrm{i}(q^2+p^2)/\hbar]$ does not correspond to $\exp_\star[\mathrm{i}(q^2+p^2)/\hbar] \equiv \exp[\mathrm{i}(\hat{q}^2+\hat{p}^2)/\hbar]$ but to the decomposition $\exp_\star\left(\frac{\mathrm{i}q^2}{2\hbar}\right) \star \exp_\star\left(\frac{\mathrm{i}p^2}{2\hbar}\right) \star \exp_\star\left(\frac{\mathrm{i}q^2}{2\hbar}\right) \equiv \exp\left(\frac{\mathrm{i}\hat{q}^2}{2\hbar}\right) \exp\left(\frac{\mathrm{i}\hat{p}^2}{2\hbar}\right) \exp\left(\frac{\mathrm{i}\hat{q}^2}{2\hbar}\right)$. In general all we can say that $\exp_\star[f(q,p)]$ is always the de-quantized form of $\exp[\hat{f}(\hat{q},\hat{p})]$ provided that $\hat{f}(\hat{q},\hat{p})$ is the Weyl transformed form of $f(q,p)$. Except the some simple cases, getting the quantized version of any ordinary exponentiated function remains an open question not answered yet.

References

[1] Weyl H 1927 *Z. Phys.* **46** 1-46, Wigner E 1932 *Phys. Rev.* **40** 749-759, Groenewold H 1946 *Physica* **12** 405-460, Moyal J 1949 *Proc. Camb. Phil. Soc.* **45** 99-124.
[2] Deenen J 1991 *J. Phys. A* **24** 3851-3858
[3] Anderson A 1994 *Ann. Phys.* **232** 292-331 (*Preprint* hep-th/9305054)
[4] Leyvraz F and Seligman T H 1989 *J. Math. Phys.* **30** 2512-2515
[5] Bayen F, Flato M, Fronsdal C, Lichnerowicz A and Sternheimer D 1978 *Ann. Phys.* **111** 61-110, *ibid* 111-151
[6] Hakioğlu T, Teğmen A and Demircioğlu B 2007 *Phys. Lett. A* **360** 501-506 (*Preprint* quant-ph/0605236)

Journal of Physics: Conference Series **462** (2013) 012055 doi:10.1088/1742-6596/462/1/012055

Holographic models with anisotropic scaling

E J Brynjolfsson[1], U H Danielsson[2], L Thorlacius[1,3] and T Zingg[1,3]

[1] University of Iceland, Science Institute, Dunhaga 3, IS-107 Reykjavik, Iceland

[2] Institutionen för fysik och astronomi, Uppsala Universitet, Box 803, SE-751 08 Uppsala, Sweden

[3] NORDITA, Roslagstullsbacken 23, SE-106 91 Stockholm, Sweden

E-mail: erlingbr@hi.is ulf.danielsson@physics.uu.se lth@nordita.org zingg@nordita.org

Abstract. We consider gravity duals to d+1 dimensional quantum critical points with anisotropic scaling. The primary motivation comes from strongly correlated electron systems in condensed matter theory but the main focus of the present paper is on the gravity models in their own right. Physics at finite temperature and fixed charge density is described in terms of charged black branes. Some exact solutions are known and can be used to obtain a maximally extended spacetime geometry, which has a null curvature singularity inside a single non-degenerate horizon, but generic black brane solutions in the model can only be obtained numerically. Charged matter gives rise to black branes with hair that are dual to the superconducting phase of a holographic superconductor. Our numerical results indicate that holographic superconductors with anisotropic scaling have vanishing zero temperature entropy when the back reaction of the hair on the brane geometry is taken into account.

1. Introduction

We consider $d+2$ dimensional gravity models that are dual to quantum critical points with anisotropic scaling in $d+1$ dimensions,

$$t \to \lambda^z t, \qquad \vec{x} \to \lambda \vec{x}\,, \tag{1}$$

with $z \geq 1$ and $\vec{x} = (x_1, \ldots, x_d)$. The $d+2$ dimensional bulk action consists of three parts,

$$S = S_{\text{Einstein-Maxwell}} + S_{\text{Lifshitz}} + S_{\text{matter}}\,. \tag{2}$$

The first term is the standard action of Einstein-Maxwell gravity with a negative cosmological constant,

$$S_{\text{Einstein-Maxwell}} = \int \mathrm{d}^{d+2}x\sqrt{-g}\left(R - 2\Lambda - \frac{1}{4}F_{\mu\nu}F^{\mu\nu}\right). \tag{3}$$

This is followed by a term involving a massive vector field,

$$S_{\text{Lifshitz}} = -\int \mathrm{d}^{d+2}x\sqrt{-g}\left(\frac{1}{4}\mathcal{F}_{\mu\nu}\mathcal{F}^{\mu\nu} + \frac{c^2}{2}\mathcal{A}_\mu\mathcal{A}^\mu\right), \tag{4}$$

whose sole purpose is to provide backgrounds with anisotropic scaling. This is a modified version of the holographic model of [1], which was formulated in four-dimensional spacetime and obtained anisotropic scaling by including a pair of coupled two- and three-form field strengths.[1] Upon integrating out the three-form field strength, the remaining two-form becomes a field strength of a massive vector [3] and in this form the model is easily extended to general dimensions. Finite temperature corresponds to having a black hole in the higher-dimensional spacetime and with the Maxwell gauge field, added in [4], the black hole can carry electric charge, which is dual to a finite charge density in the lower-dimensional theory. Finally, we consider matter in the form of a scalar field, which is charged under the Maxwell gauge field but does not couple directly to the auxiliary massive vector field,

$$S_{\text{matter}} = -\frac{1}{2}\int \mathrm{d}^{d+2}x\sqrt{-g}\,\Big(\left(\partial^\mu\phi^* + iqA^\mu\phi^*\right)\left(\partial_\mu\phi - iqA_\mu\phi\right) + m^2\phi^*\phi\Big)\,. \tag{5}$$

At low temperature there is an instability for charged black holes in the model to grow scalar hair, which corresponds to the superconducting phase of holographic superconductors with $z > 1$ asymmetric scaling [4], as described in Section 3 below.

The equations of motion obtained from the action (2) consist of the scalar field equation

$$\left(\nabla^\mu - iqA^\mu\right)\left(\nabla_\mu - iqA_\mu\right)\phi - m^2\phi = 0\,, \tag{6}$$

along with the Einstein equations, the Maxwell equations, and field equations for the auxiliary massive vector,

$$G_{\mu\nu} + \Lambda g_{\mu\nu} = T^{\text{Maxwell}}_{\mu\nu} + T^{\text{Lifshitz}}_{\mu\nu} + T^{\text{matter}}_{\mu\nu}\,, \tag{7}$$

$$\nabla_\nu F^{\nu\mu} = j^\mu_{\text{matter}}\,, \tag{8}$$

$$\nabla_\nu \mathcal{F}^{\nu\mu} = c^2\mathcal{A}^\mu\,. \tag{9}$$

The asymmetric scaling symmetry (1), sometimes referred to as Lifshitz scaling, is realized in a d+2 dimensional spacetime,

$$ds^2 = L^2\left(-r^{2z}dt^2 + r^2d\vec{x}^2 + \frac{dr^2}{r^2}\right), \tag{10}$$

whose metric is invariant under the transformation

$$t \to \lambda^z t, \quad \vec{x} \to \lambda\vec{x}, \quad r \to \frac{r}{\lambda}\,. \tag{11}$$

Length dimensions are carried by the characteristic length L while the coordinates $(t, r, \vec{x})$ are dimensionless. The scale invariant Lifshitz geometry (10) is a solution of the equations of motion when L is related to the cosmological constant Λ via

$$\Lambda = -\frac{z^2 + (d{-}1)z + d^2}{2L^2}\,, \tag{12}$$

and the mass of the auxiliary vector field is fine-tuned to $c = \sqrt{z\,d}/L$. In this solution the Maxwell field vanishes, $A_\mu = 0$, but the massive vector field has the background value

$$\mathcal{A}_t = \sqrt{\frac{2(z{-}1)}{z}}L\,r^z\,, \qquad \mathcal{A}_{x_i} = \mathcal{A}_r = 0\,. \tag{13}$$

[1] See also [2] for early work on gravitational backgrounds with anisotropic scaling.

Journal of Physics: Conference Series **462** (2013) 012055 doi:10.1088/1742-6596/462/1/012055

2. Charged black branes

In order to study finite temperature effects in the dual strongly coupled field theory we look for static black brane solutions of the equations of motion (6) - (9) which are asymptotic to the Lifshitz fixed point solution given by (10) and (13). From now on we set $L = 1$ and consider a metric of the form

$$ds^2 = -r^{2z} f(r)^2 dt^2 + r^2 d\vec{x}^2 + \frac{g(r)^2}{r^2} dr^2 \,. \tag{14}$$

An asymptotically Lifshitz black brane with a non-degenerate horizon has a simple zero of both $f(r)^2$ and $g(r)^{-2}$ at the horizon, which we take to be at $r = r_0$, and $f(r)$, $g(r) \to 1$ as $r \to \infty$. It is straightforward to generalize this ansatz to include black holes with a spherical horizon or topological black holes with a hyperbolic horizon but it is the flat horizon case (14) that is of direct interest for the gravitational dual description of strongly coupled $d + 1$ dimensional field theories. In a static electrically charged black brane background the Maxwell gauge field and the massive vector can be taken to be of the form

$$A_\mu = r^z f(r)\{\alpha(r), 0, \ldots, 0\}\,; \qquad \mathcal{A}_\mu = \sqrt{\frac{2(z-1)}{z}}\, r^z f(r)\{a(r), 0, \ldots, 0\}\,. \tag{15}$$

with $\alpha(r) \to 0$ and $a(r) \to 1$ as $r \to \infty$.

2.1. Field equations for static configurations

For static configurations the equations of motion can be expressed as a first order system of ordinary differential equations. Introducing a scale invariant radial variable $u = \log(r/r_0)$ and writing $\dot{} \equiv \frac{d}{du}$, the scalar field equation becomes

$$\dot{\phi} = \chi \,, \tag{16}$$

$$\dot{\chi} + \left(\frac{\dot{f}}{f} - \frac{\dot{g}}{g}\right)\chi = -(d+z)\,\chi + \left(m^2 - q^2\alpha^2\right) g^2 \phi \,, \tag{17}$$

while the Maxwell equations and the equations of motion for the massive vector reduce to,

$$\dot{\alpha} + \frac{\dot{f}}{f}\alpha = -z\,\alpha + g\,\beta \,, \tag{18}$$

$$\dot{\beta} = -d\,\beta + q^2 g\,\phi^2 \alpha \,, \tag{19}$$

$$\dot{a} + \frac{\dot{f}}{f} a = -z\,a + z\,g\,b \,, \tag{20}$$

$$\dot{b} = -d\,b + d\,g\,a \,. \tag{21}$$

The functions χ, β, and b are defined via (16), (18), and (20), respectively. The Einstein equations can also be written in first order form,

$$\frac{\dot{g}}{g} + \frac{\dot{f}}{f} = (z-1)\left(g^2 a^2 - 1\right) + \frac{1}{2d}\left(\chi^2 + q^2 g^2 \alpha^2 \phi^2\right), \tag{22}$$

$$2d\,\frac{\dot{f}}{f} = d(1-d-2z) + \frac{\chi^2}{2} + g^2\Big[(z-1)\left(da^2 - zb^2\right) - \frac{\beta^2}{2} + \frac{1}{2}\left(q^2\alpha^2 - m^2\right)\phi^2 + z^2 + (d-1)z + d^2\Big]\,. \tag{23}$$

The field equations (16) - (23) are manifestly invariant under the scaling (11) and the Lifshitz fixed point solution is given by $f = g = a = b = 1$ and $\alpha = \beta = \phi = \chi = 0$.

In the absence of charged matter, the Maxwell equation (19) integrates to $\beta(u) = \tilde{\rho}\, e^{-du}$. The integration constant $\tilde{\rho} = \rho/r_0^d$ is proportional to the electric charge per unit d-volume of the black brane, which in turn corresponds to the charge density in the dual field theoretic system. For given d and $z \geq 1$, the remaining field equations then have a one parameter family of black brane solutions, labelled by $\tilde{\rho}$. A neutral black brane without scalar hair has $\tilde{\rho} = 0$ while the extremal limit is given by $\tilde{\rho} \to \pm\sqrt{2(z^2 + (d-1)z + d^2)}$. A non-vanishing charged scalar field changes this picture as discussed below.

Equation (18) can easily be solved in the Lifshitz geometry (10) without matter, giving

$$\alpha(u) = \begin{cases} \tilde{\mu}\, e^{-zu} + \frac{1}{(z-d)}\tilde{\rho}\, e^{-du} & \text{if} \quad z \neq d\,, \\ \tilde{\mu}\, e^{-du} + \tilde{\rho}\, u\, e^{-du} & \text{if} \quad z = d\,, \end{cases} \tag{24}$$

where the integration constant $\tilde{\mu} = \mu/r_0^z$ corresponds to having non-vanishing chemical potential in the dual system. In general, we are not working with the Lifshitz background but with solutions that are only asymptotically Lifshitz as $u \to \infty$. However, as long as the value of z isn't too high,[2] the asymptotic behavior of the gauge potential carries over from the Lifshitz background to the more general case, and one can read off the charge density and chemical potential in the dual field theory from the leading two terms in the expansion of α at large u. Calculations in this paper refer to fixed ρ, corresponding to a fixed density of charge carriers in the dual system, but one can also work at fixed chemical potential.

2.2. Numerical solutions

Black brane solutions at generic $z > 1$ can be obtained using numerical techniques similar to those of [5]. The field equations are integrated numerically starting near the black hole, with suitable initial conditions and proceeding out towards the asymptotic region. For a regular, non-degenerate horizon we require the functions that appear in the metric ansatz (14) to behave as

$$f(u) = \sqrt{u}(f_0 + f_1 u + \ldots), \quad g(u) = \frac{1}{\sqrt{u}}(g_0 + g_1 u + \ldots), \tag{25}$$

near $u = 0$. The appropriate near-horizon behavior of the remaining field variables can be worked out and then inserted into the equations of motion to generate initial value data for the numerical integration. For any given values of d and z, we obtain a three parameter family of initial values, where for instance $\phi(0)$, $\beta(0)$, and $b(0)$ can be taken as the independent parameters. The condition that the metric and massive vector approach the Lifshitz fixed point solution (10) and (13) sufficiently rapidly as $u \to \infty$ restricts the solutions further [5, 6, 7]. As a result, $b(0)$ is fixed for given $\phi(0)$ and $\beta(0)$ and one has a two-parameter family of solutions. This means in particular that, in the absence of scalar hair, there is a unique (up to overall scale) asymptotically Lifshitz charged black brane solution for given d, z and $\tilde{\rho}$.

The Hawking temperature is determined by the near-horizon behavior of the black brane metric,

$$T_H = \frac{r_0^z}{4\pi}\frac{f_0}{g_0}\,. \tag{26}$$

The full numerical solution of the field equations is required, however, to relate the coefficients f_0 and g_0 to the charge density ρ and other physical variables of the dual field theory.

[2] At $z \geq 3d$ non-linear effects give rise to additional terms in (24) with a falloff in between that of the charge density and chemical potential terms.

2.3. Conserved charge under radial evolution

A conserved charge under radial evolution was found in [6] for electrically neutral black branes with $d = 2$ and arbitrary dynamical critical exponent z. Such a conserved charge is useful for matching solutions across the bulk geometry from the near-horizon region to the asymptotic large u region and also provides a check on numerical solutions. The charge found in [6] generalizes to charged black branes with scalar hair in general spatial dimensions,

$$\begin{aligned} D_0 &= r_0^{z+d} e^{(z+d)u} f \Big[\frac{1}{2g}(\chi^2 - 2d(d+1)) - 2d(z-1)ab - d\,\alpha\beta \\ &\quad + g\Big[(z-1)(d\,a^2 - z\,b^2) + z^2 + (d-1)z + d^2 - \frac{1}{2}(\beta^2 + m^2\phi^2 - q^2\alpha^2\phi^2)\Big]\Big]. \end{aligned} \tag{27}$$

It is straightforward to check that $\frac{d}{du}D_0 = 0$ when the field equations (16) - (23) are satisfied. The conserved charge is related to thermodynamic state variables of the dual system in a simple way. Inserting a perturbative near-horizon expansion of the fields, one finds

$$D_0 = 2r_0^{d+z}\frac{f_0}{g_0} = 32\pi S\,T, \tag{28}$$

where T is the temperature (26) and $S = r_0^d/4$ is the Bekenstein-Hawking entropy density, which are identified with the temperature and entropy of the dual field theory.

2.4. Exact solution

As always, it is useful to have explicit analytic solutions to work with. Although Lifshitz black branes at $z > 1$ can in general only be obtained numerically, it turns out that for each value of d an isolated $z = 2d$ exact solution can be found [4, 8],[3]

$$b = 1, \quad f^2 = \frac{1}{g^2} = a^2 = 1 - e^{-2du}, \quad f\alpha = \pm\sqrt{2}e^{-du}(1 - e^{-du}), \tag{29}$$

with $\phi = 0$ and $\tilde{\rho} = \pm\sqrt{2}d$. It is straightforward to continue the exact black brane metric inside the horizon and obtain the globally extended geometry [4]. Define a tortoise coordinate u_* by

$$u_* = \frac{1}{2d\,r_0^{2d}}\log\left(1 - e^{-2du}\right), \tag{30}$$

and then transform the (t, u) variables to a pair of null coordinates

$$V = \exp\left[d\,r_0^{2d}(u_* + t)\right], \quad U = -\exp\left[d\,r_0^{2d}(u_* - t)\right]. \tag{31}$$

In the new coordinate system the metric is given by

$$ds^2 = \frac{-dU\,dV}{d^2(1+UV)^2} + \frac{r_0^2\,d\vec{x}^2}{(1+UV)^{1/d}}, \tag{32}$$

and is manifestly non-singular at the horizon, which is located at $UV = 0$. There is a null curvature singularity at $UV \to \infty$, which corresponds to $r \to 0$ in the original coordinate system. The asymptotic region $r \to \infty$ corresponds to $UV \to -1$. By a further transformation

[3] The corresponding exact solutions for $d = 2$, $z = 4$ black holes with a spherical horizon and topological black holes with a hyperbolic horizon were also found in [4]. In the limit of vanishing electric charge these black hole solutions reduce to the previously discovered $z = 4$ black hole solution of [6].

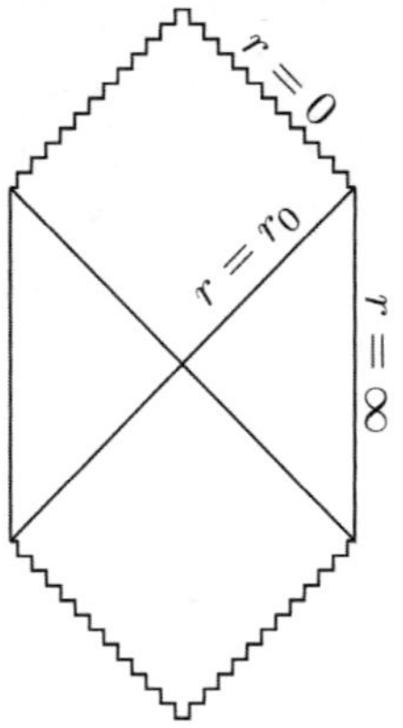

Figure 1. Global geometry of an asymptotically Lifshitz black hole. The null singularity at $r = 0$ is indicated by the jagged lines.

to new null variables P, Q, defined through $V = \tan\frac{\pi P}{2}$, $U = \tan\frac{\pi Q}{2}$, the global geometry can be represented by a simple diagram shown in Figure 1. Each point in the diagram represents an entire d-volume parametrized by $\vec{x}$. The global diagram in Figure 1 differs from standard Carter-Penrose conformal diagrams in that the boundary at $r \to \infty$ is not conformally flat. This is a consequence of the scaling asymmetry between t and $\vec{x}$ and is readily apparent in the globally extended metric (32).

When $z = 1$ the auxiliary massive vector field $\mathcal{A}_\mu$ can be consistently set to zero and the field equations then reduce to those of Einstein-Maxwell gravity with a negative cosmological constant. In this case, there is a well known exact solution, the AdS-Reissner-Nordstrm black brane, which has a timelike curvature singularity inside an inner and an outer horizon. The interior geometry of the exact $z = 2d$ black brane is markedly different with a null curvature singularity at $r = 0$ and no smooth inner horizon.

3. Holographic superconductors with asymmetric scaling

We now consider charged black branes with hair. Static spherically symmetric solutions of the scalar field equation (6) have the asymptotic form $\phi(u) \to c_-(e^{-\Delta_- u} + \ldots) + c_+(e^{-\Delta_+ u} + \ldots)$, with

$$\Delta_\pm = \frac{d+z}{2} \pm \sqrt{\left(\frac{d+z}{2}\right)^2 + m^2}, \tag{33}$$

while the asymptotic behavior of the electromagnetic field field strength is

$$\beta(u) \approx \frac{\rho}{r_0^d} e^{-du} + \ldots . \tag{34}$$

Working at fixed charge density in the dual field theory, we read the radial location of the horizon off from the asymptotic behavior of $\beta(u)$ and then the temperature can be obtained from the numerical solution for $f(u)$ and $g(u)$ using (26).

In the following we set the scalar mass squared to $m^2 = \frac{1}{4} - \left(\frac{d+z}{2}\right)^2$, which is inside the range where there is a choice of two boundary theories [9]. This choice leads to convenient values, $\Delta_\pm = \frac{d+z}{2} \pm \frac{1}{2}$, for the dimensions of the operators $\mathcal{O}_\pm$ that are dual to the scalar field in each of the two boundary theories. Non-linear descendants of the leading scalar field modes are suppressed by $O(e^{-2u})$ at $u \to \infty$ and this choice of mass squared ensures that the first descendant of ψ_- falls off faster than ψ_+.

In order to study holographic superconductivity, we first select some value for d, z and the electric charge q carried by the scalar field and then generate numerical black brane solutions for a range of initial values $\beta(0)$ and $\phi(0)$. We then investigate the asymptotic large u behavior of the scalar hair in the numerical solutions. A superconducting condensate corresponds to either

$$c_+ = 0, \quad \langle \mathcal{O}_- \rangle = c_- \neq 0, \qquad \text{or} \qquad c_- = 0, \quad \langle \mathcal{O}_+ \rangle = c_+ \neq 0, \tag{35}$$

depending on which of the two boundary theories is being considered [10, 11]. We look for a curve in the $\beta(0)$ *vs.* $\psi(0)$ plane of initial values at the horizon, for which the corresponding black brane solution has vanishing c_+ (c_-), and tabulate the value of c_- (c_+) along this curve. The temperature is found from the same numerical solutions *via* (26) and (34). Figure 2 shows a plot of c_- *vs.* T obtained by this procedure for $d = 2$, $z = 2$, and $q = 1$. The results are expressed in terms of dimensionless ratios that are insensitive to the overall scale (set by the charge density ρ, which is held fixed at some finite value throughout).

These results demonstrate that a superconducting condensate can form in systems with anisotropic scaling but we have not touched on a number of interesting topics including the electric conductivity and magnetic properties of these holographic superconductors.

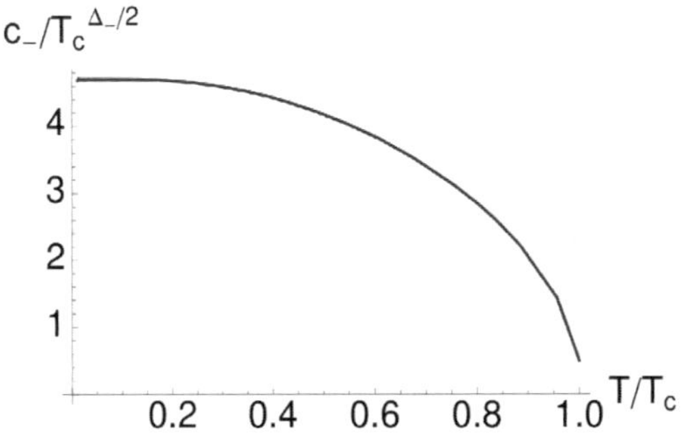

Figure 2. Scalar field condensate in a holographic superconductor at $z = d = 2$.

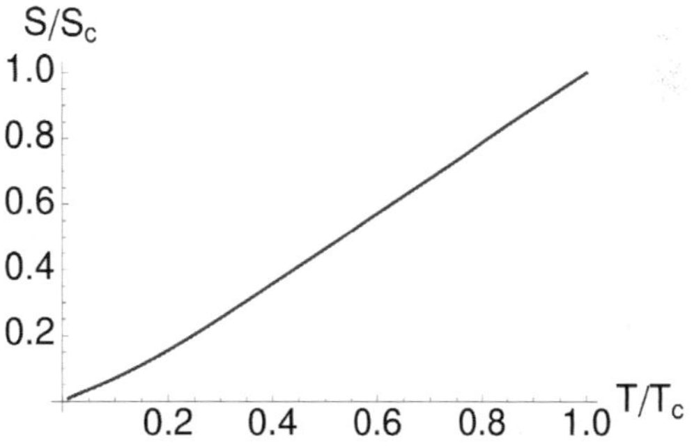

Figure 3. Entropy as a function of temperature in the superconducting phase of the holographic superconductor in Figure 2.

4. Zero temperature entropy vs. scalar hair

In the absence of charged matter, the charged black branes in our model have an extremal limit given by $\tilde{\rho} \to \pm\sqrt{2(z^2 + (d-1)z + d^2)}$. It then follows from $\tilde{\rho} = \rho/r_0^d$ that the radial location of the horizon r_0 has a finite value in the extremal limit for fixed charge density ρ. This in turn means that the entropy density $S = r_0^d/4$ remains finite in the zero temperature limit indicating a macroscopic groundstate degeneracy.

This conclusion is radically altered when the system is coupled to a charged scalar field. In this case the zero temperature limit is approached in the condensed phase and the black holes, that are dual to extreme low temperature states, have scalar hair. It is straightforward to keep track of the black brane entropy as the temperature is lowered below the critical value for forming the superconducting condensate. The result for the same $d = 2$, $z = 2$ holographic superconductor as was considered in the previous section is shown in Figure 3. Both the entropy density and the temperature are normalized to their values at the onset of condensation, S_c and T_c respectively.

Although the numerical calculations break down before absolute zero is reached, the numerical data strongly suggest that S vanishes in the $T \to 0$ limit for a generic dynamical critical exponent z, which is consistent with a non-degenerate ground state in the dual field theory. The corresponding result for conformal systems with $z = 1$ was established in [12].

5. Summary

We have presented an overview of the construction of charged black brane solutions in gravity models that realize the anisotropic scaling symmetry that is characteristic of many interesting quantum critical points. The motivation for the study of these models comes from condensed matter theory, in particular from two- and three-dimensional systems involving strongly correlated electrons. The relevance of gravitational models to real world condensed matter systems remains highly speculative, but the gravitational approach continues to produce effects that are intriguingly similar to what is seen in experiments. A recent example from our own work [13] involves non-Fermi-liquid behavior in the specific heat of anisotropic black branes of the type considered in the present paper, which turns out to be qualitatively similar to the measured specific in certain heavy fermion metals near a quantum phase transition [14, 15]. While much of the work on gravitational modeling of strongly coupled field theories to date involves asymptotically AdS spacetime and an underlying conformal symmetry, it was crucial to the success of this particular application to have a non-trivial dynamical critical exponent $z > 1$. This provides impetus for further study of gravity models with anisotropic scaling.

Acknowledgments

This work was supported in part by the Göran Gustafsson foundation, the Swedish Research Council (VR), the Icelandic Research Fund, the University of Iceland Research Fund, and the Eimskip Research Fund at the University of Iceland.

References

[1] Kachru S, Liu X, and Mulligan M 2008 Gravity duals of Lifshitz-like fixed points *Phys. Rev.* D **78** 106005 (*Preprint* arXiv:0808.1725 [hep-th])

[2] Koroteev P and Libanov M 2008 On existence of self-tuning solutions in static braneworlds without singularities *J. High Energy Phys.* JHEP02(2008)104 (*Preprint* arXiv:0712.1136 [hep-th])

[3] Taylor M 2008 Non-relativistic holography (*Preprint* arXiv:0812.0530 [hep-th])

[4] Brynjolfsson E J, Danielsson U H, Thorlacius L and Zingg T 2010 Holographic superconductors with Lifshitz scaling *J. Phys. A: Math. Theor.* **43** 065401(*Preprint* arXiv:0908.2611 [hep-th])

[5] Danielsson U H and Thorlacius L 2009 Black holes in asymptotically Lifshitz spacetime," *J. High Energy Phys.* JHEP03(2008)070 (*Preprint* arXiv:0812.5088 [hep-th])

[6] Bertoldi G, Burrington B A and Peet A (2009) Black holes in asymptotically Lifshitz spacetimes with arbitrary critical exponent *Phys. Rev.* D **80** 126003 (*Preprint* arXiv:0905.3183 [hep-th])

[7] Ross S F and Saremi O (2009) Holographic stress tensor for non-relativistic theories *J. High Energy Phys.* JHEP09(2009)009 (*Preprint* arXiv:0907.1846 [hep-th])

[8] Pang D W 2010 On charged Lifshitz black holes *J. High Energy Phys.* JHEP10(2010)116 (*Preprint* arXiv:0911.2777 [hep-th])

[9] Klebanov I R and Witten E 1999 AdS/CFT correspondence and symmetry breaking *Nucl. Phys. B* **556** 89 (*Preprint* hep-th/9905104)

[10] Gubser S S 2008 Breaking an Abelian gauge symmetry near a black hole horizon *Phys. Rev.* D **78** 065034 (*Preprint* arXiv:0801.2977 [hep-th])

[11] Hartnoll S A, Herzog C P and Horowitz G T 2008 Building a Holographic Superconductor *Phys. Rev. Lett.* **101** 031601 (*Preprint* arXiv:0803.3295 [hep-th])

[12] Horowitz G T and Roberts M M 2009 Zero temperature limit of holographic superconductors *J. High Energy Phys.* JHEP11(2009)015 (*Preprint* arXiv:0908.3677 [hep-th])

[13] Brynjolfsson E J, Danielsson U H, Thorlacius L and Zingg T 2010 Black hole thermodynamics and heavy fermion metals (*Preprint* arXiv:1003.5361 [hep-th])

[14] Löhneysen H v, Pietrus T, Portisch G, Schlager H G, Schröder A, Sieck M, and Trappmann T 1994 Non-Fermi liquid behavior in a heavy-fermion alloy at a magnetic instability *Phys. Rev. Lett.* **72** 3262

[15] Stewart G R 2001 Non-Fermi-liquid behavior in d- and f-electron metals *Rev. Mod. Phys.* **73** 797 (Addendum-ibid. 2006 **78** 743)

Geometric reduction of the string landscape

Joseph Towe
Department of Physics,
The Antelope Valley College, Lancaster, CA 93536 USA

E-Mail: jtowe@avc.edu

Abstract: The string background AdS_7XS^4 and dual CFT are adopted, and symmetric, multi-dimensional objects are identified as elements of the 7-dimensional vacuum that complements 4-spacertime in the proposed, 11-dimensional theory. Increments of the proposed, multi-dimensional elements with respect to increments of linear scale are (assuming a vacuum density) interpreted as creation operators or Higgs events that break SU(7). Each creation operator (each symmetry breaking event) and resulting mass is identified as a gauge transformation that preserves SUGRA gauge. If appropriately calibrated, the proposed model approximates the masses of the fermionic spectrum and predicts a new lepto-quark state of mass $M \cong 30$ GeV/c^2. Because the proposed model derives quantization from gauge invariance (as proposed by F. London in 1927), this model can be implemented as a large scale quantization which results in a galactic hierarchy of states that maximize Riemannian symmetry. If the number of such states is equal to the range of the quantum number which, in the micro-scale model, generates the approximate lepto-quark masses, then the large scale model generates a theoretical number of galaxies that approximates the number indicated by observation.

I. The spontaneous breaking of SU(7)

Based upon AdS/CFT correspondence [1], the string background AdS_7XS^4 [2] and dual CFT are adopted. Thus a Type IIB theory of super-strings is regarded as equivalent to a maximally super-symmetric SU(N) Yang-Mills theory in a Minkowski 4-spacetime [3]. In this context, it is postulated that symmetric, (M+1)-dimensional classes of elements: l^{M+1}: M=0,1,2,…,6 collectively constitute the 7-dimensional vacuum that complements 4-spacetime in the proposed 11-dimensional theory. It is assumed that all such elements share the ground state energy or vacuum energy of this 7-dimensional domain and that the seven postulated classes of multi-dimensional elements constitute a fundamental representation of the group SU(N): N=7.

A second postulate recalls that conformal symmetries preserve angles and admit increments of scale. In this context increments d^M of the proposed (M+1)-dimensional magnitudes (l^{M+1} : M=0, 1,2,…,6) with respect to increments, dl, of linear scale are identified as creation operators:

$$d^M(l^{M+1}) = (M+1)!(l)dl \cdots \underset{M}{\longleftrightarrow} \cdots dl : \tag{1}$$

M= 1,2,…,6 that enhance the vacuum energy (divided by c^2) which is, according to hypothesis, shared by the postulated multi-dimensional objects l^{M+1}. Consequently, the factors (M+1)! which are generated by the proposed creation operators are regarded as the factors by which mass is enhanced. The resulting masses are collectively identified as constituting a fundamental representation of broken SU(7).

The resulting masses will also be identified as the masses of known lepto-quark states plus the mass of a predicted state. Specifically, because the proposed model is calibrated by averaging two of the elements that constitute the broken fundamental representation, the symmetry that will be applied consists of only six elements. Thus the applied symmetry can be represented in terms of a projection that distinguishes lepto-quark states in terms of only two quantum numbers: isotopic spin (I_3) and generation. Each vertex of the consequent hexagon can be identified with a specific lepto-quark state and with a specific creation operator provided that the proposed hexagon is divided into three levels, that each level is associated with a value of I_3 and that increments of mass are associated with clockwise rotations about the center of symmetry of the hexagon.

This model is calibrated by association of the bottom base with I_3=0, by association of the middle level with I_3=+1/2 and by association of the top level with I_3 = -1/2. A second aspect of the proposed calibration associates the vertex at the right side of the bottom base with the average of the I_3 values that correspond to the lepto-quark states (up, $\nu^{e^-}{}_L$) and (down, $e^-{}_L$). (These are the states that will be averaged to provide a calibrating mass.) In this context, the first increment of a clockwise rotation, beginning at the vertex which is located at the left side of the bottom level of the hexagon identifies the vertex that lies at the left side of the bottom level with the lepto-quark state that contains the strange quark and with the first creation operator $d^M l^{M+1}$: M=1 of (1). A second clockwise increment of rotation associates the vertex at the left side of the middle level with the lepto-quark state that is characterized by the charmed quark and with the creation operator $d^M l^{M+1}$: M=2 of (1). A third clockwise increment of rotation associates the vertex on the left side of the top level with the lepto-quark state that is characterized by the bottom quark and with the creation operator $d^M l^{M+1}$: M=3 of (1). A fourth clockwise increment associates the vertex at the right side of the top level with a currently unknown lepto-quark state that is predicted and with the creation operator $d^M l^{M+1}$: M=4 of (1) and finally a fifth clockwise increment associates the vertex which lies on the right side of the middle level with the lepto-quark state that is characterized by the top quark and by the creation operator $d^M l^{M+1}$: M=5 of (1). To fully implement this model it is necessary to discuss details regarding observed values.

2. Comparison with observation

Since it is indicated that the proposed geometric process for the introduction of mass may approximate the masses of the fundamental fermions, it is important to discuss the observational data with which theoretical numbers are to be compared. The masses of leptons are well known, but there is some controversy regarding the masses of the quarks. The concept of quark mass is clarified however, if the constituent picture is prescribed for mesons and baryons that are made of quarks that are more massive than the up and down quarks. It is implicitly argued that the masses which emerge from the constituent quark model (where the sum of the constituent masses of valence quarks is roughly equal to the mass of a baryon or meson) are more relevant than those emerging from the Sine-Gordon Soliton model and the Chiral-Lagrange model. Based upon this argument, the masses of lepto-quark states will, in this discussion, refer to the constituent masses of quarks. These quark masses are approximately: charmed: 1.5 GeV/c^2; strange: 0.50 GeV/c^2; bottom: 4.3 GeV/c^2. The mass of the top quark is 176 GeV/c^2 (according to SLAC), but if this number is rounded to two significant digits, as are the other values, then the mass of the top quark is approximately 180 GeV/c^2 [4], [5].

The second consideration that must be clarified if theoretical and experimentally established numbers are to be compared is the aforementioned calibration of the micro-scale. Since the masses of the up and down quarks are nearly the same, the initial condition, in terms of which calibration is done involves a mean of the masses of these quarks. An average or mean of the masses of the up and down quarks will be determined from three sources. The heaviest value will be indicated by the rho vector meson; the best known moderate value will be indicated by the proton and the lightest value will be indicated by the pion. If these means are averaged, the result is approximately 0.25 GeV/c^2. This number will be assigned to the fundamental vacuum energy (divided by c^2) that was introduced in the previous section. Adopting the notation *"f"* for this vacuum "mass" which is postulated as common to the elements l^{M+1} of the proposed 7-dimensional vacuum, it will now be demonstrated that the proposed hypothesis (embodied in expression (1)) prescribes an array of lepto-quark states that includes observed particles, plus a new lepto-quark state that is not currently recognized.

If N=1, then $[(M+1)!]f$ becomes

$$(2)(0.25)GeV/c^2 = (0.5)GeV/c^2 \tag{2}$$

which approximates the observationally determined mass of the strange quark plus the relatively negligible mass of the μ^-.

If N=2, then $[(M+1)!]f$ becomes

$$(3)(2)(0.25)GeV/c^2 = (1.5)\,GeV/c^2 \tag{3}$$

which approximates the observationally determined mass of the charmed quark plus the relatively negligible mass of the $\nu^{\mu-}$.

If N=3, then $[(M+1)!]f$ becomes

$$(4)(3)(2)(0.25)GeV/c^2 = (6.0)\,GeV/c^2, \tag{4}$$

which approximates the observationally determined mass of the bottom quark plus that of the τ^- (the former of about 4.3 GeV/c^2 and the latter of about 1.7 GeV/c^2).

If N=4, then $[(M+1)!]f$ becomes

$$(5)(4)(3)(2)(0.25)GeV/c^2 = (30.0)\,GeV/c^2, \tag{5}$$

which corresponds to the mass of a lepto-quark state that is not currently recognized. This new state constitutes a prediction by which the proposed model lends itself to confirmation.

If N=5, then $[(M+1)!]f$ becomes

$$(6)(5)(4)(3)(2)(0.25)GeV/c^2 = (180.0)\,GeV/c^2, \tag{6}$$

which approximates the observationally determined mass of the top quark plus the relatively negligible mass of the tauon's neutrino.

3. Lepto-quark states as stationary super-gravitational states

A fourth postulate associates the events that break the super-conformal symmetry with gauge transformations in a theory of pure super-gravity and identifies the individual mass classes that constitute the fundamental representation of the broken SU(7) symmetry as resulting from gauge transformations that preserve SUGRA gauge. To discuss SUGRA interactions and the gauge transformations that occur to the SUGRA connections it is necessary to briefly digress. The theory of 11-dimensional super-gravity is based upon an Osp(1/.4)-symmetric Lagrangian:

$$\mathscr{L} = \sqrt{-g}R + e\overline{\psi}_{\mu}\gamma_{\nu}\gamma^{5}\nabla_{\sigma}\psi_{\rho}\varepsilon^{\mu\nu\rho\sigma}, \tag{7}$$

which is based upon the super-Poincare algebra

$$[M_A, M_B\} = f^C{}_{AB}M_C, \tag{8}$$

where the components of diagonal generators M_A are

$$M_A = \{P_a, -iM_{ab}, Q_\alpha\}. \tag{9}$$

The P_a generate the translation group, the $-iM_{ab}$ constitute the adjoint representation of the Lorentz group and the Q_α are components of the SUSY generator.

If the ω^A describe all connection fields:

$$\omega^A{}_\mu = (e^a{}_\mu, \omega^{ab}{}_\mu, \overline{\xi}^\alpha{}_\mu), \tag{10}$$

and if the $\omega^A{}_\mu$ transform under Osp(1/4) as

$$\delta\omega^A{}_\mu = f^A{}_{BC}\varepsilon^B\omega^C{}_\mu, \tag{11}$$

then the Osp(1/4) covariant derivative is

$$\nabla_\mu = \partial_\mu + M_A\omega^A{}_\mu = \partial_\mu + e^a{}_\mu P_a - i\omega^{ab}{}_\mu + \overline{\xi}^\alpha{}_\mu Q_\alpha \tag{12}$$

and the Riemannian curvature tensor is

$$[\nabla_\mu, \nabla_\nu] = R^A{}_{\mu\nu}M_A, \tag{13}$$

where

$$R^A{}_{\mu\nu} = \partial_\mu\omega^A{}_\nu - \partial_\nu\omega^A{}_\mu + \omega^B{}_\nu\omega^C{}_\mu f^A{}_{CB} \tag{14}$$

[6]. In this context the interaction Hamiltonian for pure SUGRA interactions is

$$-i(\alpha)\overline{\psi}_\mu\gamma_\nu\gamma^5 M_A\omega^A{}_\sigma\psi_\rho\varepsilon^{\mu\nu\rho\sigma}. \tag{15}$$

In this context it is assumed that each lepto-quark state that constitutes an element of the fundamental representation of the broken SU(7) symmetry is individually preserved by a characterizing class of super-gravitational interactions. Accordingly, the transitions among these super-gravitational states are associated with gauge transformations (on the SUGRA connections) that preserve SUGRA gauge:

$$\exp(i)\delta\omega^A = \exp(i)f^A{}_{BC}\varepsilon^B\omega^C = \exp[2i\pi M]: \tag{16}$$

M=1,2,3,…

An example of a SUGRA GUT interaction (a quark-lepton transition that is super-gravitationally mediated) is as follows:

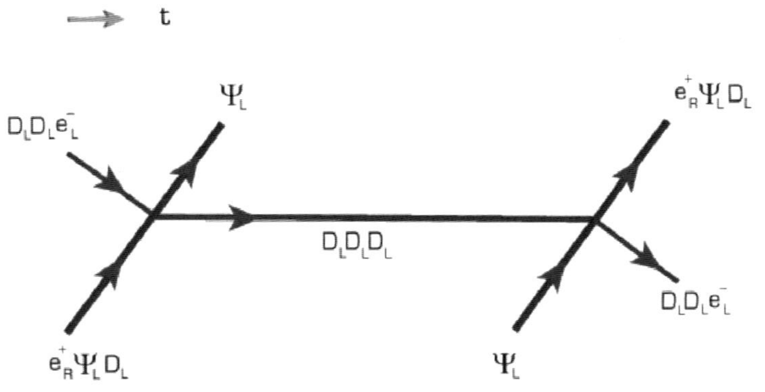

Figure 1 A First Order SUGRA Interaction

Clearly this interaction is mediated by the SUGRA pair consisting of $e^+{}_R\psi_LD_L$ and $D_LD_LD_L$. A pure SUGRA interaction that would involve a quark-lepton transition can be mediated by super-gravitational partners

$$(\omega^A{}_\mu)_L \equiv (e^-{}_Le^-{}_Le^-{}_L)^A{}_\mu \quad \text{and} \quad (\psi_\mu)_L = (e^+{}_R\psi_LD_L)_\mu \tag{17}$$

Each of the seven elements of the postulated fundamental representation of broken SU(7) is preserved by a characteristic SUGRA interaction; e.g. by an interaction that is mediated by an analogue of the SUGRA pair that is described by equation 17 (or depicted by Figure 1). Thus the 14 generators and connections of the Osp(1/4) theory are exhaustively implemented and each lepto-quark class is preserved by a characterizing super-gravitational interaction. It will now be argued that the stationary SUGRA states in terms of which the small scale is modeled can also be applied to the astrophysical scale.

4. Large scale SUGRA states

As just observed, the connection coefficients $\omega^A{}_\mu$ of local super-symmetry lend themselves to gauge transformations:

$$\exp(i\beta)\delta\omega^A = \exp(i\beta)f^A{}_{BC}\omega^A{}_\mu dx^\mu \tag{18}$$

under the local SUSY group Osp(1/4), where β represents a scale factor. The large scale model that was introduced in 2008 [7] identifies inflation events with the gauge transformations that are intrinsic to local super-symmetry. Moreover, this model imposes gauge invariance, which restricts gauge transformations to those that preserve maximal Riemannian symmetry:

$$\exp(i\beta)f^A{}_{BC}\varepsilon^B\omega^C = \exp[2i\pi N]: \tag{19}$$

Journal of Physics: Conference Series **462** (2013) 012056 doi:10.1088/1742-6596/462/1/012056

N=0,1,2,.... Since this model is applied to the large scale it is calibrated in terms of a large scale boundary condition that is established by observation: *It is observed that galaxies are typically separated by distances that are about ten times the diameter of the typical galaxy; that galactic clusters are typically separated by distances that are about ten times the diameter of the typical cluster etc.* Based upon this boundary condition, the constant β of expression (19) is chosen as

$$\frac{\beta}{2\pi} \cong 0.434294482..., \tag{20}$$

so that the equation (19) reduces to

$$\exp[f^{A}{}_{BC}\varepsilon^{B}\omega^{C}] = 10^{N}: \tag{21}$$

N=0,1,2,...

If the range of the integral values N in expression (21) is the same as the final range of the quantum number M in Section 2 (since two values of M were incorporated into a single initial condition in that discussion, the proposed range of values was reduced from 0,1,2,...,6 to 0,1,2,...,5), then the large scale model that is associated with (21) describes a galactic hierarchy that parallels observation. Specifically, observations of local conditions reveal that the typical galaxy is about $ct_0=10^5$ light years in diameter and that five galaxies populate the typical basic cluster. Based upon these boundary conditions and upon an assumption of homogeneity (up to very large scales), one can roughly associate the expression $(4/3)\pi R_1^3$, which will be regarded as approximately describing the volume of a typical basic cluster (in terms of galaxies) with the number '5.00.' Thus, rounding to three digits it is concluded that typically, the radius of the cluster which associates with N=1 is about $R_1 \cong 1.06$. Now utilizing this value of R_1 and citing the boundary condition that corresponds to the equation (21), one determines the approximate 'radius' of the N=2 state (recall that clusters are typically separated by a distance about ten times the diameter of the typical cluster): (2)(10)(1.06)=21.2, which is interpreted as the approximate 'radius' of a super-cluster. Now proceeding similarly, one can determine the 'radius' of the N=3 state: (2)(10)(21.2)=424; of the N=4 state: (2)(10)(424)=8480 and of the N=5 state: (2)(10)(8480)= 169600 ≅ 170,000.

Since gauge invariance associates with maximal Riemannian symmetry, it is inferred that the initial phase transition on a SUGRA gauge connection (referring to expression (21) where N=1) produces a spherically symmetric wave front and that the phase transitions on this radially symmetric wave are also isotropically distributed. Specifically it is argued that the second of the phase transitions that is prescribed by (21) (that corresponding to N=2) emerges from each point of the initial wave just as this wave is initially propagated. It is argued that, due to inflation, the points of propagation become instantaneously separated so that the second wave appears as an isotropic distribution of semi-spherical, secondary wavelets that are mirror images of each other about the center of radial symmetry of the expansion. Analogously it is argued that the third of the phase transitions that is prescribed by (21) (that corresponding to N=3) emerges from each point of the secondary waves (just as the primary and secondary waves are simultaneously propagated). Again, due to inflation, these points become instantaneously separated so that the second wave appears as an isotropic distribution of semi-spherical, tertiary wavelets that are mirror images of each other about the center of radial symmetry of the initial expansion etc.

This sequence of phase transitions produces a Huygens pattern along the shock front of global expansion. The predicted empty "bubbles" of instantaneous expansion are observed by radio surveys of the regions near the event horizon [8]. The distributions of galactic clusters that are, according to observation, distributed along the boundaries of the bubbles probably result from the instantaneous expansions of the bubbles, which leave the more slowly expanding distributions of mass-energy at the interstices.

It is currently argued that the observable evidence of the big bang that exists today will ultimately disappear [9]. In this vein it is argued that the event horizon, which includes the Huygens pattern that is described above will ultimately disappear beyond the event horizon. It is argued that in this limit, the galaxies that initially constitute the interstices of inflationary bubbles will approach a distribution in which they approximate the surface of a large sphere. In this limit the radius of the N=5 configuration is implemented to determine the approximate surface area of this spherical idealization as a number of galaxies. This calculation produces about $4\pi R_5^2 \cong 3.63 \text{ X } 10^{11}$, which is approximately the number of galaxies that is indicated by observation.

Conclusion

Based upon AdS/CFT correspondence, the string background AdS_7XS^4 and its dual CFT are adopted. Thus a Type IIB theory of super-strings is regarded as equivalent to a maximally super-symmetric SU(N) Yang-Mills theory in a Minkowski 4-spacetime. In this context, it is postulated that symmetric, (M+1)-dimensional classes of elements: l^{M+1}: M=1,2,…,6 collectively constitute the 7-dimensional vacuum that complements 4-spacetime in the proposed 11-dimensional theory. It is postulated that these symmetric, multi-dimensional elements share the energy of the vacuum and constitute a fundamental representation of super-conformal SU(7). Secondly it is postulated that increments of the proposed, multi-dimensional objects with respect to increments of linear scale act as creation operators that enhance the vacuum energy (divided by c^2) which is by hypothesis shared by the postulated multi-dimensional objects, thereby breaking the super-conformal symmetry of the vacuum. Calibration of this model approximates the masses of known lepto-quark states and predicts a new lepto-quark state of approximate mass $30(GeV/c^2)$.

It is argued that the elements which constitute the fundamental representation of the broken SU(7) are individually preserved by characterizing SUGRA interactions. Specifically, it is observed that mediating SUGRA connections are susceptible to gauge transformations under the local SUSY group, but it is argued that if such gauge transformations are restricted to those that preserve SUGRA gauge, then the elements of broken SU(7) that are individually preserved by characterizing SUGRA interactions are distinguished in that they preserve maximal Riemannian symmetry and thereby select small values of the cosmological constant (as opposed to the large values that seem to associate with large scale chaos).

Since quantization is derived in this model from gauge invariance, as proposed by F. London in 1927, this model is implemented as a large scale quantization, which results in a galactic hierarchy of states (each preserving maximal Riemannian symmetry). If the range of the relevant large scale quantum number is equal to the range of the quantum number which, in the micro-scale model approximates the lepto-quark spectrum, then the large scale model produces a theoretical number of galaxies that approximates the number indicated by observation.

References

[1] D'Hoker, E. and Freedman, D. "Super-symmetric gauge theories and the AdS/CFT correspondence," *arXiv:* hepthe/020123, 2002.

[2] Aharony, O., Gubser S. et. al. "Large N theories, string theory and gravity," *arXiv*: hep-th/990511f v**3**, 1999, 48-52.

[3] Zwieback, B. *A First Course in String Theory*, Cambridge University Press, New York (2004), Chap. 16, Sect. 8.

[4] *SLAC Virtual Visitor Center* (2009), SLAC National Accelerator Laboratory, Menlo Park, CA.

[5] Bock, R. at CERN, Geneva and Vasilesca, A. at IFA, Bucuresti (1999) *Particle Detector BriefBook.*

[6] Kaku, M. *Quantum Field Theory*, pp. 692-696, Oxford Universety Press, Oxford (1993).

[7] Towe, J. "The gauge-invariant spectrum of local super-symmetry as the universe that is

observed," *Int. J. Th. Phys.* 2008, **47**, 2898-2903.

[8] Huchra, J. P. and Geller, M., *Semi-Annual report: Harvard-Smithsonian Center for Astrophysics*, 1 July-31 December, 1987, 4-5.

[9] Krauss, L. and Scherrer, R. "The return to a static universe and the end of cosmology," *Journal of General Relativity and Gravitation*, 2007, **39**, No. 10, 1545-1550.

[10] London, F., "Quantenmechanische deutung der theorie von Weyl," *Z. Phys.,*1927, **42**, 375-389.

Journal of Physics: Conference Series **462** (2013) 012057 doi:10.1088/1742-6596/462/1/012057

Black Hole Bose Condensation

Cenalo Vaz[a,b] **and L.C.R. Wijewardhana**[b]

[a]RWC and [b]Department of Physics, University of Cincinnati, Cincinnati, OH 45242.

E-mail: Cenalo.Vaz@uc.edu and Rohana.Wijewardhana@uc.edu

Abstract. General consensus on the nature of the degrees of freedom responsible for the black hole entropy remains elusive despite decades of effort dedicated to the problem. Different approaches to quantum gravity disagree in their description of the microstates and, more significantly, in the statistics used to count them. In some approaches (string theory, AdS/CFT) the elementary degrees of freedom are indistinguishable, whereas they must be treated as distinguishable in other approaches to quantum gravity (eg., LQG) in order to recover the Bekenstein-Hawking area-entropy law. However, different statistics will imply different behaviors of the black hole outside the thermodynamic limit. We illustrate this point by quantizing the Bañados-Teitelboim-Zanelli (BTZ) black hole, for which we argue that Bose condensation will occur leading to a "cold", stable remnant.

By now it seems clear that most approaches to quantum gravity are able to obtain some general features of black hole thermodynamics [1] from statistical mechanics in the thermodynamic limit. However, this seeming universality of black hole thermodynamics is belied by the fact that the leading candidates for a theory of quantum gravity differ from each other both in the microstates being counted and in the statistics applied to the elementary degrees of freedom. In string theory the black hole microstates are dual to weak field D-brane states [2], in the AdS/CFT approach they are taken to be the states of a particular horizon conformal field theory (CFT) [3] and in loop quantum gravity (LQG) they are represented by punctures of a spin network on the event horizon [4]. In string theory and AdS/CFT the elementary degrees of freedom are indistinguishable whereas they must be treated as distinguishable in LQG in order to recover the Bekenstein-Hawking law. We may expect that these different models will diverge in their descriptions of the black hole at very low temperatures and outside the thermodynamic limit, where statistics plays an important role.

In this contribution we ask what behavior one may expect outside the thermodynamic limit when Bose statistics is used to count the microstates. Since such a question may only be answered in a model dependent way, we pose it within the context of a model of the quantum black hole that arises naturally from a canonical quantization of gravitational collapse. We discuss the statistical mechanics of the BTZ black hole. (The importance of the BTZ black hole lies in the fact that many black holes in string theory have near horizon geometries of the form BTZ $\times \mathcal{M}$, where $\mathcal{M}$ is a simple manifold, and their thermodynamics can be obtained directly or by duality from the entropy of the BTZ black hole [5]. This makes our discussion relevant to a very large class of interesting black hole solutions in higher dimensions.)

We will argue that Bose-Einstein condensation should occur at sufficiently low temperatures. We will determine the condensate temperature and the condensate fraction as well as the entropy. We show that the area law holds at high temperatures but breaks down below the critical temperature, where the entropy becomes dominated by a mass independent term making the black hole cold and stable even though its mass remains non-vanishing.

Both canonical quantum gravity [6] and the AdS/CFT correspondence [7, 8] predict an equispaced mass spectrum for the BTZ black hole and agree on the indistinguishability of the elementary degrees of freedom. In canonical quantum gravity one obtains a stationary state solution of the Wheeler-DeWitt equation with the mass levels

$$M_j = \frac{\hbar}{l}\left(j + \frac{1}{2}\right), \tag{1}$$

where $l^2 = -\Lambda^{-1}$ is the AdS length. The macroscopic black hole is thought of as the end state of the collapse of many, say $\mathcal{N} = \sum_j \mathcal{N}_j$, matter shells, $\mathcal{N}_j$ of which occupy level j of (1). When a boundary contribution from the origin is included [9], the total black hole mass is given as

$$M = M_0 + \sum_j \frac{\hbar}{l}\left(j + \frac{1}{2}\right)\mathcal{N}_j. \tag{2}$$

From this point of view, the elementary degrees of freedom are bosonic mass shells and a black hole microstate is a particular distribution of $\mathcal{N}$ shells between the levels in (1). We begin with the grand partition function,

$$\Xi(\mu, \beta) = e^{-\beta M_0} \prod_j (1 - e^{-\beta(\varepsilon_\alpha - \mu)})^{-1}, \tag{3}$$

where the chemical potential μ is determined from the constraint $\sum_j \langle \mathcal{N}_j \rangle = \mathcal{N}$ and ε_j is the energy of the level j. Introducing the fugacity, $\lambda^{-1} = e^{-\beta(\mu - \varepsilon_0)} > 1$, and $\Delta\varepsilon_j = \varepsilon_j - \varepsilon_0$ we then have

$$\Xi(\mu, \beta) = e^{-\beta M_0} \prod_j \frac{\lambda}{\lambda - e^{-\beta \Delta\varepsilon_j}}. \tag{4}$$

Since the average level occupation number is computed directly from $\Xi(\beta, \mu)$,

$$\langle N_j \rangle = -\frac{1}{\beta}\frac{\partial \ln \Xi}{\partial \varepsilon_j} = \frac{1}{e^{\beta\Delta\varepsilon_j} - \lambda}, \tag{5}$$

our condition for determining the chemical potential is

$$\mathcal{N} = \sum_{j=0}^{\infty} \frac{1}{e^{\beta\Delta\varepsilon_j} - \lambda} = \frac{\lambda}{1-\lambda} + \sum_{j=1}^{\infty} \frac{\lambda}{e^{\beta\Delta\varepsilon_j} - \lambda}. \tag{6}$$

The first term on the right represents the ground state occupancy, N_0 (henceforth, average values will be understood and the angular brackets will be omitted), and the second represents the number of excited shells, N_{ex}.

The condensate fraction, f_C, is the ratio of the ground state occupancy to the total number of shells, $f_C = N_0/\mathcal{N}$. We will first determine f_C as a function of the temperature. For the spectrum of the BTZ black hole in (1), $\Delta\varepsilon_j = \hbar j/l$ so replacing the last sum in (6) by its power series expansion, we can express the total number of shells as

$$\mathcal{N} = \frac{\lambda}{1-\lambda} + \sum_{j=1}^{\infty}\sum_{r=1}^{\infty} \lambda^r e^{-\beta\hbar r j/l}. \tag{7}$$

The most direct way to evaluate the double sum in (7) is to employ the Mellin-Barnes representation of the exponential function and re-express it as an integral over the complex plane [10]. This gives

$$\mathcal{N} = \frac{\lambda}{1-\lambda} + \frac{1}{2\pi i}\int_{\tau-i\infty}^{\tau+i\infty} dt\, \frac{\Gamma(t)\zeta(t)\mathrm{Li}_t(\lambda)}{(\beta\hbar/l)^t}, \tag{8}$$

where $\zeta(t)$ is the zeta function and $\mathrm{Li}_t(\lambda)$ is the polylogarithm. For $\lambda < 1$ the integral is governed by the simple pole of the $\zeta-$function at $t = 1$, therefore

$$\mathcal{N} = \frac{\lambda}{1-\lambda} + \frac{\mathrm{Li}_1(\lambda)}{\beta\hbar/l} \equiv \frac{\lambda}{1-\lambda} - \frac{1}{\beta\hbar/l}\ln(1-\lambda). \tag{9}$$

Equation (9) can be solved for λ in terms of the Lambert function, $\mathcal{W}(x)$,

$$\lambda(\beta,\mathcal{N}) = 1 - \frac{\beta\hbar/l}{\mathcal{W}(\beta\hbar/l e^{(\mathcal{N}+1)\beta\hbar/l})}, \tag{10}$$

and determines the ground state occupancy as well as the condensate fraction,

$$f_C(\beta,\mathcal{N}) = \frac{N_0}{\mathcal{N}} = \frac{\mathcal{W}(\beta\hbar/l e^{(\mathcal{N}+1)\beta\hbar/l})}{\mathcal{N}\beta\hbar/l} - \frac{1}{\mathcal{N}}, \tag{11}$$

in terms of the temperature. It is more illuminating, however, to express the condensate fraction as a function of the dimensionless ratio, $x = T/T_c$, where T_c is the critical temperature at which all the shells are excited.

To calculate the critical temperature, T_c, it is necessary to set $\lambda \approx 1$ and $N_{\text{ex}} = \mathcal{N}$, the total number of shells. Once again employing the Mellin-Barnes transformation, this time replacing the polylogarithm function $\mathrm{Li}_t(\lambda)|_{\lambda\approx 1}$ by the $\zeta-$function, we find [10]

$$N_{\text{ex}} = \mathcal{N} \approx \frac{1}{2\pi i}\int_{\tau-i\infty}^{\tau+i\infty} dt\, \frac{\Gamma(t)\zeta^2(t)}{(\beta_c\hbar/l)^t}, \tag{12}$$

but the integral now admits a double pole at $t = 1$ via the $\zeta-$function and therefore

$$\mathcal{N} \approx \left(\frac{\Gamma(t)}{(\beta\hbar/l)^t}\right)' = -\frac{l}{\beta_c\hbar}\ln(e^{\gamma}\beta_c\hbar/l), \tag{13}$$

where γ is the Euler-Mascheroni constant. Solving for the inverse critical temperature, we determine

$$\beta_c\hbar/l = \frac{\mathcal{W}(e^{-\gamma}\mathcal{N})}{\mathcal{N}}, \tag{14}$$

which is the exact version of the result in [11]. The Lambert function increases linearly for small values of its argument and slower than the log function for larger values. Thus, for a few shells the critical temperature is more or less constant whereas it increases roughly as $\mathcal{N}/\ln\mathcal{N}$ when $\mathcal{N}$ grows large. When the black hole is composed of many shells, most of them will in fact lie in the ground state. For a small number of shells the ground state occupancy can be a large percentage of the total number of shells in a significant interval of temperatures.

Using (14) to re-express the condensate fraction in terms of the variable $x = T/T_c$, we easily arrive at

$$f_C(x,\mathcal{N}) = \frac{x}{\mathcal{W}_N}\mathcal{W}\left(\frac{\mathcal{W}_N}{\mathcal{N}x}e^{\frac{(\mathcal{N}+1)\mathcal{W}_N}{\mathcal{N}x}}\right) - \frac{1}{\mathcal{N}}, \tag{15}$$

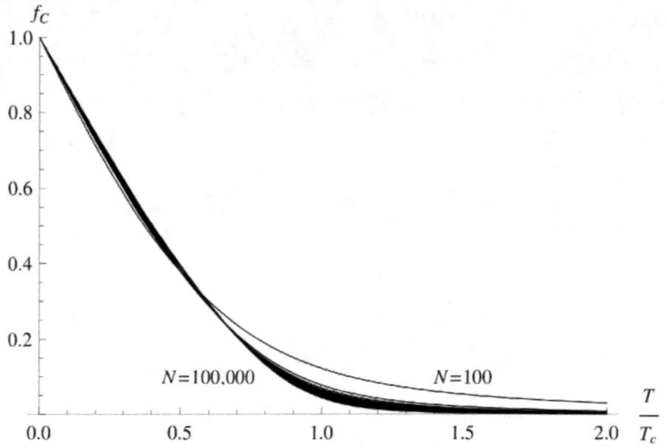

Figure 1. f_C as a function of T/T_c.

where we have defined $\mathcal{W}_N = \mathcal{W}(e^{-\gamma}\mathcal{N})$. At high temperatures,

$$f_C \overset{x\to\infty}{\approx} \frac{1}{\mathcal{N}} + \frac{\mathcal{W}_N}{\mathcal{N}x} + \mathcal{O}(x^{-2}) \tag{16}$$

whereas, at low temperatures,

$$f_C \overset{x\to 0}{\approx} 1 - \frac{\ln\mathcal{N}}{\mathcal{W}_N}x + \mathcal{O}(x^2) \tag{17}$$

Figure 1 shows the behavior of f_C as a function of x for $\mathcal{N}$ ranging from one hundred to one hundred thousand shells. It is small, increasing slowly to the left when $x \gg 1$ but significantly faster when $x \ll 1$ and approaching unity in the limit as $x \to 0$. Again, the slope of the condensate fraction, $\partial f_C/\partial x$, approaches $\ln\mathcal{N}/\mathcal{W}_N$ for small x and it is vanishingly small for large x. Thus increasing the number of shells saturates the slope of the condensate fraction. This behavior suggests that the critical temperature may be thought of as the condensation temperature.

The average energy of the system, which is associated with the black hole mass, can be expressed in terms of the level occupancy in the usual way and the sum rewritten in the integral representation we have used so far. For $\lambda < 1$ there is only a simple pole coming from the $\zeta-$function at $t = 2$, so the black hole mass can be given in terms of the temperature by

$$M(\beta, \mathcal{N}) = M_0 + \mathcal{N}\varepsilon_0 + \frac{\mathrm{Li}_2(\lambda)}{\beta^2\hbar/l} \tag{18}$$

and the heat capacity, expressed in terms of $x = T/T_c$, is

$$C(x, \mathcal{N}) = \frac{2\mathcal{N}x\mathrm{Li}_2(\lambda)}{\mathcal{W}_N} - \frac{\mathcal{N}x^2}{\lambda\mathcal{W}_N}\ln(1-\lambda)\left(\frac{\partial\lambda}{\partial x}\right)_{\mathcal{N}}, \tag{19}$$

where the prime on the polylogarithm represents a derivative with respect to λ. The specific heat, $C/\mathcal{N}$, is displayed in figure (2) for one hundred to one hundred thousand shells. The uppermost curve is for $\mathcal{N} = 100$. The specific heat is seen to decrease with increasing $\mathcal{N}$, but there is no true phase transition in this system. At low temperatures, the fugacity in (10)

Journal of Physics: Conference Series **462** (2013) 012057 doi:10.1088/1742-6596/462/1/012057

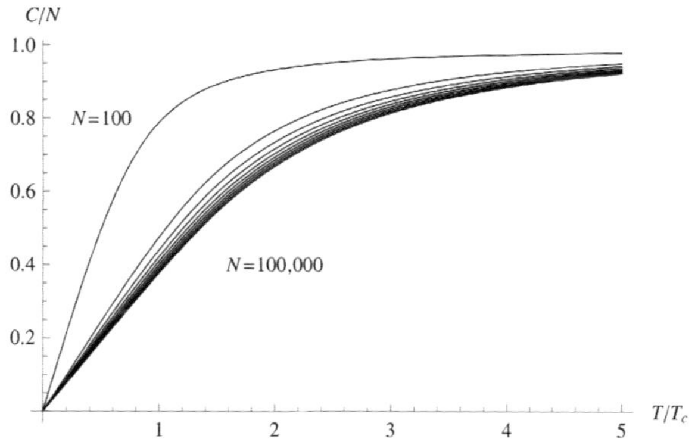

Figure 2. $C/\mathcal{N}$ as a function of T/T_c.

behaves roughly as $1-\mathcal{N}^{-1}$ and the second term in the expression for the heat capacity in (19) becomes vanishingly small. Thus in the limit as $\beta\to\infty$, using $\text{Li}_2(1)=\pi^2/6$ we find

$$\frac{C}{\mathcal{N}}\approx\frac{\pi^2 kT}{3\mathcal{N}\hbar/l}. \tag{20}$$

(see [12, 13]). On the other hand, at high temperatures the fugacity behaves as

$$\lambda = 1-e^{-N\beta\hbar/l} \tag{21}$$

and the specific heat approaches unity in the limit as $\beta\to 0$.

We now turn to the entropy, which may be computed from

$$S=\ln\Xi+\beta U-\mu\beta\mathcal{N}. \tag{22}$$

Expanding $\ln\Xi$ in a power series and applying the transformation in Mellin-Barnes transformation, we arrive at

$$\ln\Xi=-\beta M_0-\ln(1-\lambda)+\frac{\text{Li}_2(\lambda)}{\beta\hbar/l} \tag{23}$$

and combine this expression with (18). Then the entropy function can be given in terms of the ratio $x=T/T_c$ as

$$S(\beta,\mathcal{N})=\frac{2x\mathcal{N}}{\mathcal{W}_N}\text{Li}_2(\lambda)-\ln(1-\lambda)-\mathcal{N}\ln\lambda. \tag{24}$$

The specific entropy is shown in figure (3) for one hundred to one hundred thousand shells. The uppermost curve represents one hundred shells and the specific entropy also decreases with increasing $\mathcal{N}$.

So far we have considered the thermodynamic potentials as functions of the number of shells, $\mathcal{N}$, and the temperature, T, but let us now consider this function in terms of its natural variables, M and $\mathcal{N}$. For a black hole that is far below the critical temperature, *i.e.*, as $x\to 0$, (24) is approximated by

$$S\approx\frac{\pi^2}{3\beta\hbar/l}+\ln\mathcal{N}+1 \tag{25}$$

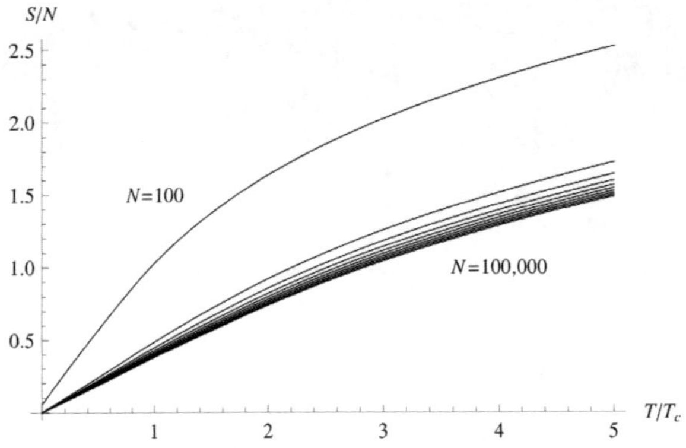

Figure 3. $S/\mathcal{N}$ as a function of T/T_c.

This is not vanishing, but the specific entropy, $S/\mathcal{N}$, does vanish in the thermodynamic limit as the temperature approaches zero and so it is consistent with a generalized version of the third law. In this same limit we find from (18) that

$$\frac{\pi}{\beta\hbar/l} \approx \sqrt{\frac{6(M - M_0 - \mathcal{N}\varepsilon_0)}{\hbar/l}} \tag{26}$$

so we can give the entropy in terms of the black hole mass and the number of shells as

$$S \approx \frac{\pi}{3}\sqrt{\frac{6(M - M_0 - \mathcal{N}\varepsilon_0)}{\hbar/l}} + \ln\mathcal{N} + 1. \tag{27}$$

This is where Bose condensation plays a dominant role. At temperatures well below the critical temperature, where a large fraction of the shells are in the ground state, the first term on the right becomes negligible and the entropy is dominated by its logarithmic behavior. Thus even though a quantization of Einstein's gravity yields the expected results in the thermodynamic limit and at high temperatures ($M, \mathcal{N} \to \infty$, $M \gg \mathcal{N}\varepsilon_0$), its behavior outside this regime presents interesting features peculiar to quantum statistics. In particular, we see that BTZ black holes turn cold once condensation sets in. This is the "small" black hole or short distance regime in 2+1 dimensions.

We suggest that this is in fact the regime in which "small" black holes and more generally quantum gravity at short distances should be studied. Here, the entropy becomes dominated by a mass independent term so that the black hole becomes cold and (hence) stable even though the mass remains non-vanishing at $M \approx M_0 + \mathcal{N}\varepsilon_0$. Furthermore, because the thermodynamics of many higher dimensional black holes are derivable from the BTZ black hole, one may reasonably expect that this feature persists in more than three dimensions. It suggests that the universe may be full of stable, Bose condensed black hole relics, which contribute to the dark matter.

LCRW was supported in part by the U.S. Department of Energy Grant No. DE-FG02-84ER40153.

References

[1] J. D. Bekenstein, Ph.D. thesis, Princeton University (1972); *ibid* Lett. Nuovo Cimento **4** (1972) 737; *ibid* Phys. Rev. D **7** (1973) 2333.

J. M. Bardeen, B. Carter, and S. W. Hawking, Comm. Math. Phys. **31** (1973) 161.
S. W. Hawking, Comm. Math. Phys. **43** (1975) 199.
T. Jacobson, Phys. Rev. Lett. **75** (1995) 1260.
T. Padmanabhan, AIPConf. Proc. 939 (2007) 114.

[2] A. Strominger and C. Vafa, Phys. Lett. B **379** (1996) 99.
G.T. Horowitz and J. Polchinski, Phys. Rev. D **55** (1997) 6189.
A. Dabholkar, Phys. Rev. Lett. **94** (2005) 241301.

[3] S. Hyun, W.T. Kim, and J. Lee, Phys. Rev. D **59** (1999) 084020.
R. Emparan, JHEP 9906 (1999) 036.
S. W. Hawking, J. Maldacena, and A. Strominger, JHEP 0105 (2001) 001.
S. A. Mukherji and S. S. Pal, JHEP 0205 (2002) 026.
S. K. Chakrabarti, K.S. Gupta, and S. Sen, Int. J. Mod. Phys. A **23** (2008) 2547.

[4] C. Rovelli, Phys. Rev. Lett. **77** (1996) 3288.
A. Ashtekar, J. Baez, A. Corichi, and K. Krasnov, Phys. Rev. Lett. **80** (1998) 904.
M. Domagala and J. Lewandowski, Class. Quant. Grav. **21** (2004) 5233.
K. A. Meissner, Class. Quant. Grav. **21** (2004) 5245.
A. Corichi, J. Diáz-Polo and E. Fernández-Borja, Phys. Rev. Lett. **98** (2007) 181301.
I. Agulló, J. F. Barbero G., J. Diáz Polo, E. Fernández-Borja and E. J. S. Villaseñor, Phys. Rev. Lett. **100** (2008) 211301.

[5] S. Nampuri, P.K. Tripathy and S.P. Trivedi, JHEP0807:072 (2008),
D. Birmingham and S. Carlip, Phys. Rev. Letts. **92** (2004) 111302,
M. Cvetic, Finn Larsen, Phys. Rev. Letts. **82** (1999) 484,
S. Carlip, Class.Quant.Grav. **15** (1998) 3609 (*and refs. therein.*)

[6] C. Vaz, S. Gutti, C. Kiefer, T.P. Singh and L.C.R. Wijewardhana, Phys. Rev. D **77** (2008) 064021.

[7] S. Carlip, Phys. Rev. D **51** (1995) 632.

[8] A. Strominger, JHEP 9802 (1998) 009.

[9] R. B. Mann and S. F. Ross, Phys. Rev. D **47**, 3319 (1993).
S. Gutti, Class. Quant. Grav. **22** (2005) 3223.

[10] Martin Holthaus, Eva Kalinowski and Klaus Kirsten, Ann. Phys. **270** 1 (1998) 198.

[11] F. Dalfovo, S. Giorgini, L.P. Pitaevskii and S. Stringari1, Rev. Mod. Phys. **71** (1999) 463.

[12] W. Ketterle and N.J. van Druten, Phys. Rev. A **54** (1996) 656.

[13] G-L. Ingold and A. Lambrecht, Eur. Phys. Jour. D **1** (1998) 29.

Strings on AdS wormholes

Mir Ali[1], Frenny Ruiz[1], Carlos Saint-Victor[1] and Justin F Vázquez-Poritz[1,2]

[1] Physics Department, New York City College of Technology
The City University of New York, 300 Jay Street, Brooklyn NY 11201, USA
[2] The Graduate School and University Center, The City University of New York
365 Fifth Avenue, New York NY 10016, USA

E-mail: jvazquez-poritz@citytech.cuny.edu

Abstract. We consider the behavior of open strings on AdS wormholes in Gauss-Bonnet theory, which are the Gauss-Bonnet gravity duals of a pair of field theories. A string with both endpoints on the same side of the wormhole describes two charges within the same field theory, which exhibit Coulomb interaction for small separation. On the other hand, a string extending through the wormhole describes two charges which live in different field theories, and they exhibit a spring-like confining potential. A transition occurs when there is a pair of charges present within each field theory: for small separation each pair of charges exhibits Coulomb interaction, while for large separation the charges in the different field theories pair up and exhibit confinement. If two charges move faster than a critical speed, then they exhibit a separation gap and energy is transferred from the leading charge to the lagging one.

1. Introduction and summary

The AdS/CFT correspondence [1] can be used to study certain strongly-coupled gauge theories. In particular, the behavior of open strings can be related to that of particles in the field theory. In the large N and large 't Hooft coupling λ limit, the string theory can be approximated by classical supergravity. A quantum theory of gravity such as string theory generally contains higher derivative corrections from stringy or quantum effects, which correspond to $1/\lambda$ or $1/N$ corrections in the field theory. However, not much is known about the precise forms of the higher derivative corrections, other than for a few maximally supersymmetric cases. The most general theory of gravity in higher dimensions that leads to second-order field equations for the metric is described by the Lovelock action [2]. The simplest such higher derivative theory of gravity is known as Einstein-Gauss-Bonnet theory, which only contains terms up to quadratic order in the curvature. Our primary motivation for considering Einstein-Gauss-Bonnet theory is that it contains Lorentzian-signature five-dimensional asymptotically locally AdS wormhole solutions which do not violate the weak energy condition, so long as the Gauss-Bonnet coupling constant is negative and bounded according to the shape of the solution [3, 4].

We consider the static wormhole solutions that were found in [5, 6] which connect two asymptotically AdS spacetimes and are described by the metric

$$ds^2 = \ell^2\Big(- \cosh^2(\rho - \rho_0)dt^2 + d\rho^2 + \cosh^2 \rho \, d\Sigma_3^2\Big) . \tag{1}$$

The base manifold Σ_3 can be either H_3/Γ or $S^1 \times H_2/\Gamma$, where H_3 and H_2 are hyperbolic spaces and Γ is a freely acting discrete subgroup of $O(3,1)$ and $O(2,1)$, respectively. The hyperbolic part of the base manifold must be quotiented so that this geometry describes a wormhole rather than a gravitational soliton with a single conformal boundary. The neck of the wormhole is located at $\rho = 0$, which connects two asymptotically locally AdS regions at $\rho \to \pm\infty$. The geometry is devoid of horizons and radial null geodesics connect the two asymptotic regions in a finite time $\Delta t = \pi$. These wormholes can evade the proof that the disconnected boundaries must be separated by black hole horizons [7], since we are dealing with a higher-derivative theory of gravity. No energy conditions are violated by these wormholes, since the stress-energy tensor vanishes everywhere [5, 6]. The stability of these wormholes against scalar field perturbations has been discussed in [13].

It has been shown that gravity pulls towards a fixed hypersurface at $\rho = \rho_0$, which lies in parallel with the neck of the wormhole [5, 6]. In particular, timelike geodesics are confined, and oscillate about this hypersurface. Although the wormhole is massless, for nonzero ρ_0, the mass of the wormhole appears to be positive for observers located at on one side and negative for the other. The parameter ρ_0 is related to the apparent mass on each side of the wormhole; for $\rho_0 = 0$, the wormhole exhibits reflection symmetry.

If we can apply the AdS/CFT correspondence to these AdS wormholes, then this is the gravity dual of two interacting field theories on $\mathbb{R} \times \Sigma_3$. Except in the UV limit, the conformal symmetry of both theories is broken by the length scale associated with Σ_3 (H_3 and H_3 must have the radius 1 and $1/\sqrt{3}$, respectively) as well as by the parameter ρ_0. However, we would like to emphasize that the AdS/CFT correspondence has not actually been tested for gravitational backgrounds of Einstein-Gauss-Bonnet theory and may not be valid. We are making the working assumption that the low-energy effective five-dimensional description of gauge theory/string theory duality has a sensible derivative expansion in which the higher curvature terms are systematically suppressed by powers of the Planck length. After making the appropriate field redefinitions, the curvature-squared terms in the action appear in the Gauss-Bonnet combination. Thus, within the limitations of the derivative expansion, one can use the AdS/CFT correspondence to determine the properties of gauge theories dual to Einstein-Gauss-Bonnet gravitational backgrounds [8]. In fact, one might expect that within the vast string landscape there are higher derivative corrections which lead to similar backgrounds with asymptotically AdS regions with multiple disconnected boundaries.

We consider the behavior of open strings on these AdS wormholes. The string endpoints lie on probe D-branes and correspond to charges in one or the other of the field theories. Technically, we should also consider the quantum fluctuations of the string worldsheet [9], which formally dominate over the higher derivative corrections of the background. However, we have checked that the corrections to the string worldsheet do not change the qualitative behavior of strings, which is the focus of our interest.

The expectation value of a rectangular Wilson loop can be computed from the proper area of the string worldsheet [10, 11], from which the energy of a pair of charges can be read off. We generalize this prescription for the case of two charges which do not live within the same field theory. Even though the string endpoints lie on opposite sides of the wormhole, there is a rectangular contour within the physical spacetime whose horizontal sides run along the distance between the two charges and whose vertical sides run along time. We find that two charges in different field theories exhibit confinement with a spring-like potential. A pair of heavy charges exhibits a weaker interaction than do lighter charges. We find that there is a rather curious transition which involves a quadruplet of charges that consists of a charge-anticharge pair within each field theory. For small charge-anticharge separations, each pair exhibits Coulomb interaction and does not interact with the pair of charges in the other field theory. However, for large separation, it is the charges of different types that interact and exhibit confinement whereas the pairs of charges of the same type no longer interact with each other. In this sense, each pair of charges of the same type exhibits the feature of effectively having a screening length, even though both field theories are at zero temperature.

We also consider the behavior of steadily-moving strings. In particular, demanding that the string configuration is timelike imposes constraints on the mass parameters of the corresponding charges, as well as on their separation. We find that the charges have an upper bound on their speed which depends on the mass parameter of one of the charges. This speed limit is generally less than the speed of light, which is a result of the fact that the proper velocity of the string endpoints is greater than the physical velocity in the field theory [12]. We find that a pair of steadily-moving charges of different types can occupy the same location with no transfer of energy or momentum between them, provided that their speed is less than a certain critical speed. This critical speed coincides with the speed limit of the charges and monotonically decreases with the mass parameters. If the charges move at this critical speed, then they exhibit a separation gap which increases with the mass parameter of the lagging charge but is not so sensitive to the mass parameter of the leading charge. Also at this critical speed, energy and momentum are transferred from the leading charge to the lagging one.

2. Static strings

We will consider a string which is localized at a point in H_2 and moves along the S^1 direction, which we label x. Choosing a static gauge, we find that the equation of motion for a string on the background metric (1) is given by

$$\frac{\partial}{\partial\rho}\Big(\frac{\cosh^2(\rho-\rho_0)\cosh^2\rho\, x'}{\sqrt{-g}}\Big) - \cosh^2\rho\,\frac{\partial}{\partial t}\Big(\frac{\dot{x}}{\sqrt{-g}}\Big) = 0\,, \tag{2}$$

where

$$-\frac{g}{\ell^4} = \cosh^2(\rho-\rho_0) + \cosh^2(\rho-\rho_0)\cosh^2\rho\, x'^2 - \cosh^2\rho\,\dot{x}^2\,. \tag{3}$$

Note that $g = \det(g_{ab})$, where g_{ab} is the induced metric. We will also make use of the fact that the string tension is given by $T_0 = \sqrt{\lambda}/(2\pi\ell^2)$, where λ is the 't Hooft coupling of the field theory.

The simplest solution is a constant $x = x_0$, for which the string is straight, as shown in Figure 1. Suppose that the endpoints are located at $\rho = -\rho_1$ and $\rho = \rho_2$, where we will always take $\rho_1, \rho_2 > 0$. We will refer to the charges associated with the endpoints at $\rho = -\rho_1$ and $\rho = \rho_2$ as a type 1 charge (blue) and a type 2 charge (red), respectively. Also, we will refer to the radial locations of the string endpoints ρ_1 and ρ_2 as the mass parameters of the charges. The D-branes (or at least the part that lies closest to the neck of the wormhole) are denoted by the aqua-colored surfaces and the grey surface represents the fixed hypersurface at $\rho = \rho_0$. The

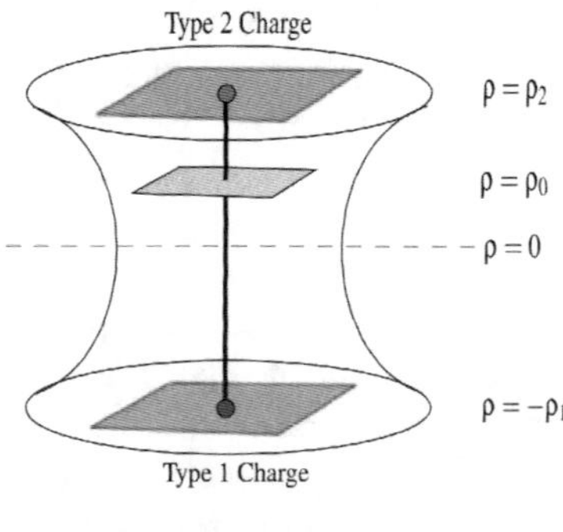

Figure 1. A static string stretching straight through a wormhole.

static string has vanishing momentum, and its energy is given by

$$E = T_0\ell^2\big(\sinh(\rho_1+\rho_0) + \sinh(\rho_2-\rho_0)\big)\,. \tag{4}$$

Curved strings can either have both endpoints on the same side of the wormhole or else on opposite sides, as shown in Figure 2. We will first consider the first case, which corresponds to

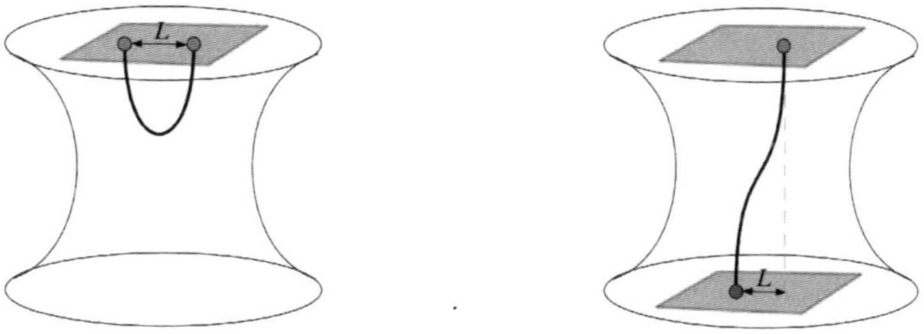

Figure 2. A string with both endpoints on the same side of the wormhole (left) and on opposite sides (right).

a pair of charges of the same type. This means that the string dips down towards the wormhole and then comes back up. We will denote the turning point of the string by ρ_t, and L is the distance between the endpoints of the string. For simplicity, we will take the endpoints to be at $\rho = \rho_2$, so that we have a pair of type 2 charges. For ρ_2, $\rho_t \gg 1$, the string is not sensitive to the presence of the wormhole and so it can be found that its energy goes as $1/L$. A Coulomb potential energy is consistent with the fact that the theory is conformal in the extreme UV regime. If the masses associated with the charges are decreased, or if the distance between the charges is increased, then the string dips closer to the wormhole. Then terms in the energy that are of higher order in L will become important.

From (2) and (3), we find the distance L between the string endpoints is given by

$$L = 2\cosh(\rho_t - \rho_0)\cosh\rho_t \int_{\rho_t}^{\rho_2} \frac{d\rho}{\cosh\rho\sqrt{\cosh^2(\rho-\rho_0)\cosh^2\rho - \cosh^2(\rho_t-\rho_0)\cosh^2\rho_t}}\,. \quad (5)$$

The above integration region assumes that $\rho_2 > \rho_t$. As we will see shortly, there are string solutions for which $\rho_2 < \rho_t$, in which case the upper and lower integration bounds must be interchanged.

For $\rho_2 > \frac{1}{2}\rho_0$, the reality of L implies that the turning point is located in the interval $\rho_2 > \rho_t \geq \frac{1}{2}\rho_0$. Similarly, a string with endpoints at $\rho_2 < \frac{1}{2}\rho_0$ has a turning point in the interval $\rho_2 < \rho_t \leq \frac{1}{2}\rho_0$. This means that if the turning points are within the interval $0 < \rho_2 < \rho_0/2$, then the string will bend *away* from the center of the wormhole. Moreover, if the endpoints are located on the opposite side of the wormhole as the hypersurface $\rho = \rho_0$, then there are strings which go through the neck of the wormhole and have a midsection on the opposite side of the wormhole as its endpoints. All of these possibilities are shown in Figure 3.

The turning point ρ_t monotonically decreases with L. In particular, as L gets large the turning point asymptotically approaches $\frac{1}{2}\rho_0$ from above. Note that, unlike the case of strings on an AdS black hole background, there is a single string configuration for each value of L. From the field theory perspective, this means that the charges do not exhibit a screening length but rather remain interacting no matter how far they are from each other. As the distance between the charges increases, the inter-charge potential becomes ever more sensitive to the IR characteristics of that sector of the theory.

We will now consider a curved string with endpoints on opposite sides of the wormhole, which corresponds to a type 1 charge and a type 2 charge. From (2) and (3), the distance between the

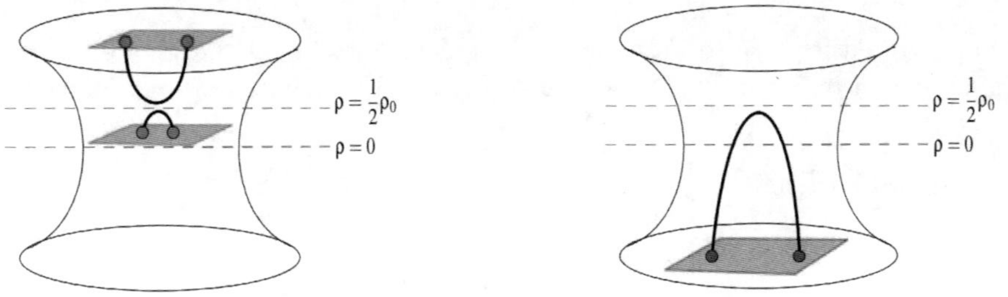

Figure 3. Strings bend towards $\rho = \frac{1}{2}\rho_0$ regardless of where the string endpoints are located, even if this means that the string bends away from the neck of the wormhole (left) or goes through the neck of the wormhole (right).

string endpoints in the x direction is

$$L = C \int_{-\rho_1}^{\rho_2} \frac{d\rho}{\cosh\rho\sqrt{\cosh^2(\rho-\rho_0)\cosh^2\rho - C^2}}\,. \tag{6}$$

For small separation L, we can expand in $C \ll 1$ so that

$$L \approx C \int_{-\rho_1}^{\rho_2} \frac{d\rho}{\cosh(\rho-\rho_0)\cosh^2\rho}\,. \tag{7}$$

We find the energy of the string to be

$$E = T_0\ell^2 \int_{-\rho_1}^{\rho_2} d\rho\, \frac{\cosh^2(\rho-\rho_0)\cosh\rho}{\sqrt{\cosh^2(\rho-\rho_0)\cosh^2\rho - C^2}}\,. \tag{8}$$

In the small separation limit $C \ll 1$,

$$E = E_{straight} + \frac{1}{2}T_0\ell^2C^2 \int_{-\rho_1}^{\rho_2} \frac{d\rho}{\cosh(\rho-\rho_0)\cosh^2\rho}\,, \tag{9}$$

where $E_{straight}$ is the energy of a straight string given by (4). Using (7) to express C in terms of L, we find

$$E = E_{straight} + \frac{1}{2}kL^2\,, \tag{10}$$

where

$$k = \frac{\sqrt{\lambda}}{2\pi}\Big(\int_{-\rho_1}^{\rho_2} \frac{d\rho}{\cosh(\rho-\rho_0)\cosh^2\rho}\Big)^{-1}\,, \tag{11}$$

and we have used $T_0 = \sqrt{\lambda}/(2\pi\ell^2)$. Thus, for small L, a pair of charges of opposite types are confined and exhibit a spring-like potential, as opposed to a pair of charges of the same type which exhibit a Coulomb potential. Since we are at zero temperature, there is no screening length. The effective force constant k monotonically decreases with the mass parameters ρ_1 and ρ_2, which means that heavy charges are not as sensitive to the confining potential as lighter charges.

For two sets of charge-anticharge pairs, an interesting transition can take place. For instance, suppose we have a pair of interacting type 1 charges and a pair of interacting type 2 charges, described by the two curved strings shown in the left plot of Figure 4. For simplicity, we will assume that both pairs of charges are a distance L apart. For small L, there is no interaction between the type 1 pair and the type 2 pair. However, as L increases, the strings bend closer to each other. Note that these strings cannot pass each other, since neither one can pass the radius $\rho = \rho_0/2$. However, even before either string hits this bound, the string configuration in the right plot of Figure 4 becomes the energetically favorable one with the same set of endpoints. Thus, for large L, each type 1 charge becomes coupled to a type 2 charge. This is similar to a screening length, in that the charges of the same type no longer interact with each other. However, since we are at zero temperature, this is really a feature of having multiple pairs of charges close to each other.

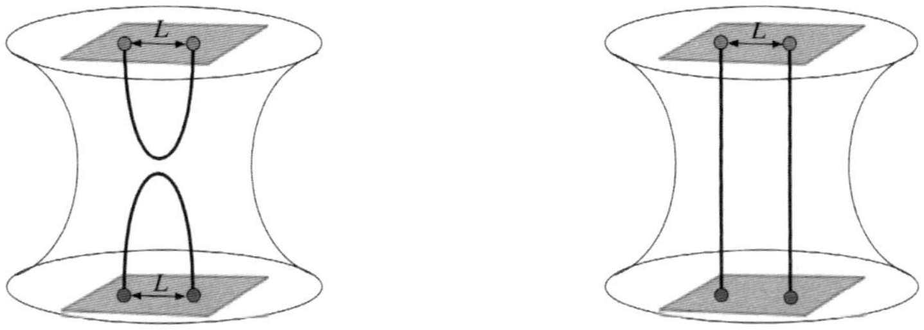

Figure 4. There is a critical distance L_{crit} associated with two sets of charge-anticharge pairs. For $L < L_{crit}$, the left configuration is energetically favorable while, for $L > L_{crit}$, the right configuration is the favored one.

3. Steadily-moving strings

3.1. Straight strings

A steadily-moving straight string is given by $x(t, \rho) = x_0 + vt$. For this solution, we find that

$$-\frac{g}{\ell^4} = \cosh^2(\rho - \rho_0) - v^2 \cosh^2 \rho \,. \tag{12}$$

For simplicity, we will take $\rho_0 \geq 0$. Then for $v < v_{crit} \equiv e^{-\rho_0}$, $-g > 0$ everywhere for $v < 1$, which implies that no parts of the string move faster than the local speed of light. On the other hand, for $v = v_{crit}$, we require that $\rho_2 < \rho_0/2$. For $v > v_{crit}$, $g = 0$ at $\rho = \rho_{crit}$ where

$$\rho_{crit} = \ln \sqrt{\frac{e^{\rho_0} - v}{v - e^{-\rho_0}}} \,, \tag{13}$$

at which point the induced metric on the string worldsheet is degenerate. We can still have steadily-moving straight strings which lie entirely within the region $\rho_{crit} < \rho$. However, for strings which have a region that lies within $\rho < \rho_{crit}$, $-g < 0$ and the action, energy and momentum are complex. This is a signal that this part of the string travels faster than the local speed of light and must be discarded.

The energy and momentum of the steadily moving string are given by the integrals

$$\begin{aligned} E &= T_0 \ell^2 \int_{-\rho_1}^{\rho_2} d\rho \, \frac{\cosh^2(\rho - \rho_0)}{\sqrt{\cosh^2(\rho - \rho_0) - v^2 \cosh^2 \rho}} \,, \\ p &= T_0 \ell^2 v \int_{-\rho_1}^{\rho_2} d\rho \, \frac{\cosh^2 \rho}{\sqrt{\cosh^2(\rho - \rho_0) - v^2 \cosh^2 \rho}} \,. \end{aligned} \tag{14}$$

For the special case of $\rho_0 = 0$, we find that

$$E = \frac{E_{rest}}{\sqrt{1-v^2}}\,, \qquad E_{rest} = T_0\ell^2(\sinh\rho_1 + \sinh\rho_2)\,. \tag{15}$$

For non-vanishing ρ_0, the expressions for E and p cannot be integrated in closed form. Note that, since π_t^1 and π_x^1 vanish, there is no energy or momentum current flowing along the string.

We will now comment on the speed limit of these co-moving charges, which stems from the fact that the proper velocity V of the string endpoints differs from the physical velocity v in the four-dimensional field theory [12]. From the metric (1), we see that

$$V = \frac{\cosh\rho_i}{\cosh(\rho_i - \rho_0)}v\,, \tag{16}$$

where $\rho_i = -\rho_1$ or ρ_2. In order to avoid a spacelike string worldsheet, we must have $V \le 1$. This corresponds to

$$v \le v_{max} = \frac{\cosh(\rho_2 - \rho_0)}{\cosh\rho_2}\,, \tag{17}$$

where we have taken $\rho_i = \rho_2$, since ρ_2 is closer to $\rho_0 \ge 0$ than is $-\rho_1$. As is the case for steadily-moving strings in AdS black hole backgrounds, the speed limit depends on the mass parameter associated to the charge. This same speed limit applies to the curved strings which we now discuss. We see that $v_{max} = v_{crit} = 1$ for $\rho_0 = 0$. On the other hand, for $\rho_0 > 0$, we find that $v_{max} < v_{crit} < 1$.

3.2. Curved strings

We will now consider steadily-moving curved strings, which are described by solutions of the form $x(t,\rho) = x(\rho) + vt$. The term with the time derivative in (2) vanishes and we are left with

$$\frac{\partial}{\partial\rho}\Big(\frac{\cosh^2(\rho-\rho_0)\cosh^2\rho\, x'}{\sqrt{-g}}\Big) = 0\,. \tag{18}$$

where

$$-\frac{g}{\ell^4} = \cosh^2(\rho-\rho_0) + \cosh^2(\rho-\rho_0)\cosh^2\rho\, x'^2 - v^2\cosh^2\rho\,. \tag{19}$$

The first integral of (18) is

$$\frac{\cosh^2(\rho-\rho_0)\cosh^2\rho\, x'}{\sqrt{-g/\ell^4}} = C\,. \tag{20}$$

where C is an integration constant. Thus,

$$x'^2 = \frac{C^2}{\cosh^2(\rho-\rho_0)\cosh^2\rho}\Big(\frac{\cosh^2(\rho-\rho_0) - v^2\cosh^2\rho}{\cosh^2(\rho-\rho_0)\cosh^2\rho - C^2}\Big)\,. \tag{21}$$

Solving for $-g$ gives

$$-\frac{g}{\ell^4} = \cosh^2(\rho-\rho_0)\cosh^2\rho\,\Big(\frac{\cosh^2(\rho-\rho_0) - v^2\cosh^2\rho}{\cosh^2(\rho-\rho_0)\cosh^2\rho - C^2}\Big)\,. \tag{22}$$

For $v \le v_{crit}$ (with the exception of $\rho_0 = 0$ and $v = 1$) then $-g > 0$ everywhere along the string provided that $|C| < C_{crit}$ where

$$C_{crit} = \frac{1}{2}\big(\cosh(2\min[\rho_2, \rho_0/2] - \rho_0) + \cosh\rho_0\big)\,. \tag{23}$$

In the limit that $\rho_0 = 0$, solutions exist with $|C| < 1$ for all $v < 1$. However, for $\rho_0 = 0$ and $v = 1$, the induced metric is degenerate everywhere, regardless of the value of C. Strings with $v > v_{crit}$ can still have $-g > 0$ everywhere provided that $\rho_2 < \rho_{crit}$ and $|C| < C_{crit}$, where ρ_{crit} and C_{crit} are given by (13) and (23), respectively. From the field theory perspective, this corresponds to an upper bound on the mass of the type 2 charge. For $v > v_{crit}$, one can also have a string that lives entirely in $\rho > \rho_{crit}$, provided that $|C| > C_{crit}$. This corresponds to a light-heavy pair of type 2 charges with a lower bound on their masses.

The distance between the charges is given by

$$L = C \int_{-\rho_1}^{\rho_2} \frac{d\rho}{\cosh(\rho - \rho_0)\cosh\rho} \sqrt{\frac{\cosh^2(\rho - \rho_0) - v^2 \cosh^2 \rho}{\cosh^2(\rho - \rho_0)\cosh^2 \rho - C^2}}\,, \tag{24}$$

For cases in which we require that $|C| < C_{crit}$, there is an upper bound on L. However, if the endpoints of the string are allowed to move freely, then the string will tend to straighten out to the $C = 0$ configuration so that it minimizes its energy. On the other hand, for the string corresponding to a light-heavy pair of type 2 charges moving with $v > v_{crit}$, we must have $|C| > C_{crit}$ and there is a corresponding lower bound on L. This string cannot straighten out without decreasing its speed.

The rate at which energy and momentum flow through the string can be written in terms of the constant C as

$$\pi_t^1 = \frac{\sqrt{\lambda}}{2\pi} C v\,, \qquad -\pi_x^1 = \frac{\sqrt{\lambda}}{2\pi} C\,, \tag{25}$$

Thus, for nonzero C, energy and momentum are transferred from the leading charge to the lagging one. If there is no external force present, then the leading charge presumably slows down and the lagging charge speeds up until an equilibrium state is reached corresponding to a steadily-moving straight string.

Acknowledgments

We would like to thank Philip Argyres, Mohammad Edalati, Hong Lü and Juan Maldacena for useful correspondence and conversations. M.A., F.R. and C.S.-V. are supported in part by the Emerging Scholars Program. F.R. and C.S.-V. are supported in part by New York City Louis Stokes Alliance for Minority Participation in Science, Mathematics, Engineering and Technology.

References

[1] Aharony O, Gubser S S, Maldacena J M, Ooguri H and Oz Y 2000 Phys. Rept. **323**, 183
[2] Lovelock D 1971 J. Math. Phys. **12**, 498
[3] Bhawal B and Kar S 1992 *Phys. Rev.* D **46** 2464
[4] Maeda H and Nozawa M 2008 *Phys. Rev.* D **78** 024005
[5] Dotti G, Oliva J and Troncoso R 2007 *Phys. Rev.* D **75** 024002
[6] Dotti G, Oliva J and Troncoso R 2007 *Phys. Rev.* D **76** 064038
[7] Galloway G J, Schleich K, Witt D and Woolgar E 2001 *Phys. Lett.* B **505** 255
[8] Buchel A, Myers R C and Sinha A 2009 *J. High Energy Phys.* JHEP03(2009)084
[9] Drukker N, Gross D J and Tseytlin A A 2000 *J. High Energy Phys.* JHEP04(2000)021
[10] Rey S J and Yee J T 2001 *Eur. Phys. J.* C **22** 379
[11] Maldacena J M 1998 *Phys. Rev. Lett.* **80** 4859
[12] Argyres P C, Edalati M and Vázquez-Poritz J F 2007 *J. High Energy Phys.* JHEP01(2007)105
[13] Correa D H, Oliva J and Troncoso R 2008 *J. High Energy Phys.* JHEP08(2008)081

Gravitational Gauge Theory and the Existence of Time

James T. Wheeler
Department of Physics, Utah State University, Logan, Utah 84322, USA
E-mail: jim.wheeler@usu.edu

Abstract. General relativity may be formulated as a gauge theory more than one way using the quotient manifold approach. We contrast the structures that arise in four gravitational gauge theories, three of which give satisfactory gauge theoris of general relativity. Of particular interest is the quotient of the conformal group of a flat space by its Weyl subgroup, which always has natural symplectic and metric structures in addition to the requisite manifold. This quotient space admits canonically conjugate, orthogonal, metric submanifolds distinct from the original space if and only if the original flat space has signature n, -n or 0. In the Euclidean cases, the resultant configuration space must be Lorentzian. This gives a 1-1 mapping between Euclidean and Lorentzian submanifolds, with induced Euclidean gravity or general relativity, respectively.

1. General relativity as a gauge theory

Direct unification of gravity with the standard model requires general relativity to be expressed as a gauge theory. Moreover, it is reasonable to expect that string or some other TOE will reduce to a gauge theory of general relativity in an appropriate limit. We therefore consider various ways to accomplish such a gauge theory[1]-[8].

While we expect the correct gauge theory limit to be supersymmetric, we restrict our attention here to the bosonic sector. The obvious choice is to gauge the Poincarï¿œ group, but it seems more natural to choose a simple group. In addition, there are metric, symplectic and complex structures required by quantum theory that a unified approach might predict. Surprisingly, one of the gauge theories we present below gives rise to these structures at a group-theoretic level.

As additional motivation, we recall the result of Ehlers, Pirani and Schild[9]. Their goal was to find the most general geometry consistent with the observable motions of particles and light. To accomplish this, they showed that the free-fall paths of particles determine a projective connection, while the free-fall paths of light determine a conformal connection. Combining these results by demanding that in the limit of high velocity, the particle geodesics coincide with the lightlike trajectories, they found that spacetime should have a Weyl connection. At this fundamental level, the local symmetry should therefore include dilatations as well as the Lorentz group. This strongly suggests that we gauge the Weyl or conformal group instead of the Poincarï¿œ group. If the dilatational gauge field, the Weyl vector, is pure gauge then we are free to make the usual definition of the second (currently, a set number of oscillations of hyperfine Cesium 133 transitions).

We examine four gauge theories of gravity, three of which lead to general relativity. Before describing these, we briefly review the quotient manifold method of gravitational gauging[3]-[5].

Let $\mathscr{G}$ be a Lie group, and $\mathscr{H}$ a Lie subgroup of $\mathscr{G}$. Then the quotient $\mathscr{G}/\mathscr{H}$ is a manifold, $\mathscr{M}$, and the Maurer-Cartan equations for the group give a connection on $\mathscr{G}$. We modify the connection so that the Maurer-Cartan equations now define the curvature 2-forms,

$$\mathbf{R}^A = d\omega^A - \frac{1}{2}c_{BC}{}^A\omega^B \wedge \omega^C$$

Once we have made the choice of $\mathscr{G}$ and $\mathscr{H}$, this construction determines

- the physical arena, $\mathscr{M}$
- the local symmetry group, $\mathscr{H}$
- the relevant tensor fields, $\mathbf{R}^A$
- any structures inherited from $\mathscr{G}$

While other structures may be imposed, we consider only those which arise directly from properties of the gauge group. To complete a gravity theory, we construct an $\mathscr{H}$-invariant action from the available tensors. While there is some arbitrariness in this step, the goal of reproducing general relativity is restrictive. Each connection form is then varied independently to find the field equations.

We illustrate the method with the Poincarï¿œ group.

The Maurer-Cartan equations for the Poincarï¿œ group are

$$\begin{aligned} \mathbf{d}\tilde{\omega}^a_{\ b} &= \tilde{\omega}^c_{\ b} \wedge \tilde{\omega}^a_{\ c} \\ \mathbf{d}\tilde{\mathbf{e}}^a &= \tilde{\mathbf{e}}^c \wedge \tilde{\omega}^a_{\ c} \end{aligned}$$

Now we take the quotient of the Poincarï¿œ group by its Lorentz subgroup. The result is a principal fiber bundle with n-dim base manifold and local Lorentz symmetry. We generalize the solder form $\tilde{\mathbf{e}}^a \to \mathbf{e}^a$ and the spin connection $\tilde{\omega}^a_{\ b} \to \omega^a_{\ b}$, leading to the addition of curvature and torsion 2-forms to the Maurer-Cartan equations,

$$\begin{aligned} \mathbf{R}^a_{\ b} &= \mathbf{d}\omega^a_{\ b} - \omega^c_{\ b} \wedge \omega^a_{\ c} \\ \mathbf{T}^a &= \mathbf{d}\mathbf{e}^a - \mathbf{e}^c \wedge \omega^a_{\ c} \end{aligned}$$

The generalization of the connection is restricted only by integrability and the condition that curvature and torsion be horizontal, i.e., $\mathbf{R}^a_{\ b} = \frac{1}{2}R^a_{\ bcd}\mathbf{e}^c \wedge \mathbf{e}^d$ and $\mathbf{T}^a = \frac{1}{2}T^a_{\ bc}\mathbf{e}^b \wedge \mathbf{e}^c$. The integrability condition, found by taking the exterior derivative of the curvature expressions, gives the Bianchi identities. Finally, the most general action linear in the curvature and torsion is

$$S = \int \left(\alpha\mathbf{R}^{ab} + \beta\mathbf{e}^a \wedge \mathbf{e}^b\right) e_{abc\ldots d} \wedge \mathbf{e}^c \wedge \ldots \wedge \mathbf{e}^d$$

Variation of the solder form and spin connection leads to vanishing torsion, and the vacuum Einstein equation with cosmological constant.

2. Four gauge theories of gravity

We repeat similar constructions for three additional quotients, giving the following four gauge theories of gravity (see also [6, 7]):

Group, $\mathcal{G}$	Subgroup, $\mathcal{H}$	dim $\mathcal{M}$	Name
Poincaré	Lorentz	n	Poincaré
Inhomogeneous Weyl	Weyl	n	Weyl geometry
Conformal	Inhomogeneous Weyl	n	Weyl conformal gravity
Conformal	Weyl	2n	Biconformal gravity

These are the only Lorentz covariant quotients with $\mathscr{G}$ no larger than conformal. There are substantial differences between these gravity theories. We compare, in turn, the following features:

(i) Number and form of the curvatures
(ii) Action and the relationship to general relativity
(iii) Induced structures
  (a) Killing metric
  (b) Symplectic form
  (c) Spacetime signature: the existence of time

2.1. Number and form of the curvatures

The number of curvatures equals the number of generators of the original group, $\mathscr{G}$, and the form of each curvature depends on the dimension of the quotient manifold, $\mathscr{M}$. In all cases we take the torsion (the curvatures associated with translations) to vanish.

As we have seen, the Poincarï¿œ gauging leads to Riemann curvature of the usual form, with the possible addition of torsion.

Weyl geometry[10]-[13] increases the number of curvatures by one – the dilatational curvature, given by the curl of the Weyl vector. The new gauge field also enters the relationship between the solder form and the spin connection, so with vanishing torsion we have

$$\begin{aligned}
\mathbf{R}^a_{\ b} &= \mathbf{d}\omega^a_{\ b} - \omega^c_{\ b} \wedge \omega^a_{\ c} \\
0 &= \mathbf{d}\mathbf{e}^a - \mathbf{e}^c \wedge \omega^a_{\ c} - \mathbf{e}^a \wedge \omega \qquad (\mathbf{T}^a = 0) \\
\Omega &= \mathbf{d}\omega
\end{aligned}$$

where ω is the Weyl vector. The form of $\mathbf{R}^a_{\ b}$ in terms of the spin connection, $\omega^a_{\ b}$, is the same as for Poincarï¿œ although the spin connection now has some dependence on the Weyl vector; the dilatational curvature is also expanded in terms of the solder form, $\Omega = \frac{1}{2}\Omega_{ab}\mathbf{e}^a \wedge \mathbf{e}^b$. This is the minimal extension of the usual Riemannian picture that is consistent with the findings of Ehlers, Pirani and Schild[9].

For both of the conformal gaugings, there are additional generators for the special conformal transformations and hence corresponding additional curvatures. There are also additional contributions to the Lorentz and dilatational curvatures because of the new special conformal gauge fields, $\mathbf{f}_a$. The curvatures are now:

$$\begin{aligned}
\Omega^a_{\ b} &= \mathbf{d}\omega^a_{\ b} - \omega^c_{\ b} \wedge \omega^a_{\ c} - \mathbf{f}_b \wedge \mathbf{e}^a + \eta_{bc}\eta^{ad}\mathbf{f}_d \wedge \mathbf{e}^c \\
0 &= \mathbf{d}\mathbf{e}^a - \mathbf{e}^c \wedge \omega^a_{\ c} - \mathbf{e}^a \wedge \omega \qquad (\mathbf{T}^a = 0) \\
\mathbf{S}_a &= \mathbf{d}\mathbf{f}_a - \omega^c_{\ a} \wedge \mathbf{f}_c - \omega \wedge \mathbf{f}_c \\
\Omega &= \mathbf{d}\omega - \mathbf{e}^a \wedge \mathbf{f}_a
\end{aligned}$$

For Weyl conformal gravity ([11], and more recently, [14] and references therein) each of the curvatures $\Omega^a_{\ b}, \mathbf{S}_a$ and Ω is expanded quadratically in the solder form as before, e.g., $\Omega^a_{\ b} = \frac{1}{2}\Omega^a_{\ bcd}\mathbf{e}^c \wedge \mathbf{e}^d$. However, since the biconformal gauging[7],[15]-[17] leaves both $\mathbf{e}^a$ and $\mathbf{f}_a$ spanning the base manifold, the curvatures of the biconformal geometry have additional terms,

$$\Omega^A = \frac{1}{2}\Omega^A_{\ ab}\mathbf{e}^a \wedge \mathbf{e}^b + \Omega^{Aa}_{\ \ b}\mathbf{f}_a \wedge \mathbf{e}^b + + \frac{1}{2}\Omega^{Aab}\mathbf{f}_a \wedge \mathbf{f}_b$$

Here, Ω^A refers to any of the curvatures, $\Omega^a_{\ b}, \mathbf{S}_a, \Omega$. The additional properties of the biconformal gauging compensate somewhat for this added complexity. Most importantly, while it is well-known that Weyl conformal gravity does not reproduce general relativity, the biconformal gauging does.

2.2. Action functionals and the relationship to general relativity

Each of these gravitational gauge theories has a different relationship to general relativity, largely because of necessary differences in the allowed action functionals.

As we have seen, the most general action linear in the curvature in Poincarï¿œ gravity is the Einstein-Hilbert action together with a cosmological constant. Palatini variation then implies vanishing torsion and the usual vacuum Einstein equation results.

Journal of Physics: Conference Series **462** (2013) 012059 doi:10.1088/1742-6596/462/1/012059

For Weyl geometry, we may write an action linear in the curvature if we introduce a scalar field. This gives the n-dim generalization of the action used by Dirac in discussing the Large Numbers Hypothesis[18]-[20],

$$S = \int \left(\kappa^2 g^{\alpha\beta} R_{\alpha\beta} - \frac{\beta}{2} \kappa^{2\left(\frac{n-4}{n-2}\right)} g^{\alpha\mu} g^{\beta\nu} \Omega_{\alpha\beta} \Omega_{\mu\nu} - \frac{\alpha}{2} g^{\alpha\beta} D_\alpha \kappa D_\beta \kappa - \lambda \kappa^{\frac{2n}{n-2}} \right) \sqrt{-g} d^n x$$

Dirac varied this action in second-order formalism, assuming a Weyl connection. However, varying the metric and connection independently, this action leads to a pure-gauge Weyl vector and hence describes a Riemannian geometry. This may also be seen by rescaling the metric by an appropriate power of the scalar field. A nontrivial Weyl geometry is obtained only with an action quadratic in the Lorentz curvature. The Dirac/Weyl action above is completely satisfactory if our goal is to write a locally scale covariant gauge theory of general relativity. The triviality of the Weyl vector is exactly the condition we need in order to allow a global definition of a time standard. With this definition, we recover general relativity.

Weyl conformal gravity, by contrast with the other theories discussed here, does not readily reproduce general relativity. Actions linear in the curvature become inconsistent, and we must write quadratic terms. Regardless of the initial form of these quadratic terms, the gauge field of special conformal transformations may be eliminated from the problem. When this is done, the action always reduces to a linear combination of the square of the Weyl curvature and the square of the dilatational curvature. In 4-dim we have

$$S = \int \left(\alpha \Omega^a_{\ b} \wedge \Omega^b_{\ a} + \beta \Omega \wedge^* \Omega \right)$$

and even this expression does not generalize to higher dimensions. We must either go to k^{th} order in the curvature in $2k$-dimensions, or we must use a Yang-Mills type of action,

$$S_{YM} = \int \left(\alpha \Omega^a_{\ b} \wedge^* \Omega^b_{\ a} + \beta \Omega \wedge^* \Omega \right)$$

In none of these cases does the theory reproduce general relativity. The Yang-Mills action gives a field equation involving the divergence of the Weyl curvature tensor, but also has an energy-momentum type source built quadratically from the curvature. The energy-momentum term arises from the variation of the metric in the Hodge dual and volume form. It can be shown that by placing a certain constraint on the curvature the resulting solutions describe conformally Ricci flat spacetimes, but the curvature constraint restricts solutions to a subset of the possible Petrov types. As we shall show below, these difficulties come with no added benefit: the additional gauge fields are auxiliary and no new structures are introduced.

Biconformal gauging was first introduced by Ivanov and Niederle[7] when they realized the drawbacks of Weyl conformal gravity. There have been numerous subsequent developments, including the introduction of an action linear in the curvature[16],

$$S = \int (\alpha \Omega^a_{\ b} + \beta \delta^a_b \Omega + \gamma \mathbf{e}^a \wedge \mathbf{f}_b) \wedge \varepsilon^{be\ldots f}_{\quad\quad ac\ldots d} \mathbf{e}^c \wedge \ldots \wedge \mathbf{e}^d \wedge \mathbf{f}_e \wedge \ldots \wedge \mathbf{f}_f$$

This action is valid in any dimension. It works because of the symplectic structure discussed below, which allows the volume element to be dimensionless. Naturally, the field equations arising from this action are complicated both by the number of equations and by the number of fields in the $2n$-dimensional base space. Nonetheless, it has been shown that the field equations generically describe general relativity on the co-tangent bundle of an n-dim Riemannian spacetime with pure-gauge Weyl vector[21]. As with the linear action for the Weyl geometry, choosing a global time standard reduces the theory to general relativity.

2.3. Additional structures

We now examine structures which arise from the original choice of gauge group, $\mathcal{G}$.

2.3.1. Killing metric The Killing metric of a semi-simple Lie group is a non-degenerate symmetric form constructible from the generators, G_A, as

$$K_{AB} = \lambda tr(G_A G_B)$$

where λ is any convenient constant. Since all of the groups considered here are subgroups of the conformal group, we can find the consequences for each gauging from the conformal Killing metric,

$$K_{AB} = \begin{pmatrix} \delta^a_b \delta^c_d - \eta^{ac}\eta_{bd} & 0 & 0 & 0 \\ 0 & 0 & \delta^a_b & 0 \\ 0 & \delta^b_a & 0 & 0 \\ 0 & 0 & 0 & 1 \end{pmatrix}$$

where the upper left corner gives the Killing metric of the Lorentz group, the off-diagonal Kronecker deltas span the translations and special conformal transformations, and the 1 in the lower right arises from the dilatations. The question we ask is whether the restriction of this metric to the base manifold induces a non-degenerate metric there.

For the Poincarï¿œ, Weyl, and Weyl conformal gaugings, the answer is no. The base manifold corresponds to the zero in the second row and second column of K_{AB}, and since that part of the Killing metric is off-diagonal, there is no induced metric in any of these gauge theories. Naturally, we may introduce the original metric by hand after gauging.

In contrast to this, the restriction of K_{AB} to the biconformal base manifold is the non-degenerate form,

$$K_{\Sigma\Delta} = \begin{pmatrix} 0 & \delta^a_b \\ \delta^b_a & 0 \end{pmatrix}$$

giving biconformal geometry a natural, conformally invariant metric.

2.3.2. Symplectic form A symplectic form is defined as a closed, nondegenerate 2-form. Such a form can exist only on even dimensional manifolds. If present it gives a way to divide fields into canonically conjugate pairs, which for convenience we call configuration/momentum pairs.

There is no reason to expect a symplectic form in any of these geometries, and certainly not in the Poincarï¿œ, Weyl or conformal Weyl cases because these theories exist in odd dimensions. Nor is there any evidence for a symplectic form in the structure equations. However, in the biconformal gauging the Maurer-Cartan equation for the Weyl vector is

$$\mathbf{d}\omega = \mathbf{e}^a \wedge \mathbf{f}_a$$

The left side shows that this 2-form is closed, since it is exact, while the right side is necessarily nondegenerate. The presence of this symplectic form survives into the curved spaces: generic solutions to the field equations have symplectic structure. This reinforces the interpretation of these solutions as co-tangent bundles, and has been used to express Hamiltonian particle dynamics[22] and quantum mechanics[23] in biconformal spaces.

2.3.3. Spacetime signature: the existence of time As we have seen, the extension to Weyl geometry offers only local dilatations as compesation for moving beyond Poincarï¿œ gauge theory, and neither the Poincarï¿œ nor the Weyl group is semisimple. Though many interpretations of Weyl geometry have been proposed, none has successfully provided new insights or become generally accepted.

The Weyl conformal gauging has also failed to introduce interesting new structures, and more importantly, fails to reproduce general relativity in any straightforward way. Even if we accept a modified theory of gravity, the Weyl conformal gauging requires a different action in every dimension.

The situation is considerably different for the biconformal gauging. As we have shown, the same linear-curvature action reproduces general relativity in any dimension, there is a natural symplectic form, and the conformal Killing metric has a nondegenerate restriction to the space. These elements

Journal of Physics: Conference Series **462** (2013) 012059 doi:10.1088/1742-6596/462/1/012059

combine to give the interpretation of a biconformal geometry as an expression of general relativity on the spacetime co-tangent bundle.

As might be expected, biconformal spaces are closely related to twistor spaces, $\mathbb{C}^n$. When the torsion vanishes, biconformal geometries have a complex structure, and may be identified with twistor spaces. However, because they arise from gauge theory, biconformal geometries have additional structure. The additional structure has an interesting consequence for our understanding of time.

Clearly, twistor space has numerous submanifolds of arbitrary signature. In particular, it has both Euclidean and Lorentzian submanifolds. However, there is no preferred map between these.

The case is different in biconformal spaces, where, using the metric and symplectic form, it becomes possible to ask for orthogonal, canonically conjugate, metric submanifolds. It turns out that these exist only in special circumstances, and the solution provides a unique map between Euclidean and Lorentzian submanifolds[24]. In the next section, we explore this construction in some detail.

3. The signature theorem

Since biconformal spaces have a symplectic form and a metric, we may ask for the existence of well-defined configuration and momentum subspaces[24]. Consider the conformal group of a compactified, n-dim space with metric η_{ab} of signature $s = p - q$, where $p + q = n$. The quotient of the conformal group by its Weyl subgroup is then a flat biconformal space of dimension $2n$.

We demand the following properties:

(i) The configuration space, $\mathscr{S}$, and the momentum space, $\mathscr{P}$ are spanned by canonically conjugate basis forms.

(ii) The spaces $\mathscr{S}$ and $\mathscr{P}$ are submanifolds.

(iii) The momentum space $\mathscr{P}$ is conformally flat.

(iv) The configuration space $\mathscr{S}$ has non-degenerate induced metric.

(v) The spaces $\mathscr{S}$ and $\mathscr{P}$ are orthogonal.

Conditions 1, 2 and 3 insure that the biconformal space is a phase space[25]. The condition 4 requires the configuration space metric to arise as the restriction of the Killing metric to $\mathscr{S}$, while condition 5 makes this induced metric unambiguous.

It is known that the solutions to the torsion-free biconformal field equations describe co-tangent bundles. Therefore, condition 3 is a consequence of the field equations. Here, we consider only flat biconformal space and impose condition 3 by hand. Since the results of this depend only on the signatures of the submanifolds, the results continue to hold for arbitrary biconformal spaces subject to the field equations.

It is straightforward to find the general form of a basis for orthogonal, canonically conjugate, subspaces. The result, up to changes of basis within the subspaces separately, is

$$\begin{aligned}\psi^a &= \frac{1}{2\alpha}\mathbf{e}^a - \alpha h^{ab}\mathbf{f}_b \\ \chi_a &= \frac{1}{2\alpha}h_{ab}\mathbf{e}^b + \alpha\mathbf{f}_a\end{aligned}$$

where $h^{ab}h_{bc} = \delta^a_c$ and h_{ab} is symmetric. With the Killing metric holding for the original $(\mathbf{e}^a, \mathbf{f}_a)$ basis, we find that h_{ab} becomes the configuration space metric. It follows from the non-degeneracy of the Killing metric that the momentum space has metric $-h^{ab}$.

We next impose conformal flatness of the subspace spanned by χ_a. It follows that h_{ab} is necessarily of one of two forms:

$$\begin{aligned}1.\ h^{ab} &= \frac{1}{nh}\eta^{ab} \\ 2.\ h^{ab} &= \frac{(n-2)}{hu^2}\left(-2u^a u^b + u^2\eta^{ab}\right)\end{aligned}$$

where $u^2 = \eta_{ab}u^a u^b$ and $h = \eta_{ab}h^{ab}$. The first solution is the expected projection from the biconformal space back to the original space. The second solution will be recognized as a signature-changing transformation of the original metric.

Before considering the conformally flat solutions in more detail, we impose the final condition. Requiring the subspaces $\mathscr{S}$ and $\mathscr{P}$ to be submanifolds places integrability conditions on the basis forms. Each set of basis forms must be involute,

$$\begin{aligned} \mathbf{d}\psi^a &\sim \psi^a \\ \mathbf{d}\chi_a &\sim \chi_a \end{aligned}$$

However, for general h_{ab}, the structure equations for ψ^a and χ_a contain terms $\left(\mathbf{D}h^{ab}\right)\chi_b$ and $\mathbf{D}h_{ab}\psi^b$, respectively. In order for these equations to be in involution, the covariant derivative $\left(\mathbf{D}h^{ab}\right)\chi_b$ must have vanishing $\psi^b\psi^c$ term, and $\mathbf{D}h_{ab}\psi^b$ must have vanishing $\chi^b\chi^c$ term. The vanishing of these quadratic terms, along with the signature-changing form of h_{ab} required by conformal flatness has the *unique* solution

$$h_{ab} = \lambda\left(y_a y_b - \frac{1}{2}y^2\eta_{ab}\right)$$

where y_a comprise half of the biconformal coordinates. A check shows that with this solution the basis forms are indeed involute.

Finally, we must consider whether this solution for h_{ab} actually does produce a consistent metric on the configuration space. Since y_a is a coordinate on a subspace with metric η_{ab}, y_a will take on both timelike and spacelike values unless η_{ab} is Euclidean. In general, the resulting signature, s', of h_{ab} will be

$$\begin{aligned} s' &= s-2 \quad y_a\,spacelike \\ s' &= -s-2 \quad y_a\,timelike \end{aligned}$$

These are consistent only if $s = 0$, or if the original space is Euclidean, $s = \pm n$. We therefore have a consistent solution for h_{ab} in only two cases. The signature of h_{ab} is $n-2$ if the original space is Euclidean, and -2 if the original space has vanishing signature.

If the original dimension is $n = 4$, both cases lead to Lorentzian spacetimes; for other dimensions, a Euclidean initial space always leads to a Lorentzian spacetime. This establishes a 1-1 correspondence between initial Euclidean spaces and Lorentzian spacetimes. Since η_{ab} also provides a solution, we can project from the biconformal space to either of these configuration spaces. In particular, we note that a gravitational solution on the biconformal space will induce Euclidean gravity on the one hand and solutions to general relativity on the other. Again, we stress that our restriction to flat biconformal spaces must generalize smoothly to curved spaces since the results depend only on signature.

As a final observation, notice that there is a particular subset of phase space coordinates, y_a, which determine the direction of the light cones in the configuration space. This will undoubtedly have consequences for the existence of tachyons – for example, if y_a is a momentumlike direction, then the momentum necessarily points inside the lightcone by definition.

4. Summary

$\mathcal{G}$	$\mathcal{H}$ = Local symmetries	General relativity	Scaling	Physical fields (tensors)	Structures on $\mathcal{M} = \mathcal{G}/\mathcal{H}$ arising from $\mathcal{G}$
Poincaré	Lorentz	Yes	No	Solder form Curvature	• Spacetime
Weyl	Lorentz Dilatations	Yes	Yes	Solder form Curvature	• Spacetime
Conformal	Lorentz Dilatations Spec. conf.	No (Weyl conformal gravity)	Yes	Solder form Weyl Curvature	• Spacetime
Conformal	Lorentz Dilatations	Yes	Yes	Solder form Curvature Symplectic form Co-torsion Dilatation	• ST cotangent bundle • Symplectic structure • Induced Killing metric • Complex structure • Time (derived Lorentz sym.)

References

[1] Utiyama R 1956 *Phys. Rev.* **101** p 1597
[2] Kibble T W B 1961 *J. Math. Phys.* **2** p 212
[3] Kobayashi S and Nomizu K 1963 *Foundations of Differential Geometry* (New York: Wiley)
[4] Ne'eman Y and Regge T. 1978 *Phys. Lett. B* **74** p 54
[5] Ne'eman Y and Regge T 1978 *Riv. Nuovo Cim.* **1N5** p 1
[6] Ivanov E A and Niederle J 1982 *Phys. Rev. D* **25** p 976
[7] Ivanov E A and Niederle J 1982 *Phys. Rev. D* **25** p 988
[8] Isham C 1999 *Modern Differential Geometry for Physicists* (Hackensack: World Scientific)
[9] Ehlers J, Pirani A E and Schild A 1972 *General Relativity* ed O'Raifeartaigh L (Oxford: Oxford University)
[10] Weyl H 1918 *Sitzung. d. Preuss. Akad. d. Wissensch.* p 465; 1923 *The Principle of Relativity, Chapter XI* (Dover)
[11] Weyl H 1952 *Space-Time-Matter* (New York: Dover)
[12] Weyl H 1921 *Nature* 106; 1918 *Math. Zeitschr.* 2 p 384; 1918 *Ann. d. Physik*, 54 p 117 ; 1919 *Ann. d. Physik*, 59 p 101; 1920 *Phys. Zeitschr.* 21 p 649; 1921 *Ann. d. Phys.* 65 p 541; 1921 *Phys. Zeitschr.* 22 p 473; 1929 *Zeit. f. Physik*, 56 p 330
[13] Adler R, Bazin M, Schiffer M 1965 *Introduction to General Relativity*, (New York: McGraw Hill) p 401
[14] Bouchami1 J and Paranjape M B 2008 *Physical Review D* **78** 044022
[15] Wheeler J T 1998 *J. Math. Phys.* **39** p 299
[16] Wehner A and Wheeler J T 1999 *Nucl. Phys. B* **557** p 380
[17] Anderson L B and Wheeler J T 2004 *Nucl. Phys. B* **686** p 285
[18] Dirac P A M 1951 *Proc. Roy. Soc.* **A209** p 291
[19] Dirac P A M 1952 *Proc. Roy. Soc.* **A212** p 330
[20] Dirac P A M 1973 *Proc. Roy. Soc.* **A333** p 403
[21] Wheeler J T *in preparation*
[22] Wheeler J T 2007 *Canadian Journal of Physics*, **85** (4) p 307
[23] Anderson L B and Wheeler J T 2006 *Int. J. Geom. Meth. Mod. Phys.* **3** p 315
[24] Spencer J A and Wheeler J T 2008 arXiv:0811.0112v1 [gr-qc]
[25] Abraham R H and Marsden J E 1978 *Foundations of Mechanics* (New York: Addison-Wesley)

Journal of Physics: Conference Series **462** (2013) 012060 doi:10.1088/1742-6596/462/1/012060

Basics of Ternary Algebras and their underlying Nambu Brackets

Cosmas K Zachos

High Energy Physics Division, Argonne National Laboratory, Argonne, IL 60439-4815, USA

E-mail: `zachos@anl.gov`

Abstract. Ternary algebras amount to closing systems of antisymmetrized trinomials of operators. The Filippov conditions (FI, which are *not identities*) for ternary algebras are contrasted to Bremner's identities dictated by associativity of operator products, and thus analogous to Jacobi identities.

Maps of the known FI-compliant ternary algebras to underlying classical Nambu brackets are constructed, which then explain this compliance: FI-compliant ternary algebras are essentially classical Nambu brackets in disguise.

In some cases involving infinite algebras, we show the classical limit may be obtained by a contraction of the quantal ternary algebra, and then explicitly realized through classical Nambu brackets. We illustrate this classical-contraction method on our Virasoro-Witt ternary algebra paradigm.

The content of the talk is in the following two references.

References

[1] Curtright T, Jin X, Mezincescu L, Fairlie D and Zachos C 2009 Classical and quantal ternary algebras *Phys. Lett.* B **675** 387–92

[2] Curtright T, Fairlie D and Zachos C 2008 Ternary Virasoro-Witt algebra *Phys. Lett.* B **666** 386–90

Self Interference of Single Electrodynamic Particle in Double Slit

J.X. Zheng-Johansson

Institute of Fundamental Physics Research, 611 93 Nyköping, Sweden

E-mail: jxzj@iofpr.org

Abstract. It is by the long established fact in experiment and theory that electromagnetic waves, here as one component of an IED particle, passing a double slit will undergo self inference each, producing at a detector plane fringed intensities. The wave generating point charge of a zero rest mass, as the other component of the particle, is maintained a constant energy and speed by a repeated radiation reabsorption/reemission scheme, and in turn steered in direction in its linear motion by the reflected radiation field, and will thereby travel to the detector along (one of) the optical path(s) of the waves leading to a bright interference fringe. We elucidate the process formally based on first principles solutions for the IED particle and known principles for wave-matter interaction.

1. Introduction

Since first hypothesised by L de Broglie in 1923[1], it has been overall experimentally demonstrated that all matter particles, like the electrons and atoms, classically pictured as point entities, manifest as also waves, referred to today as matter waves or quantum particles. These particles are capable of producing such wave effects as diffraction and interference entirely in the same way as the ordinary waves do. Overall experiments have subsequently corroborated that the matter waves formally are governed by quantum mechanics, pillared with the Schrödinger equation, or alternatively the Heisenberg equation, and the Dirac equations. Up to the present however it has remained as a basic open question as to what is waving with a matter wave (ψ). Regarding the nature of this wave, the current understanding has been formally rested on an assumption that the complex form of it, $\psi(x,t)$, specifies the physical state of a particle assumed being spatially a point, with $|\psi(x,t)|^2$ describing the probability of finding the particle at location x at time t. This in essence a "statistical point particle" picture has been successful in accounting for some of the basic properties of the quantum particles, in particular in predicting as eigen solutions the quantized dynamical quantities in direct agreement with experiments. This picture however encounters difficulties in accounting for certain other important properties, most notably the superposition, diffraction and interference, of the quantum-mechanical matter particles as well as "photons". The picture becomes paradoxical in the case of explaining the experimentally demonstrated single particle interference (see e.g. the recent reviews in [2, 3, 4]), most simply in a double slit, since a point particle is not capable of entering two slits at the same time. The incompleteness with the current understandings of the matter waves is commonly acknowledged.

From the standpoint of the general physics of particles, the (single) particle interference, or also the sometimes joined-in quantum entanglement, as we see, is only one of a range of unsolved "mysteries" in the realm of fundamental physics. By what scheme an electromagnetic radiation is upon absorption or emission converted to or from particle's mass, by a portion or as a whole? What is the origin of mass (m)? Why is mc^2, where c is the velocity of light, equal to the particle's total energy? Why do masses attract one another? And so forth. The eventual answers to these questions in a coherent theoretical framework would demand a realistic particle scheme, or model. Such basic properties of particles as mass, total energy and wave function we know today are each dependent on the particle's motion and these in turn are convertible to and from electromagnetic radiation. This is to say that, the basic particle properties and their relations are interrelated in a dynamical context. A particle scheme or model can not be an ultimate one which facilitates the predictions of some and yet not the overall fundamental properties of particles.

Stemmed from general considerations as underlined above, the author developed in [5a-l] an internally electrodynamic (IED) particle model, which briefly states that *a single-charged material particle, like the electron, proton, etc., is constituted of (i) an oscillatory point charge q (as source) with a characteristic frequency Ω and zero rest mass, and (ii) the electromagnetic waves generated by the charge.* See further a formal outline in Sec. 2. The IED process is governed by a minimal set of well established basic, or first principles laws (for a recent review see e.g. [4], [5j]). A range of basic properties and relations for particles have become predictable in terms the first principle's solutions for the IED process under corresponding conditions; solutions already obtained by the author include those reported in [5a-l]. Based on first principles solutions for the IED particle and the known principles for wave-matter interaction, we formally elucidate in this paper that the IED particle will naturally undergo self-interference in a double slit with a scheme already built-in with the IED process.

2. Wave equations, solutions for IED particle in three dimensions

Consider that a charge q of a zero rest mass and having an initial free-charge kinetic energy ε_q (hence a dynamical inertia) is inserted at position $\mathbf{R}_q$ in the vacuum to which we attach three Cartesian coordinates X, Y, Z. Owing to ε_q, the charge is spontaneously driven into an oscillation (following a mechanical scheme elucidated e.g. in [5j]) of a characteristic frequency Ω, of displacement $\mathbf{u}_q = u_q \hat{u_q}$ along a $\hat{u_q}$ axis about $\mathbf{R}_q(X_q, Y_q, Z_q)$, and in addition, a uniform translation at velocity v in the $+X$-direction. In view of the general validity of the Maxwell's equations for the resulting waves below, u_q may be judged as relatively small and thus has the usual sinusoidal solution (see e.g. [5j]), assuming zero applied force, $u_q(t) = A_q \sin(\Omega t + \alpha)$, with A_q the amplitude and α initial phase. The orientation of $\mathbf{u}_q$, $\hat{u}_q$, may change with time under the influence of random environmental and/or applied fields.

Suppose for the present that during a time δt_1, $\hat{u}_q$ is along the Z-axis; $u_q = Z'_q - Z_q$. Owing to its accelerated motion u_q, the charge generates electromagnetic waves of electric fields $\mathbf{E}^j(\mathbf{r}, t)$'s and magnetic fields $\mathbf{B}^j(\mathbf{r}, t)$'s as propagated to position $\mathbf{r}(x, y, z) = \mathbf{R}(X, Y, Z) - \mathbf{R}_q$ in time $t = r/c$, c being the velocity of light and j indicating a source motion effect. The fields are described by the Maxwell's equations given for a specified j, in regions excluding the charge q and assuming no other charge and current, as $\nabla \cdot \mathbf{E}^j = 0$, $\nabla \cdot \mathbf{B}^j = 0$, $\nabla \times \mathbf{B}^j = 0 + \frac{1}{c^2}\partial_t \mathbf{E}^j$, $\nabla \times \mathbf{E}^j = -\partial_t \mathbf{B}^j$. These lead to a wave equation for a jth component wave, denoting for brevity by a dimensionless wave displacement $\varphi^j \hat{\theta} = \mathbf{E}^j / E_q$ in the transverse $\hat{\theta}$ direction, and a corresponding equation for the total wave $\psi = \sum_j \varphi^j$,

$$\partial_t^2 \varphi^j = c^2 \nabla \varphi^j \quad (a), \quad \partial_t^2 \psi = c^2 \nabla \psi \quad (b) \qquad (1)$$

The travelling oscillatory charge q and the electromagnetic waves E^j, B^j's, or φ^j's which are continuously generated and therefore in general extend a wave train, make up our extensive IED particle, with the prefix "extensive" unless for emphasis omitted hereafter.

For $v < c$ (or $<< c$) holds ordinarily, there thus ordinarily exists a brief yet finite time interval δt in which the charge is essentially standing still, at X_q, on the X axis, for which α is a fixed value. So except for a source motion effect indicated by j, the solutions to the Maxwell's equations are of the usual form $E^j = E_q\varphi^j$, where $E_q \doteq \frac{\mu_0 q\omega^2 A_q}{4\pi}$, with $\omega \doteq \omega^j$, and

$$\varphi^\dagger = \tfrac{1}{2}\mathcal{C}_0 \sin(k_p^\dagger r - \omega_p^\dagger t + \alpha), \quad \varphi^\ddagger = -\tfrac{1}{2}\mathcal{C}_0 \sin(k_p^\ddagger r + \omega_p^\ddagger t - \alpha), \quad \text{with } \mathcal{C}_0 = C_0 \sin\theta/r, \tag{2}$$

are two opposite travelling plane waves in the $+r$- ($j = \dagger$) and $-r$- ($j = \ddagger$) directions along a certain radial path $\mathbf{r}(r,\theta,\phi)$ in spherical polar coordinates r,θ,ϕ. In these, $k_p^j = \gamma_p^j K$,$\omega_p^j = k_p^j c = \gamma_p^j \Omega$ are the Doppler displaced wavevectors and frequencies due to the source motion, with $\gamma_p^\dagger = \frac{1}{1-v_p/c}$, $\gamma_p^\ddagger = \frac{1}{1+v_p/c}$, $\Omega = Kc$, and $v_p = v\sin\theta\cos\phi$ being the component velocity of charge q in the $+r$-direction; C_0 is a normalization constant.

Let the IED particle be confined as in later applications between two massive, non absorbing vertical walls here separated at a distance L_ν along the X axis. Its waves φ^j's will travel up to the walls, be reflected, reabsorbed by the charge, partially and most probably after a two-way reflection owing to detailed resonance, and then reemitted, repeatedly, thereby maintaining a constant total energy (ε_{tot}) of the charge and wave system. If the charge has not emitted any radiation, then $\varepsilon_{tot} = \varepsilon_q$. If the charge has emitted its entire ε_q into electromagnetic waves of a mean total length $L_\varphi = \sqrt{L_\varphi^\dagger L_\varphi^\ddagger}$, of a total wave energy $\varepsilon(\equiv \varepsilon_q)$, then $\varepsilon_{tot} = \varepsilon$. In representing a realistic extensive particle able to produce interference at the scale L_ν, L_φ would wind about L_ν in a large $\frac{J_\nu}{2} = \frac{L_\varphi}{2L_\nu}$ number of loops. For this extensive IED particle, there will ordinarily be a portion a_1 of ε_{tot} conveyed by the charge and a_2 by the wave, $a_1, a_2 \le 1$, and $\varepsilon_{tot} = a_1\varepsilon_q + \alpha_2\varepsilon$. For the IED particle in stationary state mainly of our concern later, the rates of emission α^{em}, absorption α^{ab}, and transmission α^{tr} by or passing q need further be in equilibrium, $\alpha^{ab} = \alpha^{tr} + \alpha^{em}$.

At any point on a radial path r in $(0, L_\nu)$ there will thus simultaneously prevail the opposite travelling $\varphi^\dagger$ and $\varphi^\ddagger$, newly generated or reflected after round trips, of the IED particle. These superpose into a partial total wave $\psi = \varphi^\dagger + \varphi^\ddagger$, given after direct algebraic operations in complex form as

$$\psi(r,\theta,t)|_{k_{dp}} = \Phi f, \quad \Phi = e^{i[(K + (\frac{v_p}{c})k_{dp})r - (\frac{v_p}{c})\omega_p t]}, \quad f = \mathcal{C}_0 e^{i[k_{dp} r - \omega_p t + \alpha_0]}, \tag{3}$$

where $\alpha_0 = \alpha - \frac{\pi}{2}$, $\frac{1}{2}(k_p^\dagger - (-k_p^\ddagger)) = K + \gamma_p \frac{v_p}{c} k_{dp}$, $\frac{1}{2}(\omega_p^\dagger - \omega_p^\ddagger) = \gamma_p(\frac{v_p}{c})\omega_p \doteq \frac{v_p}{c}\omega_p$, $\frac{1}{2}(k_p^\dagger + (-k_p^\ddagger)) = \gamma_p k_{dp} \doteq k_{dp} = \sqrt{(k_p^\dagger - K)(K - k_p^\ddagger)}$, $\frac{1}{2}(\omega_p^\dagger + \omega_p^\ddagger) = \gamma_p\omega_p \doteq \omega_p = \sqrt{\omega_p^\dagger \omega_p^\ddagger}$, with the geometric means being alternative expressions; $\gamma_p = \sqrt{\gamma_p^\dagger \gamma_p^\ddagger} = 1/\sqrt{1 - v_p^2/c^2}$ or $\gamma_p^2 = 1 + \gamma_p^2(v_p^2/c^2)$; and

$$k_{dp} = (\gamma_p/\gamma)(v_p/v)k_d, \quad k_d = \gamma(v/c)K, \quad \omega_p = (\gamma_p/\gamma)\omega, \quad \omega = \gamma\Omega, \tag{4}$$

and $\gamma = 1/\sqrt{1 - v^2/c^2}$. ψ travels in the $+r$-direction at a phase velocity $W_{pp} = \frac{\omega_p^\dagger + \omega_p^\ddagger}{k_p^\dagger + (-k_p^\ddagger)} = \frac{\omega_p}{k_{dp}}$ and group velocity $W_{gp} = \frac{\omega_p^\dagger - \omega_p^\ddagger}{k_p^\dagger - (-k_p^\ddagger)} = \frac{(v_p/c)\omega_p}{K + (v_p/c)k_{dp}}$, or

$$W_{pp} = W_p/\sin\theta\cos\phi, \quad W_p = c^2/v, \quad W_{gp} \doteq v_p = W_g \sin\theta\cos\phi, \quad W_g = v. \tag{5}$$

The ψ due directly to the moving source above is seen each r,θ-dependent, or, "polar spherical".

The electromagnetic waves have in terms of the total field $E_q\psi$ an energy density of the usual form $\varepsilon_0 = \epsilon_0 E_q^2|\psi|^2$, which is transported at the speed of the φ^j waves, $\omega^\dagger/k^\dagger = \omega^\ddagger/k^\ddagger = c$. The intensity of the total wave in relative terms is therefore

$$I(r,\theta) = c\varepsilon_0/c\epsilon_0 E_q^2 = |\psi|^2 = \mathcal{C}_0{}^2. \tag{6}$$

The ε_0 integrated over total solid angle 4π is $\varepsilon_{0x} = \int_{4\pi} \varepsilon_0 d\tilde{\Omega} = \epsilon_0 E_q'^2 |\psi_x|^2$ where $d\tilde{\Omega} = \sin^2\theta d\theta d\phi$, $E_q' = \frac{2\sqrt{\pi}}{\sqrt{3}} E_q$, and $\psi_x = C_0 e^{i[k_d x - \omega t + \alpha_0]}$ is an apparent plane wave along the X-axis by which ε_{0x}, being a constant, is effectively transported. Along the X-axis $\theta = \frac{\pi}{2}, \phi = 0$, the $\mathcal{C}_0 (= C_0/x)$ and I are a maximum each, $\psi = \psi_x/x$, and the wave variables k_d, ω describe the dynamics of the mass center of the total ψ and hence the IED particle. The X-axis therefore represents a symmetry axis of ψ. The total wave energy across the full wave train $L_\varphi (= J_\nu L_\nu)$ is $\varepsilon = J_\nu \int_0^{L_\nu} \varepsilon_{0x} dx = L_\varphi \epsilon_0 E_q'^2$ provided with the normalisation $\int_0^{L_\nu} |\psi_x|^2 dx = 1$ and thus $C_0 = 1/\sqrt{L_\nu}$. $p = \varepsilon/c$ gives the total linear momentum.

Alternatively, employing the Planck energy equation for the ψ of a frequency ω, and in turn Newtonian kinetic energy equation for the wave train of ψ which travels rectilinearly at the finite velocity c and accordingly has a finite inertial mass m, we may write down $\varepsilon = \hbar\omega = mc^2$; and $p (= \varepsilon/c) = \hbar k = mc$ (see e.g. [5a,c,e] for a detailed account). Making the expansion $\omega (= \gamma\Omega) = \Omega + \varpi_d$ where $\varpi_d [= (1 + \frac{3}{4}\frac{v^2}{c^2} + \ldots)\frac{1}{2}(\frac{v^2}{c^2})\Omega]$, with $\mathcal{E} = \lim_{v^2/c^2 \to 0} \varepsilon = \hbar\Omega$ and $P = \lim_{v^2/c^2 \to 0} p = \hbar K$, the differences $\varepsilon_v (\doteq \frac{1}{2}mv^2) = \varepsilon - \mathcal{E}$ and $p_v (\equiv mv) = \sqrt{p^2 - P^2}$ thus give the kinetic energy and linear momentum of the IED particle. These further lead to $M^2c^4 + p_v^2 c^2 = \varepsilon^2$ and the corresponding de Broglie relations

$$\tfrac{1}{2}mv^2 = \hbar\varpi_d, \quad mv = \hbar k_d. \tag{7}$$

ψ thus follows to represent a de Broglie phase wave and reduces to a de Broglie wave Ψ (Sec. 4) at a thermal k_d scale, each being polar-spherical. More generally, at the scale k_d equation (1b) reduces to a Schrödinger equation describing the kinetic motion of the particle[5c,j].

From the above follows that $K = Mc/\hbar >> k_d$ holds ordinarily, with $M = \lim_{v^2/c^2 \to 0} m$ the particle's rest mass. The diffraction and interference effects to be dealt with later will be at the scale k_d, with $\lambda_d = 2\pi/k_d$ the de Broglie wavelength, for which $\Phi \doteq 1$, $\psi \doteq f = \mathcal{C}_0 e^{i[k_{dp} r - \omega_p t + \alpha_0]}$.

3. Optical path of ψ through a double slit. Interference fringes

Suppose that the ψ above, re-denoting now by $\psi(r', \theta', t')$, is first let strike on to an opaque screen D_1 with a narrow slit A of width b, with $b \sim \lambda_d$, at distance x' from q, see Fig. 1a. An effectively plane wave, of a fraction g_A of the integral intensity ($I_{4\pi} = \int_{4\pi} I d\tilde{\Omega}$), only will enter the narrow slit, given by setting $r' = x'$ and $\theta' = \frac{\pi}{2}$ in (3) as $\psi(x', t') = \frac{C_A}{x'} e^{i(k_d x' - \omega t' + \alpha_0)}$, where $C_A = \sqrt{g_A} C_0$. Along the b, the incident $\psi(x', t')$ may be as in the usual approach thought as if regenerated from many (N) small divisions of a width $ds = b/N$ each, into a new wave $\psi_s(r_s'', \theta_s'', t'') = \mathcal{C}_0 e^{[k_d r_s'' - \omega t'' + \alpha_A]}$ along a radial path $r_s'' = r'' + s \sin\beta''$ joining a ds at a vertical distance s from the slit center A to a point P_2 on an image plane D_2, with $r'' = AP_2$ and $\beta'' = \frac{\pi}{2} - \theta''$; $\alpha_A = \alpha_0 + k_d x_A' - \omega t_A$. The total diffracted de Broglie phase wave from b arriving at P_2 at a later time t'', for N being large and x' assumed for illustration large, is given by

$$\psi(r'', \theta'', t'')|_{k_d} = \int_{\frac{-b}{2}}^{\frac{b}{2}} \psi_s(r_s'', \theta_s'', t) ds = \mathcal{C}_A e^{i(k_d r'' - \omega t'' + \alpha_A)}, \quad \mathcal{C}_A(\theta'') = \frac{b\sqrt{g_A} C_0 \sin\eta''}{x'\eta''} \tag{8}$$

$\eta'' = \frac{\pi b \sin\beta''}{\lambda_d}$. ψ has the phase and group velocities W_p and W_g as of $\psi(x', t')$, and an intensity as given after (6) for $\mathcal{C}_A$ here, $I(\theta'') = \mathcal{C}_A^2(\theta'')$.

Let the D_2 above be an opaque screen containing two narrow slits B, C similarly of a width b each, as in Fig. 1a; and the $\psi(r'', \theta'', t'')$ given by (8), regenerated at A at time $t_0'' = 0$, is incident on D_2. Two of its partial components travelling along the equidistant radial paths $r'' = AB$ and AC at angles $\theta'' = \theta_{AB}$ and θ_{AC} to the Z axis, and of a fraction $g_B = g_C$ of total intensity each, $\psi(AB, \theta_{AB}, t'')$ and $\psi(AC, \theta_{AC}, t'')$ will at a later time t'' $(= AB/c)$ simultaneously arrive at B and C; here they have an identical amplitude $\mathcal{C}_A$ and initial phase α_A. By the standard

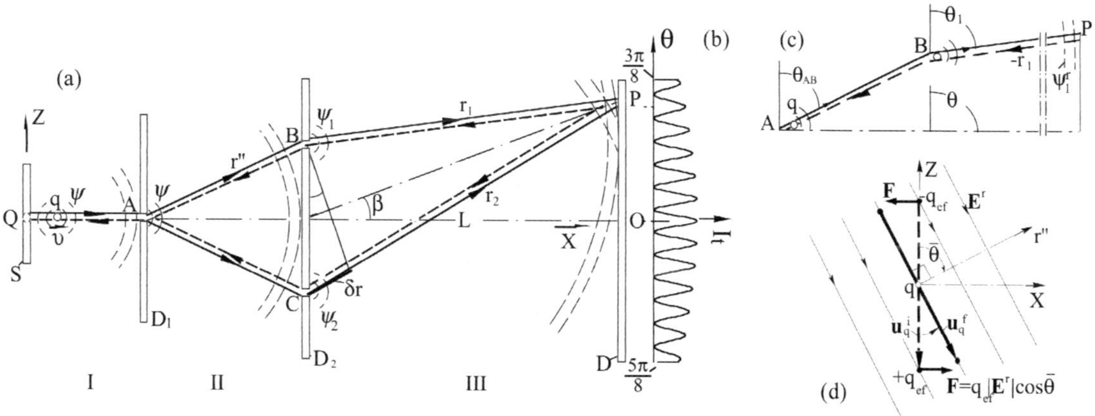

Figure 1. (a) The total electromagnetic wave ψ generated by travelling oscillatory charge q, with ψ, q making up an IED particle, travels through a single slit A and double slits B, C to detector D. (b) Intensity I_t (in units of $1/\mathcal{C}_{B0}$) of wave ψ_t on D calculated from (10) for $b = 1, \lambda_d = 0.5, \ell = 4$, showing two-slit interference fringes. (c) Charge q travels at velocity v statistically along one optical wave path $(Q)qABP$ through the double-slit up to D. (d) The effective dipole charges $+q_{\rm ef}, -q_{\rm ef}$ associated with the initial displacement $\mathbf{u}_q^{\rm i}$ of q along the Z direction, as at slit A, are acted on a torque $\mathbf{F} \times \mathbf{u}_q$ by the reflected field $\mathbf{E}^{\rm r}$ and turned into aligned with the $\mathbf{E}^{\rm r}$, in the $\bar{\theta}$ direction, with a final $\mathbf{u}_q^{\rm f}$.

notion we know that each electromagnetic wave φ^j will on passing the two slits undergo self interference. The formal conditions for this for the specific de Broglie phase wave ψ here are written down as follows.

The two slits B, C, as two new coherent sources, then regenerate at time $t_0 = 0$ the respective incident waves into two new spherical waves given at location r time t as $\psi_1 = \mathcal{C}_1 e^{i(k_d r_1 - \omega t + \alpha_B)}, \psi_2 = \mathcal{C}_2 e^{i(k_d r_2 - \omega t + \alpha_B)}$ in region III similarly as given by (8), where $\mathcal{C}_i = \mathcal{C}_{B0} \frac{\sin \eta_i}{\eta_i}$, with $i = 1, 2$, $\mathcal{C}_{B0} = \frac{b\sqrt{g_B g_A} C_0}{r''}$, $\eta_i = \frac{\pi b \sin \beta_i}{\lambda_d}$, $\beta_i = \frac{\pi}{2} - \theta_i$. Let at distance L apart in front of D_2 be placed a detector D, and P be a point on D as in Fig. 1a. Two respective partial components of ψ_1, ψ_2 travelling along paths $r_1 = BP, r_2 = CP$ at angles θ_1, θ_2 to the Z-axis, will at a later time t arrive at P, superposing linearly as $\psi_t = \psi_1 + \psi_2$. Assuming as in usual applications $L >> BC$, so for the amplitudes, the two paths r_1,r_2 can be approximated as parallel to each other, of an average distance $r = \frac{1}{2}(r_1 + r_2)$ and angle $\theta = \frac{1}{2}(\theta_1 + \theta_2)$; thus $\mathcal{C}_i \doteq \mathcal{C} = \mathcal{C}_{B0} \sin \eta / \eta$, with $\eta = \pi b \sin \beta / \lambda_d$, $\beta = \frac{\pi}{2} - \theta$. For the phase constants the difference between the optical paths, $\delta r = r_1 - r_2$, yet is nontrivial. Incorporating the foregoing, the summation for ψ_t is algebraically given as

$$\psi_t = \mathcal{C}_t(\delta r) e^{i(k_d r - \omega t + \alpha_B)}, \quad \mathcal{C}_t(\delta r) = 2\mathcal{C} \cos[\tfrac{1}{2} k_d \delta r]. \tag{9}$$

ψ_t, or the field $\mathbf{E}_t = E_q \psi_t \hat{\theta}$ has a relative intensity given similarly as (6) as $I_t = \frac{c\epsilon_0 E_q^2 |\psi_t|^2}{c\epsilon_0 E_q^2}$, or

$$I_t = |\psi_t|^2 = |\psi_1 + \psi_2|^2 = \mathcal{C}_t^2(\delta r) = 4(|\psi_1|^2 + |\psi_2|^2) \cos^2(\tfrac{1}{2} k_d \delta r) \tag{10}$$

where $|\psi_i|^2 = \frac{1}{2}\mathcal{C}^2$, $i = 1, 2$. Now,

$$\text{if} \quad \delta r = 2n\pi / k_d = n\lambda_d, \quad n = 0, 1, 2, \ldots, \quad \text{then} \quad |\mathcal{C}_t(\delta r)| = 2\mathcal{C}, \quad I_t = 4\mathcal{C}^2 \tag{11}$$

are maxima, yielding bright fringes (peaks in Fig. 1b) at distances $PO = L \tan \beta \doteq L \frac{\delta r}{\ell} = n\lambda_d L / \ell$ from O, where $\ell = BC$. Or, if $\delta r = (n + \frac{1}{2})\lambda_d$, then $\mathcal{C}_t(\delta r) = I_t = 0$ yielding dark fringes at distances $PO = (n + \frac{1}{2})\lambda_d L / \ell$ from O on D.

4. Reflection wave path

A wave ψ (as ψ_1 or ψ_2 here) striking at the detector D, assumed massive and not yet actualised a permanent absorption, will be scattered or reflected back from D, as ψ^{r}. In determining the reflection paths, the following basic principles of electrodynamics and wave mechanics (a-c) apply: (a) The scattering of ψ by a scatterer at P may be regarded as a temporary absorption of ψ by the scatterer, and then remission. (b) The amplitude of the absorption will be large if ψ and the scatterer are (nearly) in resonance, or small or zero if otherwise. (c) The orientation of the symmetry axis of ψ^{r} is determined by (c.1) the total linear momentum $p' = \hbar k' = mc'$ of ψ^{r} [after solutions to (1)] and in addition, (c.2) the path which requires the least action (the principle of least action). (d) For elucidating the mechanical scheme for the scattering of ψ in vacuum here, a physical vacuum becomes indispensable which is proposed in [5a,g-h] based on overall experiments as composed of electrically polarisable vacuuons.

For the ψ, vacuum and a massive wall D here we concretely find: (4.1): With the $\omega = \Omega + \varpi_d$ earlier, thus $\psi = \Theta\Psi$, $\Theta = \Phi e^{-i\Omega t}$, $\Psi = \mathfrak{C} e^{i[k_d r - \varpi_d t + \alpha_B]}$. This ψ may on the basis of (a)-(b) be scattered separately as Θ and Ψ by distinct scatterers. The scatterer can only be a vacuuon (Fig. 2a-b) for ψ (or Θ alone), whose frequency, ω (or Ω) of the scale of the particle's mass, is too high to be (temporarily) absorbed by a material particle. For Ψ of a thermal frequency ϖ_d, the scatterer may be a material particle, as illustrated in Fig. 2c for a free electron in the wall of a metallic surface. (4.2): The vacuuons will be polarised by the presence of q or also by the passing of ψ, and additionally, when in their proximities by the nuclear charges comprising the wall; they are thereby coupled with each other and in turn to the wall bulk (Fig. 2a). So when

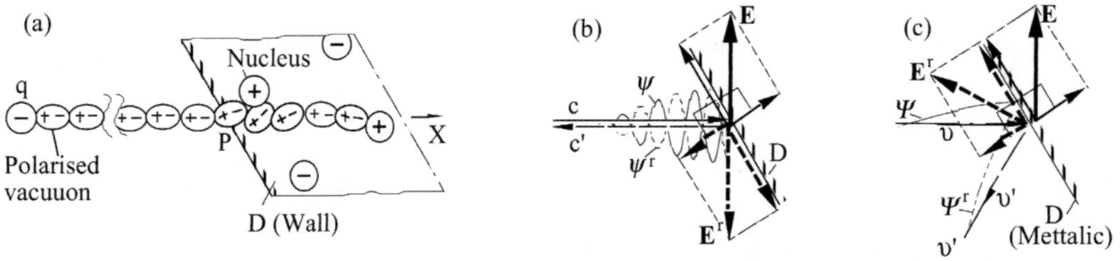

Figure 2. (a) Vacuuons comprising the vacuum are polarised about the charge q of an IED particle, and near P at a massive wall D in turn also by the mutually bound nuclear charges in the wall. (b) Reciprocal reflection of ψ by D. (c) Mirror reflection of Ψ by D.

ψ is incident onto the wall, its field $\mathbf{E}$ acts on a vacuuon scatterer *of an effective charge q_p) at the wall surface a force $\mathbf{F}(= q_p\mathbf{E})$, opposite to which the massive wall acts back a force $\mathbf{F}^{\mathrm{r}} = -\mathbf{F}$ (Newton's third law). This gives a reciprocal reflection of ψ, into ψ^{r} of a linear momentum $mc' = -mc$ (Fig. 2b). (4.3): For the ψ_t of (9) striking on D, its reflected wave ψ_t^{r}, will be determined by (4.2) to have a linear momentum in the reciprocal $-r$ direction and, while this being met, will however practically choose to retrace along the two separate reciprocal paths $-r_1, -r_2$ that are already (continuously) stretched by the incident waves and thus require a least action. The optical reflection paths for ψ_t from D are therefore for large L equal probably $PBAqQ$ and $PCAqQ$ (Fig. 1a, dashed lines). The reflection "wall" S on the left may in real applications be a particle beam injection "gun"; its mid point Q serves as a virtual source.

5. Motion of q through double slit

Provided that no permanent emission of radiation occurs on its way, the point charge q will, by the repeated radiation reabsorption/reemission scheme (Sec. 2), travel across the double slit. The charge motion in region I is straightforward, and through regions II-III (Fig. 1c) involves re-directions of its linear velocity by the reflected fields. We elucidate the latter process and the physics involved below.

On arriving at the slit A (Fig. 1c), at initial time the charge has a velocity v and symmetry axis in the $+X$-direction, and a $\mathbf{u}_q^{\rm i}$ and radiation field $\mathbf{E}^{\rm i}$ which are according to (6) invariably along the transverse, Z- direction. During a dwelling time δt here it will continuously generate radiation field and in turn be met by the reflected field $\mathbf{E}^{\rm r}$ from either side. By origin of electric field production in general, the field here immediately due to the $\mathbf{u}_q^{\rm i} = u_q\hat{u}_q'$ at A, $\mathbf{E}^{\rm i} = \frac{E_q}{A_q}\mathbf{u}_q^{\rm i}$, may be thought as produced between two effective dipole charges $+q_{\rm ef}, -q_{\rm ef}$ separated at a distance $\mathbf{u}_q^{\rm i}$ pivoted about q (Fig. 1d). So, due to a $\mathbf{E}^{\rm r}$ in a certain $\bar{\theta}$- direction the two dipole charges will be acted by two forces $+\mathbf{F}, -\mathbf{F} = q_{\rm ef}E^{\rm r}\cos\bar{\theta}$, hence a torque $\boldsymbol{\tau} = \mathbf{u}_q^{\rm i} \times \mathbf{F} = \boldsymbol{\mu}_{q_{\rm ef}} \times \mathbf{E}^{\rm r}$, with $\boldsymbol{\mu}_{q_{\rm ef}} = q_{\rm ef}\mathbf{u}_{\mathbf{q}}^{\rm i}$, until after an relaxation time δt_1, its axis $\hat{u}_q{}'$ is turned into $\mathbf{u}^{\rm f} = u_q\hat{u}_q''$ in $\bar{\theta}$-direction (Fig. 1b). (We may comprehend the microscopic physics as that, the vacuum is a dielectric[5] and is produced with a polarization $\mathbf{P}_{\rm v}$ about q for q at each fixed position, and an altered $\mathbf{P}_{\rm v}$ on deformation due to $\mathbf{u}_q^{\rm i}$, hence the $+q_{\rm ef}, -q_{\rm ef}$, and in turn due to the $\mathbf{E}^{\rm r}$; the latter passes on no energy but a turn in the dipole axis only.) Accordingly, the charge will then be (re-)directed to travel along the radial path r'' perpendicular to $\mathbf{u}_q^{\rm f}$.

In practice, $\psi^{\rm r}$ is reflected from all $\bar{r}$-directions, and thus with a $\mathbf{E}^{\rm r}$ in all $\bar{\theta}$-directions in a statistical manner; $|\mathbf{E}^{\rm r}|$ also varies with time. The probability for $\mathbf{u}_q^{\rm i}$ to be turned to a $\mathbf{u}_q^{\rm f}$ parallel with a particular $\mathbf{E}^{\rm r}(\bar{r}, \bar{\theta}, t'')$ is apparently in direct proportion to the intensity of $\mathbf{E}^{\rm r}$, $I(\bar{\theta})$ given in (6); with ψ of (8) in it here we have $\mathcal{P}(\bar{\theta}) = |\psi^{\rm r}(0, \bar{\theta}, t'')|^2 = \mathcal{C}_A^2(\bar{\theta})$. We suppose in the following that, by a probability $\mathcal{P}(\theta_{AB})$, the charge q at A has been re-directed to travel in the AB- direction and $\hat{u}_q{}''$ along the transverse θ_{AB}-direction, and at a later time t'' enters B as in Fig. 1d.

Similarly, at B the charge, with an initial $\mathbf{u}_q^{\rm i}$ in the θ_{AB} direction, is statistically subject to a torque due to a certain reflected field, which according to (4.4) is a projection of a total $\mathbf{E}_t^{\rm r}$ onto $\bar{\theta} = \theta_1$ or θ_2 perpendicular to $\bar{r} = -r_1$ or $-r_2$. Particularly here, $\mathbf{E}_t^{\rm r}$, as the incident $\mathbf{E}_t(= E_q\psi_t\hat{\theta})$, has peaked intensities $I_t^{\rm r}(\bar{\theta}) = I_t(\theta)$ given by (11); and its projection on θ_i, with $i = 1, 2$, is $I_1^{\rm r}(\bar{\theta}_i) = \frac{1}{2}I_t^{\rm r}(\bar{\theta})$ following from the effectively symmetric geometry of P relative to B, C for large L in region III, and $I(\theta''_{AB}) = I(\theta''_{AC})$ in region II. The probability for $\hat{u}_q{}''$ to be turned to $\hat{u}_q$ parallel with $\bar{\theta}_i = \theta_i$- direction perpendicular to $\bar{r}_i = -r_i$, $i = 1$ or 2, in an individual measurement is thus given by

$$\mathcal{P}_i(\theta_i) = \tfrac{1}{2}\mathcal{P}_t(\theta) = \tfrac{1}{2}(|\psi_1|^2 + |\psi_2|^2)4\cos^2(\tfrac{1}{2}k_d\delta r) = \tfrac{1}{2}|\psi_1 + \psi_2|^2, \quad i = 1, 2. \tag{12}$$

(12) implies that, in each individual measurement the point charge q has definitely, equally probably travelled through either slit B or C before striking on D. The detection signal produced directly by the striking of q however does not inform which slit is passed by q. A determination of which slit can be prevented, not in principle but in common practice. If for example one attempts to determine which slit by blocking one of the two, say C, then the intensities of the incident and reflected waves and of the so navigated q will each be of a pattern of single-slit diffraction. This feature has been well demonstrated in matter-wave double-slit experiments.

A detection signal at D may be directly due to the impingement of q or ψ_t (in the latter case the slow moving q may not have entered D), of an identical intensity I_t given by (10) or (12). Either will inform the self interference of the IED particle. From either (10) or (12), the two probabilities $|\psi_1|^2, |\psi_2|^2$ for the single particle passing two slits add up to the total I_t in the manner of $|\psi_1 + \psi_2|^2$, and not $|\psi_1|^2 + |\psi_2|^2$. This predicts an important characteristics of quantum probability as has been demonstrated by overall matter-wave interference experiments. This way how the intensities add up is completely the same as how the intensities of an elastic or light wave add up; and we have in fact just obtained (10) naturally for the IED particle for its constituent electromagnetic waves φ^j's, or "light" rays of far-above ultra violet frequencies.

An incident IED particle with a low enough energy will knock a point-like absorber in D into excitation of one energy quantum $\hbar\varpi_d$ at a time, producing a detection signal appearing as a spatial point on the D plane. When this is directly due to a (thermal) absorption of the extensive (Ψ off from) ψ), we can think of the wave as (conveying) an energy flux and its each wave front as an equi-energy surface S. An energy will first of all be transported to P longitudinally through the propagation of ψ at the speed c. In the meantime, a sudden drop of energy at P (upon onset of excitation) creates an energy potential against other points on the S intersecting with P, driving therefore a lateral energy flow toward P at a speed which may be expected to be high since no time is wasted in undulation.

The author expresses thanks to Professor H.-D. Doebner for his invaluable discussions at the QTS 6, extending over a number of previous ones, with the author over many key aspects of the IED representation and solutions, to the QTS 6 organisers for arranging the presentation of this work, and to Scientist P.-I. Johansson for his continued moral and private funding support of the research. During the meeting the author especially also enjoyed the kind reception of Professor A. Shapere, QTS6 organising chairman, and intellectual conversations with Professors M. Berry, A. Shapere, F. Verstraete, P. Devereus, J. Towe, V. Dobrev, V. Borowiec, and others.

References

[1] De Broglie L. 1923 *Comp. rend.* **177** 507-510; 1924 PhD Thesis, Univ. of Paris; 1924 *Phil. Mag.* **47** 446.

[2] Mandel L 1999 *Rev. Mod. Phys.* **71** S274-282; Greenstein G and Zajonc A G 1997 *The quantum challenge* (London: Jones and Barlett); Buzek V and Knight in P L 1995 *Prog in Optics XXXIV* ed E Wolf (Amsterdam: Elsevier) p1.

[3] Zeilinger A 1999 *Rev. Mod. Phys.* **71** S288; Rauch H and Werner S 2000 *Neutron Interferometry: Lessons in Experimental Quantum Mechanics* (Oxford: Oxford Univ. Press).

[4] Zheng-Johansson J X 2010 Internally electrodynamic particle model: its experimental basis and its predictions *Phys. Atom. Nucl.* **73** 571-581 (*Preprint* arxiv:0812.3951).

[5] Zheng-Johansson JX and Johansson P-I (a): 2006 *Unification of Classical, Quantum and Relativistic Mechanics and of the Four Forces* (New York: Nova Science); *Preprint* arxiv:physics/0412168; (b): 2010 *Inference of Basic Laws of Classical, Quantum and Relativistic Mechanics from First-Principles Classical-Mechanics Solutions* (New York: Nova Sci.); (c): 2006 Inference of Schrödinger equation from classical mechanics solution *Quantum Theory and Symmetries IV* vol 2 (*Suppl. Bulg. J. Phys.* **33**, ed Dobrev VK (Sofia: Heron Press) pp 763-770 (*Preprint* arxiv:phyiscs/0411134 v5); "Schrödinger Equation for Electrodynamic Model Particle," (internal); (d): 2006 Developing de Broglie wave *Prog. Phys.*, **4** 32-35 (*Preprint* arxiv:phyiscs/0608265); (e): 2006 Mass and mass–energy equation from classical-mechanics solution *Phys. Essay*, **19** 544 (*Preprint* arxiv:phyiscs/0501037); (f): (Zheng-Johansson J X) 2006 Spectral emission of moving atom *Prog. Phys.* **3** 78-81(*Preprint* arxiv:phyiscs/060616); (g): (Zheng-Johansson J X) Vacuum structure and potential *Preprint* arxiv:physics/0704.0131; (h): (Zheng-Johansson J X) Dielectric theory of the vacuum *Preprint* arxiv:physics/0612096; (i): 2006 Depolarisation radiation force in a dielectric medium. its analogy with gravity *Quantum Theory and Symmetries IV* vol 2, (*Suppl. Bulg. J. Phys.* **33**) ed Dobrev V K (Sofia: Heron Press), pp. 771-779, with R. Lundin; (*Preprint* arxiv:phyiscs/0411245v4); (j) (Zheng-Johansson JX) Doebner-Goldin Equation for electrodynamic model particle. the implied applications *Preprint* arxiv:0801.4279 (talk at 7th Int. Conf. Symm. in Nonl. Math. Phys., Kyiv, 2007); (k): (Zheng-Johansson JX) 2008 Dirac equation for electrodynamic model particles *J. Phys: Conf. Series* **128**, 012019, *Proc. 5th Int. Symp. Quantum Theory and Symmetries*, ed Olmo M (Valladolid, 2007); (l): (J.X. Zheng-Johansson) 2003 Unification of classical and quantum mechanics & the theory of relative motion, *Bull. Amer Phys Soc.* **G35.001**-Gen. Phys. (Austin).

AUTHOR INDEX

Institute of Physics
Dirac House, Temple Back
Bristol BS1 6BE UK

CURRAN ASSOCIATES INC
proceedings.com